AF781264

Move Into the Light:

postscript to a turbulent 2007

by The Turbulence Collective

PM Press PAMPHLET SERIES

0001: BECOMING THE MEDIA: A CRITICAL HISTORY OF CLAMOR MAGAZINE
By Jen Angel

0002: DARING TO STRUGGLE, FAILING TO WIN:
THE RED ARMY FACTION'S 1977 CAMPAIGN OF DESPERATION
By J. Smith & André Moncourt

0003: MOVE INTO THE LIGHT: POSTSCRIPT TO A TURBULENT 2007
By The Turbulence Collective

0004: PRISON-INDUSTRIAL COMPLEX
By Eve Goldberg and Linda Evans

0005: ABOLISH RESTAURANTS
By Prole

PM Press PAMPHLET SERIES No. 0003
MOVE INTO THE LIGHT: POSTSCRIPT TO A TURBULENT 2007
By The Turbulence Collective
ISBN: 978-1-60486-031-3

Copyright © 2008 Creative Commons
Published under a Creative Commons Attribution Non-Commercial Share-Alike license

This edition copyright © 2008 PM Press
All Rights Reserved

PM Press
PO Box 23912
Oakland, CA 94623
www.pmpress.org

Layout and design: Courtney Utt

Printed in Oakland, CA on recycled paper with soy ink.

It's night time and a man is crawling around on his hands and knees, looking for his car keys underneath a lamp post. A woman comes along and starts to help him. After they've been searching together for a while the woman asks the man: "Are you sure this is where you dropped them?"

The man replies: "No, I think I dropped them somewhere else."

"Then why are we looking here?" she enquires.

"Because this is where the light is."

At the beginning of 2007, the Turbulence collective commissioned 14 articles from around the global 'movement of movements', asking authors: "What would it mean to win?" We edited their responses into a newspaper and printed 7,000 copies, most of which were distributed at the mobilisation against the G8 summit in Heiligendamm, Germany, in June. A few months later, we want to return to the question of winning.

As we resume our search it's no surprise that we keep coming across the problem of visibility. When we think about winning, our eyes are drawn to things that are highly visible or easy-to-measure, such as institutional or legislative change, the opening of a social centre or an increase in membership. That's where the light is. But we also need to assess victories in the less tangible though just as real realm of possibilities. Winning in this realm may involve increased potential, changes in perception or patterns of behaviour. Yet these seem to exist at the very edge of the luminous zone.

This problem leads into another: our experiences create their own luminosity and consequently their own areas of darkness. When we think about winning we are drawn to movements, people and events that are familiar to us; and we have expectations about how things should turn out if they are to constitute a victory.

So how can we overcome our night-blindness once we move beyond the familiar? In a sense this ability to look outside ourselves was key to Heiligendamm.

HEILIGENDAMM: A DIFFERENT REPETITION?

In many ways, this year's G8 summit on Germany's Baltic coast was much as we had expected it to be: a repetition of previous counter-summit mobilisations from at least Seattle onwards (Prague, Gothenburg, Genoa, Evian, Cancun, Gleneagles...). Each of these events saw a broad constellation of actors brought together in productive cooperation. Each opened up a space and set in motion processes of contamination (often behind people's backs) that were key to the politicisation of a generation of militants. On the one hand, people launched practical challenges to the legitimacy of global command (the rejection of dialogue, the blocking of roads into the summit); on the other, commonalities and mutations were produced, in the camps and convergence centres, during debates and actions.

However, previous summit mobilisations had already shown us the limits of such events. After Seattle, in 1999, it became clear that the affect produced in mass street actions would not translate automatically into everyday practices of transformation. Two years later Gothenburg and Genoa showed the price that would be paid by a movement for entering into a logic of near-symmetrical conflict (imprisonment, injuries and death). And Gleneagles 2005 showed the extent to which the desires of a movement could be captured and turned against itself, with 300,000 marching for the G8. So if many had already seen the limits of summit mobilisations, and Heiligendamm had always promised to be a repetition, surely the last thing it was going to be was different?

Sometimes, however, what appears like mere repetition is not really a repetition at all; at least not in the sense that it is simply the same thing taking place over and over again. So rather than return to a particular point in a cycle ('Bringing Seattle to Germany', for example), the point at Heiligendamm was to start anew with an unforeseeable process of becoming—one that would hopefully go beyond the achievements and limits of the past. Less a repetition that sought to mimic, more a new experiment in the production of politics; overcoming rather than reaffirming existing identities.

In the run-up to the summit, the groups involved in the organisation of the protest underwent something of a reconfiguration. They took some significant steps towards becoming a more genuine 'movement of movements'. A common 'choreography of resistance' was built and designed by a wide spectrum of groups—from the autonomous radical left through to people organising inter-confessional prayer sessions against poverty. While more radical elements attempted to set the terms of the coalition (a rejection of the G8's legitimacy alongside a toleration of diverse forms of action), there was a willingness to compromise and come to common agreements as to which forms of action were appropriate where and when. In this way Heiligendamm moved beyond the principle of 'diversity of tactics' that had become commonplace, and returned to the earlier process of cross-pollination. Instead of different political currents engaging in different forms of action—in a spirit of solidarity but without jeopardising their own identities—the work developed in Germany was in the direction of a 'becoming-other, together'. This meant collectively devising and carrying out forms of action new to all, actions and alliances that took people beyond their comfort zones towards the practical constitution of new commons, and therefore new common potentials.

WHAT'S THE SCORE?

While Heiligendamm wasn't a quantitative high-point in the history of the counter-globalisation movement (in terms of numbers, Genoa and Gleneagles were both around four times the size), in other respects it did seem that a new qualitative high was reached. It was a 'victory' because it was a reconstitutive moment, not least for the German left. But something was missing from this affect of victory: the sense of having defeated the other side. Sure, we scored some successes against the police and summit organisers with our mass blockades. But German chancellor Merkel won legitimacy by appearing to force 'the recalcitrant Americans' into an agreement on climate change. And the G8? It is celebrating the Heiligendamm summit as one of its most successful ever. It managed to create the impression that the leaders of the world are tackling the 'global challenge' of climate change.

When the G8 first became the target of massive protests towards the end of the 1980s, it was relatively easy to point to the in-built illegitimacy of its activities. At the 1999 Cologne summit, when it clumsily responded to actions by social movements in the global South (and some Northern NGOs) by passing debt-relief programmes, hardly anyone took them seriously. But the G8 reinvented itself. It stopped being just a place for the major capitalist powers to hammer out differences and became a media-circus that presents itself as the only forum that can deal with global concerns. In other words, as the G8 came under attack, its very purpose became the re-legitimation of its global authority. And it learnt its lessons well. At Gleneagles, a big NGO operation sponsored by the UK government saw 300,000 people turn out, not to demonstrate against the G8, but to welcome and 'lobby' it in favour of debt relief and aid for Africa.

The initiative lost in Scotland—where the protests were hijacked by an efficient PR offensive—was successfully regained in Heiligendamm: the explicit goal of all major actions was the delegitimation of the G8. The problem however was that the G8 had once again moved on, now seeking to draw legitimacy by seeming to respond to widespread concern about climate change. And this is where we (got) lost. The actions carried out in Germany failed to convey a political challenge to the G8's relegitimation on the issue of climate change, which had become a new key terrain of struggle.

How did this happen? One reason is that there isn't yet an overarching 'alternative' narrative to the newly greened global capitalist agenda: however bad their story may be, there is nothing else on offer. But the problem runs deeper than that. The G8's narrative on climate change solutions is a fiction, just as it was on making poverty history. But we can't counter this with a fiction of our own: at the moment we don't know how to 'solve' climate change. None of us can see far enough or clearly enough. All we can do is move from one puddle of light to another.

WHAT'S IN A LIMIT? CAPITAL, CRISIS AND CLIMATE CHANGE

It's no coincidence that talking about the G8 should lead directly to talking about climate change. For movements, it represents the possible emergence of a new focus, as shown by the buzz in public opinion and events such as this year's climate camp in the UK which, it seems, will be repeated in Germany, the US, Sweden and elsewhere in 2008. From the perspective of governance and capital, it is becoming a key element in the management of the global system, both at the level of decision-making and of political legitimation, not to mention new market niches. In the space between movements and governance, it exemplifies the ambiguity and complexity of the question of 'winning'. If the whole emphasis of environmental activism over the last few years has been on raising awareness about the threat of climate change, then 2007 must be seen as the year when 'we won'. The issue is now everywhere, and everyone, politicians and big companies included, talk about it.

Yet it is precisely this victory that could prove to be a defeat. Global concern about climate change must be given a new form if it is to actually affect the state of things (that is, radically reduce carbon dioxide emissions in a short time-frame). In part this means constructing a new story, one that can stop the issue being turned into a huge profit-making opportunity for capital. Without this, it's easy to see climate change being used to unleash a new regime of austerity on the governed, and to excuse measures like increased 'security' and border controls as geopolitical tensions rise. But if the fight is to be more than a public opinion dispute—one where we're always on the back foot—then it has to also take place at the level of production and social reproduction.

It's common to think of climate change as a technical-environmental problem that calls for a technical-environmental solution: the problem is too much carbon dioxide going into the atmosphere, so the solution is to reduce these emissions to 'acceptable' levels via technological innovation, government legislation and the public 'doing their bit'. The difficulty with this is twofold. First, almost everything we do is bound-up with fossil fuel use and the resulting CO2 emissions: from travelling to work to phoning-in sick so we can watch DVDs. Second, the cuts required (some 60–90% before 2050) are so large they require sweeping changes, and cannot be solved simply by the world's environment ministries getting together.

An alternative way to understand climate change is in terms of metabolism. The Earth's metabolism, its ability to process carbon, runs at a slower speed than the metabolism of contemporary capitalism. The economy is on a collision course with the biosphere. Here we are talking about a limit to the expansion of capital and a possible crisis of accumulation.

For capital, limits are peculiar. Capital has an internal dynamic of expansion which must be satisfied, so limits must be ignored, subverted, side-stepped, or otherwise overcome. And the secret of capital's longevity lies precisely in its ability to use limits and the crises they engender as a launch-pad for a new round of accumulation and expansion. A good example of this dynamism is the emergence of the so-called Keynesian/Fordist phase of capitalism. The high levels of organisation of the industrial working class in the first half of the 20th century—not only the Russian Revolution but intense struggle worldwide—appeared as a limit to the expansion of capitalism, threatening not only to halt accumulation but to destroy the system once and for all. The welfare state was a direct result of these struggles, but it was also a way of neutralising this threat. And capital's greatest feat was to strike a productivity deal which actually transformed this limit into the engine of a new phase of capitalist growth.

What does an analysis of the generic response of capitalism to limit-crises tell us about likely responses to climate change? There's no doubt that climate change is a limit which presents as many opportunities as dangers to capital. Many are jumping at the chance to take this new limit, this potential crisis, and turn it into a new motor for accumulation. Look at the clamour for buying and selling rights to emit carbon: carbon credits, carbon offsets, Tradable Emissions Quotas, carbon futures. And then there's green consumerism: green cars, solar panels, green home make-overs. Could climate change inject new dynamism into the global economy? Are we looking at a new, 'green' phase of capitalism, where the atmosphere is opened up like cyberspace was in the '90s? It's possible. And it's also obvious that it's unlikely to cut carbon emissions radically!

A capitalist solution will look, well, like capitalism. Just as the effects of climate change are uneven, having a far more devastating effect on the poor—look at the impact of Hurricane Katrina in New Orleans, or the east Asian tsunami on Aceh—so almost all the current crop of solutions will also work to reinforce existing hierarchies. Most 'green' taxes will increase the price of basic goods and services, limiting mobility and access to food and heating. Access to travel, food and comfort all tied in to possession of money? No news there, of course: just the rules of the game as we know it. Except now they will be justified on the grounds that they're necessary in order to save the planet. Expect 'green capitalism' to be a new regime of austerity and discipline, imposed on the poor more than on the rich in the name of the 'greater good'.

THE EYE OF THE STORM

But capitalism is neither all-conquering nor invincible. If climate change might open up a moment of crisis, it's worth trying to understand what its dynamics might be.

One key aspect is the variable of time, understood in two different ways. First, there is a problem of time lag. The outcomes of different decisions, in climate change terms, are felt decades later. Due to the thermal inertia of the climate system there is a huge temporal mismatch between cause and effect. This means that if the impacts of climate change become 'out of control', they may stay that way for several decades. Second, all the scientific evidence points to the urgency of the problem. If we are to avoid the 'tipping points'—points at which climate change becomes potentially irreversible and catastrophic for the majority of the Earth's population (the death of the Amazon rainforest being one example)—emissions have to be drastically reduced within the next decade.

There's a positive side to this sense of urgency. A lot of the 'awareness-raising' activism of the last ten years worked with no time variable whatsoever; it addressed 'the public', a general 'other' who needed to be 'informed' of what is going on. Because of that, there were no general deadlines, no overall calendar, no sense of escalation, no particular goals; when everything is always happening 'right now', there is no time as such.

The urgency of climate change raises important questions which only exist because of the time variable; they are questions of strategy and tactics.

And here we return to the question of winning. For instance, some suggest that nothing this big can be done with such little time, and the best that can be done is start preparing now for the worst. We may as well extinguish the lights and blithely head off into the darkness. Others have said that the problem is so massive and so pressing that only a centralised body is capable of tackling it. Faced with the abyss of the unknown it's tempting to turn back to the comforting light of the state. But this harsh glare blinds as well as illuminates.

Take the example of air travel. The growth in aviation is clearly a massive environmental problem, so it's easy to get lured into supporting new taxes on flying, say, or even seeing people who fly as part of the problem. But focusing on this issue might make it harder to see some of the other dynamics at work. By restraining our autonomy or strengthening capitalist and state institutions, some climate change solutions may hinder other struggles and make it harder to tackle the larger causes of climate change. What is needed is a lens—an approach or ethic—that allows us to pose the question of how climate change politics can resonate with other struggles. Not because movements need an explicit, conscious connection in order to resonate with one another; they don't. But we do need to bring to light resonances and dissonances. Once we can see the paths, they're easier to follow.

While we have to be wary of being blinded by the glare of the state, we can't just close our eyes to it. So how can we relate to institutional forms? Perhaps recent events in Latin America can provide some clues.

CHANGING THE WORLD BY TAKING POWER?

The last few years have seen the rise and establishment of governments in different shades of red across Latin America. Chavez's Venezuela, Morales' Bolivia and Lula's Brazil have generated the most international discussion. But there's also Rafael Correa's Ecuador, Tabaré Vasquez in Uruguay, the return of Daniel Ortega in Nicaragua and, more contentiously, Michelle Bachelet in Chile and the Kirchners in Argentina. These national processes are not independent, but share two related themes: first, the neoliberal model has run out of steam in the region; and second, the movement of movements has managed to make its mark at an institutional level.

But what are we to make of this institutional success? Some people see these electoral victories as the only concrete result of those post-Seattle years. In this sense, 'winning' would also be the defeat of the 'movementism' of that period: confirmation that it is impossible to 'change the world without taking power'. By this logic, all that is left to do is to ensure that, once in power, the parties and groups that rode on that wave of resistance are able to produce change within institutional constraints. Further, these parties and groups that have ascended to power must also be forced where possible to transform institutions in ways that make them more permeable to this 'pressure from below'. It is taken for granted that such pressure can only fulfil its role if it is capable of being translated into institutional forms.

While we shouldn't underestimate the advances taking place across so much of Latin America, it's worth pausing to consider the implications of this view of social change.

First of all, it's important not to gloss over some important differences among these countries. It is only the case of Morales that directly corresponds to the picture of a growing wave of resistance leading to electoral victory. The history of Bolivia in the last ten years has been punctuated by moments of radicalisation that were always recuperated into the existing political system, only to be denied resolution again. As the episodes of resistance became more frequent and powerful, they forced the systemic rearrangement that carried Morales' MAS (Movimiento al Socialismo) to power. In Brazil's case, a similar wave took place in the 1980s, and was beaten at the polls three times until Lula was elected. By then, the Workers' Party (PT) had become the translation into party politics of a movement on the wane (with the possible exception of the landless workers movement, the MST). In Venezuela, despite a diffuse anger at the impermeability of the institutions and at the policies implemented in the 1980s and '90s, there was no movement as such. Chavez has acted as a catalyst for an intensification of mobilisation and participation which is unheard of in Venezuelan history. It remains to be seen whether he was only the catalyst, or whether he has now become the pillar without which it could all come crumbling down.

More importantly, the idea that these electoral victories are the only practical result of the last decade is flawed on two related counts. First, it assumes that 'politics' only takes place in the institutional places where we normally look for it. This discounts a whole series of networks, infrastructures, knowledge, cultures and so on—a diffuse web of collective intelligence and memory that is always active in one way or another, always producing change, and crystallising as an antagonistic force at crucial points. The escalation of resistance in Bolivia before the MAS victory is a good example of this. What 'disappeared' after each flashpoint would 'return' bigger and stronger. And it could only do so because it had never gone away.

Second, it ignores that fact that movements, as long as they keep moving, have ways of effecting and producing change that do not need to pass through, or even be recognised by, institutional politics. They can do this by, to give a few examples, transforming public discourse, by making legislation unenforceable, or simply through their power of self-management and autonomous self-constitution.

WHAT IF THERE WERE A NEW CYCLE OF STRUGGLES AND WE WEREN'T INVITED?

This question of the power of movements brings us back to where we started. As the Turbulence newspaper explained, we had three main reasons to produce a publication for the G8 summit in Heiligendamm. The first, very pragmatic, was that it would be relatively easy to distribute to a wide readership. The second was that in our experience summit mobilisations are spaces where people are more open to other ideas.

The third includes a more complex wager. Since Seattle, summit mobilisations have been the most visible face of the movement of movements, the way its role as a global force is most explicitly manifested, and also the times when its strength and orientations can be gauged. But by the same token it's at summit mobilisations that all the movement's potential limitations have been most apparent.

On one level, the wager was what everybody asked themselves on the way to Heiligendamm: how socially relevant, big, transformative is this event really going to be? Will this be a last gasp, a new beginning, or neither? And, equally important, whatever it is, how will we recognise it?

But if we criticise those who only recognise change on the institutional level, are 'movementists' not similarly guilty of looking for answers in the usual places? Here we are again writing of summit protests and counter-summits. Perhaps the impasse of the last few years has arisen precisely because people have failed to see answers in the places they searched, and did not start looking elsewhere. However hazy our image in the mirror has always been, have we not become too enamoured of it to actually have a look around? What if there were a 'new cycle of struggles', and we were not invited?

Think about what happened in the French banlieues in autumn 2005 (and appears to be re-emerging as we write). Anyone on the 'established' left—parties, trade unions, 'activists'; if you know who we are talking about, you belong in this category!—who claims that those who revolted are 'with us' in any strong sense would be guilty of appropriating someone else's struggle by misrepresenting it. Sure, they fight against many things that we oppose. But let's look at the established left's reaction to them, along its three general lines. Either the banlieues are brought into a ready-made framework, and become the 'proof' of some 'new stage of capitalism'. Or they signify the terror of a social dissolution that requires state intervention to redistribute wealth and access to opportunities in the long term (but possibly also to police them in the short term, so as to prevent civil war). Or they represent a romantic, abstract 'other' whose tough, uncompromising radicality—the poster image of revolution—is paid back with equally abstract solidarity.

If this is all 'we'—parties, unions, 'movementists'—have to offer, let's recognise that we are part of the problem. Even the most 'radical' members of the established left could only interpret the banlieues as an eruption of pure negativity, a 'force of nature' rather than the work of real people. For the mainstream politician, it is the face of fear: we're on the verge of civil war! For others, it's in itself nothing, but as an unknown quantity it can fit anywhere in the theory: 'See, it confirms our predictions!' The latter simply eliminates the event; whatever else happened would mean just the same. The first two recognise an event, but see it as something so beyond any explanation that it can only be a harbinger of the end of the world (something to celebrate or lament, depending on your taste).

All three positions ignore the fact that, if the banlieues pose a problem, it is a problem made of flesh and bones. The banlieues reveal a gap in our knowledge: for as long as that gap is not filled by banlieusards—met on their terms, introduced by their own voices—'we' feed into the game that excludes 'them'. Worse, by posing as interpreters of those to whom we don't speak, 'we' actively reproduce this game. And there is political currency to be gained, even for the most marginal leftwing groupuscules, by pretending to speak on behalf of those outside the gates. The real challenge, then, lies in effectively opening the gates to those outside. Or better, in tearing down the wall altogether. But this kind of coordination can only take place through actually working with actual people. There is little to be gained by flattering ourselves that abstract feelings of 'solidarity' matter in any real way.

Another example: for over a year now, different cities in Spain have seen a unique movement coalesce around people's frustration with the impossibility of getting 'dignified housing' in the face of rampant property speculation. The movement began when, at the height of the anti-CPE protests in France, an anonymous individual sent out an email calling for a day of protest for 'dignified housing'. The email did the rounds, and on the arranged day hundreds of people—from cab drivers to hairdressers, as well as 'activists'—took to the streets. By the second self-convoked day of protests there were thousands involved. Since then a number of local assemblies have been created, many of which are still going.

The reactions of 'activists' to this housing struggle have been interesting. They ranged from confusion ('How come there is a protest and I don't know who called it?'), to a desire to fade into the background ('In the assembly everyone is equal, people shouldn't expect us to have anything special to say'), to a recognition of how their specific knowledges could be useful ('Well, I organised a protest once, and I realise it works better if we do it like this…'). People with less of a history in politics, on the other hand, sometimes seemed to go in the opposite direction: famously, the Madrid assembly at one point discussed whether it should subscribe to a protest it had not called—'Now we're the space in which the movement is organised, so we should be the only ones to decide these things'; 'But didn't this all begin with a spontaneous demonstration in the first place? Didn't you attend that?'; 'I did, but that was before there was the assembly!…'

All of this underlines the point that, no matter what meaning we give to the label 'movement of movements', it offers no guarantees. 'Seattle' or 'Cancun' or 'Heiligendamm' don't mean 'we' are the be-all, end-all of social change. In fact, they don't even mean 'we' exist. And to pretend we do, and that history is exclusively ours to make, can only make us blind to where we fit. (And if we fit somewhere, it means logically we are not the whole.)

MOVE INTO THE LIGHT?

This 'thought travelogue' that we have tried to establish—from Heiligendamm to Latin America, from the politics of climate change to the banlieues, to the movement of movements and back again—saw us start with a question and a few thousand papers to distribute, and come across a few themes and problems that just won't go away.

We've started with the theme of visibility because it highlights the relation between movements and their dynamic of self-reproduction. It's relatively easy to think of movements grappling with institutional politics, like the electoral experiments in Latin America, or the social forum process, or recent attempts to realign social centres across Europe. Depending on your perspective, these are examples of movements 'selling out' or 'growing up' or 'being recuperated'. But all three positions make the mistake of seeing institutional forms as somehow separate from movements. Or put another way, all three see movements as discrete bodies, with an 'inside' and an 'outside', rather than as an endless moving of social relations.

As movements move, they constantly throw up new forms of organisation and practice which are constantly settling and consolidating. Of course this can be problematic: once established, identities and rituals can become huge obstacles to change. But this doesn't mean that movements die as soon as they begin to take root, or as soon as they move into the light of exposure. This process is also a way in which movements cast their own light. The 'movement of movements', for example, is an institutionalisation of a certain moment of struggles, with Seattle as one of its highlights. It has also helped to generate a whole series of other institutions, which have developed their own dynamics. Summit protests, for example, took place around the world, each building on the other with subtle and not-so-subtle modifications. As that cycle of protest seemed to wane, the social forum process took off, constructing a different kind of experiment. After the 2007 World Social Forum in Kenya, heavily sponsored and controlled by NGOs, many felt that this process itself had come to an end. But a few months later the US Social Forum showed that it's possible to organise something that would not last just a few days but would produce effects in cross-pollinating and coordinating different struggles.

In fact the recent encuentros hosted by the Zapatistas brought this point home forcefully. They offered an encounter between, on the one hand, 'movement' visions of autonomy, horizontalism and non-hierarchical practice and, on the other, a real attempt to make these visions work on the ground—under threat of attack by paramilitaries and surrounded by hostile forces. Many 'movementists' got an insight into the functioning of the 'Juntas of Good Governance', a long-term experiment in self-government by the Zapatista autonomous municipalities. A startling aspect of this was the experience of being in a space where men with guns—the EZLN—are on your side. But if we're serious about producing change in visible and tangible ways, how is that possible without creating institutions of one kind or another? How else are we really going to create other worlds?

But there's a second theme of light and luminosity. When we asked the question 'what would it mean to win?', we were deliberately not asking for a ten-point programme. We didn't want 'illumination'. Instead we wanted to assert a politics that recognises that no one has the solution, that changing the world is, at least in part, a process of 'shared investigation', and that as a first step we can begin to ask the same questions. This is a world apart from the old-school politics of certainty, which is dominated by polemical confrontations, where differing political identities and approaches are pitted against each other, recreating identitarian or ideological niches.

Of course the idea of total illumination is a fantasy. But it's a very tempting delusion, one tied up with the myth of total knowledge. If you stare at the sun for long enough, an after-image will be etched into the back of your eyelids. After the G8 seized the issue of climate change, some concluded that we just need the right narrative, one that shows only we have 'the answer'. Rather than adopt this approach, with all its overtones of dual power and counter-hegemony, it seems more productive to take another lesson from Heiligendamm. As people prepared to block the roads leading into the summit, a final piece of advice was circulated: "As you approach the police lines, don't look at the cops—look at the gaps between them."

Finally, there's a third thread that ties together the idea of light and visibility. Throughout this text we've used the idea of 'moving into the light' in the sense of making yourself visible, 'coming out' and occupying space. Yet it has a different sense in so-called near-death experiences: you see a light, and a voice invites you to move into the light. To a certain extent, change means dying. It means abandoning the comfort zone, giving up part of yourself, abandoning habits and certainties. And in a wider sense, movements need to flirt with their own death, with the possibility that they need to cease to be so that something else can be born. Dare we lay this wager? Dare we make this leap of faith? Dare we leap through those gaps, into the unknown, into the light?

December 2007

Turbulence: Ideas for movement is an irregular publishing project that we hope will become an ongoing space for thinking through the ideas and practices of our movements. Turbulence is David Harvie, Keir Milburn, Tadzio Müller, Rodrigo Nunes, Michal Osterweil, Kay Summer, Ben Trott and David Watts

www.turbulence.org.uk
editors@turbulence.org.uk

Joseph von Eichendorff

Ausgewählte Werke

Band 4

Herausgegeben
und mit einem Nachwort von
Hans A. Neunzig

nymphenburger

© Nymphenburger Verlagshandlung GmbH, München 1987
Alle Rechte, auch der photomechanischen Vervielfältigung
und des auszugsweisen Abdrucks, vorbehalten
Umschlaggestaltung: Cooperation, München,
unter Verwendung von Gemälden von Caspar David Friedrich
(Archiv für Kunst und Geschichte, Berlin)
Satz: Fotosatz Völkl, Germering
Druck und Bindung: Wiener Verlag, Himberg
Printed in Austria
ISBN 3-485-00554-1

INHALT

Dichter und ihre Gesellen

Erstes Buch

Erstes Kapitel

In den letzten Strahlen der Abendsonne wurde auf der grünen Höhe ein junger Reiter sichtbar, der zwischen dem Jauchzen der Hirten und heimkehrenden Spaziergänger fröhlich nach dem freundlichen Städtchen hinabritt, das wie in einem Blütenmeere im Grunde lag.
Er sann lange nach, was ihn hier mit so altbekannten Augen ansah, und sang immerfort ein längst verklungenes Lied leise in sich hinein, ohne zu wissen, woher der Nachhall kam. Da fiel es ihm plötzlich aufs Herz: wie in *Heidelberg* lagen die Häuser da unten zwischen den Gärten und Felsen und Abendlichtern, wie in Heidelberg rauschte der Strom aus dem Grunde und der Wald von allen Höhen! So war er als Student manchen lauen Abend sommermüde von den Bergen heimgekehrt und hatte über die Feuersäule, die das Abendrot über den Neckar warf, in die duftige Talferne gleichwie in sein künftiges, noch ungewisses Leben hinausgeschaut. –
„Mein Gott", rief er endlich, „da in dem Städtchen unten muß ja Walter wohnen, mein treuer Heidelberger Kamerad, mit dem ich manchen stillen, fröhlichen Abend auf den Bergen verlebt! Was muß der wackere Gesell nicht alles schon wissen, wenn er fortfuhr, so fleißig zu sein, wie damals!" – Er gab ungeduldig seinem Pferde die Sporen und hatte bald das dunkle Tor der Stadt erreicht. Walters Wohnung war in dem kleinen Orte leicht erfragt: ein buntes, freundliches Häuschen am Markte, mit hohen Linden vor den Fenstern, in denen unzählige Sperlinge beim letzten Abendschimmer einen gewaltigen Lärm machten. Der Reisende sprang eilig die enge, etwas dunkle Treppe hinan und riß die ihm bezeichnete Tür auf; die Abendsonne, durch das Laub vor den Fenstern zitternd, vergoldete so-

eben die ganze, stille Stube, Walter saß im Schlafrock am Schreibtische neben großen Aktenstößen, Tabaksbüchse, Kaffeekanne und eine halbgeleerte Tasse vor sich. Er sah den Hereintretenden erstaunt und ungewiß an, seine Gipspfeife langsam weglegend. „Baron Fortunat!“ rief er dann, „mein lieber Fortunat!“ und beide Freunde lagen einander in den Armen.

„Also so sieht man aus in Amt und Brot?“ sagte Fortunat nach der ersten Begrüßung, während er Walter von allen Seiten umging und betrachtete; denn es kam ihm vor, als wäre seit den zwei Jahren, daß sie einander nicht gesehen, die Zeit mit ihrem Pelzärmel seltsam über das frische Bild des Freundes dahingefahren, er schien langsamer, bleicher und gebückter. Dieser dagegen konnte sich gar nicht satt sehen an den klaren Augen und der heiteren, schlanken Gestalt Fortunats, die in der schönen Reisetracht an Studenten, Jäger, Soldaten und alles Fröhliche der unvergänglichen Jugend erinnerte. – Fragen und Gegenfragen kreuzten sich nun rasch, ohne eine Antwort abzuwarten. Walter pries vor allem sein Glück, das ihn hier so schnell eine leidliche Stelle hatte finden lassen, es fehlte nicht an größern Aussichten, und so sehe er einer heitern, sorgenlosen Zukunft entgegen. – Dazwischen hatte er in seiner freudigen Unruhe bald noch einen Brief zusammenzufalten, bald ein Paket Akten zu binden, bald draußen etwas zu bestellen, beide konnten den alten, vertraulichen Ton gar nicht wiederfinden.

Unterdes war eine alte Frau hereingetreten und fing an, eine altmodische Kaffeeserviette zierlich auszubreiten und Teller, Gläser und Weinflaschen aufzustellen, wobei sie von der Seite ehrerbietige Blicke auf den vornehmen, fremden Herrn warf, der eine solche Revolution in der einförmigen Junggesellenwirtschaft verursachte. Fortunat aber überschaute am Fenster den heitern Markt, und eine leise Wehmut flog durch seine Seele über die langsam zersetzende und zerstörende Gewalt der Verhältnisse, wie sie ihm auf Walters treues Gemüt wirksam zu sein schien.

– „Laß uns nach guter alter Art im Freien trinken!" rief er, sich schnell umwendend, aus, da er die Zurüstungen hinter sich erblickte. Walter hatte Bedenken: das sei hier nicht gewöhnlich, man werde in kleinen Städten zu sehr bemerkt. Fortunat aber hatte unterdes schon unter jeden Arm eine Flasche genommen, und wanderte damit die Treppe hinunter. Walter folgte verlegen lachend, die Alte brachte voll Verwunderung Tisch und Gläser nach, und bald war die ganze fröhliche, funkelnde Wirtschaft unter den Bäumen vor der Tür aufgeschlagen.
Die Sonne war indes untergegangen, und die Dächer und die Gipfel der Berge über der Stadt glühten noch, von denen ein erquickender Strom von Kühle durch alle Straßen und Herzen ging. Kinder jagten sich und schwärmten in den Gassen, die Vornehmen, ihre Hüte nachlässig in der Hand und sich den Schweiß abtrocknend, kehrten, von allen Seiten ehrerbietig begrüßt, von ihren Spaziergängen zurück. Andere traten in bequemen Nachtkleidern mit den Pfeifen vor die Türen, und plauderten mit dem Nachbar, während junge Mädchen, kichernd und in lebhaftem Gespräch, Arm in Arm über den Platz schlenderten und neugierig an dem Fremden vorüberstrichen.
Walter ging bei den Erinnerungen an die fröhliche Studentenzeit und bei dem langentbehrten weiteren und reichen Gespräch recht das Herz auf, er hatte gar bald alle Scheu und blöde Rücksicht abgeschüttelt. – „Wie glücklich bist du zu preisen", rief er seinem Freunde zu, „daß dir vergönnt ist, so mit den Vögeln durch den Frühling zu ziehen und die Reise nach Italien nun wirklich anzutreten, die wir in den heitersten Stunden in Heidelberg so oft miteinander besprachen. Das waren schöne Jugendträume!" – „Das verhüte Gott!" versetzte Fortunat lebhaft, „warum denn Träume? Die Ahnung war es, der erste Schauer des schönen, überreichen Lebens, das gewißlich mit aller seiner geahnten und ungeahnten herrlichen Gewalt über uns kommen wird, wenn wir nur fröhlich standhalten. Wo wären wir denn aufgewacht von den sogenannten Träu-

men? Was hätte sich denn seitdem verändert? Aurora scheint noch so jung über die Berge wie damals, die Erde blüht alljährlich wieder bis ins fernste tiefste Tal – warum sollte denn unsere unsterbliche Seele, die alle den Plunder überdauert, allein alt werden? Was hindert denn zum Exempel dich, alle den Ballast von Vor-, Neben- und Rücksichten frisch wegzuwerfen und frei mit mir in das offene Meer zu stechen? – Reise mit, alter Kumpan!"

Walter faßte lächelnd die ihm dargebotene Rechte. „Was mich eigentlich zwischen diesen Bergen festhält", sagte er, „das sollst du künftig erfahren. – Doch – du magst immerhin lachen – das kann ich außerdem ehrlich sagen: es wäre mir schwer, ja gewissermaßen unmöglich, den einmal mit Ernst und Lust begonnenen Geschäften zu entsagen, die wie ein stiller, klarer Strom in tausend unscheinbaren Nebenarmen das Land befruchten und mich so von meiner stillen Stube aus in immer wechselndem, lebendigem Verkehr mit den entferntesten Gegenden verbinden."

Fortunat sah ihn nachdenklich an. „Du meinst es immer brav", sagte er nach einer Pause, „darum glaube ich dir, wo ich dich auch nicht recht verstehe. Aber in welchem greulichen Rumor lebt ihr Beamte dabei! Keiner hat Zeit zu lesen, zu denken, zu beten. Das nennt man Pflichttreue; als hätte der Mensch nicht auch die höhere Pflicht, sich auf Erden auszumausern und die schäbigen Flügel zu putzen zum letzten, großen Fluge nach dem Himmelreich, das eben auch nicht wie ein Wirtshaus an der breiten Landstraße liegt, sondern treu und ernstlich und mit ganzer, ungeteilter Seele erstürmt sein will. Ja, ich habe schon oft nachgedacht über den Grund dieser zärtlichen Liebe so vieler zum Staatsdienst. Hunger ist es nicht immer, noch seltener Durst nach Nützlichkeit. Ich fürchte, es ist bei den meisten der Reiz der Bequemlichkeit, ohne Ideen und sonderliche Anstrengungen gewaltig und mit großem Spektakel zu arbeiten, die Satisfaktion, fast alle Stunden etwas Rundes fertig zu machen, während die Kunst und die Wissenschaften auf Erden niemals fertig werden, ja in

alle Ewigkeit kein Ende absehen." „Da rührst du", entgegnete Walter, „an den wunden Fleck, wenigstens bei mir. Daß ich, aus Mangel an Zeit, zu beiden Seiten die schönen Fernen und Tiefen, die uns sonst so wunderbar anzogen, liegen lassen muß, das ist es, was mich oft heimlich kränkt, und was ich hier nicht einmal einem Freunde klagen kann. Dazu kommt die Abgelegenheit des kleinen Orts, wo alle Gelegenheit und aller Reiz fehlt, der neuesten Literatur zu folgen."

„Ist auch nicht nötig", versetzte Fortunat. „Was willst du jedem Phantasten in seine neumodischen Parkanlagen nachschreiten! Das rechte Alte ist ewig neu, und das *rechte* Neue schafft sich doch Bahn über alle Berge und – wie ich oben bemerkt – auch in diesen Gebirgskessel. Denn wenn ich nicht irre, sah ich vorhin bei dir neben dem Corpus juris die neuesten poetischen Werke des Grafen Victor stehen." „Nun", sagte Walter, „meinen großen Landsmann muß ich doch in Ehren halten, seine Heimat liegt ja kaum eine Tagereise von hier." – Fortunat sprang überrascht auf. „Da reit' ich hin", rief er, „den muß ich sehen." – Geduld", erwiderte Walter lächelnd, „er ist schon seit mehreren Jahren auf Reisen." „Und ich reite doch hin!" entgegnete Fortunat fröhlich, „wer einen Dichter recht verstehen will, muß seine Heimat kennen. Auf ihre stillen Pätze ist der Grundton gebannt, der dann durch alle seine Bücher wie ein unaussprechliches Heimweh fortklingt." Walter schien einem Anschlage nachzudenken. „Wohlan", sagte er endlich, „wenn du durchaus willst, so begleite ich dich, ich bin dort wohlbekannt, und wir bleiben dann um so länger beisammen. Ich muß dir nur gestehen, ich hatte mich eigentlich schon selbst darauf eingerichtet, in diesen Tagen hinzugehen. Hier kann ich dir nicht viel Ergötzliches bieten, und wenn's dir recht ist, so reisen wir morgen." – Fortunat schlug freudig ein.

Walter aber fing nun an, einige Lieblingsstellen aus Victors Werken zu rezitieren, was Fortunat immer störte, weil ein gutes Gedicht keine Stellen, sondern eben nur das

ganze gute Gedicht gibt, gleich wie eine abgeschlagene Nase oder ein paar abgerissene Ohren der Mediceischen Venus für Kenner recht gut, aber sonst ganz nichtswürdig sind.

„Du kennst doch Victors Werke? Du liebst ihn doch auch?" unterbrach sich endlich Walter selbst, da Fortunat schweigend ein Glas nach dem andern hinunterstürzte. – „Ich liebe ihn", sagte dieser, „wie ich ein nächtliches Gewitter liebe, das alles Grauen und alle Wunder in der Brust regt, ich kenne ihn, weil er von den geheimnisvollsten, innersten Gedanken meiner Seele, ja ich möchte sagen, von dem Waldesrauschen meiner Kindheit wunderbaren Klang gibt. – Friede dem großen dunklen Gemüt", fuhr er, sein Glas erhebend, fort, „und freudiges Begegnen mit ihm!"

Die Freunde hatten über dem lebhaften Gespräch gar nicht bemerkt, daß unterdes der Platz allmählich öde geworden war. In der wachsenden Stille hörte man nur noch eine Geige aus einiger Entfernung und dann das einförmige Stampfen von Tanzenden dazwischen herüberschallen. Beides klappte so wenig zusammen, und die Geige wurde so unaufhörlich und entsetzlich schnell gestrichen, daß Fortunat laut auflachte und ungeachtet Walters Einwendungen sogleich dem Tanzplatze zueilte. Der verworrene Klang kam aus einem niedrigen Häuschen, über dessen Tür ein Strohbüschel als Wahrzeichen eines Weinschanks im Nachtwinde hin und her baumelte. Walter war in anständiger Ferne stehengeblieben, während Fortunat durch das Fenster in die seltsame Tanzgrube hineinblickte. Ein langes dünnes Licht, das wie ein Peitschenstiel aus einem eisernen Leuchter hervorragte, warf ungewisse Scheine über das dunkle Gewölbe eines Kellers, an dessen Seitenwänden eingeschlafene Trinker über den langen plumpen Tischen umherlagen. In der Mitte tanzten eifrig mehrere Paare lustigen Gesindels, bald mit den zierlich gebogenen Armen wie zum Fliegen ausholend, bald in den auserlesensten Figuren und Windungen sich nähernd

und wieder trennend, bevor sie einander endlich zum Walzer umfaßten. Der dicke Weinschenk ging mit aufgestreiften Hemdärmeln dazwischen herum, ahmte mit dem Munde den Wachtelschlag nach, schnitt den vorübertanzenden Frauenzimmern lächerliche Gesichter oder wagte zuweilen selbst einen künstlichen Sprung. Am auffallendsten aber war der Musikant; ein anständig gekleidetes, lebhaftes Männchen mit einem scharfen, geistreichen Gesicht, emsig in den wunderlichsten Laufern die Geige spielend, während seine Augen mit unverkennbarem Wohlbehagen die Tanzenden verfolgten. Vergebens riefen diese ihm zu, sich zu moderieren, der Unaufhaltsame drehte mit wahrem Virtuosenwahnsinn die Töne, wie einen Kreisel, immer schneller und dichter, die Tanzenden gerieten endlich ganz außer Takt und Atem, es entstand ein allgemeines Wirren und Stoßen, bis zuletzt alle zornig auf den Musikus eindrangen. Dieser erhob sich nun und retirierte besonnen in künstlichen Fechtparaden nach der Tür, immerfort mit dem Fiedelbogen in den dicksten Haufen stoßend. So kam er glücklich auf die Straße heraus, die Schlafmütze des Wirts, die er im Getümmel aufgespießt, hoch auf seinem Bogen. Der lustige Wirt folgte schimpfend und vermehrte den Lärm von Zeit zu Zeit durch das Prasseln von Feuerwerk, das er täuschend mit dem Munde nachmachte.
Jetzt bemerkte der Musikus plötzlich die beiden Freunde auf der Gasse und sah sie mit seinen klugen Augen durchdringend an, während der Wirt, mit der einen Hand seine wilden Gäste in den Keller zurückdrängend, mit der andern ruhig die ihm zugeworfene Schlafmütze wieder auf den Kopf stülpte. Walter war einen Augenblick in Verlegenheit, ob und wie er den ihm unbekannten Fremden anreden sollte, und äußerte endlich seine Verwunderung über diese heillose Fertigkeit auf der Geige. – „Kleinigkeit! Kleinigkeit!“ erwiderte der Musikus; „nichts als Taranteln, womit ich die Leute in die Waden beiße und den St. Veitstanz erfinde.“ Mit diesen Worten empfahl er sich,

nahm die Geige unter den Arm und schlenderte, noch einigemal furchtsam nach dem Keller zurückblickend, rasch durch die Nacht über den Marktplatz fort.
Fortunat, der bisher kein Auge von ihm verwendet hatte, trat nun schnell auf den Wirt zu, um etwas Näheres über das wunderbare Männchen zu erfahren. „Ein Fremder", sagte der Wirt, „ein Partikulier, wie er sich nennt, mit dem ich schon manchen Verdruß gehabt habe. Er kommt zuweilen in die Stadt, aber immer nur gerade zu mir, und wenn ich reelle Gäste habe, die nach getaner Arbeit ihr Gläschen trinken und vernünftig diskurieren wollen, setzt er sich zu ihnen, und eh' ich's mich versehe, hat er Händel unter ihnen angestiftet und hat dann keine Courage, sie auszufechten. Wenn er recht vergnügt ist, zieht er gar seine verfluchte Geige hervor und spielt tolles Zeug auf. Hol' der Teufel alle Phantasten."
Hiermit kehrte der Wirt wieder in seine Höhle zurück, und die beiden Freunde bemerkten bei dem hellen Mondenschein, wie der unbekannte Musikus soeben zum Stadttor hinauswanderte. „Ein herrlicher Narr!" rief Fortunat aus, dem Wanderer noch immer nachsehend. „Laß die Fledermäuse", erwiderte Walter, „sie geraten uns sonst noch in die Haare. Komm nun nach Haus, es ist schon spät, und ich habe noch alle Hände voll zu tun für morgen."
Auf Walters Stube ging nun ein fröhliches Rumoren an. Die alte Aufwärterin wurde herbeigerufen, Befehle wurden erteilt, Briefe versiegelt und Akten und Wäsche gepackt, wobei Fortunat, in der Vorfreude der bevorstehenden, unerwarteten Fahrt, zur Verwunderung der Alten wütend half. Der weitgestirnte Himmel sah indes durch die offenen Fenster herein, der Brunnen rauschte vom einsamen Markte, während die Nachtigallen in den Gärten schlugen, und Fortunat war es dazwischen, als ginge draußen das Geigenspiel des wunderlichen Musikanten noch einmal fern über die stillen Höhen.

Zweites Kapitel

Bei dem schönsten Frühlingswetter zogen die beiden Freunde, auf ihren Pferden fröhlich von den alten Zeiten miteinander schwatzend, in das morgenrote Land hinein. Sie hatten den weitern, aber anmutigern Weg durch das Gebirge eingeschlagen, auf welchem sie *Hohenstein*, den Sitz des Grafen *Victor*, nach Walters Versicherung noch vor Nacht bequem erreichen konnten. Das Städtchen mit seiner grünen Stille lag schon weit hinter ihnen, ein frischer Wind ging durch alle Bäume, und Walter fühlte sich recht wie ein Vogel, der aus dem Käfig entflohen. Er war fast ausgelassen heiter, schwenkte den Hut in der Luft, und stimmte alte Studentenlieder an, so daß es den beiden Reitern vorkam, als wären sie nie getrennt gewesen und zögen nur eben wieder aus dem Tor von Heidelberg den grünen Bergen zu. In dieser Stimmung ließ er sich gern von dem unruhigen Fortunat verlocken, der bald dem fremden Schall eines unbekannten Gebirgsvogels folgte, bald mit den Hirten plauderte, dann wieder einen schönen Berggipfel oder eine reizend gelegene Ruine zu erklettern hatte. So waren sie lange aufs Geratewohl umhergeschweift, als Walter endlich zu seinem Schrecken bemerkte, daß schon die Abendsonne schräg durch den Wald funkelte. Jetzt fand er auch, daß sie alle Richtung verloren hatten, er wußte nicht, wo er war. Vergebens schlug er den ersten besten Pfad ein, die Wege teilten sich bald von neuem wieder, kein Dorf war ringsumher zu sehen, je tiefer sie in den Wald kamen, je ungeduldiger wurde er; er wollte durchaus noch heut nach Hohenstein. Unterdes war die Nacht völlig hereingebrochen, sie mußten absteigen, und ihre Pferde hinter sich herführen, da der Holzweg sich nach und nach in einen verwachsenen Fußsteig verlor.

Walter war verdrießlich und sprach wenig. Fortunat aber wurde immer vergnügter, je weiter sie fortschritten, und blickte recht mit frischem Herzen in die wunderbaren

Mondlichter und die rätselhaften Abgründe, an denen sie vorüberzogen. Oft hielten sie horchend still, denn es war ihnen, als hörten sie aus weiter Ferne Hunde bellen, und den dumpfen Takt eines Pochhammers dazwischen; aber das einförmige Rauschen der Wälder verschlang immer alles wieder.

Walter schwor endlich, nicht einen Schritt mehr weiterzugehen, er band sein Pferd an und setzte sich maulend daneben. Fortunat hatte sich gleichfalls auf den Rasen hingestreckt, während sein Gefährte nun allerlei Reden über unzeitige Romantik und verlorne Zeit verlauten ließ. Fortunat antwortete nicht darauf, und da es gar nicht enden wollte, zog er seinen Mantel über den Kopf und schlummerte bald vor Ermüdung ein.

Als er wieder aufwachte, war Walter unterdes vor Ärger fest eingeschlafen. Er sah freudig rings um sich her, die tiefe Einsamkeit, die unbekannte Gegend, der Schlafende, und die Pferde im Mondschein, alles war ihm so neu und wunderbar; er ging unter den Bäumen auf und nieder und sang halb für sich:

„Wie schön, hier zu verträumen
Die Nacht im stillen Wald,
Wenn in den dunklen Bäumen
Das alte Märchen hallt.

Die Berg' im Mondesschimmer
Wie in Gedanken stehn,
Und durch verworrne Trümmer
Die Quellen klagend gehn.

Denn müd' ging auf den Matten
Die Schönheit nun zur Ruh,
Es deckt mit kühlen Schatten
Die Nacht das Liebchen zu.

Das ist das irre Klagen
In stiller Waldespracht,

Die Nachtigallen schlagen
Von ihr die ganze Nacht.

Die Stern' gehn auf und nieder –
Wann kommst du, Morgenwind,
Und hebst die Schatten wieder
Von dem verträumten Kind?

Schon rührt sich's in den Bäumen,
Die Lerche weckt sie bald –
So will ich treu verträumen
Die Nacht im stillen Wald."

Und wie er aufblickte, hörte er wirklich schon den Klang einer früh erwachten Lerche durch den Himmel schweifen. „Frisch auf!" rief er fröhlich Waltern zu, „frisch auf, ich wittre Morgenluft!" Walter erhob sich taumelnd und konnte sich lange nicht in dem wunderlichen Schlafsaal zurechtfinden. Der kurze Schlummer hatte ihn neu gestärkt und verwandelt, er schämte sich seines gestrigen Mißmuts, und bald saßen die beiden Freunde wieder rüstig zu Pferde, um womöglich noch vor Tagesanbruch aus dem Labyrinth der Wälder herauszukommen.
Nach einem kurzen Ritt hatten sie die Freude, unerwartet wieder einen ordentlichen Weg zu erreichen. „Land! Land!" rief endlich Walter vergnügt aus, „dorthin zu liegt Hohenstein!" – Sie verdoppelten nun ihre Eile und gelangten bald völlig aus dem Walde in das weite, geheimnisvolle Land hinaus. Immer tiefer und freudiger stiegen sie von den Bergen in das Blütenmeer, schon hörten sie von fern eine Turmuhr schlagen, zahllose Nachtigallen schlugen überall in den Gärten. Am Ausgang des Gebirges schien ein großes Dorf zu liegen, zerstreute Hügel, dunkle Baumgruppen und ein hohes prächtiges Schloß hoben sich nach und nach aus der verworrenen Dämmerung, alles noch unkenntlich und rätselhaft, wie in Träumen. So waren sie in eine hohe Kastanienallee gekommen, als Walter plötzlich an einem

zierlichen Gittertor stillhielt. „Sie schlafen noch alle", sagte er, „wir wollen indes hier in den gräflichen Garten gehen und die Erwachenden überraschen."
Sie banden nun ihre Pferde an den Zaun und schwangen sich von den steinernen Sphinxen, die den Eingang bewachten, über das Gitter in den Garten hinein. Da war noch alles still und duftig, einzelne Marmorbilder tauchten eben erst aus den lauen Wellen der Nacht empor. Das alte, finstere Schloß im Hintergrunde mit seinen dichtgeschlossenen Jalousien stand wie eine Gewitterwolke über einem freundlichen Nebengebäude, von dem man vor lauter Weinlaub fast nur das rote Ziegeldach sah. Unter den hohen Bäumen vor dem letzteren fanden sie einen Tisch und mehrere Stühle, als wären sie eben erst von einer Gesellschaft verlassen worden. – „Da hat sie schon wieder ihre Gitarre draußen vergessen", sagte Walter kopfschüttelnd. – „Wer denn?" fragte Fortunat – „die schöne Amtmannstochter, von der du mir erzählt hast?" – „Ja, Florentine", erwiderte Walter; „das ist des Amtmanns Wohnung, und dort oben nach dem Garten hinaus ihre Schlafstube." – „Du weißt hier gut Bescheid", entgegnete Fortunat. – Walter wurde rot und schwieg verlegen. Fortunat aber ergriff ohne weiteres die auf dem Tische liegende Gitarre, stellte sich vor das bezeichnete Fenster und sang:

„Zwei Musikanten ziehn daher
Vom Wald aus weiter Ferne,
Der eine ist verliebt gar sehr,
Der andre wär' es gerne."

„Ich bitte dich", unterbrach ihn Walter, „was singst du da für dummes Zeug!" – „Wart' nur, 's kommt gleich klüger", erwiderte Fortunat und sang weiter:

„Die stehn allhier im kalten Wind
Und singen schön und geigen:
Ob nicht ein süßverträumtes Kind
Am Fenster sich wollt' zeigen?"

Sein Wunsch ging wirklich in Erfüllung. Ein schönes Mädchen, noch ganz verschlafen, wie es schien, fuhr oben ans Fenster, schüttelte die Locken aus dem Gesichtchen und sah neugierig mit großen, frischen Augen durch die Scheiben. Als sie aber unten einen unbekannten, wohlgekleideten Mann erblickte, war sie ebenso schnell wieder verschwunden. – Walter wurde nun in der Tat unwillig, Fortunat aber griff immer lustiger in die Saiten und sang wieder:

„Mein Herz ist recht von Diamant,
Ein Blum' von Edelsteinen,
Die funkelt fröhlich übers Land
In tausend bunten Scheinen!

Und durch das Fenster steigen ein
Waldrauschen und Gesänge,
Da bricht der Sänger mit herein
Im seligen Gedränge."

Unterdes war es im Hause nach und nach lebendig geworden, Türen gingen auf und zu, im Innern hörte man dazwischen das kräftige Lachen eines Mannes, das immer näher zu kommen schien. Endlich wurde die Haustür von innen geöffnet, und, mit einer langen Pfeife im Munde, stand ein schon völlig angekleideter, großer, starker Mann vor ihnen, dessen gebräuntes, lebenslustiges Gesicht von der Morgensonne hell beschienen wurde. Es war der Amtmann selbst. Er war voller Freude, Walter so unerwartet wiederzusehen, und konnte gar nicht aufhören, über das lustige Ständchen zu lachen, durch das sich Fortunat sogleich in seine entschiedene Gunst gesetzt zu haben schien. Mit schallender Stimme rief er nun alles im Hause wach, es mußten eilig Kaffee und Pfeifen ins Freie herausgebracht werden, sie lagerten sich um den Tisch auf dem grünen Platze vor der Tür, den die beiden Gäste noch vor kurzem so einsam gesehen hatten, und

Walter mußte ausführlich ihre nächtlichen Irrfahrten vortragen.
Unterdes war auch die Frau Amtmann dazugekommen. Sie hatte sich vor dem unbekannten Gaste sorgfältig und beinahe festlich angetan und empfing Fortunat mit umständlicher, wortreicher Feierlichkeit. Fortunat, dem bei solcher Gelegenheit unwillkürlich alle Bewillkommnungskomplimente einfielen, die er in seinem ganzen Leben gehört oder auch nicht gehört hatte, konnte nicht widerstehen, mit einem unerschöpflichen Schwalle der auserlesensten Redensarten zu entgegnen, und erweckte dadurch bei der Dame eine nicht geringe Meinung von sich und seiner feinen Lebensart.
„Das ist heute ein rechter Freudentag!" sagte der Amtmann, „da soll es auch einmal hoch hergehen." Er erzählte nun, wie sie heut gegen Abend auch noch ihren jungen Neffen *Otto* hier erwarteten, der von der fernen Universität zurückkehrte, um sich zu seiner Anstellung vorzubereiten. Die Amtmännin ließ mit zufriedener Miene noch einfließen, daß Otto, der Sohn ihrer verstorbenen Schwester, aus Herrn Walters Städtchen sei, daß er schon auf der Schule immer für den Stillsten und Geschicktesten galt und nun ein wahrer Gelehrter geworden sei.
Fortunat bemerkte während dieses Gesprächs, daß sich Walter unterdes verloren hatte. Der Garten, der nun in voller Morgenpracht herüberfunkelte, lockte auch ihn schon lange, und er sagte endlich dem Amtmanne, wie er Walter vorzüglich in der Absicht hierherbegleitet habe, um die Heimat des berühmten Grafen Victor einmal in der Nähe zu sehen. Der Amtmann lächelte. „Ich weiß nicht", sagte er, „ob Sie auch solcher Meinung sind, aber wenn die andern von dem berühmten, gelehrten Grafen sprechen, denken sie sich ihn immer mit der Zipfelperücke, wie den Hilmar Curas vor seiner Grammatik. Das kann mich immer ärgern. Was da Gelehrter! Zu Pferde muß man den großen Victor sehen, im Walde auf der Jagd, auf dem Felsen, wo allen andern schwindelt – mit einem Worte: Das

ist ein rechter Mann! Das Berühmtsein und Versemachen ist nur so Lumpenzeug daneben, wie eine Schabracke auf einem schönen Roß, und er gibt selber nichts darauf. Doch wir sprechen ein andermal mehr davon." – Er stand nun auf und beschrieb Fortunat die Gänge, die er im Garten einschlagen sollte, um zu den schönsten Punkten zu gelangen, da ihn selbst die Wirtschaftsanordnungen für den anbrechenden Tag in das Haus hineinriefen.

Fortunat wandte sich nun allein in den Garten, wo er zu seinem Erstaunen ringsumher nur architektonische Formen altmodischer Gänge, hohe, feierliche Buchenalleen, Springbrunnen und künstliche Blumenbeete erblickte, von denen dunkelglühende Päonien und prächtige Kaiserkronen glänzten. Es war, als hätte ein wunderbarer Zauberer über Nacht seine bunten Signaturen über das Grün gezogen und säße nun selber eingeschlummert in dem Labyrinth beim Rauschen der Wasserkünste und träumte von der alten Zeit, die er in seine stillen Kreise gebannt.

Schon waren Schloß und Amtmannswohnung hinter Fortunat versunken, als er plötzlich einen wohlgekleideten jungen Mann bemerkte, der an den Marmorstufen eines einsamen Gartenhauses eingeschlafen war. Er wollte umkehren, aber der Schläfer, von dem Geräusch erweckt, fuhr soeben rasch auf, blickte verworren ringsumher und fragte Fortunat, wer er sei. Dieser erzählte nun sein nächtliches Abenteuer und seinen langgehegten Wunsch, diese Gegend einmal zum Angedenken des Dichtergrafen Victor zu durchstreifen. – „Vortrefflich", erwiderte der andere, „so will ich Sie sogleich herumführen!" – „Kennen Sie den Grafen Victor?" fragte Fortunat. – „Nicht sonderlich", erwiderte jener, „doch weiß ich eben genug von ihm, um Ihnen hier überall genügende Auskunft zu geben."

Fortunat nahm das unerwartete Anerbieten dankbar an und betrachtete, als sie nun miteinander weitergingen, mit freudiger Überraschung das schöne, aber etwas bleiche und wüste Gesicht des Unbekannten, über das die Mor-

genlichter durch das Laub wunderlich wechselnde Scheine warfen. Er äußerte endlich seine Verwunderung über die, wie es schien, absichtlich und sehr sorgfältig festgehaltene Altmodigkeit des Gartens. – „Der Graf", entgegnete sein Begleiter, „will es so haben. Buchsbaumene Kindlichkeit! Wie es in seiner Kindheit gewesen, so soll es hier ferner verbleiben, selbst dieselben Blumen müssen jährlich an denselben Plätzen wieder gepflanzt werden wie damals." – „Er hat recht", sagte Fortunat, „was soll ein Garten, wenn er nicht ein Gedicht von ganz bestimmtem Klange ist! In diesem einförmigen Plätschern der Wasserkünste, in dieser geisterhaften Symmetrie der Laubwände und stummen Marmorbilder ist eine Wehmut, die einen wahnsinnig machen könnte."

Jetzt standen sie an dem Abhange des Berges, dessen obere Fläche das Schloß und der eigentliche Ziergarten einnahmen. Von der mit Efeu umrankten Felswand sah man hier plötzlich in tiefe Schluchten und Wiesenplätze hinab, wo in kühlen Schatten uralter Bäume Rehe und Damhirsche weideten, die scheu die Köpfe nach ihnen emporhoben und dann pfeilschnell im tieferen Dunkel verschwanden. – „Sehen Sie da", rief Fortunats Begleiter aus, „das Großartige und Kühne dieser Komposition. Ich betrete diesen Ort nie ohne Ehrfurcht vor dem seltenen Genius dieses Dichtergrafen – oder sagen wir es nur lieber grad heraus Dichterkönigs! Besonders muß ich Sie hier auf jene leicht geschwungenen Brücken aufmerksam machen. Sie führen, wie Sie sehen, über die Wipfel der Bäume hinweg nach einzeln stehenden, hohen, abgerissenen Felsen hinüber, die, mit ihren bunten Gärtchen auf den Wipfeln, wie funkelnde Blumenzinnen über Waldeseinsamkeit emporragen. Diesen Einfall hat der liebenswürdige Graf vor dem lieben Gott voraus, er legte diese hängenden Gärten an; das waren die Blocksberge seiner Phantasie. Hier pflegte er als Knabe, wenn ein Gewitter heraufzog und im Schlosse alles ängstlich durcheinanderlief, vor der unermeßlichen Aussicht zu sitzen, mit den Beinen über dem Ab-

grunde baumelnd, bis ihm die ersten dicken Regentropfen an die seidenen Strümpfe klatschten." – „Es freut mich" – erwiderte Fortunat, der, ganz in den Anblick des wunderbaren Grundes versunken, die letzten Worte fast überhört hatte – „es freut mich recht, daß Sie Victors poetische Erscheinung so hochhalten."

Der Begleiter sah ihn aus den schönen Augen scharf und zweifelhaft an. – „Ich bedaure ihn aufrichtig", sagte er dann, „denn ich halte die Anstellung als Genie für eine der epinösesten in der Welt. Ein anderer stopft sich seine Pfeife, zieht seinen Schlafrock an, setzt sich auf dem Schreibesel zurecht, und macht seine Arbeiten ab, und geht dann zufrieden in die Ressource, wo er wieder ganz Mensch sein kann. Aber so ein Genie, zumal ein Dichter, kann das Genie gar nicht loswerden; wie ein Spaziergänger, der im Herbst über Feld gegangen, schleppt er die Sonnenfäden seiner Träume an Hut und Ärmeln bis auf die Ressource nach. Ist dort gar das Fenster offen, so sind die Nachtigallen und Lerchen draußen recht wie versessen auf ihn und rufen ihn ordentlich bei Namen, ja zuweilen spielt ihm seine kaum halbfertig gedichtete Geliebte den fatalen Streich und blickt ihn plötzlich aus den Augen irgendeiner albernen Dame an." – Hier stand er plötzlich selber überrascht still. Sie waren in das Felsental hinabgestiegen und an einen einsamen Weiher gelangt, in dessen Mitte sich eine, wie es schien, unzugängliche Insel im frischen Schmuck des Morgentaues spiegelte. Spuren ehemaliger Gänge und Blumenplätze waren von hohem Grase und Unkraut überwachsen, fremde Blütengewächse schlangen sich an den Baumstämmen empor, nur einzelne hohe Blumen funkelten noch hier und da aus der bunten Verwilderung, in der unzählige Vögel sangen. „Das war sonst Victors Lieblingsplatz", sagte der Fremde nach einem Weilchen, „hier hat er den Namen seines ersten Liebchens in die Bäume geschnitten. Das Mädchen ist tot, der Nachen zu der Insel lange zertrümmert und versenkt, und Wipfel und Zweige, Unkraut und Blüten schlingen sich drüben

verwildert durcheinander und können doch nicht in den Himmel wachsen." – Ein seltsames Leuchten flog bei diesen Worten über sein geistreiches Gesicht. Dann auf einmal zu Fortunat gewandt, sagte er: „Aber Sie sind am Ende selbst der Graf Victor – leugnen Sie nur nicht!" – Fortunat brach in lautes Lachen aus und bat den Unbekannten, der ihm wohlbehagte, zu wechselseitiger näherer Bekanntschaft sogleich mit zum Amtmanne hinaufzukommen. Der Fremde besann sich einen Augenblick und fragte dann, ob noch mehrere Gäste dort wären. Da er hörte, daß auch Walter droben sei, entschuldigte er sich, er habe zu lange am Brunnen geschlafen und müsse nun schnell wieder weiter. – „Sind Sie denn nicht hier aus dem Hause?" fragte Fortunat erstaunt. – Aber jener eilte schon fort, winkte noch einmal mit dem Hute und war bald zwischen den Bäumen verschwunden.

Drittes Kapitel

Als Fortunat wieder die Anhöhe erreichte, traute er seinen Augen kaum. Der schönste Morgenglanz blitzte jetzt über die gezirkelten Rasenfiguren und Tulpenbeete, an den Statuen hingen Mieder, Poschen und Schleier umher, ein frischer Wind ging durch den Garten, und ließ, die Zweige teilend, bald ein Paar bloße Mädchenarme, bald ein ganzes zierliches Bildchen flüchtig erblicken. Und so glich der Garten mit den bunten Tüchern, die wie Frühlingsfahnen von den Büschen flatterten, mit den funkelnden Strahlen der Wasserkünste und dem heiteren Sonnenhimmel darüber, auf einmal jenen alten Landschaften, wo alle Hecken von schwärmenden Nymphen wunderbar belebt sind. Erstaunt drang er weiter vor, da sah er eine junge Dame in wunderlichem Schmuck, mit Reifrock, Mieder und gesticktem Fächer, vor einem Springbrunnen stehen, sie bespiegelte sich, fröhlich plaudernd, im Wasser, schüt-

telte lachend die schweren blitzenden Ohrgehänge und sah wieder hinein. Auf einmal wandte sie sich, er glaubte in dem frischen Gesichtchen Florentine, die Amtmannstochter, zu erkennen, die er vorhin am Fenster gesehen. Aber nun erschallte ein lauter Schrei, und aus allen Hekken, in Taft und Seide rauschend, fuhren erschrocken fliehende Mädchengestalten durchs Grüne, als hätte der Wind Aprikosenblüten umhergestreut.
Fortunat folgte ihnen zu der Amtmannswohnung, wo sie verschlüpft waren. Aber hier hielt ihn neue Verwirrung fest, er fand auch dort alles in lebhafter Bewegung. Aus dem Mörserstampfen im Hause und dem ernstwichtigen Durcheinanderrennen der Mägde, zwischen dem man von Zeit zu Zeit die Kommandostimme der Amtmännin vernahm, schloß er sogleich auf ein großes Kuchenbacken im Innern. Draußen aber auf dem Rasen sah man große Teppiche ausbreiten, Sofas und Polsterstühle ausklopfen, überall wurden die verdunkelnden Doppelfenster ausgehoben, die Morgensonne schien lustig durch das ganze Haus, und einzelne Schwalben kreuzten jauchzend über dem Platze.
Ein langer, hagerer Mann mit dünnem Hals und hervorstehenden Augen schien besonders selig in dem Rumor, man sah ihn überall im dicksten Haufen schreiend, helfend und anordnend. Von diesem erfuhr Fortunat endlich, nicht ohne Müh' und wiederholte Fragen, daß die Pachterstöchter aus der Nachbarschaft angekommen und mit Florentinen im Garten den alten gräflichen Hofstaat anprobiert hätten, und daß alle diese Anstalten auf den feierlichen Empfang des heute erwarteten Studenten Otto zielten, der nach den eingelaufenen Nachrichten früher hier eintreffen könnte, als man anfangs glaubte. Der Mann aber war der Förster des Orts, der früher selbst das Gymnasium frequentiert und seitdem eine wütende Vorliebe für Studenten hatte. – Fortunat war diese unverhoffte Wirtschaft ein willkommenes Fest. Er mischte sich ohne Verzug in das bunte Getümmel, um den Lärm womöglich

noch größer zu machen. Dem Förster stellte er vor, wie unerläßlich es sei, den Gefeierten durch ein Triumphtor einzuführen, worauf beide sogleich voll Eifer forteilten, um die nötigen Materialien zu dem neuen Werke herbeizuschaffen. Unterwegs begegneten sie Walter, der soeben mit einem Buche in den Garten ging. „Ich muß mich ein wenig sammeln", sagte er flüchtig zu Fortunat, „ich freute mich so auf den stillen Tag im Freien, und nun bricht aller Plunder herein; es ist mir einmal nicht gegeben, mit den Leuten über nichts zu schwatzen, es ist unleidlich!"
Inzwischen verzögerte sich Ottos Ankunft von Stunde zu Stunde. Walter hatte nicht lange gelesen, sondern revidierte in seiner praktischen Lust mit dem Amtmann die Höfe, Scheunen und Ställe. Im Garten wurden die Vögel schon still, Florentine und ihre jungen Freundinnen, wieder bequem in ihren gewöhnlichen Kleidern, flüchteten vor der steigenden Sonne aus einem Schatten zum andern, die immer kürzer wurden; jede hatte ein Stück frischen Kuchen in der Hand, sie wußten nicht, was sie in der Hitze anfangen sollten mit der langen Zeit. Auch ein junger Wirtschaftsschreiber mit Sporen und neuem Frack hatte sich eingefunden. Er trug den Mädchen die Tücher nach, focht mit seiner Reitgerte galant in die Luft und wußte durch Schnalzen auf Lindenblättern und andere artige Kunststücke sich bei den Frauenzimmern angenehm zu machen.
Plötzlich versetzte der Knall eines Böllers alles in die größte Verwirrung, aus allen Hecken und Türen stürzten die Erwartenden nach der Richtung hin, wo die Explosion erfolgt war. Dort gewahrten sie schon von fern den Förster am Abhange des Gartenberges, wie er soeben durch ein altes Perspektiv, das er wütend immer länger und länger hervorschob, in die Gegend hinausblickte. Als die andern endlich atemlos und fragend anlangten, warf er auf einmal das Fernrohr fort, ergriff eine neben ihm stehende Lunte und löste zum Schrecken der lautschreienden Damen einen zweiten Böller. Und in der Tat, in demselben

Augenblick wurde durch den sich teilenden Pulverdampf zwischen den Kornfeldern am blaugewundenen Strom im Tal ein Reiter in bunter studentischer Tracht sichtbar, der nun auch seinerseits die Harrenden auf dem Berge erblickte, und, freudig seinen Hut schwenkend, die Sporen einsetzte. „Otto! Otto!" rief alles fröhlich durcheinander und winkte ihm mit den Schnupftüchern entgegen. Der Reiter hatte unterdes den Fuß des Berges erreicht, schwang sich vom Pferde, und auf dem nächsten Wege zwischen den grünen Rebengeländern aufsteigend, erschien ein schöner Jüngling von etwas kleiner, zierlich schlanker Gestalt mit einem feinen Gesicht und fast träumerischen Augen.

Aber am Eingang zur ersten Allee wurde er plötzlich durch eine seltsame Erscheinung aufgehalten. Ein schöner Tannenbaum stand dort am Abhang von alters her, wie ein dunkler Ritter auf der Wacht, und ragte mit dem Wipfel bis über die Anhöhe hinauf. Auf einmal rauschte er mit den grünen Kronen und zeigte sein Riesenhaupt mit rotbraunem Gesicht und langem Schilfbart, das Haar phantastisch von wilden Blumen und Eichenlaub umkränzt. „Salve!" redete das Haupt, die Augen sichtbar bewegend, den erstaunten Studenten an:

„Salve! Herr Doktor oder Magister!
Bin ein alter Bursch' und hass' die Philister,
Bin der Waldmann aus dem Gebirge hier,
Darf nicht näher treten zu dir,
Kann nicht zu dir kommen in Haus und Zimmer,
Trät' dort alle den Plunder in Trümmer,
Drum schau ich über die Wipfel hier hinaus;
Und bist du der Alte noch immer,
So lad' ich dich wieder in mein grünes Haus!
Da gehn, wie damals, noch mit Gefunkel
Die Quellen verworren durchs kühle Dunkel,
Waldhornsklänge und Vögelschall,
Von fern dazwischen der Wasserfall,

Und über uns rauschend die Buchen und Fichten
Erzählen dir wieder die alten Geschichten. –
Doch hast du über Pandekten und Latein
Seitdem vergessen die Sprache mein,
So magst du über deinem Buche hocken und lesen!
Das meine ist doch gescheiter gewesen!
Dann halt' ich auf ewig meinen großen Mund,
Wir sehen uns nimmermehr wieder – und –"

Und – hier blieb der Gebirgsgeist plötzlich stecken, man hörte eine andere Stimme immer lauter, aber vergeblich soufflieren. Darüber geriet das Haupt nach und nach ins Wackeln, auf einmal kollerte es zwischen den Zweigen auf die Anhöhe herunter und prasselnd hinterdrein der Förster und Fortunat zu großem Gelächter und Ergötzen der Umstehenden.
Otto stürzte dem schimpfenden Waldmann herzlich in die Arme, dann sah er mit den schönen Augen Fortunat nachdenklich an. „Gott weiß es", sagte er, „ich verstehe die Waldessprache noch immer, und was ich auch seitdem hinzugelernt habe, sie ist und bleibt doch meine rechte Muttersprache!" – Nun bemerkte er erst die andern in der Allee und fiel jubelnd dem Amtmann und seiner Frau und endlich auch den Mädchen in die Runde um den Hals, die errötend und verlegen sich des Ungestümen nicht erwehren konnten. Aber kein Mensch konnte zu Worte kommen, denn der unermüdliche Förster, der in seinem Eifer gar keine Notiz von der Rührung nahm, hatte insgeheim Pauken und Trompeten herbestellt, die jetzt furchtbar in die Ohren der Damen schmetterten, Böller auf Böller wurde dazwischen gelöst, er selbst aber rührte sehr künstlich die Pauken, auf die er zuletzt hinaufsprang und, Schlägel und Hut hoch über sich in die Luft werfend, unaufhörlich Hurra schrie. Der Amtmann wurde ganz zornig in dem Lärm, auch Otto schien verlegen und gestört. Da war der tolle Förster endlich mit senem Empfange fertig geworden, und noch ganz erhitzt von dem pappenen

Riesenkopfe, in dem er vorhin gesteckt, führte er nun mit einer wunderlichen, ungelenken Grandezza die fremden Mädchen nach der Amtmannswohnung hin.

Hier unter den Bäumen standen auf einer altmodischen Kaffeeserviette, in welche verschiedene Städte und Hirschjagden rot gewirkt waren, unzählige kleine chinesische Tassen aufgepflanzt, ein ungeheurer Kaffeekrug dampfte einladend dazwischen, die junge Dienstmagd im Sonntagsputz brachte eine Schüssel mit den in Kuchen gebackenen Namenszügen Ottos herbei, und küßte dem neu angekommenen jungen Herrn hocherrötend die Hand. Der Förster, der alte Junggesell, war inzwischen in den vollen Redestrom seiner Feiertagslaune geraten und brachte alle seine alten Jagdspäße und lateinischen Brokken wieder aufs Tapet, worüber die Pachterstöchter, die ihn insgeheim für einen gewandten Weltmann und Gelehrten hielten, jedesmal in ein unmäßiges Lachen ausbrachen. Bald aber nahm Otto die Aufmerksamkeit ausschließlich in Anspruch, noch in der vollen Heimatsfreude des ersten Wiedersehens erzählte er von seinem Studentenleben in Halle, er sprach so frisch, und als nun gar der Amtmann die funkelnden Weinflaschen auf den Tisch setzte, glitten alle Gedanken fröhlich mit dem bunten Studentenschifflein am Giebichenstein und den blühenden Kirschgärten die Saale hinab in das gelobte Land der Jugend.

So war unvermerkt der Abend herangekommen, der Förster und die Mädchen hatten sich heimlich ins Haus geschlichen, Otto erzählte noch immer, als plötzlich die Tür sich weit auftat, und bei dem Geschwirr einer Geige ein ganzer Hofstaat von Damen und Herren in Reifröcken, Haarbeuteln und altfranzösischen Fräcken sich rauschend herausbewegte. Man erkannte sogleich den Förster unter ihnen, er führte feierlich die jungen Leute vom Tisch den verlegen knicksenden Damen auf, die Geige schwirrte von neuem, und so entspann sich unversehens ein Tanz auf dem Rasen. Walter wollt' es gar nicht gelingen, er wurde

immer verlegener, je mehr die andern über ihn lachten, auch die beiden Pachterstöchter konnten sich in ihren Staat nicht finden, in dem sie sich, wie in einem Gehäuse, nur schwerfällig bewegten und alle Augenblicke verwikkelten. Jeder sprang, so gut er konnte, und als nun vom Schwung der Reifröcke die Lichter verlöschend flackerten, ergriff der Wirbel endlich auch die Alten am Weintisch, der Förster führte die sich vergebens sträubende Amtmann zu einer Sarabande, jeder der übrigen wählte gleichfalls seine Dame, und es entstand eine wundersame, künstliche Verschlingung, wobei der Förster durch kühne Schwenkungen alles in Erstaunen setzte.

Auf einmal fuhr Florentine aus dem leuchtenden Kreise wie eine Sternschnuppe in den finstern Garten hinaus. Ihre Brust flog über dem knappen seidenen Mieder, sie atmete erschöpft in der kühlen Nachtluft, dabei blickte sie immerfort nach den Bäumen zurück, als erwartete sie noch jemand. Fortunat bemerkte sie, ihn hatte unter den abenteuerlichen Gestalten nach und nach die Hofluft der alten Zeit unwiderstehlich ergriffen, er folgte rasch dem Mädchen nach, faßte sie zierlich an den äußersten Fingerspitzen und promenierte so feierlich mit ihr auf den geschnörkelten Gängen. Sie ließ ihm lachend die Finger, sah aber immer ungeduldiger zurück. So waren sie in galantem Diskurs an eine einsame Grotte gekommen, noch ein Überbleibsel jenes grillenhaften Schmuckes altmodischer Gärten. Bunte Muscheln blitzten im Mondschein von Decke und Wänden, ausgestopfte Reiher und Wasservögel standen mit weitaufgesperrten Schnäbeln auf Kristallriffen umher. – „Süßer Gott der Liebe", sagte Fortunat, „das ist recht eine Grotte zum Schnäbeln, o wären wir doch jetzt zwei Turteltäubchen!" – Sie sah ihn einen Augenblick verschmitzt an, dann drehte sie leise einen verborgenen Kran, auf einmal spritzten alle Schnäbel funkelnde Wasserstrahlen gerade auf Fortunat, und eh' er sich noch besinnen konnte, war seine wilde Taube in dem Sprühregen verflogen.

Er schüttelte sich lachend ab, und als er zu der Gesellschaft zurückkam, stand Florentine schon wieder am Tische vor der Mutter, die ihr besorglich die Locken aus der heißen Stirn strich. Sie hatte die langen Augenwimpern tief gesenkt, denn es tat ihr nun heimlich leid um Fortunats neuen Frack, die flackernden Lichter spielten auf ihrem Gesicht und dem glitzernden Mieder, so sah sie in den rauschenden Wogen von Taft und bunten Schleifen wie ein Elfchen aus, das aus einer Tulpe guckt. – Walter sah sie lange unverwandt an, dann faßte er Fortunat unter dem Arm und führte ihn rasch in den Garten. „Ist sie nicht wunderschön? o wie bin ich doch glücklich!" rief er aus und erzählte nun dem Freunde, daß er seit längerer Zeit mit Florentine verlobt sei, daß sie auf den Rat der Eltern nur noch eine bevorstehende Gehaltserhöhung Walters abwarteten und dann in dem Städtchen Haus und Garten mit der Aussicht auf Hohenstein kaufen und dort im Grünen sich für die ganze Lebenszeit miteinander einrichten wollten. –

Kaum eine Stunde darauf aber war alles verklungen, aus den Tälern schallte das Zirpen der Heimchen herauf, man hörte nur noch die Kalesche der Pachterstöchter auf dem steinigen Wege durch die Nacht fortrumpeln, in der Ferne zerplatzten einige Leuchtkugeln, die der unermüdliche Förster noch aus seinem Gärtchen warf. – O glückselige, bangsame Einsamkeit, dachte Fortunat, wer es wie Walter über sich gewönne, sich ganz darin zu versenken!

Viertes Kapitel

Schöne, stille Zeit, du liebste Heimatsgegend mit deinen frischen Morgen und mittagschwülen Tälern, und ihr rüstigen, nun nach allen Weltgegenden hin zerstreuten Jugendgesellen, die damals von den Bergen so ernst und fröhlich mit mir in das Leben hinausgesehen – ich grüß'

euch alle aus Herzensgrund! Denn alles wird mir wieder lebendig hier auf den kühlen Waldbergen, wie ich den Amtmann zwischen den Kornfeldern wandern sehe und Florentine bald oben am Fenster, beim ersten Morgenlichte singend und ihre Haare flechtend und sich strekkend und putzend um die Wette mit den erwachenden Vögeln in den Bäumen vor dem Hause, bald wieder im Garten über einer französischen Grammaire eingeschlafen, die Walter ihr gegeben, um sich für das Stadtleben auszubilden. Vor allen aber hat Fortunat, der seine Abreise von einem Tage zum andern verschiebt, sich behaglich im Garten eingerichtet. Im Grün zwischen hohen Blumen, die weite Landschaft unter sich, und über ihm die rauschenden Wipfel, setzt er sich jeden Morgen mit dem Schreibzeug an dem steinernen Fußgestell eines etwas verwitterten Apollos zurecht, um einige Novellen, die er in glücklichen Reisestunden auf seinem Pferde ersonnen, endlich einmal recht in Ruhe zu Papier zu bringen. Aber da geht es ihm wunderlich. Der lustige Morgenwind wirft ihm die Blätter ins Gras, wo sich die Hühner drum raufen, hinter ihm aber stimmen die Wipfel ihr uraltes Lied wieder an, das in keine Novelle paßt, die Waldvögel singen ganz fremde Noten dazwischen und Wolken fliegen über das Land und rufen ihm zu: Menschenkind, sei doch kein Narr! Und zog dann gar der Förster unten zur Jagd und schwenkte seinen Hut und rief Hurra hinauf, da warf er gewiß Feder und Papier fort und schwang sich auf seinem Pferde mit in den frischen, glänzenden Morgen hinaus. – Auf einem solchen Morgenritte tröstete er sich einmal mit folgendem Liedchen:

„Ich wollt' im Walde dichten
Ein Heldenlied voll Pracht,
Verwickelte Geschichten,
Recht sinnreich ausgedacht.
Da rauschten Bäume, sprangen
Vom Fels die Bäche drein,

Und tausend Stimmen klangen
Verwirrend aus und ein.
Und manches Jauchzen schallen
Ließ ich aus frischer Brust,
Doch aus den Helden allen
Ward nichts vor tiefer Lust.

Kehr ich zur Stadt erst wieder
Aus Feld und Wäldern kühl,
Da kommen all die Lieder
Von fern durchs Weltgewühl,
Es hallen Lust und Schmerzen
Noch einmal leise nach,
Und bildend wird im Herzen
Die alte Wehmut wach,
Der Winter auch derweile
Im Feld die Blumen bricht –
Dann gibt's vor Langerweile
Ein überlang Gedicht!"

Bei seiner Rückkehr fand er im Hause alles ausgeflogen, und streckte sich ermüdet im Garten an dem hohen Bogengange ins Gras. Er hatte aber noch nicht lange geruht, als er Stimmen neben sich vernahm, an denen er die Amtmann und Walter erkannte, die, ohne ihn zu bemerken, in dem Gange auf und nieder wandelnd, in lebhaftem Gespräch begriffen schienen. – „Das kommt bei dem Überstudieren heraus", sagte soeben die Amtmann, „nichts als Verse im Kopf, Reisen und dergleichen unkluges und kostspieliges Zeug." – „Ich glaube gar", rief Fortunat, „die spricht von mir!" – „Beruhigen Sie sich", hörte er nun Walter entgegnen, „ich werde versuchen, die eigentlichen Absichten dieses verschlossenen, rätselhaften Gemütes zu erforschen." – „Bei Nacht möchte er spazierengehen", fing die Amtmann wieder an, „den Tag verträumt er! Und warum verbirgt er sich vor uns?" – Hier verlor sich der Diskurs in der Ferne. Fortunat sprang hastig auf. Sie re-

den von meinem unbekannten Führer im Garten an jenem ersten Morgen, dachte er, und es fiel ihm aufs Herz, daß er ihn in der Zerstreuung so ganz vergessen hatte.
Als am Abend alle unter den Linden vor der Haustür sich wieder versammelten, beschloß er, der Sache näher auf den Grund zu kommen. Der Amtmann war der erste auf dem Platz, er erzählte ihm sogleich das ganze Begegnis, wie er damals den Unbekannten schlafend am Springbrunnen getroffen und was sie miteinander gesprochen hatten. Dieser hörte sehr aufmerksam zu, er mußte ihm Größe, Kleidung, Haare und Stimme des Fremden ausführlich beschreiben, aber der Amtmann wußte alles besser als er, alle seine Fragen trafen wunderbar ein. „So kennen Sie ihn also?" fragte Fortunat. – Der Amtmann schüttelte nachdenklich den Kopf. „Ich weiß nicht, wer es war", sagte er, „und darf nicht sagen, was ich vermute." – Unterdes war seine Frau herausgekommen, er bat Fortunat schnell, vor den Weibern nichts von der Geschichte zu erwähnen. Jetzt trat auch der Student Otto, der von einem weiten Spaziergange zurückzukommen schien, zu der Gesellschaft. Als er sich bei ihnen niederließ und in der warmen Luft seinen Rock schnell öffnete, fiel ein sauber eingebundenes Buch daraus zu Boden; es war des Grafen Victors neuestes poetisches Werk, das er bisher noch nicht gekannt und heute früh unter den zerworfenen Büchern des Amtmanns gefunden hatte. – „Ach, ich dachte, es wäre dein juristisches Handbuch", sagte die Amtmann, indem sie das Buch aufnahm und Otto zurückgab. Dann, sich gemächlich auf ihren Lehnstuhl zurücklehnend, fuhr sie nach einer kurzen Pause fort: „Hab' ich doch heute von Tagesanbruch in Haus und Hof zu schaffen gehabt, daß mir ordentlich alle Glieder wehtun. Nun, dafür schmeckt auch am Abend die Ruhe, wenn man sich wakker gerührt und seine Pflichten erfüllt hat." – Otto errötete flüchtig, ohne etwas darauf zu erwidern. – Fortunat aber fiel es bei diesen Worten erst auf, wie sonderbar allerdings Otto seit einiger Zeit erschien. Alle Morgen zog er

ganz allein in den Wald hinaus und kam selten vor Mittag wieder zum Vorschein. Dann war er einsilbig, schüchtern, zerstreut, und oft mitten in den heitersten Augenblicken flog es über sein freundliches Gesicht wie ein Wolkenschatten über eine schöne, sonnenhelle Gegend.

Man hatte unterdes das Abendessen aufgetragen, und die rüstige Amtmann, die es nun heut einmal auf Otto abgesehen zu haben schien, begann, indem sie den Braten zerschnitt und jeden reichlich davon beteilte, sich mit allerei weisen Redensarten und spitzigen Ausfällen über die teuren Zeiten zu verbreiten, und wie notwendig es sei, daß ein junger Mensch jetzt frühzeitig darauf denke, dereinst sein sicheres Brot zu haben. Da seien noch immer Toren genug in der Welt, um reichen Leuten die Zeit zu vertreiben mit schönen Bildern, Komödienspielen oder Versemachen – das sei ein bloß herrschaftliches Vergnügen, setzte sie schnell verbessernd hinzu, indem ihr dabei Graf Victor einfallen mochte. – Der Amtmann hatte die Salatschüssel vor sich geschoben und aß hastig, man konnte nicht erraten, ob er sich über Otto oder über seine Frau ärgerte. – „Da fällt mir immer mein seliger Bruder ein“, hob die letztere wieder an; „er hatte auch studiert, aber das war ein gescheiter Kopf, der ließ die Phantasten ablaufen, setzte sich auf seine Brotwissenschaften, heiratete eine gebildete vernünftige Frau, und Gott hat seinen Ehestand gesegnet. Nun, du kannst es ja selber bezeugen“ – fuhr sie zu dem Amtmann gewendet fort, empfindlich, daß er ihr gar nicht beistimmte –, „der ließ sich zu seiner Hochzeit von den besten Poeten Schäfergedichte machen, Gott weiß, wo die nun selber die Schafe hüten.“ – Hier brach Otto, der bis jetzt sichtbar mit sich selbst kämpfte, plötzlich mit verbissener Bitterkeit und einem höhnischen Stolze los, den niemand dem sanften Jünglinge zugetraut hatte. „Lieber Schweine hüten“, sagte er, „als so zeitlebens auf der Treckschuite gemeiner Glückseligkeit vom Buttermarkt zum Käsemarkt fahren. Der liebe Gott schafft noch täglich Edelleute und Pöpel, gleichviel, ob sie Adels-

diplome haben oder nicht. Und ich will ein *Herr* sein und bleiben, weil ich's *bin*, und jene Knechte sollen mich speisen und bedienen, wie es ihnen zukommt!" – Das war der bestürzten Amtmann zu toll. „Unsinniger, aufgeblasener Mensch!" rief sie hochrot vor Zorn; „so iß meinetwegen trockenes Brot, wenn du Butter und Käse verachtest! Aber wir wissen's wohl, wo du die Komödiantensprüche gelernt hast. Denke nur nicht, in unser ehrliches Haus einmal eine Theaterprinzessin heimzuführen, die nicht soviel hat, um die Löcher zu flicken, die sie in ihre Lappen gerissen, so eine von aller Welt ausgeklatschte Kreatur!"
Aber Otto hörte nicht mehr, er war rasch aufgestanden und schritt zürnend in den nächtlichen Garten hinein. Walter, in sichtbarer Verlegenheit, wollte ihm folgen, wurde aber von Fortunat aufgehalten, der ihn schnell in einen Seitengang führte. „Sage doch nur", fragte er Walter, „was gibt's denn eigentlich hier, und wo willst du hin?" – „Den Gekränkten trösten", erwiderte Walter, „und – vermag ich's sonst – ihm auch den Kopf ein wenig zurechtsetzen. Komm mit!" – „Das laß' ich wohl bleiben", rief Fortunat aus, „ich bin froh, wenn mir mein eigener Kopf zuweilen noch so leidlich sitzt." – „Mein Vorhaben", sagte Walter, „ist wahrhaftig edler, als es dir nach deinem ironischen Gesicht auf den ersten Blick vielleicht erscheinen mag. Denke dir nur recht diesen stillbeschränkten, heitern Familienkreis, dessen ganzes Trachten und Hoffen auf den einzigen Jüngling gerichtet ist, der auf der Schule immer für den aufgewecktesten und geschicktesten galt. Und nun kehrt er von der Universität zurück, verwandelt, träumerisch in sich gekehrt, unlustig zu jeder tüchtigen Arbeit und einer verworrenen Welt von ausschweifenden Gedanken und Wünschen nachhängend, um – wie ich fürchte – dereinst zu spät von der grausamsten Täuschung zu erwachen und ein verlornes Leben zu bereuen. Nein, ich will es endlich versuchen, ihn auf das Gefährliche eines Pfades aufmerksam zu machen, der einsam über die Köpfe der andern Menschen weggeht und

immer nur für sehr wenige bestimmt scheint." – Fortunat war über diese Worte ernst und nachsinnend geworden. „Du ehrliche Seele!" sagte er endlich, dem Freunde herzlich die Hand schüttelnd, „so versuche dich denn an ihm. Ist der junge Mensch ein halber Philister, so hilf ihm völlig aus dem tollen Poetenmantel heraus, und ist es rechter Ernst mit seinem Talent, so muß er ja doch weiter und rennt dich über, wärst du auch der weise Salomo selber." Alle vor dem Hause waren durch den Vorfall gestört, die kleine Gesellschaft sah stumm und kopfhängend auf die Teller. Draußen über den Tälern war es indes schon stiller und dunkler geworden, nur in weiter Ferne sah man zuweilen leichte Blitze über den Bergen schweifen. Die Amtmann blickte mit heimlicher Besorgnis, wie es schien, bald in das Wetterleuchten, bald nach der Richtung hin, wo Otto verschwunden war, und ging dann, ohne ein Wort zu sagen, in das Haus hinein. Endlich brach der Amtmann ärgerlich die umheimliche Stille. „Es geht auch alles konfus jetzt", sagte er zu Fortunat, „im Frühling Gewitter, im Sommer kalt, in der Jugend alt und im Alter närrisch. Glauben Sie mir, unsere ganze Zeit jetzt ist gerade wie dieses verrückte Frühlingswetter, die Schwüle brütet und treibt alles vorzeitig hervor, und ich fürchte, es schießt mehr ins Kraut als in die Blüte. Unsere Jungens wissen schon jetzt mehr, als wir jemals erfahren haben, und recken und sehnen sich aus allen Gelenken heraus, während wir in unserer lustigen und gesunden Jugendzeit ohne besondere Sehnsucht hinreichend dumme Streiche machten und erst die fatalen Lümmeljahre überstehen mußten. Ja, es ist recht verdrießlich! Man möchte sich gern bequem, fröhlich und auf die Dauer einrichten, wie in der guten alten Zeit, aber der ferne Donner verkündigt überall den unheimlichen Ernst, und so sitzen wir verwirrt, ungewiß und in banger Erwartung vor dem dunklen Vorhang, hinter dem fortwährend Gott weiß, was! unruhig und feurig zuckt." – Unterdes hatte Walter den verscheuchten Otto im Garten aufgefunden. Empört und

in innerster Seele verletzt, saß er wie eine Nachteule mitten im Gestrüpp. Als er Walter erblickte, sprang er rasch auf und kam ihm mit erzwungener, gleichgültiger Höflichkeit entgegen. „Die Tante", sagte er, „ist gewiß schon besorgt, daß ich draußen nicht den Schnupfen bekomme. Freilich die Nase ist ein empfindlicher Teil, da sitzt die Seele schon tiefer und wärmer, die ficht so leicht nichts an." – Walter stand einen Augenblick verblüfft, denn es war ihm, als säh' er auf einmal sich selber als Studenten vor sich stehn, er war ganz aus seinem Konzept gebracht und ergriff gerührt die Hand des aufgeregten Jünglings. „Ich komme keineswegs", sagte er endlich, „um das harte, heftige Wesen der Amtmann zu verteidigen, obgleich es auch nur eine andere, ungeschickte Form der Liebe ist. Das Angedenken meiner eigenen Jugend ist es, was mich herführt, der aufrichtige Schmerz um ein junges, heitres Gemüt, das auf diesem Wege sich immer tiefer und tiefer in der blühenden Einsamkeit verirrt und verwildert. Ich kenne diese trostlose Öde junger Seelen gar wohl, das Heimweh ohne Heimat, diese labyrinthische Selbstquälerei. Sie stehn verlassen auf der Welt, ohne Vater und Mutter – verlangt Sie in dieser Einsamkeit nach einem Freunde, und wollen Sie's mit mir versuchen, so biete ich Ihnen meine Hand bis in den Tod und will raten, schützen, helfen wo ich kann!" – Otto sah ihn erstaunt an, denn in Walters Worten war jener wunderbare Klang ernster Güte, der überall unmittelbar zum Herzen geht. – „Sie sind im Amte, angesehen, ruhig" – sagte er dann nach einer kurzen Pause. „Und wenn ich Ihnen nun auch erzählen wollte von dem zauberischen Spielmanne, der jeden Frühling, wenn der Sonnenschein sich munter über die Felder ausbreitet, aus dem Venusberge kommt mit neuen, wunderbaren Liedern und die Seelen verlockt, von dem in schwüler Mittagsstunde der einsame Vogelsang schallt, von dem die Ströme und Quellen verworren rauschen im Mondschein, und die badenden Nixen wie im Traume singen durch die stille goldne Nacht – Sie würden mich ja doch

nur für verrückt halten!" – Walter erschrak fast, so irr und fremd leuchteten die Augen des Jünglings im Streiflicht des Mondes. – „Und ich bin es ja auch in der Tat!" fuhr dieser fort, „bildete mir da ein, dem Zauberstrom von Klängen unversehrt folgen zu dürfen und ein Dichter zu sein, der die Zauber regiert! Aber nun weiß ich's besser. Alle Engel, die durch die erste Dämmerung meiner Kindheit zogen, was ich oft betend heimlich ersehnte und immer und immer vergeblich auszusprechen versuchte: ich fand es heut auf einmal mit freudigem Erschrecken in des Grafen Victors Buch, er hat es kühn, frisch und jung wie eine Zauberinsel entdeckt – und ich weiß nicht mehr, was ich will. – Aber es ist noch immer Zeit, ich bin noch jung. Und wie ich das Buch hier vom Berge in den Fluß hinunterschleudere, so entsag' ich von heut ab der fröhlichen Dichtkunst, der Metze! Und gleich den andern, die ich verachtet und die so unsäglich besser sind als ich, ich will von heut an allein und ganz der Rechtswissenschaft leben und von den Büchern nicht wieder aufsehen!" – hier brach er plötzlich in Weinen aus und stürzte wie vernichtet an Walters Brust.

Beide neuen Freunde schritten nun durch den stillen Garten, nur eine Nachtigall tönte schluchzend in der Ferne. Otto schwieg und schien gefaßter. Walter sagte: er brauche ja darum die Poesie nicht ganz aufzugeben, es bedürfe eines des andern, die Poesie des strengen, ernsten Lebens und das Leben der heitern Dichtkunst. Aber er fühlte bald, wie albern solcher Trost in solcher Stunde war, und schwieg endlich auch still.

So kamen sie an das Haus, wo sie die Amtmann in Angst und Tränen fanden. Sie hatte zuletzt gefürchtet, daß Otto in seiner Heftigkeit sich selbst ein Leids angetan, und fiel nun dem Geretteten mit großer Freude um den Hals, die dieser herzlich erwiderte. „Es ist vorbei", rief Otto in seiner seltsamen Hast, „ihr habt mich nun ganz wieder, und nächstens, will's Gott, ist Examen!" – „Du bist ein braver Junge", rief der Amtmann, „stoß an!" – Die Gläser gaben

einen hellen Klang, und so endigte der Abend noch in Freuden; die fernen Gewitter hatten sich auch verzogen, und der Himmel glänzte mit tausend Sternen über den Versöhnten.

Fünftes Kapitel

Aber es blieb nicht lange so ungestört; ein Zufall, Mißverständnis, oder wie sonst der Mensch des Himmels Führung oder sein eigenes Ungeschick benennen mag, stellte unerwartet alles anders auf Hohenstein.

Es war ein schwüler Nachmittag, die Blätter im Garten rührten sich kaum, der Amtmann war auf der Bank vor der Haustür eingeschlummert, Walter schrieb Briefe im Hause, Fortunat hatte sich mit einem Buche ins Gras gestreckt und ließ es sich vor der weiten Aussicht gern gefallen, daß die leise Luft ihm das Buch verblätterte. Florentinen wurde ganz wehe in dieser Stille, sie mußte immer etwas zu schaffen haben; so schlich sie sich heimlich nach dem Walde, um für den Abend Erdbeeren zu pflücken, die Walter für sehr gesund hielt, weil er sie gern aß. Fortunat sah sie mit ihrem Körbchen unten aus dem Dorfe gehen, er warf sein Buch weg und folgte ihr, konnte sie aber im Walde nicht wiederfinden.

Florentine war unterdes, bald sammelnd, bald haschend, von Strauch zu Strauch geschlendert und so unvermerkt an die Ruine der gräflichen Stammburg gekommen. Überrascht sah sie in der Einsamkeit an den halbzerfallenen Mauern, Toren und Fensterbogen empor; steinernes Bildwerk, das von der ehemaligen Pracht zeugte, lag im hohen Grase zerstreut, aber der Frühling hatte den verlassenen Berg wieder bestiegen und spielte fast wehmütig in dem stillen Hause. Seltsame Sagen gingen in der Gegend von diesem einsamen Ort. Die Hirten hörten oft bei Nacht fremde Stimmen in der Burg, eine wunderschöne bleiche

Frau sollte sich manchmal dort in dem ausgebrochenen Fenster sehn lassen. – Florentine war noch nie allein hier gewesen, jetzt verlockte sie der eigene Reiz des Grauens, sie betrat erst vorsichtig und zaudernd, dann immer kekker die kühlen, von oben verschatteten Hallen. Durch die Mauerlücken blickten zuweilen die Täler schillernd aus der sonnenhellen Tiefe herauf, nur hin und her sang ein Gebirgsvogel mit fremdem Schall und verstörte Eidechsen fuhren raschelnd unter das Unkraut, daß sie unwillkürlich zusammenschrak.

Jetzt kam sie in den innern Burghof, da stand ein wilder Kirschbaum in voller Blüte, dunkelrote Blumen glühten zwischen den Steinen, einzelne Schmetterlinge flatterten ungewiß in der trüben, brütenden Schwüle; und als sie plötzlich um den Pfeiler trat, sah sie eine schöne, bleiche Frau in einem seltsamen, himmelblauen Gewande mitten im Hof auf dem Rasen sitzen, die wandte sich nicht und kämmte schweigend ihr lang herabwallendes rabenschwarzes Haar. – Florentine blickte noch einmal scharf hin, dann, vom Entsetzen überwältigt, ergriff sie die Flucht.

Aber wie es oft in ängstlichen Träumen geht, sie verfehlte in der Hast die rechte Pforte; aus einem Zwinger in den andern rennend, glaubte sie sprechen zu hören, die Stimmen kamen immer näher, sie konnte den Ausgang nicht finden. Auf einmal standen zwei fremde Männer vor ihr in abgetragenen Ritterwämsern, Pickelhauben auf den Köpfen. Der eine wollte sie am Körbchen festhalten, in der Todesangst ließ sie ihm fliehend die Beeren und hörte sein schallendes Lachen hinter sich.

Wie atmete sie auf, als sie endlich Gottes freien Himmel wiedersah! Der erste, der ihr begegnete, war Fortunat. Atemlos, mit heftig klopfendem Herzen flog sie an seine Brust, er drückte das schöne Kind fester an sich und fühlte einen flüchtigen, brennenden Kuß auf seinen Lippen. – In demselben Augenblicke aber war auch *Walter,* der sie zu suchen schien, neben ihnen aus dem Gebüsch hervorge-

treten. Florentine besann sich schnell wieder, strich die Locken aus der heißen Stirn und reichte ihm die Hand hin, um ihr über die letzten Trümmer herabzuhelfen.
Nun erzählte sie in lebhafter Aufregung, und oft noch scheu zurückblickend, ihr wundersames Abenteuer. Walter war still und schien nur halb hinzuhören. Fortunat wollte sogleich in die Burg zurück, um die bleiche Frau zu sehen, aber Florentine gab es durchaus nicht zu. Während sie aber noch so stritten, stutzte sie plötzlich und wies dann ganz erstaunt nach dem Tale hinaus. Dort wurde fern am Saume des Waldes ein abenteuerlich bepackter, langsam einherziehender Wagen sichtbar, ihm folgte ein seltsam gekleidetes Mädchen zu Pferde in blauem Gewand, mit dunklem, fliegendem Haar, mehrere Männer, grüne Zweige auf ihren Hüten, schritten rüstig nebenher; unter ihnen erkannte man sogleich die beiden Burgkobolde wieder, deren Pickelhauben weit in der Sonne funkelten. Ein fröhlicher Chorgesang schallte von dem Zuge durch das Grün herauf. – „Reisende Komödianten!" rief Fortunat lachend, „nun bedarf es keiner Untersuchung weiter, das waren die Spukgeister, der Weg kommt gerade von der Burg."
So traten sie nun alle beruhigter den Rückweg nach Hohenstein an. Florentine, die sich völlig wieder erholt hatte, lachte jetzt selber mit; dann wandte sie sich noch einmal nach den Blütentälern, in die sich die künstlerischen Wandervögel gesenkt. „Es geht doch nichts übers Reisen", rief sie fröhlich aus, „wenn ich so manchmal im Sommer recht früh erwache und höre unten aus den Dörfern die Hähne krähen oder ein Posthorn von fern über den Garten herüber, da wünsch' ich mir oft, ich wäre ein Mann und könnte auch so mit in die Welt hinaus." – „Ich meine", fiel hier Walter etwas grämlich ein, „man müsse erst sich selbst und die kleine Welt um sich herum recht verstehen gelernt haben, ehe man sich weiter umsieht, und das Reisen zieme überhaupt nur dem reifern Alter." – Fortunat ärgerte der Schulmeisterton. „Gerade umgekehrt", rief er

aus, „nur die Jugend versteht recht aus Herzensgrunde die Schönheit der Welt mit ihren morgenroten Gipfeln und kühlen Abgründen und funkelnden Auen im Grün, und malt es alles fresco nach, daß das Alter sich einst daran erfrische, wenn draußen die Blätter fallen und die sinkende Herbstsonne die Schildereien noch einmal wunderbar beleuchtet. Während dein sogenanntes reifes Alter vom Schifflein sorgsam die Tiefe mit dem Senkblei mißt, sitzt die Jugend über Bord geneigt und sieht ihr eigenes weinbekränztes Haupt in der klaren Flut und hört die Glocken der versunkenen Stadt aus der Tiefe heraufklingen. Ja, glaubt nur, die Welt ist wie eine eigensinnige Schöne, die nur in jungen Augen sich mit ihrem fröhlichsten Schmukke spiegeln mag, für Klugheit und Kenntnisse gibt sie nur Brot, für Liebe und rechte Freude an ihr aber wieder Freude und Liebe."

So waren sie vor der Amtmannswohnung angelangt. Die letzten Strahlen der untergehenden Sonne vergoldeten bereits die Bäume vor dem Hause, unter denen die Amtmann schon wieder den Tisch gedeckt hatte. Ein jeder machte sich's in der Abendkühle behaglich bequem, und Florentine mußte, ausruhend, ihre Burggeschichte nochmals umständlich erzählen. Nur Walter fehlte. Auf einmal trat er, ganz reisefertig, mit dem Amtmann aus dem Hause. „Schlechte Neuigkeit", sagte der letztere, „Walter hat dringende Briefe bekommen, er muß in die Stadt und will noch heut reisen, um die nächtliche Kühle zu benutzen." – Die Amtmann machte besorgt Einwendungen gegen das gefährliche Reisen in der Nacht, Florentine ereiferte sich über die Geschäfte, die sie von jeher als eine unbekannte, feindliche Macht betrachtete, aber Walter blieb unerschütterlich und nahm, auch von Fortunat, schnell und kurz Abschied. Ganz zuletzt wandte er sich noch einmal zu diesem, als wollt' er ihm etwas sagen, schüttelte ihm aber nur rasch die Hand und ging schweigend fort. – Fortunat begleitete ihn noch heraus bis zu seinem Pferde, dem Florentine den Hals streichelte, und als es dann beim Auf-

steigen unruhig wurde, schnell nach der Haustür zurücksprang. „Herrje!“ sagte er heimlich zu Walter, „was machst du da für ein langes Gesicht! Und überhaupt, warum willst du gerade heute noch fort? die Geschäfte sind's ja doch nicht.“ – „Ich will nicht stören“, entgegnete Walter empfindlich, „du bleibst ja doch noch längere Zeit hier, ich sag' dir's vielleicht ein andermal, leb' wohl!“ – Hiermit gab er seinem Pferde die Sporen und war bald zwischen den Bäumen verschwunden. – „O, langweilige Welt!“ rief Fortunat ihm nachsehend aus, „wie glücklich könnte er sein mit seinem schlanken Reh im schönen grünen Wald, wenn er frisch vom Herzen weg liebte, anstatt den Talar von Melancholie, Eifersucht und andern hergebrachten Liebestücken durch alle Paradiese jämmerlich hinter sich nachzuschleppen!“

Als er in den Garten zurückkam, bemerkte er auf der Linde vor dem Hause zwei zierlich beschuhete Füßchen zwischen den Zweigen. Es war Florentine; sie saß im Baume, mit den Füßen baumelnd, während sie Walter nachschaute, der sich soeben in der Dämmerung zwischen Wiesen und Kornfeldern verlor. Das heitere Mädchen schien in ihrer Unbefangenheit von seinem Mißmute gar nichts zu ahnen.

Fortunat aber ging allein und unruhig durch den Garten. „Ich werde doch kein Narr sein und mich verlieben?“ sagte er zu sich selbst. „Und doch bin ich auf dem nächsten Wege dazu. Und hinter mir langsam und feierlich der abgemagerte Geist des sich selbst erschossenen Walters, und vor mir ein Zug von Tanten und Basen, und gute Wirtschaft, und Kindergeschrei, und ein Haus machen –“

Der Angstschweiß trat ihm ordentlich bei diesen Gedanken vor die Stirn. Er rannte eiligst nach dem Hause zurück und eröffnete dort ohne weiteres der erstaunten Familie, wie er zwar heute gerade keine Briefe aus der Stadt bekommen habe, aber eigentlich ebenfalls schleunigst fortreisen müsse; daß er daher für Speis' und Trank und alle die schöne, stille, herrliche Zeit aus Herzensgrund Dank

sagen und hiermit sogleich schon heut Abschied nehmen wolle, da er noch vor Tagesanbruch weiterzuziehen gedenke. Florentine wurde bei diesen Worten ganz rot, sie setzte sich schmollend auf eine entfernte Bank, und Fortunat glaubte zu bemerken, daß ihre abgewendeten Augen von Tränen glänzten. Auch die andern machten ihm durch ihre aufrichtige Trauer das Herz schwer, denn sie hatten sich alle in der kurzen Zeit schon an seine fröhliche Weise gewöhnt. Er mußte versprechen, wiederzukommen und ihnen noch ausführlich von den Ländern und Städten zu erzählen, wohin seine Reise ging; so saßen sie noch lange plaudernd vor der Haustür beisammen. Beim Schlafengehen endlich flüsterte ihm Florentine noch heimlich zu: „Und ich werde doch auf sein, eh' Sie wegreiten!" –

Er hatte alle Fenster des Schlafzimmers offen gelassen, um den Morgen nicht zu verschlafen. Da war es ihm, als gingen draußen fröhliche Stimmen unter den Fenstern auf und nieder und riefen immerfort in seinen Schlummer hinein: Frisch auf, schlafe nicht mehr! Wunderbare Berge und Gründe, schimmernde Fernen, frisch auf! und schöne, helle, fröhliche Zeit! Er sprang endlich empor und blickte durchs Fenster. Es war noch Nacht; dennoch kleidete er sich in langentbehrter Reiselust sogleich an, ging durch das stille Haus an Florentinens Schlafkammer vorüber und machte noch schnell einen Gang durch den Garten. Es war in der Nacht ein warmer Regen gefallen, die Nachtigallen schlugen überall aus den erfrischten Büschen, hin und her bellten Hunde fern in den Dörfern, sonst lag alles noch still im prächtigen Mondenschein unter dem weiten gestirnten Himmel. – Als er zurückkehrte, hörte er unten im Hause leise ein Fenster öffnen, es war Florentine, die sich in leichter Morgenkleidung hinauslehnte. „Zisch' aus! zisch' aus!" rief sie ihm entgegen, „ich bin früher wach gewesen als Sie!" Dann, sich im Garten umsehend, sagte sie: „Das ist gerade wie damals, da Sie hier das Ständchen brachten und wir Sie zum ersten Male

sahen. – Nun wird es hier wieder recht einsam sein, und ich wollte Sie eben nur noch bitten, daß Sie auf Ihrer Reise von sich hören lassen und manchmal an Walter schreiben, der Ihnen außerordentlich gut ist und gern von fremden Ländern hört." – Fortunat versprach es und bat sie um einen Kuß zum Abschiede. – „Warum nicht gar!" rief das Mädchen lachend, indem sie ihm schnell die Hand hinausreichte, dann schloß sie geschwind das Fenster, und er sah sie nicht wieder. Fortunat warf sich nun ungesäumt auf sein Pferd und ritt durch die hohe, dunkle Allee an dem Gittertor des Gartens und dem stillen Dorfe vorüber. Draußen auf dem Bege aber wandte er sich noch einmal zurück. „Gesegnet", rief er, „du schönes Waldtal, in deiner glückseligen Abgeschiedenheit, möge der Sturm der Welt dich nie verstören!"

Sechstes Kapitel

Ein schweres Gewitter zog eben an dem Gebirge hin und sandte seine Regenschauer in die Ebenen hinaus, während Fortunat, durchnäßt und lange vom Wege abgekommen, über ein weites, in Regen und Abenddunkel verhülltes Feld dahintrabte. Da hörte er unerwartet den Gesang einer schönen Männerstimme von fern herüberschallen, wovon er nur folgende Worte verstehen konnte:

„Bei dem angenehmsten Wetter
Singen alle Vögelein,
Klatscht der Regen auf die Blätter,
Sing' ich so für mich allein.

Denn mein Aug' kann nichts entdecken,
Wenn der Blitz auch grausam glüht,
Was im Wandeln könnt' erschrecken
Ein zufriedenes Gemüt."

Er gab seinem Pferde die Sporen und erreichte in kurzer Zeit ein Häufchen Wanderer, die neben einem Paar Pferde einherschritten, auf denen zwei junge Frauenzimmer saßen. Mit freudiger Überraschung erkannte er sogleich die abenteuerlichen Gestalten der Schauspieler wieder, die an Victors Stammburg vorübergezogen waren, von denen aber jetzt die Dunkelheit nur die ungefähren Umrisse erraten ließ.

Fortunats Gruß fand nur eine halbe Erwiderung, die Gesellschaft schien in üblem Humor zu sein, und langsam und schweigend, wie ein schwerer Traum, bewegte sich das Ganze weiter. Endlich unterbrach der Voranschreitende, welcher soeben gestolpert war, die Stille mit einem derben Fluche, prustete und glitt gleich wieder aus und kam gar nicht aus der Wut. – „Das haben wir davon", hob die eine Dame zu Pferde zu der anderen Reiterin an, „das haben wir nun von eurer schönen Natur. Brächen die Herren nicht ihren Flaschen auf das Wohlsein jeder alten Burg die Hälse, so wäre uns allen jetzt wohler, und wir säßen im Trocknen, denn unser Wagen ist gewiß längst in der Stadt." – Dabei breitete sie mühsam einen, wie es schien, nicht sonderlich konditionierten Regenschirm über sich aus. Aber der Wind verarbeitete ihn sogleich mit solcher Fertigkeit, daß ihre berittene Nachbarin laut auflachte und die Dame ihre Segel erbost wieder einziehen mußte. Fortunat, welcher hier heimlich auf ein ergötzliches Gezänk hoffte, ermahnte die Gesellschaft, den beiden Damen in diesem Kampfe mit den Elementen durch ein gemeinschaftliches angenehmes Gespräch galant unter die Arme zu greifen. Die Männer antworten gar nicht darauf, die Dame mit dem Regenschirme aber fragte: ob er vielleicht auch ein Künstler sei und es so gut haben wolle wie sie? „O", setzte sie spitzig nach ihrer Nachbarin gewendet hinzu, „Liebhaberrollen sind hier jederzeit zu haben." – „Bitte sehr", erwiderte die Nachbarin mit einer wohlklingenden Stimme, „bei Ihnen ist ja diese Stelle seit geraumer Zeit vakant." – Ein plötzlicher Blitz beleuchtete

hier auf einen Augenblick ein schönes, feines, aber bleiches Gesichtchen, über welches zu beiden Seiten lange schwarze Haare triefend herabhingen. – „Mein Gott, was ist das für eine Wirtschaft um das bißchen Regen!“ rief einer der jungen Männer aus, „quamquam sint sub aqua, sub aqua maledicere tentant!“ – „Sparen Sie doch Ihr Latein“, sagte die Dame mit dem Schirme. „Sie memorieren wohl eben den Bettelstudenten?“ – Sie wollte noch mehr sprechen, aber der Literatus fiel schnell in das Lied wieder ein, das Fortunat schon vorhin von fern gehört hatte, und übersang sie lustig:

„Frei von Mammon will ich schreiten
Auf dem Feld der Wissenschaft,
Sinne ernst und nehm' zuzeiten
Einen Mund voll Rebensaft.

Bin ich müde vom Studieren,
Wann der Mond tritt sanft herfür,
Pfleg' ich dann zu musizieren
Vor der Allerschönsten Tür.“

„Land! Land!“ schrie hier plötzlich der Voranschreitende dazwischen, und man erblickte zu allgemeiner Freude von weitem Mauern und Türme, die sich wie dunkle Riesen immer deutlicher aus dem trüben Grau aufrichteten. Einzelne Lichter schimmerten schon den Reisenden trostreich entgegen, ein jeder strengte neu belebt seine letzten Kräfte an, und so waren sie bald an dem Tore eines kleinen Städtchens angelangt. – Wie Zugvögel mit begossenen, hängenden Flügeln strichen sie stumm durch die engen, finstern Gassen, wo sich die Lichter aus den Fenstern blendend und verwirrend im Wasser spiegelten, während der Regen von allen Dächern aus abenteuerlich vorgestreckten Drachenköpfen auf sie herabstürzte.
So kamen sie endlich in den Hof eines Wirtshauses. Hier war der Reisewagen der Gesellschaft, den man unterwegs

umgeworfen hatte, auch soeben erst angelangt. Der Theaterprinzipal *Sorti*, ein kleines fixes Männchen, rannte eifrig hin und her, vom Wagen wurden Burgen, Drachen und lange Kamelhälse eilig über den Hof getragen, die Hofhunde bellten, überall war ein Rufen, Drängen und Schimpfen in der undurchdringlichen Finsternis, die nur von einzelnen Blitzen manchmal durchkreuzt wurde. Mitten aus diesem Rumor hob der Literatus die jüngere Reiterin schnell vom Pferde und trug sie auf seinem Arm in das Haus. Das Mädchen war arg durchnäßt, mit dem dünnen, vom Regen knapp anliegenden Kleide, mit den lang herabhängenden tröpfelnden Locken sah sie wie ein Nixchen aus, das eben den Wellen entstiegen. Sie hielt beide Hände vor das Gesicht, um sich vor dem plötzlich aus dem Hause dringenden Lichte zu schützen, aber zwischen den kleinen Fingern funkelten zwei schwarze Augen hindurch, die Fortunat im Vorüberfluge durchdringend anblickten.

Dieser konnte nur mit Mühe ein besonderes Stübchen gewinnen, wo er schnell seine Kleider wechselte, während draußen nach und nach ein gewaltiges Türzuwerfen, Streiten und Lachen, von einzelnen Operntrillern und Laufern durchschwirrt, das ganze Haus erfüllte. Unterdes hatte auch das Wetter sich wieder verzogen, und der Mond trat klar zwischen dem zerrissenen Gewölk hervor. Er verließ daher gar bald seine enge, schwüle Kammer wieder und eilte zwischen den Reifröcken, Rüstungen, Fahnen und Miedern, die über dem Treppengeländer zum Trocknen ausgehängt waren, in den Garten hinab. Ein einsames Frauenzimmer saß dort vor der Haustür auf der Bank, an dem etwas verbrauchten Federhut, dem hohen Kragen und der ganzen Haltung erkannte er die Dame mit dem Schirme wieder. – „Ich bin mir selbst noch Genugtuung schuldig", hob sie sogleich an, als sie Fortunat bemerkte, „Sie werden vielleicht eine ungünstige Meinung von mir gefaßt haben; aber Sie glauben nicht, welche Verleugnung es einem zarteren Gemüt kostet, mit den rohen

Scherzen dieser Menschen, wenn auch nur zum Schein, gleichen Schritt zu halten." – „In der Tat", erwiderte Fortunat, „der Lateiner schritt wacker und lustig aus." – „Lustig?" sagte die Dame, „Sie kennen diesen Wilden noch nicht, er hat keine Ahnung von jener geistigen Seelenlust, die schon diesseits die Gipfel der Menschheit erklimmt" – „Und jenseits rücklings wieder herunterschnurrt", fiel hier der feindliche Literatus ein, der eben mit einer Gitarre aus dem Hause tretend, das letzte Kapitel von der Lust mit angehört hatte, und, einzelne Akkorde anschlagend, sich nun weiterhin auf dem Platze im Dunkel verlor. Fortunat lachte, denn ein leiser Zornesblitz zuckte plötzlich über das Gesicht der Dame und brachte die ganze Muskeldekoration in eine augenblickliche widerliche Unordnung, zumal da gleich darauf die andere hübsche Reiterin aus der Tür guckte, ihr Näschen rümpfte, da sie die beiden beisammen erblickte, und dann gleichfalls an ihnen vorüber in den Garten schlenderte. – „Die arme Kleine! sie hat keinen ganzen Strumpf", bemerkte die Dame hämisch. Und in der Tat, auch der Mond hatte das schon bemerkt und beleuchtete wohlgefällig ein Streifchen des zierlichsten Beinchens, das blendend über dem Schuh hervorblickte, während die hochgeschürzte Kleine unbefangen unter den Linden bemüht schien, Blüten von den herabhängenden Zweigen zu streifen.

Unterdes ging ein frisches Wehen durch die Wipfel, die letzte Wolkendecke zerriß, und die alte Stadtmauer und die Waldberge darüber standen plötzlich wunderbar beglänzt. Die Dame hatte sich erhoben und unter der Linde vor der Bank eine malerische, melancholisch-heroische Stellung genommen. Das Haupt in die rechte Hand an den Baum gestützt, sah sie eine Zeitlang, wie in Gedanken verloren, nach den Höhen – „Tiedge!" – sagte sie endlich bedeutungsvoll, und drückte Fortunat leise die Hand. Fortunat, den die ganze wunderliche Wirtschaft dieses Polterabends schon lange innerlichst aufgeregt hatte, sprang rasch auf. „O Gott, wahrhaftig!" rief er, ihre Hand fest-

haltend, aus, „da schwebt er dahin als ein Veilchenduft, die Sterne scheinen ihm durch den Leib – o hören Sie nichts? – nun lispelt er mit jemand, wie gedämpfte Musik der Sphären, es ist Lafontaine, mit dem er kos't, der hat einen perlendurchwirkten Schlafrock an, aber die Perlen, alle sind Tränen – sie wandeln miteinander auf der Milchstraße – aber was ist das!" – „Wo?" sagte die Dame erschrocken und versuchte vergeblich, ihm ihre Hand zu entwinden. – „Sehen Sie die bärtige Wolke dort", fuhr er fort, „da kommt ihnen Kotzebue auf einem Ziegenbock entgegen, ach, Lafontaine weint, daß ihn der Bock stößt – o, es ist keine Tugend mehr auf der Welt!" – Hier hatte die Dame sich endlich losgemacht, sie hielt ihn längst für betrunken oder wahnsinnig, stammelte verlegen eine kurze Entschuldigung und stürzte in das Haus zurück. Er aber sprach noch immer fort, bis sie ihr Zimmer erreicht und die Tür eilfertig hinter sich abgeschlossen hatte.
Lachend warf er sich nun wieder auf die Bank hin, die Wälder rauschten in der plötzlichen Stille von den Bergen herüber, hin und her erwachten einzelne Nachtigallen, in einiger Entfernung hörte man den Literatus singen:

„Die fernen Heimathöhen,
Das stille, hohe Haus,
Der Berg, von dem ich gesehen
Jeden Frühling ins Land hinaus,
Mutter, Freunde und Brüder,
An die ich so oft gedacht,
Es grüßt mich alles wieder
In stiller Mondesnacht."

Die zierliche Reiterin hatte sich bald nach den ersten Klängen dem Sänger genähert. „Du, du" – sagte sie mit dem Finger drohend, „du hast heute wieder deine melancholische Stunde!" – „Ach", erwiderte der Literatus, halb unwillig abbrechend, „was weißt du davon, wie einem Gelehrten manchmal zumute ist!"

Ein plötzliches Getümmel an der Haustür verhinderte hier Fortunat, mehr von dieser Unterredung zu vernehmen. Ein ganzer, heller Haufen von Schauspielern kam nämlich samt einem langen, mit Weinflaschen und Gläsern besetzten Tische, den sie alle mühsam trugen, zum Hause heraus, der Gastwirt, voll Besorgnis um seine Gläser, ihnen auf dem Fuße nach. „Der liebe Gott hat hier draußen den Vorhang wieder aufgezogen", sagte der eine zum Wirte, „seht da, Menschenkind, den prächtigen Saal! Ein Reverbère, der bis auf einige verjährte Rostflecke ziemlich blank ist, eine Unzahl von Lichtern, die sich selber putzen, an allen Wänden ganze Mondlandschaften al fresco." – Die Gesellschaft hatte sich unterdes nicht ohne bedeutenden Tumult um den Tisch gelagert. Ein starker, wohlleibiger Mann von gesetzten Jahren zündete qualmend seine lange Pfeife an dem flackernden Lichte an, das in einer Glaskugel auf dem Tische stand, und in dessen Widerscheine sein vom Wein und Wetter verbranntes Gesicht sich noch dunkelroter ausnahm, es schien derselbe, der vorhin, im Regen der Gesellschaft voranschreitend, verschiedentlich gestolpert und geflucht hatte. – „Sie sollten auch Komödie spielen, mein Herr Wirt", sagte er, mit der Pfeife in breiter Behaglichkeit auf dem Stuhle zurückgelehnt. – Der Wirt äußerte Bedenklichkeiten gegen seine Geschicklichkeit. – „Ach, Flausen!" fiel ihm der Schmauchende in die Rede, „sehen Sie, so wie ich hier vor Ihnen sitze, so sitze ich auch auf dem Theater als Oberförster, als gutmütig polternder Alter und so weiter, ich rauche, ich plaudere und trinke mein Gläschen Wein so gut, wie hier." – „Das würd' ich allenfalls wohl auch treffen", meinte der Wirt. – „Nun, so seid kein Tor!" fuhr jener fort, „wollt Ihr gratis Eure Schlafmütze aufsetzen, Euer Abendpfeifchen schmauchen, Euern Kindern rührende Ermahnungen geben? Laßt's Euch bezahlen, Mensch!"
Fortunat, dem der Mann gar nicht übel dünkte, verließ hier seine Bank. „Aber, mein Bester" – sagte er, sich mit an den Tisch setzend – „wird Euch denn nicht manchmal

Angst, daß die neuere Poesie Eure Oberförstereien aufhebt und Euch Eure häuslichen Vergnügungen legt? – „Keineswegs", entgegnete der Oberförster sehr richtig, „im Gegenteil, die neuesten kurzen Dramen machen sich wieder ganz vernünftig und familiär. Und wenn ich auch in Versen spreche oder vielleicht gar ein Ritterwams anlege, ich bleibe doch der Alte. O, mein Herr, solange noch deutsche Biederkeit waltet, und Bier getrunken und Tabak geraucht wird, steht mein Charakter unerschütterlich, wie auf Elefantenfüßen." – Hier mischte sich ein junger, blasser Schauspieler mit in das Gespräch, der bisher für sich allein an dem Stümpfchen Licht in einem Buche gelesen hatte, ohne an dem Lärm der andern teilzunehmen. „Bester Herr Ruprecht", redete er den Oberförster an, „wer Sie so zum ersten Male schwatzen hört, könnte leicht an Ihnen irre werden. Ich aber weiß es wohl, wie Sie, gleich jenem Herrn, in der Kunst nur das Edlere, das Ideale schätzen." – Ruprecht, der sich nicht wenig damit wußte, daß er in seiner Jugend die Kantische Philosophie gehört hatte, räusperte sich und rückte sich soeben wohlgefällig in seinem Stuhle zurecht, als plötzlich die kleine Reiterin herbeisprang und ihm von hinten den Mund zuhielt. „Um Gottes willen", rief sie, „fangt nicht wieder von dem langweiligen Zeuge an, ihr guten Leute und schlechten Philosophen!" – „Armer Shakespeare!" entgegnete der Blasse mit einem unsäglich verachtenden Blikke. – „O", fiel ihm Kordelchen – so hieß die Reiterin – in die Rede, „der Ruprecht ist ein eingefleischter Shakespeare, hat er sich nicht schon allmählich Bardulphs feurige Nase anstudiert?" – Und in der Tat, seine stolze Nase leuchtete immer schöner, je trüber das Licht in der Glaskugel zu verlöschen begann. Er begab sich für einen Augenblick der feierlichen Gravität, in die ihn die Erinnerung an seine akademischen Studien versetzt hatte, und, täppisch Kordelchen zu sich zerrend, rief er: „So komm und gib deinem Bardulph einen Kuß, du süße Dortchen Lakenreißer!" – Da gab ihm Kordelchen, durch diese un-

zeitige Vergleichung beleidigt, geschwind eine derbe Ohrfeige, Ruprecht aber sprang zornig auf sie los, während seine nächsten Nachbarn bemüht waren, ihn festzuhalten. Bei der allgemeinen Bewegung warfen sie mit ihren Ellbogen einige Stühle und mehrere volle Gläser um, der Blasse, der ganz entrüstet sein Buch retten wollte, fiel über ein Stuhlbein, der hinzugesprungene Wirt über den Blassen, Ruprecht mit seinen Verfolgern über den Wirt, und so war auf einmal alles wie ein Rattenkönig von wundersam durcheinanderarbeitenden Armen und Beinen. In diesem Augenblicke hörte man Säbelscheiden über die Hausschwelle klirren, und zwei bärtige Polizeidiener traten in den Garten. „Was für eine skandalöse Aufführung", rief der eine die Erschrockenen an, „ist das jetzt die Zeit, durch schnöden Lärm eine gesittete Bürgerschaft zu turbieren, die, nach sauer erfüllter Berufspflicht, soeben schon den einen Fuß in das Bett gesetzt hat." – „Und die durchreisenden Herrschaften! da fährt eben eine ehrwürdige Matrone erschrocken empor", fiel sein Gefährte ein, indem er auf ein Fenster wies, wo die Dame mit dem Schirm neugierig hervorguckte, bei dieser Apostrophe aber schnell wieder verschwand. – „Nur nicht noch gar räsoniert!" – fuhr der andere zornig fort, da die Schauspieler reden wollten – „wir kennen uns, wir sind verwegene Schuldenmacher, denen kein Gläubiger mehr glauben will." – Rasch an das Licht tretend und ein Papier entfaltend, las er: „Da ist Herr *Ruprecht* – feurig von Nase, erhaben von Nase, blühend von Nase –, was? Nichts als lauter Nase! – Herr *Lothario,* dann, auch Literatus genannt. – Charakter: erster Tenor; besondere Kennzeichen: verdrehte Schleife am Halstuch, ungekämmtes Haar, spricht am vernünftigsten, wenn er betrunken ist, in Summa: großes Genie. – Aber der Teufel mag aus der Beschreibung klug werden, ich verhafte in dem Klumpen da die ersten besten Beine. – Greif' zu!" – Sein Gefährte packte nun ohne weiteres den Ruprecht an den Füßen, der in dem Gedränge vergeblich bemüht war, seine Stiefeln in

den Händen des Häschers zu lassen und sich auf die Strümpfe zu machen. Unterdes hatten sich endlich auch die andern eiligst vom Boden aufgerafft, der Direktor Sorti, schon halb entkleidet, flog in größter Bestürzung herzu, der Hofhund dicht an seinen Waden hinter ihm drein, Kordelchen lachte, der Wirt schimpfte, der Blasse deklamierte fortwährend von persönlicher Freiheit und unverletzlichen Menschenrechten.

„Seid ihr nicht rechte Narren!" rief da auf einmal der Polizeidiener dazwischen und warf Bart, Hut und Rock von sich – es war der Literatus *Lothario*. Sein Gefährte aber verwandelte sich ebenso rasch in Herrn *Fabitz,* den Komikus der Bande.

„Ich wußt' es lange" – sagte Ruprecht, der sich zuerst von dem Schreck erholt hatte –, indem er ruhig seine Pfeife ausklopfte. Die übrigen aber konnten den Scherz nicht so schnell verwinden, dem einen hatten sie auf das Hühnerauge getreten, ein anderer fuhr wütend mit dem Ellbogen aus dem Ärmel und behauptete, das Loch sei erst von jetzt, alle keiften auf Lothario los, während ihnen Fabitz unbemerkt ihr Bier austrank. Lothario aber hatte unterdes vom Reisewagen schnell eine Trommel geholt, setzte sich damit auf den Tisch und begann lustig zu wirbeln, bald piano, bald crescendo, nach der jedesmaligen Stimmung des Redenden. Kein Mensch konnte sein eigenes Wort verstehen, die Zänker schrien sich ganz heiser und verloren die Geduld, einige lachten, Lothario trommelte immerfort, bis alle nach und nach den Platz geräumt und der letzte zornig die Haustür hinter sich zugeschmissen hatte. Nur Kordelchen war zurückgeblieben. Sie setzte sich trotzig neben Lothario auf den Tisch. „Und ich bleibe gerade noch draußen", sagte sie, „mir gefällt die Nacht. Überhaupt", fuhr sie fort, „ich habe dir's schon oft gesagt, dieses stolze, herrische, hochfahrende Wesen sollst du mir endlich einmal ganz lassen!" – „Ich bitte dich", erwiderte Lothario, die Trommel weglegend, „du bist sonst gescheit, und ich kann dich wohl leiden, aber mit dem Las-

sen und Anderswerden, Kind, da ist gar nicht die Rede davon bei mir!" – Kordelchen sah ihn eine Weile an, dann brach sie plötzlich in lautes Lachen aus. „Das wollt' ich nur", sagte sie, „es steht dir gar zu schön, wenn du zornig bist. Gute Nacht!" Hiermit gab sie ihm einen Kuß und war schnell im Hause verschwunden.

Fortunat aber, der unterdes an einem entfernteren Tische sein Abendessen verzehrte, war nicht wenig erstaunt, als er in Lothario, da er vorhin seine Polizeimaske abwarf und ins volle Licht getreten war, auf einmal den wunderlichen Cicerone wiedererkannt hatte, der ihn am ersten Morgen in Hohenstein durch den Garten begleitet. Er benutzte die plötzliche Stille, um den alten Bekannten zu begrüßen, Lothario schien überrascht und sah Fortunat einen Augenblick durchdringend an. – „Hat mich sonst noch jemand dort gesehen?" fragte er endlich, und als Fortunat es verneinte, schien er noch viele Fragen auf dem Herzen zu haben, besann sich aber schnell wieder. „Ich liebe Hohenstein", sagte er nach einer kurzen Pause, „vor allen andern Orten und mache so oft wir in der Nähe vorüberziehen, einen Abstecher nach dem Garten. – Doch heut ist's schon zu spät, wir sprechen wohl noch morgen mehr davon." – Hiermit schüttelte er Fortunat die Hand und ging nach dem andern Flügel des Hauses hin.

Fortunat konnte in seiner Kammer lange nicht einschlafen. Im Hause und unter den Fenstern war alles still geworden, nur die Bäume neigten sich rauschend im Winde, während ferne Blitze zuweilen noch eine plötzliche, gespenstische Helle über den Garten warfen. Da war es ihm, als nahten sich zwei Gestalten von ferne dem Hause. Er erkannte Lothario, der mit einem fremden Manne, den er bisher in der Gesellschaft nicht bemerkt hatte, in lebhaftem Gespräche begriffen schien. Sie verloren sich bald wieder zwischen den Bäumen. Nach einem Weilchen kam Lothario allein zurück, dann wurde alles wieder still.

Siebentes Kapitel

Noch war keine Spur des Morgens am Himmel, da lagen mehrere der jüngeren Schauspieler, denen es zu schwül im Hause geworden war, in ihre Mäntel gehüllt, schlafend auf den Stühlen und Bänken unter den Linden umher. Fabitz, der Komikus, welcher sich über den langen Tisch hingestreckt hatte, erwachte zuerst. Er blickte erschrokken in den Himmel, und da er an dem Stand der Gestirne bemerkte, daß es lange nach Mitternacht war, sprang er sogleich auf den Tisch hinauf und fing wie ein Hahn zu krähen an.

Da fuhr eine dunkle Gestalt nach der andern fröhlich in die dämmernde Nacht empor, schauernd und sich schüttelnd in der kühlen Luft. Lothario aber kam, schon ganz reisefertig, tiefer aus dem Garten und pochte lustig an die Haustür. „Glück auf!" rief er, „fröhliche Botschaft! heraus da! ich habe Fortuna beim Schopf!" – Nun fuhren schlaftrunkene Mädchengesichter neugierig aus den Fenstern, immer mehr Stimmen wurden nach und nach drinnen wach, Türen flogen heftig auf und zu, und bald glich das ganze Haus einem Bienenstocke, der schwärmen will. Fortunat, von dem wachsenden Lärm aufgeschreckt, eilte gleichfalls hinab und fand schon die ganze Gesellschaft in der liebenswürdigsten Laune um Lothario versammelt. Dieser hatte nämlich in der Nacht durch einen Freund die Nachricht erhalten, daß der Fürst auf seinem, eine Tagereise von hier gelegenen, Jagdschlosse angekommen, wo er jeden Sommer einige Wochen hindurch sich den Freuden der Jagd und allerlei wunderlichen, romantischen Einfällen zu überlassen pflege. Dem Briefe lag zugleich eine Einladung des Fürsten an Herrn Sorti bei, mit seiner Truppe so schnell als möglich sich auf dem Schlosse einzufinden. – Dieser unerwartete Glücksfall verbreitete einen allgemeinen Jubel. Ein jeder schnürte eiligst sein Bündel, alle versprachen sich goldene Berge von dem reizenden Aufenthalt, die Männer Ruhm und gutes Leben, die Mäd-

chen vornehme Liebschaften und Geschenke. Fortunat selbst, den sein Weg ohnedies an dem fürstlichen Schlosse vorbeiführte, beschloß, die Fröhlichen bis in die Nähe desselben zu begleiten.

Die aufgehende Sonne traf die muntere Karawane schon draußen auf den Bergen. Kamilla – so wurde die Dame mit dem Schirm genannt – schien Fortunat ausweichen zu wollen und war daher mit Herrn Sorti auf dem Packwagen vorausgefahren. Die andern hatten in dem Städtchen einen Burschen gedungen, der sie auf den Fußsteigen durch den schönen Wald führen mußte, alle waren freudig aufgeregt und sprachen von den Festen auf dem fürstlichen Schlosse und den schönen Tagen, denen sie entgegenwanderten. Ruprecht schritt Tabak rauchend wieder voraus, und intonierte an den schönsten Waldstellen zuweilen: „In diesen heiligen Hallen" oder eine andere würdige Baßarie, während Fabitz unermüdlich die mannigfaltigsten Vögelstimmen nachahmte. Lothario schweifte unterdes, seine Flinte auf dem Rücken, allein auf den Bergen umher, von Zeit zu Zeit hörte man ihn fern im Walde schießen, was jedesmal von der Gesellschaft mit einem lauten Hurra erwidert wurde. – Fortunat aber war wunderlich zumute in der ungebundenen Freiheit. Er atmete fröhlich die kühle Waldluft, sich oft zurückwendend und des muntern Zuges erfreuend, wie die heitern Gestalten mit ihren bunten Tüchern und phantastischen Reisetrachten, bald über ihm auf überhängenden Felsen erschienen, bald tief im dunklen Grün wieder verschwanden.

Als die Sonne schon hoch stand, ruhte die Truppe auf einer schönen Waldwiese aus. Da kam plötzlich auch Lothario aus dem Walde zu ihnen. „Wer ist der fremde Herr hier in den Bergen?" fragte er rasch den Führer, – „da ist so ein Kerl im Frack, der schlüpft schon die ganze Zeit über von Strauch zu Strauch, sieht sich manchmal nach euch um und flieht dann von neuem von eurem Singsang und Geschnatter wie ein Hase auf der Klapperjagd." –

„Das ist gewiß der Doktor", erwiderte der Führer lachend, „der kam einmal mitten in einem Platzregen ins Dorf, wie vom Himmel gehagelt. Die Gegend gefiel ihm, es war grade ein Haus droben leer, da wohnt er seitdem darin, eine alte Frau aus dem Dorfe besorgt ihm das Essen. Am Abend aber, wenn die jungen Burschen und Mädchen vor den Haustüren sitzen, kommt er auch herab, und sie müssen ihm Lieder singen und Märchen erzählen, da hat er schon manche Maulschelle bekommen, wenn er die Mädchen heimlich in die Arme kniff. Aber es ist ihm nicht zu trauen", fuhr der Führer fort, „er hat droben kuriose Bücher, da ist kein christlicher Buchstabe drin, lauter Zirkumflexe, wie wenn eine Spinne übers Blatt gelaufen wäre, und so oft er aus den Büchern murmelt, zieht sich in den Bergkoppen ein Wetter zusammen, dann hört man ihn drinnen im Hause laut sprechen und schimpfen, und ist doch kein Mensch bei ihm."
In demselben Augenblicke erblickten sie auch den Zauberer selbst in der Ferne, wie er soeben hastig den Berg hinanklomm, daß die Steine hinter ihm herabkollerten. – „Den muß ich doch sprechen!" rief Lothario, dem Fliehenden sogleich rasch nachsetzend. Fortunat und noch einige andere von der Gesellschaft schlossen sich neugierig an.
So verfolgten sie rasch die Spur des Fremden, der unterdes schon den Gipfel des nächsten Hügels erreicht hatte; nur seine Rockschöße sahen sie noch manchmal zwischen den Gebüschen fliegen, bis sie ihn zuletzt ganz aus den Augen verloren. Nach mühsamem Umherirren gelangten sie endlich an ein halbverfallenes, rings von hohem Unkraut umgebenes Haus, dessen Türen und Fenster fest verschlossen waren. – „Da ist er gewiß hineingeschlüpft", sagte Lothario und klopfte an die alte Tür. Es erfolgte keine Antwort, aber im Innern des Hauses hörten sie ein gewaltiges Gepolter, als würden Tisch' und Bänke hastig an die Tür geschoben. Lothario pochte von neuem, stärker und immer stärker. Da flog plötzlich oben eine Dachluke

auf, und mit zornblitzenden Augen erschien in der Öffnung ein kleiner, lebhafter Mann, in dem Fortunat zu seinem Erstaunen sogleich den nächtlichen, seltsamen Geiger aus dem Weinkeller in Walters Städtchen wiedererkannte. – „Doktor! – Dryander!" riefen die Schauspieler überrascht aus.

„Was wollt ihr?" fuhr sie der Musikus von oben sehr heftig an. „Denkt ihr, ich werde aus den frischen Berglüften zu eurem dicken Lampendunst hinabkommen und das Volk lassen um das Publikum, und das Rauschen der Wälder um eure Triller und Sentenzen? Geht hinunter und weint um Hekuba, wenn ihr nicht über eure eigene Misere weinen könnt!" – Hier sah er erst seine Zuhörer einen nach dem andern genauer an. „Entsetzlich", sagte er nach einer kurzen Pause zu Ruprecht, „du schaust wie ein brennender Busch aus." – „Und du, idealer, blaß verwaschener Musenbräutigam, redest du jede Magd noch Jungfrau an, und forderst den Stiefelknecht in Jamben?" – „Aber dich, Barbar, der in Blut watet, und von den Tränen des Publikums lebt, dich erkannt' ich gleich an der roten, tyrannischen Stirne wieder!" – Jetzt wurde er plötzlich auch Lothario gewahr, er stutzte, und wie ein Morgenleuchten überflog es sein ganzes Gesicht, dann warf er schnell das Dachfenster zu. – Lothario aber hatte unterdes schon die morsche Tür eingerannt und über die umgeworfenen Stühle, womit sie verrammelt war, das Zimmer erreicht.

Als die übrigen eintraten, fanden sie beide in einem leisen, heftigen Gespräch, das Laub vor dem Hause verbreitete eine wunderbare, grüne Dämmerung über die kleine Stube, durchs offene Fenster hörte man den mehrstimmigen Gesang der zurückgebliebenen Schauspieler von unten heraufschallen:

„Wir wandern wohl heut noch weit.
Wie das Waldhorn schallt!
O grüner Wald,
O lustige, lustige Sommerzeit!"

Dryander war auf einmal wie verwandelt. „Das ist noch das alte Lied“, sagte er und schob ein paar Bücher in seine Rocktasche, „das hab’ ich Euch damals komponiert, um Eure Affekte von den Wirtshäusern auf die schöne, erhabene Natur zu lenken. Seid Ihr noch immer so durstig? Und lebt Kordelchen noch, den Kennern zur Freude und den Frauen zum Trotz?“ –

„O lustige, lustige Sommerzeit!“

klang es wieder herauf. Da hatte der Doktor hastig wieder ein paar Bücher eingesteckt, nahm die Geige unter den Arm und setzte seinen Hut auf. Lothario stopfte ihm schnell noch ein Bündel Wäsche nach, die andern drängten ihn schon zur Tür hinaus, und so stiegen sie eilig mit dem Doktor die Höhe hinab.

Unten auf der Waldwiese fanden sie alles soeben schon im Begriff, wieder aufzubrechen. Ein allgemeiner Jubel begrüßte die Ankommenden, und alle umringten den wiedergefundenen Doktor, der früher einmal als Musikdirektor die Gesellschaft eine Zeitlang begleitet hatte. Dieser embrassierte die alten Kameraden nach der Reihe durch, küßte dann der Dame Kamilla, die eben nicht sehr erfreut schien, ihn wiederzusehen, zierlich die Hand, und half ihr, da Herr Sorti ängstlich zur Fortsetzung der Reise trieb, mit ausnehmendem Anstande auf den Rüstwagen.

Unter diesem Bewillkommnungsgetümmel bewegte sich endlich der Zug langsam weiter. Dryander aber mit seinen dick angeschwollenen Rocktaschen setzte sich an die Spitze desselben, ergriff seine Geige und spielte und sang, daß es weit durch den Wald erschallte:

„Mich brennt’s an meinen Reiseschuhn,
Fort mit der Zeit zu schreiten –
Was wollen wir agieren nun
Vor so viel klugen Leuten?

Es hebt das Dach sich von dem Haus,
Und die Kulissen rühren
Und strecken sich zum Himmel 'raus,
Strom, Wälder musizieren!

Und aus den Wolken langt es sacht,
Stellt alles durcheinander.
Wie sich's kein Autor hat gedacht:
Volk, Fürsten und Dryander.

Da gehn die einen müde fort,
Die andern nahn behende,
Das alte Stück, man spielt's so fort
Und kriegt es nie zu Ende.

Und keiner kennt den letzten Akt
Von allen, die da spielen,
Nur der da droben schlägt den Takt,
Weiß, wo das hin will zielen."

Die Sonne stand schon tief und warf ihre letzten Strahlen zwischen den Baumstämmen schimmernd über die Wanderer, als diese durch die zierlichen Jägerhäuser und die im Walde sich kreuzenden Alleen daran erinnert wurden, daß sie dem Ziele ihrer Reise nicht mehr fern sein konnten. Von weitem vernahm man nun auch Waldhornsignale, einzelne Schüsse und Rufen dazwischen, wie das letzte Verhallen einer großen, weitverbreiteten Jagd. Die Gesellschaft wurde nun nach und nach stiller, jeder rückte sorgsam seine Kleidung zurecht und blickte erwartungsvoll vor sich in die Ferne hinaus. Fortunat aber fühlte sich unbehaglich überrascht, da nun das bisherige fröhliche Reiseleben plötzlich zum förmlichen Metier werden sollte.

Jetzt senkte sich der Weg allmählich ins Tal hinab, da sahen sie eine luftige Säulenhalle, rote Ziegeldächer und stille Wasserspiegel wechselnd aus der Tiefe aufblicken, im-

mer geheimnisvoller, je weiter sie kamen, schimmerte es bald da, bald dort zwischen dem Grün herauf, durch die Wipfel aber leuchtete ein Gewitter, das sie im Walde nicht bemerkt hatten. Auf einmal schrien die Frauenzimmer kreischend auf, denn gerade über ihnen, wie aus den Lüften, ließen sich plötzlich fremde Stimmen vernehmen, und auf der, in viele Klüfte zerspaltenen, fast unzugänglichen Felsenwand erblickte man zwei Schützen, die sich offenbar dort zwischen den Steinen verstiegen hatten. Der eine, ein kleiner, dicker runder Mann, der immer da, wo man ihn an wenigsten vermutete, wie ein Kürbis vom Felsen hing, trat beständig zu kurz, während sein überlanger, hagerer Begleiter jederzeit über sein Ziel hinausschritt. Dieser gab sich, zum Ärger des andern, das Ansehen, ihm beizustehn, obgleich er selbst jeden Augenblick das Gleichgewicht verlor und so den Dicken erst recht mit ins Unglück brachte. Endlich konnten beide weder vor, noch zurück mehr und begannen aus Leibeskräften um Hilfe zu schreien. Da erschallte vom höchsten Gipfel ein mutwilliges Lachen. Die Abendsonne warf unter der schwarzen Gewitterwolke einen dunkelroten Glanz über die ganze Gegend, und in der scharfen Beleuchtung erschien droben plötzlich eine schöne, hohe Mädchengestalt zu Pferde, ein grünsamtenes Jagdkleid umschloß die schlanken Glieder, lange, weiße Federn wogten vom Barett über ihre Schultern hinab. Während ihr Pferd ungeduldig den Boden scharrte, betrachtete sie mit großen, dunkeln Augen die Erstaunten, die unwillkürlich die Unbekannte ehrfurchtsvoll begrüßten. Sie nickte mit dem schwarzgelockten Köpfchen kaum einen flüchtigen Dank, wandte sich dann rasch und war bald in den Abendgluten wieder verschwunden.

„Herrlich!“ riefen mehrere von der Gesellschaft aus. – „Bei Gott“, sagte Lothario, die Reiterin mit durchdringenden Blicken verfolgend, „die haben gewiß heut wieder einmal ihren romantischen Tag!“ – Unterdes waren die andern schon mit langen Stangen, Stricken und Leitern

herbeigeeilt, und es gelang ihnen endlich unter größerem Lärm, als eben nötig war, die beiden verirrten Schützen glücklich auf die Ebene zu bringen. Diese waren indes übel zugerichtet, der eine hatte den Hut, der andere den Rockschoß droben gelassen, am abenteuerlichsten sah der Lange aus mit knappen, grauen Gamaschen und modernem Jagdkleid, halb Überrock, halb Frack, fast lauter Tasche. Kaum aber sahen sie sich unten in Sicherheit, als sie, Gefahr und Dank vergessend, sogleich mit spitzigen Worten aufeinander losgingen. Jeder schob dem andern die Schuld zu, es schien, als habe die schöne Jägerin, der sie in verliebter Galanterie nachgesetzt, sie absichtlich in dieses Klippenlabyrinth verlockt. – So schritten beide, ohne sich um die Schauspieler weiter zu kümmern, eilend dem Schlosse zu, und man hörte sie noch weit durch die Dämmerung zanken.

Jetzt aber fegte der Sturm alles zusammen, von allen Seiten sah man einzelne Jäger an den einsamen Waldeshängen herniedersteigen. Da begann es auch im Schlosse sich wundersam zu rühren, Türen wurden geöffnet und geschlossen, Bediente in bunten, reichen Livreen liefen die Marmortreppen auf und ab, die hellerleuchteten Fenster, hinter denen sich in prächtigen Gemächern einzelne Frauengestalten bewegten, warfen einen magischen Schein weit über den dunkeln Garten. Dann wurde auf einmal alles still in der ganzen weiten Runde, die Nacht und das Gewitter zogen immer tiefer herein; Fortunat, der keine Lust hatte, wieder naß zu werden, war bereits allein nach der Dorfschenke geritten, die Schauspieler schimpften, sie hatten zu ihrem Empfange sich Triumphbogen geträumt, einholende Kammerjunker und den Fürsten von hohem Balkon ihnen entgegenwinkend. – Endlich sahen sie vom Schlosse her Fackeln durch den Waldgrund sich bewegen und erkannten bei den wirren Scheinen mit klopfendem Herzen die bunten Livreen der fürstlichen Bedienten. „Heda, ihr Herren Komödianten!“ rief der eine, „wo Teufel steckt ihr denn?“ – „Nun Gott behüt’ uns!“ – sagte

ein anderer, im Kreise umherleuchtend – „das hängt ja wie Meltau an allen Sträuchern, als hätt' es Plunder geregnet!" – Kamilla, höchst entrüstet, rauschte mit ihrem vornehmsten Anstande daher und ließ einiges von impertinenten Domestiken fallen. Da war aber nicht lange Zeit zum Ärgern und Händelmachen. Denn der Gewitterwind wühlte schon in den Flammen der Fackeln und in den Tüchern der Damen, die Bedienten trieben zur Eile, Mäntel und Regenschirme flogen verworren durcheinander, und so wälzte sich alles in unordentlicher Flucht dem Schlosse zu.

Nur Lothario war zurückgeblieben, denn die schöne Jägerin mußte noch in den Bergen sein. Und er irrte sich nicht. Zwischen den Blitzen von Fels zu Fels, daß ihm schwindelte, lenkte sie mit kühner Gewandtheit ihr Pferd langsam den schmalen Steig hinab. Von dem letzten Abhange endlich wagte es einen verzweifelten Sprung und stürzte unten samt der Reiterin auf dem Rasen zusammen. In demselben Augenblick riß sie es gewaltsam wieder empor, beide hatten keinen Schaden genommen, nur der Zaum war entzwei. Da sprang Lothario rasch hinzu, ein langer Blitz beleuchtete plötzlich die ganze schöne Gestalt. „Wie das blendet!" rief er, während er, auf den Nakken des Pferdes gelehnt, ihr lächelnd unter dem Barett in die Augen blickte. – Sie sah ihn groß an, – „da, die Kinnkette noch", erwiderte sie kurz und stolz, dann, als er den Zaum in Ordnung gebracht, drückte sie rasch die Sporen ein, und zwischen den roten Scheinen der Windlichter sah er ihren weißen Federschmuck, wie einen Schwan, durch die finstere Nacht dahinziehn.

Achtes Kapitel

Als Fortunat erwachte, blickte er erstaunt in einem hohen, vom Morgenrot schimmernden Gemache umher. Nach und nach erst besann er sich auf alles, wie er gestern

noch vor Ausbruch des Gewitters aus der Dorfschenke in das fürstliche Schloß geladen worden, wie wunderbar da beim Widerscheine der Blitze das Schloß in der Nacht aussah, das Getümmel dann im Hofe, und wie darauf ein Bedienter ihn mitten aus dem Gewirre in dieses Gemach gewiesen. Hier hatte er durch das Fenster bemerkt, daß die übrigen Schauspieler nochmals weiterziehen mußten und beim trüben Schein einiger Windlichter einen dunklen Baumgang hinabgeführt wurden, bis zuletzt die Lichter, das Rumpeln des Reisewagens und die wohlbekannten Stimmen sich in dem Plätschern des Regens verloren, der nun plötzlich in Strömen herabstürzte.
Jetzt aber regte sich noch kein Laut, nur draußen blickten einzelne Flüsse und Landschaften mit funkelnden Kirchtürmen schon geheimnisvoll zwischen den hohen Bäumen herauf. Da kleidete Fortunat sich schnell an und eilte durch das stille Haus die breiten, dämmernden Marmortreppen hinab. Unter einer luftigen Säulenhalle, die von beiden Seiten mit hohen, ausländischen Blumen besetzt war, trat er in den prächtigen Garten. Hier war nach dem erfrischenden Regen der Morgen wie ein bunter Teppich ausgebreitet, auf dem das Schloß gleich einer schlummernden Sphinx noch rätselhaft ruhte. – Er wollte eben tiefer in das Grün hineingehen, als er überrascht in einiger Entfernung folgendes Lied singen hörte:

„Aus Wolken, eh' im nächt'gen Land
Erwacht die Kreaturen,
Langt Gottes Hand,
Zieht durch die stillen Fluren
Gewaltig die Konturen,
Strom, Wald und Felsenwand.

Wach' auf, wach' auf! die Lerche ruft,
Verträumt in Duft,
Beginnt auf Berg und Talen
Ringsum ein himmlisch Malen

In Meer und Land und Luft.
Aurora taucht die Strahlen.

Und durch die Stille, lichtgeschmückt,
Aus wunderbaren Locken
Ein Engel blickt. –
Da rauscht der Wald erschrocken,
Da gehn die Morgenglocken,
Die Gipfel stehn verzückt.

O lichte Augen, ernst und mild,
Ich kann nicht von euch lassen!
Bald wieder wild
Stürmt's her von Sorg' und Hassen –
Durch die verworrnen Gassen
Führ' mich, mein göttlich Bild!"

Fortunat folgte dem Gesang, der von einem entfernten Flügel des Schlosses herzukommen schien. Die hohe Tür war nur angelehnt, er trat herein und befand sich in einer schönen, großen Kapelle, die durch eine Kuppel erleuchtet wurde. Auf einem Gerüste stand dort ein Maler, welcher in dieser stillen, kühlen Einsamkeit, zwischen den von oben einfallenden Morgenlichtern und den halbvollendeten, betenden Gestalten mit ihren reichen, leuchtenden Gewändern wie in dem Kelch einer wunderbaren Blume schwebte. Er hörte auf zu singen, als er unten den Fremden gewahrte, und wandte schnell ein munteres Gesicht zwischen umwallenden, braunen Locken aus seinem Himmel hinab. – „Glück auf!" rief ihm Fortunat, überrascht von der ganzen, unerwarteten Erscheinung, fröhlich zu, „das ist eine herrliche Werkstatt!" – Der Maler nickte lächelnd und fuhr in seiner Arbeit fort, kehrte sich dann aber, plötzlich abbrechend, wieder zu Fortunat: „Sind Sie nicht gestern abend mit den Schauspielern gekommen?" – „Ja, und zugleich von ihnen abgekommen, ich weiß nicht wie", erwiderte Fortunat. – „O, die sind

gar nicht weit“, sagte der Maler. „Und eigentlich ist auch heute Aurora zu schön, um ihr hier ins Gesicht zu klecksen, ich will Sie lieber gleich zu Ihren Kameraden führen.“ – Bei diesen Worten hatte er rasch Pinsel und Palette weggelegt und kam die Leiter herab. Es war ein kecker, vollwangiger Jüngling mit bloßem Hals und knappem, sehr zierlichem, deutschem Rock. Er verschloß die Tür, da sie hinaustraten, und führte Fortunat eilig durch den Baumgang, in welchem gestern nacht die Schauspielergesellschaft verschwunden war. „Das muß ein glückliches Leben sein“, sagte er, „wie oft hab' ich mir schon gewünscht, so mit fröhlichen Gesellen ins Blaue hinein zu ziehen! Wir Maler sind überall an Ort und irdisches Material gebunden. Da sind die andern Künstler besser dran, zumal der Dichter. Die ganze schöne Welt ist sein Revier, und wo er singt, ist der Himmel. – Aber da sind wir schon!“ unterbrach er sich hier. „Sehn Sie dort. Es ist eigentlich ein altes Gartenpalais, das lange wüst und verlassen stand. Ich wohne auch drin, seit ich hier male, nun hat der Fürst auch die Gesellschaft mit hineinquartiert. Hören Sie doch, was für ein Rumor darin! Das ist ja wahrhaftig wie eine Menagerie, wo unzählige Loris und Papageien durcheinander kreischen und manchmal eine alte Hyäne dazwischen gähnt.“

Fortunat erblickte nun am Ende des Baumganges einen zweiten grünen Platz, wo mehrere Figuren von Buchsbaum, halbzertrümmerte Statuen und vertrocknete Wasserkünste einen ehemaligen französischen Garten andeuteten, der jetzt nur noch durch einzelne Kaiserkronen und dunkelglühende Päonien seltsam an die alte Herrlichkeit erinnerte. Im Hintergrunde stand ein alter, schwerfälliger, von der Zeit gebräunter Palast, dessen vornehme Gesimse mit Verachtung auf die aus den Fenstern flatternde Wäsche und auf Kamillas Regenschirm herabzublicken schienen, den sie vor ihrem Schlafzimmer als Markise ausgespannt hatte.

Fortunat trat mit dem Maler hinein und begrüßte seine lu-

stigen Reisegefährten, die vor Freuden auch nicht mehr schlafen konnten und sich hier nach jahrelangem, dunklem Umhertreiben in den Dachstübchen kleiner Städte sehr behaglich und laut in dem ungewohnten Glanze sonnten. Ein großer Saal mit Stuckverzierungen, verblichenen Tapeten und einem altväterlichen Billard in der Mitte, diente ihnen zum Versammlungsplatz, und wenngleich die Bursen des Billards zum Teil vom Zahne der Zeit schon abgenagt waren, so hatten die erfinderischen Geister doch sogleich ihre eigenen, ohnedies ziemlich überflüssigen Geldbeutel daran geheftet und schnitten ihre Karoline mit mehr Behagen als Geschicklichkeit. Nur Kordelchen erwies sich als Meisterin, wobei sie, in gewandten Stellungen über der grünen Tafel schwebend, ihr zierliches Figürchen zu zeigen willkommene Gelegenheit hatte.
Der enthusiastische Maler begann sogleich eine Partie mit ihr, und Fortunat wollte eben Lothario aufsuchen, den er in der Gesellschaft vermißte, als der sonst friedfertige Komiker Fabitz plötzlich mit einem seltsamen jungen Manne, mit welchem er draußen in Zank geraten, in den Saal hereinstürzte. Der junge Mensch trug die altdeutsche Tracht, deren verschossenes Schwarz aber schon bedeutend ins Grauliche spielte; lange, grobe Haare hingen ihm von beiden Seiten bis über die Schultern herab und gaben dem langen, eckigen Gesichte ein gewisses antiquitätisches Ansehen. Es ergab sich, daß es gleichfalls ein Maler, namens Albert, war, der auf seiner Rückreise von Rom hier seit einiger Zeit Beschäftigung und günstige Aufnahme gefunden. Dieser hatte nun kaum in Erfahrung gebracht, daß bei der eben angekommenen Gesellschaft ein Herr Fabitz den Kasperl zu spielen pflege, als er sogleich mit wahrem Missionarieneifer auf den Unglücklichen losging und ihm über das Unwürdige, Verkehrte und daher Unhaltbare seines Kunstgewerbes die gemessensten Vorstellungen machte. Er sprach viel vom ernsten Norden, wo die edlen Eichen höherer Bildung solch niederes Un-

kraut gar nicht aufkommen ließen, ja eine norddeutsche Zunge, wie die seinige, entsetzte sich schon vor dem barbarischen Laute: Kasperl! Fabitz dagegen meinte, er kenne zwar von den nordischen Zungen bloß die geräucherten, die langen, norddeutschen Kaspars aber seien wahrscheinlich nur zu langweilig, um auf das Theater gebracht zu werden. – Zuletzt aber, da ihm die ganze Erscheinung des Norddeutschen etwas Neues war, überwältigte ihn sein Naturell. Unwillkürlich nahm er nach und nach, Zorn und Streitpunkt vergessend, die wunderliche Haltung, Gesicht und Stimme seines Gegners, der in seinem fanatischen Eifer nichts davon merkte, selber an und focht so verzweifelt in aufgeschnappten, hochtrabenden Sentenzen, daß sein Gegner ganz konfus wurde. – Kordelchen hatte schon lange vom Billard zugehorcht. „Allerliebster Narr", rief sie nun hinzuspringend aus, und gab ihm einen herzhaften Kuß. „Pfui! wenn er nur nicht so häßlich wäre!" sagte sie dann, sich den Mund schnell abwischend.

Währenddes hatte sich, ohne von dem Streite Notiz zu nehmen, ein kurzer, runder Mann zu Fortunat gesellt, der sich ihm als den fürstlichen Schulrat vorstellte, und in welchem er sogleich den dicken Schützen wiedererkannte, dem sie gestern vom Felsen geholfen. Fortunat wußte gar nicht, wie ihm geschah, da der Kleine auf einmal sehr gelehrt von Poesie, Kunst und Religion zu sprechen anfing und sich endlich angelegentlichst erbot, ihn in den wenigen Augenblicken der Muße, die ihm blieben, mit den mancherlei Merkwürdigkeiten des Orts bekanntzumachen. Kaum hatte der kämpfende Maler Albert den Schulrat erblickt, als er vornehm den Streit abbrach und sich zu ihnen wandte. – „Vortrefflich", sagte der Schulrat, sich an Fortunats Arm hängend, „so geleite ich Sie gleich zu einem Götterfrühstück, womit ich mich jeden Morgen für meine Berufsgeschäfte zu stärken pflege." – So schritten sie eilig durch einen langen Korridor zu einer schweren, eichenen Tür, die Albert mit einer gewissen Feierlichkeit

öffnete. Es war sein Atelier, ein hohes, ritterliches Gemach, an dessen schmuckloser Hauptwand ein großes, mit der Jahreszahl 1813 bezeichnetes Schwert hing, um das sich ein verwelkter Eichenkranz wand. „Das ist mein treuer Reisegefährte“, sagte Albert zu Fortunat, „und wenn mich schlaffe Ruh’ oder weichliche Lust schleichen wollen, blick’ ich die Eisenbraut an und gedenke der ernsten, großen Zeit.“ – „Ach, das ist schon eine alte Geschichte!“ entgegnete Fortunat lachend. – „Sind Sie damals mit zu Felde gewesen?“ fragte der Maler etwas spitzig. – „Freilich“, erwiderte jener, „das versteht sich ja ganz von selbst.“

Inzwischen befand sich der Schulrat schon mitten unter Alberts Arbeiten, die in dem Gemache umherstanden und von dem erstaunenswerten Fleiße des Malers zeugten. Da waren die ungeheuersten Anstalten zur Kunst: Gliederpuppen, sorgfältig gefaltete Mäntel, Modelle und Büsten, dazwischen mehrere vollendete Bilder, Historienstücke aus der antiken Heroenzeit von sehr zusammengesetzter, studierter und nicht leicht faßlicher Komposition. „Göttlich!“ rief der Schulrat einmal über das andere aus, während er mit Kennermiene beschäftigt war, jedes Bild genau in das rechte Licht zu stellen. „Sehen Sie den ätherischen Hauch des Inkarnats, die Perspektive, diesen klassischen Ausdruck!“ – „In der Tat, ein philosophischer Pinsel“, erwiderte Fortunat. Denn diese anmaßlichen, affektierten Heldengestalten voll Männerstolz und Männerwürde wollten ihm nicht im mindesten behagen, und die Jungfrauen mit ihrer langgestreckten, anmutlosen Tugendlichkeit kamen ihm gar wie gemalte Begriffe der Jungferschaft vor.

„Nun, ich muß mich nur wieder mit Gewalt losreißen“, sagte endlich der Schulrat, seinen Hut ergreifend, „ernstere Geschäfte rufen mich.“ – „Ein Genie!“ flüsterte er, im Fortgehen auf Albert deutend, Fortunat zu. – „Ein tiefer umfassender Geist!“ sagte Albert, als der Schulrat verschwand.

Fortunat aber hatte unterdes eines von den kleineren Bildern angezogen. Man sah Rom in der Ferne mit seinen phantastischen Trümmern und Palästen in der vollen Glut des südlichen Abendhimmels. Im Vorgrunde, von Rom fort, schritt einsam durch das schon dunkelnde, öde Feld ein einzelner Mann mit antikem Faltenwurf des Mantels und feierlich ernster Miene, an dem Fortunat sogleich den Maler selbst erkannt hätte, wenn er auch nicht zum Überflusse noch mit dem obengedachten Schwerte vom Jahre 1813 umgürtet gewesen wäre. – „Aber warum in aller Welt kehren Sie dieser leuchtenden Wunderpracht hier so eilfertig den Rücken?" fragte er erstaunt. – „Dieses Bild", erwiderte Albert mit seinem allerlängsten Gesicht, „bezeichnet eigentlich die dunkle Führung überhaupt, die in meinem Leben waltet. Rom ist herrlich, und ich nahte voll Ehrfurcht den alten Heldenmalen. Aber das leichtsinnige Geschlecht und das Klingeln der Bonzen über den Gräbern versunkener Größe störte und empörte mich. Ich konnte mich den Anmutungen des Aberglaubens, auch nur zum Scheine, nicht gefällig erweisen und hatte beständig Verdruß. Dazu kam, daß das Geschick meines deutschen Vaterlandes, wo eine neue, große Zeit sich ausgebärt, heimlich an meinem Herzen fraß, ich hatte nirgends Ruhe. Meine Kameraden gefielen sich dort bald höchlichst – ich aber ermannte mich zur rechten Zeit und flüchtete vor den gleißenden Schlingen doppelter Knechtschaft nach dem ernsten heimatlichen Norden."

„Norden?!" – rief Fortunat erschrocken über dieses plötzlich wiederkehrende Lieblingsthema des Malers aus und griff hastig nach seinem Hute. Albert, welcher dies für eine Aufwallung übereinstimmender Empfindung halten mochte, drückte ihm stumm die Hand, aber mit so seltsamer Kreuzung der Finger, daß es Fortunat sogleich für das heimliche Zeichen irgendeines ihm fremden Bundes erkannte. Fortunat besann sich nicht lange, sondern erwiderte den Druck, zu Alberts Verwunderung, mit

noch abenteuerlicheren Handgriffen und stürzte dann ins Freie hinaus.
„Verdammte Wirtschaft!" rief er draußen, durch den Baumgang eilend, „überall vertreten einem solche lange Gesichter das Morgenlicht! Lassen sich da von irgendeinem kritischen Kleinmeister eine angeräucherte Brille aufheften, womit sie dann in alle Welt gehen, die Völker zu richten. So zieht das Geschmeiß, wie die Wanderraupen, durch den Glanz der Länder in stillem Wahnwitze fort, wenn es sonst Wahnsinn ist, die Dinge anders anzusehen, als sie wirklich sind!" – Zuletzt mußte er selbst laut auflachen über den wunderlichen Zorn, in den ihn das Larvenkunstkabinett des Malers versetzt hatte. Die Morgensonne spielte golden durch die Wipfel der Bäume und unzählige Vögel sangen. Er blickte fröhlich umher und fand, daß die Welt trotz allen Narren so schön und lustig blieb, wie sie war.

Neuntes Kapitel

Es war schön anzusehen, wie auf der luftigen Rampe des Schlosses, die gleich einer Blumenzinne weit über die Wälder hinausragte, schlanke Frauengestalten und bunte Uniformen zwischen den dunklen Orangenbäumen hervorschimmerten. Oben saßen die Fürsten, Herren und Damen in der heitern Morgenkühle auf buntgestickten Feldstühlen umher, die Abenteuer der gestrigen Jagd besprechend. Mehrere Bände von Shakespeare mit funkelndem Goldschnitt lagen auf einem zierlichen Tischchen, Notenhefte und eine Gitarre daneben; der Morgenwind blätterte lustig darin und ging durch die Saiten, daß es von Zeit zu Zeit zwischen dem Plaudern und Lachen einen fröhlichen Klang gab. – Weiter zurück aber standen die zur Musterung heraufbeschiedenen Schauspieler in ihren besten Feierkleidern, ganz verwirrt unter den Fürsten und

Grafen, die sie doch so oft auf ihren Brettern gespielt hatten. Vergebens suchten sie unter den fremden Gesichtern den geraden Kriegshelden, den schlauen Beichtvater, den falschen Minister; Herr Sorti vergaß darüber ganz seine wohlersonnene, altmodische Anrede, sie fanden alles anders, als sie sich's unten eingebildet hatten. Mit ehrerbietiger Neugier blickten sie zuweilen seitwärts durch die offene Tür in die prächtigen Gemächer hinein, aus denen der glatte Fußboden, hohe Spiegel und Statuen zwischen bronzenen Kandelabern geheimnisvoll glänzten. Manches junge Herz aber wünschte sich hundert Meilen von hier, denn unter der Terrasse pfiffen die Vögel lustig in der alten Freiheit, und zwischen den Wipfeln blickte die Landschaft so heiter herauf, als riefe es: Kommt nur wieder hinunter, da draußen ist's doch viel schöner!

Der Fürst, ein junger, schöner Mann in bequemer Jagdkleidung, war unterdes zu ihnen getreten und entschuldigte seine gestrige Vergeßlichkeit so leicht und vornehm, daß sie ihm für ihren schlechten Empfang noch danken mußten. Er belobte Herrn Sorti über die Eile, mit der er seiner Einladung gefolgt, und wußte in aller Geschwindigkeit durch Andeutungen seltener Belesenheit und Sachkenntnis allen zu imponieren. Dazwischen blickte er manchmal verstohlen nach Kordelchen, die das auch sogleich bemerkte und, schlau ihre Augen niederschlagend, die Verwirrte spielte. Kammerherren, junge Offiziere und Jagdjunker mischten sich nun mit in die Unterhaltung, die Schauspieler wollten in auserlesenen Redensarten ihren Weltton zeigen, die Mädchen waren naiv, die Junker charmant, zwischen ihnen und den Feldstühlen der Damen flogen häufig französische Witzworte, wie zierliche Pfeile über den glatten Boden hin und her, deren Zielscheibe eben nicht zweifelhaft war. Unter ihnen fiel der lange Schütz von gestern am meisten auf, ein reisender Lord, der überall wie ein Kamelhals mit seiner Lorgnette über die andern hervorragte. Er versicherte jeden seiner Pro-

tektion und sprach immerfort von Kunst und dramatischer Kunst und mimischer Kunst in so wunderlichem Deutsch, daß einer den andern nicht verstand.
Die Konfusion aber wurde noch immer größer. Denn seitwärts hinter einer phantastischen Palme, auf deren breiten Blättern ein Papagei linkisch auf und nieder kletterte, stand die kühne Reiterin von gestern und neckte, wie es schien recht absichtlich den Vogel, dessen durchdringendes Gekreisch jeden Augenblick den galanten Diskurs verstörte. Sie beachtete die Komödianten nicht, aber zuweilen funkelten ihre Blicke zwischen den Zweigen nach Fortunat und Lothario herüber, welcher den ersteren mit heraufgeschleppt und soeben der Fürstin als einen geistreichen, nur erst kürzlich zu ihnen gestoßenen Volontär vorgestellt hatte. Die Fürstin, eine junge, schmächtige Dame mit schwarzem Haar, bleichem Gesicht und feurigen Augen, in graziöser Lebhaftigkeit bald zu diesem, bald zu jenem Herrn ihres Gefolges plaudernd zurückgewandt, nun witzig, dann sinnig, dann wieder gelehrt, wechselte in wenigen Minuten verschwenderisch alle Farben der neuesten Bildung. Dazwischen blickte sie oft Fortunat fast lauernd an, als wollte sie prüfen, welchen Ton sie ihm gegenüber eigentlich anschlagen sollte. Sie schien es wunderbarerweise recht ausschließlich auf den Beifall des unbekannten jungen Mannes abgesehen zu haben, der sich, wie in einem plötzlichen Feuerwerk, vor den Raketen und steigenden Leuchtkugeln dieser Unterhaltung gar nicht zu fassen und nicht zu retten wußte. – Dem Fürsten aber waren die Blicke der Gräfin *Juanna* – so nannte man die schöne Jägerin – nicht entgangen, er wurde auf einmal verstimmt und entließ schnell die Schauspielergesellschaft. „Das ist ein lustiges Metier", sagte er dabei noch mit besonderem Nachdruck zu Fortunat, „sich so täglich in einen andern zu verwandeln, gestern ein Graf, heute ein Schauspieler und immer ein Poet." – „Ganz interessant", meinte die Fürstin, „die Exposition ist romantisch, die Motive lassen sich ahnen, ich bin nur

auf den letzten Akt begierig." – Fortunat war ganz verwirrt, noch mitten in dem Getümmel des Abschiednehmens konnte er bemerken, wie die Fürstin der unterdes hinzugetretenen Gräfin Juanna sehr lebhaft etwas zuflüsterte, was ihm zu gelten schien. „Also dieser?" – sagte die Gräfin, den schönen Mund spöttisch aufwerfend. – Und wie sie so fortgingen und die Terrasse hinter ihnen versank, und nur noch Juanna an dem marmornen Geländer hoch über dem schönen, weiten Kreise der Wälder stand, da war es, als sei sie die Fürstin hier, der alle andern dienten. –

Die Schauspieler schritten nun eifrig schwatzend durch den Garten, die meisten waren ganz begeistert und wie berauscht, andere, die sich zurückgesetzt glaubten, sprachen von drückender Hofluft und dem schlüpfrigen Boden der vornehmen Welt. Fortunat aber fiel nun erst alles auf: seine gestrige Aufnahme im Schloß, vorhin die Dienstfertigkeit des Schulrats, die Reden der Fürstin und Juannas letzter Ausruf. – Sollten sie den reisenden *Baron* in mir wittern? dachte er, kennen mich doch die Schauspieler selbst nicht, wie sollten die droben es wissen! –

Am Abend desselben Tages ruhete er mit Lothario auf dem grünen Abhange einer Höhe und schaute fröhlich über die Wälder in die weite, fruchtbare Gegend hinaus, in die er nun bald selbst mit dem blauen Strome hineinziehen sollte. Lothario immer rastlos umherschweifend, hatte in der kurzen Zeit alle verworrenen Verhältnisse ihres neuen Aufenthalts schnell überblickt und entwarf nun in seiner Art eine Musterkarte davon. „Der Fürst", sagte er, „ist ein erstaunlicher Virtuos, er spielt die schwierigste Romantik vom Blatte weg, ohne eben selbst zu komponieren. Die Fürstin ist ganz und gar sinniger Roman, durch viele Hände gegangen, schon sehr zerlesen; ich glaube, der lange Lord studiert sie jetzt. Diese wilde, schöne Gräfin dann, die ihnen wie ein Hirsch durch alle ihre künstlichen Gehege bricht und die Meute Liebhaber

hinter sich für Hunde hält – wahrlich, so scheues Wild weckt recht das Jagdgelüste!" – „Nimm dich in acht", entgegnete Fortunat; „was mich betrifft, so kümmert's mich wenig, wie sie sind, das Ganze zusammen macht sich doch schön, und mehr verlang' ich nicht von ihnen." – Lothario sah ihn ein Weilchen fast ärgerlich an. „Ich begreif's nicht", sagte er dann, „wie ihr Dichter es vor Langerweile aushaltet, so ein dreißig bis fünfzig Jahre auf der ästhetischen Bärenhaut rücklings über zu liegen und Kriegstrubel, Philosophie, wilde Jäger und singende Engel wie ein Wolkenspiel über euch dahinziehen zu lassen, um daraus ganz gemächlich ein paar dicke Romane zusammenzuschreiben, die am Ende niemand liest. Zum Teufel, ich bin keine Äolsharfe, die nur Klang gibt, wenn ein Poet ihr Wind vormacht! Ist das Leben schön, so will ich auch schön leben, und selber so verliebt sein wie Romeo, und so tapfer wie Götz, und so tiefsinnig wie Don Quixote. Um die Schönheit will ich freien, wo ich sie treffe, und mich mit den Philistern drum schlagen, daß die Haare davonfliegen. Warum sollte man so ein lumpiges Menschenleben nicht ganz in Poesie übersetzen können?" – „Du bist ein wunderlicher Mensch", unterbrach ihn Fortunat, „ich glaube, du könntest ein großer Dichter sein, wenn du nicht so stolz wärest." – „Ich?" – erwiderte Lothario fast betroffen und sah einen Augenblick nachdenkend vor sich hin.

Hier wurden sie auf einer weiter ins Land hinausgelegenen Anhöhe mehrere der Schauspieler gewahr, die soeben zwischen den Gebüschen emporstiegen und sich gleichfalls an der schönen Aussicht zu ergötzen schienen. Sie konnten deutlich unterscheiden, wie Herr Ruprecht sein altes Perspektiv gemächlich aus dem Futteral nahm, es wie ein Fühlhorn bald weit ausstreckte, bald wieder einzog und damit in die Ferne zielte. Bald aber schienen sie unten etwas Besonderes auf dem Korn zu haben, das Fernrohr ging eilig aus Hand in Hand, und Fortunat bemerkte nun auch seinerseits einen Fußgänger im Tal, welcher bequem

zwischen Wiesen und Büschen daherkam, zuweilen stehenblieb und sich nach den schönen, abendroten Gründen heiter zurückwandte, dann zufrieden wieder weiter schlenderte. Auf einmal erhoben die Schauspieler ein wütendes Freudengeschrei und winkten mit Perspektiv und Hüten und Schnupftüchern. Jetzt schien auch der Wanderer sie zu erkennen, er warf jubelnd seinen Hut hoch in die Luft und schritt dann eilig den Berg hinan. – „Wahrhaftig, den sollt' ich kennen!" rief Fortunat ganz erstaunt aus. – „Gott schütz', gewiß noch ein Dichter!" entgegnete Lothario, indem er aufsprang und ohne weiteres in den Wald hineinging.

Fortunat eilte sogleich zu den Schauspielern hinüber. Aber eine tiefe Kluft lag dazwischen; er verlor sie im Walde bald aus dem Gesicht und wußte nicht, wo er war, als auf einmal der Wanderer, der gleichfalls den nächsten Weg gesucht und den rechten verfehlt hatte, sich mühsam neben ihm durch das Gestrüpp hervorarbeitete. „Fortunat!" rief er höchst überrascht und sichtbar verlegen aus, da er den alten Bekannten erblickte. – „Mein Gott! Otto!" erwiderte jener, „wie kommen Sie hierher?" – „Ich?" – sagte der Student ganz verwirrt – „ist denn das nicht der fürstliche Park, wo die Schauspielergesellschaft des Herrn Sorti –"

Fortunat aber hatte keine Zeit mehr zu antworten, denn um eine Waldecke sahen sie plötzlich einen ganzen Haufen Lumpengesindel von weitem auf sich zuwanken, das sie im ersten Augenblick für Zigeuner erkannten. Sie schienen untereinander in Händel geraten zu sein und kamen in vollem Zanke daher; einige von ihnen waren bemüht, von hinten einen widerspenstigen Esel vorzuschieben, auf dem eine seltsame, phantastisch geschmückte Weibergestalt saß, die voll Zorn nach den ungestümen Treibern zurückschimpfte. Wie eine Zigeunerkönigin hatte sie ihr langes zottiges Haar mit einer Schnur von Gold und Edelsteinen oben in ein Krönchen zusammengefaßt, in den Ohren trug sie schwere Gehenke von ge-

schmelzter Arbeit, ihre Schabracke war von Scharlach, das grüne Kleid mit silbernen Posamenten verbrämt, und ihr schneeweißes Hemd an den Nähten mit schwarzer Seide nach böhmischer Art ausgenäht, woraus sie hervorschien wie eine Heidelbeere aus der Milch. – Jetzt erst erkannte Fortunat in dem Gesindel nach und nach die Gesichter der Schauspieler, ohne zu begreifen, wie sie zu dem Narrenstreiche kamen. Seitwärts bemerkte er nun auch Kamilla, welche die Rolle der Preziosa übernommen zu haben schien, wozu sie ihre große, noble Figur besonders geeignet glaubte. Sie schwärmte abgesondert von den andern, eine Gitarre im Arm, und sang: „Einsam bin ich nicht alleine." Aber sie blieb doch allein, denn alles lief einer jungen, schönen Zigeunerin nach, die plötzlich wie ein Reh aus dem Walde brach. Die pechschwarzen Haare hingen glänzend über Stirn und Wangen, ihr Gesicht war wie eine schöne Nacht. Sie blieb dicht vor Otto stehen und funkelte ihn neugierig mit den Augen von oben bis unten an. „Wußt' ich's doch", sagte sie dann, „daß es so kommen wird." – Es war Kordelchen. „Silentium!" hörte man nun auf einmal die abenteuerliche Gestalt durch das Getümmel rufen, die unterdes auf ihrem Esel herangekommen war. „Ei, mein schöner, weißer, junger Gesell", redete sie Otto an, „was machst du hier? wo kommst du so allein daher?" – Der Esel, der unterwegs ein Maul voll Gras genommen, sah die Gesellschaft, seine langen Ohren schüttelnd, ruhig an und hieb mit dem einen Hinterfuß nach den Komödianten, die ihn heimlich zwickten. Otto aber, von der allgemeinen Lust mit angesteckt, antwortete: „Meine großmächtige Frau Libuschka, ich komme von Haus und bin willens, in der Welt ein mehreres zu studieren oder einen Dienst zu bekommen, denn ich bin ein armer Schüler." – „Daß dich Gott behüte, mein Kind!" versetzte die alte Zigeunerin – „aber zum Teufel! laßt die Faxen, ich falle wahrhaftig herunter!" rief sie dazwischen den Schauspielern plötzlich mit grober Stimme zu, an der Fortunat sogleich Herrn Ruprecht erkannte. Dieser aber

ließ sich dadurch nicht irremachen. „Wenn du Lust hast, bei uns zu bleiben", fuhr er fort, „so ist der Sache bald abgeholfen." – „Ich will noch ein paar Tage mit mir selbst zu Rate gehen", erwiderte Otto, „des Studierens und Tag und Nacht über den Büchern zu hocken bin ich schon vorlängst müd' worden." – „Du hast einen weisen Menschensinn, mein Sohn", versetzte hier Ruprecht, „und kannst hierbei leicht abnehmen und probieren, was unsere Manier vor anderer Menschen Leben für einen Vorzug habe, wenn du nämlich siehst, wie wir hier in unserer Freiheit auf den alten Kaiser leben, wie die Marder und Füchse. Was ist Reichtum, was ist Geld, Habe? Wenn ich's nicht habe, acht ich's für gar nichts, und wenn ich's habe, schmeiß ich's gleich wieder weg. Man muß immer als Philosoph denken, glaube einem alten Genie, mein Sohn, und werden die Lichter ausgeputzt und es kommt die Nacht und die Schlafenszeit, so sind doch alle wieder gleich, Zigeuner und andere Leut'!"
„Oho!" riefen hier die andern darein, denen der Sermon schon zu lang wurde, „eine moralische Libuschka! eine philosophische Zigeunerin!" Ruprecht schimpfte sie ganz erbost Ignoranten, die wie Ochsen mit eingelegten Hörnern ins Blaue hineinrennten. Aber sie hörten nicht auf ihn. Ein paar rüstige Gesellen erwischten Otto bei den Beinen und schwangen ihn vor die Frau Libuschka auf den Esel, den Kordelchen unterdes mit bunten Bändern ausgeschmückt hatte; andere faßten die Zügel, und so wälzte sich der ganze tolle Zug nach dem Gartenpalaste hin.
Hier aber wurden sie selbst überrascht, die Zurückgebliebenen hatten sich schnell verkleidet und unter den Bäumen bunte Zelte aufgeschlagen, so lagen sie an lustigen Feuern umher, und zu Fortunats Verwunderung kam es nun nach und nach heraus, daß sie Otto als ein neues Mitglied ihrer Truppe heute hier erwartet hatten. Unter ihnen erwies sich *Guido* besonders geschäftig, der junge, hübsche Maler aus der Kapelle, der in seiner sorgfältigen Zi-

geunertracht sich selbst sehr hübsch zu finden schien und, von Zeit zu Zeit Kordelchen feurige Blicke zuwerfend, wohlgefällig sein Schnurrbärtchen strich. Er hatte brennende Pechkessel besorgt und war eifrig bemüht, die phantastischen Gestalten malerisch um die Flammen zu gruppieren und überall die rechten Lichteffekte anzubringen. Er mußte indes gar bald alles gehen lassen, es war schlechterdings keine Ordnung und kein künstlerisches Motiv hineinzubringen. Über dem dunklen Berge aber trat plötzlich der Mond aus einer Wolke und beschien die stillen Wälder und Gründe; da war auf einmal alles in der rechten, wunderbaren Beleuchtung: das öde Haus, der altmodische, halbverfallene Garten, die wildverwachsenen Statuen und die abenteuerlichen Gestalten, die auf den Bassins der vertrockneten Wasserkünste umhersaßen wie eine Soldatenwacht im Dreißigjährigen Kriege. – „Preziöschen!“ rief Fortunat Kordelchen zu, „bellt von fern ein Hund, liegt ein Dorf im Grund, schläft Bauer und Vieh, gibt was zu schnappen hie!“ – Kordelchen antwortete munter: „Heult der Wolf in der Heid’, ist mein Schatz nicht mehr weit; stellt aus die Wacht, gibt heut eine gute Zigeunernacht.“ – „Wilewau, wau, wau, witohu!“ riefen die andern jauchzend dazwischen. Kordelchen aber schwang plötzlich ein Tamburin, daß es schwirrte, tanzte mit ihren roten, polnischen Stiefeln auf zigeunerisch und sang dazu:

„Am Kreuzweg, da lausche ich, wenn die Stern’
Und die Feuer im Walde verglommen,
Und wo der erste Hund bellt von fern,
Da wird mein Bräut’gam herkommen.“

Fortunat antwortete lustig:

„Und als der Tag graut’ durch das Gehölz,
Sah ich eine Katze sich schlingen,
Ich schoß ihr auf den nußbraunen Pelz,
Die macht’ einmal weite Sprünge!“

Kordelchen sang wieder:

„'s ist schad' nur ums Pelzlein, du kriegst mich nit!
Mein Schatz muß sein wie die andern:
Braun und ein Stutzbart auf ungrischen Schnitt
Und ein fröhliches Herze zum Wandern."

Hier schlug sie das Tamburin dem Ruprecht, der ihrem Tanze verliebt zusah, dröhnend an den Kopf und setzte sich, in der Tat wie ein Kätzchen, dem träumerischen Otto auf den Schoß.

„Weißt du" – sagte sie, ihre Haare aus dem erhitzten Gesicht schüttelnd –, „weißt du noch, wie wir uns zum erstenmal sahen? Du kamst vom Giebichenstein herab mit einem studentischen Helm, daß der Federbusch dir in die Augen hing; damals gefielst du mir besser als jetzt so mit dem närrischen Frack." – Otto war's bei diesen Worten, als tauchte seine ganze schöne Jugendzeit wieder vor ihm auf, das Mädchen war nur so wild, das störte ihn heimlich. – „Es war in den ersten Frühlingstagen", sagte er, „überall zogen Studenten durchs Grün, du saßest auf der Bank vor dem Wirtshause unter den Linden und spieltest die Harfe." – „Ja, ja", fiel ihm Kordelchen in die Rede, „und du glaubtest, ich spielte für Geld, und setztest dich neben mich und drücktest mir einen Taler in die Hand." – „Und du", versetzte Otto, „besahst verwundert das Geld, dann stecktest du's lachend ein, gabst mir schnell einen Kuß und verschwandest im Hause, und ich sah dich nicht mehr wieder. Ach, Kordelchen! nun ist ja alles, alles wieder gut, und" – „Nun und was denn?!" rief Kordelchen lustig, sprang schnell auf und verlor sich in dem dicksten Haufen.

Kamilla, die es mit angesehen, ging eben vornehm vorüber und sprach halbleise von wilden Waldbeeren, womit man Gimpel fange. Otto aber hielt sich nun nicht länger und fiel ganz glückselig dem Fortunat um den Hals. „Ach", rief er aus, „ich bin so von Grund der Seele ver-

gnügt, wie ein Vogel in der Luft!" – Sie gingen miteinander auf den mondbeschienenen Gängen weit fort, daß sie die Stimmen der Schauspieler kaum mehr vernahmen, und Otto erzählte nun, wie entsetzlich einsam es nun auf Hohenstein geworden, nachdem Walter und Fortunat fortgezogen. Er habe sich gleich nach ihrer Abreise mit redlichem Ernst und Eifer ganz auf die Bücher geworfen, nichts anderes gedichtet und getrachtet und selbst jede Erinnerung an sein früheres Leben gewissenhaft vermieden. „Aber", fuhr er fort, „die Seele des Dichters ist wie eine Nachtigall, je tiefer man ihren Käfig verhängt, je schöner schlägt sie, und ich hörte sie oft in Träumen wunderbar klagen, aber ich hütete mich wohl, wenn ich erwachte, dem weiter nachzuhängen. Und wie nun so der Amtmann täglich um dieselbe Stunde auf das Feld hinausritt und wieder zurückkehrte, und Florentine ihre Tauben fütterte und ihre Blumen band, und ringsum in der ländlichen Stille allmählich alles wuchs und wuchs, als wollte das Grün die Menschen begraben – es war mir nicht anders, als säß' ich viele hundert Klaftern tief im Meer und hörte die Abendglocken meiner Heimat von weitem über mir. So verzehrte ich mich sichtbar selbst, der gute Amtmann sah mich oft insgeheim bedenklich an, die Amtmann steckte mir die besten Leckerbissen zu, sie dachte, wenn ich nur erst fetter wäre, so würde schon alles gut werden. – In einer schönen Nacht aber träumte mir von Halle, ich stand auf dem Giebichenstein, die Kirschgärten unten blühten wieder, und lustige Kähne mit Studenten glitten die Saale hinab, da erklang ein Lied aus dem Tale, das ich damals gehört, auf das ich mich aber seitdem durchaus nicht wieder besinnen konnte. Ich wachte vor Freude darüber auf, das Fenster stand noch offen, und als ich mich hinauslehnte, klang das Lied wirklich draußen durch die stille Nacht herüber. – Seht, ein solcher Lufthauch wendet oft das Narrenschiff des Menschen! Ohne selber recht zu wissen, was ich tat oder wollte, kleidete ich mich rasch an, schnürte mein Bündel, im Hause schliefen

noch alle, und ehe eine Viertelstunde verging, wanderte ich schon durch die dunkle Kastanienallee, das stille Dorf entlang. Als ich ins Freie kam, tönte das Lied noch immer fort, aber sehr fern." –
Hier hielt er plötzlich erschrocken inne, man hörte tief im Garten singen; die Luft kam von dort herüber; sie konnten deutlich folgende Worte vernehmen:

„Hörst du nicht die Bäume rauschen
Draußen durch die stille Rund'?
Lockt's dich nicht, hinabzulauschen
Von dem Söller in den Grund,
Wo die vielen Bäche gehen
Wunderbar im Mondenschein,
Und die stillen Schlösser sehen
In den Fluß vom hohen Stein?"

„Das ist das Lied!" rief Otto und eilte ganz verwirrt den Berg hinab. Unten aber sang es von neuem:

„Kennst du noch die irren Lieder
Aus der alten, schönen Zeit?
Sie erwachen alle wieder
Nachts in Waldeseinsamkeit,
Wenn die Bäume träumend lauschen,
Und der Flieder duftet schwül,
Und im Fluß die Nixen rauschen –
Komm herab, hier ist's so kühl."

Fortunat glaubte jetzt in dem Grunde, woher der Gesang kam, Kordelchen zwischen den mondbeglänzten Gebüschen zu erkennen. – Dann wurde auf einmal alles still, es war eine verlockende Nacht, das Wetter leuchtete von ferne, und die wechselnden Bäume schwankten verwirrend über den Steinen und Klüften.

Zehntes Kapitel

Fern von diesem Weltgetümmel, mitten zwischen den Waldbergen, lag in stiller Abgeschiedenheit ein altes Schloß mit wunderlichen, kleinen Fenstern, halbverfallenen Söllern und Türmchen, alles ganz verwildert und grün überwachsen. Zwischen den Tannenwipfeln qualmten die weißen Schornsteine des freundlichen Dorfes lustig herauf, sie schienen das Schloß schon lange einzuräuchern, denn es sah ganz braun aus, und zahllose Sperlinge lärmten und nisteten in dem Helm des steinernen Wappenschildes über dem Tore. Aus den alten Wallgräben war früher ein Garten, und aus dem Garten mit der Zeit eine grüne Wildnis von Stachelbeeren und Haselnußsträuchern geworden, in der jetzt einige Ziegen ruhig weideten.

Dort saßen an einem schwülen Nachmittage mehrere Jagdhunde unter einer Weinlaube und unter ihnen der Gutsherr, Baron Eberstein, mit dem jungen Prediger des Orts schwatzend, der zum Besuch heraufgekommen war, um dem Baron seine neuen Meerschaumköpfe anrauchen zu helfen. Sie freuten sich beide des allmählich aufsteigenden Gewitters, denn die schillernden Täler unten lechzten nach Regen, es rührte sich kein Lüftchen in der ganzen Gegend, nur die Bienen summten um die hohen Sonnenblumen vor dem Schlosse. Seitwärts aber sah man bald einen roten Schuh, bald ein zierliches Füßchen aus dem Laube eines Kirschbaums schimmern, zwischen dem manchmal ein paar schöne, dunkle Augen herausfunkelten. Es war Fräulein *Gertrud*, des Barons Tochter, die im Wipfel Kirschen naschte und die Kerne mutwillig nach den Hunden schnellte: eigentlich aber hatte sie's auf des Predigers neue, geschniegelte Weste abgesehen.

Der Prediger aber merkte nichts davon, so vertieft war er in den Diskurs. „Ja“, sagte er, „diese Gewitterschwüle ist ein bedeutungsvolles Bild der Gegenwart, alles liegt in banger Erwartung, daß man fast den leisen Schritt der Zeit

hört, Gedankenblitze spielen auf dem dunklen Grunde." – „Ah bah!" erwiderte der Baron, sich eine neue Pfeife stopfend. „Gewitter ist Gewitter und dummes Zeug ist dummes Zeug!" – Der Prediger, ein wenig pikiert, rückte sich vornehm zurecht und sprach von der unaufhaltsamen Intelligenz, von der Mündigkeit der Zeit und der unsichtbaren Gewalt unverjährbarer Wahrheit. Da wurde der Baron ganz hitzig. „Was ist wahr? was ist wahr?" rief er dicht heranrückend aus. Dem Prediger, erschrocken und verblüfft wie er war, wollte gerade in diesem kritischen Moment keine passende Antwort einfallen. – „Na seht", fuhr der Baron fort, „Ihr wißt's nicht, und ich weiß es auch nicht, das weiß der liebe Gott allein. – Aber mein Jagdrevier hier das kenn' ich ganz genau, und wer mir in meine Wildbahn bricht, mündig oder unmündig, den schieß ich vor den Kopf wie einen tollen Hund, und damit basta! Und wenn jeder so täte in seinem Revier, so hätten wir bald Ruhe vor der verjährten Intelligenz und der unsichtbaren Wahrheit und alle dem Plunder. Glaubt einem altgedienten Offiziere, Prediger, die Zeit will nur Prügel haben, weiter ist's nichts!"

„Gäste kommen! Gäste kommen!" rief hier auf einmal das Fräulein im Kirschbaume. Und in der Tat, kein Schiffer vom Mastkorbe blickt so scharf in die Ferne als ein Landfräulein in der Meeresstille ihrer einförmigen Einsamkeit, denn kaum noch schimmert' es flüchtig von dem Gipfel des gegenüberliegenden Berges herüber. Das Gewitter lag schwer über dem Berg und verdunkelte schon die ganze Gegend, nur der grüne Abhang nach dem Schlosse zu war von der Abendsonne noch hell beschienen. Da sah man auf einmal Federbüsche aus dem Grün nicken, einzelne Reiter flogen über den Plan, immer mehrere folgten, Jäger und Frauengestalten auf zierlichen Zeltern, wie wenn der Herbstwind farbige Blätter verstreut; der eine der Reiter schien eine Gitarre im Arme zu haben; man hörte seine Stimme durch die stille Luft bis herüber schallen, andere bliesen auf dem Waldhorne dazu und schossen ihre Flin-

ten ab; so bewegte sich der bunte Zug in der wunderbaren Beleuchtung heiter und eilig den Abhang hinunter – das Fräulein konnte sich nicht satt sehn daran.
„Wahrhaftig, Seine Durchlaucht mit Ihrer ganzen Literatur!“ rief der erstaunte Baron aus, indem er die Pfeife schnell weglegte. „Jetzt biegen sie in den Hohlweg, es kommt alles hierher. He, Johann! meinen Hut! meine Uniform! Was das lateinische Reiter sind! Wo bleibt der Schlingel! Das wollen Jäger sein, die Juanna, das Blitzmädel, ist noch der beste Schütz unter ihnen.“ – „Sie soll immer mitten ins Herz treffen“, versetzte der ästhetische Prediger. – „Prediger“, sagte der Baron, ihn bei der Hand festhaltend, „ich bitt' Euch um Gottes willen, lauft mir jetzt nicht davon, Ihr müßt gelehrt sprechen mit den Leuten, mir ist's immer wie chaldäisch im Halse unter ihnen.“ – „Nun, nun, wir wollen schon machen“, erwiderte der Prediger, zufrieden schmunzelnd.
Fräulein Trudchen aber war schon wie ein Reh über Wallgraben und Sträucher nach dem Schlosse gesprungen. Da gab's ein wahres Volksfest, die Türen flogen krachend auf und zu, die Hunde bellten, die alten Sofas und Stühle wurden ausgeklopft, daß es rauchte, zuweilen hörte man das lustige Lachen des Fräuleins dazwischen. Zuletzt band sie nur noch schnell ihre neue Schürze um; sie wußt' es wohl, sie war hübsch genug, so wie sie war.
Nun aber begann auch schon draußen der Lärm. In hastiger Flucht brachen Gewitter und Gäste zusammen herein; der kleine Hof füllte sich plötzlich mit Glanz und Getümmel von eleganten Uniformen, Reitern und Rossen, der Regen fiel schon in einzelnen großen Tropfen, Tücher, Mäntel und Schleier flatterten im Sturme durcheinander, und bunte Jockeis flogen von den Pferden, um in der Verwirrung den Herrschaften herabzuhelfen, während die Mägde und Knechte des Barons, ihre Mützen in der Hand, ganz verwirrt in den Türen standen. Der Fürst war der erste, der sich aus dem Knäuel herauswickelte. Er befahl seinen Leuten, mit Pferden und Hunden im Dorfe ein

Unterkommen zu suchen, so gut es gehe; dann entschuldigte er verbindlich beim Baron den plötzlichen Überfall, das Unwetter habe sie überrascht; er bat um Schutz für die Nacht; wo könne er diesen besser finden, setzte er hinzu, als bei den alten Häusern des Landes. – „Alt und wackelig in der Tat", sagte die Fürstin leise zu ihrem Nachbar, das Schloß bedenklich betrachtend. – „Es sieht aus", erwiderte dieser, „wie ein altes Rolandsbild, dem der Zahn der Zeit den Kopf abgebissen." – „Nein, wie ein einzeln stehengebliebener Backenzahn der Zeit selbst", meinte ein anderer. – Der Baron aber, in dem beim Anblick von Damen jederzeit die Ritterlichkeit seines ehemaligen Offizierslebens wieder erwachte, hatte mit scharfem Jägerblicke sogleich die Fürstin aufs Korn genommen. Er half ihr kunstgerecht aus dem Sattel, bot ihr mit altmodischer Galanterie den Arm und führte sie über den Hof, immerfort französisch mit ihr sprechend, obgleich sie ihm deutsch antwortete. Aber schon am Eingange gab es unverhofften Aufenthalt. Die fürstlichen Jagdhunde schnupperten überall vornehm umher, da gebrauchten die Hunde des Barons ihr Hausrecht, und eh' man sich's versah, gerade in der Tür, entstand plötzlich ein Balgen und Würgen, daß die Haare davonflogen. Mit gewaltiger Stimme, mit Stock und Stiefeln stiftete der Baron endlich wieder Frieden und wandte sich dann entschuldigend zur Fürstin. Die Fürstin aber kam darüber in ein unaufhaltsames Lachen, das steckte die andern mit an, und so zog alles fröhlich ein.

Dieser konfuse Anfang hatte die ganze Feierlichkeit verstört, welche der Baron im Schilde führte. Er brachte die Gesellschaft in ein großes Zimmer, das nicht zum gewöhnlichen Gebrauche bestimmt schien, wie man an der verstaubten Pracht der damastenen Gardinen abnehmen konnte. Anstatt aber Platz zu nehmen, eilte die Fürstin nach einer leichten Verbeugung sogleich mit Kennermiene zu einer alten, sehr kunstreich mit Elfenbein ausgelegten Kommode. In demselben Augenblicke fing eine ver-

goldete Spieluhr auf dem Schranke mit heisern Absätzen zu spielen an. „Mein Gott, noch aus cosa rara!“ rief die Fürstin überrascht aus. – „Ich weiß wirklich nicht“ – erwiderte der Baron, der es für Spott hielt, und zog die Augenbrauen finster zusammen. Aber er irrte sich. Cosa rara war die erste Oper, welche die Fürstin noch als Kind gehört; jetzt überwältigte sie die Erinnerung, sie hütete sich aber, es zu sagen, damit niemand die Jahre nachzählte. Unterdes hatte der Fürst auch ein Klavier entdeckt, und mit der Unbarmherzigkeit der großen Welt wurde Fräulein Trudchen ohne weiteres, wie zur Schlachtbank, zum Spielen gedrängt. Der Prediger, der sich gern bemerklich machen wollte, brachte ein Pack Noten herbei und stellte sich geschäftig hinter den Stuhl, um die Blätter umzuschlagen. Dem Fräulein ging es aber wie der Spieluhr, rot bis an die Ohrläppchen, konnte sie keinen vernünftigen Ton hervorbringen. Da warf sie plötzlich das Stutznäschen stolz in die Höh, schob die Noten zur Seite und sang herzhaft eines von den Volksliedern, wie sie damals noch auf den Bergen im Schwange waren. Da ging zur Verwunderung des erschrockenen Barons auf einmal eine freudige Bewegung durch die ganze Gesellschaft, man verglich sie einem Waldvöglein, sie mußte mehr und immer noch mehr solche Lieder singen. Dazu kam die Neuheit der ganzen Umgebung, das heimliche Gefühl der Sicherheit in der stillen Burg, während draußen schon der Sturm den Regen an die Fenster peitschte. Die Fürstin fand das altertümliche Kamin, die tiefen Fensterbogen und Erker entzückend, während der Fürst in dem einen Fenster sich nicht satt sehen konnte an dem tiefen Waldgrunde unter dem Schlosse, den die Blitze von Zeit zu Zeit seltsam erleuchteten, so daß der Baron, der lange dort nicht hinausgesehen, endlich selbst neugierig mit hinunterblickte. So war alles in der heitersten Stimmung, als nun noch in dem Kamin ein lustiges Feuer angezündet wurde; der Prediger konnte mit seiner Gelehrsamkeit gar nicht aufkommen, und der Baron fand mit Erstaunen, daß es

doch eigentlich gar nicht so übel leben sei unter diesen Leuten.

Es war noch zu früh zum Schlafengehen, die Fürstin schlug vor, Geschichten zu erzählen, jeder, was ihm eben einfiele. Der Prediger räusperte sich; eine Novelle, die er neulich für ein Taschenbuch geschrieben, steckte ihm schon im Halse. Aber zu aller Verwunderung bat der lange Lord vorweg um das Wort, der Baron brachte alten Ungarwein, wovon er ein Glas der Fürstin zierlich auf einem silbernen Teller präsentierte, alles setzte sich um das Kaminfeuer zurecht, und der Lord begann ohne weiteres folgende

Geschichte der wilden Spanierin

„In dem Kriege Napoleons gegen Spanien diente ich in der englischen Armee, welche damals den Spaniern zu Hilfe zog. Ich war Husarenoffizier, da hatt' ich vielen Ärger mit der unvernünftig hohen Bärenmütze, die alle Augenblick das Gleichgewicht verlor, während ich mich täglich ein paarmal in dem sarmatischen Gehänge und Gebommel von Säbeltasche, Dolman und Fangschnüren mit meinen langen Beinen verwickelte. Einmal waren wir versprengt und rasteten im Freien. Es regnete in einem fort, ich stand melancholisch mitten im Felde unter meinem Regenschirme, in jeder Hand, wie ich aus Vorsicht immer zu tun pflegte, eine Pistole mit gespanntem Hahn. Auf einmal heißt's: Die Franzosen! Wir waren unser nur wenige, der Feind in hellen Haufen. Meine Kameraden zerstoben im Nu nach allen Seiten. Ich aber fasse mein Pferd, fahre in der Eile mit dem Beine in den Pelzärmel des Dolmans, mit einem Arme in die Säbeltasche, mit dem andern in die verfluchte Takelage von Schnüren und Troddeln, so daß ich mich nicht rühren, viel weniger die Zügel erlangen konnte; mein Pferd erschrickt vor meiner Positur und rennt grade auf den Feind los, und so, mit ausgespreizten Armen, den Säbel zwischen den Zähnen, während meine Pi-

stolen losgehen, fliege ich, wie eine wahnsinnige Fledermaus, mitten unter die Franzosen hinein, daß ein lustiges Hussa! durch ihr ganzes Geschwader erscholl. Ich war nun in ihre Gefangenschaft geraten, sie hatten Mühe, mich aus meiner verwickelten Lage zu bringen, und nannten mich den tollsten Kerl, den sie jemals gesehen. Da ich aber französisch sprach und Gold in der Börse hatte, so wurden wir bald gute Kameraden. Sie wollten mich nach Burgos führen in ihr Depot, das war aber nicht so leicht gemacht, denn bewaffnete Banden spanischer Bauern verrannten uns überall den Weg, und so zogen wir geraume Zeit miteinander im Lande umher.

Auf diesem Zuge lagerten wir einmal in einer schönen Sommernacht an einem großen Schlosse, das schon seit langer Zeit nicht mehr bewohnt schien. Die alten, zackigen Türme warfen im Mondenschein lange Schatten über den wüsten Schloßgarten, wo wir lagen und unsere Pferde an die verwilderten Hecken angebunden hatten. Es war alles still in der ganzen Gegend, von Zeit zu Zeit nur hörte man die Pferde schnauben und die Wachen anrufen aus der Ferne, im Walde schlugen die Nachtigallen, als gäb' es keinen Krieg in der Welt. – Der Rittmeister, der den Zug führte, ein heiterer Gaskogner, lag rücklings auf seinen Mantel ausgestreckt, ich glaubte, er schliefe, er hatte aber, wie er mir nachher sagte, an seine ferne, schöne Heimat gedacht. Auch richtete er sich gleich darauf schnell und rüstig wieder auf. ‚Hier ist nicht Zeit zum Träumen', meinte er, ‚wir müssen auf unserer Hut sein heute nacht, denn das ist das Schloß der wilden Spanierin.' Und als ich fragte, wer die sei, benutzte er gern die Gelegenheit, sich munter zu erhalten, und erzählte mir alles ausführlich.

‚In diesem Schlosse', sagte er, ‚wohnte ehedem ein Graf aus uraltem Stamme, der nach und nach wohl sich zu beugen verlernt haben mochte. Wenigstens soll der Graf früher den Anforderungen des alten Hofes jederzeit trotzigen Stolz entgegengesetzt haben bis zu wechselseitiger,

bitterer Verstimmung; um so mehr durfte man voraussetzen, daß er der neuen Ordnung der Dinge geneigt sei. Auch fanden ihn die Unsrigen, als sie das Land überzogen, einsam auf seinem Schlosse, höflich, aber finster, und, wie es schien, ohne alle Teilnahme an dem, was hinter seinen Bergen vorging. Seine größte Freude war ein Töchterchen, sein einziges Kind, bei dessen Geburt die Mutter gestorben. Mit ihr pflegte er, wenn alles schon schlief, die Zinne des Schlosses zu besteigen und zeigte ihr das Land, das ehemals ihre Ahnen beherrscht, soweit der Mond die Wälder beleuchtete, und erzählte ihr halbe Nächte hindurch von der alten großen Zeit und der fürstlichen Freiheit, die sich dem Zwang der Städte nicht unterwerfe. Unter solchen Träumen wuchs das Fräulein auf, und da der Krieg alles vereinzelte, so sah sie fast kein anderes Frauenzimmer als ihre alte Amme, ein hexenhaftes Weib, das von ihrem Vater, einem Zigeuner, und ihrer Mutter, einer gefangenen Araberin, manch Zauberstückchen ererbt hatte, woran die Tradition dieser Stämme so reich ist.

Aber unseren Leuten blieb die junge Gräfin nicht lange verborgen, und die sie sahen, konnten nicht genug erzählen, wie wunderbar schön sie war: schwarze Locken, bleich mit brennend rotem Munde, die Augen wie ein dunkler Abgrund. Täglich nun flimmerte es von französischen Offizieren auf dem Schlosse. Das gefiel ihr wohl, sie ritt und focht mit ihnen und war der beste Schütze auf der Jagd, so oft aber einer näher trat mit verliebten Blicken oder Worten, sah sie ihn verwundert an und wußte nicht, was er wollte; allen gleich fern und fremd, wie ein Stern in kalter Winternacht. Das verlockte aber die lustigen Gesellen nur noch immer mehr aufs Glatteis, und ein hübscher, junger Unterleutnant – St. Val war sein Name –, der soeben erst aus der Militärschule von Paris angekommen war und davon hörte, verschwor sich mörderisch, sie müßte sein werden, oder er wollte des Teufels sein!

Unterdes wurden die Plänkeleien in der Gegend immer ernster, die Offiziere hatten vollauf zu tun und blieben aus, da konnte sich die Gräfin gar nicht wiederfinden in die alte Einsamkeit und das einförmige Rauschen der Wälder. – So stand sie auch eines Abends allein mit der Amme vor dem Schlosse. Der Krieg ging unten wie eine lustige Jagd durch die Berge, zuweilen sahen sie fern in der Abendsonne ein Geschwader von Reitern aufblitzen, einzelne Trompeten klangen herüber, dann verhallte und verdunkelte nach und nach alles wieder, nur die Flammen brennender Dörfer blieben am Horizont stehen. Die Gräfin sah lange stumm und unverwandt in das ferne Feuer, dann brach sie still in Weinen aus und sagte für sich: Wie ist das herrlich! Ach, daß ich kein Mann geworden bin! ihnen gehört alles, sie regieren die Welt. – Die kluge Amme erwiderte: Desto besser, Kind, desto besser, denn die Frauen regieren wieder die Männer. – Wieso? sagte die Gräfin und sah sie groß an, daß ihr die Tränen funkelnd in den schönen Augen stockten. – Nun, nun, antwortete die Alte, kein schlanker Tiger verwundet so tief, als wenn Ihr lacht und ihnen die weißen Zähnchen weis't oder einen beim Küssen heimlich damit beißt; keine buntgefleckte Schlange ist so schön und stark als Eure Arme, wenn Ihr einen umschlingt. – Die Gräfin hörte nur halb darauf und sagte wie in Gedanken: Darum habe ich immer in den alten Büchern meines Vaters gelesen, wie Fürsten und Könige vor Mädchen knieten und ihnen treu und gehorsam waren bis in den Tod. – Ach, liebe Amme, du weißt so viele Künste von deinem Vater, kannst du denn nicht machen, daß alle Männer, die mich sehen, in Liebe entbrennen und mir folgen müssen? – Hm, entgegnete die Amme zögernd, wenn nur – ich wüßte wohl –

Die Gräfin aber, deren Seele ganz erfüllt war von dem Gedanken, hatte sie schon am Arm gefaßt und drängte sie ungeduldig fort; die Nacht sei dunkel und schwül, alles schlafe schon im Schlosse, es sei eben die rechte Zeit. – So

gingen sie weiter den stillen Garten entlang bis ans einsamste Ende. Unterwegs sagte die Amme: Es ist nichts Geringes, dem Freier, den ich Euch zuerst zeigen werde, müßt Ihr den Ring vom Finger ziehn – aber laßt's Euch nicht anfechten, wenn er etwas bleich und wirre sieht –, den Ring drückt Ihr ans Herz, bis es blutet, dann ist Euer Herz liebefest, und Eure Augen werden schön funkeln wie der Stein im Ringe, der arme Junge aber muß sterben. – Hier waren sie an ein altes, zerfallenes Gemäuer gekommen, die Alte holte ein weißes Stäbchen aus einem hohlen Baumstamme, da schwirrten plötzlich Fledermäuse hervor und schlugen mit den Flügeln in den Zweigen, eine Schlange fuhr rasch zwischen das Gestein, unter dem sie eine dicke Kröte mit großen, rötlichen Augen ansah. Hoho, bist du auch da, Großmutter, lachte die Alte und schien lustig auf zigeunerisch mit den Tieren zu sprechen. Darauf tauchte sie Hände und Stab in einen Topf, daß sie hell leuchteten, und beschrieb unverständlich murmelnd einen feurigen Kreis, bei dessen grüngoldenem Glanz die Eidechsen neugierig im Grase hervorschlüpften. Die Gräfin stand mitten drin, es war ihr wie im Traume, als fingen die Blumen, Büsche und Wälder in der stillen Runde leise zu singen an, Johanniswürmchen zogen leuchtend um ihr Haupt, so sah sie mit tiefer, tiefer Lust vom Berge über die mondbeschienene Gegend und in den weiten, gestirnten Himmel hinein. – Die Amme aber schien in großer Unruhe, die Schweißtropfen standen auf ihrer Stirn. Siehst du noch immer nichts? fragte sie manchmal leise dazwischen. Aber nur ein Hund bellte aus dem fernen Dorfe, dann war alles wieder still, die Gräfin hielt den Atem an vor Erwartung. Auf einmal fuhren beide zusammen – ein fremder Mann, dicht im Mantel verhüllt, trat plötzlich in der Ferne zwischen den Bäumen hervor. Um Gottes willen! rief die Amme und flüsterte noch etwas in der höchsten Angst. Aber die Gräfin, wie ein Falke in den Lüften hängend, stürzte mit unmenschlicher Lust schon auf ihre Beute. Der Fremde erschrak heftig, erholte sich aber, da er ein

Weib vor sich erblickte. Sie sah ihn groß an, sie kannte ihn nicht. Auf ihre Frage, wo er hin wolle, erwiderte er zögernd und sichtbar verwirrt, er wolle der schönen, jungen Gräfin ein Ständchen bringen. Der Wind schlug ein wenig seinen Mantel auf, da fiel es ihr seltsam aufs Herz, daß es ein französischer Offizier, doch sagte sie nichts, aber ihre Blicke gingen scharf seitwärts in die Dunkelheit, denn es war ihr, als hörte sie etwas heimlich durch den Garten huschen und Pferde schnauben in der Ferne. – Kannst du mir die Fenster zeigen, wo sie schläft? sagte der Fremde wieder, und da sie ihm gefiel, umschlang er sie mit einem Arme. Die Gräfin besann sich einen Augenblick. Warum nicht! sagte sie dann schnell, wenn Ihr mir Euren schönen Ring gebt zum Lohne; aber Euren Mantel müßt Ihr mir borgen, damit man mich nicht erkennt. Der verliebte Offizier hing ihr selbst den Mantel um und meinte dabei, ihre aufgeringelten Locken sähen wie Schlangen aus bei Nacht. Sie aber hatte schon ganz andere Gedanken, und als er eben den Ring vom Finger zog, ergriff sie rasch ein Pistol, das er unter dem Rocke auf der Brust trug, und stieß ihn damit rücklings von der Rampe, auf der sie standen. – Sie ist im Garten, greift die kleine Hexe! rief jetzt eine Stimme tiefer unten. Da drückte sie schnell ihr Pistol ab und: Herr Jesus! hörte man unten dieselbe Stimme verhallen. Dann, sich in den Mantel wickelnd, rief sie hinab: Mir nach, sonst seid ihr alle verloren!

Aber es war alles schon zu spät. Die Unsrigen, die unerwartet erfahren, daß der Graf es heimlich mit dem Feinde halte, hatten die dunkle Nacht benutzt, das Schloß ohne Geräusch beschlichen und den Grafen bereits gefangen in ihrer Mitte. Dieser nun, als er die Tochter an der Stimme erkannte, glaubte sich von seinem eigenen Kinde verraten, in dieser Verblendung entriß er wütend einem der Soldaten den Degen, um sich selbst zu richten. Sein Leben war ihm nichts gegen die Ehre und Freiheit; so ward ihm schnell die letzte zuteil, indem die anderen Soldaten, da sie ihn nicht mehr aufhalten konnten, ihn von hinten mit vie-

len Stichen durchbohrten. – Unterdes aber war, wie die Gräfin vorausgesehen, durch den Schuß alles munter geworden. Gleichwie die Krähen, wenn man nachts in die Wipfel schießt, sich mit wildem Geschrei in die Lüfte stürzen, so brachen bewaffnete Jäger, Bedienten und Bauern, die damals einen leisen Schlaf hatten, plötzlich aus allen Türen, Hecken und Mauerritzen hervor. Die Unsrigen, als sie sich so umgeben sahen, folgten blindlings der Gräfin, die sie in dem Offiziersmantel für ihren Kapitän hielten. Sie wollte ihrem Vater, den sie noch im Schlosse glaubte, Zeit lassen, sich zu retten, und führte, immerfort winkend, die verstörten Soldaten bis in den äußeren Hof, wo sie dem wilden Haufen gerade in die Hände rannten. Da rangen sie still und grimmig in der Dunkelheit Mann gegen Mann, die einen ums Leben, die andern um den Leichnam ihres Herrn, die Gräfin hatte unterdes eine Meute grausamer Hunde losgelassen, welche in der Verwirrung die Fliehenden zerrissen, es war eine schreckliche Nacht. – Der Offizier aber, den die Gräfin durch den Pistolenschuß so still gemacht, war derselbe junge St. Val, der damals sie zu fangen geschworen und sich nun vermessen zu dem gefährlichen Kommando gedrängt hatte. Er war aber nur verwundet und betäubt, und als er auf dem stillen Platze einmal die Augen aufschlug, sah er, wie im Traum, zum ersten Male das Gesicht der Gräfin zwischen den schwarzen, herabwallenden Locken beim Widerschein einer Fackel über sich geneigt – er mußte die Augen wieder schließen, so furchtbar schön war der Anblick.'"

Hier wurde der Lord plötzlich von der Fürstin unterbrochen, die schon während der ganzen Erzählung eine seltsame Unruhe gezeigt und öfters ängstlich nach der Tür gesehen hatte. „Nein, das ist gar zu traurig vor dem Schlafengehen", rief sie mit einem bedeutenden Blick auf den Fürsten und schien aufbrechen zu wollen. Dieser aber, ganz vertieft in die Geschichte, merkte nicht darauf; so blutigrot also war ihr Aufgang – sagte er in Gedanken und

wollte durchaus noch das Ende wissen. Der Lord stutzte, da aber der Fürst von neuem in ihn drang und die andern mit Blicken und Kopfnicken beistimmten, erzählte er ruhig wieder weiter:
„,Seit dieser Stunde' – so fuhr mein Rittmeister fort – ,steht das Schloß wüst und verlassen, aber die wilde Gräfin geht wie ein wunderbarer Spuk durchs Gebirge. Oft nach nächtlichen Bivouaks, wenn die Sonne über der prächtigen Gegend aufgeht, erscheint sie am Saume des Waldes zu Pferd im vollen Glanze der Schönheit; da schwingt sich manch fröhlicher Reiter auf, sie zu fangen, aber keiner von allen kehrte noch jemals wieder zurück. – Seltsam, es ist ja doch nur ein Weib. Seht, ich habe mein Liebchen in Frankreich, mir soll sie nur kommen, ich spüre eine rechte Lust, ihr einmal zu begegnen! –
Dem armen St. Val aber ging es am schlimmsten. Das Bild der Gräfin stand seit jener Nacht unaufhörlich vor seiner Seele, der lustige Bursch wurde ganz schwermütig, und eines Abends war er plötzlich verschwunden, wir wußten lange nicht, wohin er gekommen. Er war aber an diesem Abende, wie er damals oft zu tun pflegte, einsam in der Gegend herumgeschweift. Da hörte er wunderschönen Klang in der Abendluft, wie eine Kriegsmusik aus der Ferne – man sagt, daß es in der Morgendämmerung vor großen Schlachten so in den Lüften musiziert – es waren die Guerillas, die im Gebirge sangen. Die Klänge verlockten ihn, er ging wie im Traume immer fort, so kam er in den Wald, wo damals die Gräfin hauste. Die Abendsonne leuchtete durchs Gebirge, als stände alles in Feuer, die Vögel sangen den funkelnden Wald entlang, dazwischen hörte er immerfort Stimmen bald da, bald dort, darunter eine wie ein Glöckchen bei Nacht, es klang ihm, als müßt' es die Gräfin selber sein. Ihm graute, und doch mußt' er der Stimme folgen. So war er schon lange gegangen, als er, plötzlich um einen Felsen tretend, auf einem stillen Rasenplatz über den Wipfeln eine weibliche Gestalt wie eingeschlummert sitzen sah, die Stirn über beiden Armen auf

die Knie gesenkt, daß die herabgefallenen reichen Locken sie wie ein dunkler Schleier umgaben. Sie hielt ein Roß am Zügel, das weidete ruhig neben ihr, von allen Seiten rauschten die Wälder herauf, sonst war's so still daneben, daß man die Quellen gehen hörte. Und wie er noch so staunend stand in dieser Einsamkeit, erblickte er seitwärts in der Ferne einen Offizier von der deutschen Legion, der unten zwischen dem Gebüsch seine Büchse angelegt hatte, er wußte nicht, ob er auf ihn oder die Schlummernde ziele, und machte erschrocken eine heftige Bewegung. Da schüttelte die Schlafende die Locken aus den Augen und richtete sich, in der Abendglut mit den Steinen ihres Gürtels leuchtend, plötzlich auf. Der Deutsche, wie geblendet, ließ seine Büchse sinken und verschwand zwischen den Bäumen; St. Val aber erkannte mit Schauern die Gräfin, denn ihm fiel die Soldatensage ein, daß es jedem den Tod bedeute, der sie unversehens im Walde erblickt. – Die Gräfin aber sah scharf nach allen Seiten, dann ihn durchdringend an. Ihr seid sehr vorwitzig, sagte sie darauf, doch es wird schon spät, ich bin so müde und verirrt, zeigt mir den Weg aus dem Walde. Da fiel es St. Val plötzlich aufs Herz; er wußte, daß die Franzosen den Wald umzingelt hatten und in welcher Gefahr sie war, er wollte sie retten, es koste, was es wolle, und dann noch diese Nacht zu seinem Regiment zurück und sich zu anderen Truppen versetzen lassen, weit von diesen Wäldern. – Während diese Gedanken verworren durch seine Seele gingen, hatte sie schon ihr Pferd gezäumt; sie befahl ihm, unterdes zu satteln, und lachte ihn aus, als er damit nicht zurechtkommen konnte, dann schwang sie sich hinauf, er mußte das Pferd am Zügel führen. Sie saß seitwärts auf einem Frauensattel, auf ihren Arm über den Hals des Pferdes gelehnt, und plauderte im Waldesgrün unbekümmert, wie ein Kind, in ihrer schönen, melodischen Sprache, daß es St. Val war, als hörte er die ferne Musik wieder in der stillen Abendluft, die ihn vorhin verlockt hatte. Auf einmal richtete sie sich lauschend auf, man hörte französisch spre-

chen dicht unter ihnen. Sie lenkte vorsichtig hin nach den Stimmen, und durch das Gebüsch sahen sie einen Trupp Reiter in ihren weißen Mänteln, die in der Dunkelheit leuchteten, langsam vorüberziehen – nur ein Laut von St. Val, und die Gräfin war verloren. – Sie aber schaute mit kühner Lust hinab, wie man nachts in ein Gewitter sieht, dann, plötzlich sich selbst unterbrechend, streckte sie den Fuß gegen St. Val: er sollt' ihr das Schuhband binden, und lächelte spöttisch, da er's tat. –

Von diesem Augenblick an war er ganz in ihrer Macht. Sie sagte: sie hätte ihn nur versuchen wollen, ob er's ehrlich meine, sie wisse den Weg besser als er, sie wolle ihn heimführen. Mit diesen Worten lenkte sie rasch herum, und in den Klüften bald hernieder, bald wieder aufwärts, an schwindelnden Abgründen vorüber, ging es immer tiefer in die Nacht und in die Wälder hinein – er konnte kaum folgen durch das Gestrüpp, wie ein getreuer Hund, und als sie endlich unerwartet ins Freie kamen, sah der Entsetzte eine Guerillabande vor sich im Waldgrund gelagert. Unzählige Röhre, da sie die französische Uniform erkannten, waren plötzlich auf ihn gerichtet, aber ein zorniger Blick der Gräfin bändigte alle; die grimmigen Bestien, ihre schwarzen Mähnen schüttelnd, zogen sich knurrend zurück und wärmten wieder ihre Tatzen an den Wachtfeuern. Nun bemerkte St. Val mit Erstaunen, wie diese wilden Männer die Gräfin, gleich einer Königin, verehrten und bedienten. Ein junger Bursch hob sie aus dem Sattel, einige breiteten einen bunten Teppich über den Rasen, während andere rasch ein lustiges Zelt darüber aufschlugen, dann war auf einmal alles wieder still und feierlich. Unterdes war auch der Mond aufgegangen und beleuchtete die Wälder. Die Gräfin saß unter ihrem Zelt und spielte auf einer Zither, St. Val lag gedankenvoll zu ihren Füßen, ihm war noch nie so himmlisch wohl gewesen. – Es war eine von den prächtigen Sommernächten jenes Landes, die alles wunderbar in Traum verwandeln. Die Gräfin hatte sich bald mit einem Teil der Bande wieder entfernt, nur

wenige bewaffnete Bauern bewachten den Gefangenen, die Luft kam von der Ebene und wehte Wohlgerüche aus den blühenden Gärten herauf, die unter den Bergen lagen. Da hörte St. Val die Trompeten seines Regiments durch die weite Stille herüberklingen, sie bliesen ein fröhliches Reiterlied aus der alten, guten Zeit. Das wandte ihm das Herz, er war wieder ganz Franzose, der die Ehre über alles stellt. Er merkte gar wohl an der geheimnisvollen Geschäftigkeit der Abenteurer, daß sie einen Hauptstreich vorhatten, da war kein Augenblick zu verlieren. So, in höchster Angst vor dem Zelte sitzend und umspähend, sann er eben, heimlich zu entfliehen und die Seinigen zu warnen, als auf einmal die ganze Bande mit Windlichtern wieder aus dem Wald zurückkehrte. Die Gräfin, mitten unter ihnen, tritt rasch hervor und, zwischen den schweifenden Lichtern mit den losgegangenen Locken wieder über ihn geneigt, wie in jener Nacht am Schloß, blickt sie ihn streng an in ihrer ganzen, furchtbaren Schönheit. Da springt er auf, entreißt einem Bauer die Fackel, und ganz verblendet und verwirrt führt er selber den Haufen zum Überfall gegen seine Landsleute! – So, rasch und schweigend, gehen sie durch den stillen Wald –'

Kaum hatte der Rittmeister diese Worte ausgesprochen, als plötzlich ein Schuß hinter uns her fiel, und bald ein zweiter und noch einer. ‚Teufel! da ist St. Val!' schrie der Rittmeister aufspringend, und ich erblickte in einem Erker des Schlosses einen schönen jungen Mann, totenbleich beim Fackelschein, ohne Hut, in einer halb zerrissenen französischen Uniform, hinter ihm im roten Widerscheine der Windlichter, der seltsam über die vergüldeten Wände der Säle schweifte, wurden wilde, trotzige Gestalten mit Dolchen und langen Vogelflinten sichtbar, wie sie der Rittmeister vorhin beschrieben. Sie schossen aus allen Fenstern auf uns, und mancher Franzose sank ins Gras eh' sich unser Häuflein nur besinnen konnte. Unterdes hatte sich das Gerücht verbreitet, die wilde Gräfin sei im

Schlosse! Der Rittmeister verlor keinen Augenblick seinen Kopf, er traute mir nicht mehr in solcher Gefahr und ließ mich tiefer in den Wald zurückbringen, dann erbrachen sie mit gewaltiger Anstrengung Tor und Riegel und drangen in die Burg hinein. Der erste, der ihnen dort begegnete, war St. Val, er focht wie ein Rasender und stürzte sich zuletzt in wildem Wahnsinne selbst in die französischen Klingen.

Über seinen Leichnam nun ging der Kampf von Treppe zu Treppe entsetzlich durch alle Gänge. Die Franzosen waren kriegsgewandter und zahlreicher, als ihre Gegner, die Gräfin und die Ihrigen wurden immer höher hinaufgetrieben – es war keine Rettung mehr für sie. Da schlug plötzlich aus dem einen Fenster ein heller Schein hervor, dann wieder aus einem andern, immer mehr rötliche Flammen züngelten schnell an allen Ecken auf, der Sturm faßte die wachsenden Lohen, und wildkühn kletterte das Feuer an den Gebälken empor, wie ein prächtiges Laubgebinde in der Nacht, mitten in der Glut sah man die dunklen Gestalten noch ringen. In dieser Not erblickte der Rittmeister auf einmal die Gräfin hoch über sich, wie den Todesengel, zwischen den Flammen. Ihm vergingen die Sinne bei dem Anblick, er vergaß Heimat, Liebchen und Ruhm, er wollte nur sie retten oder sterben. Vergebens riefen ihm die Seinigen nach, er hörte nicht mehr und drang verblendet die brennende Treppe hinan, unter sich in der wilden Beleuchtung sah er den Garten, die Schluften und den Strom, der wie eine glühende Schlange an dem Schlosse vorüberschoß – schon langte er nach ihr, sie zu umschlingen und hinabzutragen, da stieß sie ihn mächtig von der Zinne hinab, daß die Flammen wie fliegende Fahnen den braven Soldaten bedeckten.

Bald darauf stürzte der ganze Bau donnernd über Freund und Feind zusammen – man hat seitdem die Gräfin nicht wieder gesehen."

Alles schwieg, als der Lord endigte, nur der Baron, der während der Erzählung eingeschlummert war, fuhr auf

seinem Stuhle erschrocken auf über die plötzliche Stille. – „Nun – und weiter?“ sagte endlich der Fürst ganz zerstreut. – Der Lord sah ihn verwundert an. „Was wollen Sie noch weiter in der spanischen Nacht, nachdem dieser schöne Stern gesunken? Das andere lohnt nicht mehr: da der Rittmeister tot war, ergriffen die wenigen noch übriggebliebenen Franzosen voll Entsetzen die Flucht, auch meine Wächter waren verschwunden. Ich eilte nun in der neuen Freiheit sogleich zum Schlosse, um die Gräfin, von der ich so viel gehört, womöglich mit eigenen Augen zu schauen – es war zu spät. – Als ich aber an die Brandstätte kam, da war's, als wüchsen dunkle Reitergestalten aus dem feurigen Boden, die wühlten mit ihren Degen in den Trümmern, daß überall blaue Flämmchen aufschlugen. Sie ist mitverbrannt, hört' ich einen von ihnen sagen. – So war denn alles nur ein prächtiger Traum! rief ein anderer schmerzlich aus; dann stürzten sie in den Wald, den Flüchtlingen nach. – Später hörte ich, daß die schwarzen Gesellen, welche das Schloß hatten entsetzen wollen, von der englisch-deutschen Legion gewesen.“

„Und sahen Sie den Offizier nicht, der sie anführte?“ fragte der Fürst wieder. „Ich erblickte ihn nur fern und flüchtig in der wilden Nacht“, erwiderte der Lord, „bei meinem Regiment aber nannten sie nachher einen deutschen Grafen: Victor von Hohenstein.“

„Nun wahrhaftig, Ihr werdet uns am Ende gar noch überreden wollen, daß die Novelle wahr ist“, sagte hier die Fürstin, indem sie sich erhob und das Signal zum allgemeinen Aufbruch gab. Man vermißte jetzt erst die Gräfin Juanna. Der Baron sagte, sie promeniere schon seit länger als einer Stunde mit seiner Tochter durch alle Winkel des Schlosses und sei dadurch um die ganze spanische Reitergeschichte gekommen. Er ergriff nun eine seidene Klingelschnur und zog erst gelassen, dann immer heftiger, aber der Draht war durch den langen Nichtgebrauch verrostet, es wollte durchaus nicht klingeln, bis er endlich

ganz zornig zur Tür hinausschrie. Mehrere Bediente in alten, verschossenen Livreen stürzten herein und setzten sich mit massiven Armleuchtern an die Spitze des Zuges, den der Baron, die Fürstin an den Fingerspitzen haltend, feierlich eröffnete, in der Perspektive erblickte man durch die offenen Flügeltüren ein mächtiges Himmelbett mit schwerseidenen Gardinen und einem Federbusch darüber. Nun verliefen sich auch die andern mit ihren Lichtern auf den verwirrten Gängen; es sah von draußen aus wie ein verbranntes Blatt Papier, wo die Funken geschäftig, durcheinander irren, bis endlich der letzte plötzlich verlischt.

Und als nun alles ruhig geworden im ganzen Hause, stand der Fürst noch immer allein mit dem Lord am offenen Fenster eines dunkeln Saals und konnte nicht aufhören, ihn über die erzählte Begebenheit immer genauer auszufragen. Das Gewitter draußen war vorüber, es blitzte nur noch von fern, einzelne zerrissene Wolken flogen eilig über den stillen Hof. Da fuhr plötzlich der Lord auf: „Seht da, wahrhaftig, die wilde Gräfin!" – der Mond war auf einmal zwischen den Wolken hervorgetreten und beleuchtete flüchtig Juanna, die jenseits noch auf dem Balkon stand. – Der Fürst aber schloß schnell das Fenster. „Still, still", sagte er zu dem erstaunten Lord, der diesen Ausruf nur so gedankenlos hingeworfen, „verratet es niemand, daß Ihr sie kennt." –

Elftes Kapitel

Ein prächtiges Schloß über schimmernden Fernen, ein bunter, fürstlicher Hofhalt, Komödianten und ein Liebchen im Grün – was Wunder, daß *Ottos* fröhliches Studentenherz wie eine Lerche singend über dem phantastischen Herbstschmuck der Wälder hing! – Auch *Fortunat* verschob seine Abreise von einem Tag zum andern, die

geheimnisvolle Aufmerksamkeit, womit man ihn hier unbegreiflicherweise auszeichnete, wurde immer auffallender. Er glich einem Fremden, der auf der Durchreise, bevor der Postillion wieder blies, sich auf einige Minuten im Theater an einen Pfeiler gelehnt hat und nun auf einmal gewahr wird, daß droben auf den Brettern von ihm selber die Rede sei und alle Blicke sich unheimlich auf ihn heften. Das Rätsel, meinte er, müsse jeden Augenblick sich lösen, er wollte wenigstens den ersten Akt noch abwarten.

Am wunderlichsten aber war es *Dryander* ergangen. Sein Dichterruf öffnete ihm alle Flügeltüren des Schlosses, da hatte ihn aber der Hofwind so wacker gefaßt, daß er bald den Hut samt dem Kopfe darüber verloren hätte. Die unverschämte Art, mit der er sich selbst vergötterte, sein Witz und poetisches Wetterleuchten dazwischen blendete, verwirrte und belebte alles, und eh' man sich dessen versah, hatte der Fürst ihn bei Hofe angestellt, die Schauspieler meinten: als lustigen Rat. Er selbst aber nahm die Sache sehr ernst, hielt einen Bedienten, mit dem er sich täglich zankte, kleidete sich sorgfältig nach der neuesten Mode, sprach nur französisch zu den Komödianten, die es nicht verstanden, und wies Lotharios Gelächter mit gründlicher Verachtung zurück.

Währenddes hatte auch der junge schöne Maler *Guido* sich immer mehr in Kordelchens feingeschlitzte Augen vertieft und entdeckte in dem mutwilligen Mädchen täglich neue, unerhörte, nur von der Gemeinheit ihrer Umgebung verschüttete Talente, von denen sie selber nichts wisse. Strotzend von guten Vorsätzen, voll Selbstvertrauens und jugendlichen Glaubens an Tugend und Liebe, ging er mutig darauf los, sie aus ihrer Verwilderung mit sich emporzuflügeln. – Eines Nachmittags saßen beide zusammen in dem altmodischen Ziergarten, der die Wohnung der Schauspieler umgab. Sie strickte einen Strumpf, er las ihr Goethes Tasso vor. Zwischen den grünen Taxuswänden schillerten von fern die reichen Täler herauf, bun-

te Schmetterlinge flatterten auf den halbverwilderten Blumenbeeten; die feierliche Pracht der Gänge, die Hermen römischer Dichter, die in der Einsamkeit umherstanden, weiterhin über den Buchenwipfeln das heitere, fürstliche Schloß – alles versetzte ihn recht mitten in das schöne Gedicht, er las sich immer mehr ins Feuer. – „Wie schön sie ist!" rief da auf einmal Kordelchen fast traurig aus. Guido glaubte: die Prinzessin im Stück. Kordelchen aber meinte die Gräfin Juanna, die soeben, eine Laute im Arm, durch den oberen Schloßgarten ging. Er sah ihr selber nach, bis sie zwischen den Orangenbäumen wieder verschwunden war, dann fuhr er, etwas gestört, weiter fort. Aber seine Schülerin war heute ganz zerstreut. „Haben Sie gestern abend Lothario droben gesehen?" unterbrach sie ihn von neuem; „ich glaube, er wollte ein Ständchen bringen." – Guido wollte aus der Haut fahren, er nickte ihr nur flüchtig zu, er war eben an einer Lieblingsstelle und deklamierte so eifrig fort, daß ihm die Stirn davon rot wurde. Als er aber einmal über das Buch hinwegsah, hatte Kordelchen gar ihr Strickzeug weggelegt und den ganzen Schoß voll Sternblumen. – „Sie liebt ihn – sie liebt ihn nicht" – sagte sie leise in Gedanken vor sich hin, eine Blume nach der andern zerpflückend. Guido stand auf, klappte das Buch heftig zu und schob es in die Tasche, seine begeisterten Augen leuchteten im Zorne so schön unter den herabwallenden braunen Locken. „Du närrischer Junge!" rief Kordelchen, ihn mit einem herzhaften Kuß festhaltend. Da wanderte eben Otto vorüber und warf ihr einen verächtlichen Blick zu. Sie warf ihm dagegen lachend alle ihre Blumen nach und sprang dann selber schnell in den Garten fort.

Ungünstigeres aber hätte Otto in diesem Augenblick nicht begegnen können, als der unerwartete Anblick dieser Vertraulichkeit. Denn er ging soeben, das Manuskript eines Trauerspieles unter dem Arm, mit klopfendem Herzen nach dem alten Palast der Schauspieler, um es ihnen behufs einer zu verhoffenden Darstellung vorzulesen. Er

fand Herrn Sorti und die übrigen Stimmführer der Gesellschaft bereits vor dem Hause in einer Wolke von Tabaksrauch zwischen hohen Biergläsern um einen runden Tisch versammelt. Zerstreut und in Gedanken noch halb bei Kordelchen, begann er mit unsicherer, fast schüchterner Stimme die Vorlesung. Doch bald faßte ihn der rasche Strom der eigenen Dichtung, heiter glitt er an den duftigen Gestaden, Rebengeländern und Burgen hinab, und das stille Glück der Stunden, ja die Gegenden und Plätze, wo er damals gedichtet, wehten ihn wieder erfrischend an. So las er immer schöner und mächtiger und bemerkte nicht, wie die Gesichter seiner Zuhörer nach und nach immer länger wurden, dort einer heimlich durch die Nase gähnte, da ein anderer mit vornehmem Lächeln unverwandt sein Bierglas ansah. Und als er endlich schloß, erfolgte eine allgemeine Stille, daß man das Laub im Baume sich bewegen hörte – ein Zustand, wobei einem jungen Autor die Gedanken plötzlich zu Eiszapfen gefrieren können.

„Schön – recht poetisch", nahm endlich Sorti das Wort, „aber aufführen" – „Keine Drucker", platzte Ruprecht heraus. – „Zuviel Verwandlungen", meinte ein anderer. – „Kein einziger brillanter Abgang." – „Aber was hat denn alle das Teufelszeug mit meinem Gedicht zu schaffen?" fragte der erstaunte Otto in seiner poetischen Unschuld. – „Wird sich schon geben, mein Liebster", entgegnete Sorti gelassen, „wird sich nach und nach schon geben mit der zunehmenden Bühnenkenntnis." – Nun steckten alle die Nasen in das Heft, und ein jeder fing an, nach seiner Art daran zu mäkeln. Der Dialog war zu phantastisch, er sollte noch einmal überarbeitet, herabgestimmt und natürlicher gemacht werden. Der Held dagegen erschien allen zu einfach, die Dame gar zu verliebt. – Da hielt sich Otto nicht länger, diese Mädchengestalt war ihm gerade die schönste, er hatte sich, wie es jungen Dichtern wohl begegnet, nach und nach im Schreiben selber in sie verliebt. „Das Lieblichste", rief er aus, „das Heimlichste, Wahrste und Beste, was ich wußte, hab' ich gegeben, und nicht

einen Buchstaben ändere ich an dem ganzen Stück!" – Hiermit schleuderte er das Manuskript zornig auf den Tisch und ging rasch in den Garten fort, und es war ihm in einiger Entfernung, als hörte er die Schauspieler hinter sich lachen.
In diesem heftig bewegten Zustande begegnete er Lothario, der ihm sehr bald die ganze Geschichte abgefragt hatte und darauf in ein tolles Gelächter ausbrach. „Darf man erfahren, worüber Sie lachen?" fragte Otto empfindlich. „Weil Sie", erwiderte Lothario, „durch diese glückliche Begebenheit hoffentlich auf den nächsten Weg geraten sind, sich der theatralischen Flausen gänzlich zu entschlagen." Otto sah ihn verwundert an. Aber Lothario ließ sich nicht irre machen. „Überlegt doch nur selbst", fuhr er fort, „was wollen Sie denn eigentlich! Ein großer, starker Kerl, der plötzlich herausstürzt und rezitativisch schreit: Ich fürcht' mich vor dem Tode nicht! ein Posaunenstoß oder ein paar Striche über die große Baßgeige dazu – das ist ein Held. Ein zimperlich Ding, etwas verliebt und etwas tugendhaft und sehr geschnürt, das in Jamben spricht und mit den Logen kokettiert – das ist eine Jungfrau. Ein Korb voll Kaldaunen, der nach Tische zur Verdauung Poesie treibt und in Romeo und Julie eines gemalten Pomeranzenbaums bedarf, um sich nach Italien zu versetzen: das ist das Publikum." – „Und dennoch", erwiderte Otto nach einer kurzen Pause, „wenn alle so dächten, so müßte die dramatische Poesie in der Luft spielen und die Bühne zugrunde gehen." – „Ja, das hoff' ich auch!" sagte Lothario, „die Dichter müssen nur nicht nachgeben, sondern die Theater poetisch aushungern, sie an ihrer eigenen Misere und Langweiligkeit allmählich verschmachten lassen und unterdes draußen frisch und keck die Welt auf ihre eigene Hand dramatisieren. Das Publikum ist so dumm gerade nicht, wie es aussieht. Ist es erst im Buch an die ursprüngliche Schönheit wieder gewöhnt, so wird es auch die Bühnen schon zwingen, sich zu akkommodieren. Aus der alten, guten Poesie kann sich ein neues Thea-

ter bilden, nimmermehr aber eine neue Poesie aus den kranken Gelüsten des Publikums und der Pedanterei der Theatermaschinisten. Und überhaupt, junger Mensch", fuhr er fort, „wollt Ihr ein Dichter werden – und ich meine, Ihr habt die unglückliche Disposition dazu –, so müßt Ihr Euch ein für allemal daran gewöhnen, für die Handvoll Gescheiter im Lande zu dichten und nach den andern nicht zu fragen. Vor allem aber müßt Ihr Euch hier von uns Komödianten und Frauenzimmern losmachen, denn wer sich so in der Rumpelkammer des Lebens herumtreibt, dem fliegen die Fledermäuse an den Kopf, und es wäre schade um Euer weiches Flachshaar."

Otto zürnte wie ein Mädchen. Lothario aber, in seinem kühnen Wesen, griff wie ein eisiger Morgenwind durch alle Saiten seiner wunden Seele. Auch hatte es Otto ja mit eigenen Augen gesehen: Kordelchen war treulos, das Brettergerüst seines geträumten Bühnenruhms zertrümmert, er kam sich nach den heutigen Erfahrungen nun selbst hier kahl und erbärmlich vor. Und so geschah es, daß er, ehe sie noch das Ende des Gartens erreichten, dem harten Freunde mit dem Ungestüm eines frischen Entschlusses die Hand darauf gab, sogleich weiterzureisen, um ungestört und mit strengem Ernste ganz in der Dichtkunst zu leben. – Nun fehlte es aber wieder am nötigen Reisegeld zur Ausführung eines so löblichen Vorsatzes. – Lothario machte bei dieser Bemerkung eine lebhafte Bewegung und schien einen raschen Vorschlag auf dem Herzen zu haben, schwieg aber plötzlich. – Da standen sie soeben vor Dryanders Tür. „Halt!" sagte er, „hier wohnt Fortunas Hofnarr, da wollen wir anklopfen, kommen Sie nur geschwind."

Mit diesen Worten drängte er den Zögernden in das Haus hinein. Ein Bedienter empfing sie in der Vorstube und wollte anmelden. Der Schauspieler schob ihn aber lächelnd zur Seite und trat ohne weiteres in das Zimmer. Hier war durch tief herabhängende, grünseidene Gardinen ein künstliches Halblicht verbreitet, ein zierlicher

bronzener Opferaltar auf dem Mahagonitisch erfüllte das Gemach mit Wohlgerüchen, Dryander selbst, in einem feinperkalenen Negligé, ruhte mit einem Papier in der Hand nachlässig auf einer Ottomane. Er blinzelte die Eintretenden vornehm an, als könnte er sie nicht gleich erkennen, faltete und versiegelte erst den Brief und klingelte nach dem Bedienten: „An Se. Durchlaucht, aber sogleich." – Dann sprang er auf und nötigte die Gäste verbindlich auf das Sofa. – Lothario, als sie sich feierlich niedergelassen, drückte mit devoter Stimme ihre langverhaltene Freude über seinen sehr ergötzlichen Glückswechsel aus. „Mich hat es nicht im geringsten überrascht, verehrter Hofrat", sagte er, „du strebtest von jeher oben hinaus: keine Dachstube war dir zu hoch, du hattest schon damals immer die besten Aussichten." – Dryander, hofmännisch überhörend, wandte sich, ohne darauf zu antworten, zu Otto, ihn seiner besonderen Teilnahme an seinem schönen Talent versichernd, doch müsse er ihm als Freund raten, seinen Umgang sorgfältiger zu wählen. – „Eben darum", unterbrach ihn Lothario, „hat dieser junge Mann einen festen Entschluß gefaßt. Du hast gestern dein Gehalt bezogen und brauchst es nicht; wir wollten daher gehorsamst bitten, ob du vielleicht die Güte haben möchtest, ihm unter die Arme zu greifen – ein kleines Darlehen – auf kurze Frist – er will nach Italien." – „Nach Italien?" rief Dryander aus, „in das göttliche Land" – „Ja, wo, nach Goethe, die Zitronen blühn", fiel Lothario ein. – „Meine Verbindungen hier bei Hofe, ich kann Ihnen vielleicht nützlich sein", fuhr Dryander fort, „auch kenne ich mehrere Personen von Rang in Rom, Neapel, mein Freund, der Ducca" – „Degli Lazzaroni", meinte Lothario, „eine alte Familie, ich glaube, Ihr seid verwandt." – Otto stand hochrot und entrüstet auf. – „Ich bedaure nur", sagte Dryander, gleichfalls aufbrechend, „daß in diesem Augenblick dringende Amtsgeschäfte – es wird mir aber sehr erfreulich sein, Sie vor Ihrer Abreise" – „Allerliebster Hofrat!" rief hier plötzlich Lothario, seine Hand fassend,

„jetzt tanz' noch eine Menuett mit mir." – Dryander maß ihn mit verächtlichen Blicken. – „Oder soll ich dich morgen vor dem ganzen Hofe auffordern? Du kennst ja meine Kuchenreiter", sagte Lothario. – Der Hofrat wollte hastig klingeln. – „Tanz", wiederholte Lothario warnend. – Da stellte sich Dryander mit teuflischem Lächeln in Positur, Lothario sang vergnügt die Menuett à la Vigano, so führten sie auf dem bunten Teppich graziös mehrere Touren aus, und es war wunderlich anzusehen, wie Dryander seinen Gegner mit den Augen erstechen wollte, so oft sie feierlich aneinander vorüberschwebten. Dann geleitete ihn Lothario an den Fingerspitzen bis zum Sofa, machte eine tiefe Verbeugung und entfernte sich mit dem verlegenen Otto, der gar nicht wußte, wie ihm geschehen.

„Das war eine gesunde Motion", sagte Lothario lachend – als sie draußen waren –; „aber Mensch, sehen sie nicht so trübe aus! Schreiben Sie noch heut nach Hohenstein um Geld, treu, klar und aufrichtig; Sie kriegen des Plunders genug; wer ehrlich will, was er soll, der kann auch, was er will!" – Mit diesen Worten wandte er sich wieder in den Garten. Otto stand noch lange zweifelnd still, dann aber eilte er auf sein einsames Stübchen, um sogleich den guten Rat zu befolgen. – Als er oben am offenen Fenster saß, tanzte schon das Abendgold durch das Weinlaub so lustig über das reine Blatt vor ihm. Er stand oft im Schreiben auf und lehnte sich zum Fenster hinaus. Die Abendsonne beschien draußen die herbstliche Gegend, die Wandervögel zogen über das Haus fort, seine ganze Seele war voll fröhlicher Verheißung und zog mit ihnen in die schöne, wunderbare Ferne hinaus.

Währenddes kehrte unten der Fürst mit mehreren Begleitern von einem Ausfluge heim. Sie ritten zwischen den einsamen Felsenwänden den kühlen Strom entlang, die Wälder glühten im buntfarbigen Herbstschmuck. Da erblickten sie hoch über sich auf einem überhangenden Felsen die Gräfin Juanna, unter wilden Waldblumen nach dem Strome hinabgebeugt, daß die dunklen Locken Stirn

und Wangen bedeckten. – „Lurelei!“ – sagte der Fürst wie in Gedanken zu seinen Begleitern, die geblendet hinaufschauten. –

Aber er selber war schon in ihrem Bann, und als sie am Schlosse angekommen, hatte er sich unbemerkt entfernt und stieg allein hastig und verwirrt durch die schöne Einsamkeit hinauf. Er kannte von seinen Jagden den wenig betretenen Fußsteig zur Höh'. Juanna fuhr erschrocken auf, als er soeben plötzlich durch das Gebüsch brach und neben ihr auf die Knie sank, ihre Hand mit glühenden Küssen bedeckend. Sie schwieg und sah ihn lange durchdringend an. „Still, still“ – sagte sie dann, „hier kann man uns vom Schloß aus sehen.“ – Hiermit ergriff sie seine Hand und führte ihn rasch durch die Hecken, über schmale Felsrücken an jähen Abgründen vorbei. Durch seine Seele gingen wechselnd Furcht und Hoffnung, wie die Schatten im Walde. „Wo wandern wir hin?“ fragte er endlich betroffen, denn die grünen Plätze kamen ihm so bekannt vor, das Abendrot spielte, wie die alte schöne Zeit, darüber. So traten sie auf einmal zwischen den Bäumen heraus und erblickten unter einzelnen Tannen ein kleines Haus mit einem stillen, zierlichen Gärtchen davor. – Der Fürst drängte erschrocken weiter. „Hier wollen wir ausruhen“, sagte Juanna, ihn festhaltend. Er schaute nun unverwandt hinüber, wie in einen Traum. Eine alte, blinde Frau saß in der Abendsonne vor der Tür, ein schönes, bleiches Mädchen ging singend vor ihr im Garten auf und nieder. Da erblickte sie auf einmal den Fürsten und floh wie ein erschrecktes Kind zu der Mutter und setzte sich zu ihren Füßen ins Gras. – „Was hast du denn?“ fragte die Blinde. Das Mädchen sagt: es gehe ein Engel im Abendscheine durch den Wald, ein anderer stehe neben ihm, der werfe einen langen Schatten weit über den Wald und die Täler, ach, es dunkelt schon, und er kommt noch immer nicht wieder! – Sie drückte ihr Gesicht in den Schoß der Mutter und weinte bitterlich.

Der Fürst wandte sich ab. Es war das Jägermädchen, das

er so oft in früheren Jahren heimlich besucht. Ihr Herz war gebrochen, da sie in ihrem Liebsten den Fürsten erkannt, nun war sie lange wahnsinnig, er hatte sie fast vergessen. – Die Abendglut blickte noch einmal durch den Wald herauf, daß die Gegend plötzlich ganz fremd und wie verwandelt erschien. Juannas Augen funkelten beinahe tödlich, er hielt sie nicht länger aus und floh tief erschüttert von dem entsetzlichen Ort.

Sie aber war unterdes in das Gärtchen getreten und sprach trostreich zu der Blinden und ihrem armen Kind und warf ihr, ehe sie weiter ging, einige Goldstücke in den Schoß. Da betete die Alte still vor sich, denn nun glaubte sie's selbst auch, daß in der Abendstille ein Engel an ihrem Hause vorübergegangen. – Währenddes stieg der Maler Albert, bis an die Zähne bewaffnet, still und ernst den Waldberg hinan. Er hatte vorhin die Gräfin auf dem Felsen, dann den Fürsten heimlich heraufschleichen gesehen und in seiner Tugendhaftigkeit sogleich beschlossen, mit Gut und Blut die Unschuld zu beschützen. Die Nacht war schon hereingebrochen, die ganze Gegend stand wie in Gedanken im Mondglanz umher, und als Juanna wieder im Schloß an ihrem Fenster stand, hörte sie unter sich den Strom aufrauschen, wie von Ruderschlägen. Es war Lothario, der unten auf einem Nachen vorüberfuhr und sang, sie konnte durch den Nachtwind nur folgende Worte verstehen:

„Wetterleuchten fern im Dunkeln,
Wunderbar die Berge stehn,
Nur die Bäche manchmal funkeln,
Die im Grund verworren gehn,
Und ich schaue froh erschrocken
Wie in eines Traumes Pracht –
Schüttle nur die dunklen Locken,
Deine Augen sind die Nacht!"

Der Nachtwächter unter den Fenstern aber schüttelte den Kopf und sah zu seiner Verwunderung auf dem Felsen

drüben eine lange Gestalt, auf ihr Schwert gestützt, die halbe Nacht hindurch gleich einer verlornen Schildwacht stehen.

Zwölftes Kapitel

Es kann ein Mensch lange Zeit in den besten Grundsätzen wie ein Schneemann eingefroren sitzen, aber die lustigen Frühlingsbäche unterwaschen schon heimlich plaudernd und neckend den Sitz unter ihm – ein Laut, der leise Flug eines Vogels: und er stürzt kopfüber und verschüttet alle guten Vorsätze wieder. – So erging es Dryander.
Es war ein schöner, stiller Abend, da ging die Fürstin allein in einem entlegenen Teile des Gartens spazieren, sie schien unruhig, oft blieb sie stehen und hörte, wie die Schauspieler unten sangen. Aber das kluge Kordelchen hatte sie schon aus der Ferne bemerkt, Lothario fehlte heut wider seine Gewohnheit bei dem Gesange – sie hatte ihre eigenen Gedanken. So begegnete sie Dryander am Eingange des Parks, da flog ihr plötzlich ein Anschlag durch den Kopf. „Endlich finde ich Sie!“ flüsterte sie ihm geheimnisvoll zu, „die Fürstin dort, sie erwartet Sie. Aber still“ – sagte sie, den Finger auf den Mund legend, und verschlüpfte schnell wieder zwischen den Bäumen. – Eitelkeit macht dumm. Der überraschte Dryander überblätterte geschwind das Glücksbuch seiner hiesigen Anstellung, jedes Blatt rauschte ihm plötzlich wie die Schleppe der Fürstin, nun verstand er erst alles, ja, er überredete sich in allem Ernste selber, längst in die Fürstin sterblich verliebt zu sein. So, im Garten fortrennend, umspann er sich immer hitziger mit dem tollsten Roman, und als nun die schlanke Gestalt in einem dunklen Bogengange auf einmal vor ihm stand, überschüttete er sie atemlos, ohne Eingang und Vorbereitung, verworren mit der glühendsten Liebeserklärung. Die Fürstin, da er so auf sie los-

stürmte, stand erst verwundert, dann lächelte sie fein und still, es fiel ihr nicht ein, daß er sich einbilden könne, sie meine ihn. – „Tasso!“ sagte sie scherzhaft warnend, „wir sind hier nicht in Belriguardo.“ – Indem sie aber den Handschuh ausziehen wollte, um ihm ihre weiße Hand zum Kuß zu reichen, fiel ein Mondstrahl durch das Laub auf Stirn und Mund. Da kam sie Dryander schon eigentlich etwas alt vor, sie gefiel ihm auf einmal gar nicht, und seine Gedanken schlugen ihm unwillkürlich um, wie Milch beim Wetterleuchten. „O Gott, Fürstin!“ rief er aus, „die Nacht ist eine wilde, phantastische Blume, berauschenden Duft verstreuend, schöne, gefallene Engel wiegen sich auf den Blättern und singen im Traume von den Sternen, wo sie sonst gewohnt, und zwischen den träumenden Kaiserkronen und Blütenglocken flüsternd ringelt die alte Schlange sich leise empor, und von ihrem Krönlein lösen sich grüngoldene Funken und schwärmen durch das Blütengeflecht, und in ihrem streifenden Widerscheine sehen die Gesichter leichenblaß, wie Sie jetzt, Fürstin, im Mondlicht.“ – So redete er sich nach und nach in die Tugend und tragisches Wesen hinein, sprach entsetzlich von der Sünde, immer begeisterter, wilder und herzzerschneidend. Die Fürstin überlief es heimlich eiskalt dabei. Aber sie bezwang sich und unterbrach ihn lachend: „Der Duft der Nachtblume ist Ihnen zu Kopfe gestiegen, gehen Sie nach Hause und nehmen Sie ein Fußbad.“ – Dann wandte sie sich stolz nach dem Schlosse.

Dryander stand wie vom Donner gerührt. Jetzt wollte er ihr nach, sie festhalten, rannte aber in der Verwirrung mit der Stirn an einen Baum, daß er den Hut verlor. Er schimpfte sich selbst einen gefallenen Engel, der gotteslästerlich die Unschuld an die Wand male, die ihn verführt. So eilte er wie besessen quer durch den Wald, in der Ferne verklang eben noch die letzte Abendglocke, die Mädchen im Dorfe unten sangen vor den Haustüren. Und als er am Ende des Parks plötzlich heraustrat, erblickte er vor der letzten Hütte des Dorfs beim hellsten Mondenscheine

eine schöne Jungfrau, die er noch niemals gesehen, in reichem Gewand unter einer Linde sitzend. Sie hatte ein blondgelocktes Kind auf dem Schoße, ein anderes stand auf ihr Knie gestützt und sah an ihr empor, alle von einem weiten Schleier umgeben, durch den die Sterne flimmerten, als wären sie drein gewirkt. Da war's ihm, als hätte der Himmel sich barmherzig auf diesen Hügel herabgeneigt, todmüde, außer sich, warf er sich zu ihren Füßen auf den Rasen hin, vor den unschuldigen Augen. „O, heilige Jungfrau, bitte für mich!" redete er sie aus tiefstem Grund der Seele an, „beschütze mich vor der wilden Jagd – ich selber Hund und Wild – erlöse mich von der innern Lüge!" – Sie sah ihn ernsthaft an, sie konnte vor den Kindern nicht aufstehen. – Er aber achtete nicht darauf; wie ein Kranker, der einen seligen Traum hat, sprach er immerfort zu ihr und bot ihr endlich gerührt seine Hand an. Er wolle sie mit den Kindern auf einen Esel setzen, so wollten sie ziehen durchs einsame Gebirg die Klippen hinab in der schattigen Kühle, alles hinter sich lassen und vergessen, fort nach der blauen Ferne, bis in das stille Himmelreich. – „Was sind das für Bälge?" unterbrach er sich hier plötzlich selbst, das Kind hastig abwehrend, das mit den schmutzigen Händen zu ihm wollte. – „Ich brachte ihnen Speise und Medizin", erwiderte das Fräulein, „ihre Mutter liegt drin krank" – – „Krank?!" rief Dryander schnell aufspringend und bedenklich nach der Hütte blikkend, denn er hatte eine abergläubische Furcht vor Ansteckung. Ein Bedienter mit einem Handkörbchen war unterdes aus dem Hause dazugetreten, das Fräulein erhob sich, wie erlöst, von dem Rasen und entfernte sich rasch, noch öfters furchtsam zurückblickend. – In dem Gebüsch daneben aber hörte er ein feines Lachen, er glaubte ein Frauenkleid durch die Zweige schimmern zu sehen.

Es war Kordelchen, die ihm heimlich gefolgt. Aber es bekam ihr schlimm. Denn sie hatte sich kaum in ihrem Versteck zurechtgesetzt, da stürzte Dryander, wie ein Rasender, schreiend und tobend daher und fuhr mit dem Kopfe

gerade in ihre Röcke. Sie sprang erschrocken auf – eine Fledermaus, da er seinen Hut im Walde gelassen, war ihm unversehens in die Haare geflogen und blickte, dort festgenestelt, mit stieren Augen vom Kopfe des Dichters. Dieser schrie, Kordelchen schimpfte, keines mochte anfassen, darüber fuhren Köpfe, Mägde und Kinder aus allen Fenstern und Türen, die Hunde im Dorfe schlugen an, Dryander nahm ganz verblüfft Reißaus, der Nachtwächter, der eben blasen wollte, mit langen Schritten ihm nach – so kam er atemlos nach Hause, wo er, endlich von dem gespenstischen Untier befreit, sogleich zu Bett ging und sich fest einbildete, todkrank zu sein.

Feine Lebensart ist wie ein guter Firnis, den die gemeine Luft nicht angreift; so war auch die Fürstin seit jenem Abende ganz unverändert; sie erwähnte des Vorfalles mit keinem Worte, sie mochte wohl ihre Gründe dazu haben. Dryander, da es ihn nicht mehr interessierte, hatte längst alles wieder vergessen, bis auf die schöne, mildtätige Jungfrau vor der Hütte. Diese aber war niemand anders als Fräulein Trudchen von dem wüsten Schlosse des Barons. Die leichte, heitere Art der vornehmen Gäste bei dem fürstlichen Besuche hatte sie ganz verblendet; wie nach Sonnenuntergang flimmerte es noch lange in ihrer Einsamkeit nach, und sie hörte nicht auf zu bitten und zu schmollen, bis der Vater sie endlich auf mehrere Wochen zu dem fürstlichen Forstmeister, ihrem Verwandten, hinüberschickte, um sich zu bilden. – Dryander besuchte nun regelmäßig jeden Abend den Forstmeister, disputierte mit den dort häufig versammelten Gutsbesitzern, trank viel und verfolgte das Fräulein mit wahrhaft poetischer Wut. Er schleppte ihr unermüdlich Bücher zu: Goethe, Shakespeare, Calderon, Cervantes, sie mußte geschwind lesen, ihre Unwissenheit reizte ihn nur immer mehr. Es war ihr alles so neu, im Hause hatten alle großen Respekt vor seiner Gelehrsamkeit, er umstrickte sie ganz mit seinem leidenschaftlichen Wesen. Die Schauspieler hatten insgeheim ihre große Freude daran, und eines Abends kamen

die Schalksnarren Ruprecht, Kordelchen, Fabitz, eins nach dem andern, feierlich zu ihm, der eine brachte ein Gedicht, der andere einen dicken Blumenstrauß, und gratulierten zu seiner morgigen Vermählung mit dem Fräulein. Er stutzte und lief sogleich noch zum Forstmeister hinüber. – Es war schon spät, er fand einen seltsamen Rumor im Hause, Spiegel und Kronleuchter wurden geputzt, Gäste vom Lande waren angekommen, andere wurden noch erwartet. Im Garten aber sah er unter den Pflaumenbäumen ein trübes Feuer glühen, vor dem sich dunkle Gestalten seltsam hin- und herbewegten. Er eilte hin und fand sein Trudchen, eine Schürze vorgebunden und die Ärmel aufgestreift, in voller Arbeit vor dem Backofen, in welchen soeben Kuchen geschoben wurden. Neugierig und diensteifrig wollte er ihr helfen, um etwas Näheres zu erfahren. Aber sie hatte nicht viel Zeit, er war ihr überall im Wege, sie streifte ein paarmal dicht an ihn heran, daß er auf der einen Seite ganz weiß von Mehl wurde. „Nun, nun“, sagte sie, da er sich eifrig sträubte, „es ist ja nicht Ihr Hochzeitsfrack.“ – „Wahrhaftig“, rief er, „wo soll ich bis morgen einen besseren hernehmen?“ – „Kommen Sie nur in dem“, erwiderte sie, „und bringen Sie ein hübsches Gedicht mit.“ – Er wollte sie, da die Mädchen eben in den Ofen sahen, schnell haschen und küssen. Aber sie hatte gerade den Kochlöffel in einen Topf voll Pflaumenmus getunkt und fuhr ihm fix damit über den Mund. „Morgen!“ sagte sie lachend und lief nach dem Hause. Er sah ihr nach – es war ihm, als führe sie unter den Bäumen wie eine kleine Hexe auf dem Kochlöffel davon.
Am folgenden Morgen war er schon frühzeitig auf dem Platze, in Schuhen und Strümpfen, einen Klapphut unter dem Arme. In des Forstmeisters Hause schien noch alles zu schlafen; er trat unbemerkt in den stillen Gartensaal. Dort war eine lange Tafel schon festlich gedeckt, buntes Naschwerk schimmerte zwischen den künstlich gefalteten Servietten, in der Mitte ein prächtiger, altmodischer Aufsatz mit Pomeranzenbäumchen von Wachs und por-

zellanenen Götterfiguren, die sich in dem Spiegelboden, wie in einem Weiher, verdoppelten. Er schritt neugierig auf und nieder und kostete alle Teller durch. Dann ging er in den Garten, um in der Geschwindigkeit noch die Rede zu memorieren, die er an der Hochzeitstafel halten wollte. Da sangen aber die Vögel so spöttisch, und die schlanken Pappeln im Morgenwinde verneigten sich vor ihm, als wollte ihm alles gratulieren. Von einem umwachsenen Hügel konnte er gerade ins Haus seiner Liebsten sehen. Dort war es unterdes auch schon lebendig geworden, er sah, wie sich Vettern und Basen im festlichen Staate versammelten, immer neue Gestalten erschienen an den Fenstern, ein galantes Wirren, Scharren und Knicksen, flimmernd durcheinander, draußen wurden Pasteten und ein hoher Baumkuchen ins Haus getragen, vom Jubel der Dorfjugend begleitet, die eben zur Schule ging. Er hatte sich das alles noch niemals so recht voraus überlegt, jetzt aber befiel ihn, allmählich wachsend, eine unwiderstehliche Angst vor dem Heiraten, und als er eben in eine Allee hineinbiegen wollte, erblickte er am andern Ende gar zwei alte Damen, die in taftnen Kleidern feierlich auf ihn dahergerauscht kamen. Da wandte er sich schnell und entfloh in langen Sätzen unaufhaltsam durch den Garten, am Dorfe vorüber in die Berge hinein; es war ihm, als verfolgte ihn Gott Hymen und klopfte seine Fackel an seinem Kopfe aus, daß ihm die Funken knisternd um die Augen sprühten.

In dem Hause ging es unterdes schon hoch her, es war des Forstmeisters Geburtstag, kein Mensch dachte an Hochzeit, Trudchen trat oft ans Fenster und ging immer wieder ganz böse fort, daß Dryander noch nicht kam. Auch der Baron, der sich wie gewöhnlich zu dem Feste mit eingefunden, war begierig, ihn zu sehen, denn der Forstmeister hatte ihm schon von seiner Liebschaft, seiner einträglichen Stelle und seinen bedeutenden Verbindungen am Hofe erzählt, und der Baron in seinen verzweifelten Vermögensumständen dachte sogleich daran, seine Tochter

unter die Haube und sich unter Dach zu bringen, ehe sein eignes ihm über dem Kopfe zusammenstürzte. Aber vergeblich war mehrere Male nach Dryanders Wohnung geschickt worden, man hatte sich endlich zu Tisch gesetzt, die Unterhaltung wurde immer lauter, in dem Lärm flogen schon Bonbons und bedeutende Blicke zwischen den jungen Leuten hin und her, vom Knall der Champagnerflaschen salutiert, als sich auf einmal durch die Diener vom Schlosse her das Gerücht verbreitete, der Hofrat sei entsprungen und fern im Walde in vollem Staate gesehen worden. Niemand wußte sich's zu erklären, denn die Schauspieler, die einen solchen Ausgang nicht erwartet hatten, hüteten sich wohl zu verraten, was sie Dryander eingeredet. – Trudchen aber stand plötzlich auf und ging hochrot hinaus. Da wurde die Sache erst recht auffallend, alle Blicke waren auf die Fortgehende gerichtet, die Mädchen zischelten einander heimlich in die Ohren, der Baron eilte ihr nach, denn es sollte noch getanzt werden. Aber das Fräulein war wie ausgewechselt, schmollend und trotzig, und wollte durchaus nicht mehr zur Gesellschaft zurück. Sie wisse es am besten, sagte sie, die Alltäglichkeit dieser prosaischen Menschen habe den Hofrat vertrieben, sie frage gar nicht mehr nach den unwissenden Leuten, sie kenne nun eine ganz andere Welt! – Der Baron aber schalt sie eine verdrehte Närrin. Dann ließ er voller Zorn mitten in der allgemeinen Verwirrung anspannen, schob sie in den Wagen und verschwor sich: der Kerl, der Hofrat, solle sie nehmen, oder er jage ihm eine Kugel durch den Kopf!

Keinem war der Vorfall fataler als Lothario, denn der Doktor war ihm lange wie ein Blitzableiter, in den sein Witz und Ärger lustig einzuschlagen pflegte. Er ging soeben, die seltsame Flucht besprechend, mit Fortunat durch den Garten, als ihnen plötzlich *Otto* mit leuchtenden Augen entgegenkam. „Gute Nachrichten aus Hohenstein!“ rief er schon von weitem, einen Brief emporhaltend. Er hatte, über alle Erwartung, nicht nur die Zu-

stimmung des Amtmanns in seine Pläne, sondern auch eine bedeutende Summe erhalten, die mehr als zureichend schien, die Reise durch Italien behaglich zu vollenden. Auch ein Brief von Walter an Fortunat war beigeschlossen, den dieser mit großer Freude sogleich erbrach.

„Unser Otto“, schrieb der wackre Freund, „hat uns von Eurem seltsamen Zusammentreffen und dem poetischen Leben an dem Hoflager des Fürsten ausführlich Bericht erstattet. Er schreibt überaus lebendig, und es ist uns allen, als wären wir in den Palästen und grünen Gängen mitten unter Euch und sähen und hörten jeden nach seiner Weise sich bewegen und sprechen, diesen Lothario, Kordelchen und Dich selbst nicht ausgenommen. Da sitzen wir dann in Hohenstein, wenn im Feld und Haus alles besorgt ist, jeden Abend wieder unter den Linden vor der Haustür zusammen, und ich muß den Brief immer wieder von Anfang bis zu Ende laut und deutlich vorlesen, bis der Mond über uns aufgeht. So bist Du auch in der Ferne bei uns, wie denn überhaupt eine stille, mondhelle Nacht schon an sich etwas Traumhaftes hat und entfernte, geliebte Gegenden und Personen der Seele wunderbar näher bringt.

Wie glücklich seid Ihr Dichter? Eurem zauberischen Sinne erschließt sich überall, wo Ihr wandelt, wie dem Geliebten, willig und vertraulich die verborgene Schönheit der Welt, mit jedem Schritte erweitern sich die Kreise, das Entfernte, Dunkle rückt verständlich in freundliche Nähe, und neue Fernen heben sich wieder wunderbar immer weiter und schöner. Was ist Dir nicht alles wieder begegnet, seit wir uns trennten! – Mit mir geht es gerade umgekehrt. Je weiter ich komme, je enger wird der Kreis, und die Fernen, die mich in der Jugend entzückten, verbleichen und versinken mir allmählich. – Doch ich denke, das muß wohl so sein. Ruhiger als Du Dir vielleicht einbilden magst, habe ich endlich meine Stellung in der Welt erkannt und von den vornehmen Täuschungen Abschied

genommen. Ich lerne mich bescheiden und beschränken, und mir ist wohl. Eure Aufgabe ist unübersehbar, verwickelt und selten recht in Eurer eigenen Gewalt. Mein Beruf dagegen ist einfach und mir jederzeit klar, und glaube nur, es ist auch was wert, mit sich selber im reinen zu sein.

Kann ich nun nicht selbst, wie ich früher wohl träumte, mit hinaus in das schöne Land der Poesie, so will ich wenigstens den Dichtern redlich helfen, wie und wo ich's vermag. So ist es mir denn auch endlich gelungen, den Otto mit seinen Pflegeeltern zu versöhnen, denn ich meine, es stand da ein bedeutendes Talent auf dem Spiele. Glaube aber nur nicht etwa, daß das so schwer hielt. Ein rechter, fester Wille tut überall Wunder. Ottos plötzlicher Entschluß, die Heimat zu verlassen, hatdie bisherige Ansicht der Sache, ich möchte sagen, auf den Kopf gestellt und der Einbildungskraft der Hohensteiner eine ganz neue Richtung gegeben. Dem Amtmann gefällt Ottos Mut, um so mehr, je weniger er ihn dem sanften Stillen zugetraut hatte. Die gute Mutter aber freut sich nun heimlich darauf, Ottos Namen gedruckt oder gar sein Bild vor einem Buche zu sehen.

Du wirst Dich wahrscheinlich über das viele Geld wundern, das wir schicken. Aber es kommt nicht von uns. Otto hat hohe Gönner – mehr darf ich für jetzt davon nicht verraten.

Das ist jetzt eine glückliche Zeit. Kaum war diese Angelegenheit wegen Otto nach Wunsch beseitigt, so erhielt ich aus der Stadt die Nachricht, daß mir das einträchtige Amt eines Gerichtsverwalters hier in Hohenstein, das ich solange zwischen Hoffnung und Zweifeln ersehnt, zuteil geworden. Nun steht unserer Verheiratung nichts mehr im Wege. – Soeben guckt mir Florentine über die Schulter ins Blatt und hält mir schnell mit der Hand den Mund zu, damit ich nicht alles ausplaudern soll. Da ich aber unterdes fortfuhr zu schreiben, so läuft sie nun gar fort und läßt Dich nicht einmal grüßen. – Ich schreibe im Garten auf

demselben Platze mit der großen Aussicht, wo Du alle Morgen zu lesen oder zu dichten pflegtest. Aber die Felder unten sind schon leer, auf den Beeten neben mir prangen nur noch die Astern, und die Blätter auf den Bäumen färben sich und fallen. Das ängstigte mich sonst immer, diesmal ist mir gar wunderlich zumute dabei, denn im Hause durch die offenen Fenster sehe ich die Mutter emsig Federn schütten zu den Brautbetten, der Tischler hat seine muntere Werkstatt vor der Haustür aufgeschlagen und schnitzt die Doppelfenster für unsere künftige Wohnung, und ich richte mich mit innigem Behagen in Gedanken für den Winter ein – da mögen draußen Sturm und Schnee an die Fenster schlagen! doch dieses Gefühl verstehst Du wohl nicht? – Nun, Gott sei mit Dir, lieber Bruder, und führe Dich auf Deinen weiten Wegen zu solchem Glücke und solcher Herzensfreude, als ich auf dem nächsten hier gefunden habe."

Fortunat legte den Brief mit ganz eigenen Empfindungen zusammen, es war ihm, als stände er tief im stillen Abendrote. Vor ihm aber stand Otto mit Lothario an dem Abhange und schaute trunken in die Ferne, in die er nun bald hinausziehen sollte.

Dreizehntes Kapitel

„Und wo noch kein Wandrer gegangen,
Hoch über Jäger und Roß,
Die Felsen im Abendrot hangen
Als wie ein Wolkenschloß.

Dort zwischen den Zinnen und Spitzen,
Von wilden Nelken umblüht,
Die schönen Waldfrauen sitzen
Und singen im Wind ihr Lied.

Der Jäger schaut nach dem Schlosse:
Die droben, das ist mein Lieb! –
Er sprang vom scheuenden Rosse,
Weiß keiner, wo er blieb."

So sang *Lothario*, auf einer Waldhöh' auf seine Büchse gestützt. *Fortunat* trat zu ihm hinauf, da sahen sie jenseits den Wald schon von Jägern und Reitern blitzen, der Fürst hatte zum Valet noch eine große Jagd veranstaltet, bevor alles vor dem Winter wieder in die Stadt flüchte.
„Hast du die Braut nicht gesehen?" fragte Lothario unruhig umherspähend. – „Du meinst die Gräfin Juanna, so hörtest du auch davon?" erwiderte Fortunat, „sie halten's so geheim vor mir, und alle Jäger wissen's. Erst diesen Morgen hört' ich, daß der Bräutigam, ein Baron Manfred, noch heute zur Jagd erwartet wird." – „Das ist ein prächtiges Wetter zum Heiraten", sagte Lothario, „der Alteweibersommer fliegt, als hätten sich alle alten Jungfern das Haupthaar ausgerauft und in die Lüfte umhergestreut, da bleibt mancher Ritter noch mit den Sporen drin hängen. Gebt acht, es gibt eine köstliche Verwicklung!" Hiermit schüttelte er Fortunat heftig die Hand und ging schnell ins Tal hinunter.
Fortunat sah ihm verwundert nach, dann folgte er der Jagd, die jetzt immer lustiger durch die Berge ging. So verlor er sich bald in das Labyrinth der Wälder und kam zuletzt in eine grüne Schlucht, über deren Felswände von allen Seiten Efeu verwildert hinabstieg. Auf einmal brach ein Hirsch durch das Dickicht, eine Meute Hunde an seinen Fersen und hinter ihnen *Juanna*. Das edle Tier bei seinem Anblick stutzte schnaubend und stürzte sich seitwärts in den Abgrund, Hunde und Reiterin konnten ihm dorthin nicht folgen. Da hielt Juanna plötzlich über Fortunat in der wilden Einsamkeit, die Hunde streckten sich lechzend zu ihren Füßen. „Seht, der ist frei" – sagte sie, die schwarzen Locken aus dem erhitzten Gesichte schüttelnd –, „und eher fangt Ihr mit verliebten Blicken einen

Hirsch im Walde als mich! Was wollt Ihr von mir? Laßt das Werben um mich, mir ist wohl in meiner Freiheit. Was auch die Fürstin für Anschläge hat, ich werde nie die Eurige und keines Mannes Weib – hütet Euch, es wäre unser beider Tod!" – Hierauf wandte sie ihr Roß, die alten Bäume schüttelten sich und streuten ihre gelben Blätter wie einen Goldregen über die schöne Gestalt. Fortunat stand ganz verwirrt, ihm war, als sprächen ringsum die Quellen irre den Wald entlang. Unerhörteres konnte ihm nicht begegnen, als daß er nun am Ende selbst der Bräutigam sein sollte! – Unterdes hatte sich Juanna wieder höher in das Gebirge gewendet, ein plötzlicher Anschlag schien ihre ganze Seele zu bewegen. Sie kannte den Waldweg nach einem Nonnenkloster, das jenseits des Gebirges lag und dessen Äbtissin ihr verwandt war. Dort wollte sie noch heute hin und abwarten, bis der Winter Gebirge, Freier und Verliebte verschüttet. Aber mitten in diesen Gedanken erblickte sie auf einmal eine Gemse über sich, die sich hoch über den Wipfeln von Klippe zu Klippe schwang. Das war ihr ganz neu, sie konnte der gefährlichen Lust nicht widerstehen. Ein alter Jäger, der sich bis in diese Öde verstiegen hatte, arbeitete sich eben durch das Gesträuch, sie übergab ihm ihr Pferd, er sollte es hüten, bis sie wiederkäme, und eh' er sie noch warnen konnte, war sie schon zwischen den Felsen verschwunden.

Nun kletterte sie wie ein schlanker Panther über die Klippen, das scheue Wild verlockte sie immer höher hinauf, die Lust wuchs mit der Gefahr, sie hatte sich lange nicht so wohl gefühlt und erstaunte, da sie plötzlich eine Felsenwand über sich wie im Feuer erblickte, es war der Widerschein der Abendsonne, die soeben jenseits hinter den schwarzen Wäldern versank. Mit der einen Hand sich an einen Strauch haltend, sah sie über den Felsenrand hinab: die Täler unten dunkelten schon, aus weiter Ferne hörte sie noch eine Abendglocke heraufschallen, sie meinte, es komme von dem Kloster herüber. Eilig schlug sie nun die Richtung ein, aber sie konnte sich in dem wilden Gewirre

nicht zurechtfinden, wohin sie sich wandte, taten sich neue Abgründe auf; so stand sie in der entsetzlichen Einsamkeit wie einer, der nachts zwischen den Zacken und Steinbildern eines unbekannten Münsters vergessen worden. In dieser Not verfiel sie darauf, ihr Gewehr zum Signal abzuschießen. Zu ihrer Freude gab sogleich ein Schuß ganz nahe Antwort. Bald darauf hörte sie Fußtritte auf dem lockeren Gerölle, eine hohe, schlanke Gestalt trat plötzlich zwischen den Steinen hervor – es war Lothario. „Das ist ein gefährliches Revier", sagte er, „und die Nacht bricht schon herein, doch ich bin hier der Pfade kundig und meiner Richtung gewiß." – Die Gräfin aber hatte bei seinem Anblick ein seltsamer Eigensinn ergriffen, gerade ihm dachte sie hier am wenigsten zu begegnen, und eh' er's verhindern konnte, schwang sie, ihn abwehrend, sich auf einen einzelnen, senkrecht über die Tiefe hinausragenden Fels, daß ihm in innerster Seele grauste – nur ein Fehltritt, und sie glitt in den Abgrund hinunter. – Da hatte Lothario mit sicherem Blicke seinen Vorteil abgesehen. In raschem Entschlusse umfaßte er sie plötzlich und schwang die Sträubende auf seinen Arm. Erschrocken, überrascht, wußte sie nicht, wie ihr geschehe, und sah ihm verwundert und zornig in die Augen. Er aber trug sie grauenhaft an jähen Schlünden vorüber durch die Dämmerung von Klippe zu Klippe hinab, daß sie, vor Entsetzen mit dem einen Arm seinen Nacken umklammernd, ihn rings mit ihren aufgeringelten Locken umgab. So schwiegen sie beide lange Zeit.

Jetzt ging der Mond prächtig über den Wäldern auf. Lothario schaute in die wunderbare Einsamkeit und sagte halb für sich: „So hab ich's manchmal im Traume gesehen." – Juanna aber blickte spähend umher, die Gegend war ihr ganz fremd, einzelne Wolkenschatten flogen darüber, tiefer schimmerten die Gründe fast heimatlich herauf, wie die Täler in Spanien, sie gedachte der schönen Sommernächte unter den Guerillas. – Auf einmal stutzte sie, zwei gesattelte Pferde standen dicht vor ihnen im Wal-

de, und ehe sie sich besinnen und fragen konnte, hob sie Lothario schon auf das eine Roß, schwang sich selbst auf das andere, und über den mondhellen Waldgrund nun ging es rasch fort durch die stille, sternenklare Nacht. Hier blitzte plötzlich eine furchtbare Ahnung durch Juannas Seele, sie konnte kein Wort hervorbringen, dem Unglaublichen finster nachsinnend, während Büsche, Täler und ferne Dörfer geheimnisvoll an ihnen vorüberflogen. Lothario war wie verwandelt. „Juanna!" rief er ihr aus Herzensgrunde zu, „blick' um dich, die Erde ist so still und schön wie eine Brautnacht! Frei sollst du wohnen auf hohem Schloß, wo die Rehe an den Abhängen einsam grasen, dort will ich unter deinem offenen Fenster ruhen in den Sommernächten und dich in Traum singen, bis die Sterne verlöschen und die erste Lerche mich ablöst, hoch in der stillen Luft. Und fallen die Blätter, und die Vögel ziehen fort, und dich befällt Heimweh, wenn du vom Schlosse über die einsamen Wälder siehst: ich führe dich weit über die Berge fort, du arme Fremde! Auf dem Meere wollen wir fahren an glänzenden Küsten vorüber, bis die Laute deiner Muttersprache gleich bunten Wundervögeln herschweifen und deine ernste, schöne Heimat emportaucht, duftige Gärten, Gebirge und maurische Schlösser in den trunkenen Fluten spiegelnd – o Juanna, mir ist's, wie von einem hohen Berge ins Morgenrot zu sehen!" – So sprach er voll Freude, während sie ritten, Juanna war immerfort still, in der Tiefe neben ihnen rauschte ein Strom, sie horchte manchmal hinunter. Auf einmal blinkte das Wasser zwischen dunklen Bäumen hinauf, da warf sie ihr Roß gewaltsam zur Seite, setzte die Sporen ein und schwang es mit sich in den Fluß hinab. Erschrocken stürzte Lothario nach, er sah sie mit dem weitaufgelösten Haare gleich einer Nixe in klarem Mondlichte über die Flut dahinschweben, sinken und wieder emportauchen. Endlich hatte er sie gefaßt, sie ruhte an seiner Schulter, ihre feuchten Locken verdunkelten ihm Stirn und Augen. So sank er mit seiner Beute erschöpft am jenseitigen Ufer auf

den Rasen hin und lauschte in der entsetzlichen Stille kniend über ihr – aber sie atmete nicht mehr, stumm und bleich in strenger Todesschönheit.

Das hatte sich alles anders gestaltet, als die lustigen Jäger sich's dachten. Fortunat war damals noch vor Abend von der Jagd abgekommen und mehrere Tage allein im Walde umhergeschweift, um recht nach Herzenslust das schöne Gebirge zu durchforschen. Als er zurückkehrte, fand er zu seinem Erstaunen alles leer, das Abendrot schien über Schloß und Garten, aus dem einen Flügel klang eine Spieluhr noch in einzelnen, langgezogenen Tönen herüber. Bei seinen Tritten, die in dem trockenen Laube raschelten, fuhr der alte Schloßwart erschrocken empor, der auf den Marmorstufen vor dem Schlosse eingeschlummert war. Von diesem hörte er nun, die Gräfin Juanna habe sich auf der Jagd in den Klippen verstiegen, so sei sie im Fluß verunglückt, zwei Hirten hätten sie im Mondschein auf dem Strome schwimmen gesehen und mit dem Wassermann ringen. Da wäre der Fürst sogleich am andern Morgen mit seinem ganzen Gefolge nach der Residenz aufgebrochen, auch die Schauspielertruppe sei wieder weitergezogen; von Lothario wußte er nichts. – Fortunat aber befiel ein tiefes Grauen in der plötzlichen Einsamkeit, er beschloß, noch heut bis in das nächste Städtchen zu reisen und sich dann ohne weiteren Aufenthalt nach Italien zu wenden. – Als er fortritt, dunkelte es schon, fern an den Bergen sah er einen stillen Fackelzug, es war Juannas Leiche, die sie nach der Residenz brachten. So geht oft ein Schauer mahnend durch die Lust der Menschen, damit sie sich erinnern, daß ihnen die schöne Erde nur geliehen sei.

Zweites Buch

Vierzehntes Kapitel

Über einer der verborgensten Schlüfte der Schweiz rauschte leise die Nacht, nur ein Bach stieg zwischen den Felsen hernieder und plauderte, da die Menschen schliefen, heimlich mit der Wetterfahne auf der ärmlichen Waldherberge, die in dem stillen Grunde lag. Da fuhr auf dem Heuboden des Hauses ein Gesell verwirrt aus dem Schlafe empor. Es war Fortunat, der auf seiner Reise nach Italien spät des Abends das Wirtshaus erreicht und gern das luftige Nachtlager bestiegen hatte, da die wenigen Fremdenstuben schon von andern Reisenden besetzt waren. Dort hatte ihn ein Traum erweckt: es war ihm plötzlich, als hätte eine altbekannte Stimme unten seinen Namen genannt. Er lauschte hinab, es rührte sich kein Laut. Draußen aber flimmerten noch die Sterne, da setzte er sich in das offene Dachfenster auf die obersten Sprossen der Leiter und sah den weiten, stillen Kreis von Gletschern im hellsten Mondenscheine über den Wäldern, nur der dumpfe Donner einer Lawine hallte von Zeit zu Zeit durch die große Einsamkeit herüber.

Jetzt erst fiel ihm der grillenhaft verworrene Bau des Hauses auf, er betrachtete schläfrig die kleinen, hölzernen Galerien, Winkel und Erker, als auf einmal in dem alten Seitenanbau sich ein Laden öffnete und eine Dame, dicht in einen langen Schleier gehüllt, am Fenster erschien. Fortunat, scharf hinblickend, schauerte innerlichst zusammen – es war der Hut, das Reitkleid, Gestalt und Art der Gräfin Juanna! – Der Mond funkelte über ihren Gürtel, wie damals auf der Jagd, dann wurde das Fenster schnell wieder geschlossen. Gleich darauf aber sah er den Wirt zwei gesattelte Pferde auf den Hof führen, die Dame trat mit einem fremden Manne aus dem Hause, alles ganz sacht

und leise, wie Wolken in der Nacht, sie flüsterten heimlich untereinander und mit dem Wirt, der ihm auf einmal selbst gespenstisch vorkam, und eh' er sich noch besinnen konnte, war die ganze Erscheinung, wie ein Zug Verstorbener, im wechselnden Mondlichte zwischen den Felsen und Bäumen verschwunden.

Fortunat war geblendet, wie einer, der nachts in den Blitz gesehen; er eilte nun die Leiter hinab, der Hof war leer, als wäre nichts geschehen, aber zu seinem Erstaunen hörte er nun in einiger Entfernung Waffenklang durch die Stille. Fechten die Toten in der Luft? dachte er und verfolgte rasch die Richtung. Da erblickte er bald durch das auseinandergebogene Gesträuch zwei Männer, die auf einer mondhellen Wiese in heftigem Zweikampfe begriffen waren. Gestalt, Tracht und Haltung, je länger er hinsah, schien ihm nicht fremd. – „Um Gott, ihr Phantasten", rief er endlich aus, „was habt ihr wieder vor!" denn jetzt erkannte er deutlich den langen *Lord* und den Maler *Albert* von dem fürstlichen Jagdschlosse.

Als die Kämpfenden ihn bemerkten, traten sie, die Spitzen ihrer Degen senkend, jeder feierlich einen Schritt zurück und verneigten sich kurz und ernst voreinander, dann stürzte der erhitzte Lord, der vor Eifer keine Zeit zum Verwundern und Begrüßen hatte, sogleich auf Fortunat los. „Entscheiden Sie selbst", rief er, „und ich behaupt' es nochmals und tausendmal: es gibt keinen kategorischen Imperativ, die Tugend ist nur der Flügelschlag der primitiven Freiheit der Seele, die Ahnung des geistigen Urstoffs und dieser endlose Urstoff läßt sich so wenig durch Großmut, Keuschheit definieren, daß" – „Keineswegs!" entgegnete Albert ganz empört, „es gibt ein absolutes Sittengesetz, die Tugend, sie ist kein leerer Schall!" – „Aber, so sagt doch nur, was denn? was gibt's denn?" unterbrach sie endlich Fortunat höchst erstaunt, und erfuhr nun nach und nach abgebrochen in einzelnen, verworrenen Sätzen von den Heftigen, daß sie beide in der festen Überzeugung von einer Entführung Juannas durch Lothario, an je-

nem unglücklichen Abend, sobald die Gräfin vermißt wurde, die Jagd mit dem Schwur verlassen hatten, sie zurückzubringen oder niemals wiederzukehren. Sehr bald, so behaupteten sie, seien sie auch wirklich den Flüchtlingen auf die Spur gekommen, die sie bis zu diesem einsamen Wirtshause verfolgt hätten. „Und nun, da wir am Ziele sind", fuhr der Maler fort, läßt dieser Herr da plötzlich seine großmütige Larve fallen und will die Gräfin als seine eigene Beute entführen. Aber mit diesem Schwerte, das in dem großen Kriegsjahre dreizehn geweiht ist, bewahre ich die Unschuld jener Dame gegen jeden Verführer, er mag ein deutscher Komödiant oder ein englischer Lord sein!" – Und hiermit gingen sie von neuem aufeinander los und führten ihre Schulterquarten und Schlenkerprimen mit einer bewunderungswürdigen Künstlichkeit und Pedanterie aus.

Da fuhr auf einmal der dicke Wirt aus der Haustür wütend zwischen die Fechtenden hinein, er hatte einen umgekehrten Tisch über dem Kopf, wie ein Stier mit vier Hörnern, die schon gezückten Schwerter klatschten flach auf seinen rindsledernen Schlafpelz. „Tausend Parlament", schrie er, „Schändlichmens, Lordmajors oder Oberstleutnant, ich frage den Teufel danach! Ich nehme nicht tausend Pfund Sperling für den Skandal, verjagt mir da mit eurem Geklimper die besten Gäste, ist das ein Ständchen für eine schöne, ausländische Gräfin!" – „Gräfin! ist sie schon fort? wohin?" unterbrachen ihn hier die Duellanten, ihre Degen rasch einsteckend. – „Ausländisch?" stotterte Albert vor Eifer, „was für eine Sprache redete sie?" – „Wahrhaftig, mir kam's ganz spanisch vor", erwiderte der Wirt und schien nun, indem er die beiden geheimnisvoll nach dem Stalle führte, mit ihnen angelegentlichst von der Fremden zu sprechen, Fortunat konnte nur noch bemerken, daß der Schalk ihnen eine ganz andere Richtung wies, als die Dame vorhin eingeschlagen hatte. – Als er zurückkam, wollte ihn Fortunat selbst über die Gräfin näher ausfragen. Aber der dicke, schlaue Mann war nicht zu ha-

schen, er sprach von tollen Nächten, Spukgeistern und fahrenden Hexen, und brach mit solchem Lärmen den Tag an, daß der Hofhund anschlug und Knechte und Mägde aus allen Winkeln herausfuhren. Mitten in dieser Konfusion hörte Fortunat plötzlich den Lord und den Maler von der andern Seite durch die Dämmerung miteinander disputieren, und ehe er ihnen noch nachrufen konnte, hatten sie in ihren langen, bis an die Knöchel herabhängenden Wachstaftmänteln, aus denen die englischen Pferde ihre dünnen Hälse seltsam hervorstreckten, sich zwischen den fliegenden Morgennebeln schon verloren.
So stand er noch ein Weilchen ganz verwirrt, dann berichtigte auch er schnell seine Zeche, schwang sich auf sein Pferd und schlug den Waldpfad ein, den die geheimnisvolle Erscheinung vor Tagesanbruch genommen. Er ritt den ganzen Morgen fort: aber er fand sie nicht mehr wieder.

Fünfzehntes Kapitel

Die Sonne war eben über Rom untergegangen, als Fortunat von den Bergen mit der Abendkühle in die Stadt einzog. Nur ein Streifen des Meeres in der Ferne und das Kreuz der Peterskuppel brannten noch im Widerschein, dazwischen der Klang unzähliger Abendglocken, und Gärten, Paläste und einsames Gebirg unten wunderbar zerworfen – es war ihm, als zöge er in ein prächtiges Märchen hinein. „Ecco là!" rief einmal sein Vetturin und hielt still. Sie standen vor einem großen, altmodischen Palast, welcher zum Teil unbewohnt schien und in der Dämmerung melancholisch auf den einsamen Platz herniederschaute, wo hohes Gras aus dem Pflaster drang und ein Springbrunnen emsig rauschte. Es war das Haus des Marchese A., in welchem befreundete Reisende für Fortunat die Wohnung besorgt hatten.

Ein alter Diener, mit klugen, kurzen Blicken das geringe Gepäck des genügsamen Reisenden musternd, führte diesen die breiten Marmortreppen hinan, während er in großem Wortschwall die Abwesenheit des Marchese entschuldigte, welcher erst heute vom Land zurückkehrt und nicht ermangeln werde, den schuldigen Empfang morgen nachzuholen.

Die ersten Stunden in einer großen, unbekannten Stadt gehören zu den einsamsten im Leben, auch Fortunat überflog das Gefühl, als sei er jetzt erst in der Fremde. Er verlor sich ganz in den hohen Gemächern und betrachtete, als der Diener sich entfernt hatte, vor Langeweile die Stuckverzierungen an den Decken, die schweren altmodischen Stühle, die hohen Spiegel mit goldenen Rahmen sowie die umherhängenden Jagdbilder, Kavaliere in seltsamen Trachten vorstellend, halb Ritter, halb Gecken, einen Hirsch mit galanter Reiterkühnheit verfolgend, und junge, schöne Damen in Reifröcken unter einem prächtigen Zelte im Walde, Jagdhörner in den Händen, denen der glückliche Jäger seine Beute ehrfurchtsvoll zu Füßen legte. – Draußen schien ein großer Garten zu liegen, weit über den Garten her schlugen viele Uhren in der Ferne, es war ihm, als sei er schon gestorben und hörte die Totenglocke über sich.

In diesen Betrachtungen unterbrach ihn das Rasseln eines Wagens, der vor dem Schlosse zu halten schien. Er sah durchs Fenster und konnte bei dem Schein der Fackel nur noch bemerken, wie eine schlanke Mädchengestalt aus der altmodischen Karosse behende in das Haus schlüpfte. Im andern Flügel des Palastes hörte man nun Türen auf- und zuwerfen, gehen und lachen, dann war plötzlich alles wieder still. – Bald darauf aber vernahm er im Garten einzelne, langgezogene Klänge einer weiblichen Stimme, wie eine Nachtigall durch das Rauschen der Wipfel, durch welche die Glühwürmer leuchtend hinzogen. Der Mond trat eben hervor und verwandelte alles in Traum. Da öffnete Fortunat alle Flügeltüren, ergriff seine Gitarre und

schritt durch die lange Reihe der Gemächer singend auf und nieder:

„Es rauschen die Wipfel und schauern,
Als machten zu dieser Stund'
Um die halbversunkenen Mauern
Die alten Götter die Rund'.

Hinter den Myrtenbäumen
In heimlich dämmernder Pracht,
Was sprichst du wirr, wie in Träumen,
Zu mir, phantastische Nacht?

Es funkeln auf mich alle Sterne
Mit glühendem Liebesblick,
Es redet trunken die Ferne
Wie von künftigem, großem Glück!"

Der schönste Frühlingsmorgen funkelte vor dem Palast über den Garten, da grünte und sang schon alles in der reizenden Verwirrung, in den ausgetrockneten Becken der Wasserkünste jagten sich jubelnd bunte Vögel, üppig blühende Ranken umschlangen mutwillig die Marmorstatuen, als wollte der Frühling sie mit Küssen ersticken. Arglos zwischen den nackten Götterbildern stand *Fiametta,* die vierzehnjährige Tochter des Marchese, mit ihrer Kammerjungfer Lenore plaudernd, die ihr die schönen, dunklen Haarflechten aufsteckte. Sie war ihr heute ungeduldig entsprungen, beide waren neugierig, ihren Gast, den gestern angekommenen Engländer, zu sehen, wofür sie jeden reisenden Fremden hielten. „Mir träumte heut von ihm", sagte Fiametta, „er sah aus wie die jungen deutschen Maler mit den langen, blonden Locken, und stand in einer unbekannten, prächtigen Gegend, die schimmerte und blitzte, daß ich vor Blendung gar nicht hinsehen konnte. Ich wußt' es wohl, es war der Morgen, der schon durch die roten Gardinen schimmerte, aber ich drückte die Augen fest zu" – hier hielt sie ein und lachte in sich. – Lenore

sah sie fragend an. – „Nein, nein“, meinte Fiametta leicht errötend, „was er mir da ins Ohr sagte, sag' ich nicht wieder – ob er noch jung sein mag?“ – Lenore erzählte, daß sie gestern abends noch im Garten gewesen, da habe sie seinen Schatten im Zimmer auf und nieder schwanken gesehen, lang und dünn wie der Perpendikel einer Turmuhr – „Oder einer Spieluhr, denn ich hört' es wohl herüberklingen“, fiel ihr Fiametta ins Wort, während sie ihr Füßchen auf den Nacken eines umgestürzten Apollos stellte und sich die zierlichen Schuhe festband. Jetzt sahen sie auf einmal zwischen den Zweigen hindurch den besprochenen Gast selbst, sich streckend und dehnend, aus der Schloßtür treten und verschlüpften, wie Lacerten, schnell zwischen Blumen und Unkraut hinter ein halbverfallenes Gemäuer, wo er vorüber mußte, und durch dessen Ritze sie ihn ungesehen betrachten konnten. Lenore fand ihn sehr schön. Fiametta dagegen kritisierte heimlich flüsternd, sein schlichtes braunes Haar, seinen dreisten Gang und seltsamen Anzug. – Als er an die Mauer kam, sagte sie leis: „Ich schreck' ihn.“ Lenore fuhr abwehrend nach ihrer Hand, aber die kleine Marchesin hatte schon den über die Mauer herüberlangenden Ast eines blühenden Apfelbaumes gefaßt und schüttelte kurz und rasch, daß Fortunat von den Blütenflocken ganz verschneit war; dann liefen sie beide schnell davon.

Fortunat aber war heute längst über alles Verwundern hinaus. Schon beim Erwachen in den hohen Trumeaux blickend, der Himmel und Bäume abspiegelte, hatte er geglaubt, so entkleidet mitten im Garten zu liegen, und war erschrocken aufgesprungen; da hörte er draußen Lachen und Mädchenstimmen in den schönen fremden Lauten, wie Glöckchen, verlockend durch die morgenfrische Wildnis gehen. So war er die helle, stille Marmortreppe hinabgeeilt, um Rom, den Garten, den jungen Frühling und den alten Marchese zu begrüßen.

Nach allen Seiten fröhlich umschauend, wurde er in einiger Entfernung vor sich einen stattlichen Herrn mit gepu-

dertem Haar, Schnallenschuhen und einem alten, hofmäßigen Kleide gewahr, welcher ein junges Frauenzimmer am Arm führte, während ein Bedienter in verschossener Livree mit einem Sonnenschirm und in sichtbarer Langeweile ihnen langsam nachschlenderte. Seine Vermutung bestätigte sich bald, es war der alte Marchese A., welcher seinen Gast kaum bemerkt hatte, als er ihn in französischer Sprache sehr feierlich willkommen hieß und ihm in seiner Begleiterin seine Tochter Fiametta vorstellte, die errötend ihre langen schwarzen Augenwimpern senkte, da sie auf Fortunats Rock noch einige Apfelblüten erblickte. Dann lud er den Fremden ein, an ihrer Morgenpromenade teilzunehmen. Fortunat war es, da sie nun in künstlicher Verschlingung zierlicher Redensarten an den Buchsbaumwänden durch die langen Alleen mit perspektivischen Aussichten gemessen dahinschritten, als wüchse ihm langsam ein Haarbeutel im Nacken und ein Stahldegen zwischen den Rockschößen heraus, und als ginge er immer tiefer und tiefer in jene gute, alte, wunderliche Zeit hinein, wie er sie aus Büchern und Bildern wohl noch kannte. Dazwischen machten ihn die dunklen, funkelnden Augen Fiamettas recht innerlichst vergnügt, und so kam er selbst, eh' er's wußte, immer lustiger in die auserlesenste Galanterie, und es störte die Illusion kaum noch, als sich der Marchese zuletzt ganz unerwartet nach einem seiner entfernten Verwandten in Deutschland, dem Grafen Victor von Hohenstein, erkundigte. Fortunat nannte ihn einen homme de lettres, der sein Siècle mache.

MARCHESE Er ist aus einem alten Hause.

FORTUNAT Bewohnt er aber wenig, sondern ist seit geraumer Zeit auf den Parnaß verzogen, wo er sich seine eigenen Lustschlösser baut.

MARCHESE Ein barocker Einfall für einen Kavalier.

FIAMETTA Ich möchte einmal einen Dichter sehen.

FORTUNAT Ihren Augen, meine Gnädigste, kann das nicht schwer werden, wo der Frühling zaubert, muß selbst der nordische Boreas durch die Blume sprechen.

Fiametta Haben Sie auch Blumen in Deutschland?
Fortunat (mit galantem Blick). So schöne nicht. – Während dieses Diskurses hatten sie sich wieder bis an den Palast herangeschlungen, man schied mit vielen Verbeugungen am Portal unter großem Geschrei der Sperlinge in den zerbröckelten Säulenknäufen. Fortunat war es, als hätt er in aller Frühe eine Menuett getanzt, im Garten aber sangen die Vögel und rauschten die Bäume wieder, als sprächen sie noch immer von den funkelnden Augen der schönen Marchesin.

Sechzehntes Kapitel

Die ersten Tage verstrichen Fortunat wie im Rausche, alles schimmerte vor seiner Seele, er mochte in dem Glanze noch nichts Deutliches unterscheiden. Der beste Führer durch Rom und der Plan der Stadt lagen auf dem Tische aufgeschlagen, jeden Morgen ging er mit dem festen Vorsatze aus, seinen regelmäßigen, auf dem Plane im voraus rot punktierten Umlauf zu beginnen, aber eine überraschende Aussicht zog ihn an, ein Bänkelsänger, der einen Kreis von Lumpengesindel um sich versammelte, lenkte ihn von seinem Wege ab und hielt ihn lange auf, oft folgte er durch ganze Straßen ein paar seltsamen Männergestalten, deren römische Nasen und ausdrucksvolle Gebärden ihm aber besonders auffielen, und wenn er dann ermüdet von dem müßigen Umherschlendern zurückkehrte, mußte er sich dennoch eingestehen, daß er in der kurzen Zeit mehr gesehen und erfahren hatte, als sein gedruckter Führer sich träumen ließ.
Auf einem solchen Streifzuge hatte er sich eines Abends in ein entlegenes Labyrinth kleiner Gassen verirrt, die Bewohner saßen plaudernd vor den Türen, schöne, halbnackte Kinder spielten und lärmten in dem Abendschimmer. Da hörte er unerwartet weiterhin ein lautes Gezänk

in deutscher Sprache herüberschallen und eilte neugierig dem Hofe zu, woher der Lärm kam. Auf einmal sprang die Haustür hastig auf, und ein wohlgekleideter, nicht mehr ganz junger Mann kam so unsanft herausgeflogen, daß er den Hut vom Kopfe verlor. „Mein Gott! du, *Grundling!*" rief Fortunat überrascht aus – es war der deutsche Reisende, der ihm die Wohnung in dem Palaste besorgt hatte. – „Da bist du ja wie gerufen", sagte dieser, ganz ruhig seinen Hut abstäubend, „ich will dich sogleich bei Landsleuten einführen." Hiermit versuchte er den Drücker der Tür, fand sie aber hinter sich verschlossen. „Hat nichts zu sagen", meinte er nun, winkte Fortunat und führte den Erstaunten in das leerstehende Nebenhaus, im Dunkeln vorsichtig tappend, zwischen wüstem Geröll hindurch. Währenddes hörten sie im Innern nebenan eine männliche und eine weibliche Stimme immerfort lebhaft miteinander zanken. „Das sind nun meine goldenen Träume!" rief der Mann. – „Träume?" erwiderte das Mädchen schnippisch, „so zwick' dich in die Nase, damit du erwachst, ich glaube, du bist heut wirklich betrunken." – „Was wußtest du je von der Trunkenheit der göttlichen Kunst!" fiel er ihr wieder ins Wort, „ich Tor, der ich meinte, dich mit emporzuheben! Nun zerrst du mich selbst mit hinab und machst mir die Welt so gemein, daß ich ihr alle meine Farbentöpfe an den Kopf werfen möchte." – „Nun, einen deiner Pinsel wenigstens hast du schon hingeschmissen", entgegnete das Mädchen lachend. – „Da meint sie mich", sagte Grundling zu Fortunat, „fideles, geniales Volk!"
Jetzt öffnete er am Ende eines schmalen, finsteren Ganges eine Hintertür, und sie traten in ein großes wüstes, von einem Kaminfeuer zweifelhaft erleuchtetes Gemach, wo Fortunat zu nicht geringer Verwunderung in den Zankenden, *Kordelchen* und den Maler *Guido* erkannte. Die erstere saß auf einem Koffer, wo sie Wäsche zu flicken schien. Kaum hatte sie Fortunat erblickt, als sie aufspringend alles wegwarf, ihm mit großer Freude an den Hals

flog und ihn tüchtig abküßte. Guido, bleich und verstört, stand schweigend und schien einen Augenblick verlegen. Kordelchen aber erzählte in aller Geschwindigkeit: Herr Grundling, der in Rom bekannt sei, wie er in seiner eigenen Tasche, habe Guido in den Bildergalerien und bei allen Malern herumgeführt, vor kurzem seien sie in einem großen philosophischen Disput über die Kunst zurückgekehrt, da habe Grundling Guidos angefangene Bilder und Zeichnungen getadelt, daraus sei der ganze Spektakel entstanden. – „Wie du alles wieder verdrehst!" fiel ihr Guido heftig ins Wort, „es ist nicht um den Plunder auf meiner Staffelei dort! Vor den übermächtigen, alten Bildern in den Werkstätten unserer frommen, ernsten deutschen Künstler, da tat es plötzlich einen langen Blitz über mein ganzes Leben von allen Decken, Wänden herab und verbrannte, was hinter mir lag. Da wußt ich's auf einmal, wer ich bin, ein weinerlicher, erbärmlicher Wicht, der noch nichts gemalt und erdacht hat!" – Hier brach seine Stimme, er setzte schnell seinen Hut auf und stürzte, ohne jemand zu begrüßen, trotzig zur Tür hinaus.

„Es ist doch ein schöner Junge, besonders wenn er böse wird", sagte Kordelchen, ihm nachsehend. Grundling zündete unterdessen an dem Kamin, wo Überbleibsel vom Mittagessen aufgewärmt wurden, gelassen seine Zigarre an, während Fortunat Guido nacheilen wollte. Aber Kordelchen vertrat ihm den Weg. „Ich bitte Sie, lieber Baron", sagte sie, „tun Sie ihm nicht den Gefallen, denn er will doch nur bedauert und widerlegt sein. Je mehr man ihn tröstet und streichelt, je ungebärdiger wird er, wie ein ungezogenes Kind, das sich selbst in die Zunge gebissen hat."

Sie fing nun, unbekümmert um die Gegenwart der beiden Fremden, vor den Trümmern eines zerbrochenen Spiegels sich schnell zu putzen an, wobei sie Fortunat sehr lustig erzählte, wie sie nach Rom gekommen. Das fröhliche Mädchen, schon früh für die Bühne dressiert, hatte durch ihr Zusammentreffen mit Lothario zum ersten Male in

ihrem Leben Lust und Leid in ihrer tiefern Gewalt erfahren, ohne sich weiter ihren Zustand klarzumachen. Als nun aber der unbeständige Freund so plötzlich verschwunden war, wurden ihr Theater, Sorti und alle die alten Gesichter langweilig, und der enthusiastische Guido beredete sie um so leichter, ihn nach Italien zu begleiten, als sie an Lotharios Untreue sich zu rächen und insgeheim diesen hier wiederzufinden meinte, was sie aber allen verschwieg. Unterwegs hatte sie sich unzählige Male mit Guido entzweit und wieder versöhnt, sie galt für seine Frau, hier in Rom endlich zerstreute sie die neue Welt, und so führte sie gedankenlos ihr gewohntes, leichtfertiges Leben mit einer gewissen Unschuld fort, die dabei nichts Arges hatte. – „Aber wie sind Sie damals in der Schweiz den Lord und den Albert wieder losgeworden?" fragte sie plötzlich Fortunat. – „Wie!" sagte dieser erstaunt, „so war't Ihr es in jener Nacht?" – „Freilich", erwiderte sie lachend, „ich kannte ihre Einbildung und ritt und trug mich wie die arme Gräfin, um die irrenden Ritter zu foppen." –

Währenddes machte Grundling dem Mädchen bei ihrer Toilette auf seine schwerfällige Art den Hof, was sie sogleich zu benutzen wußte, indem sie beständig etwas zu kommandieren hatte, bald mußte er ihr ein Tuch holen, bald eine Nadel suchen, dann reichte sie ihm ihr Füßchen hin, um ihr den Schuh festzubinden, was der trockene Schalk mit ungeschickter Umständlichkeit ausführte. Darauf wollte sie ihre Gäste auf nordische Weise mit Tee bewirten, aber da waren die Teelöffel verlegt, die Tassen voll Farbenkleckse, zudem war es schon finster, und je mehr man suchte, je größer wurde die Verwirrung, bis Kordelchen endlich, den Einfall wieder aufgebend, die beiden Männer lustig zu einem Seitenpförtchen hinausschob, um ihnen ihren sogenannten Garten zu zeigen.

So gelangten sie durch das dunkelverbaute Hinterhaus auf einen kleinen, grünen Platz, dessen Aussicht Fortunat wunderbar überraschte. Denn hinter den Weingeländen

und duftigen Gärten, die sich terrassenartig senkten, lag plötzlich die Nacht mit ihren Trümmern und zerbrochenen Säulen wie ein Buch der Vergangenheit unter ihnen aufgeschlagen, dessen Anfangsbuchstaben der Mond rätselhaft vergoldete. Da hörten sie von fern aus den Gärten einzelne Akkorde einer Laute, bald darauf sang eine männliche Stimme:

„Jetzt wandr' ich erst gern!
Am Fenster nun lauschen
Die Mädchen, es rauschen
Die Brunnen von fern.
Aus schimmernden Büschen
Ihr Plaudern, so lieb,
Erkenn' ich dazwischen,
Ich höre mein Lieb!"

Kordelchen, die aufmerksam hinabgelauscht hatte, besann sich nicht lange und antwortete sogleich nach derselben Melodie:

„Ich höre, mein Lieb,
Beim wechselnden Scheine
Verläßt er die Seine
Und kommt wie ein Dieb.
Es hallt von den Steinen,
Die Wipfel wehn sacht
Und sagen's der Deinen –
Ja hüt' dich! bei Nacht."

Der Sänger unten schien es vernommen zu haben, er sang, immer näher und näher kommend, lustig entgegen:

„Ja, hüt' dich bei Nacht!
Pflegt Amor zu wandern,
Ruft leise die andern,
Da schreiten erwacht

Die Götter zur Halle
Ins Freie hinaus,
Es bringt sie dir alle
Der Dichter ins Haus."

Die Stimme schien Fortunat bekannt, da rauschte es in dem nächsten Gebüsch, und mit einem leichten Satze schwang sich der Sänger zwischen dem alten Gemäuer zu ihnen herauf, daß seine Laute an den Zweigen einen fröhlichen Klang gab. – „*Otto!*" rief Fortunat freudig aus, denn es war niemand anders als der poetische Student aus Hohenstein. Fast aber hätte er ihn nicht wiedererkannt, so verwandelt, von der Sonne gebräunt und rüstig erschien der Jüngling. Er hatte Fortunats Ankunft schon erfahren und erzählte ihm nun sogleich voller Entzücken von seiner Reise und dem hiesigen Aufenthalt, er war wie berauscht in den fremden Lüften. Kordelchen neckte ihn mit seinem römischen Liebchen, und Grundling schwor, das sei das schönste Frauenzimmer, das er jemals gesehen, alle Maler stiegen ihr nach, wenn sie, ihr Fruchtkörbchen auf dem Kopfe mit dem einen Arm unterstützend, schlank und zierlich über den Markt ging; einem Landsmann habe sie bei dieser Gelegenheit einmal eine Feige umsonst gereicht, nämlich hinters Ohr.
Während sie noch so sprachen, hörten sie hinter sich im Hause heftig gehen und die Türen zuschlagen. Es war Guido, der, in der ungebärdigsten Laune zurückgekehrt, nach Licht rief und im Finstern mit den Stühlen umherwarf. – „Heraus, du verstörter Poltergeist mit deinem dummen Künstlerunglück!" rief Grundling in das Haus hinein. – „Laßt mich jetzt ungeschoren, das rat' ich euch", erwiderte Guido zornig von innen, „wem sein Himmel über dem Haupte zusammenbrach, dem kommt's auf ein paar Scherben mehr oder weniger nicht an." – Hier aber verwickelte er sich unter dem alten Gerümpel des Hausflurs mit den Füßen in umherliegende Schläuche, er zuckte ungeduldig, darüber geriet ein übereinander geschich-

teter Turm von leeren Weintonnen ins Wanken und Fallen, bis auf einmal Schaff, Tonnen und Maler, unaufhaltsam übereinander kollernd, zum Hause herausgestürzt kamen. Grundling, der sich vorwitzig der Tür genähert hatte, konnte nicht so schnell entspringen, eine Tonne hüpfte ihm zwischen die Beine, er wollte sich an Otto festhalten, erwischte aber nur seine Laute, mit der er krachend niederfiel. Otto schalt auf Grundling, Grundling auf Guido, Guido auf mehrere alte Weiber, die über dem Lärm keifend aus allen Dachfenstern herausfuhren. Mitten aus dieser Verwirrung brach endlich das tiefe, weitschallende Lachen Grundlings mit solcher vehementen Herzlichkeit, daß es bald Handelnde und Zuschauer unaufhaltsam mit fortriß.
Diese unerwartete Explosion zerstreute die letzten Wölkchen an dem leicht beweglichen Horizont. Auch Guido hatte darüber seine hochmütige Zerknirschung gänzlich wieder vergessen. Man holte Wein herbei, und Kordelchen forderte Grundling auf, da sie sich eben alle wie in der Arche Noah so fröhlich zusammengefunden hätten, bei der schönen, warmen Nacht seine Lebensgeschichte zum besten zu geben, was von den andern mit großem Applaus aufgenommen wurde.
Grundling langte nun aus seinem tiefen Schubsack erst mehrere Stücke seines Pfeifenrohrs hervor, die er umständlich zusammensetzte, und einen ungeheuren Pfeifenkopf vollpfropfte, während er auf einem der umgestürzten Weinfässer Platz nahm. Die andern hatten sich, um dem Qualme des schlechten Tabaks zu entgehen, vorsichtig außer dem Winde um ihn her gesetzt, worauf derselbe endlich folgendermaßen begann:
„Du wirst dich noch erinnern, Fortunat, wie ich in Heidelberg mich so in die Wissenschaften verbissen hatte, daß ich gar nicht mehr loskommen konnte." – „Allerdings", erwiderte Fortunat, „du und dein gringräulicher Mantel hatten schon mehrere Studentengenerationen überlebt, als ich dort ankam. Du warst ein hartnäckiger Kantianer

und standst, noch immerfort nach der Aufklärung hinweisend, wie ein alter Meilenzeiger, den man mitten im Kornfelde vergessen, nachdem Fichte und Schelling längst andere Straßen gezogen hatten. Du verachtetest damals uns Jüngere unsäglich, die wir den neuen Weg eingeschlagen." – „Nun, bei Gott, das tu' ich auch jetzt noch", rief Grundling, indem er dicke Tabakswolken von sich stieß. – „Auf einmal aber warst du in Heidelberg spurlos verschwunden", sagte Fortunat. „Ein von den Ferien zurückkehrender Student hatte deinen Mantel mitten auf der Heerstraße gefunden, den wir sodann mit einem philosophischen Leichensermon feierlich zur Erde bestattet haben. Wie ging das zu?" – „Das will ich euch wohl berichten", entgegnete Grundling.

„Es trieb sich dazumal ein schlanker junger Mensch in Heidelberg herum, den niemand näher kannte, er war nicht Student, nicht Philister, aber verdammt schlau. Das kam mir gleich verdächtig vor, denn ich habe in solchen Stücken eine feine Nase. Ich fühlte dem Patron bei schicklicher Gelegenheit auf den Zahn, da sprach er von Fürsten, Ministern und Bischöfen! – versteht ihr? Bischöfen – mit denen er oft in naher Berührung stände, von Rührung, Stimmung der Seelen usw., aber alles glatt und durcheinandergeschlungen wie ein Aal. Da schoß mir endlich ganz das Blatt. Ja, liebe Freunde, es war niemand anders, als ein geheimer Jesuit, so ein verdammter proselytenmacherischer Emissär! Nun, ihr kennt mich, von Stund' an faßte ich den Kerl scharf ins Auge, sann und beobachtete ihn bei Tag und Nacht. Eines Abends sehr spät wandle ich eben in meinem Mantel vor dem Tore so für mich auf und nieder, als ich auf einmal den Emissär sacht und vorsichtig in ein dunkles Gebüsch schlüpfen sehe. Ich, nicht zu faul, lenkte sogleich meine Schritte dahin, arbeite mich durch Strauch und Dorn immer tiefer nach, und was erblick' ich?! – Unter einer hohen Linde im dämmernden Mondenschein steht der Emissär in erhabener Stellung, neben ihm ein sehr junger Mensch, der soeben, die rechte Hand zum

Himmel gereckt, einen feierlichen Schwur ablegt. Nun halt ich mich nicht länger, ich stürze hervor und donnere den Seelenverkäufer an, daß er sich unterfange, diesen Sitz der Aufklärung mit der pestilenzialischen Finsternis des Mittelalters zu verdüstern et cetera. Unterdes fing auch über meiner Rede ein Hund in der Nähe zu bellen an, einige Personen bewegten sich von fern zwischen den Bäumen, die Überraschten wurden immer verlegener, ich fuhr in meinen Ermahnungen immer nachdrücklicher fort. Aber was geschieht? Der Kerl von Jesuit packt mich auf einmal von hinten, der andere an den Füßen, daß ich die Balance verliere, so werfen sie mich in eine verfluchte Kalesche am Gebüsch, die ich vorher gar nicht bemerkt hatte, schwingen sich mit herauf, der Kutscher peitscht in die Pferde, und fort geht es über Stock und Stein in die finstere Nacht hinein. – Als ich wieder zu mir selbst kam, fand ich mit Vergnügen, daß meine Pfeife in der Konfusion nicht ausgegangen war, auch hatte ich den Tag über viel gesessen, etwas Motion konnte nicht schaden, die Nacht war schön, kein Mensch oder Dorf in der Runde – so dacht' ich denn: laß sie fahren! und setzte meine Ermahnungen ruhig wieder fort. Aber es dauerte nicht lange, so war der junge Proselyt darüber eingeschlafen. Der Jesuit dagegen, wie es die Art dieses schlauen Ordens ist, wich mir mit sophistischen Redensarten bald rechts, bald links aus, dann zog er eine Flasche guten, schweren Weines aus der Wagentasche und trank mir zu. Ich kam immer mehr ins Feuer, wir disputierten und tranken, ich verbreitete mich ausführlich über Aufklärung, Mönchtum, dicke Finsternis et cetera, aber Gott weiß, wie das zuging, es war mir, wie ich so fortsprach, als schritt' ich in der Rage unserm Säkulum um einige Jahrhunderte so unaufhaltsam vor, daß ich meine Gedanken gar nicht mehr halten konnte; vergeblich blickte ich unverwandt auf den dreieckigen Hut des Kutschers vor mir, Bäume und Dörfer und Wälder und Gedanken flogen und verwickelten sich mir im Mondenschein durcheinander, nur manchmal hört' ich

noch den Jesuiten dazwischen schnarchen, bis mir zuletzt selbst alle Sinne vergingen. – Als ich wieder aufwachte, war der Jesuit und Proselyt und Wagen und alles fort, und ich liege rücklings auf einem Rasenkanapee an der Chaussee in der angenehmsten Morgenkühle. Aber wie lieg' ich da! In einem kompletten Jesuitenroquelaure mit unzähligen Knöpfen vom Kinn bis an die Fußspitzen, und ein kleines, schwarze Barett auf dem Haupt!"

Hier brachen sämtliche Zuhörer in ein lautes Gelächter aus, nachdem Kordelchen schon während der ganzen Erzählung öfters heimlich gekichert hatte. „Dummes Zeug!" rief Grundling ärgerlich und stürzte zwei Gläser Wein hintereinander aus, „was ist da zu lachen? Das war kein Spaß. Vom Felde glotzten mich ein paar Bauern groß an, ich schämte mich in dem Aufzuge, als ob ich nackt wäre, und sprang geschwind ins Gebüsch. Aber die Bauern, wie sie das sehen, fangen an zu schreien, und hurra hinter mir drein! Ich springe und schlüpfe und duck' mich in Gräben, an Zäunen, laufe in der Verwirrung gerade ins Dorf hinein, verwickle mich mit dem langen Roquelaure im Gesträuch, da fahren euch Hunde, Kinder und Weiber aus allen Löchern, und alles schreit Mordio. – So brachten sie mich ganz atemlos zum Pastor. Da hatt ich nun gut reden, daß ich kein Jesuit, sondern eigentlich ein Philosoph sei, je mehr ich von Aufklärung sprach und auf die Jesuiten schimpfte, je schlauer und verdächtiger lächelte der Pastor dazu. Endlich gab er mir zu essen, ich hatte einen erstaunlichen Appetit. Über der Mahlzeit aber hör' ich draußen ein Pferd schnauben und scharren, der Pastor geht hinaus, ich vernehme eine feine Silberstimme, die sich voller Verwunderung und sehr eifrig nach mir erkundigt. Als ich ans Fenster trete, erblick' ich unter den alten Linden vor dem Pfarrhause ein hohes, schlankes Frauenzimmer zu Pferde, im Jagdhabit mit nickenden Federn auf dem Haupt. Sie ritt soeben wieder fort, ich konnte ihr Gesicht nicht mehr sehen, aber sie machte von hinten einen recht majestätischen Eindruck auf meine Sinne. – Nun

kam und ging der Pastor wieder hin und her und hatte immerfort das fatale Lächeln im Gesicht, ich merkte, daß Boten abgeschickt wurden, ich hörte insgeheim vom Gerichtshalter et cetera flüstern, da wurde mir zuletzt Angst, und gegen Abend schlüpfte ich unvermerkt durchs Hinterpförtchen, um die Nacht über nach Heidelberg fortzuwandern. Wie ich aber so vor dem Dorfe am Schloßpark vorüberziehe, hör' ich drin dieselbe Silberstimme sehr angenehm zur Laute singen. Das ficht mich an, ich trete in den Park, immer dreister und weiter – es war richtig die Reiterin. Sie hatte mich schon erblickt. – ‚O meine Ahnung! wußt' ich's doch, daß du kommen würdest, frommer Vater', sagte sie, zu mir tretend. Nun hätte das doch mit dem Teufel zugehen müssen, wenn ich ihr Vater hätte sein sollen, denn sie war älter als ich und häßlich, lang und vertrocknet. Sie erzählte mir nun in der Geschwindigkeit, daß sie Schriftstellerin sei unter dem Namen Blancheflour, ich würde ihre Schriften wohl kennen, sie habe diesen wichtigen Moment in ihres Herzens Herzen längst ersehnt. – Aber was wollen Sie denn eigentlich? fragte ich ganz verblüfft. – ‚Nun mein Gott! katholisch werden! Aber du kennst wohl meine geistlichen Hymnen noch nicht, ehrwürdiger Vater?' – Und nun fing sie, eh' ich's mich versah, wütend zu deklamieren an, bei jedem Vers trat sie in der Verzückung einen Schritt näher, ich einen Schritt zurück, bis an eine Laube, wo ich geschwind entwischen will. Da brechen auf einmal zwei junge Leute aus dem Buschwerk und gerade auf mich los; es war der Bruder des Fräuleins und sein akademischer Freund, ein durchreisender englischer Lord. Der Lord, der uns für verliebt hält, nimmt sich sogleich der verfolgten Unschuld der Jungfrau an, es werden Hieber angeschleppt, und ich muß mich auf der Stelle mit ihm duellieren. Ihr wißt, ich führte eine gute Klinge, der Lord ebenfalls, wir konnten einander nichts anhaben. Nun ging's drauf – das Fräulein lag in Ohnmacht – Schlenkerprimen und Schulterquarten, daß ich mein Barett vom Kopf verlor. ‚Nur noch einen

Gang!' rief der Lord entzückt aus – meinetwegen! – und wieder einen und noch einen! – Darüber wird mir endlich der Lord ganz gewogen, wirft den Hieber weg und embrassiert mich. – Nun fand sich's, daß er auch ein heller, philosophischer Kopf, und ebenso erpicht auf Menschenbildung war, als ich. Ich mußte mit ihm aufs Schloß, da hatte er alle Koffer voll neuer Konstitutionen, die er bei den verschiedenen Nationen anbringen wollte. Wir disputieren zusammen die ganze Nacht, wir werden ein Herz und ein Sinn, trinken Brüderschaft, und er proponiert mir, mit ihm zu reisen. Das Fräulein behandelte mich nun schnöde und verächtlich. Aber ich fragte nichts darnach, am folgenden Morgen saß ich mit dem Lord auf dem Wagen, und wir fuhren durch die Schweiz über Rom, Neapel, zwischen Kalabrien und Sizilien durch –"

„Halt! halt ein!" riefen hier die andern lachend dazwischen, „dein Lebenslauf kommt auf einmal so verteufelt ins Stürzen, daß einem ordentlich der Wind am Hute pfeift."

„Was da halt!" erwiderte Grundling, trinkend und wieder einschenkend. „Aber in Spanien ging's uns kurios. Das ist ein verteufelt hitziges Land, kaum hat man dort das Samenkorn der Weisheit in den Boden gelegt, so schießt's einem auch schon gleich unter den Beinen empor, Disteln und Unkraut, da ist kein Halten mehr, und eh' man sich's versieht, ist einem in dem verrückten Klima die ganze Vegetation über den Kopf gewachsen wie eine ungeheure Pelzmütze. Das haben wir dazumal wohl erfahren. Wir hatten uns durch Prozessionen, an Klöstern und Feudalsitzen vorüber, schon ziemlich tief ins Land hineingeärgert und ritten eines Abends soeben dem Gebirge zu, als sich ein paar wackere Burschen zu uns gesellten. Wem's Ernst ist, der feiert nicht gern. Wir knüpften sogleich ein Gespräch aus dem Gebiet der praktischen Philosophie mit ihnen an, bald holten wir noch ein paar Wanderer ein und wieder ein paar, bis wir zuletzt am Fuße des Berges auf einen großen, hellen Haufen stießen. Ich besinne mich

nicht lange und haranguiere das Volk. Ich sprach vom Aberglauben, von der Freiheit des Willens et cetera, ich kam immer mehr ins Feuer mit donnernder Stimme und zuckenden Gedankenblitzen, das zündet gleich rechts und links, die Kerls jauchzen, schreien Bravi und wieder Bravi, und eh' man die Hand umdreht, mitten in der Rede, heben sie mit Piken und Stangen ein altes, abgebrochenes Zelt hoch über ihre Köpfe, schwingen vor Entzücken mich und den Lord auf den Baldachin hinauf und tragen uns so im Triumphe auf ein altes, adeliges Schloß zu. Da war's doch nicht anders, als wollten sie mit unseren Köpfen die Mauern einrennen, denn in der Begeisterung fragten sie den Teufel darnach, daß das Schloßtor viel zu niedrig war für unsern Baldachin. Zum Glück erblick' ich nebst dem Lord noch zu rechter Zeit einen Balkon gerade vor uns über dem Tore, wir erfaßten schnell das Geländer, die Kerls schritten wie toll unter uns weg, und so blieben wir draußen am Balkon hängen, mit den Beinen in der Luft. Jetzt aber entstand unter uns ein Spektakel, ein Gedränge und Gewürge – denn die Kerls waren Guerillas –, die vom Schloß fielen aus, die Guerillas ein – zwischen unseren Beinen hindurch flogen die Kugeln immerfort hin und her, der Lord verwünschte unsere Philosophie, worüber wir noch heftig aneinander gerieten. Wie wir nun so bedenklich hängen und streiten, stürzt plötzlich oben im prächtigen Mondenscheine zwischen blühenden Pomeranzenbäumen das Schloßfräulein auf den Balkon heraus, dunkle Locken, Alabasterhals und Busen und eine Laute im Schwanenarm. Die sieht mich penetrant an und bleibt wie verzaubert stehen, sie sieht mich noch einmal an – und: ‚O, mein Traum!' ruft sie und läßt die Laute fallen. Darauf, schnell wieder gefaßt, erwischt sie mich hinten beim Kragen und hilft erst mir, dann dem Lord rasch übers Geländer auf den Balkon, in das Pomeranzengemach hinein. Jetzt aber war guter Rat teuer; ich unbewaffnet, kein Schwert in der Nähe, und von unten heult das Gedrossel, wie ein versessener Sturmwind, durch das alte

Haus immer höher und näher herauf. Der Lord wirft sich noch geschwind an den Sekretär des Fräuleins hin, schreibt sein Testament und setzt mich zu seinem Universalerben ein. Unterdes aber – ihr kennt die südliche Glut – verliebt sich die Prinzessin –"
„Prinzessin?" rief Fortunat, „du nanntest sie ja eben noch schlechtweg vorhin Fräulein!"
„Verliebt sich die Prinzessin", fuhr Grundling immer schneller redend und trinkend fort, „immer heftiger in mich und erzählte mir, wie sie mich schon früher einmal im Traume gesehen, mit Uniform und dreieckigem Hute durchs Morgenrot auf Wolken schwebend, et cetera. Jetzt war auch der Lord mit dem Petschieren des Testaments fertig, die Prinzessin wollte uns aus dem Schlachtgetümmel heimlich salvieren, wir retirierten durch Kammern und lange Gänge unaufhaltsam immer höher hinauf, wobei uns noch der eigensinnige Lord gefährlich wurde, der niemals seine prallen, hirschledernen Hosen ablegen mochte, die nun in dem Mondenscheine von weitem leuchteten. So kamen wir endlich auf das flache Schloßdach hinaus, da standen wieder blühende Granaten und Limonien, in der Mitte plätscherte eine Wasserkunst sehr angenehm, in der Goldfischchen bei dem klaren Mondenscheine lustig hin und her fuhren. Aber da war nicht lange Zeit zur Ergötzlichkeit. Unter uns der Kriegslärm, vor uns der nächtliche Abgrund, dazwischen die schöne Herzogin mit der südlichen Glut immer dicht hinter mir drein: ich soll katholisch werden und sie heiraten, oder ich und sie müßten auf der Stelle sterben! Ich aber kann mich in der Konfusion nicht gleich resolvieren, da zieht sie einen unvernünftig langen Dolch aus dem Gürtel, preßt mich mit dem linken Arm fest an ihre Brust, holt mit dem rechten hinter meinem Rücken aus und will mich und sich zugleich durch und durch stechen. In demselben Augenblicke platzt die Falltür neben uns mit einem ungeheuren Knall, daß die Stücke meilenweit auseinander fliegen. Sie hatten schon lange darunter gestemmt, und nun, wie

wenn ein Champagnerstöpsel unverhofft losgeht, kamen auf einmal Guerillas, Schloßsoldaten und Alguazils, die einen mit den Ellbogen, die andern mit den Stiefeln voraus, mit unglaublicher Vehemenz aus dem Loche senkrecht emporgeflogen, und so wie einer auf das Dach wieder niederfiel, fuhr er seinem Nachbar gleich wieder in die Haare, so verbissen waren sie untereinander. Die verliebte Königin, da sie nun alles verloren sieht, faßt mich beim Arme und rasch mit mir fort an den Rand der Zinne; aber ihr wißt, ich hielt niemals viel auf Kleider, mein ganzer Ärmel läßt oben in der Naht los, und die Königin stürzt sich mit meinem Ärmel in den Abgrund hinab, in der Luft noch: ‚Don Grundlinghio!' rufend. – Unterdes bekommt mein Lord plötzlich seinen englischen Spleen. Eh' ich's mich versehe, duckt er sich kopfüber in das Bassin der Wasserkunst. Das war nun aber so klein und seicht, daß ihm die Lederhosen oben trocken heraushingen. Ich schreie, die gestörten Goldfische stoßen wütend auf seinen Backenbart, alles umsonst! er stampft und stopft sich selber immer tiefer hinein und ersäuft sich so mit aller Gewalt. Es war ein kritischer Moment, Feinde ringsum, ich ziehe schnell mein Schwert und mähe mich von Etage zu Etage hinunter, ein umgefallener Alguazil beißt mich in dem Gemetzel noch in die Wade, ich spick' ihn fest an den Boden –"

„Aber was Teufel!" fuhr Grundling hier plötzlich mit sichtbarem Schrecken von seinem Sitze auf, „stehen denn die Toten wieder auf? da geht wahrhaftig der Lord vorüber!" – Und in der Tat, durch die offenen Türen des Hauses sah man draußen auf der Gasse beim hellsten Mondenscheine die gelben Lederhosen eines rasch vorübergehenden Mannes deutlich schimmern. Überrascht sprangen nun auch die andern auf, denn sie glaubten in der Figur flüchtig ihren langen Lord vom fürstlichen Hofe wiederzuerkennen. Eine schlanke Mädchengestalt, mit welcher die Eile des Fremden vielleicht in einigem Zusammenhange stehen mochte, schlüpfte unterdes, noch ein-

mal zurückblickend, schnell um die dunkle Straßenecke. Grundling aber hatte den Engländer schon erreicht, und sie sahen nun beide in der Dämmerung wie zwei Schatten im Reiche der Toten dahin schweben.

„Laßt die Phantasten laufen!“ sagte Kordelchen in der Haustür. „Wißt ihr denn nun aber auch, wer den Grundling eigentlich aus Heidelberg fortgeschafft hat? Der vermeintliche Proselytenknabe war ich selbst und der sogenannte Jesuit niemand anders, als ein junger Schauspieler, der mich damals heimlich von Heidelberg entführte. Wir mußten wohl den tollen Kauz über Hals und Kopf mit auf den Wagen packen, wenn er mit seinem Lärm nicht alles verraten sollte; mein Freund hatte in seiner kleinen Theatergarderobe zufällig eine Jesuitenkleidung, in die wir dann den Trunkenen hineinknöpften und des Nachts auf der Landstraße wieder aussetzten.“ – „Nun wahrlich“, rief Fortunat lachend aus, „das ist ja ein wahrer Sturmbeutel voll Lügen!“

Währenddes ruhte Guido, der nach den heftigen Gemütsbewegungen über Grundlings Erzählung eingeschlummert war, draußen im Gärtchen, noch im Schlafe malerisch über einen zertrümmerten Säulenknauf hingestreckt. Otto aber blickte immerfort unverwandt in die Straße hinaus, auch er hatte vorhin jene flüchtige Mädchengestalt bemerkt und schien zerstreut und unruhig. Endlich hielt er sich nicht länger und schlug Fortunat hastig noch einen Streifzug durch die Stadt vor, was dieser mit Freuden annahm. Kordelchen blickte beide listig an: „Felicissima notte!“ sagte sie dann mit einem ganz besondern, schelmischen Nachdruck, und als sich Otto darüber unwillig zu ihr wandte, war das wilde Mädchen schon im Hause und hatte die Tür laut lachend hinter sich verschlossen.

Sie eilten nun aus dem Gewirre der kleinen, engen Gäßchen ins Freie hinaus, Zithern schwirrten von fern durch die stille Luft, die Straßen waren noch voll Menschen, die fröhlich plaudernd und singend in der erquickenden Küh-

le auf und nieder schwärmten. Otto war still und schritt in Gedanken immer schneller und schneller, bis sie zuletzt an einen einsamen Platz kamen, wo er sogleich auf ein kleines, unansehnliches Haus zueilte. Er fand die Tür verschlossen und klopfte leise an; es blieb alles still drin, er klopfte noch einmal lauter. Da ließ sich eine überaus anmutige Stimme im Hause vernehmen: „Mein Herr, ich kann den Schlüssel im Dunkeln nicht finden, auch wacht die Mutter noch, aber habt die Güte, rechts die Straße hinab zu gehen, dann links um die Ecke, über die Brücke fort, dann wieder rechts, das vierte Gäßchen links hinein, so kommt Ihr in einen kleinen Hof, und wenn Ihr dort nicht auf einen Kettenhund stoßt und die Leiter findet, so könnt Ihr mir von dem Dache unseres Hinterhauses noch eine gute Nacht sagen; aber sputet Euch und fallt nicht, denn ich bin schon sehr schläfrig." Und kaum hatte sie ausgeredet, so hörten sie sie schon, leise lachend, die Treppe hinaufspringen. – „Annidi!" rief nun Otto höchst verwundert hinauf. Auf diesen Ton öffnete sich schnell ein Fenster über ihnen, und eine Mädchengestalt von überraschender Schönheit mit rabenschwarzem Haar und Augen erschien im hellsten Mondglanze. „Bist du es?" rief sie erstaunt aus, „ich meinte, es wäre der lange Engländer, der mir vorhin wie auf hohen Stelzen nachkam." Jetzt bemerkte sie auch Fortunat, stutzte und war bemüht, ihr loses Halstuch vor dem Fremden rasch in Ordnung zu bringen. Otto hatte sich unterdes auf einen Stein gestellt und reichte so bis ans Fenster. Das Mädchen legte den schönen Arm vertraulich um seinen Nacken, sich hinausbeugend, daß ihre dunklen Locken aufgingen und den Freund von allen Seiten umgaben; dabei sah sie unverwandt Fortunat an, dem sie nicht recht zu trauen schien. „Nein! nein!" rief sie endlich, nicht ohne Koketterie ihre Locken wieder aus der Stirn schüttelnd, „was fragt ihr fremden Herren nach dem Rufe eines armen römischen Mädchens! Die Nachbarn wachen noch und alle Fenster sehen im Mondenscheine wie glänzende Augen her, gute Nacht!" Hiermit

warf sie noch unversehens jedem einen frischen Blumenstrauß ins Gesicht und schloß schnell das Fenster.
Währenddes waren zwei Frauenzimmer, dicht in seidene Mäntel verhüllt, eilig über den Platz gegangen. Fortunat kam es vor, als hätten sie ihn im Vorüberstreifen scharf und verwundert angesehen. Er hörte sie darauf leise und eifrig miteinander sprechen, die eine sah noch einmal zurück, dann waren beide schnell verschwunden.
„O wie wunderschön sie ist!" rief Otto, noch immer nach dem Fenster schauend, aus, und erzählte nun begeistert, wie er sein Liebchen auf einem ländlichen Feste zum ersten Male gesehen, wie sie mit ihren armen Eltern eingezogen, aber fröhlich, lebe, wie sie von ihm Deutsch und er von ihr Poesie lerne, weil ihre Gegenwart, gleich der Morgenröte, alles verzaubere und verwandle. So gingen sie langsam durch die verlockende Nacht, die Nachtigallen schlugen aus allen Gärten, und zahllose Brunnen rauschten von fern.

Siebzehntes Kapitel

Die Villa des Marchese A. mit ihren kühlen Schatten, hohen, ausländischen Blumen und weißen Marmorbildern lag wie eine Insel in dem Weltgewühle, auf die sich Fortunat einsam verschlagen fühlte. Oft tönte es wunderlich in seine Morgenträume hinein, wie wenn eine Hochzeit in weiter Ferne schwirrend durch eine anmutige Landschaft ginge; wenn er erwachte, erkannte er Fiamettas liebliche Stimme, die treppauf treppab singend, plaudernd und lachend das ganze Haus schon mit fröhlichem Klange erfüllte. Eines Morgens fand er sogar einen frischen, vollen Blumenstrauß auf seinem Tischchen am Bett, er begriff nicht, wie er über Nacht dahin gekommen, und da er der kleinen Marchesin dafür danken wollte, schob sie's lachend auf ihre Kammerjungfer Lenore, die ihn gestern

dort vergessen, aber sie wurde über und über rot dabei. – Einmal kam er spät des Abends von einer Wanderung zurück, als er im Garten noch singen hörte, er meinte Fiamettas Stimme zu erkennen und wollte ihr noch eine Gutenacht sagen. Da war's ihm, als säh' er ihr Figürchen, verstohlen winkend und flüsternd, bald hier bald dort durch das Gebüsch schimmern, er folgte immer eifriger durch Hecken und Dorn in eine ganz unbekannte Gegend des Gartens hinein, die schadenfrohen Nesseln stichelten auf seine seidenen Strümpfe, Eidechsen schlüpften überall neugierig durch das Gestrüpp. Plötzlich stand er vor einem Gartenhause, die Tür war fest zu, durch die geschlossenen Jalousien aber glaubte er im Mondenscheine flüchtig zwei frische Augen funkeln zu sehen. Sonst war alles still im ganzen Garten, und beschämt und verdrießlich wanderte er wieder nach dem alten Schlosse zurück. – Aber es half ihm nichts, der Morgen kam doch wieder und das liebliche Stimmchen mit ihm, wie ein Zaubervogel im Walde, der ihn neckend immer tiefer in das grüne Labyrinth verlockte, von dem kein Ende abzusehen war.

So waren mehrere Wochen vergangen, Fortunat hatte, um sich alle Liebestorheit aus dem Sinne zu schlagen, sich endlich mit einer Art von Wut auf die Sehenswürdigkeiten der Stadt geworfen, mancherlei Studien und Ausflüge in die Umgebung gemacht und darüber seine deutschen Freunde fast ganz vernachlässigt. Er freute sich daher recht, als eines Tages Otto unerwartet gegen Abend zu ihm ins Zimmer trat, und bestürmte ihn sogleich mit Fragen nach Hohenstein, dessen grüne Stille mit allen ihren geliebten Personen ihm bei des Studenten Anblick wieder einmal ganz lebendig wurde. Aber zu seiner Verwunderung beantwortete Otto alles nur obenhin, ausweichend und beinahe verlegen. Dagegen schien ihn irgendeine gegenwärtige, große Freude zu drängen, seinem Herzen Luft zu machen. Gegen seine sonstige zurückhaltende Gewohnheit teilte er unaufgefordert mehrere soeben vollendete Gedichte mit, sprach voll fröhlicher Zuversicht

von seinen Plänen zu künftigen, großen Arbeiten, und entwickelte einen solchen bunten Reichtum der Seele, daß Fortunat wie in ein Kaleidoskop hineinzusehen glaubte. Draußen wehte es unterdes schon wieder kühl über die Stadt, sie machten noch einen Gang ins Freie, und Otto, sein Gespräch leidenschaftlich fortsetzend, führte den Freund zwischen kleinen Häusern und Weinbergen unvermerkt in eine schöne, abgelegene Gegend hinaus, die Fortunat noch nicht kannte. Garten stieß an Garten, ein unübersehbares, blühendes Paradies mit zierlichen Villen und Balkonen, auf denen manche schlanke Gestalt zwischen den Wipfeln erschien, alles von der untergehenden Sonne zauberhaft durchblitzt und beleuchtet. – „Wenn ich", sagte Otto, die Gegend überschauend, „wenn ich jemals aus diesem Glanze wieder in die dumpfe Enge meines deutschen Gebirgsstädtchens zurück müßte, wo sie jetzt wohl vor den Türen unter ihren hölzernen Lauben sitzen, die Hände vor Kälte fest eingewickelt, und nichts vernehmen, als das Glöcklein der Bergleute und den Schlag des Eisenhammers von fern, und die Berge sehen von allen Seiten finster auf den stillen Markt herein, und der feuchte Wind schlägt den Kohlenrauch nieder und verhüllt alles wie ein Grab – mich schauert ordentlich bei dem Gedanken!" – „Hüt' dich wohl", entgegnete Fortunat, „es ist ein wunderbares Lied in dem Waldesrauschen unserer heimatlichen Berge; wo du auch seist, es findet dich doch einmal wieder, und wär' es durchs offene Fenster im Traum, keinen Dichter noch ließ seine Heimat los." – Otto schwieg nachsinnend – es war heut fast etwas Freudeverstörtes in seinem ganzen Wesen.
Auf einmal bog er rasch mitten in das Blütenmeer von Gärten hinein. Sie kamen an ein kleines, aber wohlgebautes, reinliches Haus, von Efeu, Weinlaub und blühenden Bäumen reizend überwachsen und verdeckt; die Tauben, die sich auf dem Dache in der Abendsonne spiegelten, die offenstehenden Fenster und Türen, wo bunte Schmetterlinge flimmernd ein und aus flatterten, alles gab ein wun-

derliches Bild südlicher Häuslichkeit. Otto führte seinen Begleiter ohne weiteres gerade durch das Haus in ein dahinterliegendes, einsames Gärtchen, umgeben von Nachbargärten, die von allen Seiten blühend hereinhingen und jede Aussicht verschlossen.

„Wo sind wir denn hier?“ fragte endlich Fortunat erstaunt. Indem aber erschien ein Mädchen in der Haustür, er erkannte sogleich die schöne Annidi wieder. Sie begrüßte ihn etwas verwirrt und beschämt, dann trat sie unter eine Weinlaube und begann aus ihrem Handkörbchen einen Tisch reinlich zu decken, Gläser und Teller aufzustellen. Draußen im Nachbargarten hörten sie einen Knaben fröhlich singen:

„Es sang ein Vöglein hier jedes Jahr:
Wie schön das Kränzlein im dunklen Haar!
Heuer ist's Vöglein nicht wiederkommen;
Wer hat dir das schöne Kränzlein genommen?“

Nun hielt Otto sich nicht länger, es kam alles heraus: daß Annidis Eltern seine Besuche ohne bestimmte Erklärung nicht weiter dulden wollten, daß er seit einigen Tagen mit dem Mädchen verheiratet und sich nun samt den Ihrigen hier eingenistet habe. Fortunat erschrak über diese ganz unerwartete Entdeckung und überdachte schnell die wunderlichen Folgen, die diese Übereilung für Otto herbeiführen mußte. Doch wurde er bald durch die liebliche Erscheinung der jungen Frau wieder beschwichtigt, die sich, ihrer neuen Lage noch ungewohnt, fortwährend mehr zierlich dienend als mitgenießend erwies, als sie sich nun fröhlich unter der Laube um den Tisch setzten. Auch ihre Eltern gesellten sich jetzt zu ihnen, zu Fortunats heimlichem Unbehagen, den die gewöhnlichen, welschgekniffenen Gesichter störten. Sie mischten sich öfters ungeschickt mit in das Gespräch, redeten viel von guter Wirtschaft und dem nötigen Fleiße ihres Schwiegersohnes im Büchermachen, und Fortunat konnte wohl bemerken,

daß sie ihn selbst als einen Zeitverderber und zweideutigen Kameraden Ottos scheel ansahen. – Unbekümmert saß und schmauste unterdes das glückliche Ehepaar. Annidi auf einem Fußbänkchen, mit beiden Armen auf Ottos Knie gestützt und die gebratenen Kastanien ausschälend, die sie jede zur Hälfte miteinander teilten. Der Mond schimmerte schon durch das Weinlaub, Otto war selig still, die junge Frau überaus schön, drüben sang der Knabe wieder:

„Wer hat dir das Kränzlein genommen?“

Fortunat aber überwältigte mitten in dieser Stille eine unwiderstehliche Wehmut, als sei Otto nun hier in der Fremde märchenhaft verzaubert. Es wollte ihm das Herz zersprengen, er schützte ein dringendes Geschäft vor, ergriff schnell seinen Hut und nahm tief gerührt Abschied von dem Freunde, wie von einem Verstorbenen. Als er zurückblickte, standen Otto und Annidi noch in der Haustür. Glühwürmchen schwärmten leuchtend durch das Rebengelände, er sah von der schönen Frau nur noch die glänzenden Augen und Schultern, Otto erschien totenbleich im Mondenschein.

In wirren Gedanken war Fortunat hastig nach Hause geeilt. Der Mond schien prächtig über den alten Garten, er lauschte, ob er Fiametta nicht wieder singen hörte, doch alles blieb still. Als er aber um den Pfeiler des Schlosses trat, fuhr er heftig zusammen, denn in einer der Alleen glaubte er plötzlich sich selber zu erblicken. Unverwandt starrte er hin, die Gestalt zeigte sich noch einmal im hellsten Mondlicht, es war seine Kleidung, sein Gang, seine Haltung, und doch schien es wieder ein ganz fremder, junger Mann. Jetzt blieb der Unbekannte lauernd hinter einer Hecke stehen. Da kam auf einmal Fiametta aus dem Gebüsch hervorgesprungen, besah ihn lachend rundum, dann gingen sie Arm in Arm tiefer in den Garten hinein. Mitten im fröhlichen Plaudern aber schien sie plötzlich Fortunats Schatten auf dem Rasen zu bemerken, er sah sie

erschrocken entfliehen, und bald war die ganze Erscheinung im Dunkel wieder verschwunden.
Fortunat aber hatte sich ins Schloß gewandt und ging heftig in seinem Zimmer auf und nieder. „Also diesem galt das Abendliedchen letzthin, o ich Tor!" sagte er mit einem bittern Gefühl, das er sich selbst nicht eingestehen mochte. Es war fest beschlossen, er wollte gleich morgen weiter nach Neapel reisen, ohne Fiametta noch einmal wiederzusehen. Noch in der Nacht schrieb er sein Vorhaben dem Marchese, der eben auf dem Lande war, und packte, in geheimer Wut lustige deutsche Lieder singend, seinen Koffer. Dabei schwirrten ihm die Worte aus einem alten Liede:

„Das Kränzlein ist herausgerissen,
Ganz ohne Scheu sie mich anlacht:
Geh du vorbei: sie wird dich grüßen,
Winkt dir zu einer schönen Nacht."

immerfort durch den Sinn, daß er darüber aus Herzensgrunde hätte weinen mögen.
Am folgenden Morgen hatte er noch einige weitläufige Gänge, um das nötige Reisegeld zu erheben; so war die Mittagsstunde herangekommen, die Zeit der zauberischen Schwüle, die im Süden alles Lebendige überwältigt. Dennoch wollte er nicht abreisen, ohne vorher noch einen Streifzug durch den Garten zu machen. Da rührte sich jetzt kein Blättchen in der weiten, träumerischen Stille, die Vögel schwiegen, nur einzelne Schlangen sonnten sich ringelnd auf den einsamen Gängen, alle Menschen lagen wie tot. Es war das erstemal, daß er hier zu dieser Stunde wach war, und dieses Schlafen der Natur mit offenen Augen erschreckte ihn gespenstisch. Er flüchtete nach einem kühlen Gartenhause, blieb aber überrascht im Eingange stehen, da er Fiametta, gleichfalls schlummernd, drin erblickte. Sie ruhte auf dem rechten Arme, das Gesicht von den losgelösten Locken halb verdeckt, heiter atmend, wie

ein schönes Kind. Einige abgebrochene Worte hielten ihn fest. Sie sprach im Schlaf, immer deutlicher und zusammenhängender, aber zu seinem Erstaunen ganz in der ausländischen Weise, wie er selbst das Italienische zu sprechen pflegte. In wunderlichem Dialog hörte er nun, wie er ihr aus ihrem eigenen Munde gestand, daß er sich nur so kalt stelle, daß er sie aber eigentlich herzlich liebe. – Er erschrak, daß sie so aus seiner Seele redete. – Nun lachte sie in sich und entgegnete fröhlich: das wisse sie ja lange schon! – Dann sprach sie leise, immer leiser, als spräch sie ihm ins Ohr, er konnte nichts verstehen, bis sie zuletzt, tief aufseufzend, sich zu regen begann.
Fortunat eilte ganz verwirrt nach dem Schlosse zurück, schon rührte sich's wieder in allen Straßen, der Postillion draußen mahnte zur Abreise, er warf sich schweigend in den Wagen, und das lieblichste Rätsel, das er nicht zu lösen wußte, erfüllte seine ganze Seele.

Achtzehntes Kapitel

Mehrere Monate sind seitdem verflossen, die Sonne glüht auf den Quadern der öden Paläste, und die Reichen sind längst auf ihre Villen geflüchtet, denn auf den Trümmern der alten Stadt sitzt die Aëra cattiva schon wie ein verhülltes Gespenst, Fieber und Wahnsinn brütend. Wie ist Ottos Einsiedelei seitdem so seltsam verwildert! Die Ranken an der Haustür wuchern bis über das Dach hinaus, in dem Gärtchen hat üppiges Unkraut, in roten und gelben Blüten brennend, Beete und Gänge verschlungen. – Da kehrte Otto eines Tages ermüdet von einem weiten Spaziergang zurück, er fand im Hause alles ausgeflogen, nur die Bienen summten einförmig in dem stillen Garten, er fühlte sich unbeschreiblich verlassen, Hausflur, Stuben und Bäume kamen ihm in der ungewohnten Einsamkeit auf einmal so fremd vor, daß er erschrak. Er ging einige Male

im Garten auf und nieder, dann setzte er sich zwischen den tief herabhängenden Zweigen an den Tisch und schrieb folgende Zeilen:

Die Nachtigall schweigt, sie hat ihr Nest gefunden,
Träg' ziehn die Quellen, die so kühle sprangen,
In trüber Schwüle liegt die Welt gefangen,
So hat den Lenz der Sommer überwunden.

Noch nie hat es die Brust so tief empfunden,
Mir war's, als ob viel' Stimmen heimlich sangen:
Auch *dein* Lenz, froher Sänger, ist vergangen,
Auf welkem Laub nun liegst du selbst gebunden.

O komm, Geliebte, komm zu mir zurücke!
Daß ich in deinen Augen wieder lesen
Mein Hoffen kann, mein Singen und mein Lieben!
Doch weh! wie fremd sind plötzlich deine Blicke,
Als wärst du's, die ich meinte, nie gewesen –
Wie einsam bin ich in der Welt geblieben.

Mein Weib das schwärmt beständig,
Und Deutschland liegt so weit,
Das Dichten geht elendig
In meiner Einsamkeit.

Ich dehne alle Glieder
Aus dieser schwülen Gruft,
O Herr, gib Frühling wieder,
Luft, frische, freie Luft!

Als er von dem Blatt aufsah, hörte er draußen Vorübergehende reden in der fremden Sprache, aber ein Vogel über ihm sang wie ehemals in Hohenstein – er drückte die Stirn über beide Arme auf den Tisch und weinte aus Herzensgrunde.

Da hörte man plötzlich im Hause eine liebliche Stimme

einzelne Klänge aus Opern-Arien theatralisch anschlagen. Eine junge Dame in reicher, eleganter Kleidung trat in den Garten und hob den seidenen Hut vom Köpfchen, die reichen Locken ringelten über den schönen, vollen Nacken hinab – es war Annidi, wie war sie seitdem so prächtig geworden! Sie warf ihre Handschuhe der dienstfertig herbeieilenden Mutter nachlässig zu, während ihr Vater, der sie als Bedienter begleitet zu haben schien, im Hause Schal und Sonnenschirm niederlegte. „Der Graf Archimbaldi läßt dich grüßen", sagte sie zu Otto, „aber die ganze Noblesse wundert sich, lieber Mann, daß du so menschenscheu bist und immerfort studierst, der lustige Duca sagte: Weisheit mache weiße Köpfe. Auch die junge Malerfrau war heute dort, mein Gott, wie war die angezogen! Der junge Mensch flüsterte mir heimlich ins Ohr, sie sei wahrscheinlich, erst halb schraffiert und grundiert, ihrem Pinsel von Mann entlaufen." –

Hier aber brach sie plötzlich erschrocken ab, da Otto endlich aufsah und ihr das bleiche, wüste Gesicht zuwandte. Sie hielt ihn für krank, sie ließ es sich nicht ausreden. Die Mutter mußte sogleich nach der Küche laufen, es wurde Tee gekocht, herzstärkende Tropfen geholt und Kräuter gestampft mit großem Geräusch. – „Mir geschieht schon recht", rief Otto mit schneidender Bitterkeit aus, „ihr habt ganz recht, mit den Fingern nach mir zu weisen. Doch ich will einen Strich durch die Rechnung meines Lebens machen, o ja, ich will auch lustig sein, daß mir das Herz zerspringt!" – Aber wie es in solchen Fällen wohl geht, Annidi hatte ihn ganz mißverstanden. – „Wahrhaftig" – sagte sie, vertraulich nähertretend – „du magerst mir ganz ab bei dem Leben, und ich wollt' es dir schon lange einmal sagen: so fleißig wie du bist, es kann dir ja doch am Ende einerlei sein, was du schreibst. Da ist der junge Schreiber uns gegenüber, du schreibst eine bessere Hand, als er, das sagen alle, und was verdient der, wie lebt er gegen uns!" –

Da kam die Mutter mit dem Tee, Otto wies sie so heftig

von sich, daß Kanne und Tassen übereinanderstürzten. „Das kommt von dem ewigen Sitzen und Brüten", sagte der erstaunte Vater in der Haustür. – „Ja, und jede Henne brütet doch mehr aus fürs Haus als er", brummte die Mutter. Otto aber, um nur aus alle dem Plunder herauszukommen, war schon aus dem Garten und Hause fort, und schweifte, so müde er war, in der Abendkühle durch die Gassen und dunkelnden Felder, bis die Nacht völlig hereinbrach.

Als er zurückkehrte, war schon alles still im Hause, es ärgerte ihn heimlich, daß Annidi nicht besorgter war um ihn. Er fand sie droben eingeschlafen, der Mondenschein machte ihre Züge so mild, ach, und sie war so schön! Da blickte er durchs offene Fenster über die Dächer in die mondbeglänzten Abgründe der Stadt hinab, einzelne Wolken flogen darüber nach seiner fernen Heimat zu. – „Wunderbar", sagte er zu sich selbst, „schon in meiner Kindheit, wie oft bei stiller Nacht im Traume hört' ich der fernen Roma Glocken schallen, und nun, da ich hier bin, hör' ich sie wie damals aus weiter, weiter Ferne, als gäb' es noch eine andere Roma weit hinter diesen dunklen Hügeln." –

In dieser Zeit traf es sich, daß in der Nähe von Rom auf dem Lande eine Kirchweihe gefeiert wurde. Annidi dünkte sich zu vornehm, um an dem Feste teilzunehmen. Otto aber, den es heimlich verdroß, warf auf einmal alle Papiere und Bücher beiseite und eilte hinaus ins Freie. Es war in den ersten linden Herbsttagen, ein warmer Regen hatte die Gegend erfrischt, Otto atmete tief auf, es war ihm als wanderte er wieder nach Hohenstein. Je tiefer er ins Tal hinabstieg, je belebter wurden allmählich Busch und Felder, bunte Züge von Reitern und Spaziergängern schlangen sich wie Blumenkränze durchs Grün, von den Waldeswiesen schimmerten farbige Zelte, zwischen denen zerstreute Gruppen fröhlich lagerten, während lustige Gestalten im Ballspiel über den Rasen hin und her schwebten. Mitten in dieser Wirrung aber bemerkte Otto

einen schlanken Zitherbuben, der auf seinem geschmückten Pferde langsam über die beglänzte Au dahinritt. Ein voller Kranz von frischem Weinlaub umschloß seinen Hut, von dem bunte Bänder in der Abendluft flatterten, von Zeit zu Zeit gab er einen vollen Klang auf der Zither. – Otto folgte der zierlichen Erscheinung, erstaunte aber nicht wenig, als der Knabe auf einmal deutsch zu singen begann:

„Die Lerch', der Frühlingsbote,
Sich in die Lüfte schwingt,
Eine frische Reisenote
Durch Wald und Herz erklingt!"

„Mein Gott", rief Otto sich besinnend aus, „das ist ja das Reiselied, das ich so oft in Deutschland gesungen habe." – Er trat näher, der Zitherbube sang wieder:

„Die Wolken ziehn hernieder,
Die Lerche senkt sich gleich –
Gedanken gehn und Lieder
Ins liebe Deutsche Reich.

Aber eh' ich ihnen selbst nachreite, muß ich vorher trinken, denn ich bin beinah' erdurstet", unterbrach sich hier plötzlich der Knabe, während er vor einer Laube anhielt und lachend von seinem Pferdchen dem Otto fast in die Arme sprang. Dieser erkannte nun *Kordelchen,* die ihn schon längst in der Menge hinter sich bemerkt hatte.
Sie zog ihn in die Laube, Guido und ihre andern Begleiter, sagte sie, kauerten soeben wie Nachteulen in Ruinen und Felsenritzen, um zu zeichnen, überdies habe sie sich auch mit ihnen verzankt. – „Aber wie siehst du aus!" rief sie dann, Otto genauer betrachtend, „nüchtern und blaugrün, wie eine leere Weinflasche! Das kommt vom Ehestande. Armer Junge! bliebst du mir treu, so wärest du nicht in das Unglück geraten." – Sie bestellte nun Wein,

und sie setzten sich zusammen in die Laube. Otto hatte seit Monaten keinen Bekannten gesehen, nun war ihm nach der langen Einsamkeit wie einem Genesenen, der zum ersten Male wieder in die frische Luft kommt. „Sieh, Kordelchen", sagte er fröhlich, „gerade in solchen linden Tagen war es auch, als wir uns zum erstenmal in Deutschland sahen." – „Ganz recht", erwiderte sie mit leuchtenden Augen, „wir rasteten eben unter einer alten Burg im Grünen, da kam er aus dem Walde und sagte, er wollte mit uns ziehen." – Sie meinte Lothario, Otto dachte, sie spräche von ihm. „Wahrhaftig", fuhr er fort, „mir ist heute ganz zumute wie damals, als käme der Frühling wieder." „Ach, nein, nein", sagte sie traurig, „der kommt nicht mehr wieder." – Sie nippte schnell am Weinglase, um die Augen zu verbergen, die von Tränen glänzten, dann wandte sie das schöne, von Locken und Weinlaub verhängte Gesichtchen wieder heiter nach Otto herum. Da bemerkte sie, daß er, auf beiden Armen über den Tisch gelehnt, sie mit einem langen, wirren Blick ansah, den sie gar wohl verstand; sie schien davon überrascht, beugte sich plötzlich vor ihm und sah ihm halb fragend in die Augen. Da hielt er sich nicht länger, er drückte sie mit glühenden Küssen an sich. Sie erwiderte flüchtig den Kuß und sprang dann rasch auf. „Ei, Ehemann!" rief sie mit dem Finger drohend, schwang sich behend auf ihr Pferdchen und war im Augenblick zwischen den Zelten und Büschen verschwunden.

Otto hatte nun den Wein zu bezahlen, die Neige kam ihm jetzt schal vor, da sie die brennend roten Lippen nicht mehr darin kühlte. Draußen aber war unterdes der Abend verklungen und verblüht, nur von den Bergen sah man noch einzelne Leuchtkugeln aufsteigen. Wie im Taumel wanderte er zwischen den Gitarrenklängen, dem Singen und Plaudern der Heimschwärmenden durch die laue Nacht, als mitten in dem Jubel eine dunkle Gestalt an ihm vorüberstreifte, dann aber, plötzlich zurückgewandt, ihm fest ins Auge blickte. Mit Erstaunen sah er den Maler *Al-*

bert vor sich stehen: ganz bleich, verwildert und abgerissen. – „Mein Gott! wie kommen Sie nach Rom, und in diesem Zustande?" rief der Überraschte aus. – „Verloren, alles verloren!" erwiderte Albert finster und mit solchem Ausdruck des tiefsten Grams, daß Otto schauderte. „Aber hier belauscht uns der Mond noch, auch er ist falsch, in diesem Lande", fuhr er fort, indem er Ottos Hand faßte und ihn tiefer in den Wald hineinzog. Rasch und unzusammenhängend erfuhr nun Otto, daß sein wunderlicher Landsmann, von heimlich aufschlagenden Freiheitsflammen von neuem auf diesen vulkanischen Boden verlockt, schon seit längerer Zeit hier heimlich mit wenigen Gleichgesinnten seine Kunst, Gut und Leben an eine Tollheit gesetzt, daß aber jetzt alle Pläne gescheitert und er selbst als Carbonaro verfolgt werde. – Der gutmütige Otto bot sogleich alle seine Kräfte, Geld und Verbindungen zur Hilfe an, er wollte den Unglücklichen zunächst in seinem Hause verbergen, bis sich Gelegenheit fände, ihn heimlich aus dem Lande zu schaffen. Aber Albert schüttelte den Kopf, daß ihm die langen, struppigen Haare Augen und Wangen bedeckten. „Nicht um mich handelt sich's hier", sagte er, „sondern um die Schmach der Zeit. Horch, wie sie draußen jauchzen und mit den Sklavenketten lustig klingeln – das ist's, was mir das Herz frißt!" Hier hörte man verworrene Männerstimmen. Albert blickte wild um sich und zog einen Degen unter seinem Mantel hervor. Otto erkannte sogleich das Schwert vom großen Kriegsjahre dreizehn wieder. „Die Sbirren sind mir auf der Spur", flüsterte er, „eilen Sie fort, es ist gefährlich, die Bahn eines tragischen Geschickes zu kreuzen." Aber Otto war fest entschlossen, lieber das Äußerste zu wagen, als den Verwirrten in dieser Not zu verlassen. Rasch und geräuschlos schritten sie unterdes immer höher ins Gebirge hinauf. Albert hieb sich mit seinem Schwerte Bahn durch das Gestrüpp, aus welchem verstörte Schlangen nach den Steinritzen schlüpften. So waren sie auf einen Felsen gekommen, der schwindelerregend über

eine unermeßliche, dämmernde Tiefe hinüberhing. Albert stand am äußersten Rande und wies mit seinem Schwerte schweigend in die Ferne. – „Großer Gott, wie herrlich!" rief Otto überrascht aus – Rom lag da unten still und feierlich im Mondglanz. – Da hörte er auf einmal ein Geräusch, er sah Albert plötzlich wanken, sinken. Der Unglückliche hatte sich mit heidnischer Tugend in sein eigenes Schwert gestürzt. – „Grüße das Vaterland – – ich sterbe – frei", sagte er ohne Zeichen des Schmerzes, wehrte die Hand des hinzugesprungenen Otto kräftig ab, und glitt, eh' ihn dieser wieder fassen konnte, rettungslos in den Abgrund hinab.

Entsetzt beugte sich Otto über die Felsenwand, es war alles still unten, nur der Strom rauschte zornig herauf – da faßte ihn ein unwiderstehliches Grauen, halb bewußtlos schwang er sich von Klippe zu Klippe den Berg hinunter. Im Fliehen bemerkte er seitwärts in dem Abgrunde mehrere dunkle bewaffnete Gestalten mit Fackeln, die den Toten in ihrer Mitte gräßlich beleuchteten. Nun schlugen hin und wieder Hunde an, einzelne Stimmen wurden in dem Tale wach, der Widerschein der Windlichter spiegelte sich wild im Flusse. Otto wagte nicht mehr zurückzublicken, schaudernd flog er über die stillen Felder, durch die leeren Gassen fort zu seiner einsamen Wohnung.

Hier fiel es ihm erst ein, daß er bei den Seinigen hinterlassen, diese Nacht auf dem Lande zubringen zu wollen. Er fand nun die Türen verschlossen, alles im Hause schien längst zu schlafen. Unmutig stieg er daher über den Zaun in den Garten, wo er sich sogleich auf die Bank in der Laube hinwarf. Das leise Rauschen in den Zweigen sang gar bald den Ermüdeten ein. Da träumte ihm, er läge in dem schönen Garten zu Hohenstein und sähe die steinernen Götterbilder vor sich im hellen Mondenscheine auf den Gängen stehen. Es war, als flüsterten sie in der Stille heimlich untereinander, und als er recht hinsah, regte sich das Venusbild und stieg langsam von dem marmornen Fußgestell herab. Mit Grauen erkannte er seine Annidi, sie kam

gerade auf ihn zu, eine Marmorkälte durchdrang plötzlich alle seine Glieder, daß er erschrocken aufwachte. Als er aber noch ganz verwirrt umherblickte, stand wirklich die weiße Gestalt in der Haustür, leise flüsternd nach jemand zurückgewandt, den er nicht sehen konnte. Auf einmal schlug sie einen weiten Mantel auseinander, und Annidi trat aus den Falten hervor. Ein junger, hoher Mann umschlang und küßte sie, dann warf sie ihm lachend den Mantel zu und schlüpfte ins Haus, der Fremde schwang sich rasch über den Gartenzaun – und alles war wieder totenstill.

Otto starrte lange regungslos auf den dunklen Fleck, wo der furchtbare Spuk zerronnen. Darauf stürzte er aus dem Garten in die Nacht hinaus, ohne zu wissen, wohin – er hatte ja nun keine Heimat mehr auf Erden! – Die Straßen waren öde, die Wasserkünste im Mondenschein, die ihm sonst so bräutlich rauschten, kamen ihm jetzt gespenstisch vor, wie verschleierte Nixen, im Winde sich beugend und neigend, als flüsterten sie heimlich von ihm und seiner Schande. Unwillkürlich hatte er den Weg zu Guidos Wohnung eingeschlagen, er wollte ihn wecken, er mußte in dieser Stunde jemand haben, dem er alles sagte. Zu seinem Erstaunen fand er die Tür nur leicht angelehnt, ein Licht brannte drin. Als er in die Stube trat, sah er Kordelchen auf der Erde knien zwischen Wäsche und Kleidern, die sie eifrig in einen Mantelsack packte. Sie blickte erstaunt, fast erschrocken nach ihm herum. „Was willst du denn jetzt hier?“ sagte sie, „Guido ist noch auf dem Lande und kommt erst in einigen Tagen zurück.“ – Otto aber wollte das Herz zerspringen, er warf sich auf das Sofa und brach, sein Gesicht mit beiden Händen bedeckend, in ein unaufhaltsames Weinen aus. Da stutzte Kordelchen, sie ließ alles liegen, setzte sich zu ihm und tröstete und streichelte ihn neugierig und mit herzlicher Teilnahme, bis sie nach und nach sein ganzes Unglück erfahren. Sie hörte alles still und nachdenklich an. Als er aber schwieg, sprang sie plötzlich auf. „Wir reisen zusammen!“ rief sie

aus, „das ist eine langweilige Wirtschaft hier, und ich und Guido, wir paßten eigentlich niemals zusammen. Wenn er sich betrinkt, so ist das genial, wenn er sich verliebt, so ist's Andacht, und wenn ich ihn darüber auslache, so wird er wütend und will mich durchaus mit sich emporflügeln, wie er's nennt. Ich hab's schon seit einigen Wochen beschlossen, ich reise heimlich fort und zurück nach Deutschland, ich habe soeben Geld genug, die Pferde sind bestellt – kurz, wir reisen noch heut!" – Dabei wartete sie gar keine Antwort ab, sondern rumorte und packte dazwischen immer lustig fort, Otto wußte nicht, wie ihm geschah, durch das offene Fenster wehte frische Reiseluft herein, der Morgen dämmerte schon leise über der stillen Stadt.

Wer dem Teufel läßt ein Haar, den faßt er ganz und gar. So brannte der Kuß von gestern noch immer heimlich fort auf Ottos Lippen, über den Trümmern seines Glückes war über Nacht eine üppig blühende Wildnis schimmernder Erinnerungen und Hoffnungen giftig aufgeschossen. – Und als die ersten Streiflichter des Morgens über die Berge flogen und die früherwachten Lerchen noch halbverträumt in den Lüften hingen, da zogen Otto und Kordelchen schon durch die stillen Felder nach Deutschland zu, und sahen Rom, wie in einem Feuermeer, langsam hinter sich versinken.

Währenddes war Fortunat in Neapel und Sizilien umhergestreift. In seiner poetischen Behaglichkeit hatte er sich alles aus dem Sinn geschlagen, und machte überhaupt aus seiner Liebe gar nichts, als ein langes Gedicht in vielen Gesängen und verschiedenen Silbenmaßen, worin ein schönes, schlankes italienisches Mädchen die Hauptfigur spielte. Da begab sich's aber, daß er im Schreiben sich nach und nach in diese Figur selbst verliebte, und je verliebter er wurde, je ähnlicher wurde sie unvermerkt der kleinen Marchesin, als ob Fiametta oft plötzlich zwischen den Blütengewinden der Verse hervorguckte und, ihn

auslachend, ausrief: „Siehst du, ich hab' dich doch!" – Ja, als er in Sizilien eines Abends auf einem hohen, senkrechten Felsen über dem Meere eingeschlummert war, träumte ihm, die blaue Flut teile sich leise, und mit langem, grünem Haar und glänzenden Schultern tauche Fiametta unten empor, in irren Tönen wehmütig klagend. – Als er erwachte, war der Mond schon über dem Meere aufgegangen, in der Ferne aber sah er ein Segel schwellend durch die weite Stille nach dem jenseitigen Ufer Italiens hinübergleiten. – Da faßte ihn eine unwiderstehliche Sehnsucht, und schon die folgende Nacht segelt' er selber hinüber. Und so geschah es, daß aus demselben Morgenrot, in welchem Rom hinter Otto versank, die Gärten, Trümmer und Kuppeln vor dem glückseligen Fortunat duftig wieder emporstiegen.

Sein erster Gang war zu dem Palast des Marchese, mit klopfendem Herzen betrat er den stillen Hof. Er horchte, ob sich nicht irgendwo Fiamettas heitere Stimme vernehmen ließe, doch alles blieb lautlos, wie ausgestorben. So ging er durch die offene, luftige Säulenhalle in den Garten. Da sangen die Vögel und rauschten die Brunnen noch immer wie damals. Aber an der Hauptallee sah er Wäsche zum Trocknen aufgehängt, einzelne Ziegen weideten ungestört zwischen den verwilderten Blumenbeeten. Endlich glaubte er in einiger Entfernung Deutsch reden zu hören. Er ging dem Klange nach und begegnete einem alten, unbekannten, etwas schäbigen Diener. Hastig fragte er nach dem Marchese A. und seiner Tochter. Der Alte sah ihn von oben bis unten an und sagte dann verdrießlich, dieser Palast sei von einem deutschen Kavalier bewohnt. Fortunat war wie im Traume. – Er verlangte nun, den Herrn zu sprechen. Der Bediente wies schweigend nach einer Laube und ging fort, ohne sich weiter um den Gast zu bekümmern.

Hellen Halses aber mußte nun Fortunat auflachen, als er in die bezeichnete Laube trat und in dem deutschen Kavalier unseren Freund Grundling erkannte: in dem geblüm-

ten Schlafrock des Marchese auf einem halbzerrissenen, damastenen Sofa ausgestreckt, eine lange Tabakspfeife und ein Buch in der Hand, Talglicht, Fidibus und Kaffeekanne vor sich. Der Vielgereiste, an das wechselnde Kommen und Gehen in Rom längst gewöhnt, schien nicht im mindesten erstaunt, Fortunat wiederzusehen. „Mir ist's eben recht", sagte er, „daß der alte Marchese banquerout gemacht." – „Was! der Marchese A.?" rief Fortunat höchst überrascht aus.

„Ja, eben recht, sag' ich, daß er seinen Palast und Rom verlassen mußte, so konnte ich mich hier in der liederlichen Wirtschaft seiner Gläubiger ziemlich wohlfeil einmieten. – Wenn nur", fuhr er, seine Pfeife plötzlich grimmig wegsetzend, fort, „in der unvernünftigen Hitze der Tabak nicht so in die Zunge bisse!"

Hier verlor endlich Fortunat alle Geduld. „Nun rede zum Teufel einmal ordentlich!" rief er, Grundling rasch an der Brust fassend, „wo ist Fiametta? was macht sie?" – „In Deutschland wahrscheinlich und weint", erwiderte Grundling gelassen. – „Warum weint sie?" – „Weil sie ein junges, albernes Ding ist, dem ein konfuser Wein, der noch moussiert, lieblicher in die Nase sticht, als ein würdiges, abgelegenes Gewächs; das will heißen: die einen brutalen Phantasten, der sein Liebchen verläßt und seine Freunde drosselt, scharmanter findet als" – „Und wem gehört jetzt dieser Palast?" unterbrach ihn Fortunat ungeduldig wieder. – „Einem filzigen Kaufmanne, der ihn, seiner Entlegenheit wegen, abtragen lassen und die Steine verkaufen will." – „So führ' mich gleich zu ihm!" – Das war Grundling, der sich gern umhertrieb, eben recht. Wenige Minuten nach diesem Verhöre waren sie schon auf der Straße, und Fortunat erfuhr nun noch unterwegs, daß Fiametta unmittelbar nach seiner Abreise aus Rom bedeutend erkrankt und bald darauf mit ihrem Vater plötzlich abgereist sei. Weder er noch der Kaufmann wisse, wohin sie sich gewendet. Auch Ottos und Kordelchens Flucht hatte der Müßiggänger schon erfahren. „Der Otto", sagte

er, „war beständig in poetischem Tran, das mußte ein Ende mit Katzenjammer nehmen."

Während dieses Berichts waren sie bei dem Kaufmanne angelangt. Dieser war, gleich Grundling, nicht wenig erstaunt, als nun Fortunat den alten, verfallenen Palast und Garten des Marchese zu kaufen verlangte. Die Hast und Jugend des Fremden weckte in dem Italiener merkantilische Gelüste und abenteuerliche Forderungen, da kam er aber bei Grundling übel an, welcher sogleich ein so heftiges Gezänk darüber anfing und mit solchem Geschrei fortsetzte, daß sie in einigen Stunden, ganz erschöpft, endlich doch noch um einen leidlichen Kaufpreis einig wurden. Fortunat hatte erst kürzlich bedeutende Wechsel aus Deutschland bezogen, sie reichten eben hin, die Summe und eine genügsame Weiterreise notdürftig zu decken. Mit bewundernswürdiger Beharrlichkeit und Resignation trieb er nun das Geschäft, wie einen Kreisel, unausgesetzt zum Ausgange und endigte damit, den hocherfreuten Grundling zum Schloßwart seines neuen Besitztums einzusetzen.

Kaum aber hatten sie den Garten wieder erreicht, da erscholl im Hofe schon der fröhliche Klang eines Posthorns. Fortunat hatte seinen Wagen hierher bestellt, aus den früheren Gesprächen mit dem alten Marchese glaubte er zu ahnen, wohin er sich gewendet. Und als er nun endlich tief aufatmend draußen in den prächtigen Abend hineinfuhr, blühten alle Gärten und ein Regenbogen stand über der Gegend, als müßte nun alles, alles wieder gut werden.

Drittes Buch

Neunzehntes Kapitel

Auf dem fürstlichen Jagdschlosse, wo im vorigen Jahre alles so bunt und fröhlich war, sieht es jetzt ganz anders aus. Die Vögel picken frühmorgens auf der marmornen Treppe zwischen den Säulen, ein lässiger Gärtnerbursch dehnt sich in der Morgenkühle und schickt sich an, die verschlungenen Gänge notdürftig in Ordnung zu bringen, die überall blühend verwildern. In der alten Pracht funkeln die Sommernächte wieder über den stillen Grund, aber keine Gitarren erklingen mehr, nur die getreuen Nachtigallen schlagen wie damals in den Gebüschen, als klagten sie noch um Juannas verlorene Schönheit.
Der Fürst gedachte nicht mehr des Schlosses, er war selber lange verwildert. Zwischen Genuß und Reue, Lust und Grauen war er allmählich immer tiefer hinabgestiegen in die schimmernden Abgründe, wo mit verlockendem Gesang die Nixen im Mondenscheine auf den Klippen ihr feuchtes Haar kämmen, das ferne Wetterleuchten der Religion verwirrte ihn nur noch mehr; so hatte er sich im schönen Leben verirrt und konnte sich nicht wieder nach Hause finden. Da schlug die himmlische Liebe ihren Sternenmantel um den Todmüden. Er verfiel in eine schwere Krankheit, und als er wieder genas, war auf einmal alles vorbei. Die Leute nannten ihn wahnsinnig, er aber war vergnügt und blätterte Tag für Tag mit stiller, herzlicher Lust in den alten Bilderbüchern, die er als Kind gelesen; alles andere hatte er vergessen. Sie hatten ihn endlich in einem entlegenen Flügel des Schlosses absondern müssen von der Welt, die er nur noch wie im Traume von fern sah, nur die unschuldigen Vögel sangen alle Morgen vor seinen Fenstern von der alten Zeit, daß er oft erschrocken von seinen Bildern aufhorchte. – Aus seiner Hand aber hatte

die Fürstin rasch die Zügel des Regiments ergriffen und lenkte keck, die Rosse peitschend, in die neue Freiheit hinaus.
In dieser Zeit kam *Lothario* eines Abends einsam von dem Gebirge herab. Wir wissen nicht, wohin er wanderte, sein Weg führte ihn durch die Stadt. Der Mond trat manchmal heimlich lauernd zwischen den Wolken hervor, da lag die alte Residenz unten wie eine Ruine phantastisch in der schwülen Nacht umher, es war schon alles still, nur ein Mädchen sang noch zur Gitarre aus einem Garten drüben, und die Nachtigallen schlugen von den Bergen.
Er kehrte in einem wenig besuchten Gasthause ein, das draußen auf einer Anhöhe lag und eine weite Aussicht über die Stadt hatte. Dort mußte er lange pochen, eh' jemand erschien. Ein alter Diener sagte ihm endlich, es sei alles in die Stadt gezogen, wo heute zum Geburtstage der Fürstin ein großes Fest gegeben werde. – Lothario nahm nun im oberen Stockwerk einen Saal in Besitz und öffnete rasch alle Fenster. Die prächtige Nacht duftete fast berauschend herauf. Er ließ Licht und Wein bringen, er fühlte seit langer Zeit wieder einmal eine echte Lust zu dichten. – Als er sich aber so einsam hinsetzte und hastig trank und schrieb, da war's ihm, als riefe es durch die Stille seinen Namen, erst leise, dann lauter, und der Teufel sähe ihm beim Schreiben über die Schulter und flüsterte zu ihm: Nur zu, nur zu! die unschuldige Welt mit vornehmen Worten belogen und verführt, ich will dich dafür auf die Zinnen des Ruhms stellen, und die Welt soll dir huldigen! – Er sprang auf und erschrak, als er sich flüchtig in einem Wandspiegel erblickte, so bleich und wüst sah er aus. Da streifte der Wind klingend die Saiten einer Gitarre, die am offenen Fenster lag. Der Mond aus blassen Wolken beschien soeben wieder die stillen Bäume und unten die alte Stadt. Er trat mit der Gitarre ans Fenster und sang:

„Lieder schweigen jetzt und Klagen,
Nun will ich erst fröhlich sein,

All mein Leid will ich zerschlagen
Und Erinnern – gebt mir Wein!
Wie er mir verlockend spiegelt
Sterne und der Erde Lust,
Stillgeschäftig dann entriegelt
All die Teufel in der Brust,
Erst der Knecht und dann der Meister
Bricht er durch die Nacht herein,
Wildester der Lügengeister,
Ring mit mir, ich lache dein!
Und den Becher voll Entsetzen
Werf' ich in des Stromes Grund,
Daß sich nimmer dran soll letzen,
Wer noch fröhlich und gesund!

Lauten hör' ich ferne klingen,
Lust'ge Bursche ziehn vom Schmaus,
Ständchen sie den Liebsten bringen,
Und das lockt mich mit hinaus.
Mädchen hinterm blühnden Baume
Winkt und macht das Fenster auf,
Und ich steige wie im Traume
Durch das kleine Haus hinauf,
Schüttle nur die dunklen Locken
Aus dem schönen Angesicht!
Sieh, ich stehe ganz erschrocken:
Das sind ihre Augen licht.
Locken hatte sie wie deine,
Bleiche Wangen, Lippen rot –
Ach, du bist ja doch nicht meine,
Und mein Lieb ist lange tot!
Hättest du nur nicht gesprochen
Und so frech geblickt nach mir,
Das hat ganz den Traum zerbrochen,
Und nun grauet mir vor dir.
Da, nimm Geld, kauf' Putz und Flimmern,
Fort und lache nicht so wild!

O ich möchte dich zertrümmern,
Schönes, lügenhaftes Bild!

Spät von dem verlornen Kinde
Kam ich durch die Nacht daher,
Fahnen drehten sich im Winde,
Alle Gassen waren leer.
Oben lag noch meine Laute,
Und mein Fenster stand noch auf,
Aus dem stillen Grunde graute
Wunderbar die Stadt herauf.
Draußen aber blitzt's von weiten,
Alter Zeiten ich gedacht',
Schauernd reiß' ich in den Saiten,
Und ich sing' die halbe Nacht.
Die verschlafnen Nachbarn sprechen,
Daß ich nächtlich trunken sei –
O du mein Gott! und mir brechen
Herz und Saitenspiel entzwei! –"

Es blitzte wirklich von weitem, aber es waren nur einzelne Raketen, die von Zeit zu Zeit fern über dem dunklen, fürstlichen Parke lustig aufstiegen. Da fiel ihm das Fest wieder ein, von dem der alte Diener vorhin sprach, er beschloß, selbst noch hinzugehn.
Lässig schlenderte er durch die lange Vorstadt; bis dorthin war das Fest nicht gedrungen, die kleinen Häuser standen still und dunkel, nur wenige Laternen flackerten im Winde, der Nachtwächter schickte sich eben an, die zehnte Stunde auszurufen; von fern aber über die hellbeleuchteten Dächer und Schornsteine qualmte ihm schon der trübrote Schein der Illumination entgegen wie die aufgehende Sonne an einem nebeligen Herbstmorgen. So war er ans Theater gekommen. Durch ein hohes, verhangenes Fenster glaubte er drin die Schauspieler mit aller Gewalt der Leidenschaft pathetisch deklamieren zu hören; ihn schauerte, so kühl und nüchtern war es dagegen hier draußen.

Eine lange Reihe von Wagen, auf ihre Herrschaften wartend, standen an der finstern Mauer, die Kutscher schlummerten auf ihren hohen Kutschböcken, der eine zog gähnend seine Taschenuhr heraus und hielt sie an den ungewissen Schein der Laterne. „Was Teufel spielen sie denn heut so lange?“ fragte er einen Kerl, der eben an einem Eckpfeiler seine Fackel putzte, daß die Funken auf einen Augenblick das ganze langweilige Chaos wunderlich beleuchteten. Dieser nannte ein bekanntes Stück von Grafen Victor von Hohenstein. – Da fuhr Lothario unwillkürlich zusammen. Er ging rasch hinein, ein gutes Trinkgeld verschaffte ihm von dem verwunderten Logendiener noch einen Platz in der Fremdenloge.

Das Haus war prächtig erleuchtet und zum Erdrücken voll, aus der fürstlichen Loge zwischen den reichen Vorhängen blitzte und schimmerte es von Sternen, Lichtern und schönen Frauenaugen blendend herüber. Das Stück war fast zu Ende. Es war seltsam genug, eben Juannas frühere Geschichte in Spanien, alle wilden Waldbäche der Leidenschaft stürzten in dieser letzten Szene wie in einem mächtigen Strom zusammen. Die Schauspielerin, welche Juanna vorstellte, hatte, vielleicht bewußtlos, nach und nach das ganze Wesen der Gräfin angenommen: ihre frische Waldkühle, ihre Stimme, das strenge schöne Gesicht, so funkelte sie mit den dunklen Augen gerade auf Lothario herüber. – Lothario sprang erschüttert auf, eine Totenstille herrschte im ganzen Hause. Da auf einmal beginnt ein Flüstern unten, es wächst und steigt allmählich durch alle Reihen der Zuschauer, viele Köpfe und immer mehrere wenden sich erstaunt nach Lothario herum. – „Was gibt's da?“ frägt die Fürstin, sich weit aus ihrer Loge hervorlehnend. – Ein Kammerherr drängt sich eilig vor, auf Lothario deutend: „Dort, der Dichter selbst, sie haben ihn erkannt, Graf Victor von Hohenstein.“ – „Der?!“ – entgegnete die Fürstin und sinkt verwirrt auf ihren Sessel zurück.

Unterdes war der Vorhang gefallen, ein wütender App-

laus brach plötzlich los, sich immer wieder erneuernd. Den Grafen Victor aber – denn er war es wirklich – erfaßte ein seltsames Grauen vor dem hohlen Sturm des Beifalls, er sah noch einmal dazwischen einen sengenden Blick der Fürstin nach ihm herüberschießen, dann stürzte er entsetzt über die noch leeren Treppen ins Freie hinaus.
Mit welchen Gedanken sah er nun den weiten, gestirnten Himmel wieder! Die plötzliche Erinnerung an die Zeit, wo er das Stück geschrieben, versenkte seine ganze Seele wie in ein Meer von Wehmut. Auf dem Gebirge in Spanien, als er an jenem stillen Abend, im Walde auf den Franzosen St. Val zielend, zum ersten Male Juanna erblickte, da war's ihm, wie in die Sonne zu sehen – sie war schon lange untergegangen, aber Wald und Berge schimmerten und sprühten noch in wunderbaren Funken –, damals dichtete er das Schauspiel von der wilden Gräfin. Da dachte er nicht, daß es so kommen würde! Und als es dann Friede und alles wieder still und nüchtern wurde, kehrte auch er nach Deutschland zurück, und der Frühling und das Grün der wechselnden Landschaften breiteten sich wie ein Schleier milde über das schöne Bild im Herzen. Aber nach der ernsten, bewegten Zeit, in der er ehrlich gerungen, kam ihm zu Hause nun alles so klein und unbedeutend vor, ihm war wie einem Schiffer nach langer, stürmischer Fahrt, der den Boden unter sich noch immer wanken fühlt und aus dem Wirtshaus am Ufer sehnsüchtig wieder in den kühlen Wogenschlag hinaussieht. In solcher Laune war er nach kurzem Umhertreiben, um sich von der guten Gesellschaft zu erholen, zum Teil auch aus grillenhafter flüchtiger Neigung zu Kordelchen, unerkannt unter dem Namen Lothario mit der Schauspielerbande ausgezogen, wo wir ihn in jener regnerischen Nacht zum erstenmal trafen. – Hier hörte er plötzlich, daß die verlorengeglaubte Gräfin Juanna noch lebe und zu der ihr verwandten fürstlichen Familie geflüchtet, mit der sie auf dem nahen Jagdschlosse sich aufhalte. Da gab's auf einmal frischen Klang! Sein Plan war gleich gemacht. Durch seine

geheime Vermittlung erfolgte die Einladung der Schauspielergesellschaft nach dem Jagdschlosse, er begleitete sie in seiner Verkleidung, denn es schien ihm lächerlich, ja sinnlos, um diese märchenhafte Diana auf dem gewöhnlichen Paradepferde gräflicher Galanterie zu freien. – Bei seiner eigenen, sorglosen Unvorsichtigkeit konnte indes die Sache nicht ganz verborgen bleiben, der Fürst und seine Gemahlin wenigstens hatten unbestimmte Kunde von seinem Vorhaben, noch ehe die Truppe bei ihnen ankam. Insbesondere hatte die Fürstin, mit dem den Frauen in solchen Dingen eigentümlichen Scharfsinn, die eigentliche Absicht gar wohl erraten. Zwar erwarteten sie täglich den Baron Manfred auf dem Schlosse, den sie insgeheim zu Juannas Bräutigam ausersehen. Dennoch konnten sie's nicht lassen, die interessante Genialität einer so romantischen Maskerade um so leichtsinniger zu begünstigen, da im schlimmsten Falle Victor noch immer als eine bessere Partie für die unbemittelte Gräfin erschien als der etwas unscheinbare Manfred. So schwiegen sie recht mit innerlicher Lust und spielten die Getäuschten, täuschten aber unbewußt nur sich selbst, indem sie den zufällig dazwischengekommenen *Fortunat*, da er gleich von Anfang so rätselhaft auftrat, für den heimlich erwarteten Grafen hielten. – Victor aber verlockte indes Juannas Schönheit nach und nach immer tiefer in das wildeste Labyrinth ausschweifender Wünsche, er gab ihren herausfordernden Blicken eine Deutung, die sie selber niemals kannte. Da hörte er auf der Jagd zum erstenmal von der nahen Ankunft des unbekannten Bräutigams – es war ihm unerträglich: er entschloß sich rasch, Juanna zu entführen, nur so, meinte er, könne diese wilde Nymphennatur bezwungen werden, gleichwie eine still aufsteigende Flamme sich plötzlich entfaltet, wenn der Sturm sie zerwühlt. – „Ja, kühne, schlanke Flamme!" sagte er nun tausendmal zu sich selbst, „wie griffst du plötzlich zornig in die Waldesnacht und klettertest furchtbar schön die Felswand auf und nieder, daß alle Wipfel donnernd in die Gluten san-

ken! Die lust'gen Wälder meiner Jugend sind verbrannt."
In solchen Gedanken war *Victor* jetzt durch mehrere Straßen fortgeschritten. Die Wagen rasselten aus dem Theater, der hoffärtige Patriotismus kokettierte aus tausend geputzten Fenstern, Kinder zogen in dem magischen Lichte lärmend durch die Gassen und brachten jedem brennenden Teertopf ein Vivat. Wohin er sich wandte, immer neue Feueralleen zogen sich durch die Nacht, bis er endlich unerwartet an den fürstlichen Garten kam. Ein Feuerwerk, wie es schien, war eben abgebrannt, nur einzelne Schwärmer stiegen noch empor und erleuchteten im Zerplatzen seltsam die Gegend und die verworrene Menge, die sich nun jauchzend nach allen Seiten verlief. Bei dem flüchtigen Widerscheine glaubte Victor auf einen Augenblick sein Wirtshaus jenseits auf der stillen Anhöhe gesehen zu haben. Der Wege unkundig an dem fremden Orte, schlägt er die nächste Richtung ein und tritt durch ein Pförtchen, das er nur angelehnt findet, zwischen die Bäume hinein, verschlungene Gänge führen ihn immer weiter, auf einmal sieht er sich mitten im fürstlichen Park. Der Himmel ist schwül bezogen, zahllose Glühwürmchen schweifen in den dunklen Gängen, die weißen Statuen stehen einsam im Mondenscheine umher; da ist's, als hört' er leise seinen Namen nennen, ein Flüstern geht seitwärts durchs Gebüsch, dann alles wieder still. – Jetzt schimmern auch die hohen Schloßfenster schon herüber, drin sieht er im hellen Glanze sich Masken wundersam bewegen, die eine Saaltür öffnet sich, ein Schwall von Licht und Klängen schlägt heraus – da fährt er innerlichst zusammen, denn bei dem brennenden Streiflichte sieht er plötzlich *Juannas* Gestalt zwischen den Bäumen entschlüpfen. Außer sich folgt er nach, er erblickt sie von neuem: Reitkleid, Gürtel und Hut, wie sie in Spanien getragen, endlich erreicht er sie, sie wendet sich rasch, mit Grauen sieht er in die dunklen Augenhöhlen einer Larve.
Er steht wie angewurzelt vor ihr, während sie ihn schweigend zu betrachten scheint. – „Du fernes Wetterleuch-

ten“, sagt er endlich ganz verwirrt, „ich folge dir, und wär’ es in den Wahnsinn!“ – Da erhebt sich auf einmal tiefer im Garten ein wunderbarer Gesang, fast ohne Melodie, in wenigen herzzerreißenden Tönen. Sie schauert, als bräch’ der Tag an, ihre schwarzen Locken ringeln sich von beiden Seiten herab, er sieht die dunklen Augen aus der Larve funkeln. – „Morgen!“ flüstert sie dann kaum hörbar und verschwindet schnell zwischen den wechselnden Schatten.

Victor aber flieht entsetzt durch den Garten, der Mondenschein wiegt sich träumend auf dem Gebüsch, seitwärts schwanken Wasserkünste im Winde, wie Feen in langen, wallenden Schleiern. Plötzlich hört er den Gesang wieder erschallen. Auf dem steinernen Rande des Springbrunnens sieht er einen eingeschlummerten Mann sitzen, ohne Hut, mit dem Haupte vornübernickend, der singt im Schlaf. Bei einem flüchtigen Mondblick glaubt er den bleichen, kranken Fürsten zu erkennen.

So kommt er ganz verstört in die Stadt zurück. Dort hat sich unterdes alles verwandelt. Nur einzelne Menschen irren noch beim ungewissen Scheine der Laternen, die verlöschend flackern, zerrissene Wolken fliegen über die Dächer, die Nacht war finster und stürmisch geworden. Da schweiften zwei weibliche Gestalten eilig durch das Dunkel. „Wo schleppst du mich hin?“ fragte die eine. – „Sahst du ihn nicht vorhin?“ entgegnete die andere, „ich muß ihn haschen!“ –

„Kordelchen! du?“ rief Victor plötzlich vor ihnen stehend aus – „du siehst ja so blaß im Laternenschein, wie eine Leiche mit spielenden, funkelnden Augen.“ – „Ach, dummes Zeug’, red’ nicht so graulich“, sagte die Komödiantin. – Er wollte fort, aber sie hatte sich schon fest in seinen Mantel verwickelt.

Sie standen an der offenen Tür eines kleinen Hauses. Ihre leichtfertige Begleiterin, die zu ihrem Verdrusse noch gar nicht beachtet worden, wünschte schnippisch viel Vergnügen und verließ sie empfindlich. Kordelchen aber hat-

te ihren späten Gast bereits hineingedrängt. Ein schwüler Duft von halbvertrockneten Blumensträußen, die an den Fenstern standen, quoll ihnen aus der kleinen Stube entgegen. Das tief heruntergebrannte Licht, dem eine leere Flasche zum Leuchter diente, verbreitete eine ungewisse Dämmerung über ärmliches Hausgerät, zerbrochene Spiegel, Notenbücher und Kleidungsstücke, die überall unordentlich umherlagen. Mitten in dieser Verwirrung war ein wohlgekleideter Mann am Tische fest eingeschlafen, die Feder lag umgefallen noch zwischen seinen Fingern auf dem halbbeschriebenen Blatte vor ihm.

„Still, still, der wird ein Paar Augen machen!“ sagte Kordelchen, indem sie Victor leise an der Hand in einen entfernten Winkel führte und ihn dabei, eh' er sich's versah, herzhaft in den Finger biß. Dann setzte sie sich auf einen Reisekoffer, öffnete ihre Schürze, die voll Knackmandeln war, und fing vernügt an zu naschen und zu plaudern, man sah ihr recht die Freude aus den muntern Augen glänzen. So in aller Geschwindigkeit erzählte sie ihm, daß sie mit Otto aus dem langweiligen Italien entflohen, seit einigen Tagen hier sei und wieder aufs Theater wollte. Auf einmal sah sie Victor lange ins Gesicht. „Armer Lothario“, sagte sie, „du siehst schlecht aus. Dacht' ich's doch gleich, als du damals die Augen so hoch warfst, siehst du, wer hieß dich Gemsen jagen! – Aber so iß doch mit – und hast du die Fürstin heut gesehen? – sie ist als Gräfin Juanna maskiert.“ – Dazwischen warf sie wieder Mandelschalen nach dem Schreiber hinüber, der noch immer schlief.

Da fuhr dieser erschrocken auf – es war Otto – sie wollte sich totlachen, wie er so wild aus dem Schlafe umherstierte. Aber Victor, bisher wie in Gedanken verloren, hatte sich bei dem unerwarteten Anblick des wüsten Gesichts plötzlich aufgerichtet. „Um Gottes willen, Otto!“ rief er mit tief erschütterter Stimme, „flieh, flieh in die Nacht hinaus, in den Krieg, bau' das Feld, spalte Holz, bettle von Haus zu Haus – nur fort von hier!“ – „Geh, geh!“ sagte Kordelchen, von ihrem Koffer springend, „du bist ja so

pathetisch wie der steinerne Komtur aus dem Don Juan." – Otto, den Kopf auf beide Arme gestützt, ahnet heimlich, was jener meint, Lotharios Urteil gilt ihm alles, seine ganze Seele hängt lauschend wie an einem jähen Absturz. – Aber Victors Sinn war heut wie ein schneidendes Schwert. – „Und red' mir nicht von Poesie, von Dichterberuf", fuhr er fort, „du hast nicht mehr davon, als ein verliebtes Mädchen. Es gibt nur wenige Dichter in der Welt, und von den wenigen kaum einer steigt unversehrt in diese märchenhafte, prächt'ge Zaubernacht, wo die wilden, feurigen Blumen stehen und die Liederquellen verworren nach den Abgründen gehen, und der zauberische Spielmann zwischen dem Waldesrauschen mit herzzerreißenden Klängen nach dem Venusberg verlockt, in welchem alle Lust und Pracht der Erde entzündet, und wo die Seele wie im Traum frei wird mit ihren dunklen Gelüsten." –

Hier hielt sich Otto nicht länger. Es überlief ihn eiskalt, als zuckte ein Blitz durch die Nacht und erleuchtete auf einmal gräßlich sein ganzes verlorenes Leben. Noch ganz verwirrt, im Innersten getroffen, ergriff er wie ein Rasender einen nahegelegenen Theaterdegen und drang sinnlos auf Victor ein. Dieser schleuderte den Wütenden weit von sich, daß ihm der Degen entfiel, „Ruhig!" rief er, „und bedenke meine Worte, ehe alles zu spät! Mich aber laß, ich habe mit mir selbst zu fechten, Gott gnad' uns beiden!" – So eilte er aus dem Hause fort.

Draußen auf der leeren Gasse hörte man noch Kordelchen klagen, die ihm betroffen nachgestürzt. „Lothario!" rief sie außer sich, „lieber, schöner, verrückter Lothario! ich bitt' dich um Gottes willen, kehre um, nur noch ein einziges Mal komm zurück! Es ist ja alles nicht wahr, was die Leute sagen, ich war dir immer im Herzen treu, was kann ich dafür, daß ich arm und schön bin? Ach, verlaß mich nicht, ich habe sonst niemand auf der Welt! Wickle mich ins Schnupftuch, steck' mich in deine Rocktasche, wenn du mir nicht traust, ich will stillsitzen und dich ansehen,

wenn du mich nur wieder liebhast, du wilder, abscheulicher Kerl!" – So bat sie rührend, lachte und schimpfte, bis sie zuletzt unaufhaltsam in heftiges Weinen ausbrach.
Aber Victor hörte sie nicht mehr. Er trat aus dem dunklen Stadttore, einzelne Morgenstreifen zuckten schon über die stille Gegend. – Durch seine Seele gingen übermächtige Gedanken. Aus der tiefen Nacht seines Grams stieg allmählich Stern auf Stern, ihm war, als müßt' nun alles anders werden.

Zwanzigstes Kapitel

Zu Weinsheim klangen die Abendglocken über die anmutige Gegend, das reiche Dorf mit seinen frischen, kühlen Gärten und dem weißen herrschaftlichen Schlosse darüber lag schon vom Gebirge verschattet, während die Abendsonne weiterhin die fruchtbare Ebene und den gewundenen Strom noch heiter beleuchtete. Auf allen Feldern war ein fröhliches Erntegewimmel, bis weit hinaus hörte man singen, rufen und jauchzen und das Rasseln der Wagen dazwischen. Mitten durch die bunte Wirrung ritt ein schöner, schlanker Mann mit gebräuntem Gesichte langsam dem Schlosse zu, nach allen Seiten für den folgenden Tag Befehle erteilend und manchem scheuen, glänzenden Blick der Bauernmädchen begegnend. Es war der junge Baron Manfred, dem diese Landschaft in doppeltem Sinne angehörte, denn er hatte sie wüst geerbt und durch Umsicht und verständige Anregung in einen blühenden Garten verwandelt.
In solcher Erntezeit haben die Landschlösser etwas unbeschreiblich Einsames. Auch Manfred fand Hof und Haus noch leer, alle Diener schwärmten noch draußen im Tale, nur die gegenüberstehenden Waldberge schauten ernst durch die offenen Fenster herein. – Ermüdet setzte er sich auf das Fenstergeländer, um sich in der Abendkühle zu er-

frischen, als er auf der Straße, die vom Gebirge kam, einen wunderlichen Zug sich zwischen den Walnußbäumen langsam heranbewegen sah. Ein elegantes Kabriolett, das aber der Steinweg übel zugerichtet zu haben schien, wurde auf drei Rädern von einem Pferde mühsam fortgeschleppt. Ein Mann in seltsamer Reisetracht führte das Pferd am Zügel, eine junge Dame, mit einem Kornblumenkranz im Haar, schlenderte daneben, das Ganze gemahnte an ziehende Komödianten. Einige verspätete Jäger des Barons hatten sich dazugesellt, die Krüppelfuhre, wie es schien, mit derben Witzen gesegnend. Der Reisende aber blieb keine Antwort schuldig. Manfred konnte, da sie eben unter seinen Fenstern vorüberzogen, deutlich vernehmen, wie er den Jägern sehr eifrig demonstrierte, bei ihrer Kunst sei, außer den Frischlingen, nichts Frisches mehr, das Elend hätten sie aus den Wäldern verjagt und hegten's zu Hause, von der Blume des Ganzen dürfe man vor gebildeten Löffeln gar nicht mehr sprechen, überdies sei Diana längst eine alte Jungfer geworden, es lohne nicht mehr, Hörner zu tragen. – So kamen sie alle mit großem Rumor und Gelächter oben an.
Hier warf der Fremde dem Jäger die Zügel zu, und befahl ihm ohne weiteres, alles aufs beste unterzubringen, das Wägelchen wiederherzustellen und das Pferd reichlich zu füttern, das heute mehr die Sonne als der Hafer gestochen habe. „Das sieht hier gar nicht schlecht aus", sagte er dann, sich zufrieden nach allen Seiten umsehend, „wem gehört das Schloß?" – Die Antwort des Jägers aber schien ihn aufs höchste zu überraschen. „Was! dem Baron Manfred?" rief er aus und flog sogleich nach dem Schlosse, wo er den eben heraustretenden Baron beinahe übergerannt hätte. – „Waren Sie", sagte er hastig und ohne alle Einleitung, „waren Sie nicht vor einiger Zeit auf Reisen? So sind Sie ohne Zweifel der gewesene Bräutigam der ehemaligen Gräfin Juanna, der damals auf dem fürstlichen Schlosse erwartet wurde!" – Manfred bejahte kurz und trocken. – „Aber heiraten!" rief der Reisende aus, „wer wird eine

wildschöne Diana gleich heiraten wollen!" – „Wer sind Sie?" unterbrach ihn Manfred, den Aufdringlichen mit etwas ernsten Blicken messend. – „Ja, so!" – erwiderte dieser –, „haben Sie vielleicht schon einmal von einem gewissen Dryander gehört?" – „Dem bekannten Dichter?" – „Der bin ich, ich reise eben auf Volkslieder, und jenes Frauenzimmer dort ist meine Frau."

Nun stellte er die junge Dame mit dem Kornblumenkranze vor, die soeben an einem Eckstein noch ihre Schuhe festband und ihnen, als sie sich nennen hörte, ein munteres, etwas trotziges Gesichtchen zuwandte, in dem wir sogleich Fräulein Gertrud als alte Bekannte vom fürstlichen Schlosse wieder begrüßen. Die Kleine begann unmittelbar nach der ersten Verständigung, mit der Lebhaftigkeit eines jungen Sinnes, dem alles noch neu ist, von ihrer romantischen Fahrt durchs Gebirge, von dem Unfall mit dem Wagen und andern Abenteuern zu erzählen, wobei sie deutlich merken ließ, daß dem Baron eigentlich ein unverdientes Glück widerfahre, den berühmten Dichter Dryander bei sich beherbergen zu können. Der letztere aber, dem die Beschreibung zu schön und zu lang zu werden schien, war schnell wieder in den Hof zurückgeeilt, um Pfeife und Tabaksbeutel aus dem Wagen zu holen. – Und so sah sich denn Manfred allein mit der hübschen, jungen Frau in einer seltsamen Lage; denn wenn er sie, nach ihrer ganzen Erscheinung, als ein lebenslustiges, verliebtes Landfräulein zu nehmen geneigt war, so wandelte sie nun auf einmal die Farbe und brach, zu seiner Verwunderung, ästhetische Diskurse vom Zaun. Und je länger er schwieg, je fröhlicher geriet sie, in der sichtbaren Lust, dem Landjunker zu imponieren, wie ein munterer Wasserfall unaufhaltsam in eine plauderselige Gelehrsamkeit, unbekümmert Zeiten, Autoren und Bücher durcheinander vermengend.

Ein Lachen hinter ihnen unterbrach hier plötzlich die sonderbare Unterhaltung. Es war Dryander, der sich unterdes wieder eingefunden und eine Zeitlang ungesehen alles

mit angehört hatte. „Trudchen, Trudchen!" rief er immerfort lachend, „was geschieht dir? ich erkenne dich gar nicht wieder – dieses charmante Wesen und angenehme Klugsprechen, Attitüden und romantischer Schaltanz." – Das resolute Weibchen aber schien nicht einen Augenblick betreten. Mit veränderter Stimme, die plötzlich wie der Absatz eines Pantöffelchens klang, erwiderte sie: „Solche Faxen leid' ich nun ein für allemal nicht von dir! Willst du ein Philister sein, so ist's gut, ich werde auch sein, wie ich Lust habe!" – Dryander hatte sie unterdes umfaßt und walzte mit ihr auf dem Rasen herum. Sie aber schrie auf einmal laut auf und riß sich mit mehr Heftigkeit als Grazie von ihm los. „Du bist immer so ungeschickt", sagte sie, „du trittst mir auf den Fuß." – „Das ist nicht wahr", rief Dryander. – „Wahr oder nicht wahr!" – entgegnete sie, „ich bin todmüde von deinem Herumziehen in dem dummen Gebirge, und ich will schlafen gehn, und das jetzt gleich!" – Nun geriet Dryander seinerseits in eine wunderliche Wut. „Um Gottes willen, nur keine Launen!" rief er aus, „Weiberlaune ist mir zuwider wie das Pech am Pfropfen einer Champagnerflasche, ein ekelhafter Meltau auf Blumen, da ist offenbarer Wahnsinn noch herrlich dagegen mit seinem Abgrunde bodenloser Gedanken." – „Und ich gehe doch schlafen!" unterbrach ihn Gertrud trotzig, machte Manfred eine kurze Verbeugung und ging nach dem Schlosse, wo die alte Haushälterin des Barons, die den Spektakel in der Haustür verwundert mit angehört hatte, die Erhitzte aufnahm und in ihre Zimmer führte.

„Ist sie nicht zum Küssen schön, wenn sie böse wird?" sagte Dryander zu Manfred gewandt. Manfred, ganz entrüstet über diese verkehrte, nichtsnutzige Wirtschaft, stellte ihn ernstlich zur Rede, daß er durch solche Tollheiten die Frau geistig vernichte. – „Ganz und gar nicht", erwiderte Dryander, „faule Naturen werden erst in der Leidenschaft bedeutend und reizend, sie ist eigentlich sehr dumm." –

Unterdes war ein Tisch mit Erfrischungen im Garten aufgeschlagen worden. Dryander nahm ohne weiteres Platz, band sich eine Serviette unterm Kinne wie zum Rasieren vor, und begann so eifrig zu essen, wie Manfred noch niemals gesehen. Dazwischen erzählte er, von allen Schüsseln zugleich zulangend, wie in seinem Bräutigamsstande auf dem fürstlichen Jagdschlosse seine Aversion gegen eine feierliche Hochzeit ein unübersteigliches Hindernis geworden, wie er sodann einmal plötzlich vor dem hochaufgestapelten Hochzeitsbette erschrocken und davon gegangen, Gertrud aber bald darauf aus Melancholie gleichfalls von dem Schlosse verschwunden sei.
„Aber auf dieser außerordentlichen Flucht“, fuhr er fort, „setzte mir die Liebe nicht wenig zu, ich kam ganz herunter, ich war fast nichts als Seele. In diesem Zustand hatte ich mich einmal des Abends im Gebirge verirrt, ich wußte durchaus nicht, wo ich mich befand, und war endlich, wie es mir vorkam, über die Trümmer eines umgefallenen Zaunes in einen ehemaligen französischen Garten geraten. Durch die schnell vorüberfliegenden Wolken fielen nur einzelne Mondblicke zwischen finstern Laubwänden und künstlich verschnittenen Taxusbäumen über zerbrochene Statuen, die im hohen Grase lagen, aus dem Walde schlugen unzählige Nachtigallen. Nur eine Statue in einiger Entfernung von mir schien noch wohlerhalten, es war eine sitzende Najade an einem steinernen Bassin, dessen klare Flut ihre Füße umspülte. – Ich bin eigentlich ein Schwärmer, mit über der Brust gekreuzten Armen lehnte ich mich nachlässig an einen neben mir stehenden antiken Opferaltar, und sah eben unverwandt in den Mond, als der morsche Altar, den ich für Stein gehalten, hinter mir zusammenbrach. Daß ich mit umfiel, war das Geringste dabei. Aber denkt Euch mein Entsetzen! Über dem Gepolter wendet die Najade auf einmal den Kopf, richtet sich hoch auf, und entflieht in den dunklen Garten. Trotz meiner Gänsehaut schreite ich doch auf das Bassin los, und finde zwei der zierlichsten Pantöffelchen auf dem stei-

nernen Rande. Ich lege sie sogleich an mein Herz zwischen Frack und Weste und komme beim weiteren Vordringen an einen, von hohen Bäumen tief verschatteten Platz. Auf dem Platz war ein Schloß, und an dem Schlosse ein Altan und auf dem Altane sehe ich, wie hinter einem Schleier von Mondenschein, Blüten und Laubgewinden das weiße Gewand der Najade wieder hervorschimmert. Das kam mir auf einmal ganz spanisch vor mit dem Balkone, ich redete sie erst zierlich in Assonanzen an, sie verbarg sich halb furchtsam, halb neugierig, bald sah ich eine Locke, bald ein bloßes Füßchen, bald einen Arm, bald wieder gar nichts. Ich werde immer verliebter, die Reime flossen mir wie Lavendelwasser, ich sprach von des Mondes Zaubermacht, der das Lieben hat erdacht, von einer süßvalenzschen Nacht, vom Kosen und vom Flüstern sacht, bis daß die erste Lerche erwacht! Sie schwieg immerfort, und wie auf der Himmelsleiter meines eigenen Wohllauts stieg ich endlich ohne weiteres auf den nächsten Baum, schwang mich mit der einen Hand auf den Balkon und hielt mit der andern der Erstaunten ihre Pantöffelchen entgegen. In demselben Augenblicke aber entriß sie mir's plötzlich und schlug mir damit tüchtig um beide Ohren. ‚Also das ist deine Treue!' rief sie, ‚ich erkannte dich gleich anfangs, o ich unglückliches Mädchen!' – es war Gertrud selbst. Ich stand ganz verblüfft. Vergeblich sagte ich, daß ich sie eigentlich auch gleich anfangs erkannt hätte, und beschwor sie, nur jetzt das Maul zu halten. Aber sie glaubte und hörte nichts, sie schimpfte und weinte dazwischen immerfort. Über dem Lärm und Gezänke streckte die alte Amme, die ich noch vom fürstlichen Schlosse her kannte, ihr Gesicht aus der Schloßtür und verschwand sogleich wieder, ein großer Hund schlug im Garten ein paarmal an, und eh' ich mich noch besinnen kann, tut sich die Balkontür weit auf, und ein verworrener Haufen von Vettern, Lichtern und Dienern stürzt plötzlich hervor, voran ein großer, starker Mann in einem damaste-

nen Schlafrocke, mit kleinem dreieckigen Hute und langem Haarzopfe, in der einen Hand eine Pistole, in der andern einen bloßen Degen. Die alte Amme, der vor den Folgen ihres Verrats bange wurde, wollte den Wütenden von hinten am Zopfe aufhalten, darüber ringelte sich das Band los, und die langen Haare umflatterten ihn wunderlich wie ein phantastisches Hirngespinst. ‚Kopuliert sie in drei Teufels Namen!' donnerte er, mit der Pistole nach mir zielend, denn es war niemand anders als Gertruds Vater. Ein alter Geistlicher, der nicht wußte, wie ihm geschah, trat aus dem Gefolge, und ich und Gertrud wurden auf der Stelle kopuliert." –

Hier stand Manfred, der schon mehrere Male den beredten Dichter unterbrechen wollte, entrüstet auf. „Schändlich!" sagte er, „mich friert innerlichst bei der Geschichte." – Dryander sah ihn mit den geistvollen Augen ein Weilchen groß an, dann sprang er plötzlich auf und fiel dem Baron um den Hals. „Sie haben ganz recht", rief er aus, „das ist die verruchte Doppelgängerei in mir, ich kann nichts Großes ersinnen, ohne ihm sogleich von hinten einen Haarbeutel anzuhängen, ein tragischer, wahnsinniger König und ein Hanswurst, der ihm fix ein Bein unterstellt, die hetzen und balgen sich Tag und Nacht in mir, daß ich zuletzt nicht weiß, welcher von beiden Narren ich selber bin."

Manfred schwieg unwillig. Dryander aber war an den Abhang des Gartens getreten und schaute in das dunkle Tal hinaus; man unterschied nur noch einzelne Massen von Wald, Feldern und Dörfern, durch die weite Stille kam der dumpfe Schlag eines Eisenhammers herüber. „Das ist schön!" sagte er, „es ist mir, als hört' ich den Pendul der Zeit einförmig picken." – „Ich bleibe hier", wandte er sich dann schnell zu Manfred, „ich habe das wüste Treiben satt; Profession vom Dichten machen, das ist überhaupt lächerlich, als wenn einer beständig verliebt sein wollte und noch obendrein auf öffentlicher Straße – ich will hier bei Euch die Landwirtschaft ler-

nen!" – „Sie?" – erwiderte Manfred erstaunt, „das gäbe eine schöne Wirtschaft!" – Aber Dryander hörte nicht darauf. „Ich will mich", fuhr er fort, „ich will mich hier wie auf den Grund des Meeres versenken, daß ich von der Welt nichts mehr höre – aber Ihr müßt mir die Hand darauf geben, daß Ihr so lange kein Wort von Literatur mit mir reden wollt."

Er sprach so eifrig, daß er endlich auch den ungläubigen Manfred um so mehr mit sich fortriß, als dieser selbst überzeugt war, daß nur die Einsamkeit und eine eisern geregelte Tätigkeit den wirren Geist heilen könnte. Und mehr bedurfte es nicht, um ihn mit Leib und Seele für den Gedanken zu gewinnen.

Sie besprachen nun noch bei einer Bowle Punsch ausführlich den neuen Plan. Dryander faßte alles begeistert auf, richtete sich in Gedanken schon völlig hier ein, war beruhigt, fast weich und in diesem ungewohnten Zustande unwiderstehlich liebenswürdig; und als sie endlich schieden, begab sich Manfred mit dem Gefühle eines begonnenen guten Werkes zur Ruhe und überdachte noch lange, wie er es am besten vollführen und gestalten wollte.

Wie sehr war er daher erstaunt, als er am folgenden Morgen vernahm, daß Dryander, der von dem übermäßig genossenen Punsche vor Hitze nicht schlafen konnte, noch lange vor Tagesanbruch die Frau und den ganzen Hof aufrumort habe und soeben schon wieder abgereist sei. – In des Dichters Stube fand er mehrere vergessene Kleinigkeiten, Tücher und Strümpfe, auf allen Stühlen zerstreut, das offene Fenster klappte im Winde, auf dem Tische lag ein, wie es schien, vor kurzem von Dryander beschriebenes Blatt. Er nahm es auf und las:

Vor dem Schloß in den Bäumen es rauschend weht,
Unter den Fenstern ein Spielmann geht,
Mit irren Tönen verlockend den Sinn –
Der Spielmann aber ich selber bin.

Vorüber jag' ich an manchem Schloß,
Die Locken zerwühlet, verwildert das Roß,
Du frommes Kindlein im stillen Haus,
Schau' nicht nach mir zum Fenster hinaus.

Von Lüsten und Reue zerrissen die Brust,
Wie rasend in verzweifelter Lust,
Brech' ich im Fluge mir Blumen zum Strauß,
Wird doch ein fröhlicher Kranz nicht daraus!

Wird aus dem Schrei doch nimmer Gesang,
Herz, o mein Herz, bist ein irrer Klang,
Den der Sturm in alle Lüfte verweht –
Lebt wohl, und fragt nicht, wohin es geht!

„Sollte man nicht wirklich denken, er sei durch und durch verzweifelt", sagte Manfred, indem er das Blatt mitleidig lächelnd weglegte, „und ich wette, da hat er in der Zerstreuung alles wieder rein vergessen, war wir gestern verabredet." – Und als er hinausblickte, sah er draußen im Morgenblitzen das Wägelchen des Dichters, über dem ein durchlöcherter Sonnenschirm ausgespannt war, wie ein Schattenspiel zwischen den grünen Bäumen dahinschwanken.

Einundzwanzigstes Kapitel

Wir finden den Baron Manfred fern von seinem stillen, grünen Revier wieder, aus dem ihn eine Familienangelegenheit von besonderer Dringlichkeit verlockt hatte. Das Geschäft, das er heiter zu ordnen gedacht, war indes durch Mißverständnisse unerwartet verwickelt geworden, und unruhig, ja ernstlich besorgt verließ er soeben das Schloß einer, ihm verwandten Dame, bei der er mehrere Tage verweilt.

Schon auf dem Schlosse hatte ihn ein verworrenes Gerücht interessiert, das sich weiterhin in den Dörfern immer wunderbarer ausschmückte. Es war die fast märchenhafte Sage von der Einsamkeit eines aufgehobenen Klosters im benachbarten Gebirge und von einem Mönch, der seit kurzer Zeit dort umgehe, während andere ihn wieder für einen wahnsinnigen Einsiedler hielten. Aber auch diese wußten nicht, wann und woher er gekommen; man nannte ihn nur den Waldbruder *Vitalis*. – Da Manfreds Weg ihn durch das Gebirge führte, beschloß er endlich, den geheimnisvollen Eremiten in seiner eigenen Klause aufzusuchen.

Es war ein schöner Sommerabend, als er zwischen Wiesen und nickenden Kornfeldern den bezeichneten Bergen zuritt. Ein Gewitter war über das Gebirge fortgezogen und blitzende Tropfen hingen noch in Zweigen und Gras, aus dem ein erquickender Wohlgeruch emporstieg. Ein Holzhauer hatte ihm den Pfad nach der Einsiedelei gewiesen, die Gegend wurde immer höher, kühler und stiller, nur die Abendglocken schallten noch durch das feierliche Rauschen des Waldes aus den Tälern herauf. – In dieser kräftigen Einsamkeit konnte er sich eines zürnenden Mißtrauens gegen den Einsiedler nicht erwehren, den er soeben kennenlernen sollte. Es kam ihm kleinlich, ja verrucht vor, inmitten allgemeiner Lust und Not sich so in hochmütiger Selbstliebe abzusondern und über die andern zu stellen. „Der Mensch", sagte er zu sich selbst, „der Mensch allein verwirrt alles mit seiner Leidenschaft und Affektation!"

Durch solche Betrachtungen war er nach und nach ganz in Eifer geraten und nahm sich eben ernstlich vor, den Einsiedler durch vernünftige Überredung womöglich der Welt wieder zuzuwenden, als sein Pferd sich plötzlich scheute und heftig zur Seite sprang. Denn eine wundersame Gestalt war auf einmal zwischen den Bäumen hervorgetreten, unter denen nun auch die in den Fels gehauene, von wilden Weinranken kühl verhangene Einsiedelei

nebst einem sorgfältig umzäunten Gärtchen sich zeigte. Der Eremit trug einen breitrandigen Pilgerhut, ein ungeheurer, alter Schlafpelz, der ihm überall zu weit war, rauschte im Grase hinter ihm her, während er aus einer langen Pfeife Tabak rauchte. Manfred traute seinen Augen nicht. „Wie!“ rief er, „Herr Dryander – Sie also sind der Vitalis!?“ – „Vitalis! warum denn nicht?“ erwiderte Dryander gelassen, „aber bleiben Sie mir mit dem dummen, wilden Pferde ein wenig vom Leibe.“

Manfred band sein Pferd an einen Baum und folgte dem Doktor, der sich fast bei jedem Schritt auf den Pelz trat, zu der Klause. Dort fehlte nichts zum Hausrate eines vollkommenen Waldbruders, ein weißer Totenschädel glänzte aus der Grotte, an deren hinterer Felswand ein großes schmuckloses Kruzifix aufgerichtet war, ein Brevier lag auf der Bank vor der Klause noch aufgeschlagen. Manfred sah lange finster umher, endlich brach er los. „Das ist kein bloßer Scherz“, sagte er, „es wäre zu frevelhaft. Aber auch der bitterste Ernst ist hier ein Frevel. Armer, grillenhafter, wetterwendischer Mensch, gehe erst zu den Einfältigen in die Lehre, erkenne erst unten im Gedränge das unsichtbare Kreuz, das der Herr mitten im Leben aufgerichtet, eh du es selbst zu fassen und in Seinem Namen die Welt zu belehren und zu richten wagst!“ – „Amen, mein Sohn!“ unterbrach ihn hier Dryander mit milder Stimme, „aber nimmermehr wird es dir gelingen, durch lose Worte mir das Rauhe meines Eremitenpelzes herauszukehren, denn mich erbarmt in tiefster Seele deine Verblendung. Also von der Welt Rumor, mein Sohn, hoffst du noch immer zu lernen, was nicht von der Welt ist? Ich aber sage dir: da ist nichts zu lernen, sondern niederzustürzen auf die Knie, denn mitten in der Stille der Waldeseinsamkeit, plötzlich und von Waffen blitzend, kommt der Engel des Herrn!“ – Hier zog und qualmte der Zelot so heftig aus seiner Tabakspfeife, die ihm über dem Reden ausgehen wollte, daß Manfred mitten in seinem Ärger in ein lautes Gelächter ausbrach. Das steckte Dryander an, er stimmte

unaufhaltsam mit ein. Beide aber wandten sich erschrokken, als plötzlich hinter ihnen das herzhafte Lachen noch eines Dritten dareinschallte.

Ein großer, starkknochiger Mann mit gebräuntem Gesicht und herabhängendem Haar, eine grobe Kutte mit einem Strick um den Leib gebunden, trat aus dem Gebüsche hervor und konnte sich, noch immer lachend, gar nicht satt sehen an dem abenteuerlichen Aufzuge des Dichters. Es ergab sich nun, daß der Neuangekommene der eigentliche Besitzer der Klause sei, und daß Dryander erst vor wenigen Stunden, auf seiner Fußreise vom Gewitter überrascht und ganz durchnäßt, sich hierher geflüchtet und, während der Eremit in den Wald nach Holz gegangen, es sich in dessen trockenem Pelze bequem gemacht hatte.

Der Einsiedler machte sich nun sogleich mit Manfreds Pferde zu schaffen, er zäumte es ab, warf ihm Heu vor, streichelte und betrachtete es mit großem Wohlgefallen. „Eine saubere Kreatur!“ sagte er, „da versteh’ ich mich noch darauf aus meinen jungen Jahren, als ich bei dem löblichen Kürassierregimente stand.“ – Darauf traf er mit gleichem Eifer Anstalten, seine Gäste zu bewirten, die unterdes einige nähere Blicke in die kleine Wirtschaft tun konnten. Im Garten hatten Kartoffeln und Kohl fast alle Blumen verdrängt; am Eingange desselben aber fiel ihnen ein frisch gegrabenes Grab auf. „Das ist nur so gegen die überflüssigen Weltgedanken“, sagte der Einsiedler – „succumbit humi bos et Caesar.“ Quer über dem Grabe waren zwei große Speckschwarten auf Stangen befestigt. Der Einsiedler meinte, in der Hütte kämen ihm sonst die Ratten darüber.

Er setzte nun Weinflaschen und Gläser auf den steinernen Tisch vor der Klause, die Gäste mußten sich auf der Bank herumsetzen, er wollte auch einmal etwas Neues aus der Welt hören. Dryander, den der viele Kohl im Garten ärgerte, nannte ihn einen Canonicus in herbis und sprach wütend das tollste Küchenlatein, der Einsiedler antworte-

te ebenso und schien erst recht vergnügt in dieser barbarischen Sprachverwirrung. Dazwischen rauchte er, heftig dampfend, stinkenden Tabak aus einer kurzen, ungarischen Pfeife, im Wein aber tat er wenig Bescheid, er mache ihn, sagte er, aufgeblasen und zänkisch. Er erzählte ihnen, daß er Frater Sammler in dem Kloster oben gewesen, nach dessen Aufhebung aber sich hier angesiedelt habe und bei den Bauern in der Runde, die ihn aus alter Bekanntschaft mit allem, was er brauche, reichlich versähen, sehr gut seine Rechnung finde. Überhaupt sei es ihm im Leben immer gut gegangen. Schon als Kind habe er mit seinem alten Vater, einem blinden Geiger, so viel erbettelt, daß er die Schule besuchen konnte. Später sei er zum Kürassierregimente eingezogen, aber gleich in der ersten Bataille so übel zugerichtet worden, daß sie ihn doch wieder hätten laufen lassen müssen. Als er darauf in sein Dorf zurückgekommen, habe seine Braut unterdes einen andern geheiratet, den sie nun halb totkeife. „Laus Deo!" schloß er, mit seinem Glase lustig anstoßend.
Manfred betrachtete nicht ohne tiefe Wehmut, den fidelen Einsiedler, den das Leben mit allen seinen Stößen nicht hatte unterkriegen können, und der nun die Frömmigkeit frischweg wie ein löbliches Handwerk trieb. – „Es ist ganz unmöglich", rief er endlich nach einigem Nachsinnen aus, „auch Sie sind nicht der Vitalis!"
„Oho!" erwiderte der Waldbruder, „ich und Herr Vitalis! wo denkt Ihr hin, nicht seine Schuhriemen aufzulösen, bin ich würdig, und ich tät's ihm gern heut und allezeit, wenn er es litte! Nein, nein, der wohnt dort im ehemaligen Konvente." – „Als Nachteule", sagte Dryander, „um die Mäuse wegzuschnappen, die nach deinen Speckschwarten gehen." – „Still", fiel ihm der Einsiedler mit überfliegender Röte schnell ins Wort, „schnattert nicht so ungewaschen ins Zeug hinein, wenn Ihr nichts von der Geistlichkeit versteht. ‚Contenti estote', sagte einmal ein Kapuziner in einer Komödie, die ich noch als Soldat gesehen habe, das heißt: begnügt euch mit eurem Kommißbrote,

wenn ihr das Himmelsmanna nicht vertragen könnt!" – „Na, seid nur nicht gleich so grob", lachte Dryander, den der Vorwurf heimlich wurmte.

„Abgemacht!" rief der gutmütige Klausner. „Aber vom Herrn Vitalis muß ich Euch noch erzählen." – Er rückte voll Eifer näher und dampfte so hastig aus der ungarischen Pfeife, daß Dryander sich an das andere Ende des Tisches setzte. – „Seht", sagte er, „es war gerade eine so schöne, sternklare Sommernacht, wie Anno 1814, da wir über den Rhein rückten. Ich hatte meinen Rosenkranz eben abgebetet und stand auf und zog, wie ich alle Mitternacht zu tun pflege, die Glocke über meiner Hütte, denn den Kranken unten in den Dörfern, wenn alles schläft, ist es tröstlich, das Glöcklein von den Bergen zu hören. Auch das Wild ist's schon gewohnt, ich hab' jedesmal meine Freude daran, wie die Rehe dann im Mondenscheine dort auf die Wiese herauskommen und das Weiden vergessen und die Köpfe hoch nach dem Klange wenden, als wollten die armen Dinger auch Gott loben. Nun, jedes tut, was es kann. Aber diesmal schnaubten sie auf einmal, und eh' ich mich's versah, waren sie plötzlich nach allen Seiten zerstoben. Ich tret' heraus, da steht ein schöner, wilder Jägersmann dicht vor mir. ‚Laudetur Jesus Christus', sagte ich. Er aber, ohne Amen zu sagen: ‚Was machst du da?' – ‚Wie Ihr seht, Herr, ich bin ein Einsiedler und bete, wenn die andern schlafen.' – ‚Und schläfst, wenn die andern beten, das ist alles eins!' – ‚Gewiß, so lösen wir einander ab auf der himmlischen Schildwacht.' – Der Jäger drauf stöbert mir in der Hütte herum, sieht mein Moosbett, das Kreuz, den Totenkopf. ‚Vollständige Dekoration', sagt er, ‚bist du so faul, daß dich der Kahlkopf da mit seinen gefletschten Zähnen erst jeden Abend ins Gewissen beißen muß, um zu beten?' – ‚Herr', erwiderte ich, ‚Ihr werdet mir nichts weismachen, ich bin Soldat und Mönch in dem Kloster da droben gewesen und weiß wohl, daß es leichter ist, eine Festung als das Himmelreich zu erobern. Nun möcht' ich doch den Prahlhans sehen, der eine Festung

ohne Bajonett, Leiter und Handwerkszeug nehmen wollte! Und Ihr wollt den Himmel, der höher liegt, stürmen, nackt und erbärmlich, wie Ihr seid, ohne Wehr und Rüstung und tägliche Übung in den Waffen? Ich sage Euch: Demut ist der Anfang und das Ende, hochmütiger Mensch!' – Der Fremde sah mich groß an mit funkelnden Augen, dann stützte er auf dem Tische den Kopf in die Hand, ich meint', er betrachtete den Totenkopf, der vor ihm lag, aber er mochte wohl andere Gedanken haben. Sitz' du, solange du willst, dachte ich, ich fürcht' dich nicht, ich trau' dir nicht. Damit streckt ich mich auf meine Streu und behielt ihn in den Augen, bis sie mir am Ende zufielen.

Als ich aufwachte, waren meine Augen noch immer auf den Tisch gerichtet, aber der Jäger saß nicht mehr auf demselben Punkte. Als ich aber vor die Klause trat, sah ich ihn in der Morgendämmerung schon von dem alten Kloster herabkommen. Es war ein prächtiger Morgen, die Hähne krähten unten in den Dörfern, hin und her klang schon eine Morgenglocke durch die stille Luft. Auch der Fremde, nachdem er mich freundlich gegrüßt hatte, blieb stehen und sah lange ins Tal hinaus. ‚Sieh', sagte er, ‚das ist ein Friede Gottes überall, als zögen die Engelscharen singend über die Erde! Die armen Menschenkinder! sie hören's nur, wie im Traume. Müde da unten, verirrt in der Fremde und Nacht, wie sie weinend rufen und des Vaters Haus suchen, und wo ein Licht schimmert, klopfen sie furchtsam an die Tür, und es wird ihnen aufgetan, aber sie sollen den Fremden dienen um das tägliche Brot; darüber werden sie groß und alt und kennen die Heimat und den Vater nicht mehr. O, wer ihnen allen den Frieden bringen könnte! Aber wer das ehrlich will, muß erst Frieden stiften in sich selbst, und wenn er darüber zusammenbräche, was tut's! – Sieh, Gesell, und das ist geistliches Recht und Tagewerk.'

Ich alter Kerl stand ganz verblüfft vor ihm, denn ich verstand schon gleich damals so viel davon, daß ich bisher ei-

gentlich noch gar nichts verstanden hatte von meinem Metier. Vor meiner eigenen Tür wollt' ich kehren und die ewige Seligkeit für mich allein zusammenknickern, wie ein filziger Schuft, als wär's dem lieben Gott um mich allein zu tun in der Welt. – Und seht, von der Stund' ab blieb der Jäger hier auf den Bergen und wohnte im Kloster droben und machte sich gemein mit mir, wie ein getreuer Kamerad, und ist doch ein grundgelehrter Herr. ‚Denn du gefällst mir', sagt er, ‚du machst keine Flausen mit deiner Frömmigkeit.' Und wenn ich faste, so hungert er, und wenn ich aufwache, so hat er die ganze Nacht gewacht und gebetet, und trinkt keinen Wein und mag keinen Speck, und will ich alter Narr manchmal verzagen, so singt er ein schönes Lied, und – kurz, das ist der Herr *Vitalis,* von dem ihr unten gehört habt."
Der Einsiedler wandte sich hier und machte sich etwas mit dem Tische zu tun, denn er schämte sich, weil ihm die Tränen in den Augen standen. Manfred aber stand auf, ein überraschender Gedanke schien durch seine Seele zu fliegen. „Führt mich zu Vitalis hinauf", sagte er, „ich muß ihn durchaus sprechen!" Der Einsiedler schüttelte bedenklich den Kopf. „Ich will's wohl tun", meinte er, „aber seht Euch vor, wenn Euch bloß die Neugierde treibt. – Da war erst neulich einer, ein junges Blut, der wollte durchaus mit Einsiedler werden. – Aber ich dacht' mir's gleich – denn zum gottseligen Leben gehört eine gute, feste Natur – wenn er nachts mit mir im Walde stand, da schauerte ihn wie ein Mädchen, unsere alten Gebete waren ihm noch nicht schön genug, er setzte sie in künstliche Verse, dann weinte er auch zuviel und hatte so allerhand Sehnsuchten. Zuletzt hatte er gar ein junges, hübsches Hirtenmädchen aufgespürt, die wollt' er mit Gewalt bekehren, aber sie war schon frömmer, als er, und eh' er sich's versah, verliebt' er sich in sie, da wurde er ganz traurig – und kurz, wie ich's vorausgesagt hatte, mit dem Herrn Vitalis ist nicht zu spaßen, der jagt' ihn wieder fort –."
„Hieß der junge Mann nicht *Otto?*" fragte Dryander. –

„Wahrhaftig, so nannte er sich", erwiderte der Einsiedler verwundert. –
Die Nacht war indes völlig hereingebrochen, als sich alle drei auf den Weg nach dem Kloster machten. Der Eremit schritt mit einer Fackel auf einem schmalen, halbverwachsenen Fußsteige voran, die andern folgten schweigend und erwartungsvoll. Unterwegs fragte Manfred den Doktor, wo er denn seine kleine Frau gelassen? – „Sie ist unter die Husaren gegangen", sagte Dryander trocken und mochte durchaus nicht nähere Auskunft geben.
So waren sie, nach einem mühseligen Gange, zu der Ruine gekommen, der Widerschein der Fackel, als sie durch das Tor gingen, beleuchtete den stillen Klosterhof mit seinen alten Bäumen und dem verfallenen Brunnen in der Mitte. Ihr Führer sah sich nach allen Seiten um. „Sollte er noch im Gebirge sein?" sagte er und öffnete knarrend eine eichene Tür. Sie kamen in eine kleine Halle, aber auch dort war niemand zu finden. Nur ein Strohsack auf dem Boden, ein Kreuz auf dem Tische und einige Bücher bezeichneten Vitalis' Wohnung, durch das verfallene Fenster aber sah wunderbar die Nacht herein. Als sie an die Öffnung traten, flatterte verstörtes Nachtgevögel scheu aus den Mauerritzen empor, einzelne Mauerstücke hatten unter ihren Füßen sich abgelöst, sie lauschten, wie es schallend tiefer und immer tiefer hinabrollte. Da trat auf einmal der Mond drüben zwischen den Wolken hervor, sie sahen nichts als stille Schlünde unter sich und das dunkle Chaos uralter Wipfel. – „Entsetzlich!" rief Manfred, in Gedanken hinabschauend.
Hier aber wurden sie plötzlich durch Dryanders Geschrei unterbrochen. Er war neugierig vorgetreten, da hatte ihn der Schwindel gefaßt, er griff krampfhaft in des Einsiedlers Kutte. „Sagt ich's doch", rief dieser, „ist dir wohl, so bleibe unten, arbeite und lobe Gott und laß allen Vorwitz!" Damit packte er den Doktor beim Kragen und schleuderte ihn von dem Abgrund zurück und zur Zelle hinaus.

Indem sie aber nun ins Freie wieder hinaustraten, sahen sie auf einmal zu ihrem Erstaunen zwei fremde Gestalten erschrocken über den Klosterhof hinwegstreichen. „Er ist's, um Gottes willen, nur schnell!“ flüsterte der eine, und in demselben Augenblicke waren beide zwischen dem alten Gemäuer in der Nacht wieder verschwunden. Bei dem Klang der Stimme fuhr Manfred sichtbar zusammen, er hatte die Flüchtlinge in der scharfen Beleuchtung der Fackel unausgesetzt mit den Augen verfolgt; jetzt stürzte er ihnen selbst nach. Aber der Einsiedler schritt mit seinen langen Beinen aus, daß die Kutte rauschte, und faßte ihn mächtig am Arme. „Seid ihr toll“, rief er, „ich weiß nicht, wer es war, aber das weiß ich, daß Ihr bei Nacht im unbekannten Gebirge das Gesindel nicht fangt, sondern den Hals brecht, wenn Ihr kein Gemsbock seid!“ – Manfred mußte ihm nach kurzem Besinnen recht geben, dann aber trieb er plötzlich mit auffallender Hast zur ungesäumten Rückkehr, und blieb still und nachdenklich, während sie vorsichtig zwischen den Felsen hinabstiegen.

„Ich muß noch diese Stunde fort, suche aber bald noch einmal den Vitalis auf“, sagte er, als sie endlich bei der Einsiedelei wieder ankamen, schüttelte seinem Wirte herzlich die Hand und schwang sich sogleich auf sein Pferd. – Der Einsiedler hatte kaum die Zeit, ihm den nächsten Weg zu bezeichnen, und sah ihm dann ganz verwundert lange nach. – „Daß ich ein Narr wäre, in dieser Spuknacht weiterzuziehen“, meinte Dryander und bat sich noch eine lange Pfeife Tabak aus, er freute sich darauf, die ganze Nacht einmal das Einsiedlerleben recht gemächlich mit durchzumachen, auch wollte er noch einige von den Nachtliedern des Eremiten abschreiben.

Manfred aber ritt eifrig den Tälern zu, da hörte er nach einiger Zeit, wie im Traume, oben noch des Einsiedlers Glöcklein schallen, die Rehe weideten wieder zur Seite, seine ganze Seele fühlte sich von der Todesstille wie in ein Grab verschüttet. Die Mitternacht aber hatte unterdes den Himmel weit aufgetan und ihre wunderbaren Schleier

über die Erde geworfen. So immer tiefer und freudiger stieg er eratmend in die träumende Sommernacht hinunter, schon hörte er unten von fern die Ströme wieder rauschen, und die Nachtigallen schlugen, von einem einsamen Schlosse klang noch eine Gitarre herüber, und Düfte wehten erquickend aus den blühenden Gärten herauf. Von dem letzten Abhang des Berges rief er, wie erlöst, hinab: „Gegrüßt, du schönes Leben, ja ich spür's, ich habe dich wieder!"

Zweiundzwanzigstes Kapitel

Auf der Donau glitt bei dem heitersten Wetter ein Schiff zwischen den schönen, waldigen Bergen und Burgen hinab. Von Zeit zu Zeit erschallte ein so herzhaftes Lachen von dem Schiffe, daß die Vorübergehenden am Ufer stehenblieben und vor Lust mitlachen mußten, ohne zu wissen, warum. Es waren reisende Kaufleute, Studenten und Jäger, die auf dem Verdecke im Kreise umherlagen, in ihrer Mitte ein kleiner, stämmiger Mann mit Reisetasche und breitkrempigem Pilgerhute, der ihnen aus seinem eigenen Leben die unerhörtesten Abenteuer erzählte und jedesmal ganz entrüstet war, wenn sie lachten und ihm nicht glauben wollten. Abgesondert aber von dem lustigen Häuflein stand mitten im Schiffe ein wunderschöner Jüngling in zierlicher Jägertracht an den Mast gelehnt, er hatte eine Zither im Arme, die er in der Kajüte gefunden, ihm zu Füßen saß ein anderer hübscher Junge. Beide konnte man für Schüler halten, die zur Vakanz reisten, und es war anmutig zu sehen, wie die fröhlichen Bilder bald im kühlen Schatten der Felsen, bald von der Abendsonne hell beschienen, zwischen den wechselnden Landschaften dahinflogen. Der eine am Maste blickte frisch unter seinem Reisehut in das Grüne hinaus und sang:

„Sie stand wohl am Fensterbogen
Und flocht sich traurig ihr Haar,
Der Jäger war fortgezogen,
Der Jäger ihr Liebster war.

Und als der Frühling gekommen,
Die Welt war von Blüten verschneit,
Da hat sie ein Herz sich genommen
Und ging in die grüne Heid'.

Sie legt das Ohr an den Rasen,
Hört ferner Hufe Klang –
Das sind die Rehe, die grasen
Am schattigen Bergeshang.

Und abends die Wälder rauschen,
Von fern nur fällt noch ein Schuß,
Da steht sie stille, zu lauschen:
‚Das war meines Liebsten Gruß.'

Da sprangen vom Fels die Quellen,
Da flogen die Vöglein ins Tal,
‚Und wo ihr in trefft, ihr Gesellen,
Grüßt mir ihn tausendmal!'"

Die Gesellschaft war längst auf den schönen Gesang aufmerksam geworden; der abenteuerliche Pilger trat vor den Sänger und sang ihm sogleich nach derselben Melodie zu:

„Das klingt wie ein Waldhorn in Träumen,
Was irrst du durch das Gestein,
Mein Rehlein, unter den Bäumen?
Ich will dein Jäger sein!"

Der Sänger sah ihn einen Augenblick von der Seite an und antwortete, ohne sich lange zu besinnen:

„Sie aber lachte im Wandern:
‚Du hast einen kecken Mund,
Ich aber mein' einen andern,
Du bist mir zu kurz und rund!'"

Hier erschallte ein allgemeines Gelächter, der Sänger erschrak darüber, warf schnell die Zither fort und setzte sich zu seinem Gesellen. Der Runde aber war nicht so leicht aus dem Felde zu schlagen, er machte sich, sehr vergnügt, sogleich mit Witzen an die beiden und wollte sie ins Bockshorn jagen. „Mein zärtlicher Herr Jäger", sagte er, „mir scheint, Ihr seid viel mehr geschossen, als Ihr jemals geschossen habt." – „Und Ihr, scheint mir, habt Euch verschossen", versetzte das muntere Jägerbürschchen, „denn der Witz brennt Euch von der Pfanne." – „Wird Euch wenigstens kein Härchen über der Oberlippe versengen! Wett' ich doch, Ihr hättet gar zu gern einen Schnauzbart an Eurem Mund." – „Wenn die Schnauze darunter hübscher wär' als Eure! – „Ich bitt' Euch, schnauzt mich nur nicht so an. Aber, Bart beiseite, ich fürcht', er wird gleich grau sprossen, denn nach Eurem verliebten Liede macht Euch ein Mädchen viel Not." – „Nein, zwei, so närrisch sie sind, ich hab' sie schon ganz müde gejagt." – „Daß die Jungfern nur dabei nicht fallen! wo jagt Ihr sie hin?" – „Unter die Haube." – „Was, führt Ihr Hauben mit Euch?" – „Gewiß, da guckt her!" – Hier lüftete der Jäger ein Felleisen, das hinter ihm lag. Der Pilgrim, der etwas kurzsichtig war, fuhr neugierig mit der Nase hinzu, und eh er sich's versah, hatte ihm das Bürschchen von hinten eine schneeweiße Schlafmütze über den Kopf gestülpt.
Nun aber war der aufrechtstehende Zipfel der Nachtmütze nicht anders als wie ein Blitzableiter, in den plötzlich von allen Seiten alle Witze, matte und feurige, durcheinander einschlugen. Darüber wurde der Pilger ganz hirnschellig, man wußte bei seinem wunderlichen Wesen nicht recht, ob es ihm Ernst oder Spaß war mit der Wut. Der junge Jäger, da er unverhofft solche Wirtschaft angerich-

tet, saß unterdes mäuschenstill und blickte nur ein paarmal scheu herüber. Als er aber den Pilger so auf das allerlustigste schimpfen hörte, und unter seiner Schlafhaube wohl die Hasenohren sah, konnt' er's doch nicht lassen; er sprang von neuem auf, schnalzte mit seiner Reitgerte und parlierte immerfort keck mit drein. Die lustigen Vögel im Schiff hetzten: sie sollten sich miteinander schießen, der Abend brach auch herein und vermehrte die Verwirrung, der Pilger schwor, er wolle noch heut mit der Degenspitze aus dem schönen Jungen eine junge Schöne herauskitzeln! Das Jägerbürschchen aber flüsterte heimlich seinem Gesellen zu: „Was fangen wir nun an? Ich bitt dich, Hänschen, rat mir!" – Da stieß das Schiff ans Land.

Während die andern nun ihre Bündel, Tabakspfeifen und Feldflaschen noch zusammenrafften, eilte *Dryander* – denn niemand anders war der abenteuerliche Pilgrim – schon voraus und flog in größter Hast nach dem Wirtshaus an dem breiten Gastwirt vorüber, der das Schiff gemächlich an der Tür erwartete und ihm verwundert nachsah. In der Gaststube fand er einen jungen Mann, der auf der Brüstung des offenen Fensters saß und in das fröhliche abendliche Getümmel hinausschaute; dieser wandte sich schnell – er erkannte seinen *Fortunat.* Ohne in der Konfusion sich zu verwundern oder ihn erst zu begrüßen, rief ihm Dryander sogleich entgegen: „Verfluchte Teufelsgeschichte! hast du deine Kuchenreiter mit? So ein Mädchen von Junge! Aber ich will ihm den Bart unter der Nase wegputzen, wenn er nur einen hätt'! Da ist nichts zu lachen dabei! Er hat gut treffen, ich bin wie ein Bienenkorb gegen seine Taille, und –" „Halt ein!" unterbrach ihn Fortunat, immer heftiger lachend, „du zerplatzt ja wie eine Bombe, was gibt's denn da auf einmal?" – Aber Dryander war zu erbost, er schimpfte unaufhaltsam über die Albernheit der Ritterlichkeit, der Duelle, der Ehre, die, wie eine Regimentsfahne, erst von Kugeln zerfetzt und lumpig sein sollte, um ein Ansehen zu haben. Indem er sich aber so in Vergleichungen erschöpfte, kam das Getümmel

draußen wachsend immer näher und näher. „Dummes Zeug!“ schloß er endlich und entwischte mit solcher Geschwindigkeit aus der Tür, daß er seinen Hut im Zimmer vergaß.

Fortunat ließ ihn laufen. Was wird es sein! dachte er, die alte Posse: Sorgen ohne Not und Not ohne Sorgen. Die Rakete wird draußen verprasseln, ohne eben den Erdkreis in Brand zu stecken. – Unterdes hatte die Stube sich nach und nach lärmend gefüllt, Felleisen, Mäntel und Tabaksbeutel lagen auf Stühlen und Tischen umher, die muntere Schiffsgesellschaft machte sich behaglich breit, der eine schrie nach Wein, der andere nach Kaffee, alle waren noch ganz voll von den lustigen Händeln, und da sie vom Wirt erfuhren, daß die beiden Jäger ein eigenes Zimmer bezogen, beredeten sie sich, wie sie morgen zum Duell die Pistolen blind laden, dem Pilger Knallkugeln unter die Füße legen wollten usw. Als aber nun allmählich aus mehreren Schlünden dicker Tabaksqualm emporzuwirbeln begann, zog Fortunat, nachdem er in dem Lärm vergeblich nach einem Leuchter gerufen, auch über Dryander keine nähere Auskunft erhalten hatte, sich ohne Licht in sein Zimmer zurück, da er morgen mit Sonnenaufgang wieder aufzubrechen gedachte.

Seine Stube ging nach dem Garten hinaus, die Glastür stand noch weit offen, wie er sie vor einigen Stunden verlassen. Alle Bewohner des Hauses hatten mit den Gästen vollauf zu tun, es war so still draußen, daß man den Ruderschlag einzelner Fischer aus der Ferne hören konnte. Ermüdet setzte er sich auf die Schwelle hin. Da hörte er Stimmen im Garten, in einer fremden Sprache, wie es ihm schien. Bald bemerkte er beim hellen Mondenscheine zwei unbekannte Gestalten, die sich hier wohl für unbelauscht halten mochten. Der eine, wie ein Jäger gekleidet, saß mit untergeschlagenen Beinen auf dem Rasen, er hatte den Hut abgenommen und in der Kühle sein Wämschen gelüftet, sein wunderschönes Haar floß in reichen Locken herab; der Mond glänzte blendend auf seiner entblößten

Schulter. Der andere kniete hinter ihm und schien die Locken zu ordnen, während sie leise und lebhaft miteinander schwatzten. Ein Brunnen, den Fortunat vor dem Gebüsch nicht sehen konnte, plauderte um die Wette mit ihnen, und je nachdem die Luft sich bewegte, klang bald das Plätschern, bald die liebliche Stimme wie ein Glöcklein aus der stillen Mondnacht herüber. Die Nacht aber hatte unterdes die Gegend draußen wunderbar verwandelt, zwischen den alten Bäumen hindurch sah man weit in die Täler hinaus, da lag es verworren im Mondenschein, wie glänzende Kuppeln, Trümmer und prächtige Gärten, in dem nahen Städtchen unten sang ein Student noch vor seiner Liebsten Tür, dazwischen immerfort wieder das Rauschen des Brunnens – Fortunat saß wie im Traum, er dachte an Italien, an Rom, und unwillkürlich in Gedanken rief er – „Fiametta!"
Bei dem Klange reckten die beiden wie Rehe, wenn das Laub raschelt, plötzlich die Köpfchen in die Höh', sprangen scheu auf, und flogen dem Hause zu. Fortunat trat ihnen erstaunt entgegen, da stutzte das Jägerbürschchen plötzlich und sah ihn einen Augenblick durchdringend an, dann aber warf es sich auf einmal atemlos an seinen Hals, ihn fest umklammernd und schluchzend, er fühlte des Jünglings Tränen unaufhaltsam über seine Wangen rinnen; seine Locken rollten rings um ihn her, es war, als würde er in seinen Armen ganz und gar vergehen. Nun aber wußt er's wohl, wen er im Arme hielt. „Meine liebe, liebe *Fiametta!*" rief er aus tiefstem Herzensgrunde. Da ließ das schöne verkleidete Mädchen los, stellte sich, ihre Locken aus dem Gesicht schüttelnd, dicht vor ihn und blickte ihn aus den Tränen so fröhlich an, daß es ihm recht durch die Seele ging. Darauf schnell wieder besonnen, zog sie ihn schweigend mit sich in sein Zimmer hinein. Er sah im Vorüberschweifen dem andern Gesellen ins Gesicht und erkannte seines Liebchens Kammerjungfer, die über und über rot wurde. In der Stube aber steckte Fiametta ihr Haar wieder auf, während sie die Kammerjungfer mit

einem heimlichen Auftrage fortschickte. Dann trieb sie Fortunat, in sichtbarer Furcht, geheimnisvoll und ohne ihm Rede und Antwort zu stehen, zur unverzüglichen Abreise, half ihm unter tausend Späßen mitten in ihrer Angst und Hast seine Sachen rasch in ein Bündel schnüren und drängte ihn fort, aus dem Hause, aus dem Garten und immer weiter. Draußen auf einem abgelegenen Platz fanden sie Fortunats Diener mit seinen beiden gesattelten Pferden, die Kammerjungfer hatte ihn hergeführt. Sie sollte mit dem Diener auf dem Schiffe weiterreisen, Fiametta selbst aber schwang sich schnell auf das eine Pferd. Fortunat wußte nicht, wie ihm geschah, und ehe er sich fassen konnte, waren Kammerjungfer und Wirtshaus schon hinter ihnen verschwunden.

Als sie im Freien waren, fragte Fiametta mit tief gesenkten Augen kaum hörbar: „Was macht denn Annidi?“ Fortunat mußte sich fast auf den Namen besinnen. „Annidi?“ – sagte er, „sie hat in Rom den Studenten Otto geheiratet. Aber wie kommst du auf die?“ – Fiametta sah ihn groß an: „Ist sie denn nicht deine Liebste gewesen?“ – „Mein Gott“, erwiderte Fortunat nach einigem Nachdenken, „so warst du es wohl, die an jenem Abend im schwarzen Mäntelchen an mir vorüberstreifte, als mich Otto zu seinem Mädchen führte, das ich damals noch gar nicht kannte?“ – „Ja freilich“, entgegnete Fiametta lebhaft, „und ich spielte dann einmal des Abends die Annidi in unserm Garten, die Kammerjungfer mußte deine Kleider anziehen und so über den Gartenzaun zu mir kommen, da kamst du auf einmal selber, wir hatten dich nicht so früh zurückerwartet.“ – „O vernagelter Kopf, der ich war!“ rief Fortunat, sich vor die Stirn schlagend aus, „hätt' ich das damals gewußt!“ – Sie lachte seelenvergnügt und ihre Augen glänzten vor Tränen.

Währenddes ritten sie eilig an dem Städtchen vorüber, zwischen den schlafenden Gärten und Landhäusern immer tiefer in die weite, sternenhelle Nacht hinein. Die Nachtigallen schlugen von den waldigen Bergen, über das

stille Feld hörte man die Hunde von fern bellen, Fiametta sah sich öfters ängstlich um. „Sieh", sagte Fortunat, „mir ist wie einem Vogel in der Luft, ich folge dir über die ganze Erde! Jetzt aber sage mir auch, warum blickst du so scheu zurück? wie kamst du vorhin auf das Schiff? was in aller Welt hast du vor?" – „Ach, das ist eine lange, traurige Geschichte", entgegnete Fiametta, „die muß ich vom Anfang anfangen." – Sie ritt dicht neben ihm und, selbst wie in Träumen in der träumerischen Nacht, halb an ihn gelehnt, begann sie folgendermaßen zu erzählen:

„Als du in Rom auf einmal verschwunden warst und nun der Winter kam, und es regnete Tag und Nacht, und der Vater saß abends in dem großen Saale am Kaminfeuer und sprach kein Wort, und alles war so still im ganzen Hause, daß man die Turmuhr gehen hörte, da wurde ich plötzlich krank. Da träumte mir, ich wäre auf einer Anhöhe über Rom im Abendglanze eingeschlafen. Als ich aber erwachte, war es schon finstere Nacht, mich fror und ich kannte die Gegend nicht wieder. Da kam durch das Dunkel ein Jäger vom Berge herab. ‚Ach führ' mich zur Stadt hinunter', rief ich, ‚horch, da klingt in der Ferne noch die Glokke vom Kapitol.' – ‚Das ist die Turmuhr, die schlägt auf meinem Schloß im Walde', sagte der Jäger. – ‚Kennst du denn nicht das Schloß des Marchese A.?' fragte ich wieder. – ‚Wo Fiametta wohnt? ach, das ist lange her', sagte der Jäger, dann wandt' er sich plötzlich – du selbst warst der Jäger, aber du kanntest mich nicht mehr. – Nun stiegst du weiter den Berg hinab, ich rief voll Angst und konnte dir so schnell nicht folgen. Da ging gegenüber der Mond auf, und auf einmal, so weit ich sehen konnte, lag die ganze fremde Gegend tief verschneit und flimmerte im hellen Mondenschein, als sollt' ich sterben vor Wehmut.

Als ich mich von der Krankheit wieder erholte, stand eines Morgens der Vater vor meinem Bett, das Fenster stand offen, die Bäume draußen waren schon wieder grün, und die Vögel sangen. ‚Steh nur auf', sagte mein Vater, ‚wir reisen nach Deutschland!' – Er hatte sein Vermögen

verloren, das Haus, unser Garten sollten verkauft werden, er mochte das nicht mit ansehen. So fuhren wie in einer schönen Frühlingsnacht von Rom fort, die Brunnen rauschten auf den stillen Gassen, in unserm Garten schlugen die Nachtigallen, als wüßten sie's auch, und als die Paläste und Kuppeln allmählich hinter uns im Mondglanz versanken, sah ich meinen Vater zum erstenmal weinen."

„Wo ist der Vater jetzt?" unterbrach sie Fortunat hier. Fiametta aber ritt ein Weilchen schweigend vor sich hin, er merkte, daß sie selber weinte. Dann sah sie sich plötzlich wieder nach allen Seiten um und fuhr gefaßter fort: „Mein armer Vater fand's in Deutschland nicht so, wie er sich's gedacht. Die mächtigen Verwandten, auf die er gerechnet hatte, weil sie in der Jugend brüderlich zusammen gelebt, waren seitdem alt und anders geworden, die meisten lange tot, ihre Kinder, die ihn nicht mehr kannten, sahen ihn verwundert und neugierig an, er konnte sich in der verwandelten Welt nicht zurechtfinden und starb vor Gram. – Das war eine furchtbare Nacht, ich erinnere mich nur der schwarzbehangenen Pferde und Gestalten und des Fackelscheins zwischen den dunklen Bäumen – und als die Glockenschläge allmählich verhallten, saß ich allein mit einer alten, schwarzgekleideten Dame im Wagen, wir fuhren rasch durch unbekannte Gegenden, sie sprach immerfort französisch zu mir, aber ich hörte nur das dumpfe Rasseln des Wagens in der Nacht, mir war's, als führen wir selber ins Grab. Die Dame aber war eine reiche, kinderlose Tante, die mich nun zu sich genommen hatte. Sie wohnte auf einem großen Schloß, das einsam am Abhange des Gebirges mitten in einem prächtigen Parke lag, der wimmelte von seltsamen Tauben und Pfauen, in dem klaren Bassin vor dem Schloß spielten bunte ausländische Fische, wie Vögel in der Luft, weiterhin in einem zierlich vergitterten Wäldchen weidete ein schöner Goldfasan. Die Tante hatte ihre Freude daran, mich recht auszuputzen, obgleich wir nur selten Besuch hatten, da ging ich denn in prächtigen Kleidern, und wenn ich manchmal so

allein im Garten stand, kam ich mir selber in der Einsamkeit wie ein verzauberter Goldfasan vor. An den Sommernachmittagen aber pflegte die Tante mit mir im Garten auf einem schattigen Hügel zu sitzen, von dem man weit hinaussehen konnte, wie der Strom und die Straßen glänzend durchs Land gingen, Reiter und Wagen zogen da wie in einem Schattenspiele rasch vorbei, manchmal kam der Klang eines Posthorns aus der Ferne herüber. ‚Dort geht es nach Italien hinaus', sagte die Tante – mir war zum Sterben bange.

Eines Abends saßen wir auch dort, ich zerpflückte in Gedanken eine Sternblume; ob du kommst oder nicht kommst? ‚Er kommt!' rief ich auf einmal erschrocken aus, warf die Blume fort und flog vom Hügel, am Schlosse vorüber, immerfort ins Tal hinab. Denn zwei Reiter kamen unten vom Walde, der eine im grünen Reiserocke, gerade wie du! Als ich atemlos unten anlangte, stutzt sein Pferd – es war ein ganz fremdes Gesicht. Er mocht' es wohl erraten, wer ich bin, er schwang sich schnell vom Pferde, und indem er die Zügel seinem Bedienten zuwarf, reichte er mir höflich den Arm und führte mich wie eine Gefangene zurück. Ich glaubte, die Tante würde schmälen, aber sie besorgte nur, daß mir die Erhitzung nicht schade, strich mir die Locken aus der Stirn und nannte mich ein artiges Kind, daß ich ihren Vetter, den sie viele Jahre nicht gesehen, so freundlich empfangen. Sie nannte ihn Baron *Manfred.*"

„Manfred?" sagte Fortunat erstaunt, „den Namen habe ich oft von Lothario gehört. Doch den kennst du ja nicht." – Fiametta schüttelte das Köpfchen und fuhr weiter fort:

„Bisher hatte ich fast wie im Traume gelebt, mit dem Fremden aber kam auf einmal Hast und Unruhe in unsere ländliche Stille. Nichts war ihm recht in unserer Wirtschaft, alles wollte er gescheuter, einrichten und sah mich dabei oft so sonderbar an, daß ich erschrak, denn er schaute so klug drein, als könnte er meine Gedanken lesen. Vor

Verdruß darüber hatte ich mich eines Tages in der schwülen Mittagszeit mitten ins Gras gelegt, alle Vögel schwiegen, nur die Bienen summten, einzelne Wolken flogen über die stille Gegend fort, ich dachte an die alten Zeiten, an dich, an unsern Garten in Rom. Da kam auf einmal die Tante mit ihrem Vetter im Buchengang herunter. Ich hob mich im Grase halb empor, sie bemerkten mich nicht, ‚Ich habe auch schon daran gedacht‘, sagte die Tante, ‚so kann es mit Fiametta nicht länger bleiben, sie vergeht mir hier in der Einsamkeit wie eine Blume.‘ – ‚Abgesehen selbst von allem, was ich Ihnen eben erzählt habe‘, erwiderte der Vetter, ‚so wüßte ich in der Tat keine bessere Partie für das Fräulein als den Baron, jung, reich, unabhängig.‘ – ‚Und Sie übernehmen es also‘, fragte die Tante wieder, ‚ihn zu uns zu bringen?‘

Ich konnte seine Antwort nicht mehr verstehen. Aber wie wenn der Blitz neben mir eingeschlagen hätte, sprang ich schnell auf und flog zu meiner italienischen Kammerjungfer und erzählte ihr alles. Da war nicht lange Zeit zum Besinnen, ihr war hier so bang auf dem Schlosse wie mir, sie wollte unter dem Vorwande einer Maskerade Jägerkleider für uns beide herbeischaffen, und wir beschlossen, zu einer jungen fröhlichen Tante in Wien zu entfliehen, die ich noch aus Rom kannte und die mich vor der dummen Partie beschützen sollte.

Seitdem sahen mich die Tante und der Vetter noch häufiger geheimnisvoll und schmunzelnd an. Besonders aber ganz abscheulich war mir nun der kluge Vetter, wenn er mit seinen spitzigen Blicken, wie eine Spinne mit ihren langen Beinen, nach mir zielte. Ja, spinne und laure du nur! dachte ich. Und als er nun wirklich abreiste, um den Bräutigam zu holen, da fuhren wir, während alles schon schlief, in unsere Jägerkleider und stiegen in der schönsten Sommernacht mit klopfendem Herzen sacht die Treppen hinab durchs leere Schloß, den stillen Gang entlang, bis wir endlich im freien Felde tief aufatmeten. Da sah's draußen so frisch und waldkühl aus! – Noch dieselbe Nacht

aber hatten wir uns im Gebirge verirrt. Fragen mochten wir nicht, so kamen wir zuletzt gar an ein verfallenes Schloß. Mich schauerte und fror, die Jungfer weinte, da tat sich plötzlich eine Tür auf, drei Männer mit Windlichtern traten heraus – der eine war der Vetter, verwildert und bleich im Widerscheine der Fackeln – ich glaube, er geht um bei Nacht, was hatt' er sonst zu tun da droben? Aber erkannt hat er mich und setzt mir sicherlich nach. Wie wir da heruntergekommen, weiß ich nicht mehr, aber als der Tag endlich anbrach, sahen wir die Donau im Tale funkeln, ein Schiff wollt' eben abgehen, wir stiegen mit ein, und so fuhr ich in Lust und Angst, und bekam Händel und sollte mich duellieren und" – „Und ich", fiel Fortunat ein, „ich habe den verflogenen Goldfasan wieder eingefangen und lass' ihn nun nimmermehr los!"

Fortunat war voller Freude und doch verwirrt, er wußte gar nicht, was er mit dem lieblichen Kinde nun anfangen sollte, das sich so ganz in seine Arme geworfen, auch war die Angst vor dem Erwischen nicht gering.

Unterdessen flogen schon einzelne Streiflichter durch die stille Luft. „Wie bist du schön geworden!" sagte Fortunat, sie fast erstaunt betrachtend. Da wurde sie über und über rot, jetzt dachte sie erst daran, daß sie so ganz allein mit ihm war. Aus den fernen Dörfern aber hörte man schon einzelne Stimmen, über die wogenden Kornfelder schossen ihnen die ersten Sonnenstrahlen blitzend entgegen – so ritten sie fröhlich in den prächtigen Morgen hinein.

Dreiundzwanzigstes Kapitel

Als *Otto* – von dem strengen Vitalis verstoßen – so einsam von dem Gebirge der Einsiedler hinabstieg, weinte er sich recht von Herzensgrunde aus. Dann wurde ihm erst leichter. Er fühlte wieder einen rechten Trieb und Mut, nach dem Höchsten in der Welt zu streben, er wollte endlich

ehrlich Frieden stiften in seiner Seele, und so neugeboren zu dem Einsiedler zurückkehren, ja es kam ihm in diesen glücklichen Stunden gering vor, selbst sein Dichten zu lassen, wenn es ihn wieder in Eitelkeit verstricken wollte. Die stille Nacht sah ihn dabei von den Bergen, wie eine milde Mutter, fast wehmütig an. – Indessen verloschen nach und nach die Sterne am Himmel, und wie nun die Morgenkühle über die Felder kam und unten der Strom und von drüben die Spiegelfenster eines Schlosses lustig aufblitzten: da erschien dem Verweinten die Erde wieder so jung und frisch wie nach einem Gewitterregen, in den tröpfelnden Bäumen über ihm dehnten die Vögel erwachend die Flügel und sahen ihn neugierig an, als wollten sie fragen: „Gesell, wo bist du so lange gewesen?“ – Er wanderte fröhlich den ganzen Tag, und als er endlich auf dem letzten Berge aus dem Walde trat, erblickte er auf einmal in der Ferne mitten zwischen Gärten die alte, braune Stadt, wie eine von Efeu übergrünte Ruine. Ermüdet streckte er sich unter den Bäumen hin, er sah Handwerksbursche, Reiter und schlanke Bauernmädchen heiter durchs Grün auf dem Gebirgspfade hinabziehn, die Vögel sangen im Walde, einzelne Wolkenschatten flogen wechselnd über die glänzende Landschaft – so schlummerte er ein und träumte von der schönen, waldkühlen Jugendzeit.

Er mußte lange geschlafen haben, denn als er erschrocken wieder um sich blickte, ging die Sonne schon unter und vergoldete die Giebel und Türme der Stadt. Voll Erstaunen sah er sich ganz von Blumen bedeckt, als hätt' es Rosen geregnet. Da hörte er eine schöne Stimme lustig durch die Abendluft klingen. Ein eleganter Reisewagen stand tiefer am Saume des Waldes, zwei junge Damen, die, wie es schien, den steilen Berg zu Fuß herabgekommen, stiegen soeben wieder ein. Die eine wandte sich noch einmal und blickte nach ihm herüber, er mußte verwirrt und geblendet niedersehen, so schön war sie. „Nach der Bergvorstadt!“ rief sie dem Postillion zu – da flog der Wagen in

den duftigen Abend hinein, er hörte das Posthorn noch lange aus der Ferne schallen.
In der Stadt fand er seine Wohnung bereit: ein kleines, freundliches Stübchen im dritten Stock, alte Kupferstiche an den Wänden, der Boden neu mit Sand bestreut, ein Glas mit frischen Blumen unter dem Spiegel. Eine alte Frau empfing ihn sehr gesprächig und händigte ihm ein Briefchen ein. Sein Jugendfreund, der hier alles für ihn besorgt hatte, meldete ihm, daß ihn leider unvorhergesehene Geschäfte über Land geführt, in wenigen Wochen hoffte er wieder zurück zu sein – so befand sich denn Otto unerwartet ganz allein in der fremden Stadt. Er konnte sich nach der langen Gebirgseinsamkeit gar nicht wieder zurechtfinden, alles kam ihm neu und wunderbar vor, der heitere Reisetag hallte noch in seiner Seele nach, und als er das Fenster öffnete, dämmerte die unbekannte Gegend so seltsam über die Dächer herauf, es war ihm, als hörte er noch immer das Posthorn fern aus der Frühlingsnacht herübertönen. Er konnte nicht widerstehen, er mußte noch einen Streifzug durch die Stadt machen.
Unten erkundigte er sich nach der Bergvorstadt, er hatte sich geschämt, die Alte darnach zu fragen. Man wies ihn nach einer entfernten Anhöhe, die mit einzelnen Villen und weitläufigen Gärten geheimnisvoll in die Straße hereinsah. Das nächtliche Wandern in einer unbekannten großen Stadt hat etwas Märchenhaftes, die Häuser und Türme stehen wie im Traum im Mondenscheine, auf den Straßen schwärmt es noch laut und behaglich in der Maskenfreiheit der lauen Nacht, dann plötzlich alles wieder still im engen, dunklen Gäßchen, nur die Dachluken klappen im Winde, eine Nachtigall schlägt wehmütig am Fenster. – Otto schlenderte in Gedanken immer fort, alte Reiselieder fielen ihm ein, er sang leise vor sich hin, er wußte selbst nicht, was er draußen wollte. Endlich hatte er die Höhe erreicht, je weiter er kam, je stiller und ländlicher wurde die Straße, seitwärts schienen sich prächtige Gärten hinabzusenken. Oft blieb er stehn und sah zurück über

die Stadt hin, zwischen den vielen verworrenen Lichtern ging das dumpfe Rasseln der Wagen wie ein ferner Sturm, zuweilen brach ein Schwarm verstörter Dohlen aus einem alten Kirchendach und durchkreiste schreiend die Nacht, eine Spieluhr vom Turme sang ihr frommes Lied in der Einsamkeit der Lüfte. Von der andern Seite aber war die Gasse schon offen, ein frischer Hauch wehte herüber, er hörte eine Mühle gehen, die er nicht sah, dann Hundegebell von fern, und da und dort noch Stimmen im dunklen Feld.

Auf einmal erklang eine Gitarre und einzelne Töne eines wunderschönen Gesangs, träumerisch vom Winde verweht, wie wenn die Nachtluft durch die Saiten einer Harfe geht. Er eilte zu dem Garten, woher die Töne kamen, das Pförtchen war nur angelehnt, er trat hinein. Da stutzte er, denn es war, als flöge der Schatten einer fliehenden Gestalt heimlich zwischen den Gebüschen hin, sonst war alles still. Neugierig ging er weiter in die dunklen Schatten der alten Bäume hinein, der Mondenschein glänzte seitwärts über die Rasenplätze. Da bemerkte er einen Weiher, von Trauerweiden umhangen, eine weiße Statue schimmerte durch die Zweige herüber: eine Nymphe, die, halb abgewandt, am Weiher auf ihrem Arme ruhte, den andern verschlafen über das Haupt gelehnt. – Er wollte eben näher hinzutreten, als plötzlich tiefer aus dem Garten ein heller Lichtschimmer durch die Bäume funkelte und ebenso schnell wieder verschwand. Erschrocken, zögernd wandte er sich zurück, er suchte das Pförtchen wieder, aber die Streiflichter des Mondes und die schwankenden Schatten der Bäume dazwischen verwirrten ihn ganz, und eh' er sich besinnen konnte, stand er vor den Marmorstufen eines hohen, altertümlichen Palastes. In demselben Augenblick schüttelte sich der Fliederstrauch über ihm, daß er ganz von Tau und Blüten verschneit wird, er hört ein heimliches Kichern hinter sich, eine schlanke, weiße Mädchengestalt guckt verstohlen durch die Zweige und faßt ihn schnell an der Hand. „Siehst du, das ist der Will-

komm, weil du mich überrascht hast", flüstert sie mit der lieblichsten Stimme, „das ist ja prächtig, daß du schon heute kommst." So führt sie, vorangehend, den Erstaunten über die Stufen durch eine dunkle Halle, plötzlich treten sie in ein erleuchtetes Gemach, sie wendet sich rasch herum – er erkennt mit freudigem Schrecken die reisende Dame von heut abend im Walde.

Sie sah ihn erstaunt an, indem sie seine Hand losließ. Dann bemerkte sie eine ihrer Rosen, die er noch im Knopfloche trug, eine flüchtige Röte flog über ihr schönes Gesicht. „Aber", sagte sie kopfschüttelnd, „wie haben Sie mich denn so bald aufgefunden?" Er erzählte nun sein Erwachen auf dem Berge, seine Unruhe darauf und den Streifzug durch die schöne Nacht. Aber sie war ganz zerstreut, sie schien auf etwas zu sinnen. Dann sprang sie schnell zur Tür hinaus, er hörte sie draußen lebhaft mit jemand sprechen.

In dieser seltsamen Lage schaute er betroffen im Zimmer umher. Eine Alabasterlampe beleuchtete wunderbar das kostbare Gerät, auf dem eine Gitarre und aufgeschlagene Notenhefte unordentlich herumlagen. Hohe, ausländische Gewächse rankten sich schlangenartig an den Wänden empor und hingen mit ihren glühenden Blüten in die träumerische Dämmerung herein, als spiegelten sie sich in dem reichen Teppich am Boden.

„Armer Junge! du wirst recht müde sein", sagte jetzt die Unbekannte, indem sie fröhlich wieder hereintrat und ihn auf den Diwan niederzog. Sie setzte sich dicht neben ihn, ein Bein über das andere geschlagen, er mußte ihr erzählen, woher er gekommen, wer er sei und was er hier treibe. – „Also so sieht ein Dichter aus!" – rief sie erstaunt, als sie seinen Namen hörte, dabei wandte sie ihn an beiden Achseln zu sich herum und sah ihm mit den großen, schönen Augen gerade ins Gesicht, er mußte die seinen errötend niederschlagen. „Come è bello!" sagte sie kaum hörbar für sich. Darauf nahm sie eine Pfirsich aus der Kristallschale vor ihnen, biß mit ihren weißen Zähnen herzhaft hinein

und reichte sie ihm hin. Aber Otto war ganz verwirrt, aus ihren Augen leuchtete zuweilen eine irre, wilde Flamme, die ihn schreckte, in dieser seltsamen Verstimmung konnte er durchaus den rechten Ton nicht finden und saß blöde und unbeholfen neben der vornehmen, schönen Frau. Da lachte sie plötzlich mutwillig auf, er wußte nicht worüber, dann sprang sie auf und brachte aus einem verborgenen Wandschranke ein zierlich gebundenes Buch hervor. „Kennst du das?" fragte sie, ihm den funkelnden Goldschnitt vorhaltend; es waren seine Gedichte. – „Ich kenn' sie noch nicht", sagte sie, „lies mir was vor daraus."
Sie setzten sich wieder, er blätterte unentschlossen und begann endlich eines seiner liebsten Gedichte von der schönen Meerfei Melusina. – „Und daß du's nur weißt", unterbrach ihn die Dame, „ich bin eigentlich selbst die Melusina; du darfst nur in den Nächten vom Montag und Donnerstag in den Garten kommen. Frag' nicht nach mir, und plaudre nicht davon; wenn du mich ein einziges Mal bei Tage erblickst, sehen wir uns niemals mehr wieder." Otto sah sie verwundert an, dann las er wieder weiter. Es war ein langer Romanzenzyklus, er hatte ihn in der glücklichsten Jugendzeit gedichtet und seitdem nicht wieder gesehn; jetzt nach so langer Zeit, in der märchenhaften Umgebung, ergriff es ihn selber wunderbar, er las aus ganzer Seele fort und immer fort. Zuletzt beim Umschlagen des Blattes blickte er einmal flüchtig zur Seite – die schöne Frau lag fest eingeschlafen neben ihm. – Er schwieg, ihn schauderte heimlich, denn die schlanke Gestalt in dem weißen Nachtgewande ruhte halb abgewendet, den einen Arm nachlässig über ihr Haupt geschlagen, gerade wie die Statue vorhin am Weiher. In dieser plötzlichen Stille öffnete sich auf einmal leise die Tür, ein schwarzgelocktes Mädchenköpfchen guckte herein, überblickte spöttisch den Schauplatz dieser tiefen Ruhe und winkte ihm dann, ihr zu folgen. „Still, still", – sagte sie, als er heraustrat, ihn an der Hand schnell fortführend – „jetzt müssen Sie sacht fort, der Mond ist eben untergegangen vor Langerweile."

Draußen sang sie halb für sich:

„Ein Fink saß schlank auf grünem Reis:
Pink, pink!
Der Jäger da mit rechtem Fleiß
Zu zielen an und messen fing,
Und zielt' und dacht': Jetzt bist du mein –
Fort war das lust'ge Vögelein:
Pink, pink! mußt flinker sein!"

„Was singst du da so lustig?" fragte Otto. – „Ich pink' nur ein wenig Feuer an im Dunkeln", entgegnete das Mädchen, „wollen Sie sich vielleicht ein Pfeifchen dran anstekken und noch etwas lesen von den zwölf schlafenden Jungfrauen?" – Sie plauderte mutwillig noch vielerlei in den Wind hinein – so gingen sie rasch durch den stillen Garten. Otto blickte im Vorbeigehen noch einmal nach dem Weiher hinüber, dort ruhte die Statue wieder auf ihrem Marmorpfühl, ein eingeschlummerter Schwan fuhr bei ihren Tritten mit dem Kopf aus den Flügeldecken hervor, sah sie schlaftrunken an und träumte dann weiter. – „Gute Nacht, Herr Morpheus!" sagte das Mädchen an der Gartentür mit einem schnippischen Knicks und schob ihn lachend hinaus.

Er hörte das Pförtchen hinter sich zuklappen, es war ihm wunderbar, so plötzlich allein unter dem stillen, weiten Sternenhimmel. In der ganzen Gegend regte sich kein Laut mehr, nur die Uhren schlugen fern in der Stadt, es war lange Mitternacht vorüber.

Seit dieser Zeit war es um ihn geschehn, die schönen Mondnächte beleuchteten noch oft seinen einsamen Gang zu dem stillen Zaubergarten. Das geheimnisvolle Grauen in der Luft verlockte ihn noch mehr, er mochte nicht nach dem Namen der schönen Frau fragen, ja er hütete sich, ihr Revier bei Tage zu betreten – war sie ja doch sein mit Leib und Seele! Aber in seiner stillen Stube dann, nach solchen durchschwelgten Nächten, überkam es ihn oft wie Alp-

hornsklänge den Schweizer in der Fremde. Da befiel ihn eine tiefe Angst, er dichtete hastig oft ganze Nächte hindurch, er wollte mit Poesie sich selber beflügeln – als wäre das Talent ein Ding für sich ohne den ganzen Menschen! – So zwischen halber Lust und Reue, versank er nach und nach immer tiefer in Melancholie, Verzagen an sich selbst, in Liederlichkeit und Armut, bis zuletzt ein zehrendes Fieber die müde Seele in seinen Traummantel einhüllte: da hörte er in seinen Phantasien das Posthorn wieder durch die Frühlingsnacht, dazwischen Waldesrauschen und das Glöcklein des Einsiedlers aus der Ferne. –

Er hatte mehrere Wochen krank gelegen. Als er endlich wieder zu sich kam, konnte er sich gar nicht besinnen, wo er war. Die Sonne schien über die Dächer freundlich durch das kleine Zimmer, eine Katze nickte auf dem Fensterbrett, nebenan hörte er einen Kanarienvogel singen, dann wieder eine Wanduhr dazwischen picken, seine alte Wirtin saß auf einem Lehnstuhl neben ihm am Bett und war über ihrem Strickzeug eingeschlummert. Er sah lange verwirrt in dieser Stille umher, eh er sie weckte. Nun fuhr sie freudig empor und erzählte ihm, wie sie schon für seine Seele gebetet, wie er irre geredet im Fieber, daß sein Freund noch immer nicht zurück sei, aber ein unbekanntes junges Mädchen sei vor langer Zeit einmal ins Haus gekommen und habe nach ihm gefragt. – Da dämmerte ihm allmählich alles wieder auf. „Kam das Mädchen nicht aus der Bergvorstadt?“ fragte er, und beschrieb ausführlich Schloß und Garten. Aber die Alte schüttelte den Kopf. „Der Palast“, sagte sie, „ist ja schon seit vielen Jahren unbewohnt“ – sie glaubte, er phantasiere wieder. Otto fuhr mit der Hand über seine Stirn, er war wie im Traume.

Eines Abends aber, als die Alte ausgegangen war, hatte er sich rasch angekleidet und ging heimlich die Treppe hinab, über die wohlbekannten Gassen und Plätze in die Vorstadt hinaus. Die Abendsonne funkelte lustig durch die Straße, Kinder spielten vor den Toren, die Mädchen plauderten an den Brunnen und Lerchen hingen jubelnd hoch

im rötlichen Dufte, er taumelte, wie berauscht, in der ungewohnten Luft. So kam er an den Garten der Geliebten, das Pförtchen war zu, aber er hatte den Schlüssel noch seit dem letzten Gange in der Rocktasche. Er schloß hastig auf und trat mit klopfendem Herzen hinein. Unterdes war die Sonne untergegangen, es war schon tiefes Abendrot. In der wunderbaren Beleuchtung kam ihm alles wie verwandelt vor; die Gänge, die er bisher nur bei Nacht flüchtig gesehen, schienen wüst und verwildert, und mit Schrekken fielen ihm die Worte der Alten wieder ein, als er endlich den Palast erblickte, denn kein Laut regte sich im ganzen Hause. Das Gras wuchs aus den Ritzen der Marmorstufen, die Türen und Fenster waren alle fest verschlossen, nur der Wind klappte eben mit einem halb zerbrochenen Laden, seitwärts schlug eine Nachtigall im Gebüsch, er hatte sie oft gehört, wenn er in den schwülen Sommernächten hier zum Liebchen schlich. – „Mein Gott, wo bin ich denn so lange gewesen!" sagte er in Gedanken versunken. – Da hörte er plötzlich in einiger Entfernung ein wohlbekanntes Lied aus alter Zeit:

„Jetzt wandr' ich erst gern!
Am Fenster nun lauschen
Die Mädchen, es rauschen
Die Brunnen von fern –"

Voll Freude antwortete er sogleich mit den folgenden Worten desselben Liedes:

„Aus schimmernden Büschen
Dein Plaudern, so lieb,
Erkenn' ich dazwischen –
Ich höre mein Lieb!"

„Barmherziger Gott – Kordelchen!" rief er auf einmal erschrocken aus. Die Schauspielerin stand vor ihm, sorgfältig geschmückt, frischgepflückte Blumen im Haar. – „Ist

er noch immer nicht zu Hause?“ fragte sie, nach dem Palaste schauend. – „Wer denn?“ entgegnete Otto ganz verwirrt. – Bei dem Klange seiner Stimme horchte sie hoch auf und sah ihn lange unverwandt an. „Ich kenn’ dich recht gut“, sagte sie dann mit einem schlauen Lächeln, „weißt du noch, wie du uns in jener regnichten Nacht zum ersten Male trafst, als wir nach einem kleinen Städtchen zogen? Damals hatt’ ich ein Loch im Strumpfe, Kamilla stichelte darauf, denn Kamillen sind bitter – ach nein, du bist’s nicht!“ schloß sie traurig. Dann hing sie sich in seinen Arm und flüsterte ihm geheimnisvoll zu: „Ich weiß wohl, wie er eigentlich heißt, aber ich verrat’s nicht, sag du’s auch nicht weiter, denn die Nacht hat Ohren – Ohren –

Und Augen verstohlen,
Wenn alles im Schlaf,
Da kommt er mich holen –
’s ist ein vornehmer Graf.“

„Kordelchen! Kordelchen!“ rief jetzt eine Stimme außerhalb des Gartens. Das Mädchen riß sich schnell los und verschwand wie ein aufgescheuchtes Reh zwischen den Bäumen. – Otto sah ihr lange nach, dann, plötzlich vom Entsetzen ergriffen, floh er unaufhaltsam über die öden Gänge, aus dem Garten, durch die einsame Vorstadt fort. Es war indes schon völlig dunkel geworden, die Sterne spielten munter am Himmel, von dem fernen Turme in der Stadt sang die Spieluhr wieder ihr frommes Lied; er mußte sein Gesicht mit beiden Händen verdecken, es war, als zögen Engel über ihm singend durch die stille Nacht. Zu Hause aber schnürte er hastig sein Reisebündel; noch denselben Abend, ungeachtet der Vorstellungen der besorgten Alten, verließ er die Stadt.
Der Eilwagen rollte auf der glänzenden Straße in die schöne Sommernacht hinaus, der Postillion knallte lustig, daß es weit über die stillen Felder schallte. Vorn im Kabriolett

plauderte ein Knabe, der zum ersten Male von Hause fuhr, munter mit dem Kondukteur, dann sah er wieder lange stumm in die Gegend, wie da die dunklen Schatten der Pappeln und seitwärts Büsche, Wälder und Dörfer im Mondenschein vorüberflogen, und wenn das Posthorn erklang, stiegen allmählich prächtige Schlösser und wunderbare Gärten und Gebirge mit Wasserfällen in der dämmernden Ferne vor ihm auf. Dann dachte er nach Hause, wie die Seinigen jetzt alle ruhig schlafen, der Mond scheint durchs Fenster über die Bilder an der Wand, nur eine Fliege summt tönend durch die stille Stube – da kam er sich auf einmal so verlassen vor hier draußen, und doch so tapfer und frei in der Fremde. – So reisefrisch war auch Otto früher gar manche schöne Frühlingsnacht zumute gewesen, heute saß er still vor sich hinbrütend im dunklen Wagen, es war ihm bei dem einförmigen, schlaftrunkenen Rasseln, als ginge es immer fort bergunter, unaufhaltsam einem unbekannten Abgrunde zu. Zuweilen blitzte der Mond oder das vorüberfliegende Licht eines Bauernhauses durch den Wagen und streifte flüchtig bald eine bleiche Nase, bald einen martialischen Schnurrbart, bald die Glasaugen einer Brille. Sie schwatzten viel von einer wunderschönen Opernsängerin und einem reichen Grafen S., einem lockern Zeisig. – „Nein, ein Dompfaff", rief der eine, „denn sie hat ihn pfeifen gelehrt." – „Vogel ist Vogel", meinte ein anderer kurz; „sie hat ihn tüchtig gerupft, nun ist sie selber davongeflogen." – „Eine barocke Idee", sagte der mit der Brille, „sich da in dem verfallenen Palaste in der Vorstadt einzunisten!" – Otto, aus seinen Gedanken auffahrend, horchte plötzlich auf. „Nisten!" fiel der Schnurrbart ein, „Turteltäubchen nisten gerade am liebsten in alten Ruinen, da ist's hübsch düster und nachtigallenhaft. Ja, mein Lieber, das hatte alles seine guten Wege, nämlich so unter den Bäumen sacht fort, die plaudern nichts aus. Konnte man wohl diskreter handeln, als der Graf? er ließ ihrer Treue ein Hinterpförtchen offen. Nun, nun, er ist ein Mann von kostbaren Erfahrungen, sie war

wenigstens nicht seine *prima* Donna, und, ich denke, er hatte eben auch keine Solopartie bei ihr." – Ein schallendes Gelächter erfolgte hier. Otto schnitt es durch die Seele, sie sprachen offenbar von seiner wundersamen Melusina! Es war ihm, als hätten die Gesellen mit ihren schmutzigen Reisestiefeln auf einmal einen köstlichen Teppich umgeschlagen, und er sähe nun die groben, rohen Fäden der glühenden Traumblumen – ihm graute recht vor dieser faden Kehrseite des Lebens.

Hier hielt der Wagen plötzlich vor einem Hause mitten im Felde, ein Mann in Nachtmütze und Pelz trat verschlafen mit einer Laterne heraus, um einige Pakete zu übergeben und andere in Empfang zu nehmen. Währenddes öffnete sich hinter ihm leise der Schieber eines kleinen Fensters, der Widerschein der Laterne beleuchtete flüchtig ein wunderschönes Mädchengesicht, das schnell wieder zurückfuhr. Otto erschrak, die Züge waren ihm bekannt, er konnte sich aber durchaus nicht besinnen. Da gähnte der Mann im Pelz. „Friß mich nicht, Mauschel!" rief ihm der lustige Kondukteur vom Kutschbocke zu. – „Ich esse kein Schweinefleisch", entgegnete der Jude trocken. Die Passagiere lachten, der Postillion knallte, und rasselnd flog der Wagen wieder in die stille Nacht hinaus.

Auf der nächsten Mittagsstation verließ Otto seine Reisegesellschaft, die jetzt schlummernd in allen Winkeln der Passagierstube umherlag, während die rüstigeren, überwacht und verdrießlich, nach Kaffee, Rum und Butterbroten durcheinanderschrien. Von hier aus gingen Seitenwege nach *Hohenstein,* dort im schattigen Grün wollte er ausruhen; er hofft' es noch vor Nacht zu erreichen, so matt und krank er sich auch fühlte. Er fragte nach dem nächsten Wege, man wies ihn auf einen Fußsteig, der gerade durch die Wälder führen sollte. Einsam schritt er nun zwischen die Berge hinein. Wie so anders, dachte er, als ich vor vielen Jahren hier auswanderte! Nun ist es Schlafenszeit, und alles ist vorüber. – Die schleichende Gewalt der Krankheit, von der durchwachten Nacht und An-

strengung neu geschürt, brach und reckte und dehnte ihn heimlich in allen Gliedern, er mußte öfters rasten und verließ endlich vor Ermüdung den Fußsteig, um womöglich ein Dorf zu erlangen. Aber kein Haus wollte sich zeigen, es war so still den Wald entlang, daß man die Spechte pikken hörte. So hatte er Zeit und Weg verloren; der Abend funkelte schon durch die Wipfel, die Gegend wurde ihm immer fremder, je weiter er fortging.

Da erblickte er seitwärts ein kleines Mädchen, das im Walde Blumen pflückte. Als er hinzutrat, wandte sie sich schnell herum, es war ihm plötzlich vor den klaren, unschuldigen Augen wie in den Himmelsgrund zu sehen. Die Abendsonne schimmerte durch die blonden Locken, er streichelte und küßt' es herzlich auf die blanke Stirn. Das schien dem armen Kinde selten zu begegnen, es suchte emsig in seiner Schürze und reichte ihm eine wilde weiße Rose, und als er fragte, ob es ihm den Weg aus dem Walde weisen könne, gab es ihm vertraulich die Hand, während es mit der andern sorgfältig das Schürzchen zusammenhielt, um seine Blumen nicht zu verlieren. Wie sie so miteinander fortgingen, wurde das schöne Kind immer vergnügter und gesprächiger. Es erzählte, es wäre gar nicht mehr so lange hin, da käme wieder Weihnachten, wo die vielen Lichter in den vornehmen Häusern brannten, dann säß' es in der Kammer auf seinem Bettchen am Fenster, da flimmerten draußen die Sterne so schön über dem Schnee, und das Christkindlein flöge durch die Nacht über den stillen Garten hin und brächt' ihm von seinen Eltern viele kostbare Sachen: neue rote Schuhe und ein Mützchen. – „Wo wohnen denn deine Eltern?" fragte Otto. – Die Kleine sah ihn erstaunt an, dann wies sie nach dem Himmel. – „Aber wo führst du mich denn jetzt hin?" fragte er fast betroffen wieder. – „Nach Hause", entgegnete das Kind. – Ihn schauerte unwillkürlich bei dem Doppelsinn der Antwort.

Auf einmal traten sie an einem Abhange aus dem Walde heraus, Otto stand wie geblendet. Denn tief unter ihm lag

plötzlich seine Heimatsgegend im stillen Abendglanze ausgebreitet: das schattige Städtchen, jenseits seiner Eltern Garten und Haus, der vergoldete Strom dann im Wiesengrunde und die fernen, blauen Berge dahinter – alles, wie er's in der Fremde wohl manchmal im Traume gesehen. Ganz erschöpft sank er unter dem Baume hin. „O stille, alte Zeit", rief er aus, „wie liegst du so weit, weit von hier." – Die Kleine hatte sich zu seinen Füßen ins Gras gesetzt. „Nein, nein", sagte sie, „so ist es nicht, ich will dich's lehren." Und bei dem Vogelschall, selbst wie ein Waldvöglein, sang sie mit dem kindischen Stimmchen:

„Waldeinsamkeit,
Du grünes Revier,
Wie liegt so weit
Die Welt von hier!
Schlaf nur, wie bald
Kommt der Abend schön,
Durch den stillen Wald
die Quellen gehn,
Die Mutter Gottes wacht,
Mit ihrem Sternenkleid
Bedeckt sie dich sacht
In der Waldeinsamkeit,
Gute Nacht, gute Nacht! –"

Otto dunkelte es vor den Augen, da ging auf einmal ein Leuchten über die Gegend wie ein Blitz in der Nacht: stille Abgründe fernab, Gärten und Paläste wunderbar im Mondglanz, er erkannte unten die goldenen Kuppeln und hörte durch die stille Luft herüber die Glocken wieder gehen und die Brunnen rauschen in Rom, und das Kind sang wieder dazwischen:

„O du stille Zeit!
Kommst, eh' wir's gedacht,
Über die Berge weit.

Nun rauscht es so sacht
In der Waldeinsamkeit,
Gute Nacht –“

„Still, still“, lachte die Kleine, „er schläft“ – aber der müde Wandersmann wachte nimmermehr auf.

Vierundzwanzigstes Kapitel

Wir aber, da es nun so still geworden im Tal und auf den Höhen, lassen die Blicke weit über das schöne Land hinschweifen, um nicht in Wehmut zu vergehen. Da rauschen die Wälder so frisch über Lust und Not, als riefe es: Menschenkind! blick' auf zum weiten Sternenhimmel, da ist ja doch alles eitel und nichts dagegen! – Und fern im Gebirge, wo der Mond so hell über die Waldwiese scheint, gewahren wir plötzlich zwei Wanderer, die fröhlich niedersteigen: es sind die beiden Liebesleute auf ihrer abenteuerlichen Fahrt. *Fortunat* hat soeben die Pferde in einem Dörfchen untergebracht und wendet sich mit *Fiametta* auf einem Fußsteig zwischen die leise bewegten Kornfelder hinein, die Nacht kühlt sich am Horizont mit Wetterleuchten, eine Wachtel schlägt fern im Felde. Vor ihnen aber breiten sich dunkle Höhen aus, der Mond beleuchtet nur einzelne Abhänge, da erkennt er nach und nach Lauben und Gänge, zuweilen blitzt ein Springbrunnen auf, aus den duftigen Gebüschen hören sie schon die Nachtigallen über das Feld herübertönen. Auf einmal hält Fortunat still und schwenkt voll Freude seinen Hut. „Grüß dich Gott, du kühler Wald!“ ruft er aus Herzensgrunde. Fiametta sieht ihn einen Augenblick fragend an, dann schwenkt auch sie jubelnd ihr Hütchen, ohne zu wissen warum. – Es ist *Hohenstein*, das vor ihnen liegt.
Fortunat wußte, das Walter jetzt ganz hier wohne; zu dem Hilfreichen, Besonnenen, Ratvollen wollte er zunächst

das Mädchen bringen. Er hatte gehofft, die Berge noch am Abend zu erreichen, nun aber mochte er kein Aufsehen machen, sie beschlossen, die kurze Sommernacht im Garten zu verweilen, um sogleich am frühesten Morgen dem alten Freunde alles zu vertrauen.

Er kannte noch aus alter Zeit den Steg im Gartenzaun, sie schwangen sich hinüber und stiegen mit klopfendem Herzen den Waldberg hinan. Fortunat blickte oft seitwärts zwischen die Bäume hinein nach den stillen Gängen, wo er so oft gewandelt, es war alles so fremd und unheimlich im Mondenscheine. „Das ist Jakobs Traumleiter", sagte er fröhlich, „wie sie der liebe Gott zuweilen in solchen Frühlingsnächten herunterläßt. Nur frisch! Wir steigen ins Himmelreich, ich seh' schon die Sterne durch die Wipfel flimmern." Jetzt hatten sie die letzten Stufen erreicht, auf einmal traten sie zwischen dem dunklen Laub, wie Bergleute aus einem Schacht, ins Freie hinaus. Da sahen sie rechts das alte Schloß und vor ihm die weiten, duftigen Blumenplätze, stille Lauben und Büsche, ein Springbrunnen plätscherte schläfrig dazwischen, weiterhin dämmerte eine unermeßliche Aussicht im Mondglanze durch die wunderbare Einsamkeit herauf. Fortunat schaute schweigend in die Runde, und eh' die kleine Marchesin sich noch besinnen konnte, hatte er schon eine weitgebreitete Linde bestiegen, die am äußersten Abhange über den schimmernden Abgrund hinausging. „Fiametta!" rief er von oben, „wär's nicht um dich, ich möchte alles wachschreien vor Freude! Sieh, da unten blickt der Strom manchmal so heimlich auf, drüben grasen Damhirsche am mondbeschienenen Abhang, nun seh' ich auch das Dorf, wo die lustigen Mädchen wohnen, mit denen ich hier oben getanzt, das schläft nun alles, alles – nur eine Turmuhr schlägt dort von fern herüber, ich hörte sie damals oft bei stiller Nacht. Und Gott Vater fährt über die Saiten seiner Harfe, wie eine leise Musik zieht's gnadenreich über die stille Gegend."

Fiametta aber sah sich nach allen Seiten um wie ein scheues

Reh. In dem dunklen Buchengange, der vom Schlosse herabkam, schwankte das Mondlicht, als bewegten sich bleiche Gestalten, sie fürchtete sich so allein da unten. Fortunat bemerkte es endlich, er reichte ihr die Hand, sie stieg schnell auf die Bank, die unter dem Baume stand, und schwang sich so lachend mit hinauf. Dort setzten sie sich nun zusammen in dem dämmernden Laube bequem zwischen die Äste zurecht, vor ihnen schossen Sternschnuppen über das Land, manchmal bellte ein Hund fern in den Dörfern. Fiametta baumelte, in Erwartung der Dinge, zufrieden mit den Beinchen. „Nun erzähle was", sagte sie. Und Fortunat besann sich nicht lange, die alte phantastische Nacht flüsterte verworren durch die Zweige, er fing sogleich aus dem Stegreif an, als spräch' er im Traum:

„Es waren einmal zwei Kinder, Kasperl und Annerl, die hatten einander sehr lieb. Die saßen einmal vor dem Hause und besahen schöne Bilder in einem großen Bilderbuche, das die Annerl mitgebracht hatte, die Vögel sangen im Walde, und das Abendrot ging über die Berge vor ihnen. Auf dem Bilde war eine sehr schöne Gegend zu sehen, fruchtbare Auen, Flüsse, Dörfer und Schlösser, dahinter ein wunderbar gezacktes Gebirge mit einsamen Kapellen und Wäldern, an deren Saum eine Prozession mit bunten Fahnen dahinzog. Das Abendrot schien über das Bild, und wie sie es so mit rechtem Fleiße betrachteten, da fingen auf einmal die gemalten Bäume an leise zu rauschen, schöne, bunte Vögel flogen über die Landschaft, die Brünnlein glitzerten im Gebirge, die Fahnen wehten, sie hörten die Prozession aus weiter Ferne singen. Und eh' sich der Knabe noch besinnen konnte, sah er zu seinem Erstaunen auch das kleine Annchen schon mitten drin, sie winkte ihm fröhlich, er faßte sich endlich ein Herz und sprang ihr nach, so liefen sie beide voller Freuden in das Buch und in die Landschaft hinein. – Als Kasperl einmal zurücksah, war ihr Haus und die Gegend, wo es stand, schon hinter ihnen verschwunden, von der Prozession

hörten sie nur noch manchmal den Gesang herübertönen, die Sonne war lange untergegangen, je weiter sie kamen, je einsamer und prächtiger wurde alles. Auf einmal, da sie eben durch einen Felsbogen traten, erblickten sie ein himmelhohes Gebirge vor sich, daß es ihnen ordentlich den Atem verhielt. Auf dem höchsten Berge stand ein herrliches Schloß, das war von lauter Silber, mit Gold gedeckt, vor dem Schloßtore aber saß eine wunderschöne Frau, die war über einer Harfe eingeschlummert. Aus ihren langen Locken und Gewändern kam ein prächtiger Mondenschein und beleuchtete die Alpen und die wundersamen Klüfte, Wälder und Abgründe ringsumher. Unten, wo die Strahlen nicht mehr hinlangen konnten, sahen sie kleine bucklige Männchen in der Dämmerung lustig von den Felsenzacken Purzelbäume schießen, von fern klang das Glöcklein eines Einsiedlers, ein Jäger, der sich verirrt hatte, stand auf dem Felsen gegenüber und gab zuweilen mit seinem Waldhorne Antwort. Oben aber am Schlosse weideten weiße Schäfchen auf den Abhängen, hoch vom Turme der Burg bliesen Engel auf silbernen Zinken wunderschön über die stillen Gründe."

„Ach, da möcht' ich auch einmal hin!" rief hier Fiametta freudig aus. – „Es ist nur gar zu weit von hier", erwiderte Fortunat – „aber wackle nicht so mit den Beinchen, wir fallen sonst beide vom Baume." – Sie rückte sich nun näher zum Hören zurecht, und Fortunat fuhr wieder fort:

„‚Das ist die Göttin Luna', sagte nun Annerl, auf die Frau vom Schlosse weisend. – ‚Kennst du sie denn?' fragte Kasperl verwundert. – Sie lachte: ‚Du bist doch noch sehr dumm für dein Alter, bleib jetzt nur dicht bei mir, sonst verirrst du dich hier.' – Kasperl aber sah nun einen alten, großen, geduckten Mann seitwärts am Wege sitzen, der hatte einen Sack voll prächtiger Äpfel umhängen. Da wurde er ganz genaschig, er wollte nur geschwind noch ein paar Äpfel auf den Weg kaufen, wie er aber in den Sack hineinguckt, erwischt ihn der Mann schnell bei den Füßen, wippt ihn so hinein und schnürt den Sack über ihm

fest zu. ‚Aha, nun hab' ich dich!' sagte er und streckte zufrieden die Beine aus, um ein wenig auszuruhen."

„Pfui, der abscheuliche Kerl!" unterbrach ihn hier Fiametta von neuem, „ich möchte so einen Menschenfresser am liebsten gleich zerpflücken! Nun kommen gewiß die armen Kinder auseinander."

„Ja freilich", entgegnete Fortunat. „In der Angst und Finsternis arbeitete Kasperl wütend mit seinen Ellbogen in den Äpfeln herum. ‚Aber seien Sie doch nicht so sackgrob, Sie erdrücken mich ja', wisperte da plötzlich ein feines Stimmchen neben ihm. – ‚Bist du's?' fragte er leise. ‚Jawohl', antwortete das Stimmchen, ‚ich bin auch gefangen und nage schon lange an dem Sack, daß mir die Zähne weh tun. Jetzt ist der Alte eingeschlafen, hören Sie nur, wie er schnarcht. Sie haben so starke, dicke Finger, seien Sie doch so gütig und helfen Sie mir ein wenig reißen.' – Es war ein allerliebstes, kleinwinziges Mäuschen, das so artig sprach. Kasperl riß nun ganz vorsichtig an dem Sack, das Mäuschen wischte hinaus, biß ihn im Fortspringen noch schelmisch in den Finger und verschlüpfte dann schnell im Mondenschein, er hörte es noch fern zwischen den Steinen kichern. Jetzt kroch er selber sacht hervor, steckte noch geschwind einen hübschen Apfel in die Tasche und nahm dann eilig Reißaus. – Aber, Gott weiß, der Alte mußte einen groben Flausrock anhaben, denn Kasperl geriet auf einmal in ein verworrenes, ungebürstetes Gestrüpp, in der Eile hatte er den Weg verloren und war, anstatt herabzuklettern, an dem alten Rockärmel gerade hinaufgelaufen. Als er aber oben stand, erstaunt' er erst recht! Da war der Morgen schon angebrochen, der Menschenfresser unter ihm war nichts anderes als der alte, graue Fels vor seines Vaters Hause, und wo er das prächtige Schloß gesehen hatte und die wunderbaren Klüfte im Mondenschein, da lagen jetzt fahle, dicke Wolken übereinander und dehnten sich noch halb im Schlaf. Er sah die Schornsteine in seinem Dorfe rauchen, der Nachbar trat gähnend in die Tür. ‚Kikeriki!' rief er, ‚Kasperl, du willst wohl den

Tag auskrähen, daß du dich da so früh auf den alten Steinjürgen gestellt hast.'" –

„Aber das arme Annerl!" fiel Fiametta wieder ein. – „Wart nur, es wird gleich noch viel schöner kommen", erwiderte Fortunat. „Das schöne Annerl war fort und kam nicht wieder, und niemand wußte was von ihr, denn sie war immer nur gegen Abend heimlich aus dem Walde mit ihm zu spielen gekommen. Da war Kasperl ganz traurig, er mußte viel lernen und sehnte sich sehr und wurde darüber nach und nach groß und stark. Einmal des Nachts aber, als der Mondenschein über die Wälder glänzte, da kam es ihm vor, als säße die wunderschöne Frau draußen auf dem Berg vor dem Hause und blätterte in dem alten Bilderbuche, daß der Goldschnitt beim Umwenden zuweilen seltsam über die Bäume am Fenster funkelte. Da wurde er sehr unruhig, und als kaum noch der Morgen dämmerte, saß er schon angezogen in seiner Kammer am Tisch, den Kopf in die Hand gestützt. Da fiel es ihm erst ein, daß er den Apfel, den er damals aus dem Sacke mitgenommen hatte, noch immer in der Tasche hatte. Er nahm ihn heraus und biß vor Schwermut drein, um ihn aufzuessen. Da schreit auf einmal etwas drin, und ein Köpfchen streckt und zwingt sich hervor, und wie er endlich verwundert den Apfel aufbricht, steigt ein kleines braunes Kerlchen mit Wanderstab und Tasche aus dem Kernhause. – ,Wer bist du?' – ,Der Äpfelmann. Adieu!' – Das Männchen ging über den Tisch fort, blieb aber plötzlich am Rande stehen, weil er nicht herunterkonnte. – ,Ich will dir wohl herunterhelfen, du armer Wicht', sagte Kasperl, ,aber du mußt mir dagegen etwas versprechen. Kannst du mich zu der Göttin Luna führen?' – ,Warum nicht?' erwiderte das Kerlchen. Da nahm er es sauber zwischen die Finger und setzte es draußen auf den Rasen. Nun traten sie sogleich ihre Wanderschaft an. Der Kleine hinkte, denn Kasperl hatte ihn vorhin im Apfel in die große Zehe gebissen. Kaum aber waren sie weiter in die Heide gekommen, so humpelte das Kerlchen so ungeheuer schnell fort wie ein

Grashüpfer und lachte und rief immer zurück: ‚Komm mir doch nach, komm mir doch nach, hast ja so lange Beine!' und ehe sich's Kasperl versah, hatt' er das Kerlchen in dem hohen Grase verloren. Da war er nun wieder so klug wie vorher. – Es war aber gerade ein schöner Sonntagsmorgen. Ein Birnbaum ging eben übers Feld zur Kirche und rauschte Gottes Lob. ‚Gelobt sei Jesus Christ!' grüßte ihn Kasperl, ‚habt Ihr nicht so einen kleinen braunen Pilgrim gesehen?' – ‚In Ewigkeit', entgegnete der Birnbaum, ‚ich glaube, ich habe vorhin so was im Grase zertreten.' – ‚Ach Gott', klagte Kasperl, ‚der hat mich irregeführt, nun weiß ich nicht, wo ich bin! wenn ich nur einen Felsen oder Turm wüßte, um mich ein wenig umzusehen in der Welt.' – ‚Jetzt hab ich keine Zeit zu Narreteien', meinte der Birnbaum; da aber Kasperl betrübt weitergehen wollte, tat es ihm leid. ‚Nun, komm nur schon, komm, was man auch für Not hat mit euch Kindern', sagte er und stieg schnaufend und ächzend auf einen hohen Berg hinauf, wo er sich breit zurechtstellte und seine grünen Äste lustig in die blaue Luft hinausstreckte. Das ließ sich Kasperl nicht zweimal sagen, er kletterte schnell bis zum Wipfel hinan – da aber warf er plötzlich seinen Hut hoch in die Luft und schrie Hurra! aus Leibeskräften, denn jenseits sah er auf einmal das wunderbare Gebirge wieder, daß ihm ordentlich schwindelte vor großer Freude. – ‚Nun zaus' mich doch nicht so grob, das tut ja weh', sagte der Baum. Aber Kasperl schwang sich schon hastig wieder hinab; ‚Gott's Lohn! Gott's Lohn!' rief er einmal übers andere. Der gute Birnbaum aber schüttelte sich zum Valet im Morgenhauch, daß der ganze Rasen voll schöner, goldener Früchte lag, die kollerten und hüpften lustig über den grünen Abhang hinunter, und Kasperl sprang ihnen nach zwischen den Morgenlichtern in die prächtige Gegend hinein. – War nun das Gebirge beim Mondglanz schön gewesen, so war jetzt alles noch viel tausendmal schöner im funkelnden Morgenlicht. Das prächtige Schloß mit seinen stillen Türmen stand ganz in rosenroter

Glut, die Bäche waren von purem Gold, die Wälder rauschten und blitzten von Rubinen und Smaragden, auf den Alpen standen Engel umher und fachten mit ihren langen, regenbogenfarbenen Flügeln das Morgenrot an. Und als er endlich zum Walde kam, da erblickte er auf einmal ein wunderschönes Mädchen auf einem weißen Hirsch, die hatte ein lustiges, funkelndes Krönlein im Haar. ‚Mein Gott! die sollt' ich ja kennen', dacht' er bei sich – es war sein liebes Annerl! – Sie hielt lachend still und sagte: ‚Die schöne Frau Luna ist verwichene Nacht untergegangen, sie läßt dich noch grüßen, ich aber bin ihre Tochter Aurora, die Königin der Wälder.' – ‚So will ich König sein', rief Kasperl und schwang sich hinter sie auf den Hirsch, und hui! ging's nun durch die Waldesnacht unter einsamen Burgen, an kühlen Strömen und Gärten und schimmernden Fernen vorüber, und jedem ging das Herz auf, der sie von fern vorüberfliegen sah. – So hausten sie fortan miteinander in freudenreichem Schalle, und da sie nicht gestorben sind, so leben sie noch heute – denn ich bin der verliebte Kasperl, und du die Waldkönigin Aurora, mein liebes, liebes Dichterweibchen!"
So schloß Fortunat und küßte herzlich Fiametta auf die verschlafenen Augen. Da stieß sie ihn leise an und wies in das Land hinaus. Ein leiser Schimmer flog über die Gegend, wie wenn ein Kind im Traume lächelt, eine früherwachte Lerche hing schon liedertrunken über ihnen hoch in der Dämmerung. „Grüß dich Gott, du schöne, wunderbare Welt!" rief Fortunat; „jetzt frisch ans Werk!" – Sie schüttelten sich schauernd in der Morgenkühle, er sprang schnell vom Baum, Fiametta folgte, er fing sie unten in seine Arme auf. Dann gingen sie schweigend miteinander durch den dämmernden Garten.
Fortunat hatte sich schon im voraus alles klug ausgesonnen. Fiametta sollte fürs erste sich in der Nähe der Amtmannswohnung noch im Grünen verborgen halten, er selber wollte unterdes in der Morgenluft wie ein Falke das Haus umkreisen und auf Walter, den er als einen frühen

Vogel kannte, wo er sich blicken ließe, sogleich niederstoßen, um mit ihm das weitere zu verabreden, bevor die andern dazukämen.

Aber der Mensch denkt, und Gott lenkt. Als sie so unter den feierlich rauschenden Wipfeln des Buchenwaldes fortgingen, zupfte und rückte Fiametta mit klopfendem Herzen ihr Wämschen zurecht wie ein Vöglein, das sich im Morgenschein die Flügel putzt, und fing italienisch zu plaudern an, das klang wie ein Glöckchen durch die Stille. Fortunat aber gedachte des schönen Frühlingsmorgens, als er mit Walter zum erstenmal hier eingestiegen. Da war alles wieder so kühl und frisch wie damals. Bald erblickte er seitwärts die duftigen Blumenplätze, den Sitz unter der Linde, lauter alte Bekannte, nun guckten auch schon die weißen Schornsteine herüber – auf einmal standen sie unter den hohen Bäumen vor dem Hause. Da lag noch alles in tiefer Ruhe, durch das Weinlaub am Fenster konnte er die untere Stube übersehen, den bunten Teppich im ungewissen Schimmer und die vergoldeten Rahmen der Bilder gegenüber an der Wand, die alte Stockuhr schlug drin soeben vier. Unter den Bäumen aber stand noch der große runde Tisch mit den Stühlen umher, wie in der alten Zeit, der Amtmann hatte seine Pfeife draußen vergessen, auch Florentinens Gitarre hing wieder über dem Stuhl. Da überkam Fortunat unwiderstehlich seine alte Reiselustigkeit, der kluge Plan, Vorsicht, Geheimnis und alles war vergessen, er ergriff die Gitarre, sprang auf den Tisch hinauf und sang recht aus Herzensgrunde:

„Wer steht hier draußen? – Macht auf geschwind!
Schon funkelt das Feld wie geschliffen,
Es ist der lust'ge Morgenwind,
Der kommt durch den Wald gepfiffen.

Ein Wandervöglein, die Wolken und ich
Wir reisten um die Wette,
Und jedes dacht': Nur spute dich,
Wir treffen sie noch im Bette!

Da sind wir nun, jetzt alle heraus,
Die drin noch Küsse tauschen;
Wir brechen sonst mit der Tür ins Haus:
Klang, Duft und Waldesrauschen."

Bei den letzten Klängen öffnete sich oben leise ein Fenster. Florentine fuhr mit dem verschlafenen Köpfchen hervor, er hätte sie beinah nicht wiedererkannt, so prächtig, voll und blühend war sie geworden. „Herr Jesus! sind Sie's, Herr Baron?" rief sie ganz erschrocken, und schlug schnell das Fenster wieder zu, denn der Morgenwind wollte ihr das leichte Halstuch nehmen. Nun hörte er im Hause die Türen gehen, rufen und rumoren. Draußen aber kletterte das Morgenrot schnell über die Spaliere und Weinranken das stille Haus bis zu den Schornsteinen hinan und guckte neugierig über die Bäume und Fortunat sang von neuem:

„Ich komme aus Italien fern
Und will euch alles berichten,
Vom Berg Vesuv und Romas Stern
Die alten Wundergeschichten.

Da singt eine Fey auf blauem Meer,
Die Myrten trunken lauschen –
Mir aber gefällt doch nichts so sehr,
Als das deutsche Waldesrauschen!"

Als Walter von Italien hörte, zweifelte er nicht länger. Eilig in hohen Schmierstiefeln, die er gegen den beißenden Tau zu gebrauchen pflegte, kam er atemlos aus dem Hause gestürzt. „Mein Gott! du, Herzensbruder!" rief er schon von weitem und flog, außer sich vor Freude, in Fortunats Arme und stach ihn tüchtig mit seinem schlecht rasierten Bart. Fortunat war im ersten Augenblick ganz verblüfft, denn Walter kam ihm so verbauert vor, altmodisch, beim Reden schreiend und gebräunt im Gesicht; aber die treuen

Augen machten gleich alles wieder gut, man sah bis in den Grund der Seele, er war doch durch und durch noch der alte. Jetzt fiel plötzlich ein Schuß hinter ihnen, daß alle zusammenfuhren. Auf der Anhöhe wurde der tolle Förster sichtbar, der von dem Besuche schon Wind bekommen. Er drehte den dünnen, flechsenreichen Hals weit aus der schmalen, engen Binde, und als er nun wirklich Fortunat recht erkannt hatte, feuerte er aus seiner Doppelbüchse geschwind noch einmal über ihre Köpfe weg und stürzte mit großem Vivatgeschrei zu ihnen herab. Dann erwischte er unversehens Fiametta, die gar nicht wußte, wie ihr geschah, und walzte wütend mit ihr unter den Bäumen herum, seine langen Rockschöße, die weit im Rade umherflogen, schleuderten einen von den Gartenstühlen soeben in die Haustür, als die Frau Amtmann feierlich heraustreten wollte. „Nun Gott behüt' uns, Herr Nachbar", rief sie empfindlich, „was ist das wieder für eine Aufführung!" – „Einführung, Frau Muhme", entgegnete der Förster, „hohe Reisende, bal à la fourchette, St. Veitstanz, Apfelsinen und Italien! da hat mich so eine verfluchte Tarantul in die Füße gebissen." – Nun schwenkte auch der Amtmann seine schneeweiße Schlafmütze, hinter der das hübsche Gesichtchen Florentinens hervorguckte, alle schrien und fragten durcheinander, die Amtmann knickste unter vielen Redensarten, die niemand hörte, ein aus dem Schlafe verstörter kalekuttischer Hahn hatte sich schon während des Walzers in des Försters fliegende Schöße verbissen – man konnte gar nicht zu Worte und ins rechte Geschick kommen. – „Und der junge Herr? – mit wem hab ich die Ehre?" – sagte endlich die Frau Amtmann, mit einem halben, ungewissen Knicks gegen die hocherrötende Fiametta gewendet. – „Himmeltausend! da hab' ich nun was Schönes angerichtet!" dachte Fortunat. Er besann sich nicht lange. „Ein junger Vetter von mir aus Italien", sagte er. – „Ah" rief der Förster erstaunt, und entschuldigte nun mit abenteuerlicher Galanterie die ganz ergebenste Erdreistung seiner nichtwissenden Keckheit. Er mühte sich

sichtbar ab, in seinen überaus höflichen Diskurs eine Menge italienischer Worte zu mischen, so glaubte er verständlicher zu werden, kam wieder auf die Taranteln zu sprechen, die eigentlich in Italien ansässig seien, ging dann auf die Skorpionen über, auf die er einen ganz besondern Haß geworfen zu haben schien, und bot ihr endlich eine lange, frischgestopfte Pfeife an. – „Nicht doch, die Herren Italiener pflegen nicht zu rauchen", fiel die Amtmann vermittelnd herüber. – „Nun, so tu' ich's selbst mit Erlaubnis", erwiderte der Förster und fing in schnellen Zügen heftig an zu dampfen, während die allzeit heitere Fiametta, in dem dicken Tabaksqualm sich manchmal verhustend, ihm in aller Geschwindigkeit die ungeheuersten Geschichten erzählte von geflügelten Skorpionen und einer wahnsinnigen Tarantel, die den St. Veitstanz erfunden.

Der Amtmann, als sich endlich der erste Jubellärm ein wenig gelegt hatte, blickte vergnügt in die Runde. „Im Kalender", sagte er, „ist heute kein Feiertag angemerkt, aber der liebe Gott hat ihn draußen rot angestrichen, so weit man nur sehen kann." Und in der Tat, das alte Schloß, die Wälder, Strom und Täler glühten nun ringsum im schönsten Morgenrot. Die Frauen hatten unterdes den Tisch gedeckt, die Vögel sangen über ihnen im Walde, und die Morgenlichter funkelten lustig über die Weinflaschen und Gläser auf dem blendend weißen Tischzeug. Walter legte in seiner Fröhlichkeit die Gitarre in Florentinens Arm, sie mußte, nicht ohne häufiges Erröten, gleich zu Willkomm alle Lieblingslieder des Hauses durchsingen. Eine tiefe Wehmut flog dabei durch Fortunats Seele: es waren noch immer dieselben Lieder, die er damals hier gesungen und gedichtet – so lange hatten sie nachgeklungen in dieser Einsamkeit! – Dann mußte er selbst ihnen von seinen Reisen, von Rom und Sizilien erzählen, dazwischen kamen immer wieder hiesige Geschichten aufs Tapet von alten Bekannten und von den hübschen Mädchen, mit denen er damals im Garten getanzt, sie zeigten ihm die Dörfer in

der Ferne, wo sie nun glücklich verheiratet waren, da ein grünverschattetes Pfarrhaus, dort ein paar Schornsteine einsam über dem Wald. Der nach literarischen Neuigkeiten ausgehungerte Walter versuchte mehrere Male vergeblich, ein wissenschaftliches Gespräch mit Fortunat anzuknüpfen. Er hatte noch immer die alte Angst, mit der Bildung fortzuschreiten, und hielt eine Menge Journale, die aber meist ungelesen blieben und von seiner hübschen Frau zum Kuchenbacken verbraucht wurden. Diese hatte sich jetzt mit ihrem Kinde an der Brust vor die Haustür gesetzt, die Morgensonne spielte zwischen dem Weinlaub lieblich über Mutter und Kind. Zuweilen blickte sie unter ihren langen, dunklen Augenwimpern scharf nach Fiametta hinüber, die unterdes, das Köpfchen auf beide Arme gelegt, über dem Schwirren und Summen der Gläser, Teller und Reden am Tische eingeschlafen war.

So war es unter den muntern Gesprächen fast völlig Tag geworden, als auf einmal Walter, einen erbrochenen Brief in der Hand, eilig aus dem Hause trat. „Das ist heut ein wahrer Wundermorgen!" rief er lachend aus, „denkt euch, da schreibt mir eben unser Rechtsfreund aus der Stadt, ich möchte ihm kollegialisch beistehen, eine junge adelige Dame auszukundschaften, die mit ihrer Kammerjungfer ihrer alten Tante entflohen und deren Spur zwischen unsern Bergen verloren gegangen sein soll." – „Kurios", sagte der Amtmann, „ja, wilde Wasser lieben die Berge." – „Was!" – rief der Förster, der eben eine neue Pfeife gestopft und nur halb hingehört hatte – „eine alte, wilde Tante ist im Wasser verlorengegangen?" „Ja", fiel Fortunat ein, „und der Rechtsfreund mit ihrer Kammerjungfer entflohen." – Walter hatte Mühe, die Konfusion zu berichtigen. „Ein angesehener Mann", fuhr er dann fort, „verfolgt nun die Flüchtlinge im Auftrag der Tante und hat in der Stadt amtliche Hilfe in Anspruch genommen. Da bist du uns eben zur rechten Stunde gekommen, Fortunat." – „Ich? wieso?" fragte dieser betroffen. – „Ich meine, als Dichter in solchen romantischen Fällen." –

„Ach teurer Freund“, entgegnete Fortunat, „ich wollte, die Romantik wäre lieber gar nicht erfunden worden! Solche romantische Verliebte – und das ist die adlige Dame gewiß samt der alten Tante und dem Rechtsfreund und seiner Kammerjungfer – die machen zusammen an einem Morgen mehr dumme Streiche, als ein gesetzter Autor im letzten Kapitel jemals wieder gutmachen kann!“ – Da hatte er nun eben recht das Kapitel der Frau Amtmann getroffen. Sie nickte ihm freundlich zu, klagte über den jetzigen Leichtsinn der Jugend und schob alles auf die Poesie. Fortunat stimmte ihr in seiner Not gern bei und hetzte noch immer mehr gegen die Poeten. Der Förster aber, nachdem er endlich alles begriffen, saß währenddes wie in Konvulsionen des heftigsten Nachdenkens, bald starrte er in den Himmel, bald wieder in die dicken Tabakswolken vor sich hin. – „Topp, sie ist's“, rief er plötzlich aufspringend aus und schlug mit der Hand auf den Tisch, daß die Gläser klirrten. „Wer?!“ – wandte sich Fortunat erschrocken herum. Über dem Lärm war Fiametta aus dem Schlafe aufgefahren, Florentine sah ihr wieder scharf in die verträumten Augen – es hing alles an einem Haar.
Aber der Förster legte schnell die Pfeife hin und setzte martialisch seinen dreieckigen Hut auf. „Jetzt kommt nur mit, alle die ihr hier seid, zur Mühle dort im Wald, aber sogleich, damit wir die Vögel noch im Nest erwischen.“ – Fortunat atmete wieder leichter auf. – Vergebens drang man nun in den Geheimnisvollen, sich näher zu erklären. „Ich will die alte Tante sein“, sagte er nur, „wenn ich euch nicht das Fräulein schaffe, und sollte sie wie ein Eichhörnchen von Baum zu Baum springen.“ – Die Amtmann mochte von dem Abenteuer nichts wissen und blieb mit Florentine zurück, die andern aber wanderten erwartungsvoll dem Walde zu. In dem allgemeinen Aufruhr konnte Fortunat durchaus keinen Augenblick gewinnen, Walter auf die Seite zu nehmen, so oft er ihm auch heimlich zuwinkte.
Nach einem kurzen Gang erblickten sie die Mühle in einer

einsamen Waldschlucht. Von einem Bergeshange tief verschattet war in dem kühlen Grunde kaum noch der Tag angebrochen, die Vögel erwachten eben erst in dem stillen Gärtchen, nur die Tauben schimmerten vom Dach, das schon von der Morgensonne beleuchtet war. Hier verteilte der Förster seine Begleitung vorsichtig an allen Ausgängen, und gebot ihnen, sich still zu halten, er selbst aber ging eilig in die Mühle. Da sahen sie, wie sich im Hause ein Dachfenster halb und leise öffnete, sie glaubten oben ein junges Mädchen zu bemerken, das bei ihrem Anblick schnell den Laden wieder zuschlug. „Was ist denn das?" – flüsterte Fiametta ängstlich Fortunat zu. – „Ich glaube", erwiderte dieser, „der ganze Morgen ist toll geworden und spiegelt unsere eigene Geschichte närrisch in der Luft." – Jetzt entstand ein Tumult im Hause, der Waldbach stürzte plötzlich brausend über das Mühlrad, zwischen dem Rauschen hörten sie rennen, klappen und zanken. Auf einmal sprang die Haustür auf, und der Förster trat mit triumphierendem Anstande hervor, er führte feierlich eine fremde, wohlgekleidete Dame am Arm, der Morgenwind schlug ihren grünen Schleier zurück und zeigte ein junges, schönes Gesicht. – Da besann sich Fortunat nicht lange. „Welche Überraschung, mein Fräulein!" rief er schnell hinzuspringend aus – „als ich das Glück hatte, Sie bei Ihrer verehrungswürdigen Tante zu sehen, wer hätte da an diese verwünschte Mühle gedacht! Ich bedaure nur, wenn dieser vorwitzige Morgenwind zu früh den Schleier gelüftet und das harte Gebirg manchen Stein des Anstoßes" – Nun war auch Fiametta dazu gekommen und drückte die Hand der Dame zärtlich an ihr Herz. „Himmlisches Mädchen", sagte sie, „und das alles um mich! – Aber wie war es möglich? wie erfuhrst du, wo ich Unglückliche umherschweife? Ja, leugne nur nicht länger, ich weiß es ja doch, du Liebe, Arme! um mich verließest du Schloß und Tante – o es geht mir alles wie ein Mühlrad im Kopfe herum!" – Die Dame sah in höchster Verwirrung bald den einen, bald den andern an, und wuß-

te nicht, was sie erwidern sollte. Die beiden ließen sie aber nicht mehr los, sie führten sie in ihrer Mitte so rasch der Amtmannswohnung zu, daß die andern kaum folgen konnten, dabei sprachen sie unterwegs oft heimlich untereinander. Walter war ganz verdutzt, auch der Amtmann schüttelte bedenklich den Kopf, der Förster aber schimpfte voller Zorn. „So eine schöne Dame“, sagte er, „und einem solchen welschen Milchbart nachzulaufen, dem die Eierschalen noch am Schnabel hängen! Da ist keine Gerechtigkeit in dem Handel, ebensogut könnte sich der Herr Amtmann da in mich verlieben.“ Dann pfiff er mit großem Lärm auf dem Finger nach seinen Hunden, warf die Büchse auf den Rücken und schritt ohne Abschied in den Wald.

Unterdes waren die andern zu Hause angelangt, wo Fiametta sehr fröhlich den erstaunten Frauen ihre unverhofft wiedergefundene Freundin vorstellte. Walter wollte folgen, aber Fortunat faßte ihn am Arm und führte ihn rasch in den Garten hinein. „Nun hilf aus der Konfusion!“ rief er aus, da sie allein waren, „denn die gefangene Dame ist eigentlich die Kammerjungfer meines Vetters, und mein Vetter ist meine Geliebte, und meine Liebste ist die entsprungene Nichte der alten Tante.“ Er erzählte nun kurz den ganzen Hergang der Sache, und wie die Kammerjungfer, plötzlich so verlassen in der Fremde, heimlich ihre Spur im Gebirge verfolgt und gestern abends – was der Förster zufällig erfahren haben müsse – in der Waldmühle eingekehrt sei, um erst das Terrain auszuforschen und sich des Morgens auf eine geschickte Art wieder mit ihnen zu vereinigen.

Als er geendigt hatte, hüllte er sich in sich selbst, um den Hagelschauer freundschaftlicher Vorwürfe geduldig abzuwarten. Walter aber, aus seiner einförmigen Einsamkeit so auf einmal mitten in das romantische Abenteuer mit hineingeworfen, rief zu seinem Erstaunen: „Deine kleine Marchesin will ich mit Gut und Blut wie meinen Augapfel beschützen“, und rannte dann voll Begeisterung sogleich

nach dem Hause zu. Unterwegs begegnete ihnen Florentine und fragte, was sie vorhätten? Walter in seiner Freude erwischte sie bloß beim Kopf, küßte sie tüchtig ab und wollte weiter. Aber sie hielt ihn fest. „Tut mir nur nicht so wichtig und geheimnisvoll", sagte sie, „merkt' ich's doch längst!" – Walter sah sie groß an. – „Dieser Herr Vetter aus Italien" – fuhr sie fort –, „wie er sich gleich anfangs vorsichtig auf den Stuhl setzte, als wollt' er sich die Röcke nicht zerknittern – sein Gang, die Stimme – dann" – hier stockte sie plötzlich – „Nun?" fragte Fortunat. – „Dann sah er Sie einmal lange, lange an, als Sie eben mit den andern sprachen und niemand achtgab." – Jetzt standen sie eben auf einer freien Anhöhe. Jenseits von den Waldbergen leuchtete die alte Burg in der Morgensonne herüber, wo Florentine ihm auf jener Spazierfahrt einmal flüchtig einen Kuß gegeben hatte – sie dachten beide daran. Die schöne Frau schlug verwirrt und errötend die Augen nieder – dann reichte sie ihm freundlich lächelnd die Hand, in die er recht herzlich einschlug. Währenddes ging Walter eifrig auf und nieder und zerbrach sich den Kopf. „Wär' nur der fremde Herr nicht, der euch verfolgt!" rief er ärgerlich aus. – „Ei was!" entgegnete Fortunat lustig, „ich habe das Mädchen und er die Tante, laß ihn die heiraten!" Sie setzten sich nun auf die Bank unter der Linde und beratschlagten miteinander, was jetzt zunächst zu tun sei. Nach vielem Hin- und Hersinnen wurde endlich einmütig beschlossen, vor dem Förster und den andern das einmal eingeleitete Mißverständnis zu benutzen und die Kammerjungfer für die entflohene Geliebte des Vetters auszugeben, beide aber einstweilen im Hause zu verwahren. Fortunat dagegen sollte schleunigst zu der Tante aufbrechen und dort, bevor er ihr den Aufenthalt Fiamettas entdeckte, nach den Umständen alles selbst vorsichtig ins rechte Geleis zu bringen suchen. „Du hast Rang, Vermögen", sagte Walter, „und bist eine so gute Partie für die Marchesin als irgendeine im Lande, es müßte wahrlich mit dem Eigensinne eines Romanschrei-

bers zugehen, wenn ihr euch zuletzt nicht noch kriegtet." Währenddes hörten sie Fiametta im Hause lustig plaudern und lachen. Der Förster, den sie weit im Walde wähnten, hatte nämlich sorgfältig seinen neuen Frack und eine lange, weiße Busenkrause angelegt und wandelte unter allerlei aus der Luft gegriffenen Vorwänden um das Haus, den Hals nach den obern Fenstern verdrehend. „Ich glaube wahrhaftig", sagte die Amtmann, „der alte Narr ist in das gnädige Fräulein geschossen." Fiametta aber hatte geschwind die Kammerjungfer beredet, ans offene Fenster zu treten, warf ihr einen großen Schal um und fing hinter derselben an zu agieren und den Förster anzureden, indem sie ihm gerührt für seine Mühwaltung dankte, wodurch er ein von der Tarantel der Liebe gebissenes Herz so frühzeitig von den Holzwegen des Leichtsinns zurückgeführt. – Als er nun seinerseits sich anschickte, verbindlich zu antworten, konnte sie vor Lachen nicht weiter, winkte ihn geheimnisvoll fort, als ob sie belauscht würden, und schlug schnell das Fenster wieder zu. – Florentine schüttelte bedenklich den Kopf und konnte sich durchaus in das buntfarbige Wesen nicht finden.

Fiametta aber, da die Männer ihr jetzt ihren Plan mitteilten, war von der Aussicht einer endlichen, baldigen Entscheidung ihres verwickelten Liebeshandels wie berauscht. Und als nun Florentine noch in aller Eile anfing Kuchen zu backen, die sie morgen Fortunat auf die Reise mitgeben wollte, half sie ihr mit großer Geschäftigkeit und naschte die schönsten Rosinen weg. Zuletzt aber, da sie selbst den Teig angefaßt, mußte auf ihr klägliches Geschrei alles zu Hilfe eilen, um ihre Finger wieder rein zu machen. – Nun ließ sie das Backen ganz und gar, und zeigte Fortunat die Wohnung, die sie ihr oben angewiesen hatten. Es war die schönste Stube im ganzen Hause, sie lag nach dem Walde zu, der durch alle Fenster hereinsah. Da ging es nun lustig ans Einpacken für morgen, die Vögel sangen draußen in den Wipfeln, Fiametta kniete in der grünen Dämmerung vor Fortunats Felleisen und plauder-

te vergnügt von den schönen Bergen, über die er kommen würde, von dem prächtigen Schloß und dem Garten der Tante, dabei packte sie heimlich allerlei Kleinigkeiten von sich unter seine Wäsche und wurde über und über rot, als er's bemerkte.

So war unter munteren Verabredungen und Vorbereitungen der Tag verflossen. Walter hatte, die Müdigkeit seiner Gäste vorschützend, für den Abend jeden Besuch entfernt gehalten, die Hausgenossen selbst, nach der halbdurchwachten Nacht, waren schon früh zur Ruh gegangen. Nur Fortunat und Fiametta saßen noch vor der Haustür und hörten zu, wie die Mädchen unten im Dorfe vor dem Johannesbilde und die Heimchen von der fernen Wiese sangen. Fiametta saß zu seinen Füßen im Grase, sie hatte die Gitarre auf ihren Knien und sah still in die mondbeschienene Gegend hinaus, er hatte sie noch nie so nachdenklich gesehen. – Da erklang auf einmal weiter oben ein Waldhorn. Es war der verliebte Förster, der den Herrschaften ein Ständchen blies. Und als nun allmählich Waldhorn und Johanneslieder verklangen, und alles still geworden war im Hause und im Tal, da nahm Fiametta die Gitarre und sang:

„Es schienen so golden die Sterne,
Am Fenster ich einsam stand
Und hörte aus weiter Ferne
Ein Posthorn im stillen Land.
Das Herz mir im Leibe entbrennte,
Da hab' ich mir heimlich gedacht:
Ach, wer da mitreisen könnte
In der prächtigen Sommernacht!

Zwei junge Gesellen gingen
Vorüber am Bergeshang,
Ich hörte im Wandern sie singen
Die stille Gegend entlang:
Von schwindelnden Felsenschlüften,

Wo die Wälder rauschen so sacht,
Von Quellen, die von den Klüften
Sich stürzen in die Waldesnacht.

Sie sangen von Marmorbildern,
Von Gärten, die überm Gestein
In dämmernden Lauben verwildern,
Palästen im Mondenschein,
Wo die Mädchen am Fenster lauschen,
Wann der Lauten Klang erwacht,
Und die Brunnen verschlafen rauschen
In der prächtigen Sommernacht!"

Fiametta legte die Gitarre schnell weg, verbarg ihr Gesicht an Fortunats Knien, und weinte bitterlich. – „Wir reisen wieder hin!" flüsterte ihr Fortunat zu. Da hob sie das Köpfchen und sah ihn groß an. „Nein", sagte sie, „betrüg mich nicht!"

Fünfundzwanzigstes Kapitel

Zur Mittagszeit des folgenden Tages war Fortunat auf seiner Reise schon fern von Hohenstein und rastete eben vor der Tür einer Dorfschenke. Die Bienen summten in der blühenden Linde am Hause, vor ihm über den niedrigen Zäunen und Gärten bezeichnete ein blauer Duftstreif kaum noch die Berge, wo er sein Liebchen zurückgelassen. Ein Mädchen mit frischen Augen brachte Wein und Brot heraus, stutzte aber, da sie ihn erblickte, und sprang schnell wieder in das Haus zurück. Drin hörte er sie lebhaft reden und zu seinem Erstaunen seine Haare, Rock und Stiefeln ausführlich beschreiben. Nun trat auch der Wirt heraus, nickte ihr zu, und Fortunat erfuhr endlich, daß vor kurzem zwei fremde Herren zu Pferde hier nach einem Reiter von seinem Aussehen sich angelegentlich er-

kundigt und dann sehr eilig, der eine diese, der andere jene Straße eingeschlagen hätten. Vergebens fragte er nach Namen und näheren Kennzeichen, er konnte aus der konfusen Personenbeschreibung durchaus nicht klug werden, die eine hätte gar beinah' auf Walter gepaßt. – „Ihr fangt mich doch nicht!" dachte er, als es ihm plötzlich aufs Herz fiel, daß er jedenfalls freiwillig und aus eigenem Entschluß vor Fiamettas Tante erscheinen müsse, wenn sein ganzer Plan nicht scheitern sollte. – In dieser Unruhe trank er noch rasch des Wirts Gesundheit, schwang sich wieder auf sein Pferd und sprengte durchs Dorf den fremden Herren nach. Draußen aber nahm er sogleich die entgegengesetzte Richtung und atmete erst wieder frei auf, als ein Bauer im Felde ihm einen näheren Holzweg gerade durchs Gebirge bezeichnete, auf dem er jene Reiter zu vermeiden, ihnen wohl gar zuvorzukommen hoffen durfte.

Die Luft war schwül, er ritt lange am Rande eines waldigen Bergrückens fort, an einsamen Klüften und melancholischen Tälern vorüber. Auf einmal leuchtete seitwärts ein lustiger Grund zwischen den Bäumen herauf: rote Ziegeldächer und Gärtchen im schillernden Sonnenscheine an den Felsen hängend, unten ein glitzernder Bach mit badenden Kindern und auf der Wiese daneben fröhlich Getümmel der Heuernte, Lachen und das Klirren der Sensen dazwischen. Und wie er noch so freudig überrascht hinabschaut, erschallt jenseits plötzlich ein Peitschenknall, und um die Waldecke herum fliegt ein schöner Reisewagen über die glänzende Landschaft. Eine Dame beugt sich aus dem Wagen – Fortunat fährt erschrocken zusammen, es ist offenbar Fiametta, aber in Frauenkleidern, lustig schwatzend mit einem Unbekannten, der neben dem Schlage reitet. Jetzt senkt sich der Weg plötzlich wieder in den Wald, und zwischen den dunklen Tannen ist alles verschwunden und verklungen.

Fortunat stand wie versteinert, im ersten Augenblicke kam ihm Fiametta fast wie ein lieblicher Kobold vor, der

neckend durchs Gebirge streifte. Dann dachte er sie sich wieder in Hohenstein entdeckt und mit roher Gewalt fortgeführt; aber wie konnte sie dann noch so fröhlich plaudern! – er war ganz verwirrt. – So lenkte er rasch auf einem Fußsteige den Berg hinab, über die Wiese dem Hohlweg zu, wo die Erscheinung versunken. Bald teilten sich die Wege, auf dem einen glaubte er eine frische Wagenspur zu bemerken und setzte munter die Sporen ein.
Aber je weiter er kam, je wilder und einsamer wurde die Gegend. Sie konnten auf dem steinigen Wege unmöglich so rasch gefahren sein, als er ritt. Oft hielt er lauschend still, da glaubte er einmal wieder ihre Stimme zu hören, es war der fremde Schall eines Waldvogels aus der Ferne. Er sang laut alle Lieder, die er wußte, dann horchte er wieder, und lachte und schimpfte und ritt immerfort, bis er zuletzt mit Entsetzen bemerkte, daß ein Unwetter rasch im Anzuge war, um die Verwirrung vollkommen zu machen. Schon durchkreuzten Möwen mit ihren weißen spitzigen Flügeln pfeilschnell die schwüle Stille. Vergeblich blickte er nach einem Obdach umher, nicht einmal der Klang einer Holzaxt ließ sich im Walde vernehmen. Nur einzelne Nebelgestalten stiegen nun langsam aus den Klüften empor und setzten sich mit ihren langen, grauen Gewändern in die Wipfel der Tannen, über dem Berge vor ihm aber hatte das Gewitter allmählich sein bleifarbenes Dunkel ausgebreitet, in das die Mauerspitzen einer Ruine fast grauenhaft hineinragten. –
Indem er noch so zögernd stand und unentschlossen war, wohin er sich wenden sollte, hörte er auf einmal den Schall einer Glocke weit aus der Höhe herüberklingen. „O du göttlicher Aberglaube!“ rief er freudig aus, „was sind alle Blitzableiter der Welt gegen diesen tröstlichen Klang, der wie ein singender Engel mit gefalteten Händen über die Wälder zieht und die Wetter wendet. Ja, die Erde ist noch immer voll schöner Wunder, wir betrachten sie nur nicht mehr!“ – Er folgte nun eilig den Klängen, die bald schwächer, bald deutlicher durch den Gewitterwind von dem

Berge herabzukommen schienen, wo er vorhin die Ruine erblickt. Ein wildverwachsener, wenig betretener Fußsteig schlang sich zwischen den Klippen gerade in der Richtung hinauf. Der Pfad wurde immer enger und steiler, bald hörte er auch die Glocke nicht mehr, er mußte endlich absteigen und, sein Pferd am Zügel fassend, mühsam von Stein zu Stein hinanklimmen. Manchmal wendete er sich rastend zurück und sah durch die Wolkenrisse tief unten die Landschaft vorüberfliegen.

So war es völlig Nacht geworden, als er atemlos droben ins Freie trat. Ein Licht schimmerte ihm aus der Ferne freundlich entgegen; indem er darauf losging, glaubte er im Dunkel ein großes Schloß zu erkennen mit Türmen, Zinnen und wunderlichen Erkern. Dann, je näher er kam, verwandelte sich allmählich alles wieder, es war wildumhergeworfenes Gestein und phantastische Baumgruppen, was ihm so prächtig erschienen, und voll Erstaunen stand er auf einmal vor einer Einsiedlerklause, halb in den Felsen gehauen, ein Türmchen mit einer Glocke darüber. Eine Lampe von der Decke warf ungewiß flackernde Scheine über die leeren Wände und den hölzernen Tisch und Stuhl in der Mitte. Plötzlich fuhr sein Pferd schnaubend zusammen, aus einem Winkel der Halle blinkte ihnen ein hochaufgerichtetes weißes Totengerippe entgegen. – „Schauerlicher Gesell!“ sagte Fortunat, „bist du der Einsiedler hier und ziehst bei Nacht heimlich die Glokke?“ – Er rief nun laut nach allen Seiten, aber nur seine eigene Stimme gab zwischen den Klüften Antwort. Da faßt er sich ein Herz, band sein Pferd vor der Hütte an und trat hinein.

Er fand sie wohnlicher, als er erwartet hatte. Ein großes Buch lag auf dem Tische, er schlug es auf, es war ein altes Brevier, zu seiner Verwunderung fand er eine kurze, ungarische Tabakspfeife drin als Zeichen eingelegt. Nun, die Toten schmauchen doch nicht, dachte er und spähte eifriger umher. Da entdeckte er in einer Ecke einen Vorrat köstlichen Heues, weiterhin auch einen vollen Weinkrug

und Gläser daneben. Erfreut über den unverhofften Fund, zäumte er vor allem sein müdes Pferd ab und versah es reichlich mit Futter. Das ungewohnte Hantieren in dieser Abgeschiedenheit, das Brausen der Wipfel, die ganze unerhörte Lage, in der er sich hier befand, versetzte ihn in eine seltsame Heiterkeit. „Gute Nacht!“ rief er fröhlich vom Berge hinab, „wie hat der Herr nun alles untergetaucht in den wunderbaren Strom der Träume! Was ist das für ein Traumlied in den Wäldern, gleichwie die Saiten einer Harfe, die der Finger Gottes gestreift. Wahrlich, wen Gott lieb hat, den stellt er einmal über allen Plunder auf die einsame Zinne der Nacht, daß er nichts als die Glocken von der Erde und vom Jenseits zusammenschlagen hört und schauernd nicht weiß, ob es Abend bedeute oder schon Morgen.“ –
Darauf setzte sich Fortunat zufrieden vor die Klause, doch so, daß er seitwärts die eine Wand im Auge behielt; er traute dem dürren Gesellen im Winkel doch nicht recht, daß er sich nicht unversehens erhübe und murmelnd am Tische aus dem Buche zu lesen anfinge. Draußen aber war es so endlos still, er hörte nur manchmal das Schnauben des Pferdes und den Schrei des Wildes tiefer im Walde, vor ihm streiften durchsichtige Wolken gespensterhaft leise den Rasen wie Schleppen fliehender Feen.
In dieser Einsamkeit überwältigte endlich der Schlaf den Erschöpften, und als er mitten in der Nacht plötzlich wieder aufwachte, waren die Wetter unterdes verzogen, der Mond schien prächtig über die Wälder. Da war's ihm, als hörte er in einiger Entfernung zwei Männer eifrig miteinander sprechen, und im zitternden Mondlichte unter den Bäumen bemerkte er einen riesengroßen Mönch, der mit einem Unbekannten schnell durch den Wald fortging. Vor dem Rauschen der Wipfel konnte er nur einzelne abgebrochene Laute vernehmen, er hörte aber deutlich, wie sie im Gespräche mehrmals seinen und Fiamettas Namen nannten. – „Träume ich denn, oder träumt diese phantastische Nacht von mir?“ – rief er erschrocken aufsprin-

gend aus, aber die Stimmen waren schon weit, und auf der stillen Höh', wo sie sich endlich im Dunkel ganz verloren hatten, sah er nun plötzlich eine Fackel aufleuchten. Mehrere dunkle Gestalten folgten, sie trugen lautlos einen Sarg. Die roten Widerscheine schweiften wunderbar zwischen den Tannen über ein Felsentor, in welchem auf einmal alles wieder verschwand. – Da war's ihm, als trügen sie Fiametta fort, er stürzte hastig nach in den Wald. Aber vergebens suchte er einen Steg durch die Wildnis, in der flimmernden Dämmerung des Mondenscheins starrten ihm überall zackige Klüfte entgegen, er mußte wieder umkehren. „Nur zu", sagte er ganz verstört, „nur immer zu! der Spuk und die Nacht müssen doch einmal ein Ende nehmen!" – Dann lehnte er sich über den Hals seines schlummernden Pferdes und starrte gedankenvoll in die weite Einsamkeit hinaus.

So hatte er lange halb im Traume gestanden, als er auf einmal von fern den lustigen Schrei eines Waldvogels zu hören glaubte. Erfreut blickte er umher, da schweifte wirklich schon ein ungewisser Morgenschein leise über den Himmel, wie ein Hauch über den Spiegel, seitwärts, als er sich bewegte, fuhr ein Reh auf und flog scheu durch die Dämmerung. Nun dacht' er dran, daß heute Sonntag war. Da rannte er schnell in die Klause. „Schau nicht so grämlich in dieser gnadenreichen Stunde", rief er dem knöchernen Klausner zu, „jetzt ist's ja Licht und alles, alles wieder gut!" Dann zog er fröhlich die Glocke, als wollt' er den Tag anbrechen, und das Herz wurde ihm still und weit, als der Schall so hell durch die Waldesnacht ging, er hatte schon lange nicht so fromm in Gedanken gebetet.

Jetzt fiel ihm erst ein, daß der Glockenklang wohl die rätselhaften Nachtwandler herbeigelockt haben könnte. Er trat hinaus und spähte nach allen Seiten umher. Aber es rührte sich nichts, der Wind hatte die Klänge nach den Tälern geweht, die noch im tiefen Schatten lagen. Auf dem Gipfel des Berges aber, an dessen Lehne die Klause sich befand, bemerkte er jetzt im falben Zwielichte die Mauer-

trümmer wieder, die er gestern aus dem Tale gesehen. Dort zogen sie hinauf, dachte er und schwang sich eilig auf sein Pferd. Bald hatte er nun auch den verschlungenen Pfad und das Felsentor entdeckt, das von der andern Seite nach der Höhe führte, und verfolgte unterdessen die Spur, um droben, wo möglich, nähere Auskunft über die Vorgänge der Nacht und die einzuschlagende Reiserichtung zu erhalten.

So ritt er wohlgemut in den wachsenden Morgen hinein, auf dem Berge vor ihm trat allmählich das alte Gemäuer immer deutlicher zwischen den Tannen hervor. Schon unterschied er eine halbverfallene Kirche, leere Fensterbogen und einzelnstehende Pfeiler, von Efeu üppig umrankt, Ziegen kletterten in der grünen Wildnis, alles von der Morgensonne wunderbar beleuchtet. Da erschien auf einmal ein hoher, schlanker Jäger auf der Wand, der Morgen funkelte glutrot darüber, es war, als stünd' er ganz im Feuer. Auf seine Büchse gelehnt, schaute er von der andern Seite in die Täler hinab, er hörte ihn oben singen:

„Hier steh' ich wie auf treuer Wacht,
Vergangen ist die dunkle Nacht,
Wie blitzt nun auf der Länder Pracht!
Du schöne Welt, nimm dich in acht!"

Jetzt wandt' er sich herum – es war Lothario! Auch er hatte nun den Ankommenden bemerkt, sprang rasch herab, und die beiden Freunde lagen einander in den Armen. Der wilde Jäger sah bleich, gebräunt und dennoch schöner aus als ehemals, Fortunat erschrak fast vor der wunderbaren Tiefe der dunklen Augen, in die er so lange nicht gesehen. – „Aber wie kommst du hier herauf?" fragte er endlich aufs höchste überrascht. – „Ich spiele den letzten Akt", erwiderte jener lächelnd, „Gräber, Hochzeit, Gottes grüne Zinnen und die aufgehende Sonne als Schlußdekoration." – Hier waren sie am Gipfel bei den Trümmern angelangt, er band Fortunats Pferd an einen Baum. „Laß un-

terdes hier alles stehen, und komm nur schnell mit mir." – „Du bist nicht allein hier oben", meinte Fortunat, „wen habt ihr heute nacht im Walde begraben?" – „Den armen Otto." – „O Gott! du fröhliches Liederherz, so frisch wie eine Lerche singend aus der Luft zu fallen! mir ist's, als hört' ich's noch im Ohre klingen." – „Wohl ihm", entgegnete der Begleiter, „er hatte rasch gelebt und stand schon müd' und schlaftrunken im tiefen Abendrote, dort ruht er aus."

Sie traten durch ein halbverfallenes Bogentor auf einen freien, grünen Platz, es schien ein ehemaliger Klosterkirchhof zu sein. Ein neues Grab, soeben erst mit schönem Rasen belegt, schimmerte ihnen taufrisch entgegen. Ein Mönch kniete betend daneben zwischen wilden, bunten Blumen, und Vögel flatterten und sangen lustig in dem Grün, das aus allen Mauerritzen rankte, über die Gräber aber leuchtete auf einmal eine unermeßliche, prächtige Aussicht aus der rauschenden Tiefe herauf. – „Gott gebe jedem Dichter solch ein Grab!" rief Fortunat freudig überrascht.

Bei dem Klange seiner Stimme aber hob sich's plötzlich unter den Blumen, er stand wie im Traum – es war Fiametta. „Ist er da!" rief sie emporfahrend aus, schüttelte die Locken aus dem Gesicht und sprang fröhlich zu ihm. Nun kam zu seinem Erstaunen auch Walter eilig zwischen den Steinen hervor mit einem Einsiedler und einem Fremden, der Fortunat mit den klugen, scharfen Augen freundlich betrachtete. „Wie haben wir dich gesucht", rief Walter schon von weitem, „wer von uns hätte das gedacht!" – Aber Fortunat konnte sich noch gar nichts denken, er blickte verwirrt in dem Kreise umher. Da glänzten unten die Täler in der schönen Sonntagsstille, und die Morgenglocken klangen von fern herauf. – „Nun lobet alle Gott!" sagte Lothario, faßte Fortunat und Fiametta bei der Hand und führte sie in die alte Kirche neben dem grünen Platze, die andern folgten schweigend. Der Mönch stand schon vor dem Altar, zu dem Lothario sie brachte. Die Morgen-

sonne schien seltsam durch das hohe, gemalte Bogenfenster, die Pfeiler waren mit frischen Birken verziert, durch die offene Tür rauschten die Bäume herein. Jetzt bemerkte Fortunat erst, daß Fiametta festlich geschmückt war und ein Myrtenkränzchen im Haare hatte, er wußte nicht, wie ihm geschah. Und als nun der Mönch sich zu ihnen wandte und fragte, ob sie als getreue Eheleute einander lieben wollten bis in den Tod, sagte Fiametta errötend aus Herzensgrunde: „Ja", und er legte segnend ihre Hände zusammen.

Sechsundzwanzigstes Kapitel

Jungen Eheleuten kommt am Hochzeitsmorgen die Welt wie verwandelt vor, als wäre über Nacht alles schöner und jünger geworden, denn die Erde putzt und spiegelt sich gern in fröhlichen Augen. Wieviel lustiger unserm Paare, gleich Zugvögeln über der prächtigsten Gegend, da war des grünen Waldlebens genug, schattige Gründe, fliegende Schimmer über das Land und unabsehliche, selige Fernen! – Allmählich erst tauchte Fortunat alles aus dem Morgenglanze auf. Er erfuhr nun, daß der seltsame Lothario Graf Victor selber war und seit geraumer Zeit hier oben als Vitalis lebe, heiter und streng, ein Einsiedler ohne Kutte, ein Jäger, nach höherm Wild gestellt. – Jetzt gab sich auch jener Fremde als Baron Manfred kund, denselben Vetter, der damals Fiamettas Tante auf ihrem Schlosse besucht. Er hatte von ihrer Liebe und ihrem Heimweh gehört und für *Fortunat* um ihre Hand bei der Tante geworben. Als aber darauf die scheue Marchesin vor dem vermeintlich unbekannten Bräutigam so plötzlich die Flucht ergriffen, verfolgte er unausgesetzt ihre Spur bis Hohenstein, wo er unmittelbar nach Fortunats Abreise eintraf. Dort erfuhr Walter von ihm den ganzen Zusammenhang,

sowie den gegenwärtigen Aufenthalt des Grafen Victor, und voll Freude waren sie nun alle noch denselben Morgen aufgebrochen, um Fortunat eiligst einzuholen. So hatte also Fortunat sein Liebchen vor sich selber entführt und ein jeder vor lauter Klugheit die möglichst größte Konfusion angerichtet, der liebe Gott aber unversehens alles wieder gescheiter gemacht.

Morgen wollten die Gäste wieder weiterziehen. Unerwartet waren sie hier auf einer jener Zinnen des Lebens zusammengekommen, die immer nur für wenige Raum hat – das fühlten sie wohl. Was hatten die Freunde nun alles einander zu erzählen in der kurzen Zeit, Lust und Leid, Vergangenes und Künftiges! So war ihnen der Tag in der schönen Einsamkeit schnell verflossen. Als es aber schon wieder abendkühl wurde, saßen sie alle zusammen vor der großen Aussicht unter den hohen Buchen, welche den Abhang an der Klosterruine beschatteten. Spuren von Kiesgängen, sorgfältig mit Buchsbaum umzäunt, Lauben und halbzerworfene Rasenbänke bezeichneten ringsumher den ehemaligen Klostergarten, nur einzeln zerstreute Blumen, wie verlorene Kinder, schimmerten noch aus der alten Zeit durch das wuchernde Unkraut. Hier hatte der geschäftige Einsiedler einen Tisch gedeckt und Stühle gesetzt, er ließ es sich durchaus nicht nehmen, die Herrschaften aufs beste zu traktieren mit Wein, Obst, Honig und Nüssen, was er nur hatte. Für Fiametta aber hatte er einen Kranz von lustigen Waldnelken besorgt. Er blätterte emsig in seinem Brevier und schenkte ihr die schönsten Heiligenbilder daraus, dabei steckte er ihr immerfort das Beste von dem Naschwerk zu und hatte seine herzinnige Freude, wie sie so schön mit dem Kränzlein aussah und fröhlich plaudernd die Nüsse knackte. – Da hörten sie auf einmal in geringer Entfernung einige Saitenklänge. „Dacht ich's doch“, fuhr der Einsiedler auf, „da hat er mir doch meine alte Zither in der Klause aufgestöbert!“ – Im Gebüsch aber hörten sie singen:

„Wir zogen manchen Wald entlang,
Viel fröhliche Gesellen,
Und salutierten mit Gesang
Die Burgen und die Quellen.

Nun sang man den zu Grabe still,
Dem sie zur Hochzeit geigen,
Der andre in den Himmel will
Auf wilden Felsen steigen.

Von den einsamen Felsensteigen
Schau' ich ins Land so weit,
Da dunkelt und rauscht's so eigen
Von der alten, schönen Zeit."

„Da kriegen wir alle was ab", sagte Fiametta. – „Nun, nun, wir wollens ihm schon zurückzahlen", meinte der Einsiedler, „aber er singt eine schöne Note, es ist mir ganz wie in meiner Soldatenzeit, wenn ich so bei stiller Nacht mit der Zither im Biwak lag." – Er sang wieder:

„Was das für ein Gezwitscher ist!
Durchs Blau die Schwalben zucken
Und schrei'n: „Sie haben sich geküßt!"
Vom Baum Rotkehlchen gucken.

Der Storch stolziert von Bein zu Bein:
„Da muß ich fischen gehen –"
Der Abend wie im Traum darein
Schaut von den stillen Höhen.

Und wie im Traum von den Höhen
Seh ich nachts meiner Liebsten Haus,
Die Wolken darüber gehen
Und löschen die Sterne aus."

Fiametta flüsterte wieder: „Ist ihm denn seine Liebste gestorben?"
„Ach, das ist eine dumme Geschichte mit seiner Amour", erwiderte der Einsiedler, „tut mir den Gefallen und bedauert ihn nicht lange, das will er nur, sonst macht er noch immer mehr Flausen davon." „Wer ist's denn?" fragte Fiametta. Aber der Spielmann sang von neuem:

„Im Schloß ihr wohl am Fenster steht
Und herzt euch nach Gefallen,
Der Herbst schon durch die Felder geht,
Da hört ihr's unten schallen.

„Das klingt ja wie vom Felsenrand
Einst bei des Klausners Buchen,
Ich glaub', das ist der Musikant,
Der kommt zum Kindtaufskuchen."

Und die Vögel ziehn über die Buchen,
Der Sommer der ist vorbei,
Ich aber muß wandern und suchen,
Wo der ewige Frühling sei."

Hier entstand plötzlich ein heftiges Geräusch, und ehe sie sich's versahen, kam der Sänger in hastiger Flucht durch Laub und Äste geradezu über die alte Gartenmauer dahergeflogen, daß die losen Steine hinter ihm dreinkollerten. Fiametta drängte sich scheu an Fortunat, dieser erkannte zu seinem Erstaunen in dem Flüchtling Dryander. Der Doktor aber blickte noch immer wild zurück, setzte seinen Hut, der vor Eile ganz schief saß, auf dem Kopf zurecht und schimpfte, außer sich vor Zorn und Schreck, über die dumme Romantik: kaum beträte man das Revier eines Poeten, so schössen verstorbene Doppelgänger, gleich wahnsinnigen Pilzen, aus dem unvernünftigen Boden und säßen auf den Klippen umher und wackelten mit den Köpfen. – Da erkannte er auf einmal in Fiamettas Au-

gen das hübsche Jägerbürschchen vom Donauschiff, und seine ganze Gedankenfolge bekam dadurch plötzlich einen anderen Zug. Fiametta errötete und fragte ihn lächelnd, ob er sich noch mit ihr schlagen wolle? Er aber besann sich nicht lange. „O“, entgegnete er tapfer, „ich habe damals auf dem Schiffe alles recht gut gewußt, und wollte nur die Damen ein wenig schrecken.“ – „Ja, ja, das hat die Schiffsgesellschaft wohl gemerkt“, sagte Fortunat, „denn sie haben deinen zurückgelassenen Hut über der Tür des Wirtshauses angenagelt zum ewigen Gedächtnis eines verwegenen Duellanten, der vor Zorn und Wut plötzlich die Verschwindsucht bekommen.“

Unterdes hatte der Einsiedler das Gebüsch hinter der Mauer untersucht und kam nun mit großem Gelächter zurück. Gerade in dem wildverwachsenen Versteck, wo Dryander das Ständchen gebracht, befand sich der zertrümmerte Eingang zur Klostergruft; dort saß seit alter Zeit ein Totengerippe wie ein Wächter zwischen den Steinen, dem der Einsiedler, als er vorhin Tisch und Stühle abräumte, in der Eile des Doktors Schlafpelz umgehangen. Mitten im Gesange nun sich umwendend, hatte Dryander plötzlich sich selbst zu erblicken geglaubt und so mit größter Behendigkeit die Flucht ergriffen.

Jetzt erfuhr Fortunat auch, daß der Doktor schon seit längerer Zeit in einem angeblichen Bußanfall bei dem Einsiedler sich aufgehalten, der ihm sehr gut war und immer tausend Spaß und Händel hatte mit dem kuriosen Gesellen. Heut noch vor Tagesanbruch aber war Dryander gleichfalls voll Eifer ausgezogen, um Fortunat aufzusuchen, ohne in seiner Zerstreuung vorher erst die Braut zu betrachten. Unterwegs aber hatte er bald die ganze Geschichte wieder vergessen, und schlenderte wohlgemut nach dem nächsten Städtchen, wo er sich im Gasthause tüchtig restaurierte. Das gefiel ihm so wohl, daß er unverzüglich einen großen Einkauf an Wein, Braten und Kuchen machte und einen Burschen zum Tragen mietete, der soeben zu allgemeinem Ergötzen aus seinem Korbe den

Markt hervorlangte und sich dann ermüdet neben sie ins Gras setzte.
Wer Dryander genau kannte, konnte bald bemerken, daß er sich wieder einmal in jener phantastischen Faselei befand, wo er sich und andere überredete, ganz besonders unglücklich zu sein. Victor sah ihn scharf an. „Nun beichte nur gleich", sagte er, „was ist wieder passiert?" – Der Doktor zögerte erst, dann begann er mit einer gewissen weichen Feierlichkeit: „Ihr wißt alle, daß meine liebe kleine Frau mich verlassen." – „Sie mußte wohl", fiel ihm Victor ins Wort, „du wolltest ihre gesunde, herbe, klare Prosa durchaus auf die poetische Lyra spannen, was Wunder, wenn endlich die Saite sprang!" – „Und einem Husarenleutnant an den Schnurrbart flog", sagte der Doktor, ärgerlich über die Unterbrechung. „Kurz, ich wußte wohl ein Jahr lang nicht, wohin sie gekommen. Heute nun, als ich mit diesem guten Jungen da soeben zu den Bergen zurückkehren will, sehen wir ein rotes Ziegeldach durchs Grün schimmern. Wir treten näher, da steht ein Brunnen unter einem blühenden Apfelbaum, die Bienen summen drin in der schwülen Mittagsstille, an dem Brunnen aber sitzt ein junges Weib, ihr Kindlein auf dem Arm – es war mein liebes Trudchen. ‚Gott grüß dich, schöne Frau', sag' ich und bitt' um einen frischen Trunk. Da blickt sie erschrocken auf – sie kannte mich nicht mehr." – „Nein, Herr", fiel hier der Bursch mit dem Korbe ein, „sie erkannte Euch gleich und schrie: ‚Herrje, Fritz, komm geschwind, da ist mein alter Mann!'" – „Ganz recht", fuhr Dryander fort, „und da kommt ihr neuer Mann, der verabschiedete Husarenleutnant, in hohen Schmierstiefeln und Hemdsärmeln, Heu und Häcksel in den Haaren, und fährt in der Eile in seinem alten Flauschrock mit der Faust zum Ellbogen heraus, ein Kernwirt, sonst ein guter Kerl. Wir gingen nun miteinander ins Haus, ich lobte alles nach Kräften." – „Ihr erzählt alles so konfus", sagte der Bursch wieder, „Ihr fragtet zuerst, was in der Stadt der Spieß Lerchen koste, die draußen so hübsch sängen?" – „Kann

sein!" – „Nein, ich weiß noch alles ganz genau. ‚Und, einmal als Philosoph gesprochen', sagtet Ihr dann, ‚was braucht ein fühlendes Herz mehr: ein ländliches Schloß mit wackligen Mansarden, ein sanfter, unter dreijährigen Dünger gesetzter Hügel daneben, ein schlängelnder Bach aus dem Kuhstall nach der lachenden Wiese'" – „Halt das Maul", fuhr ihn Dryander an. „Ich stand in der Haustür, mit tiefer Wehmut überblickte ich noch einmal den Apfelbaum, das stille Gärtchen und Trudchens Gestalt – dann wandt' ich mich" – Hier konnte der Bursche das Lachen nicht halten. – „Was hast du?" fragten die andern. „Mit Erlaubnis", sagte er, „und als der Herr so von dem schlängelnden Bach sprach, erwischte ihn der Herr Leutnant am Flügel und schmiß ihn zum Hause heraus, daß er mir bald in den Korb gefallen wäre." „Nun, wenn ihr's besser wißt, so ist mir's auch recht", entgegnete der Doktor, ergriff eine Flasche und wollte fort, kehrte aber wieder um, nahm noch eine zweite unter den Arm und ging eilig in die Ruine. „Wahrhaftig", sagte Fortunat lachend, „da ist Lug und Einbildung, Wahrheit und Dichtung so durcheinander gefilzt und gewickelt, daß er selber nicht mehr heraus kann! Ich wette, er ist nun auf acht Tage in allem Ernst wieder in seine kleine Frau verliebt!"

Während dieser Gespräche war es völlig dunkel geworden. Für Fiametta hatte man unterdes zwischen den Trümmern eine Lagerstatt von duftendem Heu bereitet, und ihre müden Augen waren schon untergegangen, als der Mond über die stillen Wälder aufging. Der Einsiedler, über seinem Rosenkranze nickend, bewachte sie von fern, die andern saßen noch zusammen bis tief in die Nacht. – Dryander aber hatte mit großer Umständlichkeit Papier, Federn, Wein und gestopfte Pfeifen in eine Zelle zusammengeschleppt, wo man ihn öfters eifrig auf und nieder gehen sah. Er wollte die schöne Nacht benutzen, um ein großes Gedicht, mit dem er sich schon lange in Gedanken herumtrug, endlich recht mit Ruhe niederzuschreiben. Da hatte er aber lauter Störungen.

Erst nickte ihn aus irgendeinem Mauerloch unaufhörlich ein melancholischer Schuhu an, gegen den er sich sehr erboste, weil er es für eine üble Vorbedeutung hielt. Dann erwachte eine Nachtigall und schmetterte gerade vor seinem Fenster. Er wollte sie mit dem Schnupftuche verjagen, darüber verlor er seine beste Feder hinterm Ohr, die Zugluft durchs offene Fenster fuhr in die beschriebenen Blätter, und als er um sich griff, schimpfte und haschte, löschte ihm gar der Wind das Licht aus. Da ballte er voller Zorn alle Papiere in seine Tasche zusammen, setzte den Hut auf den Kopf und nahm draußen, da alles schon schlief, mit wenigen Worten nur von dem Einsiedler Abschied, der, halb im Traum, nicht wußte, was geschah. Dann raffte er noch geschwind die Viktualien vom Tische in den Korb und rüttelte den Burschen auf. Der mußte ohne weiteres voraus, und so wanderte er mit langen Schritten den Wald hinab, um nie mehr auf diesen Berg zurückzukehren, wo ihm die ungeheure Tugendwirtschaft auf einmal unglaublich langweilig vorkam.

Wir aber lassen das Irrlicht wandern, und überschauen noch einmal das nächtliche Gebirge. Die Wälder und Abgründe liegen noch geheimnisvoll umher in der tiefen Stille, nur das ungewisse Flimmern der Sterne verkündet die Nähe des Morgens. Durch die weite Einsamkeit aber tönt ein Gesang, es ist Victors Stimme:

„Nächtlich macht der Herr die Rund',
Sucht die Seinen unverdrossen,
Aber überall verschlossen
Trifft er Tür und Herzensgrund.

Und er wendet sich voll Trauer:
Niemand ist, der mit mir wacht –
Nur der Wald vernimmt's mit Schauer,
Rauschet fromm die ganze Nacht.

Waldwärts durch die Einsamkeit
Hört' ich über Tal und Klüften
Glocken in den stillen Lüften,
Wie aus fernem Morgen weit. –

An die Tore will ich schlagen,
An Palast und Hütten: Auf!
Flammend schon die Gipfel ragen,
Wachet auf, wacht auf, wacht auf!"

Da regt sich's nach und nach immer lauter und lauter vor dem verfallenen Kloster, gesattelte Pferde wiehern durch die Dämmerung. Walter treibt geschäftig zur Eile, um noch vor der Mittagshitze ins Tal zu kommen, Fiametta sitzt schon auf ihrem Zelter, und schüttelt sich und plaudert reiselustig in der Kühle. Jetzt tritt zu aller Erstaunen auch Victor mit dem Einsiedler ganz wanderfertig aus dem Kloster. „Glück auf!" ruft er ihnen fröhlich entgegen, indem er Fiamettas Pferd am Zügel faßt und den Zug beginnt, der wegekundige Einsiedler, eine Reisetasche umgehängt und einen dicken Wanderstab in der Hand, schreitet im Zwielicht rüstig voran.
„Nun, das ist einmal ein Wort!" rief Fortunat freudig aus, während sie langsam den Wald hinabzogen, „du wanderst also mit?" – „Was hast du vor?" fragte Manfred fast betroffen. – „Beschlossen war es längst", sagte Victor, „und heute leuchten schöne, gute Sterne. Ihr wißt's noch nicht: ich bin auch Bräutigam." – Hier öffnete er den Reiserock, unter dem die Kleidung eines katholischen Priesters sichtbar wurde. – „Mein Lieb ist streng und ernst", fuhr er lächelnd fort, „drum wollt' ich hier oben mich erst zusammenraffen und innerlich besinnen. Glaubt mir, ein herrlich Ding ist's um die Einsamkeit auf hohen Bergen; das Buch des Lebens versteht doch nur, wer um Gottes willen lernt und nicht um der Welt Gunst." – Manfred sah ihn lange schweigend an. „Nun wahrlich", sagte er dann, „wenn ich dich auf dem Schlachtfelde wiedergefunden

hätte, hoch zu Roß mit der Fahne voran!" – „Du sprichst ja wie ein Mädchen davon", entgegnete Victor, „wie wenn es keinen Krieg gäbe, als den die schmucken Leutnants führen." – „Und dein großes poetisches Talent", unterbrach ihn Manfred wieder, „du wirfst es fort wie ein Verschwender?" – „Was wär' denn Poesie", meinte Victor unwillig, „wenn sie in feinem Goldschnitt auf einer Morgentoilette durchzublättern wäre? Talent! das ist nur ein Blitz, den der Herr fortschleudert in die Nacht, um zu leuchten, und der sich selbst verzehrt, indem er zündet. Nein, Freunde, genug endlich ist des weiblichen Sehnens, wer gibt uns das Recht zu klagen, wenn niemand helfen mag! Nicht morsche Mönche, Quäker und alte Weiber; die Morgenfrischen, Kühnen will ich werben, die recht aus Herzensgrund nach Krieg verlangt. Auch nicht übers Meer hinüber blick' ich, wo unschuldige Völker unter Palmen vom künftigen Morgenrot träumen, mitten auf den alten, schwülen, staubigen Markt von Europa will ich hinuntersteigen, die selbstgemachten Götzen, um die das Volk der Renegaten tanzt, gelüstet's mich umzustürzen und Luft zu hauen durch den dicken Qualm, daß sie schauernd das treue Auge Gottes wiedersehen im tiefen Himmelsgrund." – Manfred konnte sich lange nicht erholen. „Ist mir's doch", sagte er endlich, „wie von einem hohen Berg ins Meer zu sehen, wo mir dein Schiff in der Morgenglut verschwindet. Der Anblick schreckt und blendet mich, ich muß den festen Boden fühlen unter mir, ein nahes Ziel von Tag zu Tag im Auge haben." – „Geht, geht", fiel Fortunat hier ein, „über euren Reden verlier' ich mich selber ganz. Du Victor zumal, verwirrst mir schon seit gestern, wie ein nächtliches Wetterleuchten, der Seele Grund: tiefe Klüfte mit kühnen Stegen darüber und manche alte, geliebte Gegend fernab, aber alles so fremd und wunderbar wie in Träumen. Zuletzt ist's doch dasselbe, was ich eigentlich auch meine in der Welt, ich habe nur kein anderes Metier dafür als meine Dichtkunst, und bei der will ich leben und sterben!"

Jetzt standen sie auf einem Abhang, von dem verschiedene Pfade auseinandergingen. Hier hielt Victor plötzlich an, sein Weg führte ihn noch weiter über den Gebirgskamm nach der Stadt, wo die neuen Gefährten seiner harrten. Er schien tief bewegt. – „Wie's da unten nebelhaft sich durcheinanderschlingt" – sagte er, in die Täler schauend –, „man hört schon Stimmen da und dort verworren aus dem Grund, Kommandoruf und Trompetenklänge durch die stille Luft und Morgenglocken dazwischen und den Gesang verirrter Wanderer. Und wo die Nebel auf einen Augenblick sich teilen, sieht man Engel ernst mit blanken Schwertern auf den Bergen stehen, und unten weite Geschwader still kampfbereit aufblitzend, und der Teufel in funkelndem Ritterschmuck reitet die Reihen entlang und zeigt den Völkern durch den Wolkenriß die Herrlichkeit der Länder und ruft ihnen zu: Seid frei, und alles ist euer! – O Freunde, das ist eine Zeit! glücklich, wer drin geboren ward, sie auszufechten!" – Hier reichte er ihnen noch einmal die Hand und wandte sich schnell zum Walde.

„Ade, du geistliches Soldatenherz!" rief Fortunat erschüttert aus. Sie sahen ihm alle noch lange schweigend nach, dann schieden auch sie voneinander. Manfred wollte dem Ruf zu einem bedeutenden Staatsdienste folgen, da hoffte er, wenn auch auf anderer Bahn, auf dem frischen Gipfel des Lebens mit Victor wieder zusammenzutreffen. Walter aber begleitete das junge Ehepaar zunächst noch nach Hohenstein; ihm war's, als sei seit seiner Jugendzeit die Welt zu groß und weit geworden für ihn, er sehnte sich recht aus Herzensgrunde nach seinem stillen, schattigen Gärtchen zurück. – Und so sehen wir denn die rüstigen Gesellen auf verschiedenen Wegen das Gebirge langsam hinabreiten, und eine tiefe Wehmut befällt uns unter den leise rauschenden Bäumen, da nun alle die lieben, langgewohnten Stimmen nach und nach verhallen, wie wenn wir im Herbst die bunten Wandervögel über uns fortziehen hören.

Fiametta aber ritt voll stiller Freude und Erwartung neben Fortunat in den dämmernden Morgen hinein, denn er hatte ihr nun entdeckt, daß er ihren Palast in Rom angekauft habe, dort wollten sie wieder hin. – Vor ihnen glänzte schon manchmal die Landstraße unermeßlich herauf, alle Ströme zogen da hinaus, Wolken und Vögel schwangen sich durchs heitere Blau ihnen nach, und die Wälder neigten sich im Morgenwind nach der prächtigen Ferne. – „Weißt du noch dein Märchen im Baum?" sagte Fiametta lachend, „nun bin ich wirklich Aurora." –

Und als Victor sich noch einmal auf der Höhe zurückwandte, waren schon alle im Morgenrot verschwunden. Durch eine Waldschlucht nur sah er unten einen schwerbepackten Rüstwagen und ein Häuflein Wanderer zu Fuß und zu Roß am Walde vorüberziehen, er erkannte seine alten Komödianten, Dryander schritt mit der Geige wieder voran. – So stand er noch lange in Gedanken oben – da ging die Sonne prächtig auf, die Morgenglocken klangen über die stille Gegend, und der Einsiedler sang:

„Wir ziehen treulich auf die Wacht,
Wie bald kommt nicht die ew'ge Nacht
Und löschet aus der Länder Pracht,
Du schöne Welt, nimm dich in acht!"

Die Freier

Lustspiel in drei Aufzügen

PERSONEN

GRÄFIN ADELE
FLORA, ihr Kammermädchen
GRAF LEONARD
HOFRAT FLEDER
FLITT, ein Schauspieler
SCHLENDER, ein Musikant
VICTOR, Jäger
FRIEDMANN, Gärtner
MARIE, Friedmanns Nichte
KNOLL, ein Weinschenk
Ein BOTE

Erster Aufzug

Erste Szene

Studierstube, Akten liegen auf den Stühlen, Hofrat FLEDER *sitzt auf einem Reitsessel vor dem Pult und schreibt.*

FLEDER. Vorwärts! vorwärts! Morgenstunde ist die stille Saatzeit der Gedanken. *Er schreibt.* „Das Amt hat dem Mäusefraß mit Energie und gebührendem Ernste entgegenzutreten." – Ja, es ist des Mannes edelster Beruf, so unmittelbar in das Leben, in die Welt zu greifen. Vaterland – deutsches Vaterland – ein erhabener Gedanke! *Er schreibt.* „Mäusefallen aber sind nicht etatsmäßig und können nicht passieren, wonach sich zu achten." – So diese wichtige Angelegenheit wäre auch reguliert. Frisch weiter! Was kommt nun! *Er steht auf und wühlt in den Akten.* Das ist fertig – hier ebenfalls – das auch. Weiß Gott! ich habe alles schon abgemacht! Ach, das ist ärgerlich, die kostbare Morgenzeit! Was fange ich nun an? – Nur nicht lange besonnen, die Zeit entflieht, man muß jede Minute benützen, sich menschlich ausbilden. Harmonie der Kräfte ist das erste Gesetz.

Er hat den Schreibesel mitten in die Stube gestellt und macht gymnastische Übungen. Währenddes tritt ein BOTE *herein und bleibt verwundert stehen.*

BOTE. Ih, Herr Hofrat! was machen Sie denn da?

FLEDER *erschrocken.* Ich? Wahrhaftig – ei, so früh schon vom Bureau? –

BOTE. Ich habe in meinen jungen Jahren immer soviel vom Turnieren gelesen –

FLEDER. Turnen, turnen heißt das, guter Mann.

BOTE. Wahrhaftig, wenn Sie nichts dawider hätten, ich möcht das Ding einmal selbst probieren.

FLEDER. Ich bitte Sie, tun Sie Ihrer edlen Aufwallung keinen Zwang an! Versuchen Sie, das Volk ist ja eben um dieser Gesittungsanstalten willen da.

BOTE. Nun, in Gottes Namen. *Er springt.* Hopp!

FLEDER. Ah – nicht doch diesen barbarischen Laut dazu! Das will mit Würde, mit geziemendem Ernst getrieben sein. Sehn Sie, so – der Schwerpunkt ruht auf den beiden Ballen, der Körper bildet einen stumpfen Winkel. – *Er springt.*

BOTE. Nun treff ich's auch. *Er macht einen Sprung und fällt jenseits nieder.*

FLEDER *ihn aufhebend.* Sie haben doch keinen Schaden genommen?

BOTE *aufstehend.* Bloß eine sanfte Empfindung hinten. – Es geschieht mir schon recht, das soll so ein alter Esel den Jungen überlassen. – Aber über dem Turnieren da wäre ich bald retourniert, ohne – *Er sucht in seinen Taschen.* Ich soll – der Präsident – o, er ist verloren!

FLEDER. Was, um Gotteswillen! verloren? Und das sagen Sie mir erst jetzt? was ist geschehen? wie verließen Sie ihn, wie sah er aus?

BOTE. Blau, mit einem eingekniffenen Eselsohr.

FLEDER. Blau? Liegt er?

BOTE. Nein, Gott sei Dank, er sitzt –

FLEDER. Fest?

BOTE. Ja, im Futter.

FLEDER. Wer?

BOTE. Was? *Er zieht einen Brief in blauem Umschlage hervor und übergibt ihn dem Fleder.* Da ist er.

FLEDER. Aber ich begreife nicht –

BOTE. Na, na, ich hab nicht länger Zeit. Guten Morgen! *Ab.*

FLEDER. Hat mich der einfältige Mensch doch erschreckt! Er ist immer so dumm zu verstehen. – Das Schreiben ist vom Präsidenten selbst, seine gewöhnliche vergnügte Hand. *Er erbricht den Brief.* Ach – eine konfidentielle Mitteilung! Laß sehen, was gibt es da?

Er liest. „Bester Hofrat Fleder! Sie kennen meinen alten Lieblingswunsch, daß mein Neffe, der junge Graf Leonard, der soeben von Reisen zurückgekehrt ist, endlich das ewige, zwecklose Umherschweifen lasse, sich vermähle und seine Kenntnisse und Talente der Welt zum Besten gebe."
Gott, jawohl, des Würdigen beut die ernste Zeit soviel! - „Ich habe daher meinem Neffen zugeredet, um die Hand der unabhängigen, liebenswürdigen und reichen Gräfin Adele zu werben. Sie lebt, wie Sie wissen, jetzt auf ihrem einsamen Waldschloß, weil ihr die Residenz zu langweilig ist. Er hat sich vor der Langweiligkeit der Welt in die Residenz geflüchtet, um unsere Kurzweil gründlich zu studieren. Sie verachtet die Männer, er die Weiber. Sie ist launisch, phantastisch; er verfolgt wütend die Phantasten und ist doch selbst der größte – in Summa: die beiden passen zusammen wie zwei scharfgezahnte Räder, wenn's auch anfangs tüchtig knarrt. Das wäre nun alles gut. Aber kaum erzähle ich ihm von allen diesen Kuriositäten der Gräfin, so springt er zerstreut vom Stuhle auf – ich füge noch hinzu, daß sie oft reisende Künstler auf ihrem Schlosse sieht – da greift er nach seinem Hut – daß sie sie aber immer am folgenden Morgen wieder fortjagt, weil sie ihr zu ordentlich sind – da ist der Narr schon zur Türe hinaus. – Nun hör' ich soeben, daß er gar schon hingereist ist, aber, wie ich von sicherer Hand weiß, wieder ganz auf seine Weise: er will nämlich – denken Sie nur! – unerkannt, als reisender Schauspieler das Schloß der Gräfin besuchen."
O irregeleiteter Jüngling! o armer, würdiger Oheim! – „Ich fürchte da wieder große Konfusion und wünschte sehr, daß ein dritter auf dem Schlosse ein wenig zum Rechten sähe, um im Falle sehr wahrscheinlicher dummer Streiche alles möglichst ins Geleise zu bringen oder doch jedenfalls mich von dem Ausgange schnell in Kenntnis zu setzen; aber das müßte jemand sein, der allen unbekannt ist, denn merken die lustigen Vögel et-

was davon, so sitze ich mit meinem Kuppelpelz wie der Schuhu auf der Stange, sie stoßen alle nach mir. – Ich weiß, lieber Fleder, wie Sie immer recht erpicht sind auf Menschheitsveredlung und nach nützlichen Taten dürsten; hier gibt es einmal einen rechten Humpen für solchen Durst. – Die Gräfin Adele kennt Sie nicht persönlich, Leonard ebensowenig, Sie aber haben ihn schon einmal bei mir aus dem Kabinett gesehen und werden ihn leicht wiedererkennen. Daneben blasen Sie auch scharmant die Flöte."

In der Tat, wenn der seelenvolle Blick manches gebildeten weiblichen Wesens bei den schwierigen Passagen nicht log –

„Wie wäre es, wenn Sie selbst sich nach dem Schlosse aufmachten, allenfalls unter erdichtetem Namen, als reisender Musikus, gleichsam so hingeblasen. – Aber was geschehen soll, geschehe schnell, recht schnell! Denn mein Neffe, wie gesagt, ist schon fort, darum wählte ich, der Kürze wegen, die schriftliche Mitteilung in der frühen Morgenzeit. –

Postskript: Sie haben die junge Gräfin noch niemals gesehen. Um des Himmels willen nehmen Sie sich in acht, daß die Verwirrung nicht noch größer wird!"

In acht? – Diese Nachschrift ist ein Dintenklecks auf die reine Seele des Präsidenten! Hält er mich für einen Jungfernknecht, wie den gemeinen Haufen? – *Indem er den Brief schnell einsteckt.* Nun, des Mannes Antwort ist die Tat! ich reise als Flötenspieler zur Gräfin, noch in dieser Stunde. Heilige Pflicht, zwei verwilderte Herzen für die allgemeine Sache der Menschheit zu erwärmen durch ernster Rede Kraft – die jüngste Regierungsratsstelle wird nächstens erledigt – die gebührende Achtung vor der Familie meines würdigen Chefs – sie wirft jährlich tausend Taler ab – ja, ein edler Vorwurf, so recht unmittelbar in das Leben hinein zu nützen – und freie Wohnung nebstbei – fort, hinaus, zu Pferde! *Eilig ab.*

Zweite Szene

Das Innere einer Dorfschenke. *Flitt* und *Schlender sitzen schlafend auf den Stühlen.*

Flitt *erwacht und sieht an sich herunter, nach einer Pause.* Eine längliche Ansicht, viel Einsicht, manche Durchsicht. – Ich glaube, ich werde von dem über Stühl- und Bänke-Hängen immer länger und länger, wie ein Perspektiv, wo will das am Ende hinaus? – Laß sehn, ob meine Stimme über Nacht nicht erdurstet ist. *Er singt schläfrig.*

Wenn die Nacht vor Altersschwächen
Drückt ihr letztes Auge zu:
Geht der Hirt den Tag anbrechen
Und bläst Tu, tu, tu, tu, tu –

Schlender *auffahrend.* Was ist's?

Flitt. Ich weck' nur das liebe Vieh auf.

Schlender *sich umsehend.* Wahrhaftig, schon wieder Tag, und beißt auch gleich so in die Augen! Daß der Sonne auch nicht die Zeit lang wird, so vom frühen Morgen bis in den späten Abend in ein solches Wirtshaus hineinzuscheinen.

Flitt. Aurora musis amica. Sie weiß, daß dieses sperlingstichige, gewesene Schindeldach zwei Künstler beherbergt, sie begrüßt das Handwerk. Aber wie geht's dir, alte Muse?

Schlender. Es geht vielmehr noch gar nicht, es hängt noch alles so miserabel an mir, mir tun alle Haare weh.

Flitt. Geduld, guter Schlender, Geduld! Dein Wandel wird dich, wie eine herzhafte Magd, bald so kahl scheuern, daß ich keinen Groschen für die leere Glatze geben möchte.

Schlender. Groschen? – Sag, lieber Flitt, hast du noch was in der Tasche?

Flitt. Bloß ein großes Loch – da, ich komme mit der gan-

zen Hand durch, und wenn sie zwanzig Finger hätte. Aber du? Untersuch' doch einmal –

SCHLENDER *nach seiner Rocktasche greifend, erschrocken.* Bei Gott, ich habe gar keine Tasche mehr!

FLITT *lachend.* Du warst außer dir, Schlender, nun weißt du's nicht einmal. Ja, eine lustige Nacht! Aber aus dem guten Humor wurde ein böser Rumor, da rissen dir die einen im Getümmel die Rockschöße vom Leibe und die anderen flickten ihn wieder mit blauen Flecken.

SCHLENDER. Mir? Aber weshalb denn?

FLITT. Beim Spiele, Freundchen. Der Wein eröffnet das Herz. Du warst zu aufrichtig mit deinen falschen Würfeln, und darüber entstand die lebhafteste Teilnahme in der ganzen Gesellschaft.

SCHLENDER *sich in seinem halben Frack von allen Seiten besehend.* Das ist gottserbärmlich, wie ich ausschau'. Nein, das muß anders werden, bei dem Leben kommt nichts heraus –

FLITT. Als der Ellbogen aus dem Ärmel.

SCHLENDER. Ich hab mir's ernstlich überlegt –

FLITT. Leg dich noch einmal über, Schlender, und schlaf besser aus, du bekommst schon wieder deinen Katzenjammer.

SCHLENDER. Nein, guter Flitt, diesmal ist's mein unerschütterlicher Vorsatz, ich mache einen neuen Absatz in meinem Leben!

FLITT. Was! Noch einen neuen Absatz an solchen alten, ausgetretenen Schuh?

SCHLENDER. Wir sind auf Umwegen zur Tugend begriffen – ich bin noch jung, talentvoll, schön gebaut – ich will mich ganz umwenden.

FLITT. Lohnt nicht, lohnt nicht, bist schon von beiden Seiten verteufelt abgetragen. – Frisch, laß uns lieber etwas auf deinen neuen Absatz trinken. – Heda , Wirt, dicker Ölgötze, schrote dich heraus, altes Weinfaß!

KNOLL *tritt ein.* Bon schnur, Bestiös! Was macht Ihr da

schon wieder für Lärm, bei so frühem, nachtschwärmendem Tage?

FLITT. Bon schnur? – Häng dich daran, alter Galimathias! Vorher aber gib uns noch von deinem schlechten französischen Weine auf unser gutes deutsches Wort.

KNOLL. Paraßol d'honneur! ein Wort ist ein Wind, und ein Wind ist nichts, und auf nichts geb' ich nichts mehr.

FLITT. Oho, mein Wort und dein Wein sind einander gerade wert. Also du willst wirklich nicht länger mehr auf mein ehrlich Gesicht? –

KNOLL. Kupfer, eitel Kupfer!

FLITT. Aber kostbares Kupfer, ich hätte eine Nase von Gold dafür haben können.

KNOLL. Ihr hättet sie doch wieder im Weinkruge sitzenlassen.

FLITT. Hör', Knoll, ich habe einen durstigen Einfall. Hör' einmal, wenn du durch deinen Speck noch Vernunft hören kannst.

KNOLL. Nun, kretipleti, Bestiös?

FLITT. Mein Kamerad da ist ein Virtuos' auf der Geige, wie du weißt, und ich bin ein reisender Schauspieler; meiner Treu, wir wollen unsere Talente hier nicht länger verschimmeln lassen!

KNOLL. Hat keine Not wegen des Verschimmelns, sie haben Euch erst heute nacht wieder tüchtig gescheuert.

FLITT. Ich bitt' dich, schluck' jetzt deinen Witz hinunter und wenn du daran erwürgen solltest! – Also ich sage: ich und mein Kamerad da, wir haben beide einen devoranten guten Geschmack; was gibt's nun hier in der Runde für Herrschaften, die auf ihre Ausbildung was draufgehen lassen? Wer wohnt zum Beispiel in dem großen Schlosse, das man dort vom Berge sieht?

KNOLL. Die Gräfin Adele – wollt' sagen: Krudele, denn sie stößt alle Liebhaber vor den Kopf, schlägt viel Geld tot, schlägt die Laute, schlägt Triller –

FLITT. Genug des Gemetzels, Barbar! Da wollen wir hin und frisch mit dreinschlagen, ich denk', es soll einen gu-

ten Silberklang geben. Schlender! – Der Kerl ist wahrhaftig vor Schwermut wieder eingeschlafen. – Halt, da blitzt's mir eben durch den ganzen Kopf! – Still, still, stör' ihn nicht. – Sag' mir einmal, Knoll, wie haben dir heute nacht seine Würfel gefallen?

KNOLL. Sie sind gut gefallen; sie hatten alleweil so viele Augen, daß sie den feinsten Silberstrich aus der tiefsten Tasche herauslasen.

FLITT. Möchtest du die Dinger wohl haben, mein Lieber?

KNOLL. Quinten, pure Quinten! Was ist eigentlich Eure Intentation dabei?

FLITT. Ich bin einmal so ein ehrlicher Bursche, daß ich's niemals in der Welt zu etwas Rechtschaffenem bringen kann; ich möchte gern uns allen dreien helfen, bevor dich der Teufel ganz in deinem Talg erstickt hat. – Sieh, erstens: Mein Kollege da hat beschlossen, sich ganz umzuwenden; wir wollen ihm helfen auf dem Wege der Besserung – erst die Taschen! *Er wendet vorsichtig Schlenders Tasche um.* Er hat von seinen Nachtsünden Zahnschmerzen in den Haaren, wir wollen ihm die schadhaften Knochen ausziehen – *Er bringt Würfel aus Schlenders Tasche hervor.* Da sind sie! – Nunmehr zweitens –

KNOLL. Eure Intentation! Eure Intentation!

FLITT. Also zweitens: Wir haben dir unsre guten feinen Kleider zum Pfande gegeben, da wir mehr Durst als Geld hatten. Wieviel haben wir davon noch zugute, wenn du die doppelten Kreidestriche abrechnest?

KNOLL. Jeder einen halben Ärmel etwa.

FLITT. Eine lumpige Sicherheit! Sei kein Narr, gib mir die Kleider heraus, und du sollst die Würfel haben. – Nun?

KNOLL. Knifftheologie, lauter Knifftheologie! Pfiffe tragen Püffe! – Nein, behaltet Ihr Eure Knochen da auf neue Kleider, und ich behalte Eure neuen Kleider für meine eigenen Knochen.

FLITT. Bist du gescheut? Unsere Kleider für dich? – Einen Frack über den Bauch, den andern über den Rücken, in

jede Pumphose einen Arm, da bleibt dir nichts für die Beine; willst du herumgehen wie ein Bergschotte? – Aber mir auch recht, wie du willst – eigentlich – wie ich mir's recht überlege. – Wie waren doch heut nacht die beiden Hauptwürfe Schlenders? – Laß einmal sehen! *Er wirft.* Nummer eins!

KNOLL *gespannt zusehend, für sich.* Sei kein Narr, – Knoll, ich bitt' dich, sei kein Narr. –

FLITT *wirft wieder.* Numero zwei: dix huit!

KNOLL. Nimm's mit – nimm's mit! *Er geht zögernd in den Hintergrund und schließt einen Schrank auf.*

FLITT *für sich.* Wahrhaftig, der Walfisch beißt an! Gott gesegne seinen breiten Rücken! Jetzt geschwind gewechselt, die falschen Köder in die Tasche und meine aufrichtige ordinäre Würfel auf den Tisch. *Er steckt die falschen Würfel ein und legt andere hin.* So – und dann rasch fort von hier, sonst bekomme ich noch Agio aufgezählt in dem Wechselnegoz.

KNOLL. Da ist der Plunder! *Er wirft einen Pack auf den Tisch, während er mit der andern Hand hastig die Würfel ergreift und einsteckt.* Nun weiß ich, was ich habe, und nies' Euch was auf Eure Pfiffe, ha, ha, ha!

FLITT *den Pack unter den Arm nehmend.* Prosit! Ha, ha, ha!

SCHLENDER *erwachend.* Was ist denn das für ein Spektakel, daß man sein eignes Nachdenken nicht hören kann. –

FLITT. Munter, Schlender! Frisch auf, zur Gräfin Fortuna! Ha, ha, ha! – Du schaust ja so bedenklich und verschoben wie ein Paraplü im Sturmwind.

SCHLENDER. Paraplü? – regnet's schon wieder ein?

KNOLL. Muß doch, denn Ihr hängt ja nach allen Seiten über wie eine nasse Pelzmütze. – Ha, ha, ha! – die Ratten haben Euch über Nacht beknappert, Ihr seid so ohne Umschweife. –

SCHLENDER *springt wütend auf.* So darf mir niemand kommen, und wenn er noch so dick wäre! Was! hab' ich

nicht Anno dreizehn den ganzen deutschen Patriotismus mitgemacht und bin als freiwilliger Janitschar vor dem Feinde ausgezogen und habe neben der Leipziger Schlacht die Becken geschlagen und bin –

KNOLL. Kaldaunen, Katabomben und Sonaten!

SCHLENDER. O, ich fürchte mich nicht, ich fürcht' mich gar nicht! Ich geb' nicht soviel auf so ein Ochsenmaul voll Worte; es hat schon manch dickerer Kerl, als du bist, mich durchgeprügelt.

FLITT *ihn nach der Tür drängend.* Ich bitt dich, Schlender, sei doch vergnügt, wenn du zur Türe hinausgeschmissen wirst. Willst du hier sitzen und alle Striche bezahlen, die wir da an die Wand getrunken haben?

SCHLENDER. Betrunken? Was? wer ist betrunken?

FLITT *schiebt ihn zur Türe hinaus.* Ach, frag' nicht soviel! Fort!

KNOLL. Bon Pakage! Behüt' Euch Gott, daß Ihr in Euren Unternehmungen nicht wieder zurückkommt!

FLITT. Möchte uns unser Weg recht bald wieder weit an dir vorbeiführen!

Alle ab.

Dritte Szene

Wald. Man hört in weiter Ferne Waldhörner. Graf LEONARD *in etwas phantastischer Reisekleidung tritt auf.*

LEONARD.

Hallo, poet'sche Wirtschaft da! Von allen Gipfeln
Rings Hörnerklänge, und wie Meereswogen
Hoch über mir frisch' Rauschen in den Wipfeln,
Als käm' durch's Holz geflogen
Die Zaubergräfin auf phantast'schem Besen
Hier zum Walburgistanze des Parnasses. –
Wer kein Genie ist, lass' es!

Mir aber schwillt mein Busen so verwogen!
Wo kühn hin über das gemeine Wesen
Ein Weib das Seil der Poesie gezogen,
Um drauf mit keckem Schwung und Neigen
Sich vor dem Volk zu zeigen:
Da will mich der Bajazzo in mir zwingen,
Mich mit auf's Seil zu schwingen
In Entrechats und unerhörten Sprüngen.
Hopp! Nochmals hopp! Das Volk ist außer sich,
Je höher die Gräfin, je höher ich –
„Sie brechen den Hals, mein Lieber!"
Hilft nichts! Dreimal noch in der Luft kopfüber
Stürz' ich zuletzt mich ins Hurra der Menge,
Verschlüpfend wieder ins gemeine Wesen,
Und keiner im Gedränge
Weiß, daß der Narr Graf Leonard gewesen. –
Das ist das Flügelpferd mit Silberschellen,
Das heitere Gesellen
Emporhebt über Heidekraut und Klüfte,
Daß durch den Strom der Lüfte,
Die um den Reisehut melodisch pfeifen,
Des Ernstes Gewalt und Torenlärm der Schlüfte
Als Frühlingsjauchzen nur die Brust mag streifen;
Und so im Flug belauschen
Des trunknen Liedergottes rüst'ge Söhne,
Wenn alle Höhn und Täler blühn und rauschen,
Im Morgenbad des Lebens ew'ge Schöne,
Die in dem Glanz erschrocken
Sie glühend anblickt aus den dunklen Locken. –
So rauscht nur, Wälder frisch! Ich wandre,
Die Schöne sei mein Lieb und keine andre!
Er will weitergehen, bleibt aber plötzlich, in die Ferne sehend, stehen.
Was gibt's denn dort? – Ich glaub, das Flügelpferd
Hat eben zwei Genies da abgesetzt!
Ein Roß, im Winde Mähn' und Zügel flatternd,
Fliegt lustig waldwärts übern grünen Grund,

Zwei Herren hinterdrein – nun fällt der eine,
Der andre über ihn – das Pferd ist fort,
Sie stehn und wundern sich – *Er ruft.* Heda! Hup hup!

SCHLENDER *von der entgegengesetzten Seite, hinter der Szene.* Hup hup!

LEONARD *sich schnell umwendend.* Was Tausend, noch ein Dritter! Die ich rief, hören mich nicht, und den ich höre, rief ich nicht. Hoho, guter Freund, sind Sie bloß ein Echo?

SCHLENDER *hinter der Szene.* Laß die Narrenspossen und hilf mir lieber aus der Verwickelung! Hagebutten, Pfingstrosen, Stachelbeeren, alles niederträchtige Gewächs stichelt auf meine neue Kleidung.

LEONARD. Wie soll ich Ihnen helfen? Ich sehe nichts von Ihnen als Ihre Stimme.

SCHLENDER *in eleganten Kleidern, eine Geige unter dem Arme, tritt auf.* Was! Sind Sie nicht Flitt?

LEONARD. O, ein Troubadour! Bei Gott, ein Troubadour!

SCHLENDER *seine Geige in die Rocktasche steckend.* Erlauben Sie, mein Herr, haben Sie nicht ein gesatteltes Pferd gesehen?

LEONARD. Jawohl, und zwei abgesattelte Herren hinterdrein. – Aber sagen Sie mir doch nur –

SCHLENDER. Die Landstürzer, die Schnappsackspringer die! – Man hat seine Not mit vielen Domestiken, je mehr Gefolge, je weniger folgt es, je größer die Suite, je mehr Suiten werden gerissen.

LEONARD. Wie, und alle diese Suitiers ritten auf *einem* Pferde?

SCHLENDER. O, das hätte noch zwanzig mehr ertragen! Was denken Sie? Es war recht mein Augapfelschimmel!

LEONARD. Nein, wahrhaftig, es war ein Brauner.

SCHLENDER Ja, oder mein Augenbrauner. Tausend Dukaten, auf Ehre! dafür hätt' ich ihn gegeben. Ich ritt einmal mit einem arabischen Lord, die bekanntlich die be-

sten Pferde sind, um die Wette drauf – das hab' ich nun verloren!

LEONARD. Meiner Treu, ich meine, das Pferd hat Sie verloren.

SCHLENDER. Ja, ich muß nun zu Fuß gehn, leider! Aber es tut nichts, besser schlecht gegangen, als gut gehangen, wer reitet, kann fallen, und wer zu Fuß geht, kann auch fallen, das ist mir alles egal.

LEONARD *der unterdessen um ihn herumgegangen und ihn mit Vergnügen betrachtet hat.* Bei Gott, wie eine Trödelbude! Der Frack zu lang, die Weste zu kurz – o, ein köstlicher Fund!

SCHLENDER. Fund? – Wieso, Fund? *Für sich.* Das ist eine gute Geschichte. Ich glaube, der Kerl ist gar einer von den Glücksrittern, die allezeit finden, was andern ehrlichen Leuten verlorengeht.

LEONARD. Wahrhaftig, ich lass' Sie nicht wieder los.

SCHLENDER. Mich, mein Herr? Was wollen Sie von mir, mein Herr? Weil ich ein vornehmes Äußeres habe? Ja, gehorsamster Diener!

LEONARD. Ganz ergebenster! – Aber wie kamen Sie denn eigentlich in diesen Wald?

SCHLENDER. Ja, wüßt' ich lieber, wie ich wieder herauskäme! – Aber das will ich Ihnen wohl sagen. Sehn Sie – treten Sie mir nur nicht so nahe, ich bin sehr empfindsam an den Hühneraugen – also: wir waren eben auf der Wanderschaft, und mein Kamerad Flitt lehrt mich unterwegs Galliarde springen und Pirouetten, kurz, das ganze Solo mit den Füßen. – Aber bleiben Sie dort stehen, so – indem kommt ein fremder Herr zu Pferde des Weges daher. Ich denk' an nichts und exerzier' mich so für mich in den Sprüngen, da kommt's mit einemmal dem Pferde auch in die Beine, ich sage Ihnen, wie ein Bock auf der grünen Wiese. Ich nicht zu faul, streng' mich an mit Sätzen, das Pferd aber immer noch höher, mein Kamerad schreit, der Herr schimpft, das war ein Spektakel, vier Pferdebeine, zwei Stiefel, ein Paar Arme,

alles wie Windmühlenflügel in der Luft durcheinander. Endlich kommt der Hut auf die Erde geflogen, und gleich darauf der ganze Herr, leider! Er aber geschwind wieder auf und läuft hinter dem Pferde drein, mein Kamerad hinter ihm, ich hinter meinem Kameraden –

LEONARD. Halt, halt wieder! –

SCHLENDER. Was Halt! Da war kein Halten mehr! Die beiden hatten gut rennen mit ihren vier Beinen, ich aber hatt' nur zwei, da mag der Teufel nachkommen mit meiner Violine. So bin ich in diese Einsamkeit und Konfusion geraten und weiß nun nicht den Weg auf das Schloß der Gräfin Adele.

LEONARD. Wie, Sie wollen auf das Schloß der Gräfin?

SCHLENDER. Ja, das will ich! Haben Sie was an sie zu bestellen? Ich und mein Kamerad, der ein berühmter Schauspieler –

LEONARD. O auserlesene Wirtschaft! O kommen Sie geschwind, ich geh' auch zur Gräfin!

SCHLENDER *zögernd.* Verzeihen Sie, mein Herr, wer sind Sie eigentlich?

LEONARD. Ja so! – eigentlich – Sagten Sie nicht vorhin, daß Ihr Kamerad ein Schauspieler? – *für sich* Zwei Komödianten, das könnte auffallen, ich muß die Farbe wechseln. *Laut.* Wer sind denn Sie?

SCHLENDER. Ich bin eigentlich ein Sänger, aber ich habe meine Gurgel viel strapaziert und unterwegs die Stimme verloren, da akkompagniere ich jetzt meinen gewesenen Gesang auf der Violine.

LEONARD. Gut, so bin ich eigentlich bloß ein Violinspieler, der aber bloß singen kann, und Sie sollen mich begleiten. Mein Name ist – Florestin.

SCHLENDER. Also ein Sänger? – Nun, wahrhaftig, dacht' ich doch Wunder! *Wieder keck vortretend.* O mein Herr Florestin, ich will nichts voraussagen, aber Sie werden mich schon kennenlernen.

LEONARD. Ist gar nicht nötig, ich sah es Ihnen gleich an, Sie haben einen schönen Strich.

SCHLENDER. Ich sag' Ihnen, in dem Hotel, wo ich heute über Nacht war, haben sie sich ordentlich um mich gerissen, das war ein Furore und ein Klatschen von allen Fäusten!

LEONARD. Das ist alles noch nichts, das soll noch besser kommen!

SCHLENDER. Nein, ich dank' Ihnen sehr. –

LEONARD. Nachher, nachher! Jetzt nur fort zur Gräfin! Bei Gott, Sie sind gerade das rechte Akkompagnement zu meiner Reise-Note. Frisch dort hinaus! Sie können sich unterwegs noch in Ihren Solosprüngen üben. *Schlender in Pirouetten voraus.* So, sassa! *Beide ab.*

Vierte Szene

Garten der Gräfin Adele. Rechts das Schloß, über dessen Tür ein mit hohen Blumen besetzter Balkon. Auf der andern Seite ein Bassin mit einem Springbrunnen, im Hintergrunde ein zierlicher Gartenzaun.

FLORA *in reicher Kleidung, steht, ihre Haarflechten zurechtsteckend, am Springbrunnen und singt.*

Offen stehen Fenster, Türen,
Und ich bin im Schloß allein,
Draußen sich die Bäume rühren,
Und der Abend rauscht herein.
Überm Fluß, weit durch die Haine,
Ziehet draußen muntre Lieb',
Ruft mir zu: Was stehst alleine? –
Ja, wer da zu Hause blieb'! –

Sie setzt sich auf den Rand des Bassins und besieht sich im Wasser.

Ich sehe doch hübsch aus in den prächtigen Kleidern. Wenn ich eine Prinzessin wäre, legte ich mich alle Abend zwischen die Blumen auf dem Balkon und sä-

he den Wolken zu, wenn die Wasserkünste unten im Garten rauschten, und so oft die Nachtigallen schwiegen, säng' ich da oben, und wo der erste Stern herunterfiele, da käm' mein Bräutigam her. –

VICTOR *von der andern Seite aus dem Gebüsch hervortretend.* Da sitzt das Singvögelchen und putzt sich die Federn. –

FLORA *ohne ihn zu bemerken, lacht für sich.* Ich fürchte mich ordentlich vor mir selbst im Wasser dort. – Wie eine Sirene – das sieht mich so starr an, als wollt' mir's in der Stille etwas erzählen.

VICTOR *ist hinter ihr herangeschlichen und hält ihr mit den Händen beide Augen zu. Flora schreit laut auf.*

FLORA. Ach, du bist's ja wieder, Victor! Pfui, deine Finger riechen nach Tabakrauch, laß mich los, du bist unausstehlich, du verdirbst mir das schöne Kleid!

VICTOR. Und soll ich denn lassen deine Augen licht, so reich erst dein Schnäblein – aber pick mich nicht!

FLORA. Ach, dummes Zeug, ich kann ja deinen Mund nicht sehn. – Da, nun hast du was Schönes angerichtet, meine schöne goldene Brustnadel ist ins Wasser gefallen.

VICTOR *sie loslassend.* Was? wo? –

FLORA *entspringend.* Zisch aus, zisch aus! –

VICTOR *sie erstaunt betrachtend.* Ei der Tausend! wie siehst du heut aus? Laß dich doch besehn, Florchen, laß dich besehn. Der Anzug ist passend für eine Jungfer, in lauter Spitzen gewickelt wie ein Igel. Wen willst du eigentlich stacheln?

FLORA. Jeden Narren, der mir zu nah kommt. Willst du mein Hofnarr sein, Victor?

VICTOR. Unbedenklich! Der lebt von der Narrheit seines Hofes, da stünd' ich hier in der besten Nahrung. – Wahrhaftig, Flora, wir beide sind doch *reizende* Personen, wir können eins das andere nicht ansehn, ohne sogleich miteinander zu fechten.

FLORA. Ja, wirklich, von den vielen Gefechten ist dein

Witz schon ganz unscheinlich geworden. Ich bitt' dich, schone dich, du wirst ihn brauchen in diesen Tagen.

VICTOR. Ja, ohne dich zu schonen. Aber warum gerade in diesen Tagen?

FLORA. Ach du kommst eben aus dem Walde, du weißt wohl noch gar nicht – Noch heute treffen hier zwei fremde Herren ein, um die Gräfin zu befreien, Graf Leonard, ein Neffe des Präsidenten, als reisender Schauspieler, und der Hofrat Fleder als Flötenspieler.

VICTOR. Was? – Nein, wenn ihr hier anfangen wollt zu heiraten, so nehm' ich meinen Abschied! Aber woher wißt ihr denn alles so genau? Haben Sie einen Komödienzettel vorausgeschickt?

FLORA. Der Präsident vertraute es unter dem Siegel der Verschwiegenheit der Gräfin Jolante, und sie versiegelte es noch einmal und schickte es an unsre Gräfin.

VICTOR. Und was sagt denn die Gräfin zu der versiegelten Geschichte?

FLORA. Erst zerriß sie den Brief und wollte augenblicklich zu Pferde und in die Türkei reiten, wo es weder Grafen noch Präsidenten gäbe. Dann lachte sie aber auf einmal sehr und sagte, wir alle sollten uns stellen, als kennten wir die Fremden nicht und wüßten nichts von der ganzen Maskerade; ich aber sollte mir ihre besten Kleider anziehen und mich solange die Gäste hier sind, für die Gräfin ausgeben, sie selbst will unterdes die Jungfer spielen.

VICTOR. O vortrefflich! – Aber ich möcht' nicht Jungfer spielen in einem Lustspiele, sie bekommt im letzten Akt eine Haube weg, sie weiß nicht wie.

FLORA. Pfui, so ein Haubenstock. – Das ist's ja eben, warum wir uns maskieren. Laß nur, ich will als Gräfin so viel dummes Zeug machen, daß sie lieber einer den andern heiraten möchten, als länger hier bleiben.

VICTOR. Höre, Flora, das ist nicht anders, als hättest du mir eine Prise Nieswurz gereicht, so licht und vergnügt wird mir's auf einmal im Haupte! Nein, da frag' ich

nichts danach, da mag draus werden, was da will, ich spiel' auch mit! Frisch, jeder auf seine eigene Hand, soweit sein Witz reicht, wir wollen sehn, wo das zuletzt zusammenklappt!

FLORA. Auf deinem Rücken, hoffe ich.

FRIEDMANN *der unterdes mit einem Blumenkörbchen eingetreten ist, bleibt unentschlossen stehen.* Der Victor! – Ich möchte ihm lieber ausweichen, er ist ein Phantast und Händelmacher.

VICTOR *ihn erblickend, stürzt auf ihn los und umarmt ihn.* Gärtner! Maskerade, Konfusion, Komödie! Ihr wißt doch schon? es geht bald los, mir spielt schon die ganze Ouvertüre im Kopfe mit erstaunlichen Fugen und Sätzen!

FRIEDMANN *sich entrüstet losmachend.* Was ist denn das? Lassen Sie mich ungeschoren mit Ihren Fugen! Ich habe keine Zeit, wie gewisse andere Leute, an solche Narrenspossen zu denken!

VICTOR. Aber Mensch! wollt Ihr denn wirklich vor lauter Vernunft zerplatzen? Ficht's Euch denn nicht auch einmal an, Euch auf den Kopf zu stellen und Vivat Friedmann! mit den Beinen in die Luft zu schreiben, oder auf der großen Gartenschere über die Blumenbeete zu reiten, oder –

FRIEDMANN. Ach Gott behüte, da müßt' ich ja toll im Kopfe sein! *Er erblickt Flora.* Sieh da, Mamsell Florchen! – Ei, ei, das steht Ihnen allerliebst!

FLORA. Ja? *In dem Blumenkorbe wühlend.* Aber was haben Sie da für schöne Blumen, Herr Friedmann, das möchte mich putzen!

FRIEDMANN. Gern, gern, mein Herzchen. – Hi, hi, hi, das flimmert einem recht vor den Augen da im Korbe, man weiß nicht, was Händchen sind oder Blumen, eins so hübsch wie das andere.

FLORA. Da, stecken Sie mir sie vor, hier an die Seite, aber nehmen Sie sich in acht, ich habe Nadeln.

FRIEDMANN *indem er Blumen ordnet und sie ihr vorsteckt.*

Ja, ja, je frischer das Röslein, je herzhafter sticht's. –

FLORA. Sie sprechen ja ordentlich durch die Blume.

FRIEDMANN. Selber Blume, Florchen, selber Blume.

VICTOR. Ihr habt ein recht angenehmes Lächeln unter der Nase, Gärtner; ich hätte nicht geglaubt, daß Ihr noch so lächerlich sein könnt, Ihr alter Tausendsassa.

FRIEDMANN. Ich bitte Sie, Herr Victor, wenn Sie nur nicht immer solche gemeine Redensarten –

FLORA. Was macht denn die liebe Frau?

FRIEDMANN. Danke, danke – so, nun ist's fest.

VICTOR. Sitzt Ihr noch alle Abend mit Eurer Familie in der Jelängerjelieber-Laube und denkt über das Vergnügen nach, Papa zu sein?

FLORA *wieder bei dem Bassin, sich die Blumen anders steckend.* Und die lieben Kinderchens bringen Ihnen die Pantoffeln heraus und Ihre lange Pfeife.

FRIEDMANN. Ja, Florchen, so nach erfüllter Berufspflicht, mein Pfeifchen im Munde und ein zufriedenes Herz in der Brust! –

VICTOR. Da seht, das kommt bei eurem langweiligen Metier heraus! Blumen und Kohl, das wächst so still und wohlgezogen in den Tag hinein –

FLORA. Und Sie murmeln, als ein zufriedener Bach, zwischen lauter Veilchen und Vergißmeinnicht dahin.

FRIEDMANN. In der Tat, die rohen, lärmenden Vergnügungen der Jagd sind niemals nach meinem Geschmack gewesen. – Aber ich habe mehr zu tun, als hier zu plaudern. *Für sich.* Die Narren sollen mich nicht aus meinen moralischen Grundsätzen herausdisputieren. *Ab.*
Man hört Hörnerklänge.

VICTOR. Horch! da kommt die Gräfin von der Jagd zurück.

FLORA *springt zu dem Gartenzaun und sieht hinaus.* Sieh nur, wie hoch und herrlich sie zu Pferde sitzt und zwischen den Abendlichtern da die Kastanienallee heraufkommt! – Sie steigt am Gartenpförtchen ab – wahrhaftig, sie kommt grad hier herauf.

VICTOR. Es ist eine rechte Freude, einer solchen Gräfin Diana zu dienen!

GRÄFIN ADELE *in einem grünen Jagdkleide, eine Reitgerte in der Hand, tritt auf.* Nun, Fastnachtsbräutchen, sieh dich vor – sie kommen!

FLORA. Wo, wo?

ADELE. Wie ich den hohen Waldweg ritt, erblickte ich zwei Fremde fern im Grunde. Sie sind's gewiß, ich sprengte schnell voraus.

Sie wirft ihr Jagdkleid ab und nebst Gerte dem Victor zu, in einfachem Hauskleide erscheinend.

Da trag' das fort, ich hoff', auf kurze Frist.
Victor ab; zu Flora.
Ich bitt' dich, deine lustigsten Gedanken
Nimm all zusammen nun! Des muntern Grafen
Geheimnisvoll verschlungne Redeblumen
Gesegne mit so vollem Maienschein
Von Freundlichkeit, als, ohne zu erröten,
Ein Mädchen darf. Und wenn er dann, mich meinend,
Dich schlau umstellt, mich will, dich freit – was lachst du?

FLORA.

Und wenn er nun den Komödiantenmantel
Auf einmal aufschlägt und ein schöner Ritter –

ADELE.

Ja, ja, Ich seh' ihn vor mir stehn: mehr Milch
Als Blut, von glattem Haar und blonder Seele,
Ein guter Neffe so und schlechter Vogel,
Der mitten in der Jugend frischer Zugzeit
Vom Onkel sich aus seinem Reisehimmel
Herunterschießen läßt und zahm sich stellen
Auf den gemeinen Freiersfuß. – Doch hör' nun,
Wirst du den Hofrat auch erkennen, Flora?

FLORA.

Gewiß, obgleich er mich nicht kennt. Wie oft
Sah ich ihn, wenn sie alle promenierten,
In selbstgefälliger Glückseligkeit,

Wie einen Weißfisch schnalzen auf dem bunten Strom,
Bald ernst, bald stolz und immer sehr zufrieden
Mit seinem neuen Frack – dort, dort! da sind sie! –
Der dort am Halstuch rückt, das ist der Hofrat!
Jetzt sehn sie uns – ei, wie sie zierlich grüßen!
Macht Verbeugungen.

ADELE.
Der Victor weist sie grad hierher zum Garten.
Frisch denn, die lustige Rakete steigt!

FLORA.
Nun weiß der Himmel, wem sie im Zerplatzen
Den Bart versengen wird. – Still, still, sie kommen. –
FLITT *in eleganter, etwas theatralischer Kleidung, und* FLEDER *treten auf.*

FLITT *zu Adele, die ihnen bis in den Hintergrund entgegengegangen.* Sieh da, süßer, bescheidener Mond! Wir sind Adler, wir fliegen nach der Sonne – wo ist sie?

ADELE *auf Flora zeigend.* Sie steht eben im ersten April, sie sticht heut ein wenig.

FLITT. Unser Teint kann schon etwas vertragen, mein Kind, um so besser schlagen die Gedanken aus, um so besser schießt der Witz – ja, ich hoffe, hier wieder ganz in Blüte zu kommen.
Sich Flora nähernd.
Gnädigste Gräfin, wo eine Muse wohnt, ist der Parnaß, und dem Parnaß fliegt unaufhaltsam der Pegasus zu, und auf dem Pegasus reisen die Künstler – so folgten wir nur dem höhern Zuge, um unsere schuldige Ehrfurcht – Erlauben Sie, daß ich mir die Ehre gebe –
Er will Fleder der Flora vorstellen.

FLEDER *mit Anstand vortretend.* Arthur, ein Flötenspieler *mit einem verächtlichen Blick auf Flitt,* welcher das Vergnügen hatte, erst heute durch einen Zufall die Bekanntschaft dieses Herrn zu machen. –

FLITT. Bitte sehr! das Vergnügen war bei diesem Falle ganz auf meiner Seite. –

ADELE *leise zu Flora.* Wie kunstreich sie einander vor uns verleugnen!

FLORA *ebenso.* Ach, der Graf ist eben auch gar nicht schön! *Laut zu Flitt.* Und Sie selbst, mein Herr?

FLITT. Flitt, ein untertänigster reisender Bühnenkünstler.

FLORA. Flitt? – Also einer, der etwas anderes vorzustellen pflegt, als er ist? Nun, wir treiben diese Kunst hier auch zuweilen.

FLITT. O, Sie versetzen mich in Entzücken! Vortrefflich, so könnten wir hier vielleicht ein Liebhabertheater einrichten?

FLORA. Wahrhaftig, an närrischen Liebhabern wenigstens wird es nicht fehlen. In der Tat ein herrlicher Gedanke! ist's mir doch, als wären wir alle schon mitten im ersten Akt. O meine Herren, Sie sind eine rechte Bereicherung unserer ländlichen Langenweile. – Kommen Sie, das muß mit mehr Muße verhandelt werden, lassen Sie uns hineingehen, einige Erfrischungen –

FLITT. Erfrischungen? – Erlauben Sie – *Er reicht Flora den Arm.*

FLORA. Ich denke, die ungewohnte Fußreise hat Sie ein wenig heruntergebracht, Sie bedürfen der Restauration.

FLITT. *Flora fortführend.* Kleinigkeit, Kleinigkeit, meine Gnädigste, wir haben's wohl schlimmer gehabt. Seit ich das Flügelpferd reite, habe ich, außer dem Flügelschlag, noch manche andere Schläge – des Schicksals erfahren. *Ab.*

ADELE *zu Fleder.* Nun, Herr Arthur – ein schöner Name! – Sie werden doch mitspielen in der Komödie?

FLEDER. Soweit es sich mit der Würde eines gebildeten Mannes verträgt. Der Komische, liebe Jungfer, war niemals mein Fach.

ADELE. Ei der Tausend! Sehn Sie wohl? Vortrefflich gleich zum Anfange, ganz vortrefflich!

FLEDER. Wie? was denn? –

ADELE. Ja, ja, verstellen Sie sich nur nicht, Sie Loser! – O

kommen Sie nur, Sie werden uns schon zu lachen geben.

Beide ab.

MARIE *tritt auf.* Also das sind die Bräutigams, um die sie so viel Aufhebens und Heimlichkeiten machen? – Hm, wenn ich nicht einmal einen schöner'n Bräutigam bekäme, das lohnte auch Jungfer zu sein! Ob sie jetzt im Gartensaale sein mögen? – Hier sieht mich niemand. – *Sie steigt vorsichtig auf die Blumentöpfe und sieht in das Fenster des Schlosses.*

VICTOR *eintretend.* Sieh da, Marie, des Gärtners Mühmchen, ich glaube, das Kätzchen ist eben auf der Lauer. *Er nähert sich ihr leise und hebt sie dann plötzlich in die Höhe.* Wart, ich will dir's bequemer machen, Kleine.

MARIE *sich erzürnt von ihm losmachend.* O geben Sie sich keine Mühe, Herr Victor! – Kleine! Zur Klugheit hilft nicht lang sein, sonst wären gewisse Personen verständiger.

VICTOR. Gut, so laß einmal hören, ob du so klug als kurz bist. Weißt du, wie man die Rotkehlchen fängt?

MARIE. Wie die Gimpel, mit roten Beeren.

VICTOR. Ganz recht – und da sind denn die kleinen Dinger vorwitzig. –

MARIE. Ja, und die Gimpel sind gar nicht witzig, weder vor- noch nachher.

VICTOR. Ich bitte dich, Mariechen, da du noch, sozusagen, die Eierschalen auf dem Kopfe trägst, so benutze die Gelegenheit, wenn der Mund des reiferen Alters von guten Lehren überfließt und würdige Männer Erfahrungssätze machen und einen Rat schlagen wollen.

MARIE. Gehn Sie, gehn Sie, Herr Victor, mit Ihren altmodischen Jagdspäßen; die sind mir schon ganz langweilig. Und was ich Ihnen schon längst sagen wollte: ich komme nun auch gerade schon in die Jahre, daß Sie mich *Sie* nennen könnten.

VICTOR. Gut. Also: sei Sie nicht so vorwitzig und guck' Sie nicht in die Fenster nach fremden jungen Herrn.

MARIE. Wenn Sie mich immer kränken wollen, Herr Victor, so ist es sehr schlecht von Ihnen, das kann ich Ihnen sagen, Herr Victor. Wahrhaftig, und da können Sie lange warten, eh Sie dasjenige von mir erlangen, weshalb Sie mich neulich so baten.

VICTOR. Ich? –

MARIE. O ja, an dem schönen Sonntagsmorgen, wie ich allein am Springbrunnen stand und mir das Gesicht wusch, da sagten Sie: ich sei ein hübsches Mühmchen, Sie wollten mein lieber Vetter sein, und ich sollte Ihnen ein Küßchen darauf geben.

VICTOR. Still, still! Da kommt jemand auf den Balkon.

MARIE. Das haben wir von Ihren guten Lehren! nun kann ich noch in schlechten Ruf kommen mit Ihnen.

VICTOR. Hör' Sie, Mühmchen, wenn du mir hier noch lange mit dem roten Mündchen so vorparlierst, so entschließ' ich mich kurz und gebe Ihnen noch einen Kuß auf Borg zu deinem zukünftigen Küßchen.

MARIE. Wenn Sie hübsch artig sein wollen –

Beide ab.

Gräfin Adele und Flora treten oben auf den Balkon heraus. Man hört von Zeit zu Zeit Hörnerklang in der Ferne.

ADELE.

Nur einen Augenblick muß ich eratmen
Hier in der Kühle! Fast gereut mich schon
Der fade Schwank. – Laß sie drin schwatzen,
Wir ruhn indes hier in der lauen Luft. –
Der Harfe gleicht in solcher stillen Zeit
Der Seele Grund, da haucht ein leises Tönen
Durch alle Seiten in der Einsamkeit,
Und niemand weiß, woher, wohin es geht.

MARIE.

Die Jäger kehren heim schon von den Bergen,
Wie lustig geht der Widerhall durchs Tal.

ADELE.

Aus der Verwirrung dieser Töne taucht

Ein langversunknes Bild mir wieder auf. –
Denkst du des Abends noch in Heidelberg?
So standen auf dem Söller wir der Burg,
Bis alles still, und nur die Wälder rauschten
Noch über uns und unter uns der Neckar.
Da kam ein Schifflein auf dem Strom gezogen
Mit Waldhornsklang und Fackelschein, der seltsam
Sich spiegelt' rings am Fels und in der Flut –

FLORA.
Und auf des Schiffes Spitze, über alle
Hochragend, stand ein fröhlicher Gesell.

ADELE.
Und als das Schiff vorrüberrauscht' am Schlosse,
Da wandt' er plötzlich sich und grüßt' die Burg,
Sein Windlicht schwingend, daß gleich einem Feuer
Die herrlichste Gestalt im Dunkel aufstieg.
So ist das wunderbare Bild uns schnell
Erschienen und verschwunden, wie bei Nacht
Ein schöner, wilder Grund im Wetterleuchten.

FLORA.
Wir stiegen voll Gedanken drauf hinab,
Wo alles wieder dunkel war und still –
Und bei dem ersten Strahl sahn wir schon weit
Vom Reisewagen wieder Strom und Burg
Im Morgenrote hinter uns versinken. –

ADELE *nach einer Pause.*
Oft ist's, als hättst du recht: es mag mehr Glück
Und Lust da unten sein, als ich hier oben
Mir träumen ließ, wer sich nur dran gewöhnte –
Was kümmert's mich, ich bleibe auf der Höh! –
Horch, wer naht dort, so spät?

FLORA. Es scheint ein Fremder.

LEONARD *erscheint auf dem Gartenzaun, sich nach allen Seiten umsehend.*
Hier also ist das Feenreich? – Nicht übel!
Die Büsche flüstern heimlich miteinander
Und Wasserkünste singen Zauberlieder –

Auf lust'ger Höhe zwischen schlanken Wipfeln,
Das ist ein guter Sitz für einen Sänger!
Doch wo blieb mein Kumpan? He, Schlender, hier!

ADELE *zu Flora.*
Blick einmal recht ihn an – fällt dir nichts ein?

FLORA.
Ich wüßte nicht –

ADELE. Und doch – nein! Laß uns gehen,
Verlockend weht die Abendluft herauf
Und dieses Dämmerlicht verwirrt das Auge. –

FLORA.
O bleiben wir – bei Gott, da kommt noch einer!

SCHLENDER *gleichfalls auf den Zaun heraufsteigend.*
Ich wollt, wir fänden erst die Tür. –

LEONARD Geduld,
Sie werden sie uns bald genug wohl zeigen.
Er schwingt sich in den Garten, Schlender folgt zögernd.

ADELE *zu Flora.*
Wie – sagt er nicht vorhin, er sei ein Sänger?

FLORA.
Ja, so verstand ich. – Nun, fürwahr, die haben
Den rechten Schwung, zu mindesten über Zäune;
Die sind vom Metier, das sieht man gleich.

LEONARD.
Horch, flüstert' dort nicht was? Gewänder schimmern –
Bei Gott, die Damen stehn auf dem Balkon!
Galantes Glück! 's ist just die rechte Stunde
Zu Serenaden. – *Zu Schlender.* Frisch! heraus die Geige!
Wir nahn in Tönen. – So, hierher – nur näher.
Sie stellen sich in einiger Entfernung dem Balkon gegenüber.

LEONARD *singt, während Schlender akkompagniert.*
O ihr Güt'gen und Scharmanten!
Auf der späten Wanderung

Nehmt der lust'gen Musikanten
Ganz ergebne Huldigung!

FLORA.

Das ist unleidlich! *Laut vom Balkon herab.*
Unverschämt' Gesindel!
Hier ist kein Wirtshaus, um den Durst zu löschen!

LEONARD *singt wie oben.*

Jeder schlürft und denkt die Seine,
Und wer nichts Besondres weiß:
Nun – der trinkt ins Allgemeine
Frisch zu aller Schönen Preis.
Da bricht der schlaue Mond hervor – laß sehen,
Was er bescheint. *Er tritt näher an den Balkon heran.*

ADELE *plötzlich an Floras Brust sinkend.*

Das ist Er! – Fort, nur fort!

Die Damen verschwinden oben.

LEONARD *singt.*

Ob sie schmähn, ob sie sich zieren,
Musikanten spielen drein,
Brechen so im Musizieren
Kühn in Tür und Herzen ein!
Sie gehen singend und spielend in das Schloß.

Zweiter Aufzug

Erste Szene

Garten der Gräfin, wie am Ende des ersten Aktes. LEONARD *ruht zum Teil von Zweigen verdeckt, schlummernd auf einer Bank.* ADELE *tritt oben auf den Balkon heraus.*

ADELE.

Noch ist es einsam rings auf Tal und Höhn,
Und wie ein scheues Reh schweift nur die Dämmrung
Leis durch die tau'ge Stille –

LEONARD *erwachend.*

Horch, was klang
So lieblich durch die Luft! – Schlief ich so lange?
Adele erblickend.
Wie! geht Aurora heut so zeitig auf,
Eh' noch die Stern' am Firmament erblichen? –
Was hat sie vor in dieser Einsamkeit?
Laß sehn – ich tu, als schlief' ich.

ADELE *ohne ihn zu bemerken.*

Wie so anders
Sahn Berg' und Garten aus bei dunkler Nacht!
Ich lag im Fenster lange noch und schaute,
Wie da der Mond durch finstre Wipfel ging,
Die flüsternd sich im leisen Winde neigten.
Da war's als trät' ein Fremder aus dem Schloß
Und wandelt' schweigend durch die stillen Gänge,
Gleichwie in Ringen weit das Haus umkreisend.
Mich schauerte – und schnell trat ich zurück. –
Doch durch die offnen Fenster rauscht's herauf
Die ganze, schwüle Nacht, es blitzt' von ferne,
Und Nachtigallen schlugen – oft dazwischen
War mir's, als hörte ich fernab im Garten
Den fremden Sänger singen – und so träumt' ich
Gar wunderbar, ich weiß nicht mehr wovon,

Doch war es wie ein unermeßlich Glück
Im Herzen mir, daß ich vor Lust erwachte. –
Wohl sind die Träume nur ein Widerhall
Von himmlischer Musik, die wir nicht kennen
Und die wie ferner Morgenglockenklang
Im roten Duft zerrinnt. – Ja, rausch' nur wieder,
Du morgenfrischer Grund, mit allen Wipfeln,
Ich komme! – Wer ist dort! – Er scheint zu schlummern –
Der fremde Sänger ist's, – ich geh hinein.
Sie bleibt zögernd stehen.
Er schläft – nur einmal möcht ich recht ihn ansehn –
Ich kann es nicht, wenn er die Augen aufschlägt –
Als ruhte träumend unter stillen Bäumen
Der frische Morgen noch –

LEONARD. O süßes Plaudern!
Wie früher Lerchen Lieder über mir
Weckst du hell Morgenrot mir in der Seele!

ADELE.
Er spricht im Schlafe – von was mag er träumen?
Nein, nein, ich weil' nicht länger hier allein.
Sie will gehen.

LEONARD *rasch aufspringend.*
Hör', Mädchen, hör' doch! Bleib, nur auf ein Wort!

ADELE *verwirrt.*
Ich? – Ich wollt' nur – ich sah nur nach dem Gärtner –
Was trieb so früh ins Freie Sie hinaus?

LEONARD.
Ich schlaf' am liebsten unterm Himmelsbette,
Leicht mit dem Sternenmantel zugedeckt.
Es ist so schwül im Schloß. – Komm auch herab!

ADELE.
Hinab? – Mich hütet eine strenge Herrin,
Wenn *ich* auch möcht', die Gräfin möchte schelten.

LEONARD.
Die Gräfin, Jäger, Laufer, alle schlafen.
Das ist die rechte Zeit, just, wenn die Rehe

Noch sorglos grasen in dem stillen Grund
Und alle Wälder, wie in Träumen, rauschen.

ADELE *ihr Gesicht mit beiden Händen bedeckend.*
O daß ein Mann so zu mir reden darf! –

LEONARD.
Kind, nimm doch nur die Hände fort! – Die Augen
Sind's gerade, die ich meine. Nur ein Weilchen
Will ich dicht vor dir stehen und hineinsehen. –
Wahrhaftig, dort stehn Leitern an der Hecke –
Bleib' nur, ich hol' mir eine schnell und steige
Zu dir hinauf!

ADELE.
Eh' stürzt ich beide uns
Den Felsenhang hinab – zurück, Verwegner!
Sie wendet sich rasch, bleibt aber plötzlich in der Tür stehen und kehrt dann zurück.
Ich möchte nicht, daß wir im Zorne schieden –
Es ist so meine heft'ge Art nur – ach,
Ich weiß nicht, was ich spreche!

LEONARD.
Wär' die Gräfin
So süß wie du! –

ADELE.
Wie?

LEONARD. Nichts. Nenn' mich auch du!
Willst du?

ADELE *über das Geländer gebeugt, leise.*
Leb wohl, du unbescheidner Mann! –

LEONARD *ihr in Gedanken nachsehend, nach einer Pause.*
Ich werd' doch toll nicht sein und mich verlieben? ––
Dort schleicht der Komödiant, was will die Eule
In diesem Frühlingsglanz? Ich prügelt' ihn,
Wär's eben nicht so fröhlich mir im Herzen!
Ab.

FLITT *kommt lauernd, mit leisen langen Schritten hervor.*
Alles wieder still – wie Eidechsen fix entschlüpft in Laub und Ritzen vor dem Sonnenblick.

VICTOR *zwischen dem Gebüsch den Kopf vorstreckend.* Der schleicht um den Taubenschlag. Mir gerade recht! Ich kann keinen alten Fuchs sehen, ohne daß es mich gleich anficht, ihn zu prellen.

FLITT *ohne Victor zu bemerken.* Verliebte Kammerjungfern und Musikanten, kuriose Gräfinnen, heimliche Winke; Flüstern und Geheimnisse überall – gib acht, Flitt, gib acht, hier ist was zu machen – wenn ich nur erst wüßte, wo eigentlich Fortunas Haarzopf flattert in dieser Konfusion.

VICTOR. Ich will kein Jäger sein, wenn ich nicht alle diese mausigen Stoßvögel von Freiern an *Einem* Narrenspieß stecke, um sie langsam am Feuer der Liebe zu braten.

FLITT *wie oben.* Diese falbe, nüchterne Morgenzeit ist mir die liebste für Gedanken, so an den stillen Erkern und Pfeilern hinzustreichen in der mausefarbenen Dämmerung.

Er will weitergehen.

VICTOR. Nun frisch, ans Werk! sonst geht er mir durch die Lappen! *Hervortretend.* Herr Flitt! Herr Schauspieldirekteur!

FLITT *erschrocken.* Was macht er denn da für einen unverhofften, rasenden Lärm? Man könnte den Tod haben vor Schreck so auf nüchternen Magen.

VICTOR. Desto besser, Herr Direkteur, desto besser! Ich bringe eben einen Schnaps, der soll ihnen wohl bekommen.

FLITT. Recht so! Er ist ein verständiger Mann, ich möcht' nicht, daß uns die Damen dabei erwischten. Mädchen verstehn nichts vom Trinken – so einen Mund voll gemeines Wasser für den Durst, wie das liebe Vieh.

VICTOR *heimlich.* Sie haben doch gestern nach Mittag die Gräfin unter dem Balkon gesehen?

FLITT. Schnaps trinken? – Nein, sie zeichnete mit der Reitgerte im Sande –

VICTOR. Aber was zeichnete sie? – Lauter Fl's und einen

Haufen Gedankenstriche dahinter! – Merken Sie etwas, Herr Flitt? –

FLITT. Ich glaube wahrhaftig, das soll da der ganze Schnaps sein!

VICTOR. Ja, und als Sie hinzutraten, verwischte sie alles schnell wieder mit ihren Füßchen.

FLITT. Ah pah! Ich hätte ihn auch für gescheiter gehalten. Mich deshalb in meinem Nachdenken aufzuhalten! In der Morgenstunde, mein Freund, da rekapitulier' ich mir den ganzen kommenden Tag und übe mir ein wenig die Zukunft ein.

VICTOR. Da hören Sie doch nur weiter! Die Gräfin weiß gar wohl, daß Sie alle Abend einsam dort an der letzten Laube des Gartens spazierengehen.

FLITT. Sie weiß? – Fataler Streich! Die Damen im Schloß können den Tabaksrauch nicht vertragen, da pflege ich denn dort abends an der Laube meine Pfeife ins Gesicht zu stecken, um die Mücken zu vertreiben.

VICTOR. Aber die Liebesmücken sind gerade wie versessen auf Tabak. *Leise.* Die Gräfin – es bleibt unter uns – die Gräfin wird heute abend um zehn Uhr, mit einem Schleier über dem Gesicht und einem Nelkenstrauß an der Brust, allein an derselben letzten Laube lustwandeln. – Nun, Sie verstehen mich –

FLITT. So? – *Mit einem durchdringenden Blick auf Victor, nach einer kurzen Pause:* Hör' Er, guter Mensch, Er sieht verteufelt pfiffig aus. –

VICTOR. Bitte recht sehr! Bloß als Futteral für die Ehrlichkeit, damit sie sich nicht so schnell abnutzt.

FLITT. So? – Aber wie ist Er den zu allen diesen Neuigkeiten gekommen? –

VICTOR *heimlich.* Herr Flitt – aber verraten Sie mich nicht, sprechen Sie kein Wort mit ihr darüber – die Kammerjungfer hat mir alles vertraut.

FLITT. Die? – In der Tat, das läßt sich hören. Sie ist eine Jungfer von Erfahrungen, ein vertrauliches Geschöpf. – Nun, mein Freund *er sucht in den Taschen,* Verschwie-

genheit – also um zehn Uhr? – Schuldige Dankbarkeit – ich werde unvergessen sein. *Er drückt dem Victor ein Geldstück in die Hand.*

VICTOR *in die Hand blickend, für sich.* Bei Gott, ein alter, abgescheuerter Kupferkreuzer! *Laut.* Nicht doch! Was denken Sie von mir, Herr Flitt? Nicht um schnöden Lohn –

FLITT. Mache er doch keine Umstände.

VICTOR. Aber –

FLITT *ihn fortdrängend.* Nur keinen weitläuftigen Dank! behalt' er doch nur, ich geb' es gern, in der Freude des Herzens greift man aufs Geratewohl ins Volle und berechnet nicht erst lange. –

Victor ab.

Besser konnte ich den Knopf von meinem alten Rock nicht losschlagen; er wird denken, daß ich mich in der Wut der Großmut vergriffen habe! – *auf und nieder gehend.* Hm – Laube, Schleier, Nelkenstrauß – das gibt einen Vers, aber es reimt sich nicht. – Und doch – sie scheinen mich hier für einen andern zu halten; es ist das Los großer Geister, verkannt zu werden! – Und diese Aufmerksamkeit der Gräfin gegen mich, diese heimlichen kurzen Blicke unter den langen Augenwimpern wie Schlagtriller! – Was ist denn das? bin ich denn so schön? Das hat mir doch noch niemand nachgesagt. – Nun, Amor ist blind, so kann ich selbst wohl auch ein Auge zudrücken über meiner roten Nase und mit dem andern über meine schmalen Beine hinwegsehen, das ist Geschmackssache. – Ja – ich komme zu der angenehmen Laube!

Er will gehen, bleibt aber plötzlich erstaunt stehen.

Was hat das zu bedeuten? Da steigt leibhaftig der Knoll auf wie ein Orkan! – Was ist das für ein fruchtbares Jahr, treibt unversehens solche fette Schwämme hier aus dem Boden! Ich kann ihm nicht mehr entwischen, eher weicht man einer Windmühle aus.

KNOLL *im Sonntagsstaate, tritt mit Bücklingen auf.*

Flitt. Aber sag' mir, Knoll, welches unverhoffte Wiedersehen – das ist ja ein Vergnügen zum Teufelholen!

Knoll. Ich bitt' um Exkommunikation – ich konnte nicht unterlassen, mich nach Ihrer gnädigen Gesundheit zu erkundigen.

Flitt. Gesundheit? – Hm, man muß eben zufrieden sein – schlaflose Nächte, eine gewisse Unlust zum Arbeiten, brennender Durst – das alte Übel. – Aber nun, allerliebstes kleines Spaßvögelchen, nur heraus, was willst du eigentlich hier? Brennen dich die Würfel in der Talggrube deiner Faust? Bringst du die Kreidetafel im Schubsack? –

Knoll. Gott behüt mich! Solche Lumperei! Alles in den besten Händen!

Flitt. In der Tat, du bist ein höfliches, ein honnettes Ungeheuer.

Knoll. Oh, ich möchte mich selbst in die Nase beißen, daß ich so ein Esel war, Sie für meinesgleichen zu halten. – Aber das geringe Gefolge, Ihr ganzer Paraplü –

Flitt. In Summa, Knoll, du hast heute schon am frühesten Morgen zu schwer geladen, tue mir daher den Gefallen und schieß' dich hier bald wieder ab, alte Fleischbombe.

Knoll. Geruhen Sie sich doch nicht länger mit Verstellung abzustrapazieren, ich weiß es ja doch. – Mein alter Freund, der Gärtner hier, hat mir alles erzählt.

Flitt. So? – von mir? – der Gärtner? – Höre, guter Knoll, sag' mir doch einmal aufrichtig, wie hat dir denn der Gärtner eigentlich die ganze Geschichte erzählt? – Es interessiert mich, wie die Leute hier die Sache nehmen.

Knoll. Euer hochgräflichen Gnaden wollen untertänigst verzeihen – ich darf nicht soviel aus der Schule schwatzen. – Euer Gnaden wissen's ja doch am besten.

Flitt. Ja – leider wahrhaftig – *vornehm.* In der Tat, mein guter Knoll – ich sehe wohl – was soll ich's länger leugnen – meine Stellung hier unter diesen Jungfrauen –

Knoll. Will's Gott, bald junge Frauen.

Flitt. Ja – die Gräfin liebt mich. –

Knoll. Der gute Gärtner sagte –

Flitt *rasch.* Was sagt er? – *sich verbessernd.* Wollt' sagen: inwieweit hat dir der Gärtner auch über diesen Punkt nähere Mitteilung gemacht?

Knoll. Der Gärtner sagte schon oft: Es ist hier nicht länger mehr auszuhalten so unter uns Mädchen, es fehlt ein Herr, damit endlich auch auf dem Schlosse einmal eine soliderische Wirtschaft anfängt.

Flitt. Man kann das dem Manne nicht verdenken – ja, es wäre eine schöne Wirtschaft! –

Knoll. Der alte Gärtner kennt seine Gräfin, sie liebt das Rumorantische – er meint, wenn Ew. hochgräflichen Gnaden da schnell zufahren wollten wie der Ochs ins Heubündel – so eine Überraschung, eine kleine Entführung –

Flitt. Wahrhaftig, der Gärtner ist nicht so dumm, wie er aussieht! *auf und nieder gehend, für sich.* Laube – hochgräfliche Gnaden – Nelkenstrauß – *laut zu Knoll.* Hör, Knoll, wir rechnen auf deine Anhänglichkeit –

Knoll. Wie Pech, gnädiger Herr. – Ich hoffe, Ew. Gnaden werden auch dann den armen ehrlichen Knoll nicht vergessen, wenn einmal –

Flitt. Man wird deine vergangenen Gefälligkeiten nach deinen zukünftigen Diensten zu belohnen wissen. – Gold wie Treue, Knoll, aber vorher: treu wie Gold!

Knoll. Gold, Gold!

Flitt. Gut also – kennst du die letzte Laube dort an der Abendseite dieses Parks?

Knoll. Ja, wo das kleine Pförtchen aus dem Garten in den Wald führt.

Flitt. Ganz recht. *Nachdem er sich nach allen Seiten umgesehen hat, leise.* Dort wird die Gräfin heut abend um zehn Uhr heimlich mit mir – du verstehst mich –

Knoll. Versteh', verstehe, ein heimliches bete a bete.

Flitt. Der Platz ist einsam, abgelegen – auf was soll ich warten? Durch Verschieben kann leicht alles wieder

verschoben werden. – Wenn du um zehn Uhr im Walde deine Pferde und Wagen bereit hieltest –

KNOLL. Und die Gräfin daraufgesetzt und zickzack mit ihr fort durch die Nacht und den Wald –

FLITT. Nach deinem Hotel, daß sie nicht weiß, wie und wo sie hingekommen! – Nun willst du?

KNOLL. Paraßol d'honneur, ich bin dabei! Solche lustige Einfälle und Fasanerien sind mir gerade recht bei Tag und bei Nacht! Ich will am Pförtchen zum Signal wie ein Uhu pfeifen oder wie eine Rohrdrommel – das soll einmal ein Jubiläum geben diese Nacht!

FLITT. Still, ums Himmels willen still! – geheimer Anschlag steht auf hohlem Boden, da darf man nicht so massiv drauf herumtrampeln, sonst gibt es eine Resonanz. – Kurz: die Sache ist abgemacht, und nun gehe, sie dürfen uns hier nicht zusammen sehen.

KNOLL. Ich verstehe – Punkt zehn Uhr im Walde – empfehle mich in ihre hohe Profession. *Ab.*

FLITT. Das ist die Beutelschneiderei, denk' ich; hat allerdings ein hohes Ziel, daß der längste Kerl mit den Fußspitzen den Boden nicht erreicht. – Aber wozu halten sie mich hier gerade für einen Grafen? und wozu will man mit aller Gewalt gerade diesen Grafen heiraten? – Gleichviel! Wer viel grübelt, fällt zuletzt selbst in die Grube. Die Grafenkrone sitzt mir verdammt schief und wacklig, ich muß die Zeit benutzen, ehe der Wind umschlägt und sie mir wieder vom Kopfe reißt. – Was kann mir geschehen? Will mich die Gräfin wirklich, so ist es gut. Will sie mich nicht, so – ist es auch gut; so muß ich das geheime Rendezvous samt der Entführung mit dem Mantel der Verschwiegenheit bedecken, wenn ich Lust habe; und dieser Mantel muß, wenn ich dazu Lust haben soll, so reich mit Gold gestickt und so ungeheuer weit sein, daß ich mich und Knolls dicken Bauch bequem dareinwickeln kann. – Aber nur sachte, Flitt, nur vorsichtig! – Diese vorwitzige, schnippische Kammerjungfer darf nicht in der Nähe sein. Alle das müßige Ge-

sindel, welches hier, wie die Fledermäuse, durch die Dämmerung schwärmt, muß ich um die zehnte Stunde auf die entgegengesetzte Seite dirigieren. – Laß sehen, da kommt gleich einer.

LEONARD *tritt auf.*

FLITT. Guten Morgen, Herr Florestin.

LEONARD. Morgen, morgen!

FLITT. Nein, guter Rat darf nie zu morgen verschoben werden; noch ein paar solche Morgen – wie vorhin da vom Balkon – geben sonst wahrhaftig gar bald eine gute Nacht. –

LEONARD. Also: gute Nacht! – wie Sie wollen. *Er will gehen.*

FLITT *ihn zurückhaltend.* Nicht doch, mein Bester! ich meine es wahrlich gut mit Ihnen, und mein Rang, meine Verhältnisse berechtigen, ja verpflichten mich einigermaßen – doch, was rede ich da in meinem freundschaftlichen Eifer! – mein Herr, ich rechne auf Ihre Diskretion –

LEONARD *für sich.* Der Kerl ist wahrhaftig imstande, sich selbst für mich zu halten! – *Laut.* Genieren Sie sich nicht, wenn Ihnen bei Ihrem Inkognito die Haut zu enge wird, fahren Sie immer ein wenig mit dem Ellbogen heraus.

FLITT *erschrocken nach seinem Ärmel sehend, für sich.* Ellbogen heraus? – dacht' ich doch wirklich schon wieder – *Laut.* Nun, ich wollte nur sagen, daß ich hier im Schlosse, wie Sie wissen, auf ziemlich gutem und festem Fuße stehe.

LEONARD. Ja, vor der Mahlzeit.

FLITT. Wir haben uns da unverhofft ganz aus unserm ursprünglichen Diskurs herauskomplimentiert. – Was war es doch? – Ja – Sie unterhielten sich, wenn ich nicht irre, vorhin mit der Kammerjungfer. Schätzen Sie dieses Frauenzimmer?

LEONARD. Nein, denn ich will sie nicht kaufen, wenn sie nicht umsonst mein Schatz werden mag.

FLITT. Freundchen, sein Sie auf Ihrer Hut – Freundchen! Sie ist eine falsche Münze – entre nous – auf jeder Seite das Bild eines andern Potentaten.

LEONARD. Entre nous? – Allerdings, wenn sie zwischen uns beiden steht – und bei dem einen, deucht mir, kommt schon das Kupfer aus der Nase. –

FLITT. Möchten Sie nicht die Güte haben, mich gelassen anzuhören? und, wenn Sie mir nicht glauben, sich heut' abend um zehn Uhr gefälligst dort an die Morgenseite des Parks, unter die hohen Linden zu verfügen? –

LEONARD. Nun geschwind, was soll's?

FLITT. Was es nicht sollte. – Ich saß zufällig im Gebüsch – sie merkten mich nicht – da verabredete die Kammerjungfer auf heut abend zehn Uhr ein Stelldichein unter jenen Linden mit dem Flötenspieler Arthur.

LEONARD *ihn an der Brust fassend.* Das lügst du, feiger Schleicher!

FLITT. Ei was der Tausend! Lassen Sie mich los, lassen Sie mich gleich los! – Und just! ich bleibe dabei, ich schrei' es durch den ganzen Garten: sie hat ein gutes Herz, ein großes Herz, wie ein Wirtshaus!

LEONARD. Mensch, ich erwürge dich, wenn du nicht gleich auf deinen Knien bekennst, daß du ein schimmlichter schäbiger Schuft bist.

FLITT. Auf den Knien? – Nimmermehr! Stehend – will ich –

LEONARD. Nun, möchtest du wohl die Güte haben? –

FLITT *noch immer von Leonard festgehalten, kniet nieder.* Ich bekenne, daß ich ein Schuft –

LEONARD. Nein, ein schimmlichter –

FLITT. Ein schimmlichter, schäbiger Schuft bin –

LEONARD *ihn loslassend.* Amen! denn da kommt eben jemand – ich empfehle mich in Ihr gütiges Andenken! *Ab.*

FLITT *aufspringend und hinter Leonard ein Schnippchen schlagend.* Und er geht doch heut' abend unter die Linden! – Seht doch! – Ja, schüttle nur, droßle nur, du tapferer auf der Gurgel spielender Musikant, du! – O ja,

Tapferkeit! eine ochsige, eine stoßige Tugend, wenn's Hirn ins Horn geschossen ist!

ADELE *rasch auftretend.* Was für ein wüster Lärm! Ist alle Zucht denn hier entflohen? Von dem widerwärtigen Schalle werden rings die Echos wach wie Janhagel, wenn Trunkne durch die Gassen ziehn!

FLITT. Nur zu! nur immer zu, mein muntrer Stier! – Laß Sie mich jetzt, Jungfer, und menge sie sich nicht drein. – O ich will dir einen Popanz vorhalten, von roter Liebe und gelber Eifersucht – stoß nur drauf zu, daß dir dein Notenkasten von Schädel wackelt!

ADELE. Aber, bester Herr Flitt, was ist Ihnen denn widerfahren, warum sind Sie auf einmal so wütend geworden? – Und wie Sie aussehn! Der Hut auf der Erde – was gibt es denn? Herr Florestin eilte eben ganz verstört von Ihnen.

FLITT *sich sammelnd, indem er den Hut aufsetzt.* Tat er das? – Ich kann es ihm nicht verdenken, gar nicht, ja das kann eine schlimme Geschichte werden, aber ich frage nichts darnach.

ADELE. So sprechen Sie doch nur.

FLITT. Man spricht nicht gern von so etwas, es ist nicht meine Art, zu prahlen. – Pah, eine Kleinigkeit – Herausforderung –

ADELE. Wie! Mit Florestin?

FLITT. Ja, er tut mir leid, aber er wollte es nicht besser haben.

ADELE. Aber was brachte Sie denn so aneinander?

FLITT. Aneinander? – Eine Jungfer darf nicht alles wissen. – Es wird sich abends alles ausweisen, dort unter den hohen Linden – o ich wollte, die zehnte Stunde wäre schon da!

ADELE. Was, ein Duell um zehn Uhr; da ist es ja schon finster.

FLITT. Finster oder nicht, mir alles gleich! Glaubt Sie denn, daß die Tapferkeit sich vor Gespenstern fürchtet? Ich brauche nicht mehr Mondschein, als auf meiner De-

genspitze Platz hat, um sie meinem Gegner in das Herz zu stoßen. – Es sind noch keine vier Wochen her, denk' ich, da hatte ich auch einen solchen Milchbart vor der Klinge, Locken an den Schläfen wie ein Merino und Ringe in den Ohren. – Auf Ehre, dachte ich, das lohnt nicht, so in Milch und Blut zu stoßen, das hält keinen Stich. – Es war im Walde, er legt sich in die Parade, er fällt aus, sticht rechts und links um sich. Ich lasse das gut sein, vigiliere immerfort auf sein Ohrläppchen, dränge ihn endlich an einen Baum und stoße plötzlich durch seinen Ohrring in den Baum hinein, daß die Degenspitze von der andern Seite des Stammes wieder herauskommt. – So spickte ich ihn an wie einen Schmetterling, und wenn ihn nicht jemand losgemacht hat, so hängt er noch heute.

ADELE *für sich.* O ich wünschte, er hinge lieber selbst an beiden Ohren! *Laut.* Aber, Herr Flitt, das sind gewiß nur kleine Mißverständnisse mit dem Florestin, gibt es denn gar kein anderes Mittel? –

FLITT. Zu spät, gute Jungfer, alles zu spät, auf Ehre, er hat einmal mein Wort! Blut will ich sehn, Blut, sag' ich, Blut! – Schade in der Tat um den jungen Menschen, er sähe nicht übel aus, wenn er nicht so einen gewissen sentimentalen, einfältig hängenden Zug am rechten Mundwinkel –

ADELE *rasch.* Das habe ich nie bemerkt, Herr Flitt, das ist nicht wahr, das – *schlägt, plötzlich abbrechend, die Augen nieder.*

FLITT. Nun, es kann auch vielleicht der linke Mundwinkel sein. – Was geht denn aber Sie Herrn Florestins Mund an? Sie wird ja über und über rot. –

ADELE. Ich? – o nein – ich wollte nur – o Gott, ich weiß vor Verwirrung nicht mehr, was ich rede! verwünschte Maskerade! *Sie eilt fort.*

FLITT. Na, die geht auch zu den Linden! Verliebte und Verrückte sind leicht zu betrügen, sie gehen beide geradeaus auf ihre fixierten Ideale los – dem Himmel sei

Dank, ich bin niemals sonderlich verliebt gewesen! – Nun muß ich noch den Flötenspieler und den Schlender haranguieren, damit sie mir nicht etwa wie die wilden Gänse in mein Wachtelnetz hineinfliegen und mir die Maschen verwirren. *Ab.*

VICTOR *tritt auf.* Aha, da streicht der alte Fuchs hinter seiner spitzigen Nase her ins Garn hinein. – Den Schlender hab' ich auch schon um zehn Uhr zur Laube bestellt, und gleich darauf werfe ich die nämliche spanische Fliege dem Schreibesel Fleder hinter die langen Ohren, das zog vortrefflich lauter philosophische Blasen. Ich sah ihn soeben voller Nachdenken vorüberschreiten. – „Aber der Auftrag, das Vertrauen des Herrn Präsidenten!" sagte er zu sich selbst und blieb mit verschränkten Armen stehn, „wie, und ist es nicht des Mannes höchste, älteste Pflicht, sich Raum zu schaffen in der Welt zu edlem, großem Wirken!" Der kalekutische Hahn im nahen Hühnerhofe kollerte eben dazwischen – da schritt er stolz weiter, stand wieder still und sprach: „Warst du nicht eher Mensch als Hofrat, Fleder? – Die Liebe kennt Präsidenten nicht und ist sich selber König!" – Ja, Euer Majestät, es werden noch ein paar solche Monarchen im Dunkel eintreffen, und blutige Kronen dürften nicht rar sein. *Ab.*

Zweite Szene

Anderer Teil des gräflichen Gartens. FLITT *und* SCHLENDER *sitzen an einem Tisch, auf welchem Weinflaschen und Gläser stehen.*

FLITT. Nunc est bibendum, nunc pede libero pulsanda tellus! – Aber du verstehst nichts von der klassischen Literatur als bibendum.

SCHLENDER. Was heißt das andere zu deutsch?

FLITT. Das heißt: Wir sind bei Hofe hier, so müssen wir denn auch die Hofetikette beobachten und uns einen Haarbeutel anhängen. *Er trinkt.*

SCHLENDER *trinkend.* Die sind hier wohlfeil. – Aber höre, Flitt, wenn du so früh anfängst, siehst du nachher den ganzen Tag aus wie ein feuriges Gewitter.

FLITT. Desto besser! Du bist mein Blitzableiter, in den alle Witze einschlagen. – *Trinkt.* Hol' der Teufel das vornehme Leben, die Langeweile trocknet aus und macht durstig; ich glaube, mir wachsen schon schnöde Pilze auf meinem guten Humor.

SCHLENDER. Ja, war das heut früh nicht ein Gekicher und Spektakel unter den Damen, als ich mein Kompliment machte und mir ein paar lumpige Federn im Haar und Rock hingen, weil ich mich gestern abend aus Versehen im Frack zu Bett gelegt hatte!

FLITT. Ja, und ist es nicht gescheuter, wenn dir ein Bauer mit allen fünf Fingern durch den Schopf fährt und dir den Rock am Leibe ausklopft, ohne sich erst lange nach einem Trinkgeld dafür umzusehen?

SCHLENDER. Ach was, Bauer! So ein Mensch ist ein Esel wie ich, da geht alles natürlich zu, aber so eine Dame –

FLITT. Die sich schämt, ein Mensch zu sein. – *Er trinkt.* Hol's der Teufel! Nichts als Finten und Schwermut in der Welt!

Er singt.

Einstens, da ich Lust bekam,
Mir zu freien eine Dam –

Wie geht es doch weiter? Stimme mit ein, Schlender!

SCHLENDER. Ja, ja, das Lied von des guten Kerls Freierei! So ein Rundgesang muß recht vornehm klingen hier in dem Park.

Beide singen.

Einstens, da ich Lust bekam,
Mir zu freien eine Dam –
Und sie freundlich fragte,
Ob ich ihr auch wohlgefiel;

Wahrlich nicht besonders viel!
Sie gar spöttisch sagte.

FRIEDMANN *eilig eintretend.* Aber um Gottes willen, meine Herren, wo denken Sie hin? So ein Spektakel, hier im Garten, bei hellem lichtem Tage?

FLITT. Ei, ei, ei, Friedmann, das hätt' ich nicht geglaubt von Euch. – Bei hellem Tage? – Wann soll man denn singen? – O nein, guter Gärtner, das Nachtschwärmen ist nicht für moralische junge Leute, laßt solche liederliche Gedanken, geht in Euch, alter Mann, geht in Euch!

SCHLENDER. Ja, geht!

FRIEDMANN. Wahrhaftig, ein ehrbarer Lebenswandel würde Sie besser kleiden.

FLITT. Kleiden! Seht Ihr wohl, schon wieder! –

SCHLENDER. Ja, in der Tat schon wieder!

FRIEDMANN *rasch und empfindlich zu Schlender.* Was denn schon wieder, Herr Musikant?

SCHLENDER *zusammenfahrend.* Na, was weiß ich denn! – Erschreckt einen doch nicht so, unvernünftiger Mensch!

FLITT. Ruhe, Ruhe da! – Besser kleiden, sagtet Ihr, Friedmann? Schämt Euch, eitler Greis! Ein Philosoph gibt nichts auf Kleider, große Männer haben große Blößen.

FRIEDMANN. Nun, ich will nicht prophezei'n – aber wenn die Gräfin erfährt – sie könnte sich wohl besinnen –

FLITT. Oho! *Er singt.*

Daß im Walde finster ist,
Das machen die Birken,
Daß mich mein Schatz nicht mag,
Das kann ich nicht merken.

SCHLENDER *singt.* Nein, das kann ich nicht merken!

FRIEDMANN. Das wird nun eine gute Geschichte, da kommt sie selbst dazu.

ADELE *rasch auftretend.* Sie, Herr Flitt – Ich dürfte wenigstens bei Ihnen mehr gute Lebensart voraussetzen.

FLITT. Gute Lebensart? Wir leben hier ganz gut.
Er singt, während Schlender mit einstimmt.

Ich sprach wieder: bin ich nicht
Ein gut Kerle, gebt Bericht!
Drauf fragt' sie mich wieder:
Was denn ein gut Kerle wär?
Ich sprach: Setzt Euch unbeschwert
Etwas zu mir nieder.

ADELE. Ich mich noch zu Euch hinsetzen? Nein, wahrhaftig, ich weiß nun genau, was so ein guter Kerle ist, ich will ein Loblied auf ihn singen.

FLITT. Ja, tunk' da erst dein Schnäbelchen mit ein, dann singe.

SCHLENDER. Ja, tunke Sie!

ADELE *zu Friedmann.* Was fangen wir mit ihnen an?

FRIEDMANN. Ich habe die Herren schon gebeten, sich zu moderieren, aber –

ADELE. O, ich möchte lachen, wenn ich vor Ärger könnte! *Zu Flitt.* Nun, das ist wahrlich der Anfang vom Ende, es soll rasch alles klarwerden!

Ab, während Friedmann achselzuckend an ein Beet tritt und sich mit den Blumen beschäftigt.

FLITT und SCHLENDER *singen.*

Wollt Ihr nun, so ist es klar,
Und wir werden bald ein Paar.
Drauf spricht sie gar sachte:
Ihr mögt mir nach allem Schein
Gar ein guter Kerle sein;
Schmunzelt drauf und lachte.

FLITT *plötzlich aufspringend.* Silentium! Dort kommt wahrhaftig die Gräfin her! – Geschwind, decke die Batterie, Schlender, unterhalte die Gräfin, bis die Flaschen unter den Tisch gebracht sind!

SCHLENDER. Laß du mich nur machen.

Er geht ab, während Flitt den Tisch abräumt, und kehrt dann mit Flora im Gespräch zurück.

FLORA. In der Tat, ich verstehe nicht recht, was Sie meinen.

SCHLENDER. Ha, ich verstehe, ich verstehe! – Dort nahen

kalte Menschen – o niemals soll das zudringliche Maul des Tages belauschen, was das plauderhafte, gefühlvolle Ohr der Nacht –

FLORA. Wie?

FLEDER *tritt von der andern Seite auf.* Guten Morgen, meine Gnädigste! *Geheimnisvoll, mit zärtlichem Nachdruck!* O könnt' ich guten Abend sagen schon!

FLORA. Guten Abend schon? Da fingen wir ja den Tag von hinten an.

FLEDER. Gleichviel, schöne Gräfin, gleichviel – es gibt keine Zeit, wo ein glückliches Herz schlägt.

SCHLENDER. Ja, da fängt die unsterbliche Ewigkeit an!

FLORA *welche beide verwundert angesehen, Flitt bemerkend.* O bester Herr Flitt, warum stellen Sie sich dort unter den Scheffel der Bescheidenheit? Ich bitte Sie mir die beiden Herren hier aufzuhalten, sie nehmen eben einen Anlauf, über diesen Tag hinwegzusetzen.

FLITT. So plumpen sie jenseits in die Nacht. – Ich für meine Person begnüge mich, nur den Saum der Nacht leise umzuschlagen, wie ein Carbonaro seinen Mantel, wenn er auf heimliche Verbindungen ausgeht.

FLORA. Wie, auch Sie? Aber weshalb sind Sie den alle gegen diesen Tag so erbost?

SCHLENDER *in Ekstase.* Tag? – O der Tag, der zerstreuungsvolle Tag – er fährt –

FLITT *dicht vor ihn tretend, leise.* Du bekommst wieder deinen poetischen Stich, schweig oder ich trete dir die große Zehe platt.

SCHLENDER *wie oben.* Laß mich! – Der verschwenderische Tag, er fährt mit seinem balsamischen Staubmantel über die tränenduftende Stirne der Nacht und läßt das Veilchen unzerknickt, weil noch das Lämpchen glüht, und wandle auf Tuberosen und Vergißmeinnichts!

FLORA. Vortrefflich! Wollen Sie sich nicht mit der Violine dazu accompagnieren? Das ist ja wie aus einem Stammbuch, das der Wind durcheinanderblättert, von jedem Blatt eine Zeile.

FLITT *leise zu Flora.* Lassen Sie ihn schwatzen! Ich puppe mich unterdes in einen Schleier, bis mich der Mondschein ausbrütet wie einen Nachtfalter.

FLORA. Schleier? – Was denn?

FLEDER *heimlich zu Flora.* Ich lege die Hand aufs Herz und sage nichts als – Nelke!

FLORA. Nelke? – *Für sich.* Träume ich denn oder sind sie alle hier vor lauter Müßiggang toll geworden?

SCHLENDER *leise zu Flora.* Ha, meine Gedanken – alle stehn in jenem grünen Holz!

FLORA *ihm furchtsam ausweichend.* So lassen Sie sie nur stehen! *Mit einer leichten Verbeugung gegen alle.* Der Herr behüte Sie auf Ihren Holzwegen.
Ab.

VICTOR *zwischen dem Gebüsch hervorsehend.* Hoho! Hier ist, wie mir scheint, die Konfusion soeben im besten Humor. Laßt sehen, ob ich sie aneinanderbringen kann. *Er tritt vor.* Was ist hier geschehen, meine Herren? Soeben eilte die Gräfin fort und prustete und nieste, blickte in die Sonne und nieste noch einmal.

SCHLENDER. Ja, Gott helf ihr! Es gibt nächtliche Tageszeiten, wo keine Sonne scheint, und im Finstern geht der tugendhafteste Mensch darauf los wie blind!

FLITT. Aber, Schlenderchen, du bist wahrhaftig zu voll, die Gedanken laufen dir schon über.

SCHLENDER. Was! ich mach mir nicht so viel aus ein paar lumpigen Gedanken! – O, meine Wege sind nicht die deinigen!

FLITT. Nein, also geh du nur jetzt deiner Wege.

VICTOR *leise zu Schlender.* Wollen Sie das leiden? Er zeigt Ihnen –

SCHLENDER. Was zeigt er mir? – Das sagen Sie mir einmal! – Seht doch – ob ich leide oder nicht leide – und das geht Sie gar nichts an! *Er schreitet zornig über die Bühne auf und nieder.*

FLITT *sich vornehm zu Fleder wendend.* Apropos, mein Lieber, werden Sie uns nicht einmal auf Ihrer Flöte et-

was zum besten geben? Ich bin ein großer Gönner der Musik.

FLEDER. Ja, ich hoffe nächstens einen Ton anzugeben, der sie in Erstaunen setzen dürfte. –

VICTOR *leise zu Fleder.* Besser, stärker angesetzt! Blasen Sie die Backen auf wie bei der Drommete! –

FRIEDMANN *sich von seiner Arbeit aufrichtend, zu Victor.* Was ist denn das, wie sprechen denn die heute alle so verwirrt?

VICTOR *auf die Stirn deutend.* Hier – wißt Ihr noch nicht? – manchmal Anfälle –

FRIEDMANN. Wie! Eines unglücklichen Deliriums? –

SCHLENDER *plötzlich vor Flitt stehenbleibend.* Und nun fängt es mich erst recht an zu wurmen – und wenn ich einmal in die Courage hineinkomme –

FLITT. Hör', Schlender, bei der Gelegenheit muß es endlich heraus. Ja, dieser gemeine Ton – diese gewisse Familiarität zwischen uns, das muß künftig ganz aufhören.

SCHLENDER. Was?

FLITT. Es gibt Lagen in menschlichen Verhältnissen, wo man aus gebührender Rücksicht auf seine Familie –

SCHLENDER. I Herrje! Lauter Bärenführer und Puppenspieler! – O ja, ich gehe, wer in Pech tritt, läßt leicht den Absatz drin stecken. Ich empfehle mich, Herr Don Tragödio dellas Comedias! *Ab.*

FLITT. Ich habe es lange prophezeit, er wird so lange schnapsen, bis er überschnappt. – *Zu Fleder.* Wie gesagt, wenn ich Ihnen vielleicht dienen kann – es geschieht nichts weniger als gern – meine Reisen – bedeutende Konnexionen – nun, wir sprechen weiter davon.

FLEDER. O ja, ich bitte, wenn es beliebt, so spät und weit als möglich. *Für sich.* Ich fürchte mich vor ihm – ganz der gläserne Blick des stillen Wahnsinns –

FLITT. Adieu, Gärtner! *Ab.*

FRIEDMANN *mit einer tiefen Verbeugung.* Untertänigsten guten Morgen! – Bei alledem doch ein recht humaner, herablassender Herr! –

FLEDER *erstaunt.* Wie ist Ihnen denn, mein Guter? Warum erweisen Sie diesem Manne so unverhoffte Ehrerbietung?

FRIEDMANN. Ich – diesem Herrn? Schauspieler, wollt' ich sagen. – In der Tat: man kann bei solchen Schauspielern niemals aus dem ersten Auftritt ihren letzten Abtritt voraussehen – heute mitten unter uns, morgen über uns.

FLEDER. Abtritt? über uns? *Zu Victor leise.* Was ist denn das? – Sie müssen den Mann ja genau kennen – ich hoffe doch nicht?

VICTOR *ebenso.* Wahrhaftig, ich kann nicht leugnen – sein Verstand fuselt bisweilen.

FLEDER *zu Friedmann.* Nun, nun, beruhigen Sie sich. Schon gut, schon gut, es kann hier bald, ja über Nacht manches anders werden, man wird gern Rücksicht nehmen – ich werde nie aufhören, Mensch zu sein.

FRIEDMANN *für sich.* O Gott, er hat wahrhaftig soeben seinen Anfall! *Laut.* Möchten Sie nicht Ihre Flöte ein wenig vornehmen, vielleicht wird Ihnen besser?

FLEDER. Besser? – Armer Mann!

FRIEDMANN. I! Schade um den feinen Verstand!

Beide, sich voreinander scheuend, von verschiedenen Seiten ab.

VICTOR *ihnen mit dem Hute nachschwenkend.* Hetzoh! Halali! Das ist ein herrliches Jagdwetter heut! Und ist noch irgend Wahrheit in der Wahrscheinlichkeit, so erleben wir heut abend die kostbarste Prügelei, wenn die drei bei der Laube unversehens zusammenstoßen! *Ab.*

Dritter Aufzug

Erste Szene

Zimmer im Hause des Gärtners. FLORA, *als Offizier verkleidet, sitzt auf einem Stuhl,* MARIE *kniet vor ihr und schnallt ihr die Sporen an.*

FLORA. Aber, Mühmchen, hast du recht vernommen?

MARIE.

Ich saß ja dicht hinter einem Strauch.
Sie wollen alle zur Laube kommen,
Graf Flitt und Schlender und Fleder auch,
Sie bilden wahrhaftig sich ein, die Narren,
Die Gräfin würd' ihrer dort abends harren.
Pfui, das war ein Strecken und Blähn und Geschniegel –
Nichts langweiliger als ein zärtlicher Mann!

FLORA.

Ei nun, das kommt noch gar sehr drauf an.
Verliebte Augen sind lustige Spiegel,
Man sieht doch noch drin, daß man jung und schön.

MARIE. Hm – da brauch' ich bloß in den Brunnen zu sehn.

FLORA. Aber hör', mit dem Victor – wie war das doch?

MARIE.

Mit dem? – Nun ich saß in dem Strauche noch,
Da schlich er durch den Garten ganz leise, leise
Und lacht' in sich und blickt' im Kreise,
Und ein Bündel trug er unter dem Arm,
Hauben und Bänder und Mäntel von Seiden,
Als wollt' er seine Braut zur Hochzeit kleiden,
Alles zerdrückt, daß Gott erbarm! –

FLORA *rasch aufspringend.* Es ist aber auch hier unausstehlich schwül!

MARIE. Ei, 's ist gleich Nacht, ja, es wird schon hübsch kühl.

Flora *an das offene Fenster tretend.*
Ich bitt' dich, Herzensmühmchen, so mach' nur geschwind
Und putze mich aus aufs allerbeste. –
Draußen rührt schon der Abendwind
Vor dem Fenster die dunkeln Äste, –
Der Springbrunnen rauscht, und die Sterne scheinen –
Was mag er nur mit dem Brautstaat meinen? –

Marie *mit Floras Anzug beschäftigt.*
Der Victor? – Das blitzt wie von Edelsteinen!
Ein Offizier müßt's sein, wenn ich einmal freite,
All andern sehn aus wie gemeine Leute.

Flora. Ich weiß doch ein Mühmchen, die auch Jäger gern mag.

Marie. Nein, die schmauchen mir gar zu viel Tabak!

Flora. So wend'st du das Näschen und reichst ihm die Wange.

Marie.
Wenns der Victor just wär – ich besänn' mich nicht lange,
Er hat einen recht hübschen Schnitt um den Mund.
Aber sein Schnurrbart spielt alle Farben,
Das zeigt auf Falschheit im Herzensgrund,
Ich möcht' ihn nicht zum Liebhaber haben.

Flora.
Nun weiß ich doch, wo seine Feinde zu sprechen.
Weil er kein Narrengesicht kann sehn,
Ohne eine Nase daran zu drehn,
So möchten sie ihn nun alle mit den Nasen erstechen.
O, die ärgsten sind die Stillen im Lande,
Die schleppen so schwer an ihrem Verstande,
Daß sie nicht kommen zu Lust und Witz,
Sie möchten gern donnern, aber es fehlt der Blitz –

Marie *in Lachen ausbrechend.* Gefangen! Zisch aus! – Du wirst ja ganz wilde. –

Flora. Ach! Was nur führst du da wieder im Schilde?

MARIE. Hm, ich denke nur so, was mir eben gefällt. –
FLORA. Du bist recht kindisch. –
Sie schnallt den Degen um und setzt den Hut auf.
Nun frisch ins Feld,
Ich bin jetzt gerade im rechten Humor!
MARIE. Aber was hast du denn eigentlich da draußen vor?
FLORA.
Rettung, Entlarvung, List, Händel, Rumor. –
Erst stell' ich mich heimlich hinter die Laube,
Bis sich die Bande versammelt zum Raube,
Dann brech' ich als weitläuftiger Vetter vom Haus,
Der eben von Reisen kommt, plötzlich heraus,
Staun', frage, drück' in die Augen den Hut,
Und wenn alles konfus recht, ruf' ich voll Wut
Diener und Fackeln – und spieße verwegen
In der Luft so die Haube mit meinem Degen,
Die über dem Scheitel der Gräfin tut zücken.
Komm auch mit, lieb Mühmchen.
MARIE. Das möchte sich schicken!
So unter Männern – in dem Gewirr!
FLORA. Hüt' dich! Kluge Vögel werden am ersten kirr.
Beide ab.

Zweite Szene

Ein vom Schloß entlegener Teil des Gartens mit einer Laube zur Seite. Es ist schon dunkel. Die Gräfin ADELE *tritt rasch auf, sich nach allen Seiten umsehend.*

ADELE.
Hier auch kein Laut rings – nur die Bäume rauschen;
Das Schloß steht leer; wem ich begegne, weicht
Geheimnisvoll mir aus, und Büsch' und Marmorbilder

Sehn in der Einsamkeit mich seltsam an,
Als schauert alles heimlich vor den Schatten,
In die vielleicht ein fröhlich atmend Herz
Auf ewig bald versinkt! – Dort an den Linden
Wo sie sich treffen wollten, ist's totstill. –
Er täuschte mich wohl nur, um ungestört
Ihn anderswo zu morden. – O, schon wächst
Und wächst das Schweigen rings; was soll ich tun,
Was lassen in der Not? – Viel tausendmal
Rief seinen Namen ich im Herzen leise
Und wag's nicht, laut zu schreien, wie ich möcht'!
O Nacht, verhülle ihn – und meine Schande! –
Sie sinkt, das Gesicht verhüllend, auf einen Rasensitz, dann sich wieder aufrichtend.
Seit wann kenn' ich die Angst? – Es ist vorüber! –
Nie sag' die Welt: die Tochter hoher Helden
Sei so entartet aller Heldenhoheit,
Daß sie verlernt', sich selbst zu überwinden.
Jetzt gleich, eh' blut'ges Unheil sich begibt,
Brech' ich die Rätsel dieser Mummerei,
Die dem gemeinen Trieb der Welt mich gleichstellt.
Noch einmal, als die freie Herrin, tret' ich
Vor Florestin – und seh ihn nie mehr wieder!
Ab.

FLORA *als Offizier, tritt vorsichtig auf.* Horch – nein, es war der Wind im Laube. – Da schlägt es zehn im Dorfe unten. *Sich schüttelnd.* Es ist auch so entsetzlich still im ganzen Garten, daß man alle Bäche aus der Ferne rauschen hört. – Da – Fußtritte – es kommt näher – geschwind in mein Versteck!

Sie verbirgt sich hinter der Laube.

VICTOR *führt* SCHLENDER *an der Hand herein. Letzterer ist als Dame verkleidet, hat einen Schleier auf dem Hut und einen Nelkenstrauß an der Brust.*

FLORA *leise.* Was ist das? – Victor – mit einer Dame!

SCHLENDER. Also die Gräfin wollte durchaus, daß ich als Frauenzimmer –

VICTOR. Scht, scht! Das versteht sich und der Nelkenstrauß ist das Feldzeichen, damit sie Sie gleich erkennt.

SCHLENDER. Aber die Gräfin selbst?

VICTOR. Wie ich Ihnen schon sagte, sie kommt als Mann, ganz wie Herr Arthur gekleidet. Merken Sie wohl? – Verdammt schlau!

SCHLENDER. Ja wahrhaftig, verdammt pfiffig!

VICTOR. Aber die Nacht hat lange Ohren – die Gräfin wird daher, um alles Aufsehen zu vermeiden, auch ganz die Stimme des Herrn Arthur nachahmen; lassen Sie sich dadurch nicht irremachen.

SCHLENDER. Ich! Da kennen Sie mich schlecht – ich irremachen! Da bin ich gerade der rechte Mann danach.

VICTOR. Das mein' ich selbst. Aber Sie müssen auch Ihre Stimme etwas dünner machen.

SCHLENDER. O, das versteh' ich aus dem Grunde! Wenn mir zuweilen auf meiner Geige die Quinte springt, spiel' ich sie mit dem Munde durch die ganze Sonate weiter fort.

VICTOR. Vortrefflich! So spielen Sie nur, Gott weiß, wer den Takt dazu schlagen wird! – Und nun geschwind den Schleier herab! *Er bedeckt damit Schlenders Gesicht.* So – jetzt lasse ich Sie mit Ihrem Glück allein. *Für sich.* Ich muß nun das andere Wild stellen. *Ab.*

SCHLENDER. He, Herr Victor, bleiben Sie, bleiben Sie doch noch ein wenig! – Da streicht er wie eine Fledermaus durch die Gebüsche. –

FLORA. Wo sie nur so auf einmal hergelaufen ist? Wie ein in Gedanken stehengebliebener Haubenstock! – O, ein auserlesener Geschmack, den der Victor hat! – Ob ich ihr lieber gleich die Augen auskratze?

SCHLENDER. Das ist das Allerdümmste bei der Liebe, daß sie immer nachtwandeln will, wenn alle andern vernünftigen Kreaturen schlafen. – Ich fürcht' mich grade nicht – was regt sich da? – Ich will nur etwas singen, das zerteilt die Gedanken. *Er singt.*

Der Himmel ist so trübe,

Es scheint nicht Mond noch –
Heda, wer ist da? Wer ist da?

FLEDER *tritt auf, für sich.* Das ist die Gräfin – der Schleier – der Nelkenstrauß. *Laut zu Schlender.* Wie! – Ich erstaune – hier in Nacht und Tau –

SCHLENDER *galant und immer mit verstellter Stimme.* O, es dürften wohl vielmehr die perlenden Tränen sein, die mein gefühlvolles Herz so im Promenadieren verloren hat.

FLORA *für sich.* Was! Ihr Herz hat doch nicht Löcher? Nun, ich glaub' es wohl, Amors Pfeile mögen es schon übel zugerichtet haben.

FLEDER *zu Schlender.* Ich würde nie gewagt haben, dieses stille Gedankenheiligtum zu betreten – aber jener kleine Nelkenstrauß – sein Duft zieht Zauberkreise durch den stillen Garten rings, ich taumelte berauscht – nun richten Sie, verdammen Sie, wenn ich zu kühn –

SCHLENDER. Au contraire! Ich bin Ihnen außerordentlich verbunden für Ihre lange Nase – wollt sagen wohlriechende Nase – wollt sagen, daß Sie so weit riechen können –

FLEDER. O, wie unwiderstehlich kleidet Sie dieser natürliche Schmuck süßen Gekoses!

SCHLENDER. Ja, ich kann es nicht leugnen, mein Bau ist nicht übel, ich bin von jeher so ein dünner Schlankel gewesen.

FLEDER. Ja, das ist sie, das ist die Sprache der Liebe! –

SCHLENDER. O, ich kann's manchmal noch feiner, aber es greift die Lunge an.

FLEDER. Die Lunge? – So schweigen wir. Der Liebe Schweigen ist Gesang! – O, ich wüßte einen Mann, den ein einziger Druck dieses Händchens –

SCHLENDER. Mann? – Nein, wahrhaftig, wenn ich jemanden drücken sollte, so wäre es nur eine Dame. –

FLEDER. Eine Dame?

SCHLENDER. Eine gewisse schöne Gräfin –

FLEDER. Die hier nicht weit ist –

SCHLENDER. Ja, und deren kirschrote Hand ich gern an meine Lilienlippen –

FLEDER. Wie! Sie wollen sich selbst die Hand küssen?

SCHLENDER. Mir selbst? – Das ist ja heute lauter Konfusion. Es ist auch so finster hier, daß man sein eigen Wort nicht versteht. – Aber es wundert mich nur, wie Sie so grob reden können.

FLEDER. Ich – grob?

SCHLENDER. Ja, wirklich ganz wie der Herr Arthur.

FLEDER. Nun, das wäre eben keine große Kunst.

SCHLENDER. Nein, wahrhaftig nicht – so ein verlaufener Stadtpfeifer! – O werfen Sie diese Verstellung ab, wir sind unter uns, sein Sie wieder, wie Sie die Natur geschaffen hat.

FLEDER. In der Tat, ich fühle mich wunderbar überrascht – ich kann nicht in Abrede stellen – es könnte sein, daß hinter dieser bescheidenen Maske des armen Flötenspielers ein Mann –

SCHLENDER. Schon wieder ein Mann? – Aber warum wollen Sie denn gerade ein Mann sein?

FLEDER. Ich? – In der Tat – ich verstehe Sie nicht. – O lassen Sie uns diesen ernsten, heiligen Moment nicht in leichter Tändelei verscherzen!

SCHLENDER *in steigender Begeisterung nach und nach in seine natürliche Stimme übergehend und auf Fledern eindringend.* Ja, ich will kein Narr sein! – O diese pechschwarze Nacht, sie ist der Lichtpunkt in dem Silberblicke meines Seins – oder Nichtsein! – Ha, das ist keine Frage, daß mein Herz von dem nagenden Wurm der heimlichen Liebe zerfressen sei – aber ein tapferes Herz frägt nichts darnach, ob ihm die Seele zum Ellbogen heraushängt, wenn es in jene elysäischen Felder –

FLEDER *der währenddes erstaunt vor Schlender zurückgewichen.* Au, meine Zehen! – Mein Gott! Ich glaube gar, Sie sind –

SCHLENDER *zu Fleders Füßen sinkend.* Der selige Schlender!

FLEDER. Ha, abscheulich! Jetzt seh' ich –

SCHLENDER *aufspringend und sich rasch umsehend.* Wo? wen?

FLEDER. Verrat, heimtückischer Verrat! –

SCHLENDER. In der Tat, ich sah ihn schon vorher dort hinter der Laube –

FLEDER *erschrocken.* Wen?

SCHLENDER. Was? – Ich salvier mich!

FLEDER. Wenn ich nur wüßte, was es gibt, – mein Gott, mein guter Ruf – ich eile dort hinaus!

SCHLENDER. Und ich dorthin!

Sie entlaufen nach verschiedenen Richtungen.

FLORA *die währenddes, mit ihrem Säbel klirrend, rasch vorgetreten ist.*

Heda, halt, halt! – O liederliche Wirtschaft,
Wie Spreu im Wind! Wer kommt denn da schon wieder?

LEONARD *bewaffnet, bleibt im Hintergrunde, zurücksehend, stehen.*

Die Kammerjungfer war's, die dort entfloh –
Auch nicht ein Mäuschen fand ich an der Linde.
Hieher sah ich vorhin sie gehn – von hier
Entwich sie jetzt. – Frisch auf! an jedem Strauch
Und Zaune will ich rütteln! Dann laß sehn,
Wo der Galan die Hasenohren andrückt!
Flora erblickend.
Wer schleicht da? Steh! Ich denk', dich such' ich eben!
Warst du hier mit der Gräfin Kammerjungfer?

FLORA.

Gewiß. – Es wär' unmöglich, uns zu trennen.

LEONARD.

So liebt ihr euch?

FLORA. Nicht sonderlich. Sie macht
Viel Unruh mir, bei Nacht durch lose Träume
Und abends wohl durch manchen dummen Streich!

LEONARD.

Ei seht, wie klug und fein das Zünglein schwatzt. –

Wer gab das Recht dir, ohne rechte Liebe
Mit güldner Sporn und glatter Worte Klang
Ein fröhlich stilles Herz schnöd zu betören?

FLORA.
Fürwahr, gäb's so ein Recht, ich hätt's zunächst.
Doch könnt' das, glaub' ich, andern besser glücken.

LEONARD.
Du schmähst sie noch! – Sieh, ich bin just kein Raufer,
Doch rauft's mich bei den Haaren recht dazu,
Mit meinem Degen hier zu untersuchen,
Ob du auch Manns genug zu einem Freier.

FLORA.
I, Gott behüt mich!

LEONARD. Ja, wie's Gott gefällt.
Ich weiß nicht, wer von uns der größte Narr hier.
Doch sind Verliebte, wie man sagt, Genies,
Die Hohes tun und Tolles, weil sie müssen!
So genial ist eben mir zumute –
Kurz: Zieh vom Leder, Fant, wehr dich! Zieh aus!

FLORA.
Sehr gern. – *Für sich.* Ich kenn hier jeden Schlich. So'n Mannsbild
Im tölpschen Eifer stolpert tausendmal. – *Sie entflieht.*

LEONARD.
Heisa, bunt' Vögelchen, entfliegst mir nicht!
Ich spür so was vom Habicht heut in mir! *Ihr nach.*

Dritte Szene

Einsamer Waldplatz. FLITT *tritt auf, sieht sich nach allen Seiten um und pfeift dann auf dem Finger.* KNOLL *erscheint im Hintergrunde und beantwortet das Signal auf dieselbe Weise.*

FLITT. Bist du toll? Wozu pfeifst du erst, Hanswurst, wenn du selber schon da bist?

KNOLL. Damegogen und Karbonaden! Das ist der Komment so, wo's was Verschworenes gibt.

FLITT. Ich bitt' dich, liebstes Ungeheuer, möchtest du nicht die Güte haben und hier deinen Rachen ein wenig zähmen? Du kommst schon wieder wie ein Haifisch dahergefahren und schnappst nach den Trümmern deiner in Tertia gestrandeten Gelehrsamkeit. – Sprich, alter bemooster Tertianer, sind die Pferde am Walde bereit?

KNOLL. Rosse und Karosse und alle Madrigalien nebst Impertinenzen sind zu der Entführung prädesteniert.

FLITT. Ich glaube gar, du hast deine alte fliegende Kalesche mitgebracht, in der du immer zum Ochsenmarkt fährst, mit deinem Brustbild von gelben Zwecken an der Tür. Die rumpelt ja in der Nacht, daß man uns bis zum andern Morgen hört, und hängt von der einen Seite hinten über; ich hoffe, sie sollen dich, wenn's einmal flink über die Steine geht, draus verlieren wie einen Sack voll Kaldaunen.

KNOLL. Labet egal! Es wird mich niemand in die Tasche stecken. – Aber warum treten Euer Gnaden hier Ihrem eigenen Schatten auf die Fersen im Mondscheine? Warum sind Sie nicht schon im Garten, etwa auf dem Place de Grandefuß? Sie haben doch dort nicht etwa ein Haar drin gefunden?

FLITT. Nein, sondern einen ganzen Schopf von Haaren; ich weiß nicht, welchem Querkopf er angehören mochte. Dort ist es nicht geheuer. Aber es tut nichts. Wir wollen die Gräfin hier erwarten, hier muß sie durchaus vorüberkommen, wenn sie zu dem verabredeten Platze geht. Gib acht! sobald sie kommt, stellst du dich als chinesisches Lusthaus oder umgestülpte Baumwurzel zur Seite – für den Notfall – ich aber bringe die Gräfin über die Rosen und Vergißmeinnicht lieblicher Redensarten zu den Pferden, du begleitest uns immerfort in einiger Entfernung wie eine Batterie von schweren Stücken

und triffst mit uns zugleich am Wagen ein. – Gelingt es, Knoll, so will ich dich mit kalekutischen Hähnen und Rinderkeulen vollstopfen und dann in goldene Reifen schlagen –

KNOLL. Schlagen? – Ja, ich hoffe, wir sollen heute hier die schönsten Prügel bekommen. Aber es verschlägt nichts. So ein Spaß ist schon ein Dutzend Beulen und Rippenstöße wert; mir kommen sie nicht so leicht durch den Speck.

FLITT. O du gemeine Seele! – O du sanguinische, materialistische Talggrube, in der der Teufel nichts tut, als Schmalz kochen, den er nächstens als Speckschwarte in seinem Schornstein aufhängen wird, den –

KNOLL. Silentium! Da ist sie, da ist sie! Ich nehme meinen Abtritt.

Er tritt hinter die Kulissen.

FLITT. Wahrhaftig, da kommt die Gräfin wie die bleiche Furcht dahergeflattert. Sie wird noch den Nelkenstrauß verlieren – wie der grüne Schleier im Winde fliegt! –

SCHLENDER *noch in der vorigen Damenkleidung mit herabgelassenem Schleier atemlos hereinstürzend.* Flitt, rette mich! – Verraten – Offiziere und zehn oder hundert Gemeine und blanke Schwerter und Dolche und alles hinter mir drein! –

FLITT. Wo? Ich hoffe doch nicht gar. – Nun gilt es rasch sein! Schnelle Pferde sind bereit, ergebene Treue denkt an alles. –

SCHLENDER. Was? –

FLITT. Der starke Gott der Liebe steht voll Schwermut uns zur Seite, und bei dem ersten Morgenstrahl beschirmt ein Strohdach zwei beglückte Herzen!

SCHLENDER. Wie?

FLITT. Um Gottes willen! Nur keinen Laut jetzt, sonst sind wir alle verloren –

SCHLENDER. Aber –

FLITT. Nur still, nur fort! – *Schlender mit einem Arm umfassend und fortführend, für sich.* Süße Beute! Nun laß

die Narren mit ihren Wünschelruten angeln, ich hab' den Schatz gehoben. – Schlau muß man sein!
Ab mit Schlender.

KNOLL *aus seinem Hinterhalte den andern rasch folgend.* Katabomben und Sonaten, ich deck das champs de canaille!

Vierte Szene

Wieder der Teil des Gartens mit der Laube, wie in der zweiten Szene dieses Aufzugs. FLORA *tritt eilig auf, von* LEONARD *und* VICTOR *von verschiedenen Seiten verfolgt.*

FLORA.
Was tu ich nur? Mondsüchtig sind sie alle,
Verwünschte Hosen und verwunschner Victor!
Wo nur ein Narr fehlt, ist er gleich zur Hand.

VICTOR *für sich.*
Sie denkt, ich kenn' sie nicht. – Die will ich ängst'gen!
Laut.
Hallo, trara! Mein Hirschlein, springst recht artig!
Ist denn kein Zaun mehr, drüber wegzusetzen? –
Ein leichtes Wild und liebt bei Nacht zu wechseln.

FLORA.
Ich wünscht' du wärst ein Hirsch und hätt'st Geweihe,
So hoch, daß du im Wald dran hängen bliebst!

LEONARD
Nun, Bürschchen, hast du deinen Mut erjagt?
Steh' still und zieh! Die Hälfte deines Schnurrbarts
Mäh' ich dir sauber ab – mehr will ich nicht.

FLORA.
Ich bitt' Euch, seid gescheit! Mit zweien fechten!

VICTOR.
Mit ihm allein nähmst du's wohl auf?

LEONARD *zu Viktor.*
Geh, halt die Scheide fest, hilf ihm herausziehn.
VICTOR *faßt das untere Ende von Floras Degen.*
So: – Nun beleidigt ihn noch etwas wen'ges,
Sonst kriegen wir ihn nicht heraus!
FLORA *zieht den Degen, läßt ihn aber schnell wieder sinken.*
Nein, nein,
Mir graut zu sehr! – Hört doch! –
VICTOR. Nachher!
LEONARD. Den Bart!
FLORA.
Ich bin kein Mann –
LEONARD.
Ach, das könnt' jeder sagen.
Sie fechten.
Die Gräfin ADELE *in veränderter, ihren Stand bezeichnender Kleidung, rasch auftretend.*
ADELE.
Von hier kam Waffenklang –
Sie erblickt die Fechtenden und stürzt sich zwischen die Degen an Leonards Brust.
Mein Florestin! –
LEONARD.
Was ist dir, scheues Kind? Welch' prächt'ge Kleider! –
Du hast dich so verändert diesen Abend –
Ich kenn' dich kaum mehr – nur der schöne Klang
Der Stimme dringt wie ehemals noch zum Herzen
Und weiß von Falschheit nicht. –
ADELE *sich von Leonards Brust erhebend mit Hoheit.*
Das falsche Spiel
Komm' ich zu enden. – *Auf Flora deutend.*
Wer ist dieser Mann?
FLORA *vor der Gräfin niederknieend.*
Ach! Einer, der gern Federhut, Schwert, Schnurrbart
Und alle Manneswürde jetzt dahingäb'

Um einen einz'gen Weiberrock – ein Mutwill',
Der seinen Willen eben hart gebüßt,
Weil er den Mut dazu nicht konnt' erschwingen. –

LEONARD *sich vor die Stirn schlagend.*
O Tor! – ein Weib wollt' ich zum Ritter schlagen! –
Man sieht in der Ferne den Widerschein von Fackeln.

ADELE *die bisher in Gedanken versunken dastand, plötzlich aufsehend.*
Welch wilder Haufe naht dort? – was bedeutet
Der Fackelschein?

LEONARD *freudig.*
Bei Gott, ich hoffe, Hochzeit! –

FLORA *die noch kniete, aufspringend.*
Und Hochzeit: Freude, und wer froh, verzeiht gern.
O sein Sie wieder meine gnädge Gräfin!

LEONARD.
Was ist denn das? – Sind Sie denn nicht? –

ADELE *Leonarden mit der Hand abwehrend.*
O fort,
Fort, eh die anderen kommen!

LEONARD.
Nein, um Gott!
Wer von den beiden ist die Gräfin?

ADELE.
Ich!
Sie verbirgt ihr Gesicht an Floras Brust und bleibt in dieser Stellung. Unterdessen treten FRIEDMANN *und* BEDIENTEN *mit Fackeln auf.*

FLORA *zur Gräfin.*
Was ist geschehn? – Gehn wir nach Haus – Sie
weinen –

FRIEDMANN.
Dem alten treuen Diener sei's vergönnt –
Ein Wort nur: – Wunderliche Reden kreuzten
Wie Fledermäuse durch die Dämmerung,
Ein Flüstern, Schlüpfen, heimlich her und hin –
Auch Pferde wollt' man sehn am Saum des Gartens

Und einen fremden, furchtbar dicken Mann. –
Ein treuer Diener vigiliert auf alles;
Ja, rief ich, alter Gärtner exponier' dich!
Was kann es kosten als dein graues Haupt!
Ich rief sogleich die Burschen hier zusammen,
Bei Fackelschein durchstreichen wir den Wald;
Da hören wirs noch fern im Grunde rumpeln,
Dazwischen Stimmen zankend durch die Nacht,
Dann alles wieder still. – Nur einen einz'gen
Erwischten wir dort auf verdächt'ger Flucht;
Ich fürchte sehr, sein Kopf sitzt ihm was schief –
Er spricht ganz wirr: von Damen, die nicht Damen –

FLEDER *sich von den Dienern losmachend, tritt stolz hervor.*
Ich sag's Euch, hütet Euch vor meinem Zorn!
Es träf' sich leicht – die Gräfin hält mich hoch –

FRIEDMANN.
Schon wieder! – Ei, besinnen Sie sich doch!

FLEDER.
Fort! Mummerei! Ich bin der Hofrat Fleder!

FRIEDMANN.
Oho! – Sanft, sanft! gemach! – Nein, packt ihn nicht!

FLEDER *Leonarden bemerkend.*
Ein Wörtchen nur mit jenem Herren dort –
Graf Leonard!

ADELE *die sich bei diesen Worten aufgerichtet hat, Leonarden erstaunt ansehend.*
Mein Gott! – Sie sind? –

LEONARD *sich vor ihr auf ein Knie senkend.*
Der Graf. –

FLEDER.
Dem Präsidenten eil' ich's zu berichten,
Wie hier sein Neffe, seines Hauses Stolz,
Bei Nacht sich wegwirft an ein Kammermädchen,
Die freilich lieber Gräfin wär' als Jungfer. *Ab.*

LEONARD *zu Adele.*
Mir träumte einst von einem süßen Kinde –

O sprich noch einmal aus dem schönen Traum
Zu Florestin: Bedeutet's Hochzeit? –

ADELE *zögernd und leise.*

Ja! –

LEONARD *rasch aufspringend und ihre Hand fassend.*

Musik! Nun, Jäger, laßt die Hörner klingen!
Voran die Fackeln, daß die Ström' im Grunde
Und alle Fenster in dem stillen Schloß
Aufblitzen in dem lust'gen Widerscheine!
Tag soll es sein – mir ist so licht im Herzen!

Er führt die Gräfin schnell fort. Hörnerklang, nach und nach immer ferner und leiser.

VICTOR.

Da mag der Teufel Junggeselle bleiben!

Er hebt Flora hoch auf seinen Arm.

Wild warst du stets, und Wild gehört für Jäger!
Hier ist der Jäger in das Wild geschossen,
Ich bin just recht im Schuß – Hoch, lust'ge Braut!

Er trägt sie fort.

FRIEDMANN *schreit.*

Vivat! und Feuerwerk und Hochzeitscarmen!

Alle ab.

Autobiographische Schriften

Kapitel von meiner Geburt

Der Winter des Jahres 1788 war so streng, daß die Schindelnägel auf den Dächern krachten, die armen Vögel im Schlaf von den Bäumen fielen, und Rehe, Hasen und Wölfe ganz verwirrt bis in die Dörfer flüchteten. In einer Märznacht desselben Winters gewahrte man auf dem einsamen Landschloß zu L. ein wunderbares, geheimnisvolles Treiben und Durcheinanderrennen, treppauf, treppab, Lichter irrten und verschwanden an den Fenstern, aber alles still und lautlos, als schweiften Geister durch das alte Haus. Mein Vater ging in dem großen, von einer Wachskerze ungewiß beleuchteten Tafelzimmer auf und nieder, von Zeit zu Zeit horchte er bald in die Nebenstube, bald in den tiefverschneiten Hof hinaus; dann trat er unruhig ans Fenster, hauchte die prächtigen Eisblumen von den Scheiben und betrachtete den weiten gestirnten Himmel. Die *Konstellation* war überaus günstig. Jupiter und Venus blinkten freundlich auf die weißen Dächer, der Mond stand im Zeichen der Jungfrau und mußte jeden Augenblick kulminieren. Da schlug plötzlich ein Hund an tief unten im Dorf, drauf wieder einer, immer mehrere und näher, eine Peitsche knallte und Pferdegetrappel ließ sich im Hofe vernehmen. Endlich! – rief mein Vater, eilig vor die Haustür hinausstürzend. Eine auf Kufen gesetzte, festverschlossene, altmodische Karosse dunkelte aus dem dicken Dampf der Pferde, wie aus einem Zauberrauch, in welchem der Kutscher seine erstarrten Arme gleich Windmühlenflügeln hin und her bewegte. Bitte, Herr Doktor, – sagte mein Vater, selbst den Kutschenschlag öffnend – Sie sind wohl gar drin eingeschlafen? – Auf Ehre, ein klein wenig! war die Antwort, und aus dem Wagen erstaunlich fix sprang zu aller Verwunderung, anstatt des erwarteten Doktors, ein langer, schmaler Kerl, den niemand kannte, in einer ganz knappen, verschossenen Livrey, aus welcher

beim hellen Mondschein sein Ellbogen glänzte, daß einem innerlich fror, wenn man ihn ansah. Mein Vater betrachtete ihn voller Erstaunen, der Fremde nahm schnell eine Handvoll Schnee und rieb sich damit die halberfrorne Nase, der Kutscher fluchte, der Schnee knirschte unter den Tritten, der Hofhund bellte – da wurde ich in der Stube neben dem Tafelzimmer geboren. Mein Vater, da er einen Kindsschrei hörte, blickte erschrocken nach dem Himmel: der Mond hatte soeben kulminiert! um ein Haar wäre ich zur glücklichen Stunde geboren worden, ich kam gerade nur um anderthalb Minuten zu spät, und *zwar in der Konfusion mit den Füßen zuerst, man sagt, ich habe damit ein Entrechat gemacht.*

Daß ich aber, trotz der vortrefflichen Aspekten, die rechte Konstellation verpaßt, verdrießt mich noch bis auf den heutigen Tag, *wie jenen armen Jungen, der bei der Hochzeit beinah einen Kuchen bekommen hätte.* Es wäre ja sonst für mich ein wahres Kinderspiel gewesen, eine reiche Frau, einen Orden, vortreffliche Konnexionen und Protektionen, anstatt meiner dürren Figur einen vornehmen à plomb, oder gar im Morgenblatt einen Lorbeerkranz zu bekommen. Die heutige Welt will Gleichheit haben, so wills die Natur. Oho! meint der Kavalier, die Natur ist grade recht unsinnig aristokratisch, stellt den Ochsen über das Kalb, den Hund über die Katze, die Katze über die Ratze, und unter den Menschen den *echten* hohen Geburtsadel des Talents über das andere gemeine Pack. Ich habe immer gefunden, es gibt nur ein Mittel, die Gleichheit herzustellen, das ist die Liebe, womit unser Herr Jesus Christus alle gleich gemacht vor dem Vater, indem er selbst dem durch eine fatale Konstellation Unterdrückten Macht gibt über das Talent durch Tugend und die Kraft eines heiligen Willens. – Amen!

Erlebtes

Vorwort

An einem schönen warmen Herbstmorgen kam ich auf der Eisenbahn vom andern Ende Deutschlands mit einer Vehemenz dahergefahren, als käme es bei Lebensstrafe darauf an, dem Reisen, das doch mein alleiniger Zweck war, auf das allerschleunigste ein Ende zu machen. Diese Dampffahrten rütteln die Welt, die eigentlich nur noch aus Bahnhöfen besteht, unermüdlich durcheinander wie ein Kaleidoskop, wo die vorüberjagenden Landschaften, ehe man noch irgendeine Physiognomie gefaßt, immer neue Gesichter schneiden, der fliegende Salon immer andere Sozietäten bildet, bevor man noch die alten recht überwunden. Diesmal blieb indessen eine Ruine rechts überm Walde ganz ungewöhnlich lange in Sicht. Europamüde vor Langerweile fragte ich, ohne daß es mir grade um eine Antwort sonderlich zu tun gewesen wäre, nach Namen, Herkunft und Bedeutung des alten Baues; erfuhr aber zu meiner größten Verwunderung weiter nichts als gerade das Unerwartetste, daß nämlich dort oben ein Einsiedler hause. – „Was! so ein wirklicher Eremit mit langem Bart, Rosenkranz, Kutte und Sandalen?“ – Keiner von der Gesellschaft im fliegenden Kasten konnte mir jedoch über diesen impertinenten Rückschritt genügende Auskunft erteilen, niemand hatte den Einsiedel selbst gesehen. Einer der Herren erklärte ihn schlechtweg für einen hochmütigen Sonderling, da er, wie er erfahren, bei der gebildeten Nachbarschaft nirgends Besuch gemacht, ja nicht einmal Visitenkarten umhergeschickt habe. Ein zweiter meinte, da stecke wohl etwas ganz anderes, eine dunkle Tat, ein großes politisches Verbrechen dahinter. – „Ja, diese heimlichen Jesuiten!“ fiel ihm da ein dritter mit einem wichtigen Augenzwick in die Rede, und sprach

nichts weiter. Eine Berliner Dame dagegen, die eben ihre Zigarre angeraucht, versicherte lachend, das sei ohne Zweifel der letzte Romantiker, der sich vor dem Fortschritt der wachsenden Bildung in den mittelalterlichen Urwald geflüchtet. Alle stimmten endlich darin überein, daß besagter Einsiedler etwas verdreht im Kopfe sein müsse.

Diese Notwendigkeit wollte mir zwar keineswegs so unbedingt einleuchten, doch war das wenige, das ich gehört, abenteuerlich genug, um mich neugierig zu machen. Ich beschloß daher, auf der nächsten Station zurückzubleiben, und den seltsamen Kauz womöglich in seinem eignen Neste aufzusuchen.

Das war aber nicht so leicht, wie ich's mir vorgestellt hatte. In den Bahnhöfen ist eine so große Eilfertigkeit, daß man vor lauter Eile mit nichts fertig werden kann. Die Leute wußten genau, in welcher Stunde und Minute ich in Paris oder Triest oder Königsberg, wohin ich nicht wollte, sein könne, über Zugang und Entfernung des geheimnisvollen Waldes aber, wohin ich eben wollte, konnte ich nichts Gewisses erfahren; ja der Befragte blickte verwundert nach der bezeichneten Richtung hin, ich glaube, er hatte die Ruine bisher noch gar nicht bemerkt. Desto besser! dachte ich, schnürte mein Ränzel und schritt wieder einmal mit lang entbehrter Reiselust in die unbestimmte Abenteuerlichkeit des altmodischen Wanderlebens hinein.

Schon war die Rauchschlange des Bahnzuges weit hinter mir in den versinkenden Tälern verschlüpft, statt der Lokomotive pfiffen die Waldvögel grade ebenso wie vor vielen, vielen Jahren, da ich mir als Student zum erstenmal die Welt besehen, als wollten sie fragen, wo ich denn so lange gewesen? So kletterte ich unter dem feierlichen Waldesrauschen auf dem steilen Fußsteig, den mir die Hirten verraten, an einsamen Wiesen vorüber, wo die weidenden Kühe scheu und neugierig nach mir aufsahen, zwischen Weißdorn und Berberitzen, die im vollen Blütenstaat jugendlichen Übermuts auf meine grauen Haare und abge-

tragene Wandertasche stichelten, siegreich immer höher und höher hinan, bis ich mich endlich durch das Dickicht auf die letzte Höhe herausgearbeitet hatte.
Da lag plötzlich, wie in einem Nest von hohem Gras und Unkraut und die Tatzen weit nach mir vorgestreckt, eine riesenhafte Sphinx neben mir, die mich mit ihren steinernen Augen fragend anglotzte. Und in der Tat, das unverhoffte Ungeheuer gab mir ein Rätsel auf, das mich ganz verwirrte. Denn statt der erwarteten Klüfte, wilden Quellen und Felsenklausen nebst Zubehör erblickte ich einen, freilich arg verwilderten, altfranzösischen Garten: hohe Alleen und gradlinige Kiesgänge; rechts und links einzelne Päonien und Kaiserkronen, über denen bunte Schmetterlinge wie verwehte Blüten dahinschwebten, und in der Mitte eine Fontäne, die einförmig fortplätscherte in der großen Stille; nur ein Pfau spazierte stolz zwischen den Kaiserkronen. Es war aber eben Mittagszeit und eine fast gespenstische Beleuchtung ohne Schatten, die Sonne brannte, die Vögel schwiegen, der Wald rauschte kaum noch wie im Traume. Mir war's, als ginge ich durch irgendeine Verzauberung mitten in die gute alte Zeit, ich schüttelte mehrmal mit dem Kopf, ob mir nicht etwa unversehens ein Haarbeutel im Nacken gewachsen.
So kam ich an die Ruine, oder vielmehr an ein Schloß, das allerdings ruiniert genug war, aber offenbar weniger durch sein Alter, als durch einen gewaltsamen Brand. Der eine Teil lag malerisch verfallen, und fraternisierte längst mit dem Frühling, der mit seinen blühenden Ranken überall an Pfeilern und Wänden lustig hinaufkletterte. Nur der nach dem Garten hin gelegene Flügel, alle Fenster künstlich umschnörkelt und durch steinerne Blumengirlanden miteinander verbunden, sah noch sehr vornehm aus, wie eine Residenz des Prinzen Rokoko. Ein Fenster unten stand offen. Ich blickte hinein und übersah eine lange Reihe großer und hoher Gemächer mit reichen Tapeten, parkettierten Fußböden und prächtigen Stuckverzierungen an den Decken, überall samtene Kanapees und

Sessel, die Lehnen weißlackiert mit goldenen Leisten, große Spiegel und Marmortische darunter. Es war so kühl da drinnen in der feierlichen Einsamkeit, aus einem der entfernteren Gemächer flötete soeben eine unsichtbare Spieluhr eine Menuett herüber, die ich noch aus meiner Kindheit zu kennen glaubte.

Jetzt hörte ich kleine, feine Stimmen hinter mir: es war ein Knabe und ein Mädchen, die einander gejagt hatten und stutzig stillstanden, da sie mich erblickten. „Wo ist der Zauberer – der Herr Einsiedler?" fragte ich, mich selbst verbessernd. Der erhitzte Knabe schüttelte die Locken aus dem hübschen Gesichtchen und sah mich schweigend und fast trotzig an. Das etwas ältere Mädchen aber wies nach einer Laube hin und sagte mit einem zierlichen Knicks: „Er betet." – Also am Ende doch wirklich ein Eremit im alten Stil, dachte ich und eilte der bezeichneten Geißblattlaube zu. Dort saß ein Mann, den Rücken nach mir gekehrt und, wie es schien, eifrig in einen schweinsledernen Quartanten vertieft, der auf dem steinernen Tische vor ihm lag. Auf einmal aber, als ich schon ziemlich nahe war, fuhr ein zahmer Storch, den ich bisher gar nicht bemerkt hatte, erschrocken neben mir aus seinen Gedanken, legte den Hals hintenüber, sperrte den langen Schnabel weit auf und klapperte aus Leibeskräften. Da wandte sich der Einsiedler. – „Arthur!" rief ich ganz erstaunt – es war mein liebster Kriegskamerad vom Lützowschen Korps!

„Um des Himmels willen, was machst denn du hier?" – „Ich lese Calderons ‚Autos'." – „Aber just in dieser seltsamen Abgeschiedenheit!" – „Das sind die Trümmer meiner Heimat", entgegnete er ruhig, „und das dort die Enkel meiner Spielgesellen aus der Kinderzeit", fügte er lächelnd hinzu, auf die beiden Kinder weisend, die unterdes neugierig mir gefolgt waren.

Er hatte sich inzwischen hoch aufgerichtet. Er trug nichts weniger als eine korrekte Einsiedleruniform, sondern einen grünen kurzen Jagdrock und nur einen schönen vollen Bart, wie ihn unsere modernen Einsiedler in den Kaf-

feehäusern und Lesekabinetten tragen. Wir hatten uns seit den Kriegsjahren nicht mehr gesehen; nun beschauten wir einander eine Zeitlang stillschweigend, bis wir zuletzt beide in ein lautes Lachen ausbrachen: so uralt und ehrwürdig waren wir beide seitdem geworden; nur seine Augen waren noch immer die alten, treuen, ich hatte ihn sogleich an dem ganz eigentümlichen Blicke wiedererkannt.

I. Der Adel und die Revolution

Sehr alte Leute wissen sich wohl noch einigermaßen der sogenannten guten alten Zeit zu erinnern. Sie war aber eigentlich weder gut noch alt, sondern nur noch eine Karikatur des alten Guten. Das Schwert war zum Galanteriedegen, der Helm zur Zipfelperücke, aus dem Burgherrn ein pensionierter Husarenoberst geworden, der auf seinem öden Landsitz, von welchem seine Vorfahren einst die vorüberziehenden Kaufleute gebrandschatzt hatten, nun seinerseits von den Industriellen belagert und immer enger eingeschlossen wurde. Es war mit einem Wort die mürb und müde gewordene Ritterzeit, die sich puderte, um den bedeutenden Schimmel der Haare zu verkleiden; einem alten Gecken vergleichbar, der noch immer selbstzufrieden die Schönen umtänzelt, und nicht begreifen kann und höchst empfindlich darüber ist, daß ihn die Welt nicht mehr für jung halten will.
Der Adel in seiner bisherigen Gestalt war ganz und gar ein mittelalterliches Institut. Er stand durchaus auf der Lehenseinrichtung, wo, wie ein Planetensystem, die Zentralsonne des Kaisertums von den Fürsten und Grafen und diese wiederum von ihren Monden und Trabanten umkreist wurden. Die wechselseitige religiöse Treue zwischen Vasall und Lehnsherrn war die bewegende Seele aller damaligen Weltbegebenheiten und folglich die welthistorische Macht und Bedeutung des Adels. Aber der Dreißigjährige Krieg, diese große Tragödie des Mittelalters,

hatte den letzteren, der ohnedem schon längst an menschlicher Altersschwäche litt, völlig gebrochen und beschlossen. Indem er die Idee des Kaisers, wenigstens faktisch, aus der Mitte nahm oder doch wesentlich verschob, mußte notwendig der ganze strenggegliederte Bau aus seinen Fugen geraten. Die Stelle der idealen Treue wurde sofort von der materiellen Geldkraft eingenommen; die mächtigeren Vasallen kauften Landsknechte und wurden Raubritter im großen, die kleinern, die in der allgemeinen Verwirrung oft selbst nicht mehr wußten, wem sie verpflichtet, folgten dem größeren Glücke oder besserem Solde. Und als endlich die Wogen sich wieder verlaufen, bemerkte der erstaunte Adel zu spät, daß er sich selbst aus dem großen Staatsverbande heraus, auf den ewig beweglichen Triebsand gesetzt hatte; aus dem freien Lehensadel war unversehens ein Dienstadel geworden, der zu Hofe ging oder bei den stehenden Herren sich einschreiben ließ. So war denn namentlich auch die Ritterlichkeit zuletzt fast ausschließlich an die modernen Offizierkorps gekommen. Auf diese warf nun der Siebenjährige Krieg noch einmal einen wunderbaren Glanz, Ruhmbegier, kecke Lust am Abenteuer, Tapferkeit, aufopfernde Treue und manche der anderen Tugenden, die das Mittelalter groß gemacht, schienen von neuem aufzuleben. Allein es war kein in sich geschlossenes Rittertum im alten Sinne mehr, sondern nur das Aufleuchten einzelner bedeutender Persönlichkeiten, die eben deshalb wohl ihre Namen, nicht aber den Geist des Ganzen unsterblich machen konnten. Auch hier gibt schon das Kostüm, das niemals willkürlich oder zufällig ist, ein charakteristisches Signalement dieses neuen Ritters. Die Eisenrüstung war ihm allmählich zum Küraß, der Küraß zum bloßen Brustharnisch und dieser endlich gar zu einem handbreiten Blechschildchen zusammengeschrumpft, das er gleichsam zum Andenken an die entschwundene Rüstung wie etwa jetzt der Orden zweiter Klasse, dicht unter dem Halse trug, die Rechte, der die Manschette nicht fehlen durfte, ruhte auf einem stattli-

chen spanischen Rohr, das gepuderte Haupt umschwebten zu beiden Seiten, anstatt der alten Geierflügel, zwei wurfähnlich aufgerollte Locken und „der Zopf der hing ihm hinten“. Ein Ritter mit dem Zopf ist aber durchaus eine undenkbare Mißgeburt, was die armen Bildhauer, welche die Helden des Siebenjährigen Krieges darstellen sollen, am schmerzlichsten empfinden. Und dieser fatale Zopf war in der Tat das mystische Symbol der verwandelten Zeit: alles Naturwüchsige, als störend und abgemacht, hinter sich geworfen und mumienhaft zusammengewikkelt, bedeutete er zugleich den *Stock*, die damalige Zentripetalkraft der Heere.

Die jungen Kavaliere jener Zeit dienten in der Regel nicht um einen Krieg, sondern um einen galanten Feldzug gegen die Damen so lange mitzumachen, bis sie die Verwaltung ihrer Güter antreten konnten, oder, wenn sie keine hatten, bis sie mit der glänzenden Uniform eine Schöne oder auch Häßliche erobert, die ihre vielen Schulden zu bezahlen bereit und imstande war. Vom Ritterwesen hatten sie einige verworrene Reminiszenzen ererbt und auf ihre Weise sich zurechtgemacht: vom ehemaligen Frauendienst die fade Liebelei, von der altdeutschen Ehre einen französischen, höchst kapriziösen point d'honneur, vom strengen Lehnsverbande einen kapriziösen Esprit de corps, der nur selten über den ordinärsten Standesegoismus hinauslangte. Es war die hohe Schule des Junkertums, an die selbst Fouqués Recken mit ihren Gardereiterpositionen und ausbündig galanten Redensarten noch zuweilen erinnern.

Der Adel überhaupt aber zerfiel damals in drei sehr verschiedene Hauptrichtungen. Die zahlreichste, gesündeste und bei weitem ergötzlichste Gruppe bildeten die, von den großen Städten abgelegenen kleineren Gutsbesitzer in ihrer fast insularischen Abgeschiedenheit, von der man sich heutzutage, wo Chausseen und Eisenbahnen Menschen und Länder zusammengerückt haben und zahllose Journale wie Schmetterlinge, den Blütenstaub der Zivili-

sation in alle Welt vertragen, kaum mehr eine deutliche Vorstellung machen kann. Die fernen blauen Berge über den Waldesgipfeln waren damals wirklich noch ein unerreichbarer Gegenstand der Sehnsucht und Neugier, das Leben der großen Welt, von der wohl zuweilen die Zeitungen Nachricht brachten, erschien wie ein wunderbares Märchen. Die große Einförmigkeit wurde nur durch häufige Jagden, die gewöhnlich mit ungeheurem Lärm, Freudenschüssen und abenteuerlichen Jägerlügen endigten, sowie durch die unvermeidlichen Fahrten zum Jahrmarkt der nächsten Landstadt unterbrochen. Die letzteren insbesondere waren seltsam genug und könnten sich jetzt wohl in einem Karnevalszuge mit Glück sehen lassen. Vorauf fuhren die Damen im besten Sonntagsstaate, bei den schlechten Wegen nicht ohne Lebensgefahr, unter beständigem Peitschenknall in einer mit vier dicken Rappen bespannten altmodischen Karosse, die über dem unförmlichen Balkengestell in ledernen Riemen hängend, bedenklich hin und her schwankte. – Die Herren dagegen folgten auf einer sogenannten „Wurst", einem langen gepolsterten Koffer, auf welchem diese Haimonskinder dicht hintereinander und einer dem andern auf den Zopf sehend, rittlings balancierten. – Am liebenswürdigsten aber waren sie unstreitig auf ihren Winterbällen, die die Nachbarn auf ihren verschneiten Landsitzen wechselweise einander ausrichteten. Hier zeigte es sich, wie wenig Apparat zur Lust gehört, die überall am liebsten improvisiert sein will und jetzt so häufig von lauter Anstalten dazu erdrückt wird. Das größte, schnell ausgeräumte Wohnzimmer mit oft bedrohlich elastischem Fußboden stellte den Saal vor, der Schulmeister mit seiner Bande das Orchester, wenige Lichter in den verschiedenartigsten Leuchtern warfen eine ungewisse Dämmerung in die entfernteren Winkel umher, und über die Gruppe von Verwalter- und Jägerfrauen, die in der offenen Nebentüre Kopf an Kopf dem Tanze der Herrschaften ehrerbietig zusahen. Desto strahlender aber leuchteten die frischen

Augen der vergnügten Landfräulein, die beständig untereinander etwas zu flüstern, zu kichern und zu necken hatten. Ihre unschuldige Koketterie wußte noch nichts von jener fatalen Prüderie, die immer nur ein Symptom von sittlicher Befangenheit ist. Man konnte sie füglich mit jungen Kätzchen vergleichen, die sorglos in wilden und doch graziös-anmutigen Sprüngen und Windungen im Frühlingssonnenscheine spielen. Denn hübsch waren sie meist, bis auf wenige dunkelrote Exemplare, die in ihrem knappen Festkleide, wie Päonien, von allzu massiver Gesundheit strotzten. – Der Ball wurde jederzeit noch mit dem herkömmlichen Initialschnörkel einer ziemlich ungeschickt ausgeführten Menuett eröffnet, und gleichsam parodisch mit dem graden Gegenteil, dem tollen „Kehraus" beschlossen. Ein besonders gutgeschultes Paar gab wohl auch, von einem Kreise bewundernder Zuschauer umringt, den „Kosakischen" zum besten, wo nur ein Herr und eine Dame ohne alle Touren, sie in heiter zierlichen Bewegungen, er mit grotesker Kühnheit wie ein am Schnürchen gezogener Hampelmann, abwechselnd gegeneinander tanzten. Überhaupt wurde damals, weil mit Leib und Seele, noch mit einer aufopfernden Todesverachtung und Kunstbeflissenheit getanzt, gegen die das heutige vornehm nachlässige Schlendern ein ermüdendes Bild allgemeiner Blasiertheit darbietet. Dabei schwirrten die Geigen und schmetterten die Trompeten und klirrten unaufhörlich die Gläser im Nebengemach, ja zuweilen, wenn der Punsch stark genug gewesen, stürzten selbst die alten Herren, zum sichtbaren Verdruß ihrer Ehefrauen, sich mit den ungeheuerlichsten Kapriolen mit in den Tanz; es war eine wahrhaft ansteckende Lustigkeit. Und zuletzt dann noch auf der nächtlichen Heimfahrt durch die gespensterhafte Stille der Winterlandschaft unter dem klaren Sternenhimmel das selige Nachträumen der schönen Kinder.

Die Glücklichen hausten mit genügsamem Behagen großenteils in ganz unansehnlichen Häusern (unvermeidlich

„Schlösser" geheißen), die selbst in der reizendsten Gegend nicht etwa nach ästhetischem Bedürfnis schöner Fernsichten angelegt waren, sondern um aus allen Fenstern Ställe und Scheunen bequem überschauen zu können. Denn ein guter Ökonom war das Ideal der Herren, der Ruf einer „Kernwirtin" der Stolz der Dame. Sie hatten weder Zeit noch Sinn für die Schönheit der Natur, sie waren selbst noch Naturprodukte. Das bißchen Poesie des Lebens war als nutzloser Luxus lediglich den jungen Töchtern überlassen, die denn auch nicht verfehlten, in den wenigen müßigen Stunden längst veraltete Arien und Sonaten auf einem schlechten Klaviere zu klimpern und den hinter dem Hause gelegenen Obst- und Gemüsegarten mit auserlesenen Blumenbeeten zu schmücken. Gleich mit Tagesanbruch entstand ein gewaltiges Rumoren in Haus und Hof, vor dem der erschrockene Fremde, um nicht etwa umgerannt zu werden, eilig in den Garten zu flüchten suchte. Da flogen überall die Türen krachend auf und zu, da wurde unter vielem Gezänk und vergeblichem Rufen gefegt, gemolken, und gebuttert, und die Schwalben, als ob sie bei der Wirtschaft mit beteiligt wären, kreuzten jubelnd über dem Gewirr, und durch die offenen Fenster schien die Morgensonne so heiter durchs ganze Haus über die vergilbten Familienbilder und die Messingbeschläge der alten Möbel, die jetzt als Rokoko wieder für jung gelten würden. An schönen Sommernachmittagen aber kam häufig Besuch aus der Nachbarschaft. Nach den geräuschvollen Empfangskomplimenten und höflichen Fragen nach dem werten Befinden, ließ man sich dann gewöhnlich in der desolaten Gartenlaube nieder, auf deren Schindeldache der buntübermalte hölzerne Cupido bereits Pfeil und Bogen eingebüßt hatte. Hier wurde mit hergebrachten Späßen und Neckereien gegen die Damen scharmützíert, hier wurde viel Kaffee getrunken, sehr viel Tabak verraucht, und dabei von den Getreidepreisen, von dem zu verhoffenden Erntewetter, von Prozessen und schweren Abgaben verhandelt; während die ungezogenen

kleinen Schloßjunker auf dem Kirschbaum saßen und mit den Kernen nach ihren gelangweilten Schwestern feuerten, die über den Gartenzaun ins Land schauten, ob nicht der Federbusch eines insgeheim erwarteten Reiteroffiziers der nahen Garnison aus dem fernen Grün emportauche. Und dazwischen tönte vom Hofe herüber immerfort der Lärm der Sperlinge, die sich in der Linde tummelten, das Gollern der Truthähne, der einförmige Takt der Drescher und all jene wunderliche Musik des ländlichen Stillebens, die den Landbürtigen in der Fremde, wie das Alphorn den Schweizer oft unversehens in Heimweh versenkt. In den Tälern unten aber schlugen die Kornfelder leise Wellen, überall eine fast unheimlich schwüle Gewitterstille, und niemand merkte oder beachtete es, daß das Wetter von Westen bereits aufstieg und einzelne Blitze schon über dem dunklen Waldeskranze prophetisch hin und her zuckten.

Man sieht, das Ganze war ein etwas ins Derbe gefertigtes Idyll, nicht von Geßner, sondern etwa wie das „Nußkernen" vom Maler Müller. Da fehlte es nicht an manchem höchst ergötzlichen Junker Tobias oder Junker Christoph von Bleichenwang, aber ebensowenig auch an tüchtigen Charakteren und patriarchalischen Zügen. Denn diese Edelleute standen in der Bildung nur wenig über ihren „Untertanen", sie verstanden daher noch das Volk und wurden vom Volke wieder begriffen. Es war zugleich der eigentliche Tummelplatz der jetzt völlig ausgestorbenen Originale, jener halb eigensinnigen, halb humoristischen Ausnahmenaturen, die den stagnierenden Strom des alltäglichen Philisteriums mit großem Geräusch in Bewegung setzten, indem sie, gleich wilden Hummeln, das konventionelle Spinnengewebe beständig durchbrachen. Unter ihnen sah man noch häufig bramarbasierende Haudegen des Siebenjährigen Krieges und wieder andre, die mit einer unnachahmlich lächerlichen Manneswürde von einer gewissen Biederbigkeit Profession machten. Die fruchtbarsten in diesem Genre aber waren die sogenann-

ten „Krippenreiter", ganz verarmte und verkommene Edelleute, die, wie die alten Schalksnarren, von Schloß zu Schloß ritten und, als Erholung von dem ewigen Einerlei, überall willkommen waren. Sie waren zugleich Urheber und Zielscheibe der tollsten Schwänke, Maskeraden und Mystifikationen, denn sie hatten, wie Falstaff, die Gabe, nicht nur selbst witzig zu sein, sondern auch bei anderen Witz zu erzeugen.

Unser deutscher Lafontaine ist, bei aller sentimentalen Abschwächung, nicht ohne einige historische Bedeutung, indem er uns oft einen recht anschaulichen Prospekt in jene gute alte Zeit eröffnet, deren adeliger Zopf sich noch fühlbar durch alle seine Romane hindurchzieht.

In der zweiten Reihe des Adels dagegen standen die Exklusiven, Prätentiösen, die sich und andere mit übermäßigem Anstande langweilten. Sie verachteten die erstere Gruppe und wurden von dieser ebenso gründlich verachtet; beides sehr natürlich, denn diese hatten die frischere Lebenskraft, die jene als plebejisches Krautjunkertum bemitleideten, die Exklusiven aber eine zeitgemäßere Bildung voraus, welche von ersteren nicht verstanden oder als affektierte Vornehmtuerei zurückgewiesen wurde. Bei diesen Vornehmen war nun die ganze Szenerie eine andere. Sie bewohnten wirkliche Schlösser, der Wirtschaftshof, dessen gemeine Atmosphäre besonders den Damen ganz unerträglich schien, war in möglichste Ferne zurückgeschoben, der Garten trat unmittelbar in den Vordergrund. Und diese Gärten müssen wir uns hier notwendig etwas genauer ansehen. Denn diese Adelsklasse, wie bereits erwähnt, ambitionierte sich durchaus, mit der Zeitbildung fortzuschreiten; und obgleich sie in der Regel nichts weniger als Literaten waren, so konnten sie doch nicht umhin, den Geist der jedesmaligen Literatur wenigstens äußerlich, als Mode, in ihrem Luxus abzuspiegeln. Die Gartenkunst aber, wie alle Künste untereinander, hängt mit den wechselnden Phasen namentlich der eben

herrschenden *poetischen* Literatur jederzeit wesentlich zusammen.
Es ist leider hinreichend bekannt, daß wir einst das große poetische Pensum, das uns der Himmel aufgegeben, ungeschickterweise vergessen hatten und daher zu gerechter Strafe lange Zeit in der französischen Schule nachsitzen mußten, wo die Muse, sie mochte nun mutwillig oder tragisch sein, nur in Schnürleib und Reifrock erscheinen durfte. Und der abgemessenen Architektonik dieser Schule entspricht denn auch zunächst der feierliche Kurialstil unserer damaligen geradlinigen Ziergärten:

Es glänzt der Tulpenflor, durchschnitten von Alleen,
Wo zwischen Taxus still die weißen Statuen stehen,
Mit goldnen Kugeln spielt die Wasserkunst im Becken,
Im Laube lauert Sphinx, anmutig zu erschrecken.

Die schöne Chloe da spazieret in dem Garten,
Zur Seit' ein Kavalier, ihr höflich aufzuwarten,
Und hinter ihnen leis' Cupido kommt gezogen,
Bald duckend sich im Grün, bald zielend mit dem Bogen.

Es neigt der Kavalier sich in galantem Kosen,
Mit ihrem Fächer schlägt sie manchmal nach dem Losen,
Es rauscht der taftne Rock, es blitzen seine Schnallen,
Dazwischen hört man oft ein art'ges Lachen schallen.

Jetzt aber hebt vom Schloß, da sich's im West will röten
Die Spieluhr schmachtend an, ein Menuett zu flöten.
Die Laube ist so still, er wirft sein Tuch zur Erde
Und stürzet auf ein Knie mit zärtlicher Gebärde.

„Wie wird mir, ach, ach, ach, es fängt schon an zu dunkeln –"
„So angenehmer nur seh' ich zwei Sterne funkeln –"
„Verwegner Kavalier!" – „Ha, Chloe, darf ich hoffen? –"
Da schießt Cupido los und hat sie gut getroffen.

So ungefähr sind uns diese, ganz bezeichnend *französisch* benannten, Lust- und Ziergärten jederzeit vorgekommen. Wir konnten uns dieselben niemals ohne solche Staffage, diese Chloes und galanten Kavaliere nicht ohne solchen Garten denken; und insofern hatten diese Paradegärten allerdings ihre vollkommene Berechtigung und Bedeutung: Sie sollten eben nur eine Fortsetzung und Erweiterung des Konversationssalons vorstellen. Daher mußte die zudringlich störende Natur durch hohe Laubwände und Bogengänge in einer gewissen ehrerbietigen Form gehalten werden, daher mußten Götterbilder in Allongeperücken überall an den Salon und die französierte Antike erinnern; und es ist nicht zu leugnen, daß in dieser exklusiven Einsamkeit, wo anstatt der gemeinen Waldvögel nur der Pfau courfähig war, die einzigen Naturlaute: die Tag und Nacht einförmig fortrauschenden Wasserkünste, einen um so gewaltigeren, fast tragischen Eindruck machten. Allein solche wesentlich architektonische Effekte sind immer nur durch große würdige Dimensionen erreichbar, wozu es bei den deutschen Landschlössern gewöhnlich an Raum und Mitteln fehlte. Überdies war das Ganze im Grunde nichts weniger als national, sondern nur eine Nachahmung der Versailler Gartenpracht; jede Nachahmung aber, weil sie denn doch immer etwas Neues und Apartes aufweisen will, gerät unfehlbar in das Übertreiben und Überbieten des Vorbildes. Und so erblicken wir denn auch hier, besonders von Holland her, sehr bald und nicht ohne Entsetzen die Mosaikbeete von bunten Scherben, die Pyramiden und abgeschmackten Tiergestalten von Buxbaum, die vielen schlechten, zum Teil hölzernen Götterbilder, mit einem Wort: die Karikatur; und auf diesen Plätzen promenierte der alte Gottsched als Prinz Rokoko mit seinem Gefolge.

Aber dem feierlichen Professor trat fast schon auf die Ferse die bekannte literarische Rebellion gegen das französische Regime, zum Teil durch Franzosen selbst. Rousseau, Diderot, Lessing, jeder in seiner Art, vindizierten der Na-

tur wieder ihr angeborenes Recht. Da brach auf einmal auch das Prachtgerüste jener alten Gärten zusammen, die lang abgesperrte Wildnis kletterte hurtig von allen Seiten über die Buxwände und Scherbenbeete herein, die Natur selbst war ihnen noch nicht natürlich genug, man wollte womöglich bis in den Urwald zurück, und ein wüstes Gehölz mit wenigen Blumen und vielen ärgerlichen Schlangenpfaden, auf denen man nicht vom Fleck und zum Ziele gelangen konnte, mußte den neuen Park bedeuten. Dazu kam noch die in Deutschland unsterbliche Sentimentalität, in beständigem Handgemenge mit dem Terrorismus einer groben Vaterländerei, Lafontaine und Iffland gegen Spieß und Cramer, und über alle hinweg schritt der stolze, kein Vaterland anerkennende Kosmopolitismus. Und sofort finden wir denn dieselbe Anarchie auch in dem neuen Garten wieder: idyllische Hütten und Tränenurnen für imaginäre Tote neben schauerlichen Burgruinen, Heiligenkapellen neben japanischen Tempeln und chinesischen Kiosks; und damit in der totalen Konfusion doch jeder wisse, wie und was er eigentlich zu empfinden habe, wurden an den Bäumen als gefühlvolle Wegweiser, Tafeln mit Sprüchen und sogenannten schönen Stellen aus Dichtern und Philosophen ausgehängt. – Jeder wahre Garten aber, sagt Tieck irgendwo ganz richtig, ist von seiner eigentümlichen Lage und Umgebung bedingt, er muß ein schönes *Individuum* sein, und kann also nur einmal existieren.

Und eben dies war auch das Geschick oder vielmehr Ungeschick der damaligen Bewohner jener Schlösser. Sie waren, wie ihre Gärten, nicht eigentümlich ausgeprägte Individuen, hatten auch keine Nationalgesichter, sondern nur eine ganz allgemeine Standesphysiognomie; überall bis zur tödlichsten Langweiligkeit, dieselbe Courtoisie, dieselben banalen Redensarten, Liebhabereien und Abneigungen. Sie waren die Akteurs der großen Weltbühne, die nicht den Zeitgeist *machten,* sondern den Zeitgeist *spielten;* das Dekorationswesen der Repräsentation war daher ihr eigentliches Fach und Studium, und bühnengerecht zu

sein ihr Stolz. Die alten Kavaliere nebst Haarbeutel und Stahldegen waren nun freilich von der Bühne verschwunden, die neuen hatten aber von ihnen die pedantische Kultur des *Anstandes* als heiligstes Familienerbstück überkommen. Allein der an sich löbliche Anstand ist doch nur der *Schein* dessen, was er eigentlich bedeuten soll, und so ging ihnen denn auch ihr Dasein lediglich in einer traditionellen Ästhetik des Lebens auf. Ihre Ställe verwandelten sich in Prachttempel, wo mit schönen Pferden und glänzenden Schweizerkühen ein fast abgöttischer Kultus getrieben wurde, im Innern des Schlosses schillerte ein blendender Dilettantismus in allen Künsten und Farben, die Fräuleins musizierten, malten oder spielten mit theatralischer Grazie Federball, die Hausfrau fütterte seltene Hühner und Tauben oder zupfte Goldborten, und alle taten eigentlich gar nichts. Sie hatten sich gleichsam die Prosa des Lebensdramas in ein prächtiges Metrum transferiert, und das ist ihre große negative Bedeutsamkeit, daß sie dadurch allerdings langehin das absolut Gemeine und Rohe unterdrückten und abwehrten. Aber Metrik ist noch keine Poesie, und den Gehalt des Lebens konnten sie dadurch nicht veredeln.
Die dritte und bei weitem brillanteste Gruppe endlich war die extreme. Hier figurierten die ganz gedankenlosen Verschwender, jene „im Irrgarten der Liebe herumtaumelnden Kavaliere“, welche ziemlich den Zug frivoler Libertinage repräsentierten, der sich wie eine narkotische Liane durch die damalige Literatur schlang. Zu diesem Berufe wurden die jungen Herren schon frühzeitig mit der sogenannten „guten Konduite“ ausgerüstet, d. h. sie mußten bei meist sehr zweideutigen und abenteuernden Strolchen tanzen, fechten, reiten und Französisch sprechen lernen. Die Eltern hatten vor lauter feiner Lebensart und gesellschaftlichen Pflichten weder Zeit noch Lust, sich um die langweilige Pädagogik zu kümmern, die eigentliche Erziehung war vielmehr gewöhnlich gewissenlosen oder unwissenden Ausländern von armer und geringer „Extraktion“ überlassen; die natürlich von ihren vornehmen Zög-

lingen in aller Weise düpiert wurden. Eine Anekdote aus dem Leben mag vielleicht am anschaulichsten andeuten, wie cavalièrement sich dieses Verhältnis oft gestaltete. Einer dieser Jünglinge hatte einen zwar gewissenhaften, aber sehr pedantischen Mentor, der wohl nicht ohne Grund nächtliche Ausflüge argwöhnen mochte und daher, wenn er, nachts im Garten eine ungewöhnliche Bewegung wahrnahm, jedesmal sich vorsichtig zum Fenster hinauszulehnen pflegte, um seinen Zögling zu belauern. Das war dem letztern schon längst störend und verdrießlich gewesen; er machte daher einmal in seinem nächtlichen Versteck absichtlich ein verdächtiges Geräusch. Kaum aber hatte der Mentor den Kopf wieder aus dem Fenster gestreckt, als zwei unten bereitstehende, als Spukgeister vermummte Lakaien ihm ihrer Instruktion gemäß einen hölzernen Bogen über den Nacken warfen und den Erschrockenen damit am Fensterbrett festklemmten, während ein dritter ihm, zum großen Ergötzen der Schalke, mit einem langen Pinsel das ganze Gesicht einseifte.

Nach dergleichen Studien wurden dann die „jungen Herrschaften" mit ihrem autonomen Hofmeister auf Reisen geschickt, um insbesondere auf der hohen Schule zu Paris sich in der Praxis der Galanterie zu vervollkommnen. Da sie jedoch, bei Strafe der sozialen Exkommunikation, nirgends mit dem Volke, sondern wieder nur in den Kreisen von ihresgleichen verkehren durften, die sich damals überall zum Erschrecken ähnlich sahen, so ist es leicht begreiflich, daß sie auf allen ihren Fahrten nichts erfuhren und lernten, und regelmäßig ziemlich blasiert zurückkehrten. Und ebenso natürlich machten sie nun zu Hause, um nur die unerträgliche Langeweile loszuwerden, die verzweifeltsten Anstrengungen, fuhren mit Heiducken, Laufern und Kammerhusaren zum Besuch, rissen ihre alten Schlösser ein und bauten sich lustig moderne Trianons. Allein das forcierte Lustspiel nahm gewöhnlich ein tragisches Ende, dem kurzen Rausche folgte der moralische und finanzielle Katzenjammer. So ein Lebenslauf

verpuffte rasch wie ein prächtiges Feuerwerk mit Geprassel, leuchtenden Raketen und sprühenden Feuerrädern, bis zuletzt plötzlich nur noch die halbverbrannten dunklen Gerüste dastanden; und das verblüffte Volk rieb sich die Blendung aus den Augen und lachte auseinanderlaufend über den närrischen Spaß. – Der Spaß hatte jedoch auch seine ernste Kehrseite, und grade diese Gruppe hat dem Adel am empfindlichsten geschadet, wie denn überall liebenswürdiger Leichtsinn und Unverstand gefährlicher ist als abstoßende Bosheit. Denn sie waren es vorzüglich, die nicht nur ihren eigenen Stand in schlimmen Ruf brachten, sondern auch in den unteren Schichten der Gesellschaft, die damals noch gläubig und bewundernd zum Adel aufblickten, die Seuche der Glanz- und Genußsucht verbreiteten. Sie haben zuerst die schöne Pietät des von Generation zu Generation fortgeerbten Grundbesitzes untergraben, indem sie denselben in ihrer beständigen Geldnot durch verzweifelte Güterspekulationen zur gemeinen Ware machten. Und so legten sie unwillkürlich mit ihrem eigenen Erbe den Goldgrund zu der von ihnen höchst verachteten Geldaristokratie, die sie verschlang und ihre Trianons in Fabriken verwandelte.

Glücklicherweise aber läßt sich das menschliche Walten nicht in einzelne Kapitel und Paragraphen einfangen. Es versteht sich daher von selbst, daß die Grenzen aller jener Gruppen, die hier nur des klareren Überblicks wegen so konzentriert und scharf gesondert wurden, im Leben häufig ineinanderliefen. Am isoliertesten standen wohl die Prätentiösen durch ihre außerordentliche Langweiligkeit, die sie aller Welt als guten Geschmack aufdringen wollten. Am leichtesten dagegen sympathisierten die erste und dritte Gruppe miteinander, denn die unbefangenen Landjunker besaßen eben noch hinreichenden Humor, um sich an dem Mutwillen und den tollen Luftsprüngen ihrer extremen Standesgenossen zu ergötzen, während die letzteren beständig das Bedürfnis immer neuer und frappanterer Amüsements verspürten, und sich von dem ewigen

Nektar nach derberer Hausmannskost sehnten; es bestand zwischen beiden ein stillschweigender Pakt wechselseitiger Erfrischung. In allen Klassen aber gab es noch Familien genug, die gleichsam mit einem traditionellen Instinkt, den alten Stammbaum frommer Zucht und Ehrenhaftigkeit in den Stürmen und Staubwirbeln der neuen Überbildung, wenn auch nicht zu regenerieren, doch wacker aufrechtzuhalten wußten; sowie einzelne merkwürdige und alle Standesschranken hoch überragende Charaktere, auf die wir weiterhin noch besonders zurückkommen wollen.

So ungefähr standen die Sachen in den letzten Dezennien des vorigen Jahrhunderts. Es brütete, wie schon gesagt, eine unheimliche Gewitterluft über dem ganzen Lande, jeder fühlte, daß irgend etwas Großes im Anzuge sei, ein unausgesprochenes, banges Erwarten, man wußte nicht von was, hatte mehr oder minder alle Gemüter beschlichen. In dieser Schwüle erschienen, wie immer vor nahenden Katastrophen, seltsame Gestalten und unerhörte Abenteurer, wie der Graf St. Germain, Cagliostro u. a., gleichsam als Emissäre der Zukunft. Die ungewisse Unruhe, da sie nach außen nichts zu tun und zu bilden fand, fraß immer weiter und tiefer nach innen; es kamen die Rosenkreuzer, die Illuminaten, man improvisierte allerlei private Geheimbünde für Beglückung und Erziehung der Menschheit, wie wir sie z. B. in Goethes „Wilhelm Meister" auf Lotharios Schlosse sehen; albern und kindisch, aber als Symptome der Zeit von prophetischer Vorbedeutung. Denn der Boden war längst von heimlichen Minen, welche die Vergangenheit und Gegenwart in die Luft sprengen sollten, gründlich unterwühlt, man hörte überall ein spukhaftes unterirdisches Hämmern und Klopfen, darüber aber wuchs noch lustig der Rasen, auf dem die fetten Herden ruhig weideten. Vorsichtige Grübler wollten zwar schon manchmal gelinde Erdstöße verspürt haben, ja die Kirchen bekamen hin und wieder bedenkliche Risse, allein die Nachbarn, da ihre Häuser und Krämerbuden

noch ganz unversehrt standen, lachten darüber, den guten Leuten im „Faust“ vergleichbar, die beim Glase Bier vom fernen Kriege, weit draußen in der Türkei behaglich diskurrieren.

Man kann sich daher heutzutage schwer noch einen Begriff machen von dem Schreck und der ungeheueren Verwirrung, die der plötzliche Knalleffekt durch das ganze Philisterium verbreitete, als nun die Mine in Frankreich wirklich explodierte. Die Landjunker wollten gleich aus der Haut fahren und den Pariser Drachen ohne Barmherzigkeit spießen und hängen. Die Prätentiösen lächelten vornehm und ungläubig und ignorierten den impertinenten Pöbelversuch, Weltgeschichte machen zu wollen; ja es galt eine geraume Zeit unter ihnen für plebejisch, nur davon zu sprechen. Die Extremen dagegen, die ohnedem zu Hause damals nicht viel mehr zu verlieren hatten, erfaßten die Revolution als ein ganz neues und höchst pikantes Amüsement und stürzten sich häufig kopfüber in den flammenden Krater. – Es ist überhaupt ein Irrtum, wenn man den Adel jener Zeit als die ausschließlich konservative Partei bezeichnen will. Er hatte, wie wir gesehen, damals nur noch ein schwaches Gefühl und Bewußtsein seiner ursprünglichen Bedeutung und Bestimmung, eigentlich nur noch eine vage Tradition zufälliger Äußerlichkeiten und folglich selbst keinen rechten Glauben mehr daran. Überdies war das Neue in Deutschland noch keineswegs bis zum Volke gedrungen, es war lediglich eine Geheimwissenschaft der sogenannten gebildeten Klassen, und daher häufig von Adeligen vertreten. Unter ihnen befanden sich viele ernste und hochgestimmte Naturen, die überall zuletzt den Ausschlag geben; aber grade diese, da sie die Unrettbarkeit des Alten einsahen, waren dem Neuen zugewandt. Und diese hatten den schlimmsten Stand. Den Landjunkern waren sie zu gelehrt und durchaus unverständlich, den Prätentiösen zu bürgerlich, den Extremen zu schulmeisterlich; sie wurden von allen ihren Standesgenossen als Renegaten desavouiert, was sie denn frei-

lich in gewissem Sinne auch wirklich waren. Aus diesen Sonderbündlern sind später, als die Revolution zur Tat geworden, einige höchst denkwürdige Charaktere hervorgegangen. So der rastlos unruhige Freiheitsfanatiker Baron Grimm, unablässig wie der Sturmwind die Flammen schürend und wendend, bis sie über ihm zusammenschlugen und ihn selbst verzehrten. So auch der berühmte Pariser Einsiedler Graf Schlabrendorf, der in seiner Klause die ganze soziale Umwälzung wie eine große Welttragödie unangefochten, betrachtend, richtend und häufig lenkend, an sich vorübergehen ließ. Denn er stand so hoch über allen Parteien, daß er Sinn und Gang der Geisterschlacht jederzeit klar überschauen konnte, ohne von ihrem wirren Lärm erreicht zu werden. Dieser prophetische Magier trat noch jugendlich vor die große Bühne, und als kaum die Katastrophe abgelaufen, war ihm der greise Bart bis an den Gürtel gewachsen.
Wenn auf den unwirtbaren Eisgipfeln der Theorie die Lawine fertig und gehörig unterwaschen ist, so reicht der Flug eines Vogels, der Schall eines Wortes hin, um, Felsen und Wälder entwurzelnd, das Land zu verschütten; und dieses Wort hieß: Freiheit und Gleichheit. Das Alte war in der allgemeinen Meinung auf einmal zertrümmert, der goldene Faden aus der Vergangenheit gewaltsam abgerissen. Aber unter Trümmern kann niemand wohnen, es mußte notwendig auf anderen Fundamenten neu gebaut werden, und von da ab begann das verzweifelte Experimentieren der vermeintlichen Staatskünstler, das noch bis heut die Gesellschaft in beständiger fieberhafter Bewegung erhält. Es wiederholte sich abermals der uralte Bau des Babylonischen Turmes mit seiner ungeheueren Sprachenverwirrung, und die Menschheit ging fortan in die verschiedenen Stämme der Konservativen, Liberalen und Radikalen auseinander. Es waren aber vorerst eigentlich nur die Leidenschaften, die unter der Maske der Philosophie, Humanität oder sogenannten Untertanentreue, wie Drachen mit Lindwürmen auf Tod und Leben gegenein-

ander kämpften; denn die Ideen waren plötzlich Fleisch geworden und wußten sich in dem ungeschlachten Leibe durchaus noch nicht zurechtzufinden.

Fassen wir jedoch diesen Kampf der entfesselten und gärenden Elemente schärfer ins Auge, so bemerken wir den der Religion gegen die Freigeisterei, als das eigentlich bewegende Grundprinzip, offenbar im Vordertreffen, denn die Veränderungen der religiösen Weltansicht machen überall die Geschichte. Hier aber war der Kampf zunächst ein sehr ungleicher. Der kleine Landadel trieb großenteils die Religion nur noch wie ein löbliches Handwerk, und blamierte sich damit nicht wenig vor den weit ausgreifenden Fortschrittsmännern. Die vermeintlich gebildeten Adelsklassen dagegen, denen die Lächerlichkeit jederzeit als die unverzeihlichste Todsünde erschien, hatten, schon längst mit den freigeisterischen französischen Autoren heimlich fraternisierend, die neue Aufklärung als notwendige Mode- und Anstandssache, gleichsam als moderne Gasbeleuchtung ihrer Salons stillschweigend bei sich aufgenommen, und erschraken jetzt zu spät vor den ganz unanständigen Konsequenzen, da ihre Franzosen plötzlich Gott abschafften und die nackte Vernunft leibhaftig auf den Altar stellten. Wie aber sollten sie so halbherzig und nachdem sie die rechte Waffe selbst aus der Hand gegeben, sich nun den ungestümen Drängern entgegenstemmen? Es konnte nicht anders sein: die neue Welt schritt über ihre ganz verblüfften Köpfe hinweg, ohne nach ihnen zu fragen. Christus galt fortan für einen ganz guten, nur leider etwas überspannten Mann, dem sich jeder Gebildete wenigstens vollkommen ebenbürtig dünkte. Es war eine allgemeine Seligsprechung der Menschheit, die durch ihre eigene Kraft und Geistreichigkeit kurzweg sich selbst zu erlösen unternahm; mit einem Wort: der vor lauter Hochmut endlich toll gewordene Rationalismus, welcher in seiner praktischen Anwendung eine Religion des Egoismus proklamierte.

Hatte man aber hiermit alles auf die subjektive Eigen-

macht gestellt, so kam es natürlich nun darauf an, diese Eigenmacht auch wirklich zu einer Weltkraft zu entwickeln; und daraus folgte von selbst der gewaltige Stoß der neuen Pädagogik gegen die alte Edukation. Diese war bisher wesentlich eine partikuläre Standeserziehung gewesen, das Individuum ging in seinem bestimmten Stande, alle Stände aber in der allgemeinen Idee des Christentums auf. Jetzt dagegen sollte auch hier die bloße Natur frei walten, jeder Knabe sollte seine subjektive Art oder Unart ungeniert herausbilden, gleichsam spielend sich selbst erziehn, man wollte lauter Rousseausche Emile, das Endziel war der „starke Mensch". Diese Emanzipation der Jugend vom alten Schulzwange hatte zunächst Basedow in die derbe Faust genommen, von dessen Dessauer Philantropie Herder sagte: „Mir kommt alles schrecklich vor; man erzählte mir neulich von einer Methode, Eichwälder in zehn Jahren zu machen; wenn man den jungen Eichen unter der Erde die Herzwurzeln nähme, so schieße alles über der Erde in Stamm und Äste. Das ganze Arkanum Basedows liegt, glaub ich, darin, und ihm möchte ich keine Kälber zu erziehen geben, geschweige Menschen." – Basedow war ein revolutionärer Renommist, sein Nachfolger Campe ein zahmer Philister; jener hat diesen Realismus aufgebracht, Campe hat ihn für die Gebildeten zurechtgemacht und Goethe das ganze Treiben in seinen „Wanderjahren" köstlich parodiert.

Allein solcher Umschwung macht sich nirgend so plötzlich, als die sich überstürzenden Pädagogen es wollten und erwarteten. Namentlich die Gymnasien waren noch keineswegs nach der neuen Schablone zugeschnitten, und es dauerte eine geraume Zeit, ehe hier der moderne Realismus, neben dem alten Klassizismus freundnachbarlich Platz nehmen konnte. Sie waren noch weit davon entfernt, jene Musterkarte von Vielwisserei zu bieten, die nur das eingebildete Halbwissen erzeugt, indem sie das fröhliche Argonautenschiff der Jugend über seine natürliche Tragfähigkeit, mit einer ganz disparaten Ausrüstung bela-

stet, von der dann gewöhnlich die Hälfte als unnützer Ballast wieder über Bord geworfen wird. Die protestantischen Gymnasien jener Zeit basierten noch wesentlich auf der Reformation, welche die Philologie als eine Weltmacht hingestellt hatte. Sie litten daher allerdings an einer, fast nur für künftige Professoren oder Theologen berechneten philologischen Starrheit; haben aber in dieser einseitigen Gründlichkeit Außerordentliches geleistet und eine Menge namhafter Gelehrten in die Welt gesandt. – Dasselbe kann man von den damaligen katholischen Gymnasien nicht rühmen. Diese befanden sich früher größtenteils in den Händen der Jesuiten, die eine mehr allgemeine Bildung mit einer gewissen klösterlichen Zucht und Strenge gar wohl zu vereinigen wußten. Jetzt aber, nach Aufhebung des Ordens, sahen sie sich plötzlich von allen Seiten den Anfechtungen des tumultuarischen Zeitgeistes, und zwar wehrlos, ausgesetzt. Denn die übriggebliebenen Exjesuiten und mit ihnen ihre alten Erziehungstraditionen waren allmählich ausgestorben, und die neuen Lehrkräfte, wie sie die veränderte Zeit durchaus erforderte, noch keineswegs herangebildet. Es entstand aber, bevor man sich nur erst einigermaßen orientiert hatte, notwendig ein augenblicklicher Stillstand, eine sehr fühlbare hin und her schwankende Unsicherheit und schüchterne Nachahmung des protestantischen Wesens, die natürlich anfangs ziemlich ungeschickt ausfallen mußte. Nur das fortdauernde Bedürfnis eines feierlichen Gottesdienstes erhielt hier noch lange Zeit eine ernste und gründliche musikalische Schule, aus der mancher berühmte Künstler hervorgegangen ist. Die Schüler veranstalteten zwar noch immer zur Weihnachtszeit theatralische Vorstellungen, aber statt der früheren, mit aller würdigen Pracht ausgestatteten Aufführung geistlicher Schauspiele, wo man nicht selten kühn auf die Meisterwerke Calderons zurückgegriffen hatte, wurden jetzt alberne Stücke aus dem „Kinderfreund", ja sogar Kotzebueaden gegeben. Auch ihre sogenannten Konvikte bestanden noch, wirkten jedoch häufig

störend durch den aristokratischen Unterschied zwischen den armen Freischülern (Fundatisten) und den reichen Pensionärs, die fast ausschließlich dem Adel angehörten. Denn auch der Adel mußte nun, wenn er nicht von der Zukunft exkludiert sein wollte, dem allgemeinen Zuge folgen. Das nach dem neuen Maßstabe durchaus unzureichende Hauslehrerunwesen, sowie die Pariser Reisestudien hatten fast ganz aufgehört, der Offiziersdienst reduzierte sich immer mehr erblich von Generation zu Generation auf bestimmte unbegüterte Militärfamilien, die jungen Kavaliere gingen auf die Gymnasien wie die andern. Ihre Erziehung war also keine spezifisch adelige mehr, sondern mehr oder minder in die Volksschule aufgegangen.
Fast noch unmittelbarer berührte jedoch den Adel der gleichzeitig zur Herrschaft gelangte Kosmopolitismus, jener seltsame „Überall und Nirgends“, der in aller Welt und also recht eigentlich nirgend zu Hause war. Aus allen möglichen und unmöglichen Tugenden hatte man für das gesamte Menschengeschlecht eine prächtige Bürgerkrone verfertiget, die auf alle Köpfe passen sollte, als sei die Menschheit ein bloßes Abstraktum und nicht vielmehr ein lebendiger Föderativstaat der verschiedensten Völkerindividuen. Alle Geschichte, alles Nationale und Eigentümliche wurde sorgfältig verwischt, die Schulbücher, die Romane und Schauspiele predigten davon; was Wunder, daß die Welt es endlich glaubte! Der Adel aber war durchaus historisch, seine Stammbäume wurzelten grade in dem Boden ihres speziellen Vaterlandes, der ihnen nun plötzlich unter den Füßen hinwegphilosophiert wurde. Diese barbarische Gleichmacherei, dieses Verschneiden des frischen Lebensbaumes nach einem eingebildeten Maße war die größte Sklaverei; denn was wäre denn die Freiheit anderes, als eben die möglichst ungehinderte Entwickelung der geistigen Eigentümlichkeit?
Hiermit hing wesentlich auch das politische Dogma zusammen, wonach alle Laster, wie etwa jetzt den Jesuiten, dem Adel, alle Tugenden den niederen Ständen zugewie-

sen wurden. Wer erinnert sich nicht noch aus den damaligen Leihbibliotheken und Theatern der falschen Minister, der abgefeimten Kammerherren, der Scharen unglücklicher Liebender, die vom Ahnenstolz unbarmherzig unter die Füße getreten werden, sowie andrerseits der edelmütigen Essighändler, biederen Förster usw., wovon z. B. Schillers „Kabale und Liebe" ein geistreiches Resumee gibt. Allein in der Wirklichkeit verhielt es sich anders als in den Leihbibliotheken; es war, nur unter verschiedenen Formen und Richtungen, der eine eben nicht besser und nicht schlimmer als der andre. Der Bauernstolz ist sprichwörtlich geworden, und die Bauern sind noch heutzutage die letzten Aristokraten vom alten Stil. Der Bürgerstand aber hatte längst dieselbe retrograde Bewegung gemacht, wie der Adel. Seine ursprüngliche Bedeutung und Aufgabe war die Wiederbelebung der allmählich stagnierenden Gesellschaft durch neue bewegende Elemente, mit einem Wort: die Opposition gegen den verknöcherten Aristokratismus. In seiner frischen Jugend daher, da er noch mit dem Rittertum um die Weltherrschaft gerungen, atmete er wesentlich einen republikanischen Geist. Die Städte regierten und verteidigten sich selbst, ihre streng gegliederten Handwerkerinnungen waren zugleich eine kriegerische Verbrüderung zu Schutz und Trutz, und die Handelsfahrten in die ferne Fremde erweiterten ihr geistiges Gebiet weit über den beschränkten Gesichtskreis der einsam lebenden Ritter hinaus. Da war überall ein rüstiges Treiben, Erfinden, Wagen, Bauen und Bilden, wovon ihre Münster, sowie ihre welthistorische Hansa ein ewig denkwürdiges Zeugnis geben. Nachdem aber draußen die Burgen gebrochen und somit die bewegenden Ideen der zu erobernden Reichsfreiheit abgenutzt und verbraucht waren, fingen sie nach menschlicher Weise an, die materiellen Mittel, womit ihre jugendliche Begeisterung so Großes geleistet, als Selbstzweck zu betrachten; gleichwie sie ja auch in der Kunst nun die handwerksmäßigen Reimtabulaturen ihres Meistergesanges für Poesie nah-

men. Und mit dieser gemeinen Herabstimmung hatten sie auch sich selbst schon aufgegeben, denn ihre Stärke war die Korporation, die Korporation aber ist nur stark durch den beseelenden Geist, der alle dem Ganzen unterordnet und keinen Egoismus duldet. Da aber, wie gesagt, dieser strenge Geist ihnen im Siegesrausch abhanden gekommen, so mußten nun wohl ihre großartigen Vereine in ihre einzelnen Bestandteile auseinanderfallen und jeder Teil in seinen bloßen Schein umschlagen; von ihrer lebendigen Gliederung blieb nur die pedantische Schablone, von ihrem fröhlichen Volksliede nur die Reimtabulatur übrig, ihre Stadtwehr wurde zur geputzten Schützengilde, die nach gemalten Feinden schoß, der alte Welthandel zur Kleinkrämerei. In ihrer schönen Jugendzeit hatten sie die Buchdruckerkunst um der Wissenschaft willen ersonnen und um Gottes willen Kirchen gebaut, an deren kühnen Pfeilern und Türmen die heutigen Geschlechter schwindelnd emporschauen. Jetzt bauten sie Fabriken und Arbeiterkasernen, erfanden klappernde Maschinen zum Spinnen und Weben, und es ist offenbar, die Industrie wuchs zusehends weit und breit. Aber wir dürfen uns keine Illusionen machen. Die Industrie an sich ist eine ganz gleichgültige Sache, sie erhält nur durch die Art ihrer Verwendung und Beziehung auf höhere Lebenszwecke Wert und Bedeutung.

So hatte also der Bürgerstand – dessen Seele die geistige Bewegung, oder wie wir es jetzt nennen würden: das Prinzip des beständigen Fortschritts war – sich kampfesmüde auf den goldenen Boden des Handwerks gelegt, und die Städte waren allmählich aus einer *Welt*macht eine *Geld*macht geworden. Allein hierin war ihnen der Adel im allgemeinen durch seinen großen Landbesitz noch immer bedeutend überlegen; sie hatten sich mit ihm auf denselben materiellen Boden gestellt, auf dem sie ihn unmöglich innerlich bewältigen konnten. Sie suchten daher nun äußerlich mit ihm zu rivalisieren, sie wollten nicht bloß frei und reich, sondern auch *vornehm* sein. Das ist aber jeder-

zeit ein höchst mißliches Unternehmen, denn um vornehm zu erscheinen, muß man, wie Goethe irgendwo sagt, wirklich vornehm, d. h. durch die allgemeine Meinung irgendwie bereits geadelt sein. Das forcierte Vornehm*tun* macht gerade den entgegengesetzten Effekt: „man merkt die Absicht und ist verstimmt"; wogegen das wirklich Vornehme sich durchaus bequem und passiv zeigt, als ein natürliches bloßes Ablehnen des Gemeinen bei völliger Unbekümmertheit um eine höhere Geltung, die sich ja schon ganz von selbst versteht. Es ist demnach sehr begreiflich, daß jene kleinliche Rivalität der Bürgerlichen, da sie auf der neuen Bühne die ihnen noch mangelnde Routine durch feierlichen Pathos zu ersetzen strebten, anfangs noch ziemlich ungeschickt ausfallen mußte, und daß der Adel seinerseits diese gewaltsamen und pompösen Anstrengungen der „Ellenreiter" mit einer gewissen Schadenfreude belächelte.

Beides indes, dieses Lächeln sowie jenes Großtun, nahm plötzlich ein Ende mit Schrecken, als gegen Schluß des vorigen Jahrhunderts auf einmal die ganze Aufklärung, die echte und die falsche, aus den Bücherschränken in alle Welt ausgefahren. Es handelte sich nun nicht mehr um dies und jenes, sondern um die gesamte Existenz, Satan sollte durch Beelzebub ausgetrieben werden, es war ein Krieg aller gegen alle. Der grobe Materialismus rang mit körperlosen Abstrakten, die zärtliche Humanität fraternisierte mit der Bestialität des Freiheitspöbels, die dickköpfige Menschheit wurde mit Bluthunden zu ihrer neuen Glückseligkeit gehetzt, und Philosophie und Aberglauben und Atheismus rannten wild gegeneinander, so daß zuletzt in dem rasenden Getümmel niemand mehr wußte, wer Freund oder Feind. – Und in dieser ungeheueren Konfusion tat der Adel grade das Allerungeschickteste. Anstatt die im Sturm umherflatternden Zügel kraft höherer Intelligenz kühn zu erfassen, isolierte er sich stolz grollend und meinte durch Haß und Verachtung die eilfertige Zeit zu bezwingen, die ihn natürlich in seinem

Schmollwinkel sitzenließ. Aber nur die völlige Barbarei kann ohne Adel bestehen. In jedem Stadium der Zivilisation wird es, gleichviel unter welchen Namen und Formen, immer wieder Aristokraten geben, d. h. eine bevorzugte Klasse, die sich über die Massen erhebt, um sie zu lenken. Denn der Adel (um ihn bei dem einmal traditionell gewordenen Namen zu nennen) ist seiner unvergänglichen Natur nach das ideale Element der Gesellschaft; er hat die Aufgabe, alles Große, Edle und Schöne, wie und wo es auch im Volke auftauchen mag, ritterlich zu wahren, das ewig wandelbare Neue mit dem ewig Bestehenden zu vermitteln und somit erst wirklich lebensfähig zu machen. Mit romantischen Illusionen und dem bloßen eigensinnigen Festhalten des längst Verjährten ist also hierbei gar nichts getan. Dahin aber scheint der heutige Aristokratismus allerdings zu ziehen, dem wir daher zum Valet wohlmeinend zurufen möchten:

Prinz Rokoko, hast dir Gassen
Abgezirkelt fein von Bäumen,
Und die Bäume scheren lassen,
Daß sie nicht vom Wald mehr träumen.

Wo sonst nur gemein' Gefieder
Ließ sein bäurisch Lied erschallen,
Muß ein Papagei jetzt bieder:
„Vivat Prinz Rokoko!“ lallen.

Quellen, die sich unterfingen,
Durch die Waldesnacht zu tosen,
Läßt du als Fontänen springen
Und mit goldnen Bällen kosen.

Und bei ihrem sanften Rauschen
Geht Damöt bebändert flöten
Und in Rosenhecken lauschen
Daphnen fromm entzückt Damöten.

Prinz Rokoko, Prinz Rokoko,
Laß dir raten, sei nicht dumm!
In den Bäumen, wie in Träumen,
Gehen Frühlingsstimmen um.

Springbrunn' in dem Marmorbecken
Singt ein wunderbares Lied,
Deine Taxusbäume recken
Sehnend sich aus Reih' und Glied.

Daphne will nicht weiter schweifen
Und Damöt erschrocken schmält,
Können beide nicht begreifen,
Was sich da der Wald erzählt.

Laß die Wälder ungeschoren,
Anders rauscht's, als du gedacht,
Sie sind mit dem Lenz verschworen,
Und der Lenz kommt über Nacht.

II. Halle und Heidelberg

Das vorige Jahrhundert wird mit Recht als das Zeitalter der Geisterrevolution bezeichnet. Allein damals wurden nur erst Parole und Feldgeschrei ausgeteilt, es war nur der erste Ausbruch des großen Kampfes, der sich unter wechselnden Evolutionen an das neunzehnte Jahrhundert vererbt hat, und noch bis heute nicht ausgefochten ist. Die deutschen Universitäten aber sind die Werbeplätze und Übungslager dieses von Generation zu Generation sich erneuernden Kriegsheeres. Von Wittenberg ging einst die Reformation aus, von Halle die Wolffsche Lehre, von Königsberg die Kantsche, von Jena die Fichtesche und Schellingsche Philosophie; lauter unsichtbare Gedankenkatastrophen, die einen wesentlichen und entscheidenderen

Einfluß auf das Gesamtleben ausgeübt haben, als sich die Staatskünstler träumen ließen.
Bekanntlich ist unser Jahrhundert unter dem Gestirn der Aufklärung geboren. Kant hatte soeben die philosophische Arbeit seiner Vorgänger streng geordnet und, da er dieselbe in seiner großartigen Wahrheitsliebe für das Ganze als unzureichend erkannte, die Welt lieber sogleich in zwei Provinzen geteilt: in die durch menschliche Erfahrung wahrnehmbare, die er sich glorreich erobert, und in die terra incognita des Unsichtbaren, die er mit der nur dem Genie eigenen heiligen Scheu auf sich beruhen ließ. Seine Schüler aber wollten klüger sein als der Meister und *alles* aufklären; eine Art chinesischer Schönmalerei ohne allen Schatten, der doch das Bild erst wahrhaft lebendig macht. Sie setzten daher nun ihren lichtseligen Verstand ganz allgemein als alleinigen Weltbeherrscher ein; es sollte fortan nur noch einen Vernunftstaat, nur Vernunftreligion, Vernunftpoesie usw. geben. Da jedoch jene zweite dunkle Provinz höchst unvernünftig mit ihrer Phantasie, mit ihrem Glauben, ihren Volksgefühlen und Traditionen gegen dieses unerhörte Regiment zu rebellieren unternahm, so machten sie sich's bequem, indem sie das Geheimnisvolle und Unerforschliche, das sich durch das ganze menschliche Dasein hindurchzieht, ohne weiteres als störend und überflüssig negierten. Kein Wunder demnach, daß das deutsche Leben und das deutsche Reich, das grade auf diesen unsichtbaren Fundamenten vorzugsweise geruht, sich nun nach allen Seiten hin bedenklich senkte und zuletzt so lebensgefährliche Risse bekam, daß es von Polizei wegen abgetragen werden mußte. Und so war denn in der Tat der ganze alte Bau schon im Anfange unseres Jahrhunderts in sich zusammengebrochen; der Sturm der Französischen Revolution und der nachfolgenden Fremdherrschaft hat nur den unnützen Schutt auseinandergefegt.
Allein auf freiem Felde können dauernd nur Wilde wohnen, über die man sich bei aller Naturvergötterung doch

so unendlich erhaben fühlte. Das begriffen alle, und so entstand damals sofort ein unerhörtes Treiben, Klopfen, Hämmern und Richten, als wäre alle Welt plötzlich Freimaurer geworden. Aber der Neubau förderte nicht, weil sie über Fundament, Grund- und Aufriß fortwährend untereinander zankten. Am geschäftigsten und vergnügtesten nämlich zeigten sich zunächst die alten zähen Enzyklopädisten, die jetzt auf dem völlig kahlgefegten Bauplatze endlich ganz freie Hand hatten. Diese wußten wirklich nicht, daß seit Erschaffung der Erde schon mancherlei Bemerkenswertes darauf sich zugetragen; sie wollten daher schlechterdings die Welt ganz von neuem anfangen und abstrakt konstruieren. Als Material hierzu trockneten sie vorerst alle Seelenkräfte auf, um sie in ihren philosophischen Herbarien gehörig zu klassifizieren, und daraus gingen damals die zahllosen neuen Gesetzbücher mit ihren Urrechten und Menschenveredelungen hervor. Sie waren, was sie freilich am wenigsten sein wollten, eigentlich gutmütige Phantasten, wie ja jederzeit grade bei den Nüchternsten das bißchen defekte Phantasie am häufigsten überschnappt, welches der gesunden nicht leicht begegnet. Es ist hiernach auch sehr begreiflich, daß in dieser alles verwischenden Gleichmacherei ohne Nationalität und Geschichte ein kühner Geist, wie Napoleon, den Gedanken einer ganz gleichförmigen europäischen Universalmonarchie fassen konnte.

Aber diesen Transzendentalen gegenüber oder vielmehr direkt entgegen arbeiteten gleichzeitig ganz andere Bauleute: die Freischar der Romantiker, die in Religion, Haus und Staat auf die Vergangenheit wieder zurückgingen; also eigentlich die historische Schule. Das deutsche Leben sollte aus seinen verschütteten geheimnisvollen Wurzeln wieder frisch ausschlagen, das ewig Alte und Neue wieder zu Bewußtsein und Ehren kommen. – Da jedoch beide Parteien einander keineswegs hinreichend gewachsen waren, so nahm bei solchem Stoß und Gegenstoß späterhin die ganze Sache eine diagonale Richtung. Es entstand die

aus beiden widerstrebenden Elementen wunderlich komponierte moderne Vaterländerei; ein imaginäres Deutschland, das weder recht vernünftig, noch recht historisch war.

Alle diese verschiedenen Richtungen waren natürlich vorzugsweise und in möglichster Konzentration auch auf den deutschen Universitäten repräsentiert. Namentlich in dem ersten Dezennium unseres Jahrhunderts bildeten dort die obenerwähnten Abstrakten, meist halbverkommene Kantianer, durchaus noch die tonangebende Majorität. Die Philosophen setzten in ihrer Logik, wie wenn man beim Lesen erst wieder buchstabieren sollte, umständlich auseinander, was sich ganz von selbst verstand; die Theologen lehrten eine elegante Aufklärungsreligion; die Juristen ein sogenanntes Naturrecht, das nirgends galt und niemals gelten konnte. Nur etwa die Lehrer des römischen Rechts machten hie und da eine auffallende Ausnahme, weil der Gegenstand sie zwang, sich in das Positive einer großartigen Vergangenheit zu vertiefen. Es ist bekannt, wie Bedeutendes *Thibaut* auf diesem Felde geleistet und wie der mild-ernste *Savigny,* der überdies niemals in dieser Reihe gestanden, grade damals sich überall neue Bahnen gebrochen hat. Jene halbinvaliden und philosophischen Handwerker dagegen, da sie an sich so wenig Anziehungskraft besaßen, suchten nun mit allerlei schlauen Kunststücken zu werben; die Derbsten unter ihnen durch zum Teil sehr schmutzige Witze und Späße, die alljährlich bei demselben Paragraphen wiederkehrten; die vornehmeren, zumal wenn sie heiratslustige Töchter hatten, durch intime Soireen und Plaudertees, um die bärtigen Burschen zu zivilisieren. Und das gelang auch ganz vortrefflich, denn zu ihnen hielt in der Tat bei weitem die Mehrzahl der jungen Leute, nämlich alle die unsterblichen Bettelstudenten, wie man sie billigerweise nennen sollte, da sie bloß auf Brot studieren. Es war wahrhaft rührend anzusehen, wie da in den überfüllten Auditorien in der schwülen Atmosphäre der entsetzlichsten Langenweile

Lehrer und Schüler um die Wette verzweiflungsvoll mit dem Schlummer rangen, und dennoch überall die Federn unermüdlich fortschwirrten, um die verschlafene Wissenschaft zu Papier zu bringen und in sauberen Heften gewissenhaft heimzutragen.
Allein nebenher ging auch noch ein anderer geharnischter Geist durch diese Universitäten. Sie hatten vom Mittelalter noch ein gut Stück Romantik ererbt, was freilich in der veränderten Welt wunderlich und seltsam genug, fast wie Don Quijote, sich ausnahm. Der durchgreifende Grundgedanke war dennoch ein kerngesunder: der Gegensatz von Ritter und Philister. Stets schlagfertige Tapferkeit war die Kardinaltugend des Studenten, die Muse, die er oft gar nicht kannte, war seine Dame, der Philister der tausendköpfige Drache, der sie schmählich gebunden hielt, und gegen den er daher, wie der Malteser gegen die Ungläubigen, mit Faust, List und Spott beständig zu Felde lag; denn die Jugend kapituliert nicht und kennt noch keine Konzessionen. Und gleichwie überall grade unter Verwandten – weil sie durch gleichartige Gewohnheiten und Prätensionen einander wechselseitig in den Weg treten – oft die grimmigste Feindschaft ausbricht, so wurde auch hier aller Philisterhaß ganz besonders auf die Handwerksburschen (Knoten) gerichtet. Wo diese etwa auf dem sogenannten breiten Steine (dem bescheidenen Vorläufer des jetzigen Trottoirs) sich betreten ließen, oder gar Studentenlieder anzustimmen wagten, wurden sie sofort in die Flucht geschlagen. Waren sie aber vielleicht in allzu bedeutender Mehrzahl, so erscholl das allgemeine Feldgeschrei: „Burschen heraus!“ Da stürzten, ohne nach Grund und Veranlassung zu fragen, halbentkleidete Studenten mit Rapieren und Knütteln aus allen Türen, durch den herbeieilenden Sukkurs des nicht minder rauflustigen Gegenparts wuchs das improvisierte Handgemenge von Schritt zu Schritt, dichte Staubwirbel verhüllten Freund und Feind, die Hunde bellten, die Häscher warfen ihre Bleistifte (mit Fangeisen versehene Stangen) in den ver-

wickelten Knäuel; so wälzte sich der Kampf oft mitten in der Nacht durch Straßen und Gäßchen fort, daß überall Schlafmützen erschrocken aus den Fenstern fuhren und hie und da wohl auch ein gelocktes Mädchenköpfchen in scheuer Neugier hinter den Scheiben sichtbar wurde.

Die damaligen Universitäten hatten überhaupt noch ein durchaus fremdes Aussehen, als lägen sie außer der Welt. Man konnte kaum etwas Malerischeres sehen, als diese phantastischen Studententrachten, ihre sangreichen Wanderzüge in der Umgebung, die nächtlichen Ständchen unter den Fenstern imaginärer Liebchen; dazu das beständige Klirren von Sporen und Rapieren auf allen Straßen, die schönen jugendlichen Gestalten zu Roß, und alles bewaffnet und kampfbereit wie ein lustiges Kriegslager oder ein permanenter Mummenschanz. Alles dies aber kam erst zu rechter Blüte und Bedeutsamkeit, wo die Natur, die ewig jung, auch am getreusten zu der Jugend hält, selber mitdichtend studieren half. Wo, wie z. B. in Heidelberg, der Waldhauch von den Bergen erfrischend durch die Straßen ging und nachts die Brunnen auf den stillen Plätzen rauschten, und in dem Blütenmeer der Gärten rings die Nachtigallen schlugen, mitten zwischen Burgen und Erinnerungen einer großen Vergangenheit; da atmete auch der Student freier auf und schämte vor der ernsten Sagenwelt sich der kleinlichen Brotjägerei und der kindischen Brutalität. Wie großartig im Vergleich mit anderen Studentengelagen war namentlich der Heidelberger Kommers, hoch über der Stadt auf der Altane des halbverfallenen Burgschlosses, wenn rings die Täler abendlich versunken, und von dem Schlosse nun der Widerschein der Fackeln die Stadt, den Neckar und die drauf hingleitenden Nachen beleuchtete, die freudigen Burschenlieder dann wie ein Frühlingsgruß durch die träumerische Stille hinzogen und Wald und Neckar wunderbar mitsangen. – So war das ganze Studentenwesen eigentlich ein wildschönes Märchen, dem gegenüber die übrige Menschheit, die alt-

klug den Maßstab des gewöhnlichen Lebens daran legte, notwendig, wie Sancho Pansa neben Don Quijote, philisterhaft und lächerlich erscheinen mußte.
In jener Zeit brütete äußerlich noch ein unheimlicher Frieden über Deutschland, aber die prophetischen Gedanken, die den Krieg bedeuten, arbeiteten gebunden in jeder Brust, und suchten sich überall in wunderlichen Geheimbünden Luft zu machen. Auch auf den Universitäten bestanden dergleichen Ordensverbindungen, noch ohne speziell politischen Beischmack, bloß auf allgemeine humanistische Zwecke gerichtet, mit allerlei abenteuerlichen Symbolen, furchtbaren Eiden und rasselndem Heldenschmuck, wie man es damals in den vielen Ritterromanen fand. Bestand auch ihr Hauptreiz eben nur in ihrer Heimlichkeit, die Sache war doch ehrlich, bitterernst und für die ganze Lebenszeit gemeint. Als aber jene humanistischen Ideen nach und nach abgenutzt, und alle Lebensverhältnisse immer matter wurden, da trat auch hier an die Stelle der strengen Orden die laxere Observanz der Landsmannschaften. Wie man draußen in der Philisterwelt nun mit dem Anstand statt der Tugend sich begnügte, so gingen auch diese Landsmannschaften eigentlich nur auf den Schein des Seins, auf den bloßen „Komment". Gegen eine nähere Verbrüderung der speziellen Landsleute, obgleich im allgemeinen beengend und einseitig, ließ sich im Grunde nicht viel einwenden. Allein dies war nicht einmal der Fall bei ihnen, sie warben eifersüchtig auch aus anderen Provinzen und verfolgten die eigenen Landsleute, wenn sie sich ihrem Zwange nicht unterwerfen mochten. Und da mithin hier die rechte sittliche Grundlage fehlte, dieses Treiben vielmehr, wie schon der selbstgewählte fade Name „Kränzchen" andeutet, sich lediglich auf der Oberfläche geselliger Verhältnisse bewegte; so artete das Ganze sehr bald in bloßes Dekorationswesen, in ein pedantisches Systematisieren der Jugendlust aus; Mut, Fröhlichkeit, Tracht, Trinken, Singen, alles hatte seine handwerksmäßige Tabulatur, das unwürdige

Prellen und Pressen der „Füchse" war ein löbliches Geschäft, Sittenlosigkeit und affektierte Roheit eine besondere Auszeichnung, und es ist hiernach leicht erklärlich, daß grade ihre Matadore im späteren Leben oft die stattlichsten Philister wurden. Mit der inneren Hohlheit aber wuchs die Prätension, sie knechteten die akademische Freiheit, indem jeder nur auf ihre Weise frei sein sollte, und so währte noch langehin ein gewaltiges Ringen zwischen ihnen und den alternden Orden; ein Kampf, der in einzelnen Fällen mit einer heroischen Aufopferung geführt wurde, die wohl eines größeren Zieles würdig gewesen wäre. So faßte z. B. einst ein hervorragendes Ordensmitglied den kühnen Gedanken sich unerkannt mitten in das feindliche Lager zu begeben, um durch Überredung, Rat und Tat die Gegenpartei zu den Seinigen herüberzuführen. Er hatte sich auch wirklich bereits zum Senior einer Landsmannschaft heraufgeschwungen, und der abenteuerliche Plan wäre fast geglückt, als feiger Verrat alles zu früh aufdeckte, und er nun in zahllosen Zweikämpfen sich durch sämtliche Landsmannschaften wieder herausschlagen mußte, was allerdings ein Kampf auf Tod und Leben war. Das mag uns in gesetzteren Jahren jetzt unnütz und kindisch erscheinen; es war aber immerhin eine Vorschule bedeutender Charaktere, die, wie wir wissen, zur Zeit der Not und als es höhere Dinge galt, sich als tüchtig bewährt haben.

So war in der Tat auf den Universitäten eine gewisse mittelalterliche Ritterlichkeit niemals völlig ausgegangen und selbst in jener Verzerrung und Profanation noch erkennbar. Unter allen diesen Jünglingen aber bildeten die eigentlichen, die *literarischen* Romantiker wiederum eine ganz besondere Sekte. – Die allgemeine Stimmung oder vielmehr Verstimmung war schon seit langer Zeit so prosaisch geworden, daß jeder romantische Anflug für ein Sakrilegium gegen den gesunden Menschenverstand gehalten und höchstens als ein barocker Jugendstreich noch toleriert wurde. Der schwere Proviantwagen der Brotwis-

senschaften bewegte sich langsam in dem hergebrachten Geleise eines hölzernen Schematismus, die Religion mußte Vernunft annehmen und beim Rationalismus in die Schule gehn, die Natur wurde atomistisch wie ein toter Leichnam zerlegt, die Philologie vergnügte sich gleich einem kindisch gewordenen Greise mit Silbenstechen und endlosen Variationen über ein Thema, das sie längst vergessen, die bildende Kunst endlich brüstete sich mit einer sklavischen Nachahmung der sogenannten Natur. Die Kraftgenies in den achtziger Jahren des vorigen Jahrhunderts hatten durch ihre Übertreibung und lärmende Renommisterei das Übel eigentlich nur noch schlimmer und unheilbarer gemacht, indem sie in vollem Burschenwichs ohne weiteres aus der Universität in die Welt hinaussprengten und Leben und Literatur burschikos einrichten wollten, was natürlicherweise einen allgemeinen Landsturm der Gelehrten gegen die Freibeuter auf die Beine brachte. Zwar hatten Lessing, Hamann und Herder nach den verschiedensten Richtungen hin schon Blitze und Leuchtkugeln dazwischengeschleudert. Allein Lessings kritische Blitze waren nur kalte Schläge, und da sie nicht zündeten, meinte jeder, es gelte den Nachbar, und hielt ihn getrost für den Seinigen; Herder dagegen trug aus aller Welt herrliche Bausteine zusammen, als es aber ans Bauen kam, war er inzwischen alt und müde geworden, sein Leben und Wirken blieb ein großartiges Fragment; und Hamanns Geisterstimme verklang unverstanden in den Wolken. Auch in der Poesie hatten Goethe und Schiller bereits den neuen Tag angebrochen, aber sie hatten noch keine Gemeinde. Das Wetterleuchten dieser Genien, obgleich den Frühling andeutend und vorbereitend, blendete und erschreckte vielmehr im ersten Augenblick die Menge; man hörte überall die Sturmglocken gehn, niemand aber wußte, ob und wo es brennt, die einen wollten löschen, die anderen schüren, und so entstand die allgemeine Konfusion, womit das neunzehnte Jahrhundert debütierte.

Da standen unerwartet und fast gleichzeitig mehrere ge-

waltige Geister in bisher ganz unerhörter Rüstung auf: Schelling, Novalis, die Schlegels, Görres, Steffens und Tieck. *Schelling* mit seiner kleinen Schrift über das akademische Studium, worin er den geheimnisvollen Zusammenhang in den Erscheinungen der Natur sowie in den Wissenschaften andeutete, warf den ersten Feuerbrand in die Jugend; gleich darauf suchten andere diese pulsierende Weltseele in den einzelnen Doktrinen nachzuweisen: Werner in der Geologie, Creuzer im Altertum und dessen Götterlehre, Novalis in der Poesie. Es war, als sei überall, ohne Verabredung und sichtbaren Verein, eine Verschwörung der Gelehrten ausgebrochen, die auf einmal eine ganz neue wunderbare Welt aufdeckte.

Am auffallendsten wohl zeigte sich die Verwirrung, welche diese plötzliche Revolution anrichtete, auf der damals frequentesten Universität: in *Halle*, weil dort das heterogenste Material auch den entschiedensten Kampf provozierte. Hier trennte sich alles in zwei Hauptlager: in das stabile der Halbinvaliden, und das bewegliche des neuen Freikorps, während das letztere wieder in mehrere verschiedenartige Gruppen zerfiel, welche aber von der Jugend, die noch nicht so ängstlich sondert, unter den Begriff der Romantik zusammengefaßt wurden. An der Spitze der Romantiker stand *Steffens.* Jung, schlank, von edler Gesichtsbildung und feurigem Auge, in begeisterter Rede kühn und wunderbar mit der ihm noch fremden Sprache ringend, so war seine Persönlichkeit selbst schon eine romantische Erscheinung, und zum Führer einer begeisterungsfähigen Jugend vorzüglich geeignet. Sein freier Vortrag hatte durchaus etwas Hinreißendes durch die dichterische Improvisation, womit er in allen Erscheinungen des Lebens die verhüllte Poesie mehr divinierte, als wirklich nachwies. Am unmittelbarsten mußte diese Naturphilosophie begreiflicherweise die Mediziner berühren, unter denen die besseren Köpfe sich jetzt von der bisherigen Empirie zu dem ritterlichen *Reil* und zu *Froriep* wandten, die überall auf das geheimnisvolle Walten höherer Natur-

kräfte hindeuteten. – Eine andere Gruppe wieder bildeten die jungen Theologen, welche sich um *Schleiermacher* scharten. Dieser merkwürdig komponierte Geist schien, seiner ursprünglichen stachelichten Anlage nach zum Antipoden der Romantik geeignet; und doch hielt er wacker zu ihr, und hat auf demselben platonischen Wege der Theologie, die damals zum Teil in toten Formeln, zum Teil in fader Erfahrungsseelenlehre sich erging, wieder Gemüt erobert; eine Art von geharnischtem Pietismus, der mit scharfer Dialektik alle Sentimentalität männlich zurückwies. – Am entferntesten wären vielleicht die Philologen geblieben, hätte nicht *Wolf,* obgleich persönlich nichts weniger als Romantiker, hier wider Wissen und Willen die Vermittelung übernommen durch den divinatorischen Geist, womit er das ganze Altertum wieder lebendig zu machen wußte, sowie durch eine geniale Humoristik und den schneidenden Witz, mit dem der stets Streitlustige gegen *Schütz* und andere, welche die Alten noch immer mumienhaft einzubalsamieren fortfuhren, fast in dramatischer Weise beständig zu Felde lag. – Zwischen diese Gruppen klemmte sich endlich noch eine ganz besondere Spezies von Philosophen herein, die den unmöglichen Versuch machte, die Kantsche Lehre ins Romantische zu übersetzen. Hierher gehörte Professor *Kayßler,* ein ehemaliger katholischer Priester, der geheiratet, und nun, gleichsam zur Rechtfertigung dieses abenteuerlichen Schrittes, sich eine noch abenteuerlichere Philosophie erfunden hatte. Er hatte es indes als doppelter Renegat mit den Kantianern wie mit den Romantikern verdorben; seine trockenen, abstrusen Vorträge fanden fast nur unter seinen schlesischen Landsleuten geringen Anklang, und wir wollten ihn hier bloß nennen, um das Bild der damaligen elementarischen Gärung möglichst zu vervollständigen. – Gegenüber allen diesen neuen Bestrebungen lag aber die breite schwere Masse der Kantschen Orthodoxen und der Stockjuristen, sämtlich von dem wohlfeilen Kunststück vornehmen Ignorierens fleißig

Gebrauch machend; unter den letzteren einerseits *Schmaltz*, der nachherige Geheimrat der Demagogenjäger, der die Kantsche Philosophie, die er vor kurzem sich in Königsberg geholt, auf seine faselige Weise elegant zu machen suchte; andrerseits Dabelow, König, Woltaer u. a., die von der Philosophie überhaupt nichts wußten. Übrigens stand Halle, so unfreundlich auch die Stadt und ein großer Teil ihrer Umgebung ist, in jener Zeit noch in mancherlei lokalem Rapport mit der romantischen Stimmung. Der nahe Giebichenstein mit seiner Burgruine, an die sich die Sage von Ludwig dem Springer knüpft, war damals noch nicht modern englisiert und eingehegt, wie jetzt, und bot in seiner verwilderten Einsamkeit eine ganz artige Werkstatt für ein junges Dichterherz. Wer als Jüngling von dieser Höhe hinabgeblickt, und sie im Alter nach vielen Jahren wiedersieht, dem wird vielleicht dabei ungefähr zumute sein, wie dem Autor nachstehenden Liedchens:

Da steht eine Burg überm Tale
Und schaut in den Strom hinein,
Das ist die fröhliche Saale,
Das ist der Giebichenstein.

Da hab' ich so oft gestanden,
Es blühten Täler und Höhn,
Und seitdem in allen Landen
Sah ich nimmer die Welt so schön!

Durchs Grün da Gesänge schallten,
Von Rossen, zu Lust und Streit,
Schauten viel schlanke Gestalten
Gleichwie in der Ritterzeit.

Wir waren die fahrenden Ritter,
Eine Burg war noch jedes Haus,
Es schaute durchs Blumengitter
Manch' schönes Fräulein heraus.

Das Fräulein ist alt geworden,
Und unter Philistern umher
Zerstreut ist der Ritterorden,
Kennt keiner den andern mehr.

Auf dem verfallenen Schlosse,
Wie der Burggeist, halb im Traum,
Steh' ich jetzt ohne Genossen
Und kenne die Gegend kaum.

Und Lieder und Lust und Schmerzen,
Wie liegen sie nun so weit –
O Jugend, wie tut im Herzen
Mir deine Schönheit so leid.

Völlig mystisch dagegen erschien gar vielen der am Giebichenstein belegene Reichhardsche Garten mit seinen geistreichen und schönen Töchtern, von denen die eine Goethesche Lieder komponierte, die andere sogar Steffens' Braut war. Dort aus den geheimnisvollen Bosketts schallten oft in lauen Sommernächten, wie von einer unnahbaren Zauberinsel, Gesang und Gitarrenklänge herüber; und wie mancher junge Poet blickte da vergeblich durch das Gittertor, oder saß auf der Gartenmauer zwischen den blühenden Zweigen die halbe Nacht, künftige Romane vorausträumend. – Nicht allzu fern davon aber, um auch in dieser Beziehung die Gegensätze zu vervollständigen, bewohnte *Lafontaine* ein idyllisches Landhaus. Man erzählte von ihm, daß er selbst an seinen schlechten Romanen eigentlich am wenigsten schuld sei, daß ihn vielmehr seine Verleger von Zeit zu Zeit nach Berlin verlockten und dort so lange gleichsam eingesperrt hielten, bis er einen neuen dicken Roman fertiggemacht; was er denn, um nur wieder freizukommen, jedesmal mit unglaublicher Geschwindigkeit besorgt habe. Und hiemit stimmte in der Tat auch seine ganze äußere Erscheinung. Es war ein bequemer, freundlicher, lebensfroher Mann,

der jetzt, da die Zeit seine Sentimentalität quiesziert hatte, sich getrost auf das Übersetzen alter Klassiker verlegte, und wie ein harmloser Revenant unter der verwandelten Generation umherging.
Von nicht geringer Bedeutsamkeit war auch die Nähe von Lauchstädt, wo die Weimarschen Schauspieler während der Badesaison Vorstellungen gaben. Diese Truppe war damals in der Tat ein merkwürdiges Phänomen, und hatte unter Goethes und Schillers persönlicher Leitung wirklich erreicht, was späterhin andere, z. B. Immermann in Düsseldorf, vergeblich anstrebten, nämlich das Theater zu einer höheren Kunstanstalt und poetischen Schule des Publikums emporzuheben. Sie hatten allerdings, und wir möchten fast hinzufügen: glücklicherweise, keine eminent hervorragenden Talente, die durch das Hervortreten einer übermächtigen Persönlichkeit so oft die Harmonie des *Ganzen* mehr stören als fördern, gleichwie die sogenannten schönen Stellen noch lange kein Gedicht machen. Aber sie hatten, was damals überall fehlte, ein künstlerisches Zusammenspiel. Denn eben jener höhere Aufschwung der waltenden Intentionen hob alle gleichmäßig über das Gewöhnliche und schloß das Gemeine oder Mittelmäßige von selbst aus; jeder hatte ein intimeres Verständnis seiner Kunst und seiner jedesmaligen Aufgabe, und ging daher mit Lust und Begeisterung ans Werk. Und so durften sie wagen, was den berühmtesten Hoftheatern bei unverhältmäßig größeren Kräften damals noch gar nicht in den Sinn kam. Mitten in der allgemeinen Misere der Kotzebueaden und Iffländerei eroberten sie sich kühn ganz neue Provinzen; gleichsam die Tragweite der Kunstwerke und des Publikums nach allen Seiten hin prüfend, brachten sie Calderon auf die Bühne, gaben den „Alarcos" und den „Jon" der Schlegel, Brentanos „Ponce de Leon" usw. – Man kann leicht denken, wie sehr dieses Verfahren grade das empfänglichste und dankbarste Publikum der Studenten enthusiasmieren mußte. Die Komödienzettel kamen des Morgens schon, gleich Götterboten, nach Hal-

le herüber, und wurden, wie später etwa die politischen Zeitungen und Kriegsbulletins, beim „Kuchenprofessor" eifrigst studiert. War nun eines jener literarischen Meteore oder ein Stück von Goethe oder Schiller angekündigt, so begann sofort eine wahre Völkerwanderung zu Pferde, zu Fuß, oder in einspännigen Kabrioletts, nicht selten einer großen Retirade mit lahmen Gäulen und umgeworfenen Wägen vergleichbar, niemand wollte zurückbleiben, die Reicheren griffen den Unbemittelten mit Entree und sonstiger Ausrüstung willig unter die Arme, denn die Sache wurde ganz richtig als eine Nationalangelegenheit betrachtet. In Lauchstädt selbst aber konnte man, wenn es sich glücklich fügte, Goethe und Schiller oft leibhaftig erblicken, als ob die olympischen Götter wieder unter den Sterblichen umherwandelten. Und außerdem gab es dort auch vor und nach der Theatervorstellung, in der großen Promenade noch eine kleine Weltkomödie, in welcher, wenigstens in den Augen der jüngeren Damen, die Studenten selbst die Heldenrollen spielten. Diese fühlten sich hier überhaupt wahrhaft als Musensöhne, es war ihnen zumute, als sei dies alles eigentlich nur ihretwegen veranstaltet; und sie hatten im Grunde recht, da sie vor allen andern das rechte Herz dazu mitbrachten.

Dieses althallesche Leben aber wurde im Jahre 1806 beim Zusammensturz der preußischen Monarchie unter ihren Trümmern mit begraben. Die Studenten hatten unzweideutig Miene gemacht, sich in ein bewaffnetes Freikorps zusammenzutun. Napoleon, dem hier zum ersten Male ein Symptom ernsteren Volkswillens gleichsam prophetisch warnend entgegentrat, hob daher zornentbrannt die Universität auf, die Studenten wurden mit unerhörtem Vandalismus plötzlich und unter großem Wehgeschrei der Bürger nach allen Weltgegenden auseinandergetrieben und mußten, ausgeplündert und zum Teil selbst der nötigen Kleidungsstücke beraubt, sich einzeln nach Hause betteln. – Wunderbarer Gang der Weltgerichte! Dieselben

vom übermütigen Sieger in den Staub getretenen Jünglinge sollten einst siegreich in Paris einziehen.

Der Geist einer bestimmten Bildungsphase läßt sich nicht aufheben, wie eine Universität. Was wir vorhin als das Charakteristische jener Periode bezeichnet: die Opposition der jungen Romantik gegen die alte Prosa war keineswegs auf Halle beschränkt, sondern ging wie ein unsichtbarer Frühlingssturm allmählich wachsend durch ganz Deutschland. Insbesondere aber gab es dazumal in Heidelberg einen tiefen, nachhaltenden Klang. Heidelberg ist selbst eine prächtige Romantik; da umschlingt der Frühling Haus und Hof und alles Gewöhnliche mit Reben und Blumen, und erzählen Burgen und Wälder ein wunderbares Märchen der Vorzeit, als gäb es nichts Gemeines auf der Welt. Solch gewaltige Szenerie konnte zu allen Zeiten nicht verfehlen, die Stimmung der Jugend zu erhöhen und von den Fesseln eines pedantischen Komments zu befrein; die Studenten tranken leichten Wein anstatt des schweren Bieres, und waren fröhlicher und gesitteter zugleich als in Halle. Aber es trat grade damals in Heidelberg noch eine ganz besondere Macht hinzu, um jene glückliche Stimmung zu vertiefen. Es hauste dort ein einsiedlerischer Zauberer, Himmel und Erde, Vergangenheit und Zukunft mit seinen magischen Kreisen umschreibend – das war *Görres*.

Es ist unglaublich, welche Gewalt dieser Mann, damals selbst noch jung und unberühmt, über alle Jugend, die irgend geistig mit ihm in Berührung kam, nach allen Richtungen hin ausübte. Und diese geheimnisvolle Gewalt lag lediglich in der Großartigkeit seines Charakters, in der wahrhaft brennenden Liebe zur Wahrheit und einem unverwüstlichen Freiheitsgefühl, womit er die einmal erkannte Wahrheit gegen offene und verkappte Feinde und falsche Freunde rücksichtslos auf Tod und Leben verteidigte; denn alles Halbe war ihm tödlich verhaßt, ja unmöglich, er wollte die *ganze* Wahrheit. Wenn Gott noch

in unserer Zeit einzelne mit prophetischer Gabe begnadigt, so war Görres ein Prophet, in Bildern denkend und überall auf den höchsten Zinnen der wildbewegten Zeit weissagend, mahnend und züchtigend, auch darin den Propheten vergleichbar, daß das „Steiniget ihn!" häufig genug über ihm ausgerufen wurde. Drüben in Frankreich hatte er bei den Banketten der bluttriefenden Revolution, hier in den Kongreßsälen der politischen Weltweisen das Menetekel kühn an die Wand geschrieben, und konnte sich nur durch rasche Flucht vor Kerker und Banden retten, oft monatelang arm und heimatlos umherirrend. – Seine äußere Erscheinung erinnerte einigermaßen an Steffens und war doch wieder grundverschieden. Steffens hatte bei aller Tüchtigkeit, etwas Theatralisches, während Görres, ohne es zu wollen oder auch nur zu wissen, schlicht und bis zum Extrem selbst die unschuldigsten Mittel des Effekts verschmähte. Sein durchaus freier Vortrag war monoton, fast wie fernes Meeresrauschen schwellend und sinkend, aber durch dieses einförmige Gemurmel leuchteten zwei wunderbare Augen und zuckten Gedankenblitze beständig hin und wider; es war wie ein prächtiges nächtliches Gewitter, hier verhüllte Abgründe, dort neue ungeahnte Landschaften plötzlich aufdeckend, und überall gewaltig, weckend und zündend fürs ganze Leben.
Neben ihm standen zwei Freunde und Kampfgenossen: *Achim von Arnim* und *Clemens Brentano*, welche sich zur selben Zeit nach mancherlei Wanderzügen in Heidelberg niedergelassen hatten. Sie bewohnten im „Faulpelz", einer ehrbaren aber obskuren Kneipe am Schloßberg, einen großen luftigen Saal, dessen sechs Fenster mit der Aussicht über Stadt und Land die herrlichsten Wandgemälde, das herüberfunkelnde Zifferblatt des Kirchturms ihre Stockuhr vorstellte; sonst war wenig von Pracht oder Hausgerät darin zu bemerken. Beide verhielten sich zu Görres eigentlich wie fahrende Schüler zum Meister, untereinander aber wie ein seltsames Ehepaar, wovon der ru-

hige mild-ernste Arnim den Mann, der ewig bewegliche Brentano den weiblichen Part machte. Arnim gehörte zu den seltenen Dichternaturen, die, wie Goethe, ihre poetische Weltansicht jederzeit von der Wirklichkeit zu sondern wissen, und daher besonnen *über* dem Leben stehen und dieses frei als ein Kunstwerk behandeln. Den lebhafteren Brentano dagegen riß eine übermächtige Phantasie beständig hin, die Poesie ins Leben zu mischen, was denn häufig eine Konfusion und Verwickelungen gab, aus welchen Arnim den unruhigen Freund durch Rat und Tat zu lösen hatte. Auch äußerlich zeigte sich der große Unterschied. Achim von Arnim war von hohem Wuchs und so auffallender männlicher Schönheit, daß eine geistreiche Dame einst bei seinem Anblick und Namen in das begeisterte Wortspiel: „Ach im Arm ihm" ausbrach; während Bettina, welcher, wie sie selber sagt, eigentlich alle Menschen närrisch vorkamen, damals an ihren Bruder Clemens schrieb: „Der Arnim sieht doch königlich aus, er ist nicht in der Welt zum zweitenmal." – Das letztere konnte man zwar auch von Brentano, nur in ganz anderer Beziehung sagen. Während Arnims Wesen etwas wohltuend Beschwichtigendes hatte, war Brentano durchaus aufregend; jener erschien im vollsten Sinne des Worts wie ein Dichter, Brentano dagegen selber wie ein Gedicht, das, nach Art der Volkslieder, oft unbeschreiblich rührend, plötzlich und ohne sichtbaren Übergang in sein Gegenteil umschlug und sich beständig in überraschenden Sprüngen bewegte. Der Grundton war eigentlich eine tiefe, fast weiche Sentimentalität, die er aber gründlich verachtete, eine eingeborene Genialität, die er selbst keineswegs respektierte und auch von andern nicht respektiert wissen wollte. Und dieser unversöhnliche Kampf mit dem eigenen Dämon war die eigentliche Geschichte seines Lebens und Dichtens, und erzeugte in ihm jenen unbändigen Witz, der jede verborgene Narrheit der Welt instinktartig aufspürte und niemals unterlassen konnte, jedem Toren, der sich weise dünkte, die ihm gebührende Schellenkappe auf-

zustülpen, und sich somit überall ingrimmige Feinde zu erwecken. Klein, gewandt und südlichen Ausdrucks, mit wunderbar schönen, fast geisterhaften Augen, war er wahrhaft zauberisch, wenn er selbstkomponierte Lieder oft aus dem Stegreif zur Gitarre sang. Dies tat er am liebsten in Görres' einsamer Klause, wo die Freunde allabendlich einzusprechen pflegten; und man könnte schwerlich einen ergötzlicheren Gegensatz der damals florierenden ästhetischen Tees ersinnen, als diese Abendunterhaltungen, häufig ohne Licht und brauchbare Stühle, bis tief in die Nacht hinein: wie da die dreie alles Große und Bedeutende, das je die Welt bewegt hat, in ihre belebenden Kreise zogen, und mitten in dem Wetterleuchten tiefsinniger Gespräche Brentano mit seinem witzsprühenden Feuerwerk dazwischenfuhr, das dann gewöhnlich in ein schallendes Gelächter zerplatzte.

Das nächste Resultat dieser Abende war die „Einsiedlerzeitung", welche damals Arnim und Brentano in Heidelberg herausgaben. Das selten gewordene Blatt war eigentlich ein Programm der Romantik; einerseits die Kriegserklärung an das philisterhafte Publikum, dem es feierlich gewidmet und mit dessen wohlgetroffenen Porträt es verziert war; andrerseits eine Probe- und Musterkarte der neuen Bestrebungen: Beleuchtung des vergessenen Mittelalters und seiner poetischen Meisterwerke, sowie die ersten Lieder von Uhland, Justinus Kerner u. a. Die merkwürdige Zeitung hat nicht lange gelebt, aber ihren Zweck als Leuchtkugel und Feuersignal vollkommen erfüllt. Übrigens standen ihre Verfasser in der Tat einsiedlerisch genug über dem großen Treiben und Arnim und Brentano, obgleich sie neben Tieck, die einzigen *Produzenten* der Romantiker waren, wurden doch von der Schule niemals als vollkommen zünftig anerkannt. Sie strebten vielmehr, die Schule, die schon damals in überkünstlichen Formen üppig zu luxurieren anfing, auf die ursprüngliche Reinheit und Einfachheit des Naturlauts zurückzuweisen. In diesem Sinne sammelten sie selbst auf

ihren Fahrten und durch gleichgestimmte Studenten überall die halbverschollenen Volkslieder für „Des Knaben Wunderhorn“, das, wie einst Herders „Stimmen der Völker“, durch ganz Deutschland einen erfrischenden Klang gab.

Auch *Creuzer* lebte damals in Heidelberg und gehörte, wiewohl dem genannten Triumvirat persönlich ziemlich fernstehend, durch seine Bestrebungen diesem Kreise an. Seine mystische Lehre hat, z. B. später in Lobeck, sehr tüchtige Gegner gefunden, und wir wollen keineswegs in Abrede stellen, daß die phantastische Weise, womit er die alte Götterlehre als ein bloßes Symbolum christlich umzudeuten sucht, gar oft an den mittelalterlichen Neuplatonismus erinnert und am Ende zu einer gänzlichen Auflösung des Altertums führt. Allein in Kriegszeiten bedarf ein grober Feind auch eines gewaltsamen Gegenstoßes. Erwägt man, wie geistlos dazumal die Mythologie als ein bloßes Schulpensum getrieben wurde, so wird man Creuzers Tat billigerweise wenigstens als eine sehr zeitgemäße und heilsame Aufregung anerkennen müssen. – Noch zwei andere, höchst verschiedene Heidelberger Zeitgenossen dürfen hier nicht unerwähnt bleiben; wir meinen: *Thibaut* und *Gries.* In solchen Übergangsperioden ist die sanguinische Jugend gern bereit, den Spruch: „Wer nicht mit uns ist, ist gegen uns“ gelegentlich auch umzukehren und jeden für den Ihrigen zu nehmen, der nicht zum Gegenpart hält; und in dieser Lage befand sich *Thibaut.* Schon seine äußere Erscheinung mit den lang herabwallenden, damals noch dunkelen Locken, was ihm ein gewisses apostolisches Ansehen gab, noch mehr der eingeborene Widerwillen gegen alles Kleinliche und Gemeine unterschied ihn sehr fühlbar von dem Troß seiner eigentlichen Zunftgenossen, und mit seiner propagandistischen Liebe und Kenntnis von der Musik der alten tiefsinnigen Meister berührte er in der Tat den Kreis der Romantiker. – Bei weitem unmittelbarer indes wirkte *Gries.* Wilhelm Schlegel hatte soeben durch das dicke Gewölk verjährter

Vorurteile auf das Zauberland der südlichen Poesie hingewiesen. Gries hat es uns wirklich erobert. Seine meisterhaften Übersetzungen von Ariost, Tasso und Calderons Schauspielen treffen, ohne philologische Pedanterie und Wortängstlichkeit, überall den eigentümlichen Sinn und Klang dieser Wunderwelt; sie haben den poetischen Gesichtskreis unendlich erweitert und jene glückliche Formfertigkeit erzeugt, deren sich unsere jüngeren Poeten noch bis heut erfreuen. Auch war Gries sehr geeignet, für den Ritt in das alte romantische Land Proselyten zu machen. Er verkehrte gern und viel mit den Studenten, die Abendtafel im Gasthofe „Zum Prinzen Karl" war sein Katheder, und es war, da er sehr schwerhörig, oft wahrhaft komisch, wie da die leichten Scherze und Witze gleichsam aus der Trompete gestoßen wurden, so daß die heitere Konversation sich nicht selten wie ein heftiges Gezänke ausnahm. Man sieht, die Romantik war dort reich vertreten. Allein sie hatte auch damals schon ihren sehr bedenklichen Afterkultus. *Graf von Löben* war in Heidelberg der Hohepriester dieser Winkelkirche. Der alte Goethe soll ihn einst den vorzüglichsten Dichter jener Zeit genannt haben. Und in der Tat, er besaß eine ganz unglaubliche Formengewandtheit und alles äußere Rüstzeug des Dichters, aber nicht die Kraft, es gehörig zu brauchen und zu schwingen. Er hatte ein durchaus weibliches Gemüt mit unendlich feinem Gefühl für den salonmäßigen Anstand der Poesie, eine überzarte empfängliche Weichheit, die nichts Schönes selbständig gestaltete, sondern von allem Schönen wechselnd umgestaltet wurde. So durchwandelte er in seiner kurzen Lebenszeit ziemlich fast alle Zonen und Regionen der Romantik; – bald erschien er als begeisterungswütiger Seher, bald als arkadischer Schäfer, dann plötzlich wieder als aszetischer Mönch, ohne sich jemals ein eigentümliches Revier schaffen zu können. In Heidelberg war er gerade „Isidorus Orientalis" und novalisierte, nur leider ohne den Tiefsinn und den dichterischen Verstand von Novalis. In dieser Periode entstand sein frühe-

ster Roman „Guido“, sowie die „Blätter aus dem Reisebüchlein eines andächtigen Pilgrims“; jener durch seine mystische Überschwenglichkeit, diese durch ein unkatholisches Katholisieren, ganz wider Wissen und Willen, die erstaunlichste Karikatur der Romantik darstellend.
Er hatte in Heidelberg nur wenige sehr junge Jünger, die ihn gehörig bewunderten; aber die Gemeinde dieser Gleichgestimmten war damals sehr zahlreich durch ganz Deutschland verbreitet. Es wäre eine schwierige, ja fast unmögliche Aufgabe, jenes wunderliche Gewirr von Talent und Zopf, Lüge und Wahrheit mit wenigen Worten in einen Begriff zusammenzufassen; und doch ist dieses Treiben insofern von literarhistorischer Wichtigkeit, als dasselbe den schmählichen Verfall der Romantik vorzüglich verschuldet hat. Es sei uns daher lieber vergönnt, aus unserer frühesten Schrift („Ahnung und Gegenwart“) die aus dem Leben gegriffene Darstellung der damaligen Salonwirtschaft hier einzuschalten, da sie, obgleich erfunden, und doch vielleicht unmittelbarer als eine Definition, in den Zirkel einführen dürfte.
Es ist nämlich dort von einer Soiree in der Residenz die Rede, wobei die Gesellschaft über die soeben beendigte Darstellung eines lebenden Bildes in große Bewegung geraten. „Mitten in dieser Entzückung fiel der Vorhang plötzlich wieder, das Ganze verdeckend, herab, der Kronleuchter wurde heruntergelassen und ein schnatterndes Gewühl und Lachen erfüllte auf einmal wieder den Saal. Der größte Teil der Gesellschaft brach nun von allen Sitzen auf und zerstreute sich. Nur ein kleiner Teil von Auserwählten blieb im Saale zurück. Graf Friedrich“ (der Held des Romans) „wurde währenddessen vom Minister, der auch zugegen war, bemerkt und sogleich der Frau vom Hause vorgestellt. Es war eine fast durchsichtig schlanke, schmächtige Gestalt, gleichsam im Nachsommer ihrer Blüte und Schönheit. Sie bat ihn mit so überaus sanften, leisen, lispelnden Worten, daß er Mühe hatte, sie zu verstehen, ihre künstlerischen ‚Abendandachten‘, wie

sie sich ausdrückte, mit seiner Gegenwart zu beehren, und sah ihn dabei mit blinzelnden, fast zugedrückten Augen an, von denen es zweifelhaft war, ob sie ausforschend, gelehrt, sanft, verliebt, oder nur interessant sein sollten.

Die Gesellschaft zog sich nun in eine kleinere Stube zusammen. Die Zimmer waren durchaus prachtvoll und im neuesten Geschmacke dekoriert, nur hin und wieder bemerkte man einige auffallende Besonderheiten und Nachlässigkeiten, unsymmetrische Spiegel, Gitarren, aufgeschlagene Musikalien und Bücher, die auf den Ottomanen zerstreut umherlagen. Friedrich kam es vor, als hätte es der Frau vom Hause vorher einige Stunden mühsamen Studiums gekostet, um in das Ganze eine gewisse unordentliche Genialität hineinzubringen.

Es hatte sich unterdes ein niedliches, etwa zehnjähriges Mädchen eingefunden, die in einer reizenden Kleidung mit langen Beinkleidern und kurzem schleiernen Röckchen darüber, keck im Zimmer herumsprang. Es war die Tochter vom Hause. Ein Herr aus der Gesellschaft reichte ihr ein Tamburin, das in einer Ecke auf dem Fußboden gelegen hatte. Alle schlossen bald einen Kreis um sie, und das zierliche Mädchen tanzte mit einer wirklich bewundeрungswürdigen Anmut und Geschicklichkeit, während sie das Tamburin auf mannigfache Weise schwang und berührte und ein niedliches italienisches Liedchen dazu sang. Jeder war begeistert, erschöpfte sich in Lobsprüchen und wünschte der Mutter Glück, die sehr zufrieden lächelte. Nur Friedrich schwieg still. Denn einmal war ihm schon die moderne Knabentracht bei Mädchen zuwider, ganz abscheulich aber war ihm diese gottlose Art, unschuldige Kinder durch Eitelkeit zu dressieren. Er fühlte vielmehr ein tiefes Mitleid mit der schönen kleinen Bajadere. Sein Ärger und das Lobpreisen der anderen stieg, als nachher das Wunderkind sich unter die Gesellschaft mischte, nach allen Seiten hin in fertigem Französisch schnippische Antworten erteilte, die eine Klugheit weit

über ihr Alter zeigten, und überhaupt jede Unart als genial genommen wurde.
Die Damen, welche sämtlich sehr ästhetische Mienen machten, setzten sich darauf nebst mehreren Herren unter dem Vorsitz der Frau vom Hause, die mit vieler Grazie den Tee einzuschenken wußte, förmlich in Schlachtordnung und fingen an, von Ohrenschmäusen zu reden. Der Minister entfernte sich in die Nebenstube, um zu spielen. – Friedrich erstaunte, wie diese Weiber geläufig mit den neuesten Erscheinungen der Literatur umzuspringen wußten, von denen er selber manche kaum dem Namen nach kannte; wie leicht sie mit Namen herumwarfen, die er nie ohne heilige tiefe Ehrfurcht auszusprechen gewohnt war. Unter ihnen schien besonders ein junger Mann mit einer verachtenden Miene in einem gewissen Glauben und Ansehen zu stehen. Die Frauenzimmer sahen ihn beständig an, wenn es darauf ankam, ein Urteil zu sagen, und suchten in seinem Gesichte seinen Beifall oder Tadel im voraus herauszulesen, um sich nicht etwa mit etwas Abgeschmacktem zu prostituieren. Er hatte viele genialische Reisen gemacht, in den meisten Hauptstädten auf seine eigene Faust Ball gespielt, Kotzebue einmal in einer Gesellschaft in den Sack gesprochen, fast mit allen berühmten Schriftstellern zu Mittag gegessen oder kleine Fußreisen gemacht. Übrigens gehörte er eigentlich zu keiner Partei, er übersah alle weit und belächelte die entgegengesetzten Gesinnungen und Bestrebungen, den eifrigen Streit unter den Philosophen oder Dichtern: Er war sich der Lichtpunkt dieser verschiedenen Reflexe. Seine Urteile waren alle nur wie zum Spiele flüchtig hingeworfen mit einem nachlässig mystischen Anstrich, und die Frauenzimmer erstaunten nicht über das, was er sagte, sondern was er, in der Überzeugung nicht verstanden zu werden, zu verschweigen schien.
Wenn dieser heimlich die Meinung zu regieren schien, so führte dagegen ein anderer fast einzig das hohe Wort. Es war ein junger voller Mensch mit strotzender Gesundheit,

ein Antlitz, das vor wohlbehaglicher Selbstgefälligkeit glänzte und strahlte. Er wußte für jedes Ding ein hohes Schwungwort, lobte und tadelte ohne Maß und sprach hastig mit einer durchdringenden gellenden Stimme. Er schien ein wütend Begeisterter von Profession und ließ sich von den Frauenzimmern, denen er sehr gewogen schien, gern den heiligen Thyrsusschwinger nennen. Es fehlte ihm dabei nicht an einer gewissen schlauen Miene, womit er niedrere, nicht so saftige Naturen seiner Ironie preiszugeben pflegte. Friedrich wußte gar nicht, wohin dieser während seiner Deklamationen so viel Liebesblicke verschwende, bis er endlich ihm gerade gegenüber einen großen Wandspiegel entdeckte. Der Begeisterte ließ sich übrigens nicht lange bitten, etwas von seinen Poesien mitzuteilen. Er las eine lange Dithyrambe von Gott, Himmel, Hölle, Erde und dem Karfunkelstein mit angestrengtester Heftigkeit vor, und schloß mit solchem Schrei und Nachdruck, daß er ganz blau im Gesicht wurde. Die Damen waren ganz außer sich über die heroische Kraft des Gedichts, sowie des Vortrags.
Ein anderer junger Dichter von mehr schmachtendem Ansehen, der neben der Frau vom Hause seinen Wohnsitz aufgeschlagen hatte, lobte zwar auch mit, warf aber dabei einige durchbohrende neidische Blicke auf den vom Lesen erschöpften Begeisterten. Überhaupt war dieser Friedrich schon vom Anfang an durch seinen großen Unterschied von jenen beiden Flausenmachern aufgefallen. Er hatte sich während der ganzen Zeit, ohne sich um die Verhandlungen der andern zu bekümmern, ausschließlich mit der Frau vom Hause unterhalten, mit der er eine Seele zu sein schien, wie man von dem süßen zugespitzten Munde beider abnehmen konnte, und Friedrich hörte nur manchmal einzelne Laute, wie: ‚Mein ganzes Leben wird zum Roman' – ‚überschwengliches Gemüt' – ‚Priesterleben' – herüberschallen. Endlich zog auch dieser ein ungeheures Paket aus der Tasche, und begann vorzulesen, unter andern folgendes Assonanzenlied:

‚Hat nun Lenz die silbern'n Bronnen
Losgebunden:
Knie ich nieder, süßbeklommen,
In die Wunder.

Himmelreich, so kommt geschwommen
Auf die Wunden
Hast du einzig mich erkoren
Zu den Wundern?

In die Ferne süß verloren
Lieder fluten,
Daß sie, rückwärts sanft erschollen,
Bringen Kunde.

Was die andern sorgen, wollen,
Ist mir dunkel,
Mir will ew'ger Durst nur frommen
Nach dem Durste.

Was ich liebte und vernommen,
Was geklungen,
Ist den eignen tiefen Wonnen
Selig Wunder!'

Er las noch einen Haufen Sonette mit einer Art von priesterlicher Feierlichkeit. Keinem derselben fehlte es an irgendeinem wirklich aufrichtigen kleinen Gefühlchen, an großen Ausdrücken und lieblichen Bildern. Alle hatten einen einzigen, bis ins Unendliche breit auseinandergeschlagenen Gedanken, sie bezogen sich alle auf den Beruf des Dichters und die Göttlichkeit der Poesie; aber die Poesie selber, das ursprüngliche, freie, tüchtige Leben, das uns ergreift ehe wir darüber sprechen, kam nicht zum Vorschein vor lauter Komplimenten davor und Anstalten dazu. Friedrich kamen diese Poesien in ihrer durchaus polierten, glänzenden, wohlerzogenen Weichlichkeit wie

der fade unerquickliche Teedampf, die zierliche Teekanne mit ihrem lodernden Spiritus auf dem Tische wie der Opferaltar dieser Musen vor. – Es ist aber eigentlich nichts künstlicher und lustiger, als die Unterhaltung einer solchen Gesellschaft. Was das Ganze noch so leidlich zusammenhält, sind tausend feine, fast unsichtbare Fäden von Eitelkeit, Lob und Gegenlob usw., und sie nennen es dann gar zu gern ein Liebesnetz. Arbeitet aber unverhofft einmal einer, der davon nichts weiß, tüchtig darin herum, so geht die ganze Spinnewebe von ewiger Freundschaft und heiligem Bunde auseinander.
So hatte auch heute Friedrich den ganzen Tee versalzen. Keiner konnte das künstlerische Weberschiffchen, das sonst fein im Takte so zarte ästhetische Abende wob, wieder recht in Gang bringen. Die meisten wurden mißlaunisch, keiner konnte oder mochte, wie beim babylonischen Baue, des anderen Wortgepräng verstehen, und so beleidigte einer den andern in der gänzlichen Verwirrung. Mehrere Herren nahmen endlich unwillig Abschied, die Gesellschaft wurde kleiner und vereinzelter. Die Damen gruppierten sich hin und wieder auf den Ottomanen in malerischen und ziemlich unanständigen Stellungen. Friedrich bemerkte bald ein heimliches Verständnis zwischen der Frau vom Hause und dem Schmachtenden. Doch glaubte er zugleich an ihr ein feines Liebäugeln zu entdecken, das ihm selber zu gelten schien. Er fand sie überhaupt viel schlauer, als man anfänglich ihrer lispelnden Sanftmut hätte zutrauen mögen; sie schien ihren schmachtenden Liebhaber bei weitem zu übersehen und selber nicht so viel von ihm zu halten, als sie vorgab und er aus ganzer Seele glaubte."
Als aber Friedrich späterhin, noch ganz entrüstet, dieses Abenteuer einem Freunde erzählt, erwidert dieser: „Ich kann dir im Gegenteil versichern, daß ich nicht bald so lustig war, als an jenem Abende, da ich zum ersten Male in diese Teetaufe oder Traufe geriet. Aller Augen waren prüfend und in erwartungsvoller Stille auf mich neuen Jünger

gerichtet. Da ich die ganze heilige Synode, gleich den Freimaurern mit Schurz und Kelle, so feierlich im poetischen Ornate dasitzen sah, konnt ich mich nicht enthalten, despektierlich von der Poesie zu sprechen und mit unermüdlichem Eifer ein Gespräch von der Landwirtschaft, von Runkelrüben usw. anzuspinnen, so daß die Damen wie über den Dampf von Kuhmist die Nasen rümpften und mich bald für verloren hielten. Mit dem Schmachtenden unterhielt ich mich besonders viel. Er ist ein guter Kerl, aber er hat keine Mannesmuskel im Leibe. Ich weiß nicht, was er gerade damals für eine fixe Idee von der Dichtkunst im Kopfe hatte, aber er las ein Gedicht vor, wovon ich trotz der größten Anstrengung nichts verstand und wobei mir unaufhörlich des simplizianisch-deutschen Michels verstümmeltes Sprachgepränge im Sinne lag. Denn es waren deutsche Worte, spanische Konstruktionen, welsche Bilder, altteutsche Redensarten, doch alles mit überaus feinem Firnis von Sanftmut verschmiert. Ich gab ihm ernsthaft den Rat, alle Morgen gepfefferten Schnaps zu nehmen, denn der ewige Nektar erschlaffe nur den Magen, worüber er sich entrüstet von mir wandte. – Mit dem vom Hochmutsteufel besessenen Dithyrambisten aber bestand ich den schönsten Strauß. Er hatte mit pfiffiger Miene alle Segel seines Witzes aufgespannt und kam mit vollem Winde der Eitelkeit auf mich losgefahren, um mich Unpoetischen vor den Augen der Damen in den Grund zu bugsieren. Um mich zu retten, fing ich zum Beweise meiner poetischen Belesenheit an, aus Shakespeares ‚Was ihr wollt‘ wo Junker Tobias den Malvolio peinigt, zu rezitieren: ‚Und besäße ihn eine Legion selbst, so will ich ihn doch anreden.‘ Er stutzte und fragte mich mit herablassender Genügsamkeit und kniffigem Gesichte, ob vielleicht gar Shakespeare mein Lieblingsautor sei? Ich ließ mich aber nicht stören, sondern fuhr mit Junker Tobias fort: ‚Ei Freund, leistet dem Teufel Widerstand, er ist der Erbfeind der Menschenkinder.‘ Er fing nun an, sehr salbungsvolle, genialische Worte über

Shakespeare ergehen zu lassen, ich aber, da ich ihn sich so aufblasen sah, sagte weiter: ‚Sanftmütig, sanftmütig! Ei, was machst du, mein Täubchen? Wie geht's, mein Puthühnchen? Ei, sieh doch, komm, tuck tuck!' – Er schien nun mit Malvolio zu bemerken, daß er nicht in meine Sphäre gehöre, und kehrte sich mit einem unsäglich stolzen Blicke, wie von einem unerhört Tollen, von mir. Das Schlimmste war aber nun, daß ich dadurch demaskiert war, ich konnte nicht länger für einen Ignoranten gelten; und die Frauenzimmer merkten dies nicht so bald, als sie mit allerhand Phrasen, die sie da und dort ernascht, über mich herfielen. In der Angst fing ich daher nun an, wütend mit gelehrten Redensarten und poetischen Paradoxen nach allen Seiten um mich herumzuwerfen, bis sie mich, ich sie, und ich mich selber nicht mehr verstand und alles verwirrt wurde. Seit dieser Zeit haßte mich der ganze Zirkel und hat mich als eine Pest der Poesie förmlich exkommuniziert." —
Es ist sehr begreiflich, daß dieses prätentiöse Unwesen von den Gedankenlosen und Schwachmütigen für die wirkliche Romantik gehalten, von den Hämischen aber gern benutzt wurde, den neuen Aufschwung überhaupt zu verketzern. Vergebens verspottete Tieck selbst in den wenigen Nummern seines „Poetischen Journals" jene falsche Romantik, vergebens zogen Arnim und Görres mitten durch den Lärm neue leuchtende Bahnen; das Gekläff der Wächter des guten Geschmacks, die den Mond anbellen und bei Musik heulen, war einmal unaufhaltsam erwacht. Es erschien ein „Klingkling-Almanach", der die Lyrik der Romantiker parodisch lächerlich machen sollte, aber durch ein stupides Mißverständnis des Parodierten nur sich selbst blamierte. Der Däne Baggesen schrieb einen „Faust", eine Komödie, worin Fichte, Schelling, Schlegel und Tieck die lächerlichen Personen spielen; an Witzlosigkeit, Bosheit und Langweiligkeit, etwa Nicolais „Werthers Leiden" vergleichbar. Garlieb Merkel endlich trommelte in seinem „Freimütigen" ein wahres Falstaffs-

heer zusammen, allerdings freimütig genug, denn die armutselige Gemeinheit lag ganz offen zutage. In Heidelberg selbst aber saß der alte Voß, der sich bereits überlebt hatte, und darüber ganz grämlich geworden war. Mitten in dem staubigen Gewebe seiner Gelehrsamkeit lauerte er wie eine ungesellige Spinne, tückisch auf alles Junge und Neue zufahrend, das sich unvorsichtig dem Gespinste zu nähern unterfing. Besonders waren ihm, nebst dem Katholizismus, die Sonette verhaßt. Daher konnte Arnim, obgleich er anfangs aus großmütiger Pietät mit dem vereinsamten Greise friedlich zu verkehren suchte, dennoch zuletzt nicht umhin, ihm zu Ehren in der „Einsiedlerzeitung" in hundert Sonetten den Kampf des Sonetts mit dem alten Drachen zu beschreiben.
Und auf ähnliche Weise hatte sich die Romantik überhaupt ihren Gegnern gegenübergestellt, indem sie – wie in Tiecks „Verkehrter Welt", im „Zerbino" und „Gestiefelten Kater", in Schlegels „Triumphpforte für den Theaterpräsidenten Kotzebue", in Mahlmanns „Hussiten vor Naumburg" – jenes hämische Treiben heiter als bloßes Material nahm und humoristisch der Poesie selbst dienstbar zu machen wußte.
Aber die Romantik war keine bloß literarische Erscheinung, sie unternahm vielmehr eine innere Regeneration des Gesamtlebens, wie sie Novalis angekündigt hat; und was man später die romantische Schule nannte, war eben nur ein literarisch abgesonderter Zweig des schon kränkelnden Baumes. Ihre ursprüngliche Intentionen, alles Irdische auf ein Höheres, das Diesseits auf ein größeres Jenseits zu beziehen, mußten daher insbesondere auch das ganze Gebiet der Kunst gleichmäßig umfassen und durchdringen. Die Revolution, die sie in der Poesie bewirkt, ist schon zu vielfach besprochen, um hier noch besonders erörtert zu werden. Der Malerei vindizierte sie die Schönheit der Religion als höchste Aufgabe, und begründete durch deutsche Jünglinge in Rom die bekannte Malerschule, deren Führer Overbeck, Philipp Veit und Corne-

lius waren. Derselbe ernstere Sinn führte die Tonkunst vom frivolen Sinnenkitzel zur Kirche, zu den altitalienischen Meistern, zu Sebastian Bach, Gluck und Händel zurück; er weckte auch in der Profanmusik das geheimnisvolle wunderbare Lied, das verborgen in allen Dingen schlummert, und Mozart, Beethoven und Karl Maria von Weber sind echte Romantiker. Die Baukunst endlich, diese hieroglyphische Lapidarschrift der wechselnden Nationalbildung, war grade in das allgemeine Stadium der damaligen Literatur mit eingerückt: kaserniertes Bürgerwohl mit heidnischen Substruktionen, die Antike im Schlafrock des häuslichen Familienglücks. Da erfaßte plötzlich die erstaunten Deutschen wieder eine Ahnung von der Schönheit und symbolischen Bedeutung ihrer alten Bauwerke, an denen sie so lange gleichgültig vorübergegangen. Der junge Goethe hatte zuerst vom Straßburger Münster den neuen Tag ausgerufen, sich aber leider dabei so bedeutend überschrien, daß er seitdem ziemlich heiser blieb. Besonnener und gründlicher wies Sulpice Boisserée auf den Riesengeist des Kölner Domes hin, der bekanntlich noch bis heut sein mühseliges Auferstehungsfest feiert. – Das augenfälligste Bild dieser Umwandlung aber gibt die Geschichte der Marienburg, des Haupthauses des deutschen Ritterordens in Preußen. Dieser merkwürdige Bau hatte nicht einmal die Genugtuung, in malerische Trümmer zerfallen zu dürfen, er wurde methodisch für den neuen Orden der Industrieritter verstümmelt und zugerichtet. Die kühnen Gewölbe wurden mit unsäglicher Mühe eingeschlagen, in den hohen luftigen Sälen drei niedrige Stockwerke schmutziger Weberwerkstätten eingeklebt; ja um den letzten Prachtgiebel des Schlosses waren bereits die Stricke geschlungen, um ihn niederzureißen, als ein Romantiker, Max von Schenkendorf, ganz unerwartet in einer vielgelesenen Zeitschrift Protest einlegte gegen diesen modernen Vandalismus, den der damalige Minister von Schrötter, ein sonst geistvoller und für alles Große empfänglicher Mann, im Namen der Aufklä-

rung als ein löblich Unternehmen trieb. Jetzt veränderte sich plötzlich die Szene. Schrötter, da er seinen wohlgemeinten Mißverstand begriff, hieß, fast erschrocken darüber, sofort alle weitere Zerstörung einstellen, die Weber wurden ausgetrieben, Künstler, Altertumsfreunde und Techniker stiegen verwundert in den rätselhaft gewordenen Bau hinab, wie in einem Bergwerke dort ein Fenster, hier einen verborgnen Gang oder Remter entdeckend, und je mehr allmählich von der alten Pracht zutage kam, je mehr wuchs, erst in der Provinz, dann in immer weiteren Kreisen der Enthusiasmus, und erweckte, soviel davon noch zu retten war, das wunderbare Bauwerk aus seinem jahrhundertelangen Zauberschlaf.
Ein ähnliches Bewandtnis beinah hatte es mit dem Einfluß der Romantik auf die religiöse Stimmung der Jugend, indem sie gleichfalls den halbvergessenen Wunderbau der alten Kirche aus seinem Schutte wieder emporzuheben strebte. Allein was dort genügte, konnte hier unmöglich ausreichen, denn die Romantiker, wenn wir Novalis, Görres und Friedrich Schlegel ausnehmen, taten es nicht um der Religion, sondern um der Kunst willen, für die ihnen der Protestantismus allzu geringe Ausbeute bot; ein Grundthema, das in „Sternbalds Wanderungen", in Tiecks „Phantasien" und in den „Herzensergießungen eines kunstliebenden Klosterbruders" durch die ganze Klaviatur der Künste hindurch auf das anmutigste variiert ist. Wir wollen daher auf die Konversion einiger, durch die Musik, die Pracht des äußeren Gottesdienstes u. dgl. m. bekehrter protestantischer Jünglinge keineswegs ein besonderes Gewicht legen. Der ganze Hergang aber erinnert lebhaft an Schillers Grundsatz von der ästhetischen Erziehung des Menschengeschlechts; wir meinen die indirekte Macht, welche diese katholisierende Ästhetik auf die katholische Jugend selber ausgeübt. Es ist nicht zu leugnen, ein großer Teil dieser, fast überall protestantisch geschulten Jugend ist in der Tat durch die Vorhalle der Romantik zur Kirche zurückgekehrt. Die katholi-

schen Studenten, die überhaupt etwas wollten und konnten, erstaunten nicht wenig, als sie in jenen Schriften auf einmal die Schönheit ihrer Religion erkannten, die sie bisher nur geschmäht oder mitleidig belächelt gesehen. Der Widerspruch, in den sie durch diese Entdeckung mit der gemeinen Menge gerieten, entzündete ihren Eifer, voll Begeisterung brachten sie die alt-neue Lehre von der Universität mit nach Hause, ja sie kokettierten zum Teil damit in der Philisterwelt, wo man über die jungen Zeloten verwundert den Kopf schüttelte; mit einem Wort: Das Katholische wurde förmlich Mode. Die Mode ging nach Art aller Mode bald vorüber, aber der einmal angeschlagene Ton blieb und hallte in immer weiteren Kreisen nach, und daraus entstand im Verlauf der immer ernster werdenden Zeiten endlich wieder eine starke katholische Gesinnung, die der Romantik nicht mehr bedarf.

So war die Romantik bei ihrem Aufgange ein Frühlingshauch, der alle verborgenen Keime belebte, eine schöne Zeit des Erwachens, der Erwartung und Verheißung. Allein sie hat die Verheißung nicht erfüllt, und weil sie sie nicht erfüllte, ging sie unter, und wie und warum dies geschehen mußte, haben wir bereits an einem anderen Orte ausführlich nachzuweisen versucht. Als jedoch auf solche Weise die Ebbe kam und jene Springfluten zurücktobten, wurde auch der alte Boden wieder trockengelegt, den man für neuentdecktes Land hielt. Der zähe Rationalismus, die altkluge Verachtung des Mittelalters, die Lehre von der alleinseligmachenden Nützlichkeit, wozu die sublime Wissenschaft nicht sonderlich nötig sei; all das vorromantische Ungeziefer, das sich unterdes im Sande eingewühlt, kam jetzt wieder zum Vorschein und heckte erstaunlich. Dennoch war aber der bloßgelegte Boden nicht mehr ganz derselbe. Die Romantik hatte einige unvertilgbare Spuren darauf hinterlassen; sie hatte durch ihr beständiges Hinweisen auf die nationale Vergangenheit die Vaterlandsliebe, durch ihren Experimental-Katholizismus ein religiöses Bedürfnis erweckt. Allein diese Vaterlandsliebe war

durch die abermalige Trennung vom Mittelalter ihres historischen Bodens und aller nationalen Färbung beraubt, und so entstand aus dem alten abstrakten Weltbürgertum die ebenso abstrakte Deutschtümelei. Andrerseits konnte das wiederangeregte religiöse Gefühl natürlicherweise weder von dem romantischen Katholisieren, noch von dem wiedererstandenen Rationalismus befriediget werden, und flüchtete sich daher bei den Protestanten zu dem neuesten Pietismus.

Von diesen veränderten Zuständen mußten denn auch zunächst die Universitäten wieder berührt werden; sie verloren allmählich ihr mittelalterliches Kostüm und suchten sich der modernen Gegenwart möglichst zu akkommodieren. Das deutsche Universitätsleben war bis dahin im Grunde ein lustiger Mummenschanz, in exzeptioneller Maskenfreiheit die übrige Welt neckend, herausfordernd und parodierend; eine Art harmloser Humoristik, die der Jugend, weil sie ihr natürlich ist, großenteils gar wohl anstand. Jetzt dagegen, durch die halbe Schulweisheit und Vielwisserei aufgeblasen, und von der epidemischen neuen Altklugheit mit fortgerissen, begnügten sie sich nicht mehr, sich an den dünkelhaften Torheiten der Philisterwelt lachend zu ergötzen; sie wollten sich *über* die Welt stellen, sie meistern und vernünftiger einrichten. Dazu kam, daß sie in den Befreiungskriegen wirklich auf dem Welttheater rühmlich mitagiert hatten, und nun auch das Recht beanspruchten, die übrigen Akte des großen Weltdramas mit fortzuspielen, mit einem Worte: Politik zu machen. Das war aber höchst unpolitisch, denn auf dieser komplizierten Bühne fehlte es glücklicherweise der Jugend durchaus an der unerläßlichen Kenntnis, Erfahrung und Routine. Die Burschenschaften, die zunächst aus jener inneren Umwandlung der Universitäten hervorgingen, waren ohne allen Zweifel ursprünglich gut und ernst gemeint und mit einem nicht genug zu würdigenden moralischen Stoizismus gegen die alte Roheit und Sittenlosigkeit gerichtet. Anstatt aber nur erst sich selbst gehörig zu

befestigen, wollten sie sehr bald im leicht erklärlichen Eifer des guten Gewissens auch die kranken Staaten durch utopische Weltverbesserungspläne regenerieren, die man am füglichsten als unschädliche Donquijotiaden hätte übersehen sollen, wenn sich nicht, wie es scheint, nun die wirklichen Politiker mit dareingemischt, und die jugendliche Unbefangenheit für ihre ehrgeizigen und unlauteren Zwecke gemißbraucht hätten. Und so wurden die Studenten, die so lange heiter die Welt düpiert hatten, nun selber von der undankbaren Welt düpiert.

Als ein anderes Symptom der neuesten Zeit haben wir vorhin den bei den Protestanten wiedererwachten Pietismus bezeichnet. Man könnte ihn, da er wesentlich auf der subjektiven Gefühlsauffassung beruht, füglich die Sentimentalität der Religion nennen. Daher der absonderliche Haß der Pietisten gegen das strenge positive Prinzip der Kirche, die von einem subjektiven Dafürhalten und Umdeuten der Glaubenswahrheiten nichts weiß. Dieser moderne Pietismus ist jetzt auf den deutschen Universitäten sehr zahlreich vertreten, nicht eben zum sonderlichen Heile der Jugend. Denn der nackte Rationalismus war an sich so arm, trocken und trostlos, daß er ein tüchtiges Gemüt von selbst zur resoluten Umkehr trieb. Der weichliche, sanft einschmeichelnde Pietismus dagegen, zumal wenn er Mode wird und zeitliche Vorteile in Aussicht stellt, erzeugt gar leicht heuchlerische Tartüffe, oder, wo er tiefer gegriffen, einen geistlichen Dünkel und Fanatismus, der das ganze folgende Leben vergiftet. Eine Sekte dieser Pietisten gefällt sich darin, grundsätzlich allen Zweikampf abzulehnen, und sich dies als einen Akt besonderen Mutes anzurechnen. Allein dieser passive Mut, die gemeine Meinung zu verachten und gelassen über sich ergehen zu lassen, ist noch sehr verschieden von der persönlichen Tapferkeit, die jeden Jüngling ziert. Es ist ganz löblich, aber noch lange nicht genug, das Unrechte hinter dem breiten Schilde der vortrefflichsten Grundsätze von sich selber abzuwehren; das Böse soll direkt *bekämpft*

werden. Überhaupt aber darf hierbei nicht übersehen werden, daß dem Zweikampf ein an sich sehr ehrenwertes Motiv zum Grunde liegt: das der gesunden Jugend eigentümliche, spartanische Gerechtigkeitsgefühl, das sich ohne innere Einbuße nicht unterdrücken läßt. Es gibt fast unsichtbare Kränkungen, infam, perfid und boshaft, die bis in das innerste Mark verwunden, und doch, eben weil sie juridisch ungreifbar sind, vom Gesetz nicht vorgesehen werden können. Dies ist der eigentliche Sitz des Übels, der Kampfplatz, wo der Zweikampf, wie früher die Gottesgerichte, ausgleichend eintritt. Dasselbe gilt im großen auch von den Kriegen, diesen barbarischen Völkerduellen um Güter, die das materielle Staatsrecht nicht zu würdigen und zu schützen vermag, und zu denen wir namentlich die Nationalehre rechnen. – Demungeachtet sind wir weit entfernt, die ganz unchristliche Selbsthilfe des Zweikampfs irgendwie verteidigen zu wollen, wünschen vielmehr vorerst nur eine genügende Vermittelung und Beseitigung seines tieferen Grundes, ohne welche, nach menschlichem Ermessen, alle Verbotsgesetze dagegen stets illusorisch bleiben werden.
Mit der neuen Umwandelung des Zeitgeistes hängt auch der Grundsatz wesentlich zusammen, die Universitäten möglichst in die großen Residenzstädte zu verlegen. Wir wollen keineswegs in Abrede stellen, daß die großen Städte mit ihrem geselligen Verkehr, mit ihren Kunstschätzen, Bibliotheken, Museen und industriellen Anstalten eine sehr bequeme Umschau, eine wahre Universitas alles Wissenswürdigen bieten. Allein es frägt sich nur, ob dieser Vorteil nicht etwa durch Nachteile anderer Art wieder neutralisiert, ja überwogen wird? Uns wenigstens scheint das alles mehr für die Professoren, als für die Studenten geeignet zu sein. Es kommt für die letzteren auf der Universität doch vorzüglich nur auf eine Orientierung in dem Labyrinth der neuen Bildung an. Auf jenen großen Stapelplätzen der Kunst und Wissenschaft aber erdrückt und verwirrt die überwältigende Masse des Verschiedenartig-

sten, gleichwie schon jeder Reisende, wenn er eine reiche Bildergalerie hastig durchlaufen hat, zuletzt selbst nicht mehr weiß, was er gesehen; und namentlich die großen Bibliotheken kann nur der Gelehrte, der sich bereits für ein bestimmtes Studium entschieden und gehörig vorbereitet hat, mit Nutzen gebrauchen. Wie aber soll der, für alles gleich empfängliche Jüngling mitten zwischen den nach allen Seiten auslaufenden Bahnen sich wahrhaft entscheiden, wo jedes natürliche Verhältnis zwischen Lehrer und Schüler, wie es in kleinen Universitätsstädten stattfindet, durch den betäubenden Lärm und die allgemeine Zerfahrenheit der Residenz ganz unmöglich wird? Auch hier also droht abermals ein vager Dilettantismus und der lähmende Dünkel der Vielwisserei. Bei der Jugend ist eine kecke Wanderlust, sie ahnt hinter dem Morgenduft die wunderbare Schönheit der Welt; sie sich selbsttätig zu erobern ist ihre Freude. In den großen Städten aber fängt die Jugend gleich mit dem Ende an: aller Reichtum der Welt liegt in der staubigen Mittagschwüle schon wohlgeordnet um sie her, sie braucht ihren Fauteuil nur gähnend da- oder dorthin zu wenden, sie hat nichts mehr zu wünschen und zu ahnen – und ist blasiert. Und auch in sittlicher Hinsicht ist der Gewinn nur illusorisch. In den kleinen Universitätsstädten herrscht allerdings oft eine arge Verwilderung, und die Studenten werden in den großen Städten gewiß ruhiger und manierlicher sein. Allein dort erscheint die Liederlichkeit in der Regel so handgreiflich, bestialisch roh und abschreckend, daß jedes gesunde Gemüt von selbst ein Ekel davor überkommt, während hier die schön übertünchten und ästhetisierten Pestgruben wohl auch die Besseren mit ihrem Gifthauch betäuben. – Unsere Universitäten sind endlich bisher eine Art von Republik gewesen, die einzigen noch übriggebliebenen Trümmer deutscher Einheit, ein brüderlicher Verein ohne Rücksicht auf die Unterschiede der Provinz, des Ranges oder Reichtums, wo den Niedriggeborenen die Überlegenheit des Geistes und Charakters zum Senior über Für-

sten und Grafen erhob. Diese uralte Bedeutung der Universitäten wird von der, in ganz andern Bahnen kreisenden Großstädterei notwendig verwischt, die Studenten werden immer mehr in das allgemeine Philisterium eingefangen und frühzeitig gewöhnt, die Welt diplomatisch mit Glacéhandschuhen anzufassen.

Dies halten wir aber, zumal in unserer materialistischen Zeit, für ein bedeutendes Unglück. Denn was ist denn eigentlich die Jugend? Doch im Grunde nichts anderes, als das noch gesunde und unzerknitterte, vom kleinlichen Treiben der Welt noch unberührte Gefühl der ursprünglichen Freiheit und der Unendlichkeit der Lebensaufgabe. Daher ist die Jugend jederzeit fähiger zu entscheidenden Entschlüssen und Aufopferungen, und steht in der Tat dem Himmel näher, als das müde und abgenutzte Alter; daher legt sie so gern den ungeheuersten Maßstab großer Gedanken und Taten an ihre Zukunft. Ganz recht! denn die geschäftige Welt wird schon dafür sorgen, daß die Bäume nicht in den Himmel wachsen und ihnen die kleine Krämerelle aufdrängen. Die Jugend ist die Poesie des Lebens, und die äußerlich ungebundene und sorgenlose Freiheit der Studenten auf der Universität die bedeutendste Schule dieser Poesie, und man möchte ihr beständig zurufen: sei nur vor allen Dingen *jung!* Denn ohne Blüte keine Frucht.

Zeittafel

1788 *10. März:* Joseph Karl Benedikt Freiherr von Eichendorff auf Schloß Lubowitz bei Ratibor geboren. Sein um zwei Jahre älterer Bruder Wilhelm (geb. am 14. September 1786) wird der enge Gefährte von Eichendorffs Jugend bis zum Abschluß der Studienzeit. Die Brüder werden auf Schloß Lubowitz von einem Hofmeister, dem Pfarrer Bernhard Heinke, unterrichtet.

1791 *27. Dezember:* Geburt der Schwester Henriette Sophie.

1793 *22. Oktober:* Geburt des Bruders August Adolf.

1797 *26. März:* August Adolf stirbt.
7. April: Henriette Sophie stirbt.

1798 Erste Tagebuchaufzeichnungen Eichendorffs. (Viele der nachfolgend aufgeführten genauen Daten entstammen den Tagebüchern, sie sind nur fragmentarisch erhalten.)

1799 *13. Juni:* Geburt der Schwester Luise Antonie Sophie. Die Familie reist nach Dresden, Karlsbad und Prag.

1800 *7. September:* Geburt des Bruders Gustav.
11. November: Eichendorff beginnt eine Naturgeschichte zu schreiben, die er auch illustriert.

1801 *5. Oktober:* Joseph und Wilhelm von Eichen-

dorff reisen mit Pfarrer Heinke nach Breslau. Drei Jahre lang werden sie das Matthias-Gymnasium besuchen und im St.-Josephs-Konvikt wohnen.
27. Dezember: Besuch einer Aufführung von Mozarts „Zauberflöte“.

1802 Schüleraufführungen unter der Beteiligung Eichendorffs.
Erste Gedichte.

1803 *25. April:* Gustav von Eichendorff stirbt. Am 20. Mai erscheint das Gedicht „Am Grabe unseres Bruders“ in den *Schlesischen Provinzialblättern.*
10. September: Luise Antonie Sophie stirbt.

1804 *13. April:* Geburt der Schwester Luise Antonie Nepomucene Johanna. Sie wurde später die Freundin Adalbert Stifters.

1805 *25. März:* Beendigung der Gymnasialzeit.
20. April: Von Lubowitz nach Halle. Beginn des Jurastudiums. Teilnahme an Galls Vorlesung über Schädellehre, die auch Goethe hörte.
3. August: „Götz von Berlichingen“ im nahen Lauchstädt in Anwesenheit Goethes.
20. August: Strafaktion auf dem Marktplatz von Halle. Spießrutenlaufen sog. Revolutionäre.
Im Herbst besuchen die Brüder Hamburg und Lübeck und sehen zum erstenmal das Meer.

1806 *1. August:* Ferienreise nach Lubowitz.
30. Oktober: Nachricht von der Schließung der Universität Halle durch Napoleon.

1807 *4. Mai:* Abreise der Brüder nach Heidelberg, wo sie ihre Studien fortsetzen wollen.
19. Mai: Erstes Zusammentreffen mit Görres.
15. November: Bekanntschaft mit Otto Heinrich Graf von Loeben (Isidorus Orientalis), der großen Anteil an Eichendorffs schriftstellerischen Versuchen nimmt.
Lektüre von „Des Knaben Wunderhorn".

1808 Unter dem Pseudonym „Florens" erscheinen acht Gedichte Eichendorffs in der *Zeitschrift für Wissenschaft und Kunst.* Vermittler war Loeben.
5. April: Die Brüder reisen nach Paris.
4. Mai: Rückkehr der Brüder nach Heidelberg.
12. Mai: Abreise nach Lubowitz.

1809 In Lubowitz schreibt Eichendorff Lieder und Gedichte. Er löst sich vom Stil Loebens.
9.–19. November: Fahrt auf der Oder bis Frankfurt.
20. November: Ankunft der Brüder in Berlin. Bekanntschaft mit Adam Müller, Arnim und Brentano.
3. Dezember: Erkrankung an Nervenfieber.
23. Dezember: Einzug des Königspaares in Berlin. Eichendorff in patriotischem Überschwang.
24. Dezember: Das Nervenfieber verstärkt sich.

1810 *1. Februar:* Brentano bei Eichendorff.
15. Februar: Erster Ausgang als Rekonvaleszent.
2. März: Besuch bei Arnim und Brentano.
In diesem Jahr entstehen sowohl das Fragment eines patriotischen Dramas „Hermann und Thusnelda" als auch mit einiger Wahrscheinlichkeit die Gedichte „Abschied" (O Täler weit, o

Höhen ...) und „Der Jäger Abschied“ (Wer hat dich, du schöner Wald ...).
Seit Mitte März sind die Brüder wieder in Lubowitz, im Oktober reisen sie nach Wien. Hier haben sie Umgang mit Friedrich und Dorothea Schlegel, mit Dorotheas Sohn Philipp Veit und

1811 wiederum mit Adam Müller.

1812 Eichendorff besteht die juristischen Examina durchweg mit Auszeichnung.
Friedrich Schlegels Vorlesung „Über die Geschichte der alten und neuen Literatur“.
Im Spätherbst schließt Eichendorff die Niederschrift von „Ahnung und Gegenwart“ ab.

1813 *17. März:* König Friedrich Wilhelm III. von Preußen erläßt in Breslau den Aufruf „An mein Volk“. Im April geht Eichendorff mit Philipp Veit zum Lützowschen Freikorps. Wilhelm, der Bruder, bleibt in Wien. Hier vollzieht sich die Trennung. Zu seinem Ärger nimmt Eichendorff kaum an kriegerischen Handlungen teil.
7. Oktober: Ernennung zum Leutnant im 17. schlesischen Landwehrregiment.
Freundschaft mit Friedrich de la Motte-Fouqué.

1814 Garnisonsdienst in der Festung Torgau.
März: Immediatgesuch des Offizierskorps, das Regiment an die Front zu beordern.
Mai: Urlaub in Lubowitz;
schließlich Entlassung aus dem Kriegsdienst.
Fouqué schreibt das Vorwort zu „Ahnung und Gegenwart“.

1815 Anstellung als Expedient (Sekretär) beim Oberkriegskommissariat in Berlin auf Empfehlung Gneisenaus.

7. April: Vermählung mit Luise Anna Viktoria von Larisch in Breslau, gegen den Willen der Mutter, die eine vermögende Schwiegertochter vorgezogen hätte.
22. April: Eichendorff will Soldat bleiben. Er reist zum Blücherschen Hauptquartier nach Lüttich.
11. Mai: In Aachen Mitwirkung bei der Aufstellung von Landwehrregimentern in den neuerworbenen rheinischen Provinzen.
7. Juli: Als Ordonnanzoffizier bei Gneisenau in Paris.
Ohne sein Wissen ist inzwischen „Ahnung und Gegenwart" im Druck erschienen, bei Johann Leonhard Schrag in Nürnberg.
30. August: Geburt des Sohnes Hermann.

1816 *12. Januar:* Auflösung des Regiments und Rückkehr nach Lubowitz.
24. Dezember: Anstellung als Referendar bei der Breslauer Regierung.

1817 Abschluß der Niederschrift von „Das Marmorbild" und Arbeit am „Taugenichts".
Wiedersehen mit dem Bruder Wilhelm in Lubowitz.
Geburt des Sohnes Rudolf.

1818 *27. April:* Tod des Vaters.
Die verschuldeten Familiengüter werden verkauft, vorläufig ohne Lubowitz und ohne Sedlnitz.

1819 Geburt der Tochter Therese.
Fouqué bringt „Das Marmorbild" in seinem *Frauentaschenbuch für das Jahr 1819.*
16. Oktober: Große Staatsprüfung in Berlin.

24. November: Ernennung zum Regierungsassessor.

1820 Wiedersehen mit dem Bruder Wilhelm in Wien.
14. Oktober: Berufung als sog. Hilfsarbeiter beim Kultusministerium in Berlin.
22. Dezember: Übertragung der kommissarischen Verwaltung der Geschäfte eines katholischen Konsistorial- und Schulrats beim Oberpräsidium und Konsistorium der Provinz Westpreußen sowie bei den Regierungen zu Danzig und Marienwerder.

1821 *21. Januar:* Abreise nach Danzig.
Freundschaftliches, wenn auch stets distanziertes Verhältnis zu Theodor von Schön, dem preußischen Oberpräsidenten.
Der „Taugenichts" ist abgeschlossen.
5. September: Ernennung zum Regierungsrat.
Geburt der Tochter Agnes. Sie stirbt im nächsten Jahr.

1822 *15. April:* Tod der Mutter.
Nun wird auch Lubowitz verkauft. Nur Sedlnitz bleibt in der Familie.

1823 In Berlin zur Vertretung ins Kultusministerium berufen.
Das erste Kapitel des „Taugenichts" erscheint in den *Deutschen Blättern für Poesie, Literatur, Kunst und Theater* ebenso wie das erste Abenteuer aus „Krieg den Philistern". Auf Anregung des Bischofs von Ermland Prinz Joseph von Hohenzollern entsteht das Gedicht „O Maria, meine Liebe".

1824 Als Oberpräsidialrat und Mitglied der Ostpreußischen Regierung (mit einem Gehalt von nun-

mehr 1500 Reichstalern) nach Königsberg, zusammen mit Theodor von Schön.

1826 „Das Marmorbild“ und „Aus dem Leben eines Taugenichts“ erscheinen mit einem Anhang von 48 Gedichten.

1828 „Ezelin von Romano“ erscheint.

1830 Geburt der Tochter Anna Hedwig Josephine, die bereits im zweiten Lebensjahr stirbt („Auf meines Kindes Tod“).
Das Schauspiel „Der letzte Held von Marienburg“ erscheint im Druck.

1831 *6. Juni:* Zur Vertretung im Kultusministerium in Berlin.
Eichendorff bemüht sich um seine Versetzung aus Königsberg.
Besuch der Berliner Mittwochsgesellschaft, einem literarisch-musischen Club, den auch Chamisso, Fouqué, Holtei und andere besuchen.

1832–1837 Mehrere Prosarbeiten werden fertig und erscheinen: „Viel Lärmen um Nichts“ in Gubitzens *Gesellschafter* (1832), „Dichter und ihre Gesellen“ bei Duncker und Humblot (1834), „Das Schloß Dürande“ (1837) in Brockhausens *Urania.* Abgeschlossen wurden in diesen Jahren auch das Theaterstück „Die Freier“ (1833), die Satire „Auch ich war in Arkadien“ (1832) und die Erzählung „Die Entführung“ (1839 in der *Urania* abgedruckt).
1837 erscheint, wiederum bei Duncker und Humblot in Berlin, die erste Gesamtausgabe der Gedichte. Bereits in diesen Jahren begann

Eichendorffs Beschäftigung mit der spanischen Literatur.

1838 Reise nach München und Wien.

1841 *12. Januar:* Ernennung zum Geheimen Regierungsrat. „Die Glücksritter“ erscheinen im *Rhein. Jahrbuch für Kunst und Poesie.*
Die erste Werkausgabe.

1843 Eichendorff erkrankt an einer Lungenentzündung. Nach seiner Genesung nimmt er Urlaub für historische Studien im Zusammenhang mit dem Auftrag des Königs, eine Geschichte der Marienburg zu schreiben.
Gleichzeitig sucht Eichendorff um seine Entlassung aus dem Staatsdienst nach.

1844 *30. Juni:* Entlassung aus dem Staatsdienst. In Königsberg erscheint „Die Wiederherstellung des Schlosses der deutschen Ordensritter zu Marienburg“.

1845 Zusammenkunft mit dem Bruder Wilhelm auf Gut Sedlnitz.
Calderon-Übersetzungen.

1846 Eichendorff zieht von Danzig nach Wien.
Lebhafter Austausch mit der Wiener Literaten-Gesellschaft. Er trifft unter anderen auch Robert und Clara Schumann hier an, neben Grillparzer und Bauernfeld und Stifter und wie sie alle heißen. Anregend vor allem der Umgang mit Karl Ernst Jarcke, für dessen *Historisch-politische Blätter* Eichendorff fortan schreibt. Zunächst: „Zur Geschichte der neuern romantischen Poesie in Deutschland“.

1847 Im Herbst Übersiedlung nach Berlin (auf dem Umweg über Danzig).
Eichendorff schreibt weiter an seinen literarhistorischen Schriften.

1848 Die Unruhen und Straßenkämpfe in Berlin, die am 18. März ausbrechen, veranlassen Eichendorff, sich mit seiner Familie zunächst nach Köthen und von dort nach Dresden zurückzuziehen.
Bekanntschaft mit dem Spät-Romantiker Leberecht Dreves, dessen Werke er ein Jahr später herausgibt.

1849 Tod des Bruders Wilhelm in Innsbruck.
Die Revolutionssatire „Libertas und ihre Freier" entsteht.

1850 Rückkehr nach Berlin.

1851 „Der deutsche Roman des 18. Jahrhunderts in seinem Verhältnis zum Christentum" erscheint.

1852 Beginn der Ausarbeitung des „Julian".

1853 Verleihung des bayerischen Maximiliansordens für Wissenschaft und Kunst.
„Julian" und der zweite Band der „Geistlichen Schauspiele" von Calderon erscheinen.

1854 „Zur Geschichte des Dramas" erscheint.

1855 Eichendorffs Frau erkrankt schwer. Erholungsaufenthalt in Köthen.
Von Anfang Juni bis Mitte Juli Kuraufenthalt in Karlsbad.
„Robert und Guiscard" erscheint.
Übersiedlung nach Neiße.

Joseph von Eichendorff

Ausgewählte Werke
Band 1

Joseph von Eichendorff

Ausgewählte Werke

Band 1

Herausgegeben
und mit einem Nachwort von
Hans A. Neunzig

nymphenburger

© Nymphenburger Verlagshandlung GmbH, München 1987
Alle Rechte, auch der photomechanischen Vervielfältigung
und des auszugsweisen Abdrucks, vorbehalten
Umschlaggestaltung: Cooperation, München,
unter Verwendung von Gemälden von Caspar David Friedrich
(Archiv für Kunst und Geschichte, Berlin)
Satz: Fotosatz Völkl, Germering
Druck und Bindung: Wiener Verlag, Himberg
Printed in Austria
ISBN 3-485-00554-1

INHALT

I. Wanderlieder

Viele Boten gehn und gingen
Zwischen Erd' und Himmelslust
Solchen Gruß kann keiner bringen,
Als ein Lied aus frischer Brust.

Frische Fahrt

Laue Luft kommt blau geflossen,
Frühling, Frühling soll es sein!
Waldwärts Hörnerklang geschossen,
Mut'ger Augen lichter Schein;
Und das Wirren bunt und bunter
Wird ein magisch wilder Fluß,
In die schöne Welt hinunter
Lockt dich dieses Stromes Gruß.

Und ich mag mich nicht bewahren!
Weit von euch treibt mich der Wind,
Auf dem Strome will ich fahren,
Von dem Glanze selig blind!
Tausend Stimmen lockend schlagen,
Hoch Aurora flammend weht,
Fahre zu! Ich mag nicht fragen,
Wo die Fahrt zu Ende geht!

Allgemeines Wandern

Vom Grund bis zu den Gipfeln,
Soweit man sehen kann,
Jetzt blüht's in allen Wipfeln,
Nun geht das Wandern an:

Die Quellen von den Klüften,
Die Ström' auf grünem Plan,
Die Lerchen hoch in Lüften,
Der Dichter frisch voran.

Und die im Tal verderben
In trüber Sorgen Haft,
Er möcht' sie alle werben
Zu dieser Wanderschaft.

Und von den Bergen nieder
Erschallt sein Lied ins Tal,
Und die zerstreuten Brüder
Faßt Heimweh allzumal.

Da wird die Welt so munter
Und nimmt die Reiseschuh',
Sein Liebchen mitten drunter
Die nickt ihm heimlich zu.

Und über Felsenwände
Und auf dem grünen Plan
Das wirrt und jauchzt ohn' Ende –
Nun geht das Wandern an!

Der frohe Wandersmann

Wem Gott will rechte Gunst erweisen,
Den schickt er in die weite Welt;
Dem will er seine Wunder weisen
In Berg und Wald und Strom und Feld.

Die Trägen, die zu Hause liegen,
Erquicket nicht das Morgenrot,
Sie wissen nur von Kinderwiegen,
Von Sorgen, Last und Not und Brot.

Die Bächlein von den Bergen springen,
Die Lerchen schwirren hoch vor Lust,
Was sollt' ich nicht mit ihnen singen
Aus voller Kehl' und frischer Brust?

Den lieben Gott laß' ich nur walten;
Der Bächlein, Lerchen, Wald und Feld
Und Erd und Himmel will erhalten,
Hat auch mein' Sach' aufs best' bestellt!

Im Walde

Es zog eine Hochzeit den Berg entlang,
Ich hörte die Vögel schlagen,
Da blitzten viel Reiter, das Waldhorn klang,
Das war ein lustiges Jagen!

Und eh' ich's gedacht, war alles verhallt,
Die Nacht bedecket die Runde,
Nur von den Bergen noch rauschet der Wald
Und mich schauert im Herzensgrunde.

Zwielicht

Dämmrung will die Flügel spreiten,
Schaurig rühren sich die Bäume,
Wolken ziehn wie schwere Träume –
Was will dieses Graun bedeuten?

Hast ein Reh du lieb vor andern,
Laß es nicht alleine grasen,
Jäger ziehn im Wald und blasen,
Stimmen hin und wieder wandern.

Hast du einen Freund hienieden,
Trau ihm nicht zu dieser Stunde,
Freundlich wohl mit Aug und Munde,
Sinnt er Krieg im tückschen Frieden.

Was heut müde gehet unter,
Hebt sich morgen neugeboren.
Manches bleibt in Nacht verloren –
Hüte dich, bleib wach und munter!

Nachts

Ich wandre durch die stille Nacht,
Da schleicht der Mond so heimlich sacht
Oft aus der dunklen Wolkenhülle,
Und hin und her im Tal
Erwacht die Nachtigall,
Dann wieder alles grau und stille.

O wunderbarer Nachtgesang:
Von fern im Land der Ströme Gang,
Leis Schauern in den dunklen Bäumen –
Wirrst die Gedanken mir,
Mein irres Singen hier
Ist wie ein Rufen nur aus Träumen.

Der wandernde Musikant

1

Wandern lieb' ich für mein Leben,
Lebe eben wie ich kann,
Wollt' ich mir auch Mühe geben,
Paßt es mir doch gar nicht an.

Schöne alte Lieder weiß ich,
In der Kälte, ohne Schuh'
Draußen in die Saiten reiß ich,
Weiß nicht, wo ich abends ruh'.

Manche Schöne macht wohl Augen,
Meinet, ich gefiel' ihr sehr,
Wenn ich nur was wollte taugen,
So ein armer Lump nicht wär. –

Mag dir Gott ein'n Mann bescheren,
Wohl mit Haus und Hof versehn!
Wenn wir zwei zusammen wären,
Möcht' mein Singen mir vergehn.

2

Wenn die Sonne lieblich schiene
Wie in Welschland lau und blau,
Ging' ich mit der Mandoline
Durch die überglänzte Au.

In der Nacht dann Liebchen lauschte
An dem Fenster süß verwacht,
Wünschte mir und ihr, uns beiden,
Heimlich eine schöne Nacht.

Wenn die Sonne lieblich schiene
Wie in Welschland lau und blau,
Ging' ich mit der Mandoline
Durch die überglänzte Au.

3

Ich reise übers grüne Land,
Der Winter ist vergangen,
Hab' um den Hals ein gülden Band,
Daran die Laute hangen.

Der Morgen tut ein'n roten Schein,
Den recht mein Herze spüret,
Da greif' ich in die Saiten ein,
Der liebe Gott mich führet.

So silbern geht der Ströme Lauf,
Fernüber schallt Geläute,
Die Seele ruft in sich: Glück auf!
Rings grüßen frohe Leute.

Mein Herz ist recht von Diamant,
Ein' Blum' von Edelsteinen,
Die funkelt lustig übers Land
In tausend schönen Scheinen.

Vom Schlosse in die weite Welt
Schaut eine Jungfrau runter,
Der Liebste sie im Arme hält,
Die sehn nach mir herunter.

Wie bist du schön! Hinaus, im Wald
Gehn Wasser auf und unter,
Im grünen Wald sing, daß es schallt,
Mein Herz, bleib frei und munter!

Die Sonne uns im Dunklen läßt,
Im Meere sich zu spülen,
Da ruh' ich aus vom Tagesfest
Fromm in der roten Kühle.

Hoch führet durch die stille Nacht
Der Mond die goldnen Schafe,
Den Kreis der Erden Gott bewacht,
Wo tief ich unten schlafe.

Wie liegt all' falsche Pracht so weit!
Schlaf wohl auf stiller Erde,
Gott schütz dein Herz in Ewigkeit,
Daß es nie traurig werde!

4

Bist du manchmal auch verstimmt,
Drück' dich zärtlich an mein Herze,
Daß mirs fast den Atem nimmt,
Streich' und kneif' in süßem Scherze,
Wie ein rechter Liebestor

Lehn' ich sanft an dich die Wange
Und du singst mir fein ins Ohr.
Wohl im Hofe bei dem Klange
Katze miaut, Hund heult und bellt,
Nachbar schimpft mit wilder Miene –
Doch was kümmert uns die Welt,
Süße, traute Violine!

5

Mürrisch sitzen sie und maulen
Auf den Bänken stumm und breit,
Gähnend strecken sich die Faulen,
Und die Kecken suchen Streit.

Da komm' ich durchs Dorf geschritten,
Fernher durch den Abend kühl,
Stell' mich in des Kreises Mitten,
Grüß' und zieh' mein Geigenspiel.

Und wie ich den Bogen schwenke,
Ziehn die Klänge in der Rund'
Allen recht durch die Gelenke
Bis zum tiefsten Herzensgrund.

Und nun geht's ans Gläserklingen,
An ein Walzen um und um,
Je mehr ich streich', je mehr sie springen,
Keiner fragt erst lang: warum? –

Jeder will dem Geiger reichen
Nun sein Scherflein auf die Hand –
Da vergeht ihm gleich sein Streichen,
Und fort ist der Musikant.

Und sie sehn ihn fröhlich steigen
Nach den Waldeshöh'n hinaus,

Hören ihn von fern noch geigen,
Und gehn all vergnügt nach Haus.

Doch in Waldes grünen Hallen,
Rast ich dann noch manche Stund,
Nur die fernen Nachtigallen
Schlagen tief aus nächtgem Grund.

Und es rauscht die Nacht so leise
Durch die Waldeseinsamkeit,
Und ich sinn' auf neue Weise,
Die der Menschen Herz erfreut.

6

Durch Feld und Buchenhallen
Bald singend, bald fröhlich still,
Recht lustig sei vor allen,
Wer's Reisen wählen will!

Wenns kaum im Osten glühte,
Die Welt noch still und weit:
Da weht recht durchs Gemüte
Die schöne Blütenzeit!

Die Lerch' als Morgenbote
Sich in die Lüfte schwingt,
Eine frische Reisenote
Durch Wald und Herz erklingt.

O Lust, vom Berg zu schauen
Weit über Wald und Strom,
Hoch über sich den blauen
Tiefklaren Himmelsdom!

Vom Berge Vöglein fliegen
Und Wolken so geschwind,

Gedanken überfliegen
Die Vögel und den Wind.

Die Wolken ziehn hernieder,
Das Vöglein senkt sich gleich,
Gedanken gehn und Lieder
Fort bis ins Himmelreich.

Die Zigeunerin

Am Kreuzweg, da lausche ich, wenn die Stern'
Und die Feuer im Walde verglommen,
Und wo der erste Hund bellt von fern,
Da wird mein Bräut'gam herkommen.

„Und als der Tag graut', durch das Gehölz
Sah ich eine Katze sich schlingen,
Ich schoß ihr auf den nußbraunen Pelz,
Wie tat die weitüber springen!" –

's ist schad nur ums Pelzlein, du kriegst mich nit!
Mein Schatz muß sein wie die andern:
Braun und ein Stutzbart auf ungrischen Schnitt
Und ein fröhliches Herze zum Wandern.

Der wandernde Student

Bei dem angenehmsten Wetter
Singen alle Vögelein,
Klatscht der Regen auf die Blätter,
Sing' ich so für mich allein.

Denn mein Aug' kann nicht entdecken,
Wenn der Blitz auch grausam glüht,
Was im Wandern könnt' erschrecken
Ein zufriedenes Gemüt.

Frei von Mammon will ich schreiten
Auf dem Feld der Wissenschaft,
Sinne ernst und nehm zu Zeiten
Einen Mund voll Rebensaft.

Bin ich müde vom Studieren,
Wann der Mond tritt sanft herfür,
Pfleg' ich dann zu musizieren
Vor der Allerschönsten Tür.

Der Maler

Aus Wolken, eh' im nächtgen Land
Erwacht die Kreaturen,
Langt Gottes Hand,
Zieht durch die stillen Fluren
Gewaltig die Konturen,
Strom, Wald und Felsenwand.

Wach auf, wach auf! Die Lerche ruft,
Aurora taucht die Strahlen
Verträumt in Duft,
Beginnt auf Berg und Talen
Ringsum ein himmlisch Malen
In Meer und Land und Luft.

Und durch die Stille, lichtgeschmückt,
Aus wunderbaren Locken
Ein Engel blickt. –
Da rauscht der Wald erschrocken,
Da gehn die Morgenglocken,
Die Gipfel stehn verzückt.

O lichte Augen, ernst und mild,
Ich kann nicht von euch lassen!
Bald wieder wild

Stürmt's her von Sorg' und Hassen –
Durch die verworrnen Gassen
Führ mich, mein göttlich Bild!

Der Soldat

1

Ist auch schmuck nicht mein Rößlein,
So ists doch recht klug,
Trägt im Finstern zu 'nem Schlößlein
Mich rasch noch genug.

Ist das Schloß auch nicht prächtig:
Zum Garten aus der Tür
Tritt ein Mädchen doch allnächtig
Dort freundlich herfür.

Und ist auch die Kleine
Nicht die Schönst' auf der Welt,
So gibt's doch just keine,
Die mir besser gefällt.

Und spricht sie vom Freien:
So schwing' ich mich auf mein Roß –
Ich bleibe im Freien,
Und sie auf dem Schloß.

2

Wagen mußt du und flüchtig erbeuten,
Hinter uns schon durch die Nacht hör ich's schreiten,
Schwing auf mein Roß dich nur schnell
Und küss' noch im Flug mich, wildschönes Kind,
Geschwind,
Denn der Tod ist ein rascher Gesell.

Seemanns Abschied

Ade, mein Schatz, du mocht'st mich nicht,
Ich war dir zu geringe.
Einst wandelst du bei Mondenlicht
Und hörst ein süßes Klingen,
Ein Meerweib singt, die Nacht ist lau,
Die stillen Wolken wandern,
Da denk an mich, 's ist meine Frau,
Nun such dir einen andern!

Ade, ihr Landsknecht', Musketier'!
Wir ziehn auf wildem Rosse,
Das bäumt und überschlägt sich schier
Vor manchem Felsenschlosse,
Der Wassermann bei Blitzesschein
Taucht auf in dunklen Nächten,
Der Haifisch schnappt, die Möwen schrein –
Das ist ein lustges Fechten!

Streckt nur auf eurer Bärenhaut
Daheim die faulen Glieder,
Gott Vater aus dem Fenster schaut,
Schickt seine Sündflut wieder,
Feldwebel, Reiter, Musketier,
Sie müssen all' ersaufen,
Derweil mit frischem Winde wir
Im Paradies einlaufen.

Die Spielleute

Frühmorgens durch die Klüfte
Wir blasen Viktoria!
Eine Lerche fährt durch die Lüfte:
„Die Spielleut' sind schon da!"
Da dehnt ein Turm und reckt sich

Verschlafen im Morgengrau,
Wie aus dem Traume streckt sich
Der Strom durch die stille Au,
Und ihre Äuglein balde
Tun auf die Bächlein all'
Im Wald, im grünen Walde,
Das ist ein lust'ger Schall!

Das ist ein lust'ges Reisen,
Der Eichbaum kühl und frisch
Mit Schatten, wo wir speisen,
Deckt uns den grünen Tisch,
Zum Frühstück musizieren
Die muntern Vögelein,
Der Wald, wenn sie pausieren,
Stimmt wunderbar mit ein,
Die Wipfel tut er neigen,
Als gesegnet' er uns das Mahl,
Und zeigt uns zwischen den Zweigen
Tief unten das weite Tal.

Tief unten da ist ein Garten,
Da wohnt eine schöne Frau,
Wir können nicht lange warten,
Durchs Gittertor wir schaun,
Wo die weißen Statuen stehen,
Da ist's so still und kühl,
Die Wasserkünste gehen,
Der Flieder duftet schwül.
Wir ziehn vorbei und singen
In der stillen Morgenzeit,
Sie hört's im Traume klingen,
Wir aber sind schon weit.

Vor der Stadt

Zwei Musikanten ziehn daher
Vom Wald aus weiter Ferne,
Der eine ist verliebt gar sehr,
Der andre wär' es gerne.

Die stehn allhier im kalten Wind
Und singen schön und geigen:
Ob nicht ein süßverträumtes Kind
Am Fenster sich wollt' zeigen?

Dryander mit der Komödiantenbande

Mich brennt's an meinen Reiseschuhn,
Fort mit der Zeit zu schreiten –
Was wollen wir agieren nun
Vor so viel klugen Leuten?

Es hebt das Dach sich von dem Haus
Und die Kulissen rühren
Und strecken sich zum Himmel 'raus,
Strom, Wälder musizieren!

Und aus den Wolken langt es sacht,
Stellt alles durcheinander,
Wie sich's kein Autor hat gedacht:
Volk, Fürsten und Dryander.

Da gehn die einen müde fort,
Die andern nahn behende,
Das alte Stück, man spielt's so fort
Und kriegt es nie zu Ende.

Und keiner kennt den letzten Akt
Von allen, die da spielen,
Nur der da droben schlägt den Takt,
Weiß, wo das hin will zielen.

Der verliebte Reisende

1

Da fahr' ich still im Wagen,
Du bist so weit von mir,
Wohin er mich mag tragen,
Ich bleibe doch bei dir.

Da fliegen Wälder, Klüfte
Und schöne Täler tief,
Und Lerchen hoch in den Lüften,
Als ob dein' Stimme rief'.

Die Sonne lustig scheinet
Weit über das Revier,
Ich bin so froh verweinet
Und singe still in mir.

Vom Berge geht's hinunter,
Das Posthorn schallt im Grund,
Mein' Seel wird mir so munter,
Grüß dich aus Herzensgrund.

2

Ich geh' durch die dunklen Gassen
Und wandre von Haus zu Haus,
Ich kann mich noch immer nicht fassen,
Sieht alles so trübe aus.

Da gehen viel' Männer und Frauen,
Die alle so lustig sehn,
Die fahren und lachen und bauen,
Daß mir die Sinne vergehn.

Oft wenn ich bläuliche Streifen
Seh' über die Dächer fliehn,
Sonnenschein draußen schweifen,
Wolken am Himmel ziehn:

Da treten mitten im Scherze
Die Tränen ins Auge mir,
Denn die mich lieben von Herzen
Sind alle so weit von hier.

3

Lied, mit Tränen halb geschrieben,
Dorthin über Berg und Kluft,
Wo die Liebste mein geblieben,
Schwing dich durch die blaue Luft!

Ist sie rot und lustig, sage:
Ich sei krank von Herzensgrund;
Weint sie nachts, sinnt still bei Tage,
Ja, dann sag: ich sei gesund!

Ist vorbei ihr treues Lieben,
Nun, so end' auch Lust und Not,
Und zu allen, die mich lieben,
Flieg und sage: ich sei tot!

4

Ach Liebchen, dich ließ ich zurücke,
Mein liebes, herziges Kind,
Da lauern viel Menschen voll Tücke,
Die sind dir so feindlich gesinnt.

Die möchten so gerne zerstören
Auf Erden das schöne Fest,
Ach, könnte das Lieben aufhören,
So mögen sie nehmen den Rest.

Und alle die grünen Orte,
Wo wir gegangen im Wald,
Die sind nun wohl anders geworden,
Da ist's nun so still und kalt.

Da sind nun am kalten Himmel
Viel tausend Sterne gestellt,
Es scheint ihr goldnes Gewimmel
Weit übers beschneite Feld.

Mein' Seele ist so beklommen,
Die Gassen sind leer und tot,
Da hab ich die Laute genommen
Und singe in meiner Not.

Ach, wär' ich im stillen Hafen!
Kalte Winde am Fenster gehn,
Schlaf ruhig, mein Liebchen, schlafe,
Treu' Liebe wird ewig bestehn!

5

Grün war die Weide,
Der Himmel blau,
Wir saßen beide
Auf glänzender Au.

Sind's Nachtigallen
Wieder, was ruft,
Lerchen, die schallen
Aus warmer Luft?

Ich hör die Lieder,
Fern, ohne dich,
Lenz ist's wohl wieder,
Doch nicht für mich.

6

Wolken, wälderwärts gegangen,
Wolken, fliegend übers Haus,
Könnt' ich an euch fest mich hangen,
Mit euch fliegen weit hinaus!

Tag'lang durch die Wälder schweif' ich,
Voll Gedanken sitz' ich still,
In die Saiten flüchtig greif' ich,
Wieder dann auf einmal still.

Schöne, rührende Geschichten
Fallen ein mir, wo ich steh',
Lustig muß ich schreiben, dichten,
Ist mir selber gleich so weh.

Manches Lied, das ich geschrieben
Wohl vor manchem langen Jahr,
Da die Welt vom treuen Lieben
Schön mir überglänzet war;

Find' ichs wieder jetzt voll Bangen:
Werd' ich wunderbar gerührt,
Denn so lang ist das vergangen,
Was mich zu dem Lied verführt.

Diese Wolken ziehen weiter,
Alle Vögel sind erweckt,
Und die Gegend glänzet heiter,
Weit und fröhlich aufgedeckt.

Regen flüchtig abwärts gehen,
Scheint die Sonne zwischendrein,
Und dein Haus, dein Garten stehen
Überm Wald im stillen Schein.

Doch du harrst nicht mehr mit Schmerzen,
Wo so lang dein Liebster sei –
Und mich tötet noch im Herzen
Dieser Schmerzen Zauberei.

Rückkehr

Mit meinem Saitenspiele,
Das schön geklungen hat,
Komm' ich durch Länder viele
Zurück in diese Stadt.

Ich ziehe durch die Gassen,
So finster ist die Nacht,
Und alles so verlassen,
Hatt's anders mir gedacht.

Am Brunnen steh' ich lange,
Der rauscht fort, wie vorher,
Kommt mancher wohl gegangen,
Es kennt mich keiner mehr.

Da hört' ich geigen, pfeifen,
Die Fenster glänzten weit,
Dazwischen drehn und schleifen
Viel' fremde, fröhliche Leut'.

Und Herz und Sinne mir brannten,
Mich trieb's in die weite Welt,
Es spielten die Musikanten,
Da fiel ich hin im Feld.

Auf einer Burg

Eingeschlafen auf der Lauer
Oben ist der alte Ritter;
Drüber gehen Regenschauer,
Und der Wald rauscht durch das Gitter.

Eingewachsen Bart und Haare,
Und versteinert Brust und Krause,
Sitzt er viele hundert Jahre
Oben in der stillen Klause.

Draußen ist es still und friedlich,
Alle sind ins Tal gezogen,
Waldesvögel einsam singen
In den leeren Fensterbogen.

Eine Hochzeit fährt da unten
Auf dem Rhein im Sonnenscheine,
Musikanten spielen munter,
Und die schöne Braut die weinet.

Jahrmarkt

Sind's die Häuser, sind's die Gassen?
Ach, ich weiß nicht wo ich bin!
Hab' ein Liebchen hier gelassen,
Und manch Jahr ging seitdem hin.

Aus den Fenstern schöne Frauen
Sehn mir freundlich ins Gesicht,
Keine kann so frischlich schauen,
Als mein liebes Liebchen sicht.

An dem Hause poch' ich bange –
Doch die Fenster stehen leer,

Ausgezogen ist sie lange,
Und es kennt mich keiner mehr.

Und ringsum ein Rufen, Handeln,
Schmucke Waren, bunter Schein,
Herr'n und Damen gehn und wandeln
Zwischendurch in bunten Reih'n.

Zierlich Bücken, freundlich Blicken,
Manches flücht'ge Liebeswort,
Händedrücken, heimlich Nicken –
Nimmt sie all' der Strom mit fort.

Und mein Liebchen sah ich eben
Traurig in dem lust'gen Schwarm,
Und ein schöner Herr daneben
Führt' sie stolz und ernst am Arm.

Doch verblaßt war Mund und Wange,
Und gebrochen war ihr Blick,
Seltsam schaut' sie stumm und lange,
Lange noch auf mich zurück. –

Und es endet Tag und Scherzen,
Durch die Gassen pfeift der Wind –
Keiner weiß, wie unsre Herzen
Tief von Schmerz zerrissen sind.

In der Fremde

Ich hör' die Bächlein rauschen
Im Walde her und hin,
Im Walde in dem Rauschen
Ich weiß nicht, wo ich bin.

Die Nachtigallen schlagen
Hier in der Einsamkeit,

Als wollten sie was sagen
Von der alten, schönen Zeit.

Die Mondesschimmer fliegen,
Als säh' ich unter mir
Das Schloß im Tale liegen,
Und ist doch so weit von hier!

Als müßte in dem Garten,
Voll Rosen weiß und rot,
Meine Liebste auf mich warten,
Und ist doch lange tot.

Sehnsucht

Es schienen so golden die Sterne,
Am Fenster ich einsam stand
Und hörte aus weiter Ferne
Ein Posthorn im stillen Land.
Das Herz mir im Leib entbrennte,
Da hab' ich mir heimlich gedacht:
Ach, wer da mitreisen könnte
In der prächtigen Sommernacht!

Zwei junge Gesellen gingen
Vorüber am Bergeshang,
Ich hörte im Wandern sie singen
Die stille Gegend entlang:
Von schwindelnden Felsenschlüften,
Wo die Wälder rauschen so sacht,
Von Quellen, die von den Klüften
Sich stürzen in die Waldesnacht.

Sie sangen von Marmorbildern,
Von Gärten, die überm Gestein
In dämmernden Lauben verwildern,

Palästen im Mondenschein,
Wo die Mädchen am Fenster lauschen,
Wann der Lauten Klang erwacht
Und die Brunnen verschlafen rauschen
In der prächtigen Sommernacht. –

Abschied

O Täler weit, o Höhen,
O schöner, grüner Wald,
Du meiner Lust und Wehen
Andächt'ger Aufenthalt!
Da draußen, stets betrogen,
Saust die geschäft'ge Welt,
Schlag noch einmal die Bogen
Um mich, du grünes Zelt!

Wenn es beginnt zu tagen,
Die Erde dampft und blinkt,
Die Vögel lustig schlagen,
Daß dir dein Herz erklingt:
Da mag vergehn, verwehen
Das trübe Erdenleid,
Da sollst du auferstehen
In junger Herrlichkeit!

Da steht im Wald geschrieben
Ein stilles, ernstes Wort
Von rechtem Tun und Lieben,
Und was des Menschen Hort,
Ich habe treu gelesen
Die Worte schlicht und wahr,
Und durch mein ganzes Wesen
Ward's unaussprechlich klar.

Bald werd' ich dich verlassen,
Fremd in der Fremde gehn,
Auf buntbewegten Gassen
Des Lebens Schauspiel sehn;
Und mitten in dem Leben
Wird deines Ernsts Gewalt
Mich Einsamen erheben,
So wird mein Herz nicht alt.

Wann der Hahn kräht

Wann der Hahn kräht auf dem Dache,
Putzt der Mond die Lampe aus,
Und die Stern' ziehn von der Wache,
Gott behüte Land und Haus!

Der Morgen

Fliegt der erste Morgenstrahl
Durch das stille Nebeltal,
Rauscht erwachend Wald und Hügel:
Wer da fliegen kann, nimmt Flügel!

Und sein Hütlein in die Luft
Wirft der Mensch vor Lust und ruft:
Hat Gesang doch auch noch Schwingen,
Nun, so will ich fröhlich singen!

Hinaus, o Mensch, weit in die Welt,
Bangt dir das Herz in krankem Mut;
Nichts ist so trüb in Nacht gestellt,
Der Morgen leicht machts' wieder gut.

Mittagsruh

Über Bergen, Fluß und Talen,
Stiller Lust und tiefen Qualen
Webet heimlich, schillert, Strahlen!
Sinnend ruht des Tags Gewühle
In der dunkelblauen Schwüle,
Und die ewigen Gefühle,
Was dir selber unbewußt,
Treten heimlich, groß und leise
Aus der Wirrung fester Gleise,
Aus der unbewachten Brust,
In die stillen, weiten Kreise.

Der Abend

Schweigt der Menschen laute Lust:
Rauscht die Erde wie in Träumen
Wunderbar mit allen Bäumen,
Was dem Herzen kaum bewußt,
Alte Zeiten, linde Trauer,
Und es schweifen leise Schauer
Wetterleuchtend durch die Brust.

Die Nacht

Wie schön, hier zu verträumen
Die Nacht im stillen Wald,
Wenn in den dunklen Bäumen
Das alte Märchen hallt.

Die Berg' im Mondesschimmer
Wie in Gedanken stehn,
Und durch verworrne Trümmer
Die Quellen klagend gehn.

Denn müd' ging auf den Matten
Die Schönheit nun zur Ruh
Es deckt mit kühlen Schatten
Die Nacht das Liebchen zu.

Das ist das irre Klagen
In stiller Waldespracht,
Die Nachtigallen schlagen
Von ihr die ganze Nacht.

Die Stern' gehn auf und nieder –
Wann kommst du, Morgenwind,
Und hebst die Schatten wieder
Von dem verträumten Kind?

Schon rührt sich's in den Bäumen,
Die Lerche weckt sie bald –
So will ich treu verträumen
Die Nacht im stillen Wald.

Wegweiser

„Jetzt mußt du rechts dich schlagen,
Schleich dort und lausche hier,
Dann schnell drauf los im Jagen –
So wird noch was aus dir."

Dank! doch durchs Weltgewimmel,
Sagt mir, ihr weisen Herrn,
Wo geht der Weg zum Himmel?
Das Eine wüßt' ich gern.

Täuschung

Ich ruhte aus vom Wandern,
Der Mond ging eben auf,

Da sah ich fern im Lande
Der alten Tiber Lauf,
Im Walde lagen Trümmer,
Paläste auf stillen Höh'n
Und Gärten im Mondesschimmer –
O Welschland, wie bist du schön!

Und als die Nacht vergangen,
Die Erde blitzte so weit,
Einen Hirten sah ich hangen
Am Fels in der Einsamkeit.
Den fragt' ich ganz geblendet:
Komm' ich nach Rom noch heut?
Er dehnt' sich halbgewendet:
Ihr seid nicht recht gescheut!
Eine Winzerin lacht' herüber,
Man sah sie vor Weinlaub kaum,
Mir aber ging's Herze über –
Es war ja alles nur Traum.

Schöne Fremde

Es rauschen die Wipfel und schauern,
Als machten zu dieser Stund'
Um die halbversunkenen Mauern
Die alten Götter die Rund'.

Hier hinter den Myrtenbäumen
In heimlich dämmernder Pracht,
Was sprichst du wirr wie in Träumen
Zu mir, phantastische Nacht?

Es funkeln auf mich alle Sterne
Mit glühendem Liebesblick,
Es redet trunken die Ferne
Wie von künftigem, großem Glück!

Liebe in der Fremde

1

Jeder nennet froh die Seine,
Ich nur stehe hier alleine,
Denn was früge wohl die Eine:
Wen der Fremdling eben meine?
Und so muß ich, wie im Strome dort die Welle,
Ungehört verrauschen an des Frühlings Schwelle.

2

Wie kühl schweift sich's bei nächt'ger Stunde,
Die Zither treulich in der Hand!
Vom Hügel grüß' ich in die Runde
Den Himmel und das stille Land.

Wie ist da alles so verwandelt,
Wo ich so fröhlich war, im Tal,
Im Wald wie still! Der Mond nur wandelt
Nun durch den hohen Buchensaal.

Der Winzer Jauchzen ist verklungen
Und all der bunte Lebenslauf,
Die Ströme nur, im Tal geschlungen,
Sie blicken manchmal silbern auf.

Und Nachtigallen wie aus Träumen
Erwachen oft mit süßem Schall,
Erinnernd rührt sich in den Bäumen
Ein heimlich Flüstern überall.

Die Freude kann nicht gleich verklingen,
Und von des Tages Glanz und Lust
Ist so auch mir ein heimlich Singen
Geblieben in der tiefsten Brust.

Und fröhlich greif' ich in die Saiten,
O Mädchen, jenseits überm Fluß,
Du lauschest wohl und hörst's vom weiten
Und kennst den Sänger an dem Gruß!

3

Über die beglänzten Gipfel
Fernher kommt es wie ein Grüßen,
flüsternd neigen sich die Wipfel,
Als ob sie sich wollten küssen.

Ist er doch so schön und milde!
Stimmen gehen durch die Nacht,
Singen heimlich von dem Bilde –
Ach, ich bin so froh verwacht!

Plaudert nicht so laut, ihr Quellen!
Wissen darf es nicht der Morgen!
In der Mondnacht linde Wellen
Senk' ich still mein Glück und Sorgen.

4

Jetzt wandr' ich erst gern!
Am Fenster nun lauschen
Die Mädchen, es rauschen
Die Brunnen von fern.
Aus schimmernden Büschen
Ihr Plaudern, so lieb,
Erkenn' ich dazwischen,
Ich höre mein Lieb!

Kind, hüt' dich! Bei Nacht
Pflegt Amor zu wandern,
Ruft leise die andern,
Da schreiten erwacht

Die Götter zur Halle
Ins Freie hinaus,
Es bringt sie dir alle
Der Dichter ins Haus.

Lustige Musikanten

Der Wald, der Wald! daß Gott ihn grün erhalt',
Gibt gut Quartier und nimmt doch nichts dafür.

Zum grünen Wald wir Herberg' halten,
Denn Hoffart ist nicht unser Ziel,
Im Wirtshaus, wo wir nicht bezahlten,
Es war der Ehre gar zu viel.
Der Wirt, er wollt' uns gar nicht lassen,
Sie ließen Kann' und Kartenspiel,
Die ganze Stadt war in den Gassen,
Und von den Bänken mit Gebraus
Stürzt' die Schule heraus,
Wuchs der Haufe von Haus zu Haus,
Schwenkt' die Mützen und jubelt' und wogt',
Der Hatschier, die Stadtwacht, der Bettelvogt,
Wie wenn ein Prinz zieht auf die Freit',
Gab alles, alles uns fürstlich Geleit.
Wir aber schlugen den Markt hinab
Und durch die Leut' mit dem Wanderstab,
Und hoch mit dem Tamburin, daß es schallt', –

Zum Wald, zum Wald, zum schönen, grünen Wald!

Und da nun alle schlafen gingen,
Der Wald steckt' seine Irrlicht' an,
Die Frösche tapfer Ständchen bringen,
Die Fledermaus schwirrt leis voran,
Und in dem Fluß auf feuchtem Steine
Gähnt laut der alte Wassermann,

Strählt sich den Bart im Mondenscheine,
Und fragt ein Irrlicht, wer wir sind?
Das aber duckt sich geschwind;
Denn über ihn weg im Wind
Durch die Wipfel der wilde Jäger geht,
Und auf dem alten Turm sich dreht
Und kräht der Wetterhahn uns nach:
Ob wir nicht einkehr'n unter sein Dach?
O Gockel, verfallen ist ja dein Haus,
Es sieht die Eule zum Fenster heraus,
Und aus allen Toren rauschet der Wald.

Der Wald, der Wald, der schöne, grüne Wald!

Und wenn wir müd' einst, sehn wir blinken
Eine goldne Stadt still überm Land,
Am Tor Sankt Peter schon tut winken:
„Nur hier herein, Herr Musikant!"
Die Engel von den Zinnen fragen,
Und wie sie uns erst recht erkannt,
Sie gleich die silbernen Pauken schlagen,
Sankt Peter selbst die Becken schwenkt,
Und voll Geigen hängt
Der Himmel, Cäcilia an zu streichen fängt,
Dazwischen Hoch vivat! daß es prasselt und pufft,
Werfen die andern vom Wall in die Luft
Sternschnuppen, Kometen,
Gar prächt'ge Raketen,
Versengen Sankt Peter den Bart, daß er lacht,
Und wir ziehen heim, schöner Wald, gute Nacht!

Wandersprüche

1

Es geht wohl anders, als du meinst:
Derweil du rot und fröhlich scheinst,

Ist Lenz und Sonnenschein verflogen,
Die liebe Gegend schwarz umzogen;
Und kaum hast du dich ausgeweint,
Lacht alles wieder, die Sonne scheint –
Es geht wohl anders, als man meint.

2

Herz, in deinen sonnenhellen
Tagen halt nicht karg zurück!
Allwärts fröhliche Gesellen
Trifft der Frohe und sein Glück.

Sinkt der Stern: alleine wandern
Magst du bis ans End' der Welt –
Bau du nur auf keinen andern
Als auf Gott, der Treue hält.

3

Was willst auf dieser Station
So breit dich niederlassen?
Wie bald nicht bläst der Postillon,
Du mußt doch alles lassen.

4

Die Lerche grüßt den ersten Strahl,
Daß er die Brust ihr zünde,
Wenn träge Nacht noch überall
Durchschleicht die tiefen Gründe.

Und du willst, Menschenkind, der Zeit
Verzagend unterliegen?
Was ist dein kleines Erdenleid?
Du mußt es überfliegen!

5

Der Sturm geht lärmend um das Haus,
Ich bin kein Narr und geh' hinaus,
Aber bin ich eben draußen,
Will ich mich wacker mit ihm zausen.

6

Ewig muntres Spiel der Wogen!
Viele hast du schon belogen,
Mancher kehrt nicht mehr zurück.
Und doch weckt das Wellenschlagen
Immer wieder frisches Wagen,
Falsch und lustig wie das Glück.

Der Wandrer, von der Heimat weit,
Wenn rings die Gründe schweigen,
Der Schiffer in Meeres Einsamkeit,
Wenn die Stern' aus den Fluten steigen:

Die beide schauern und lesen
In stiller Nacht,
Was sie nicht gedacht,
Da es noch fröhlicher Tag gewesen.

Wandernder Dichter

Ich weiß nicht, was das sagen will!
Kaum tret' ich von der Schwelle still,
Gleich schwingt sich eine Lerche auf
Und jubiliert durchs Blau vorauf.

Das Gras ringsum, die Blumen gar
Stehn mit Juwelen und Perl'n im Haar,
Die schlanken Pappeln, Busch und Saat
Verneigen sich im größten Staat.

Als Bot' voraus das Bächlein eilt,
Und wo der Wind die Wipfel teilt,
Die Au verstohlen nach mir schaut,
Als wär' sie meine liebe Braut.

Ja, komm' ich müd ins Nachtquartier,
Die Nachtigall noch vor der Tür
Mir Ständchen bringt, Glühwürmchen bald
Illuminieren rings den Wald.

Umsonst! Das ist nun einmal so,
Kein Dichter reist inkognito,
Der lust'ge Frühling merkt es gleich,
Wer König ist in seinem Reich.

Erinnerung

1

Lindes Rauschen in den Wipfeln,
Vöglein, die ihr fernab fliegt,
Bronnen von den stillen Gipfeln,
Sagt, wo meine Heimat liegt?

Heut im Traum sah ich sie wieder,
Und von allen Bergen ging
Solches Grüßen zu mir nieder,
Daß ich an zu weinen fing.

Ach, hier auf den fremden Gipfeln:
Menschen, Quellen, Fels und Baum,
Wirres Rauschen in den Wipfeln, –
Alles ist mir wie ein Traum.

2

Die fernen Heimathöhen,
Das stille, hohe Haus,
Der Berg, von dem ich gesehen
Jeden Frühling ins Land hinaus,
Mutter, Freunde und Brüder,
An die ich so oft gedacht,
Es grüßt mich alles wieder
In stiller Mondesnacht.

Heimat

Wer in die Fremde will wandern,
Der muß mit der Liebsten gehn,
Es jubeln und lassen die andern
Den Fremden alleine stehn.

Was wisset ihr, dunkele Wipfel,
Von der alten, schönen Zeit?
Ach, die Heimat hinter den Gipfeln,
Wie liegt sie von hier so weit!

Am liebsten betracht' ich die Sterne,
Die schienen, wie ich ging zu ihr,
Die Nachtigall hör' ich so gerne,
Sie sang vor der Liebsten Tür.

Der Morgen, das ist meine Freude!
Da steig' ich in stiller Stund
Auf den höchsten Berg in die Weite,
Grüß' dich, Deutschland, aus Herzensgrund!

An der Grenze

Die treuen Berg' stehn auf der Wacht:
„Wer streicht bei stiller Morgenzeit
Da aus der Fremde durch die Heid'?" –
Ich aber mir die Berg' betracht'
Und lach in mich vor großer Lust,
Und rufe recht aus frischer Brust
Parol' und Feldgeschrei sogleich:
Vivat Östreich!

Da kennt mich erst die ganze Rund',
Nun grüßen Bach und Vöglein zart
Und Wälder rings nach Landesart,
Die Donau blitzt aus tiefem Grund,
Der Stephansturm auch ganz von fern
Guckt übern Berg und säh' mich gern,
Und ist er's nicht, so kommt er doch gleich,
Vivat Östreich!

Wanderlied der Prager Studenten

Nach Süden nun sich lenken
Die Vöglein allzumal,
Viel' Wandrer lustig schwenken
Die Hüt' im Morgenstrahl,
Das sind die Herrn Studenten,
Zum Tor hinaus es geht,
Auf ihren Instrumenten
Sie blasen zum Valet:
Ade in die Läng' und Breite
O Prag, wir ziehn in die Weite:
Et habeat bonam pacem,
Qui sedet post fornacem!

Nachts wir durchs Städtlein schweifen,
Die Fenster schimmern weit,

Am Fenster drehn und schleifen
Viel' schön geputzte Leut'.
Wir blasen vor den Türen
Und haben Durst genung,
Das kommt vom Musizieren,
Herr Wirt, einen frischen Trunk!
Und siehe über ein kleines
Mit einer Kanne Weines
Venit ex sua domo –
Beatus ille homo!

Nun weht schon durch die Wälder
Der kalte Boreas,
Wir streichen durch die Felder,
Von Schnee und Regen naß,
Der Mantel fliegt im Winde,
Zerrissen sind die Schuh'.
Da blasen wir geschwinde
Und singen noch dazu:
Beatus ille homo
qui sedet in sua domo
Et sedet post fornacem
Er habet bonam pacem!

Rückkehr

Wer steht hier draußen? – Macht auf geschwind!
Schon funkelt das Feld wie geschliffen,
Es ist der lustige Morgenwind,
Der kommt durch den Wald gepfiffen.

Ein Wandervöglein, die Wolken und ich
Wir reisten um die Wette,
Und jedes dacht': Nun spute dich,
Wir treffen sie noch im Bette!

Da sind wir nun, jetzt alle heraus,
Die drin noch Küsse tauschen!
Wir brechen sonst mit der Tür ins Haus:
Klang, Duft und Waldesrauschen.

Ich komme aus Italien fern
Und will euch alles berichten,
Vom Berg Vesuv und Romas Stern
Die alten Wundergeschichten.

Da singt eine Fei auf blauem Meer,
Die Myrten trunken lauschen –
Mir aber gefällt doch nichts so sehr
Als das deutsche Waldesrauschen!

Zur Hochzeit

Was das für ein Gezwitscher ist!
Durchs Blau die Schwalben zucken
Und schrein: „Sie haben sich geküßt!"
Vom Baum Rotkehlchen gucken.

Der Storch stolziert von Bein zu Bein;
„Da muß ich fischen gehen –"
Der Abend wie im Traum darein
Schaut von den stillen Höhen.

Und wie im Traume von den Höhen
Seh' ich nachts meiner Liebsten Haus,
Die Wolken darüber gehen
Und löschen die Sterne aus.

Der irre Spielmann

Aus stiller Kindheit unschuldiger Hut
Trieb mich der tolle, frevelnde Mut.

Seit ich da draußen so frei nun bin,
Find' ich nicht wieder nach Hause mich hin.

Durchs Leben jag' ich manch trüg'risch Bild,
Wer ist der Jäger da? Wer ist das Wild?
Es pfeift der Wind mir schneidend durchs Haar,
Ach Welt, wie bist du so kalt und klar!

Du frommes Kindlein im stillen Haus,
Schau nicht so lüstern zum Fenster hinaus!
Frag mich nicht, Kindlein, woher und wohin?
Weiß ich doch selber nicht, wo ich bin!

Von Sünde und Reue zerrissen die Brust,
Wie rasend in verzweifelter Lust,
Brech' ich im Fluge mir Blumen zum Strauß,
Wird doch kein fröhlicher Kranz daraus! –

Ich möcht' in den tiefsten Wald wohl hinein,
Recht aus der Brust den Jammer zu schrein,
Ich möchte reiten ans Ende der Welt,
Wo der Mond und die Sonne hinunterfällt.

Wo schwindelnd beginnt die Ewigkeit,
Wie ein Meer, so erschrecklich still und weit,
Da sinken all' Ström' und Segel hinein,
Da wird es wohl endlich auch ruhig sein.

Letzte Heimkehr

Der Wintermorgen glänzt so klar,
Ein Wandrer kommt von ferne,
Ihn schüttelt Frost, es starrt sein Haar,
Ihm log die schöne Ferne,
Nun endlich will er rasten hier,
Er klopft an seines Vaters Tür.

Doch tot sind, die sonst aufgetan,
Verwandelt Hof und Habe,
Und fremde Leute sehn ihn an,
Als käm' er aus dem Grabe;
Ihn schauert tief im Herzensgrund,
Ins Feld eilt er zur selben Stund'.

Da sang kein Vöglein weit und breit,
Es lehnt' an einem Baume,
Der schöne Garten lag verschneit,
Es war ihm wie im Traume,
Und wie die Morgenglocke klingt,
Im stillen Feld er niedersinkt.

Und als er aufsteht vom Gebet,
Nicht weiß, wohin sich wenden,
Ein schöner Jüngling bei ihm steht,
Faßt mild ihn bei den Händen:
„Komm mit, sollst ruhn nach kurzem Gang." –
Er folgt, ihn rührt der Stimme Klang.

Nun durch die Bergeseinsamkeit
Sie wie zum Himmel steigen,
Kein Glockenklang mehr reicht so weit,
Sie sehn im öden Schweigen
Die Länder hinter sich verblühn,
Schon Sterne durch die Wipfel glühn.

Der Führer jetzt die Fackel sacht
Erhebt und schweigend schreitet,
Bei ihrem Schein die stille Nacht
Gleichwie ein Dom sich weitet,
Wo unsichtbare Hände baun –
Den Wandrer faßt ein heimlich Graun.

Er sprach: Was bringt der Wind herauf
So fremden Laut getragen,
Als hört' ich ferner Ströme Lauf,

Dazwischen Glocken schlagen?
„Das ist des Nachtgesanges Wehn,
Sie loben Gott in stillen Höh'n."

Der Wandrer drauf: Ich kann nicht mehr –
Ist's Morgen, der so blendet?
Was leuchten dort für Länder her? –
Sein Freund die Fackel wendet:
„Nun ruh' zum letzten Male aus,
Wenn du erwachst, sind wir zu Haus."

II. Sängerleben

Singen kann ich nicht wie du,
Und wie ich nicht der und jener,
Kannst du's besser, sing frisch zu!
Andre singen wieder schöner,
Droben an dem Himmelstor
Wird's *ein* wunderbarer Chor.

Schlimme Wahl

Du sahst die Fei ihr goldnes Haar sich strählen,
Wenn morgens früh noch alle Wälder schweigen,
Gar viele da im Felsgrund sich versteigen,
Und weiß doch keiner, wen sie wird erwählen.

Von einer andern Dam' hört ich erzählen
Im platten Land, die Bauern rings dir zeigen
Ihr Schloß, Park, Weiler – alles ist dein eigen,
Freist du das Weib – wer möcht' im Wald sich quälen!

Sie werden dich auf einen Phaeton heben,
Das Hochzeitscarmen tönt, es blinkt die Flasche,
Weitrauschend hinterdrein viel vornehm Wesen.

Doch streift beim Zug dich aus dem Walde eben
Der Feie Blick, und brennt dich nicht zu Asche:
Fahr wohl, bist nimmer ein Poet gewesen!

Anklänge

1

Vöglein in den sonn'gen Tagen!
Lüfte blau, die mich verführen!
Könnt' ich bunte Flügel rühren,
Über Berg und Wald sie schlagen!

Ach! es spricht des Frühlings Schöne,
Und die Vögel alle singen:
Sind die Farben denn nicht Töne,
Und die Töne bunte Schwingen?

Vöglein, ja, ich lass' das Zagen!
Winde sanft die Segel rühren,
Und ich lasse mich entführen,
Ach! wohin? mag ich nicht fragen.

2

Ach! Wie ist es doch gekommen,
Daß die ferne Waldespracht
So mein ganzes Herz genommen,
Mich um alle Ruh' gebracht!

Wenn von drüben Lieder wehen,
Waldhorn gar nicht enden will,
Weiß ich nicht, wie mir geschehen,
Und im Herzen bet' ich still.

Könnt' ich zu den Wäldern flüchten,
Mit dem Grün in frischer Lust
Mich zum Himmelsglanz aufrichten –
Stark und frei wär' da die Brust!

Hörnerklang und Lieder kämen
Nicht so schmerzlich an mein Herz,
Fröhlich wollt' ich Abschied nehmen,
Zög' auf ewig wälderwärts.

Intermezzo

Wie so leichte läßt sich's leben!
Blond und rot und etwas feist,
Tue wie die andern eben,
Daß dich jeder Bruder heißt,
Speise, was dir Zeiten geben,
Bis die Zeit auch dich verspeist!

3

Wenn die Klänge nahn und fliehen
In den Wogen süßer Lust,
Ach! nach tiefern Melodien
Sehnt sich einsam oft die Brust!

Wenn auf Bergen blüht die Frühe,
Wieder buntbewegt die Straßen,
Freut sich alles, wie es glühe,
Himmelwärts die Erde blühe:
Einer doch muß tief erblassen,
Goldne Träume, Sternenlust
Wollten ewig ihn nicht lassen –
Sehnt sich einsam oft die Brust.

Und aus solcher Schmerzen Schwellen,
Was so lange dürstend rang,
Will ans Licht nun rastlos quellen,
Stürzend mit den Wasserfällen,
Himmelstäubend, jubelnd, bang,
Nach der Ferne sanft zu ziehen,
Wo so himmlisch Rufen sang,
Ach! nach tiefern Melodien.

Blüten licht nun Blüten drängen,
Daß er möcht' vor Glanz erblinden;
In den dunklen Zaubergängen,
Von den eigenen Gesängen
Hold gelockt, kann er nicht finden
Aus dem Labyrinth der Brust.
Alles, alles will's verkünden
In den Wogen süßer Lust.

Doch durch dieses Rauschen wieder
Hört er heimlich Stimmen ziehen,
Wie ein Fall verlorner Lieder

Und er schaut betroffen nieder:
„Wenn die Klänge nahn und fliehen
In den Wogen süßer Lust,
Ach! nach tiefern Melodien
Sehnt sich einsam oft die Brust!"

4

Ewig's Träumen von den Fernen!
Endlich ist das Herz erwacht
Unter Blumen, Klang und Sternen
In der dunkelgrünen Nacht.

Schlummernd unter blauen Wellen
Ruht der Knabe unbewußt,
Engel ziehen durch die Brust;
Oben hört er in den Wellen
Ein unendlich Wort zerrinnen,
Und das Herze weint und lacht,
Doch er kann sich nicht besinnen
In der dunkelgrünen Nacht.

Frühling will das Blau befreien,
Aus der Grüne, aus dem Schein
Ruft es lockend: Ewig dein –
Aus der Minne Zaubereien
Muß er sehnen sich nach Fernen,
Denkend alter Wunderpracht,
Unter Blumen, Klang und Sternen
In der dunkelgrünen Nacht.

Heil'ger Kampf nach langem Säumen,
Wenn süßschauernd an das Licht
Lieb' in dunkle Klagen bricht!
Aus der Schmerzen Sturz und Schäumen
Steigt Geliebte, Himmel, Fernen –
Endlich ist das Herz erwacht

Unter Blumen, Klang und Sternen
In der dunkelgrünen Nacht.

Und der Streit muß sich versöhnen,
Und die Wonne und den Schmerz
Muß er ewig himmelwärts
Schlagen nun in vollen Tönen:
Ewig's Träumen von den Fernen!
Endlich ist das Herz erwacht
Unter Blumen, Klang und Sternen
In der dunkelgrünen Nacht.

Rettung

Ich spielt', ein frohes Kind, im Morgenscheine,
Der Frühling schlug die Augen auf so helle,
Hinunter reisten Ström' und Wolken schnelle,
Ich streckt' die Arme nach ins Blaue, Reine.

Noch wußt' ich's selbst nicht, was das alles meine:
Die Lerch', der Wald, der Lüfte blaue Welle,
Und träumend stand ich an des Frühlings Schwelle,
Von fern rief's immer fort: Ich bin die Deine!

Da kam ein alter Mann gegangen,
Mit hohlen Augen und bleichen Wangen,
Er schlich gebogen und schien so krank;
Ich grüßt' ihn schön, doch für den Dank
Faßt' er mich tückisch schnell von hinten,
Schlang um die Arme mir dreifache Binden,
Und wie ich rang und um Hilfe rief,
Geschwind noch ein andrer zum Alten lief,
Und von allen Seiten kamen Menschen gelaufen,
Ein dunkelverworrner, trübseliger Haufen.
Die drängten mich gar tückisch in ihre Mitte,
Führten durchs Land mich mit eiligem Schritte.

Wie wandt' ich sehnend mich oft zurücke!
Die Heimat schickte mir Abschiedsblicke;
Die Büsche langten nach mir mit grünen Armen,
Es schrien alle Vöglein recht zum Erbarmen.
Doch die Alten hörten nicht die fernen Lieder,
Sumsten düstere Worte nur hin und wieder,
Führten mich endlich in ein altes Haus,
Da wogt' es unten in Nacht und Graus,
Da war ein Hämmern, ein Schachern und Rumoren,
Als hätte das Chaos noch nicht ausgegoren.
Hier hielt der Alte würdig und breit:
Mein Sohn, sprach er zu mir, das ist die Nützlichkeit!
Die haben wir so zum gemeinen Besten erfunden.
Das betrachte hübsch fleißig und sei gescheit. –
So ließen sie mich Armen allein und gebunden.

Da schaut' ich weinend aus meinem Kerker
Hinaus in das Leben durch düstern Erker,
Und unten sah ich den Lenz sich breiten,
Blühende Träume über die Berge schreiten,
Drüber die blauen, unendlichen Weiten.
Durchs farbige Land auf blauen Flüssen
Zogen bunte Schifflein, die wollten mich grüßen.
Vorüber kamen die Wolken gezogen,
Vorüber singende Vöglein geflogen;
Es wollt' der große Zug mich fassen,
Ach, Menschen, wann werd't ihr mich wieder hinunter lassen!
Und im dunkelgrünen Walde munter
Schallte die Jagd hinauf und hinunter,
Eine Jungfrau zu Roß und blitzende Reiter –
Über die Berge immer weiter und weiter
Rief Waldhorn immerfort dazwischen:
Mir nach in den Wald, den frischen!

Ach! weiß denn niemand, niemand um mein Trauern?
Wie alle Fernen mir prophetisch singen
Von meinem künft'gen wundervollen Leben!

Von innen fühlt' ich blaue Schwingen ringen,
Die Hände konnt' ich innigst betend heben –
Da sprengt' ein großer Klang so Band wie Mauern.

Da ward ich im innersten Herzen so munter,
Schwindelten alle Sinne in den Lenz hinunter,
Weit waren kleinliche Mühen und Sorgen,
Ich sprang hinaus in den farbigen Morgen.

Hippogryph

Das ist ein Flügelpferd mit Silberschellen,
Das heitere Gesellen
Empor hebt über Heidekraut und Klüfte,
Daß durch den Strom der Lüfte,
Die um den Reisehut melodisch pfeifen,
Des Ernsts Gewalt und Torenlärm der Schlüfte
Als Frühlingsjauchzen nur die Brust mag streifen;
Und so im Flug belauschen
Des trunknen Liedergottes rüst'ge Söhne,
Wenn alle Höhn und Täler blühn und rauschen,
Im Morgenbad des Lebens ew'ge Schöne,
Die, in dem Glanz erschrocken,
Sie glühend anblickt aus den dunklen Locken.

Die zwei Gesellen

Es zogen zwei rüst'ge Gesellen
Zum erstenmal von Haus,
So jubelnd recht in die hellen,
Klingenden, singenden Wellen
Des vollen Frühlings hinaus.

Die strebten nach hohen Dingen,
Die wollten, trotz Lust und Schmerz,

Was Recht's in der Welt vollbringen,
Und wem sie vorüber gingen,
Dem lachten Sinnen und Herz. –

Der erste, der fand ein Liebchen,
Die Schwieger kauft' Hof und Haus;
Der wiegte gar bald ein Bübchen,
Und sah aus heimlichem Stübchen
Behaglich ins Feld hinaus.

Dem zweiten sangen und logen
Die tausend Stimmen im Grund,
Verlockend' Sirenen, und zogen
Ihn in der buhlenden Wogen
Farbig klingenden Schlund.

Und wie er auftaucht' vom Schlunde,
Da war er müde und alt,
Sein Schifflein das lag im Grunde,
So still war's rings in die Runde,
Und über die Wasser weht's kalt.

Es singen und klingen die Wellen
Des Frühlings wohl über mir;
Und seh' ich so kecke Gesellen,
Die Tränen im Auge mir schwellen –
Ach Gott, führ' uns liebreich zu Dir!

Das Bilderbuch

Von der Poesie sucht Kunde
Mancher im gelehrten Buch,
Nur des Lebens schöne Runde
Lehret dich den Zauberspruch;
Doch in stillgeweihter Stunde
Will das Buch erschlossen sein,

Und so blick' ich heut hinein,
Wie ein Kind im Frühlingswetter
Fröhlich Bilderbücher blättert,
Und es schweift der Sonnenschein
Auf den buntgemalten Lettern,
Und gelinde weht der Wind
Durch die Blumen, durch das Herz
Alte Freuden, alten Schmerz –
Weinen möcht' ich wie ein Kind!

Mandelkerngedicht

Zwischen Akten, dunkeln Wänden
Bannt mich, Freiheitsbegehr*enden*,
Nun des Lebens strenge Pflicht,
Und aus Schränken, Aktenschichten
Lachen mir die beleid*igten*
Musen in das Amtsgesicht.

Als an Lenz und Morgenröte
Noch das Herz sich erlab*ete*,
O du stilles, heitres Glück!
Wie ich nun auch heiß mich sehne,
Ach, aus dieser Sand*ebene*
Führt kein Weg dahin zurück.

Als der letzte Balkentreter
Steh' ich armer Enter*beter*
In des Staates Symphonie,
Ach, in diesem Schwall von Tönen
Wo fänd' ich da des *eig*enen
Herzens süße Melodie?

Ein Gedicht soll ich euch spenden:
Nun, so geht mit dem Leid*enden*
Nicht zu strenge ins Gericht!

Nehmt den Willen für Gewährung,
Kühnen Reim für Begeist*erung*,
Diesen Unsinn als Gedicht!

Der Unverbesserliche

Ihr habt den Vogel gefangen,
Der war so frank und frei,
Nun ist ihm's Fliegen vergangen,
Der Sommer ist lange vorbei.

Es liegen wohl Federn neben
Und unter und über mir,
Sie können mich alle nicht heben
Aus diesem Meer von Papier.

Papier! wie hör' ich dich schreien,
Da alles die Federn schwenkt
In langen, emsigen Reihen –
So wird der Staat nun gelenkt.

Mein Fenster am Pulte steht offen,
Der Sonnenschein schweift übers Dach,
Da wird so uraltes Hoffen
Und Wünschen im Herzen wach.

Die lustigen Kameraden,
Lerchen, Quellen und Wald,
Sie rauschen schon wieder und laden:
Geselle, kommst du nicht bald?

Und wie ich durch die Gardinen
Hinaussah in keckem Mut,
Da hört' ich Lachen im Grünen,
Ich kannte das Stimmlein recht gut.

Und wie ich hinaustrat zur Schwelle,
Da blühten die Bäume schon all'
Ein Liebchen, so frühlingshelle,
Saß drunter beim Vogelschall.

Und eh' wir uns beide besannen,
Da wiehert' das Flügelroß –
Wir flogen selbander von dannen,
Daß es unten die Schreiber verdroß.

Die Werber

„O Frühling, wie bist du helle!
Ade nun Hof und Haus!"
Und jubelnd auf den Schwellen
Mit fröhlichen Gesellen
Wandert der Dichter aus.

Doch ihre Lieder wecken
Rings leises Zischeln bald,
Kobold' aus allen Hecken
Erweisen sich mit Necken
Gar wunderbar im Wald.

Zu Roß, so schön und wüste,
Ein hohes Weib fliegt her,
Behelmt, entblößt die Brüste,
Ihr Aug' weckt wild Gelüste,
Sie heißt Soldatenehr'.

Ihr nach aus Felsenritzen
Schaun graue Wichte klein,
Verstreun von ihren Mützen
Dukaten rings, die blitzen
Blutrot ins Land herein.

Der Schlauste gar durchs Blaue
Als Flügelbübchen schwirrt,
Führt über Berg und Aue
Daher die schönste Fraue –
Die macht erst all' verwirrt.

Und der Dichter in dem Toben
Steht einsam auf der Höh',
Die andern sind zerstoben,
So still nun ist's da oben,
Sein Herz tut ihm so weh.

Er hört der Quellen Gänge
Durch die Waldeinsamkeit,
Da sinnt er auf Gesänge,
Die Welt gibt volle Klänge,
Sein Herz wird ihm so weit.

Und jeden Frühling wieder
Von der schönen Jugendzeit
Singt er vom Berg hernieder,
Und Heimweh faßt die Brüder,
Die in dem Tal zerstreut.

Sonette

1

So viele Quellen von den Bergen rauschen,
Die brechen zornig aus der Felsenhalle,
Die andern plaudern in melod'schem Falle
Mit Nymphen, die im Grün vertraulich lauschen.

Doch wie sie irrend auch die Bahn vertauschen,
Sie treffen endlich doch zusammen alle,

Ein Strom, mit brüderlicher Wogen Schwalle
Erfrischend durch das schöne Land zu rauschen.

An Burgen, die vom Felsen einsam grollen,
Aus Waldesdunkel, zwischen Rebenhügeln
Vorübergleitend in die duft'ge Ferne,

Entwandelt er zum Meer, dem wundervollen,
Wo träumend sich die sel'gen Inseln spiegeln
Und auf den Fluten ruhn die ew'gen Sterne.

2

So eitel künstlich haben sie verwoben
Die Kunst, die selber sie nicht gläubig achten,
Daß sie die Sünd' in diese Unschuld brachten:
Wer unterscheidet, was noch stammt von oben?

Doch wer mag würdig jene Reinen loben,
Die in der Zeit hochmüt'gem Trieb und Trachten
Die heil'ge Flamme treu in sich bewachten,
Aus ihr die alte Schönheit neu erhoben!

O Herr! Gib Demut denen, die da irren,
Daß, wenn ihr' Künste all' zuschanden werden,
Sie töricht nicht den Gott in sich verfluchen!

Begeisterung, was falsch ist, zu entwirren,
Und Freudigkeit, wo's öde wird auf Erden,
Verleihe denen, die Dich redlich suchen!

3

Ein Wunderland ist oben aufgeschlagen,
Wo goldne Ströme gehn und dunkel schallen,
Gesänge durch das Rauschen tief verhallen,
Die möchten gern ein hohes Wort dir sagen.

Viel' goldne Brücken sind dort kühn geschlagen,
Darüber alte Brüder sinnend wallen –
Wenn Töne wie im Frühlingsregen fallen.
Befreite Sehnsucht will dorthin dich tragen.

Wie bald läg' unten alles Bange, Trübe,
Du strebtest lauschend, blicktest nicht mehr nieder,
Und höher winkte stets der Brüder Liebe:

Wen einmal so berührt die heil'gen Lieder,
Sein Leben taucht in die Musik der Sterne,
Ein ewig Ziehn in wunderbare Ferne!

4

Wer einmal tief und durstig hat getrunken,
Den zieht zu sich hinab die Wunderquelle,
Daß er melodisch mit zieht selbst als Welle,
Auf der die Welt sich bricht in tausend Funken.

Es wächst sehnsüchtig, stürzt und leuchtet trunken
Jauchzend im Innersten die heil'ge Quelle,
Bald Bahn sich brechend durch die Kluft zur Helle,
Bald kühle rauschend dann in Nacht versunken.

So laß es ungeduldig brausen, drängen!
Hoch schwebt der Dichter drauf in goldnem Nachen,
Sich selber heilig opfernd in Gesängen.

Die alten Felsen spalten sich mit Krachen,
Von drüben grüßen schon verwandte Lieder,
Zum ew'gen Meere führt er alle wieder.

5

Nicht Träume sind's und leere Wahngesichte,
Was von dem Volk den Dichter unterscheidet.

Was er inbrünstig bildet, liebt und leidet,
Es ist des Lebens wahrhafte Geschichte.

Er fragt nicht viel, wie ihn die Menge richte,
Der eignen Ehr' nur in der Brust vereidet;
Denn wo begeistert er die Blicke weidet,
Grüßt ihn der Weltkreis mit verwandtem Lichte.

Die schöne Mutter, die ihn hat geboren,
Den Himmel liebt er, der ihn auserkoren,
Läßt beide Haupt und Brust sich heiter schmücken.

Die Menge selbst, die herbraust, ihn zu fragen
Nach seinem Recht, muß den Beglückten tragen,
Als Element ihm bietend ihren Rücken.

6

Ihm ist's verliehn, aus den verworrnen Tagen,
Die um die andern sich wie Kerker dichten,
Zum blauen Himmel sich empor zu richten,
In Freudigkeit: Hie bin ich, Herr! zu sagen.

Das Leben hat zum Ritter ihn geschlagen,
Er soll der Schönheit neid'sche Kerker lichten;
Daß nicht sich alle götterlos vernichten,
Soll er die Götter zu beschwören wagen.

Tritt erst die Lieb' auf seine blühnden Hügel,
Fühlt er die reichen Kränze in den Haaren,
Mit Morgenrot muß sich die Erde schmücken;

Süßschauernd dehnt der Geist die großen Flügel,
Es glänzt das Meer – die mut'gen Schiffe fahren,
Da ist nichts mehr, was ihm nicht sollte glücken!

Wehmut

1

Ich kann wohl manchmal singen,
Als ob ich fröhlich sei,
Doch heimlich Tränen dringen,
Da wird das Herz mir frei.

So lassen Nachtigallen,
Spielt draußen Frühlingsluft,
Der Sehnsucht Lied erschallen
Aus ihres Käfigs Gruft.

Da lauschen alle Herzen,
Und alles ist erfreut,
Doch keiner fühlt die Schmerzen,
Im Lied das tiefe Leid.

2

Sage mir mein Herz, was willst du?
Unstät schweift dein bunter Will';
Manches andre Herz wohl stillst du,
Nur du selbst wirst niemals still.

„Eben, wenn ich munter singe,
Um die Angst mir zu zerstreun,
Ruh' und Frieden manchen bringe,
Daß sich viele still erfreun:

Faßt mich erst recht tief Verlangen
Nach viel andrer, beßrer Lust,
Die die Töne nicht erlangen –
Ach, wer sprengt die müde Brust?"

3

Es waren zwei junge Grafen
Verliebt bis in den Tod,
Die konnten nicht ruhn, noch schlafen
Bis an den Morgen rot.

O trau den zwei Gesellen,
Mein Liebchen, nimmermehr,
Die gehn wie Wind und Wellen,
Gott weiß: wohin, woher. –

Wir grüßen Land und Sterne
Mit wunderbarem Klang
Und wer uns spürt von ferne,
Dem wird so wohl und bang.

Wir haben wohl hienieden
Kein Haus an keinem Ort,
Es reisen die Gedanken
Zur Heimat ewig fort.

Wie eines Stromes Dringen
Geht unser Lebenslauf,
Gesanges Macht und Ringen
Tut helle Augen auf.

Und Ufer, Wolkenflügel,
Die Liebe hoch und mild –
Es wird in diesem Spiegel
Die ganze Welt zum Bild.

Dich rührt die frische Helle,
Das Rauschen heimlich kühl,
Das lockt dich zu der Welle,
Weil's draußen leer und schwül.

Doch wolle nie dir halten
Der Bilder Wunder fest,
Tot wird ihr freies Walten,
Hältst du es weltlich fest.

Kein Bett darf es hier finden.
Wohl in den Tälern schön
Siehst du sein Gold sich winden,
Dann plötzlich meerwärts drehn.

Intermezzo

Dein Bildnis wunderselig
Hab' ich im Herzensgrund,
Das sieht so frisch und fröhlich
Mich an zu jeder Stund'.

Mein Herz still in sich singet
Ein altes, schönes Lied,
Das in die Luft sich schwinget
Und zu dir eilig zieht.

Laß das Trauern

Laß, mein Herz, das bange Trauern
Um vergangnes Erdenglück,
Ach, von diesen Felsenmauern
Schweifet nur umsonst der Blick.

Sind denn alle fortgegangen:
Jugend, Sang und Frühlingslust?
Lassen, scheidend, nur Verlangen
Einsam mir in meiner Brust?

Vöglein hoch in Lüften reisen,
Schiffe fahren auf der See,
Ihre Segel, ihre Weisen
Mehren nur des Herzens Weh.

Ist vorbei das bunte Ziehen,
Lustig über Berg und Kluft,
Wenn die Bilder wechselnd fliehen,
Waldhorn immer weiter ruft?

Soll die Lieb' auf sonn'gen Matten
Nicht mehr baun ihr prächtig Zelt,
Übergolden Wald und Schatten
Und die weite, schöne Welt? –

Laß das Bangen, laß das Trauern,
Helle wieder nur den Blick!
Fern von dieser Felsen Mauern
Blüht dir noch gar manches Glück!

Dichterfrühling

Wenn die Bäume lieblich rauschen,
An den Bergen, an den Seen,
Die im Sonnenscheine stehen,
Warme Regen niederrauschen,
Mag ich gern begeistert lauschen.
Denn um die erfrischten Hügel
Auf und nieder sich bewegen
Fühl' ich Winde, Gottes Flügel,
Und mir selber wachsen Flügel,
Atm' ich still den neuen Segen.

Wie der Kranke von der Schwelle
Endlich wieder in die warme
Luft hinausstreckt Brust und Arme,

Und es spült des Lebens Welle
Fort die Glieder in das Helle:
Also kommt ein neues Leben
Oft auf mich herab vom Himmel,
Und ich seh' vor mir mein Streben
Licht und unvergänglich schweben
Durch des Lebens bunt Gewimmel.

Will erquickt nun alles prangen,
Irrt der Dichter durch die Schatten,
Durch die blumenreichen Matten,
Denkt der Zeiten, die vergangen,
Ferner Freunde voll Verlangen,
Und es weben sich die Träume
Wie von selbst zum Werk der Musen,
Und rings Berge, Blumen, Bäume
Wachsen in die heitern Räume
Nach der Melodie im Busen.

Intermezzo

Wohl vor lauter Sinnen, Singen
Kommen wir nicht recht zum Leben;
Wieder ohne rechtes Leben
Muß zu Ende gehn das Singen;
Ging zu Ende dann das Singen:
Mögen wir auch nicht länger leben.

Aufgebot

Waldhorn bringt Kund' getragen,
Es hab' nun aufgeschlagen
Auf Berg und Tal und Feld
Der Lenz seine bunten Zelt'!

Ins Grün ziehn Sänger, Reiter,
Ein jeglich Herz wird weiter,
Möcht' jauchzend übers Grün –
Mit den Lerchen ins Blaue ziehn.

Was stehst du so alleine,
Pilgrim, im grünen Scheine?
Lockt dich der Wunderlaut
Nicht auch zur fernen Braut?

„Ach! diese tausendfachen
Heilig verschlungnen Sprachen,
So lockend Lust, wie Schmerz,
Zerreißen mir das Herz."

„Ein Wort will mir's verkünden,
Oft ist's, als müßt ich's finden,
Und wieder ist's nicht so,
Und ewig frag ich: Wo?" –

So stürz' dich einmal, Geselle,
Nur frisch in die Frühlingswelle!
Da spürst du's im Innersten gleich,
Wo's rechte Himmelreich.

Und wer dann noch mag fragen:
Freudlos in blauen Tagen
Der wandern und fragen mag
Bis an den jüngsten Tag!

Intermezzo

Der Bürgermeister

Hochweiser Rat, geehrte Kollegen!
Bevor wir uns heut aufs Raten legen,

Bitt' ich, erst reiflich zu erwägen:
Ob wir vielleicht, um Zeit zu gewinnen,
Heut sogleich mit dem Raten beginnen,
Oder ob wir erst proponieren müssen,
Was uns versammelt und was wir alle wissen? –
Ich muß pflichtmäßig voranschicken hierbei,
Daß die Art der Geschäfte zweierlei sei:
Die einen sind die eiligen,
Die andern die langweiligen,
Auf jene pfleg' ich *Cito* zu schreiben,
Die andern können liegen bleiben.
Die liegenden aber, geehrte Brüder,
Zerfallen in wicht'ge und in höchstwicht'ge wieder.
Bei jenen – nun – man wird verwegen,
Man schreibt nach amtlichem Überlegen
More solito hier, und dort *ad acta,*
Die Diener rennen, man flucht, verpackt da,
Der Staat floriert und bleibt im Takt da,
Doch werden die Zeiten so ungeschliffen,
Wild umzuspringen mit den Begriffen,
Kommt gar, wie heute, ein Fal, der eilig
Und doch höchstwichtig zugleich – dann freilich
Muß man von neuem unterscheiden:
Ob er mehr eilig oder mehr wichtig. –
Ich bitte, meine Herrn, verstehn Sie mich richtig!
Der Punkt ist von Einfluß. Denn wir vermeiden
Die *species facti* wie billig, sofort,
Find't sich der Fall mehr eilig als liegend,
Ist aber das Wichtige überwiegend,
Wäre die Eile am unrechten Ort,
Meine Herren, Sie haben nun die Prämissen,
Sie werden den Beschluß zu finden wissen.

Terzett

Hirt

Wenn sich der Sommermorgen still erhebt,
Kein Wölkchen in den blauen Lüften schwebt,
Mit Wonneschauern naht das Licht der Welt,
Daß sich die Ährenfelder leise neigen,
Da sink' ich auf die Knie im stillen Feld,
Und bete, wenn noch alle Stimmen schweigen.

Jäger

Doch keiner atmet so den Strom von Lüften,
Als wie der Jäger in den grünen Klüften!
Wo euch der Atem schwindelnd schon vergangen,
Hat seine rechte Lust erst angefangen,
Wenn tief das Tal auffunkelt durch die Bäume,
Der Aar sich aufschwingt in die klaren Räume.

Hirt

Und sinkt der Mittag müde auf die Matten,
Rast' ich am Bächlein in den kühlsten Schatten,
Ein leises Flüstern geht in allen Bäumen,
Das Bächlein plaudert wirre wie in Träumen,
Die Erde säuselt kaum, als ob sie schliefe,
Und mit den Wolken in den stillen Räumen
Schiff' ich still fort zur unbekannten Tiefe.

Jäger

Und wenn die Tiefe schwül und träumend ruht,
Steh' ich am Berg wie auf des Landes Hut,
Seh' fern am Horizont die Wetter steigen,
Und durch die Wipfel, die sich leise neigen,
Rauscht droben schwellend ein gewaltig Lied,
Das ewig frisch mir durch die Seele zieht.

Hirt

Es blitzt von fern, die Heimchen Ständchen bringen,
Und unter Blüten, die im Wind sich rühren,
Die Mädchen plaudernd sitzen vor den Türen;
Da lass' ich meine Flöte drein erklingen,
Daß ringsum durch die laue Sommernacht
In Fels und Brust der Widerhall erwacht.

Jäger

Doch wenn die Täler unten längst schon dunkeln,
Seh' ich vom Berge noch die Sonne funkeln,
Der Adler stürzt sich jauchzend in die Gluten,
Es bricht der Strom mit feuertrunknen Fluten
Durchs enge Steingeklüft, wie er sich rette
Zum ew'gen Meer – ach, wer da Flügel hätte!

Angela

Wenn von den Auen
Die Flöte singt,
Aus Waldesrauschen
Das Horn erklingt,
Da steh' ich sinnend
Im Morgenlicht –
Wem ich soll folgen,
Ich weiß es nicht.

Doch kehrt ihr beide
Im letzten Strahl
Der Sonne wieder
Zurück ins Tal,
Schaut mir so freudig
Ins Angesicht:
Da weiß ich's plötzlich –
Doch sag' ich's nicht.

Intermezzo

Chor der Schmiede

Bis zum künft'gen Holmgang
Nun gehämmert, Nordmann!
Schlängelt sich im Todkampf
Glutrot einst dein Schwertblitz –
Sehr weint da die Heldbraut –
Denk! der Waffenmeister
Hämmert, singet! Ist's auch
Ungereimt, so klappt's doch!

Morgenlied

Ein Stern still nach dem andern fällt
Tief in des Himmels Kluft,
Schon zucken Strahlen durch die Welt,
Ich wittre Morgenluft.

In Qualmen steigt und sinkt das Tal;
Verödet noch vom Fest
Liegt still der weite Freudensaal,
Und tot noch alle Gäst'.

Da hebt die Sonne aus dem Meer
Eratmend ihren Lauf;
Zur Erde geht, was feucht und schwer,
Was klar, zu ihr hinauf.

Hebt grüner Wälder Trieb und Macht
Neurauschend in die Luft,
Zieht hinten Städte, eitel Pracht,
Blau' Berge durch den Duft.

Spannt aus die grünen Tepp'che weich,
Von Strömen hell durchrankt,
Und schallend glänzt das frische Reich,
So weit das Auge langt.

Der Mensch nun aus der tiefen Welt
Der Träume tritt heraus,
Freut sich, daß alles noch so hält,
Daß noch das Spiel nicht aus.

Und nun geht's an ein Fleißigsein!
Umsumsend Berg und Tal,
Agieret lustig groß und klein
Den Plunder allzumal.

Die Sonne steiget einsam auf,
Ernst über Lust und Weh
Lenkt sie den ungestörten Lauf
Zu stiller Glorie. –

Und *wie* er dehnt die Flügel aus,
Und *wie* er auch sich stellt,
Der Mensch kann nimmermehr hinaus
Aus dieser Narrenwelt.

Intermezzo

Chor der Schneider

Nur vom Ganzen frisch gerissen,
Eh' die Ware ganz verschlissen,
Hier ein uralt gülden Stück,
Gibt so 'n gewissen frommen Blick,
Hier ein bunter welscher Flick,
Drauf ein Stück Hausleinewand,
Macht das Welsche erst pikant,

Hie 'nen Fetzen Bärenhaut,
Daß man auch das Deutsche schaut,
Drüber einen span'schen Kragen,
Das Erhabne wird behagen,
Frisch gestichelt, fein zum Werke,
Und wird auch nichts Ganzes draus,
Sieht es doch gar niedlich aus.

Guter Rat

Springer, der in luft'gem Schreiten
Über die gemeine Welt,
Kokettieret mit den Leuten,
Sicherlich vom Seile fällt.

Schiffer, der nach jedem Winde,
Blas' er witzig oder dumm,
Seine Segel stellt geschwinde,
Kommt im Wasser schmählich um.

Weisen Sterne doch die Richtung,
Hörst du nachts doch fernen Klang,
Dorthin liegt das Land der Dichtung,
Fahre zu und frag nicht lang.

Umkehr

Leben kann man nicht von Tönen,
Poesie geht ohne Schuh',
Und so wandt' ich denn der Schönen
Endlich auch den Rücken zu.

Lange durch die Welt getrieben
Hat mich nun die irre Hast,
Immer doch bin ich geblieben
Nur ein ungeschickter Gast.

Überall zu spät zum Schmause
Kam ich, wenn die andern voll,
Trank die Neigen vor dem Hause,
Wußt' nicht, wem ich's trinken soll.

Mußt' mich vor Fortuna bücken
Ehrfurchtsvoll bis auf die Zeh'n
Vornehm wandt' sie mir den Rücken,
Ließ mich so gebogen stehn.

Und als ich mich aufgerichtet
Wieder frisch und frei und stolz,
Sah ich Berg' und Tal gelichtet,
Blühen jedes dürre Holz.

Welt hat eine plumpe Pfote,
Wandern kann man ohne Schuh' –
Deck mit deinem Morgenrote
Wieder nur den Wandrer zu!

Intermezzo

Blonder Ritter

Blonder Ritter, blonder Ritter,
Deine Blicke, weltschmerzdunkel,
Statt durch Helmes Eisengitter,
Durch die Brille gläsern funkeln.

Hinterm Ohre, statt vom Leder,
Zornig mit verwegner Finte
Ziehst du statt des Schwerts die Feder,
Und statt Blutes fließet Dinte.

Federspritzeln, Ehr' beklecken,
Ungeheueres Geschnatter!

Wilde Recken, wilde Recken,
Trampelt nicht die Welt noch platter.

Liedesmut

Was Lorbeerkranz und Lobestand!
Es duftet still die Frühlingsnacht
Und rauscht der Wald vom Felsenrand,
Ob's jemand hört, ob niemand wacht.

Es schläft noch alles Menschenkind,
Da pfeift sein lust'ges Wanderlied
Schon übers Feld der Morgenwind
Und fragt nicht erst, wer mit ihm zieht.

Und ob ihr all' zu Hause faßt,
Der Frühling blüht doch, weil er muß,
Und ob ihr's lest oder bleiben laßt,
Ich singe doch aus frischer Brust.

Entgegnung

„Sei antik doch, sei teutonisch,
Lern, skandiere unverdrossen,
Freundchen, aber nur ironisch!
Und vor allem laß die Possen,
Die man sonst genannt: romantisch." –
Also hört man's ringsher schallen;
Aber mich bedünkt: pedantisch
Sei das Schlimmste doch von allen.

Wem der Herr den Kranz gewunden,
Wird nach alle dem nicht fragen,
Sondern muß, wie er's befunden,
Auf die eigne Weise sagen,

Stets aufs neu' mit freud'gem Schrecken,
Ist sie auch die alte blieben,
Sich die schöne Welt entdecken,
Ewig jung ist, was wir lieben!

Oft durch des Theaters Ritzen
Bricht's mit wunderbarem Lichte,
Wenn der Herr in feur'gen Blitzen
Dichtend schreibt die Weltgeschichte,
Und das ist der Klang der Wehmut,
Der durch alle Dichtergeister
Schauernd geht, wenn sie in Demut
Über sich erkannt den Meister.

Der Isegrimm

Aktenstöße nachts verschlingen,
Schwatzen nach der Welt Gebrauch,
Und das große Tretrad schwingen
Wie ein Ochs, das kann ich auch.

Aber glauben, daß der Plunder
Eben nicht der Plunder wär,
Sondern ein hochwichtig Wunder,
Das gelang mir nimmermehr.

Aber andre überwitzen,
Daß ich mit dem Federkiel
Könnt' den morschen Weltbau stützen,
Schien mir immer Narrenspiel.

Und so, weil ich in dem Drehen
Da steh' oft wie ein Pasquill,
Läßt die Welt mich eben stehen –
Mag sie's halten, wie sie will!

Tafellieder

1

(Damenliedertafel in Danzig)

Die Frauen

Gleich wie Echo frohen Liedern
Fröhlich Antwort geben muß,
So auch nahn wir und erwidern
Dankend den galanten Gruß.

Die Männer

O, ihr Güt'gen und Charmanten!
Für des Echos holden Schwung
Nehmt der lust'gen Musikanten
Ganz ergebne Huldigung!

Frauen

Doch ihr huldigt, will's uns dünken,
Andern Göttern nebenbei.
Rot und golden sehn wir's blinken –
Sagt, wie das zu nehmen sei?

Männer

Teure! zierlich, mit drei Fingern,
Sichrer, mit der ganzen Hand –
Und so füllt man aus den Dingern
's Glas nicht halb, nein, bis zum Rand.

Frauen

Nun, wir sehen, ihr seid Meister.
Doch wir sind heut liberal;
Hoffentlich, als schöne Geister,
Treibt ihr's etwas ideal.

Männer

Jeder nippt und denkt die Seine,
Und wer nichts *Besondres* weiß:
Nun – der trinkt ins allgemeine
Frisch zu *aller* Schönen Preis!

Alle

Recht so! Klingt denn in die Runde
An zu Dank und Gegendank!
Sänger, Fraun, wo die im Bunde,
Da gibt's einen hellen Klang!

2

Trinken und Singen

Viel Essen macht viel breiter
Und hilft zum Himmel nicht,
Es kracht die Himmelsleiter,
Kommt so ein schwerer Wicht.
Das Trinken ist gescheiter,
Das schmeckt schon nach Idee,
Da braucht man keine Leiter,
Das geht gleich in die Höh'.

Chor

Da braucht man keine Leiter,
Das geht gleich in die Höh.

Viel Reden ist manierlich:
„Wohlauf?“ – Ein wenig flau. –
„Das Wetter ist spazierlich.“ –
Was macht die liebe Frau? –
„Ich danke“ – und so weiter,
Und breiter als ein See –
Das Singen ist gescheiter,
Das geht gleich in die Höh’.

Chor

Das Singen ist gescheiter,
Das geht gleich in die Höh’.

Die Fisch’ und Musikanten
Die trinken beide frisch,
Die Wein, die andern Wasser –
Drum hat der dumme Fisch
Statt Flügel Flederwische
Und liegt elend im See –
Doch wir sind keine Fische,
Das geht gleich in die Höh’.

Chor

Doch wir sind keine Fische,
Das geht gleich in die Höh’.

Ja, Trinken frisch und Singen
Das bricht durch alles Weh,
Das sind zwei gute Schwingen,
Gemeine Welt, ade!

Du Erd' mit deinem Plunder,
Ihr Fische samt der See,
's geht alles, alles unter,
Wir aber in die Höh'!

Chor

's geht alles, alles unter,
Wir aber in die Höh!

3

Zum Abschied

Horcht! Die Stunde hat geschlagen,
Und ein Schiffer steht am Bord,
Grüßt noch einmal, und es tragen
Ihn die Wellen rauschend fort.

Sturm wühlt und die Zeiten bäumen
Sehnsüchtig sich himmelan,
Hoch in solcher Wellen Schäumen
Segle, kühner Steuermann!

Und den letzten Becher, Brüder,
Eh' wir hier verlassen stehn,
Und den letzten Klang der Lieder
Auf ein freudig Wiedersehn!

4

Berliner Tafel

Viele Lerchen hellerwacht,
Die zum Himmel steigen,

Viele Sterne in der Nacht,
Vieler Wipfel Neigen,
Viele frische Herzen dann,
Die begeistert lauschen –
Da bricht erst der Lenz recht an,
Klang und Waldesrauschen.

So sind viele hier gesellt:
Rüstige Gesellen,
Die ihr' Sach' auf Klang gestellt,
Schauspiel und Novellen,
Viele dann, die recht sich freun,
Wenn wir's löblich machen,
Und, greift einer falsch darein,
Auch von Herzen lachen.

Und wo *solche* Resonanz,
Klingt das Lied erst helle,
Wie wir hier vereint zum Kranz,
Blüht die sand'ge Schelle,
Kuckuck ruft und Nachtigall
Und von Lust und Schmerzen
Weckt der Schall den Widerhall
Rings in tausend Herzen.

Ein Land, das ihr schweigend meint
Und wir freudig singen,
Und *ein* Meer, das uns vereint
Soll hinüberbringen.
Frische Fahrt denn, nah und fern,
Allen mut'gen Seglern,
Die getreu dem rechten Stern,
Schleglern oder Heglern!

5

Die Heimonskinder

Auf feur'gem Rosse kommt Bachus daher,
Den Becher hoch in der Hand,
Sein Rößlein wird wild, sein Kopf ist ihm schwer,
Er verschüttet den Wein auf das Land.

Den Dichter erbarmet der Rebensaft,
In den Bügel er kühn sich stellt
Und trinkt mit dem Gotte Brüderschaft –
Nun geht's erst, als ging's aus der Welt!

„Ei, sieh da, so einsam, Herr Komponist!
Steig auf mit, 's ist schad um die Schuh',
Du lös'st erst die Schwinge – und wo keine ist,
Da mach uns die Flügel dazu!"

Und was sie ersonnen nun, singen die drei.
„O weh!" ruft ein Sänger herauf,
„Ihr schreit ja die köstlichen Noten entzwei!"
Und schwingt zu den dreien sich auf.

Nun setzt der Tonkünstler, skandiert der Poet,
Der Sänger gibt himmlischen Schall,
Es lächelt Herr Bachus: „Wahrhaftig, das geht,
Und's Trinken verstehen sie all'."

Und wie sie nun alle beisammen sind,
Hebt's sachte die seligen Leut,
Es wachsen dem Rosse zwei Schwingen geschwind
Und überfliegen die Zeit.

6

Der alte Held

(Tafellied zu Goethes Geburtstag 1831)

„Ich habe gewagt und gesungen,
Da die Welt noch stumm lag und bleich,
Ich habe den Bann bezwungen,
Der die schöne Braut hielt umschlungen,
Ich habe erobert das Reich."

„Ich habe geforscht und ergründet
Und tat es euch treulich kund:
Was das Leben dunkel verkündet,
Die heilige Schrift, die entzündet
Der Herr in der Seelen Grund."

„Wie rauschen nun Wälder und Quellen
Und singen vom ewigen Port:
Schon seh' ich Morgenrot schwellen,
Und ihr dort, ihr jungen Gesellen,
Fahrt immer immerfort?"

Und so, wenn es still geworden,
Schaut er vom Turm bei Nacht
Und segnet den Sängerorden,
Der an den blühenden Borden
Das schöne Reich bewacht.

Dort hat er nach Lust und Streiten
Das Panner aufgestellt,
Und die auf dem Strome der Zeiten
Am Felsen vorübergleiten,
Sie grüßen den alten Held.

7

Toast

Auf das Wohlsein der Poeten,
Die nicht schillern und nicht göthen,
Durch die Welt in Lust und Nöten
Segelnd frisch auf eignen Böten.

Treue

Frisch auf, mein Herz! Wie heiß auch das Gedränge,
Bewahr' ich doch mir kühl und frei die Brust!
Schickt Wald und Flut doch noch die alten Klänge,
Erschütternd mich mit wunderbarer Lust,
Und ob die Woge feindlich mit mir ränge:
So frömmer nur sing' ich aus treuer Brust;
Da bleicht das Wetter, Himmelblau scheint helle,
Das Meer wird still und zum Delphin die Welle.

„Was wollt ihr doch mit eurem Liederspaße!
Des Würd'gern beut die große Zeit so viel!"
So schallt's hoffärtig nun auf jeder Gasse,
Und jeder steckt sich dreist sein glänzend Ziel.
Die Lieder, die ich stammelnd hören lasse,
Ew'ger Gefühle schwaches Widerspiel, –
Sie sind es wahrlich auch nicht, was ich meine,
Denn ewig unerreichbar ist das Eine.

Doch lieben oft, der Sehnsucht Glut zu mildern,
Gefangne wohl, das ferne Vaterland
An ihres Kerkers Mauern abzuschildern.
Ein Himmelsstrahl fällt schweifend auf die Wand,
Da rührt's lebendig sich in allen Bildern. –
Dem Auge scheint's ein lieblich bunter Tand –
Doch wer der lichten Heimat recht zu eigen,
Dem wird der Bilder ernster Geist sich zeigen.

So wachse denn und treibe fröhlich Blüte,
Du kräftig grüner, deutscher Sangesbaum!
Rausch nur erfrischend fort mir ins Gemüte
Aus deiner Wipfel klarem Himmelsraum!
Du aber, wunderbare, ew'ge Güte,
Die mir den Himmel wies im schönen Traum,
Erhalt' auf Erden rüstig mir die Seele,
Daß ich, wo's immer ehrlich gilt, nicht fehle!

Heimweh

An meinen Bruder

Du weißt's, dort in den Bäumen
Schlummert ein Zauberbann,
Und nachts oft, wie in Träumen,
Fängt der Garten zu singen an.

Nachts durch die stille Runde
Weht's manchmal bis zu mir,
Da ruf' ich aus Herzensgrunde,
O Bruderherz, nach dir.

So fremde sind die andern,
Mir graut im fremden Land,
Wir wollen zusammen wandern,
Reich treulich mir die Hand!

Wir wollen zusammen ziehen,
Bis daß wir wandermüd'
Auf des Vaters Grabe knien
Bei dem alten Zauberlied.

Dichterlos

Für alle muß vor Freuden
Mein treues Herze glühn,

Für alle muß ich leiden,
Für alle muß ich blühn,
Und wenn die Blüten Früchte haben,
Da haben sie mich längst begraben.

Spruch

Bau nur auf Weltgunst recht
Und pass' auf jeden Wink und Gruß,
Wirst dabei nimmer fröhlich werden!
Es hat's kein Hund so schlecht,
Der hinter seinen Herren muß,
Nicht frei spazieren kann auf Erden.

Lockung

Hörst du nicht die Bäume rauschen
Draußen durch die stille Rund'?
Lockt's dich nicht, hinabzulauschen
Von dem Söller in den Grund,
Wo die vielen Bäche gehen
Wunderbar im Mondenschein
Und die stillen Schlösser sehen
In den Fluß vom hohen Stein?

Kennst du noch die irren Lieder
Aus der alten, schönen Zeit?
Sie erwachen alle wieder
Nachts in Waldeseinsamkeit,
Wenn die Bäume träumend lauschen
Und der Flieder duftet schwül
Und im Fluß die Nixen rauschen –
Komm herab, hier ist's so kühl.

Rückblick

Ich wollt' im Walde dichten
Ein Heldenlied voll Pracht,
Verwickelte Geschichten,
Recht sinnreich ausgedacht.
Da rauschten Bäume, sprangen
Vom Fels die Bäche drein,
Und tausend Stimmen klangen
Verwirrend aus und ein.
Und manches Jauchzen schallen
Ließ ich aus frischer Brust,
Doch aus den Helden allen
Ward nichts vor tiefer Lust.

Kehr' ich zur Stadt erst wieder
Aus Feld und Wäldern kühl,
Da kommen all' die Lieder
Von fern durchs Weltgewühl,
Es hallen Lust und Schmerzen
Noch einmal leise nach,
Und bildend wird im Herzen
Die alte Wehmut wach,
Der Winter auch derweile
Im Feld die Blumen bricht –
Dann gibts vor Langerweile
Ein überlang Gedicht!

Zweifel

Könnt' es jemals denn verblühen,
Dieses Glänzen, dieses Licht,
Das durch Arbeit, Sorgen, Mühen
Wie der Tag durch Wolken bricht,
Blumen, die so farbig glühen,
Um das öde Leben flicht?

Golden sind des Himmels Säume,
Abwärts ziehen Furcht und Nacht,
Rüstig rauschen Ström' und Bäume
Und die heitre Runde lacht,
Ach, das sind nicht leere Träume,
Was im Busen da erwacht!

Bunt verschlingen sich die Gänge,
Tost die Menge hin und her,
Schallen zwischendrein Gesänge,
Die durchs Ganze golden ziehn,
Still begegnet im Gedränge
Dir des Lebens ernster Sinn.

Und das Herz denkt sich verloren,
Besser andrer Tun und Wust,
Fühlt sich wieder dann erkoren,
Ewig einsam doch die Brust.
O des Wechsels, o des Toren,
O der Schmerzen, o der Lust!

Dichterglück

O Welt, bin dein Kind nicht von Hause,
Du hast mir nichts geschenkt,
So hab' ich denn frisch meine Klause
Im Morgenrot mir versenkt.

Fortuna, streif nur die Höhen
Und wende dein Angesicht,
Ich bleibe im Wald bei den Rehen,
Flieg zu, wir brauchen dich nicht.

Und ob auf Höh'n und im Grunde
Kein Streifchen auch meine blieb,
Ich segne dich, schöne Runde,
Ich habe dich dennoch so lieb!

Glückliche Fahrt

Wünsche sich mit Wünschen schlagen,
Und die Gier wird nie gestillt.
Wer ist in dem wüsten Jagen
Da der Jäger, wer das Wild?
Selig, wer es fromm mag wagen,
Durch das Treiben dumpf und wild
In der festen Brust zu tragen
Heil'ger Schönheit hohes Bild!

Sieh, da brechen tausend Quellen
Durch die felsenharte Welt,
Und zum Strome wird ihr Schwellen,
Der melodisch steigt und fällt.
Ringsum sich die Fernen hellen,
Gottes Hauch die Segel schwellt –
Rettend spülen sich die Wellen
In des Herzens stille Welt.

Sommerschwüle

1

Ich klimm' zum Berg und schau' zur niedern Erde,
Ich klimm' hinab und schau' die Berge an,
Süß-melancholisch spitzt sich die Geberde
Und gift'ge Weltverachtung ficht mich an;
Doch will aus Schmerz und Haß nichts Rechtes werden.
Ermanne dich! – Ich bin doch wohl ein Mann? –
Und ach! wie träge Silb' aus Silbe schleichet,
Mit Not hab ich den letzten Reim erreichet.

O weg mit Reim und Leierklang und Singen!
Faß, Leben, wieder mich lebendig an!

Mit deiner Woge will ich freudig ringen,
Die tief mich stürzt, hebt mich auch himmelan.
Im Sturme spannt der Adler seine Schwingen –
Blas zu! Da spür' ich wieder, daß ich Mann!
Viel lieber will ich raschen Tod erwerben,
Als, so verschmachtend, lebenslang zu sterben.

2

Die Nachtigall schweigt, sie hat ihr Nest gefunden,
Träg ziehn die Quellen, die so kühle sprangen,
In trüber Schwüle liegt die Welt umfangen,
So hat den Lenz der Sommer überwunden.

Noch nie hat es die Brust so tief empfunden,
Es ist, als ob viel' Stimmen heimlich sangen:
„Auch *dein* Lenz, froher Sänger, ist vergangen!
An Weib und Kind ist nun der Sinn gebunden!"

O komm, Geliebte, komm zu mir zurücke!
Kann ich nur deine hellen Augen schauen,
Fröhlich Gestirn in dem verworrnen Treiben:

Wölbt hoch sich wieder des Gesanges Brücke,
Und kühn darf ich der alten Lust vertrauen,
Denn ew'ger Frühling will bei *Liebe* bleiben.

Frisch auf!

Ich saß am Schreibtisch bleich und krumm,
Es war mir in meinem Kopf ganz dumm
Vor Dichten, wie ich alle die Sachen
Sollte aufs allerbeste machen.
Da guckt am Fenster im Morgenlicht
Durchs Weinlaub ein wunderschönes Gesicht,
Guckt und lacht, kommt ganz herein

Und kramt mir unter den Blättern mein.
Ich, ganz verwundert: „Ich sollt' dich kennen" –
Sie aber, statt ihren Namen zu nennen:
„Pfui, in dem Schlafrock siehst ja aus
Wie ein verfallenes Schilderhaus!
Willst du denn hier in der Dinte sitzen,
Schau, wie die Felder da draußen blitzen!"
So drängt sie mich fort unter Lachen und Streit,
Mir tat's um die schöne Zeit nur leid.
Drunten aber unter den Bäumen
Stand ein Roß mit funkelnden Zäumen.
Sie schwang sich lustig mit mir hinauf,
Die Sonne draußen ging eben auf,
Und eh' ich mich konnte bedenken und fassen,
Ritten wir rasch durch die stillen Gassen,
Und als wir kamen vor die Stadt,
Das Roß auf einmal zwei Flügel hatt',
Mir schauerte es recht durch alle Glieder:
„Mein Gott, ist's denn schon Frühling wieder?" –
Sie aber wies mir, wie wir so zogen,
Die Länder, die unten vorüberflogen,
Und hoch über dem allerschönsten Wald
Machte sie lächelnd auf einmal halt.
Da sah ich erschrocken zwischen den Bäumen
Meine Heimat unten, wie in Träumen,
Das Schloß, den Garten und die stille Luft,
Die blauen Berge dahinter im Duft,
Und alle die schöne, alte Zeit
In der wundersamen Einsamkeit.
Und als ich mich wandte, war ich allein,
Das Roß nur wiehert' in den Morgen hinein,
Mir aber war's, als wär' ich wieder jung,
Und wußte der Lieder noch genung!

Kriegslied

Nicht mehr in Waldesschauern
An jäher Klüfte Rand,
Wo dunkle Tannen trauern,
Siehst du die Brut mehr lauern
Auf wüster Felsenwand.

Die Greifen nicht mehr fliegen,
Lindwürm' auf heißem Sand
Nicht mehr mit Löwen kriegen,
Auf ihren Bäuchen liegen
Die Drachen im platten Land.

Doch wo das Leben schimmelt,
So weit man reisen kann,
Von Würmern es noch wimmelt,
Und was auf Erden himmelt,
Sie hauchen's giftig an.

Noch halten sie in Schlingen
Die wunderschöne Braut,
Bei Nacht hört man ihr Singen
Die stille Luft durchdringen
Mit tiefem Klagelaut.

Das ist die Brut der Natter,
Die immer neu entstand:
Philister und ihre Gevatter,
Die machen groß Geschnatter
Im deutschen Vaterland.

Sankt Georg, du blanker Streiter,
Leg' deine Lanze ein,
Und wo ein wackrer Reiter,
Dem noch das Herz wird weiter,
Der steche frisch mit drein!

Eldorado

Es ist von Klang und Düften
Ein wunderbarer Ort,
Umrankt von stillen Klüften,
Wir alle spielten dort.

Wir alle sind verirret,
Seitdem so weit hinaus,
Unkraut die Welt verwirret,
Find't keiner mehr nach Haus.

Doch manchmal taucht's aus Träumen,
Als läg' es weit im Meer,
Und früh noch in den Bäumen
Rauscht's wie ein Grüßen her.

Ich hört' den Gruß verfliegen,
Ich folgt' ihm über Land,
Und hatte mich verstiegen
Auf hoher Felsenwand.

Mein Herz ward mir so munter,
Weit hinten alle Not,
Als ginge jenseits unter
Die Welt im Morgenrot.

Der Wind spielt' in den Locken,
Da blitzt' es drunten weit,
Und ich erkannt' erschrocken
Die alte Einsamkeit.

Nun jeden Morgenschimmer
Steig' ich ins Blütenmeer,
Bis ich Glücksel'ger nimmer
Von dorten wiederkehr'.

Frühlingsklage

Ach, was frommt das Wehen, Sprossen,
In der schönen Frühlingszeit:
Ist des Liedes Born verschlossen
Und der Seele Freudigkeit,
Die erst Blüten bringt den Sprossen
Und den Frühling in die Zeit.

Gib den alten Frieden wieder,
In der Brust den Sonnenschein,
Gib die Laute mir und Lieder,
Dann laß blühen oder schnein,
Selbst weck' ich den Lenz mir wieder,
Sollt' es auch der letzte sein!

An die Waldvögel

Konnt' mich auch sonst mit schwingen
Übers grüne Revier,
Hatt' ein Herze zum Singen
Und Flügel wie ihr.

Flog über die Felder,
Da blüht' es wie Schnee,
Und herauf durch die Wälder
Spiegelt' die See.

Ein Schiff sah ich gehen
Fort über das Meer,
Meinen Liebsten drin stehen –
Dacht' meiner nicht mehr.

Und die Segel verzogen,
Und es dämmert' das Feld,
Und ich hab mich verflogen
In der weiten, weiten Welt.

Vorwärts!

Wie der Strom sich schwingt
Aus den Wolken, die ihn tränken,
Alle Bäche verschlingt,
Sie ins Meer zu lenken –
Drein möcht' ich versenken,
Was in mir ringt!

Tritt nur mit in mein Schiff!
Wo wir landen oder stranden,
Erklinget das Riff,
Bricht der Lenz aus dem Sande,
hinter uns dann ins Branden
Versenk' ich das Schiff!

Frühe

Im Osten graut's, der Nebel fällt,
Wer weiß, wie bald sich's rühret!
Doch schwer im Schlaf noch ruht die Welt,
Von allem nichts verspüret.

Nur eine frühe Lerche steigt,
Es hat ihr was geträumet
Vom Lichte, wenn noch alles schweigt,
Das kaum die Höhen säumet.

Zum Abschied

Der Herbstwind schüttelt die Linde,
Wie geht die Welt so geschwinde!
Halte dein Kindlein warm.
Der Sommer ist hingefahren,

Da wir zusammen waren –
Ach, die sich lieben, wie arm!

Wie arm, die sich lieben und scheiden!
Das haben erfahren wir beiden,
Mir graut vor dem stillen Haus.
Dein Tüchlein noch läßt du wehen,
Ich kann's vor Tränen kaum sehen,
Schau still in die Gasse hinaus.

Die Gassen schauen noch nächtig,
Es rasselt der Wagen bedächtig –
Nun plötzlich rascher der Trott
Durchs Tor in die Stille der Felder,
Da grüßen so mutig die Wälder,
Lieb Töchterlein, fahre mit Gott!

Vergebner Ärger

Im alten Hause steh' ich in Gedanken;
Es ist das Haus nicht mehr, der Wind mit Schauern
Geht durch das Gras im Hof, und Eulen lauern
In leeren Fenstern, die schon halb versanken.

Mich ärgern nur die jungen, kecken Ranken,
Die wie zum Spott noch schmücken Tor und Mauern,
Die grünen Birken, die mit falschem Trauern
Leicht überm Grabe meiner Lieben schwanken.

So, Nachteul' selber, auf dem öden Gipfel
Saß ich in meines Jugendglücks Ruinen,
Dumpfbrütend über unerhörten Sorgen;

Da blitzten Frühlingslichter durch die Wipfel,
Die leuchtend unter mir das Land beschienen,
Und nichts nach Eulen fragt der junge Morgen.

Der Wegelagerer

Es ist ein Land, wo die Philister thronen,
Die Krämer fahren und das Grün verstauben,
Die Liebe selber altklug feilscht mit Hauben –
Herr Gott, wie lang willst du die Brut verschonen!

Es ist ein Wald, der rauscht mit grünen Kronen,
Wo frei die Adler horsten, und die Tauben
Unschuldig girren in den grünen Lauben,
Die noch kein Fuß betrat – dort will ich wohnen!

Dort will ich nächtlich auf die Krämer lauern
Und kühn zerhaun der armen Schönheit Bande,
Die sie als niedre Magd zu Markte führen.

Hoch soll sie stehn auf grünen Felsenmauern,
Daß mahnend über alle stillen Lande
Die Lüfte nachts ihr Zauberlied verführen.

Der Glücksritter

Wenn Fortuna spröde tut,
Lass' ich sie in Ruh,
Singe recht und trinke gut,
Und Fortuna kriegt auch Mut,
Setzt sich mit dazu.

Doch ich geb' mir keine Müh':
„He, noch eine her!"
Kehr' den Rücken gegen sie,
Lass' hoch leben die und die –
Das verdrießt sie sehr.

Und bald rückt sie sacht zu mir:
„Hast du deren mehr?"

Wie Sie sehn. – „Drei Kannen schier,
Und das lauter Klebebier!“ –
’s wird mir gar nicht schwer.

Drauf sie zu mir lächelt fein:
„Bist ein ganzer Kerl!“
Ruft den Kellner, schreit nach Wein,
Trinkt mir zu und schenkt mir ein,
Echte Blum’ und Perl’.

Sie bezahlet Wein und Bier,
Und ich, wieder gut,
Führe sie am Arm mit mir
Aus dem Haus, wie’n Kavalier,
Alles zieht den Hut.

Der Schreckenberger

Aufs Wohlsein meiner Dame,
Eine Windfahn’ ist ihr Panier,
Fortuna ist ihr Name,
Das Lager ihr Quartier!

Und wendet sie sich weiter,
Ich kümmre mich nicht drum,
Da draußen ohne Reiter,
Da geht die Welt so dumm.

Statt Pulverblitz und Knattern
Aus jedem wüsten Haus
Gevattern sehn und schnattern
Alle Lust zum Land hinaus.

Fortuna weint vor Ärger,
Es rinnet Perl’ auf Perl’.

„Wo ist der Schreckenberger?
Das war ein andrer Kerl.“

Sie tut den Arm mir reichen,
Fama bläst das Geleit,
So zu dem Tempel steigen
Wir der Unsterblichkeit.

Trost

Es haben viel' Dichter gesungen
Im schönen deutschen Land,
Nun sind ihre Lieder verklungen,
Die Sänger ruhen im Sand.

Aber so lange noch kreisen
Die Stern' um die Erde rund,
Tun Herzen in neuen Weisen
Die alte Schönheit kund.

Im Walde da liegt verfallen
Der alten Helden Haus,
Doch aus den Toren und Hallen
Bricht jährlich der Frühling aus.

Und wo immer müde Fechter
Sinken im mutigen Strauß,
Es kommen frische Geschlechter
Und fechten es ehrlich aus.

An die Dichter

Wo treues Wollen, redlich Streben
Und rechten Sinn der Rechte spürt,

Das muß die Seele ihm erheben,
Das hat mich jedesmal gerührt.

Das Reich des Glaubens ist geendet,
Zerstört die alte Herrlichkeit,
Die Schönheit weinend abgewendet,
So gnadenlos ist unsre Zeit.

O Einfalt gut in frommen Herzen,
Du züchtig schöne Gottesbraut!
Dich schlugen sie mit frechen Scherzen,
Weil dir vor ihrer Klugheit graut.

Wo find'st du nun ein Haus, vertrieben,
Wo man dir deine Wunder läßt,
Das treue Tun, das schöne Lieben,
Des Lebens fromm vergnüglich Fest?

Wo findest du den alten Garten,
Dein Spielzeug, wunderbares Kind,
Der Sterne heil'ge Redensarten,
Das Morgenrot, den frischen Wind?

Wie hat die Sonne schön geschienen!
Nun ist so alt und schwach die Zeit;
Wie stehst so jung du unter ihnen,
Wie wird mein Herz mir stark und weit!

Der Dichter kann nicht mit verarmen;
Wenn alles um ihn her zerfällt,
Hebt ihn ein göttliches Erbarmen –
Der Dichter ist das Herz der Welt.

Den blöden Willen aller Wesen,
Im Irdischen des Herren Spur,
Soll er durch Liebeskraft erlösen,
Der schöne Liebling der Natur.

Drum hat ihm Gott das Wort gegeben,
Das kühn das Dunkelste benennt,
Den frommen Ernst im reichen Leben,
Die Freudigkeit, die keiner kennt,

Da soll er singen frei auf Erden,
In Lust und Not auf Gott vertraun,
Daß aller Herzen freier werden,
Eratmend in die Klänge schaun.

Der Ehre sei er recht zum Horte,
Der Schande leucht' er ins Gesicht!
Viel Wunderkraft ist in dem Worte,
Das hell aus reinem Herzen bricht.

Vor Eitelkeit soll er vor allen
Streng hüten sein unschuld'ges Herz,
Im Falschen nimmer sich gefallen,
Um eitel Witz und blanken Scherz.

Oh, laßt unedle Mühe fahren,
O klingelt, gleißt und spielet nicht
Mit Licht und Gnad', so ihr erfahren,
Zur Sünde macht ihr das Gedicht!

Den lieben Gott laß in dir walten,
Aus frischer Brust nur treulich sing!
Was *wahr* in dir, wird sich gestalten,
Das andre ist erbärmlich Ding. –

Den Morgen seh' ich ferne scheinen,
Die Ströme ziehn im grünen Grund,
Mir ist so wohl! – Die's ehrlich meinen,
Die grüß' ich all' aus Herzensgrund!

Wünschelrute

Schläft ein Lied in allen Dingen,
Die da träumen fort und fort,
Und die Welt hebt an zu singen,
Triffst du nur das Zauberwort.

III. Zeitlieder

Wo ruhig sich und wilder
Unstäte Wellen teilen,
Des Lebens schöne Bilder
Und Kläng' verworren eilen,
Wo ist der sichre Halt? –
So ferne, was wir sollen,
So dunkel, was wir wollen,
Faßt alle die Gewalt.

Die Freunde

I

Wer auf den Wogen schliefe,
Ein sanft gewiegtes Kind,
Kennt nicht des Lebens Tiefe,
Vor süßem Träumen blind.

Doch wen die Stürme fassen
Zu wildem Tanz und Fest,
Wen hoch auf dunklen Straßen
Die falsche Welt verläßt:

Der lernt sich wacker rühren,
Durch Nacht und Klippen hin
Lernt der das Steuer führen
Mit sichrem, ernstem Sinn.

Der ist vom echten Kerne,
Erprobt zu Lust und Pein,
Der glaubt an Gott und Sterne,
Der soll mein Schiffmann sein!

II

An L ...

Vor mir liegen deine Zeilen,
Sind nicht Worte, Schriften nicht,
Pfeile, die verwundend heilen,
Freundesaugen, treu und schlicht.

Niemals konnte so mich rühren
Noch der Liebsten Angesicht,
Wenn uns Augen süß verführen,
Und die Welt voll Glanz und Licht:

Als in Freundesaugen lesen
Meiner eignen Seele Wort,
Fester Treue männlich Wesen,
In Betrübnis Trost und Hort.

So verschlingen in Gedanken
Sich zwei Stämme wundertreu,
Andre dran sich mutig ranken
Kron' an Krone immer neu.

Prächt'ger Wald, wo's kühl zu wohnen,
Stille wachsend Baum an Baum,
Mit den brüderlichen Kronen
Rauschend in dem Himmelsraum!

III

An L …

Mit vielem will die Heimat mich erfreuen,
Ein heitres Schloß an blaugewundnem Flusse,
Gesell'ge Lust, Mutwill' und frohe Muße,
Der Liebe heitres Spiel, süß zu zerstreuen.

Doch wie die Tage freundlich sich erneuen,
Fehlt doch des Freundes Brust in Tag und Muße,
Der Ernst, der herrlich schwelget im Genusse,
Des reichen Blicks sich wahr und recht zu freuen.

Wo zwei sich treulich nehmen und ergänzen,
Wächst unvermerkt das freud'ge Werk der Musen.
Drum laß mich wieder, Freund, ans Herz dich drücken!

Uns beide will noch schön das Leben schmücken
Mit seinen reichen, heitern, vollen Kränzen,
Der Morgenwind wühlt um den offnen Busen!

IV

An Fräulein ...

Schalkhafte Augen reizend aufgeschlagen,
Die Brust empört, die Wünsche zu verschweigen,
Sieht man den leichten Zelter dich besteigen,
Nach Lust und Scherzen durch den Lenz zu jagen.

Zu jung, des Lebens Ernste zu entsagen –
Kann ich nicht länger spielen nun und schweigen,
Wer Herrlich's fühlt, der muß sich herrlich zeigen,
Mein Ruhen ist ein ewig frisches Wagen.

Laß mich, so lang noch trunken unsre Augen,
Ein'n blühnden Kranz aus den vergangnen Stunden
Dir heiter um die weiße Stirne winden;

Frag nicht dann, was mich deinem Arm entwunden,
Drück fest den Kranz nur in die muntern Augen,
Mein Haupt will auch und soll den seinen finden!

V

An Fouqué

1

Seh' ich des Tages wirrendes Beginnen,
Die bunten Bilder fliehn und sich vereinen,
Möcht' ich das schöne Schattenspiel beweinen,
Denn eitel ist, was jeder will gewinnen.

Doch wenn die Straßen leer, einsam die Zinnen
Im Morgenglanze wie Kometen scheinen,
Ein stiller Geist steht auf den dunklen Steine,
Als wollt' er sich auf alte Zeit besinnen:

Da nimmt die Seele rüstig sich zusammen,
An Gott gedenkend und an alles Hohe,
Was rings gedeihet auf der Erden Runde.

Und aus dem Herzen lang verhaltne Flammen,
Sie brechen fröhlich in des Morgens Lohe,
Da grüß' ich, Sänger, dich aus Herzensgrunde!

2

Von Seen und Wäldern eine nächt'ge Runde
Sah ich, und Drachen ziehn mit glühnden Schweifen,
In Eicheswipfeln einen Horst von Greifen,
Das Nordlicht schräge leuchtend überm Grunde.

Durch Qualm dann klingend brach die Morgenstunde,
Da schweiften Ritter blank durch Nebelstreifen,
Durch Winde scharf, die auf der Heide pfeifen,
Ein Harfner sang, lobt' Gott aus Herzensgrunde.

Tiefatmend stand ich über diesen Klüften,
Des Lebens Mark rührt' schauernd an das meine,
Wie ein geharn'schter Riese da erhoben.

Kein ird'scher Laut mehr reichte durch die Lüfte,
Mir war's, als stünde ich mit Gott alleine,
So einsam, weit und sternhell war's da oben.

3

In Stein gehaun, zwei Löwen stehen draußen,
Bewachen ewig stumm die heil'ge Pforte.

Wer sich, die Brust voll Weltlust, naht dem Orte,
Den füllt ihr steinern Blicken bald mit Grausen.

Dir wächst dein Herz noch bei der Wälder Sausen,
Dich rühren noch die wilden Riesenworte.
Nur Gott vertraund, dem höchsten Schirm und Horte –
So magst du bei den alten Wundern hausen.

Ob auch die andern deines Lieds nicht achten,
Der Heldenlust und zarten Liebesblüte,
Gedanken treulos wechseln mit der Mode:

So felsenfester sei dein großes Trachten,
Hau klingend Luft dir, ritterlich Gemüte!
Wir wollen bei dir bleiben bis zum Tode.

Der Riese

Es saß ein Mann gefangen,
Auf einem hohen Turm,
Die Wetterfähnlein klangen
Gar seltsam in den Sturm.

Und draußen hört' er ringen
Verworrner Ströme Gang,
Dazwischen Vöglein singen
Und heller Waffen Klang.

Ein Liedlein scholl gar lustig:
Heisa, so lang Gott will!
Und wilder Menge Tosen;
Dann wieder totenstill.

So tausend Stimmen irren,
Wie Wind' im Meere gehn,

Sich teilen und verwirren,
Er konnte nichts verstehn.

Doch spürt' er, wer ihn grüße,
Mit Schaudern und mit Lust,
Es rührt' ihm wie ein Riese
Das Leben an die Brust.

Sängerfahrt

Kühlrauschend unterm hellen
Tiefblauen Himmelsdom
Treibt seine klaren Wellen
Der ew'gen Jugend Strom.

Viel' rüstige Gesellen,
Den Argonauten gleich,
Sie fahren auf den Wellen
Ins duft'ge Frühlingsreich.

Ich aber fass' den Becher,
Daß es durchs Schiff erklingt,
Am Mast steh' ich als Sprecher,
Der für euch alle singt.

Wie stehn wir hier so helle!
Wird mancher bald schlafen gehn,
O Leben, wie bist du schnelle,
O Leben, wie bist du schön!

Gegrüßt, du weite Runde,
Burg auf der Felsenwand,
Du Land voll großer Kunde,
Mein grünes Vaterland!

Euch möcht' ich alles geben,
Und ich bin fürstlich reich,

Mein Herzblut und mein Leben,
Ihr Brüder, alles für euch!

So fahrt im Morgenschimmer!
Sei's Donau oder Rhein,
Ein rechter Strom bricht immer
Ins ew'ge Meer hinein.

In das Stammbuch der M. H.

Akrostichon mit aufgegebenen Endreimen

Ist hell der Himmel, heiter alle *Wellen,*
Betritt der Schiffer wieder seine *Wogen,*
Vorüber Wald und Berge schnell *geflogen,*
Er muß, wohin die vollen Segel schwellen.
In Duft versinken bald all' liebe *Stellen,*
Cypressen nur noch ragen aus den *Wogen,*
Herüber kommt manch süßer Laut *geflogen,*
Es trinkt das Meer der Klagen sanfte *Quellen,*
Nichts weilt. – Doch zaubern Treue und *Verlangen,*
Da muß sich blühnder alte Zeit *erneuern,*
Oeffnet die Ferne drauf die Wunder*lichtung,*
Ruht dein Bild drin, bekränzt in heil'ger *Dichtung.* –
Fern laß den Freund nach Ost und West nur *steuern,*
Frei scheint er wohl – du hältst ihn doch *gefangen!*

In E...s Stammbuch

Mit einem Blatte, ein Bergschloß vorstellend

In klaren Ebenmaßen, schön gefugt,
Gleich dem Palaste freundlich sich erhebend,
Stark wie die Burg, die von dem Fels dort lugt,
In ernster Höh' der alten Freiheit lebend,

Gleich jenem Turm stets nach dem Höchsten strebend,
Schloß, Burg und was da irdisch, überflügelnd –
Dabei, still wie die See dort, im Gemüt
Des Himmels Blau und was auf Erden blüht,
In frommer Klarheit ewig heiter spielend;
Vor allem dann fern über Strom und Land
Den alten Freunden treulich zugewandt!

Auf dem Schwedenberge

Da hoben bunt und bunter
Sich Zelte in die Luft,
Und Fähnlein wehten munter
Herunter von der Kluft.

Und um die leichten Tische,
An jenem Bächlein klar,
Saß in der kühlen Frische
Der lust'gen Reiter Schar.

Eilt' durch die rüst'gen Zecher
Die Marketenderin,
Reicht' flüchtig ihre Becher,
Nimmt flücht'ge Küsse hin.

Da war ein Toben, Lachen,
Weit in den Wald hinein,
Die Trommel ging, es brachen
Die lust'gen Pfeifen drein.

Durch die verworrnen Klänge
Stürmt' fort manch wilde Brust,
Da schallten noch Gesänge
Von Freiheit und von Lust.

Fort ist das bunte Toben,
Verklungen Sang und Klang,

Und stille ist's hier oben
Viel hundert Jahre lang.

Du Wald, so dunkelschaurig,
Waldhorn, du Jägerlust!
Wie lustig und wie traurig
Rührst du mir an die Brust!

Lieber alles

Soldat sein ist gefährlich,
Studieren sehr beschwerlich,
Das Dichten süß und zierlich,
Der Dichter gar possierlich
In diesen wilden Zeiten.
Ich möcht' am liebsten reiten,
Ein gutes Schwert zur Seiten,
Die Laute in der Rechten,
Studentenherz zum Fechten.
Ein wildes Roß ist's Leben
Die Hufe Funken geben,
Wer's ehrlich wagt, bezwingt es,
Und wo es tritt, da klingt es!

Sonette

An A ...

1

Die Klugen, die nach Gott nicht wollten fragen,
Den heil'gen Kampf gern irdisch möchten schlichten,
Zum Tod kein Herz, nicht Lieb', sich aufzurichten,
Verzehren sich nur selbst in eitlen Klagen.

Sind alle eure Schiffe denn zerschlagen:
Sieht man die heil'ge Flagge *dich* aufrichten,
Vom Liebessturm, der jene mußt' vernichten,
Dein junges Schiff siegreich hinweggetragen.

Südwinde spielen lau um Laut' und Locken,
Im Morgenrot des Hutes Federn schwanken,
Und Gottes Atem macht die Segel schwellen.

Wen noch die alten Heimatklänge locken,
Dem füllt der Segel wie der Töne Schwellen
Die Brust mit jungen, ewigen Gedanken.

2

Wir sind so tief betrübt, wenn wir auch scherzen,
Die armen Menschen mühn sich ab und reisen,
Die Welt zieht ernst und streng in ihren Gleisen,
Ein feuchter Wind verlöscht die lust'gen Kerzen.

Du hast so schöne Worte tief im Herzen,
Du weißt so wunderbare, alte Weisen,
Und wie die Stern' am Firmamente kreisen,
Ziehn durch die Brust dir ewig Lust und Schmerzen.

So laß dein' Stimme hell im Wald erscheinen!
Das Waldhorn fromm wird auf und nieder wehen,
Die Wasser gehn und einsam Rehe weiden.

Wir wollen stille sitzen und nicht weinen,
Wir wollen in den Rhein hinuntersehen,
Und, wird er finster, nicht von sammen scheiden.

3

Es will die Zeit mit ihrem Schutt verdecken
Den hellen Quell, der meiner Brust entsprungen,

Umsonst Gebete himmelan geschwungen,
Sie mögen nicht das Ohr der Gnade wecken.

So laß die Nacht die grausen Flügel strecken,
Nur immerzu, mein tapfres Schiff, gedrungen!
Wer einmal mit den Wogen hat gerungen,
Fühlt sich das Herz gehoben in den Schrecken.

Schießt zu, trefft, Pfeile, die durchs Dunkel schwirren!
Ruhvoll um Klippen überm tück'schen Grunde
Lenk' ich mein Schiff, wohin die Sterne winken.

Mag dann der Steuermann nach langem Irren,
Rasch ziehend alle Pfeile aus der Wunde,
Tot an der Heimatküste niedersinken!

Der Geist

Nächtlich dehnen sich die Stunden,
Unschuld schläft in stiller Bucht,
Fernab ist die Welt verschwunden,
Die das Herz in Träumen sucht.

Und der Geist tritt auf die Zinne,
Und noch stiller wird's umher,
Schauet mit dem starren Sinne
In das wesenlose Meer.

Wer ihn sah bei Wetterblicken
Stehn in seiner Rüstung blank:
Den mag nimmermehr erquicken
Reichen Lebens frischer Drang –

Fröhlich an den öden Mauern
Schweift der Morgensonne Blick,
Da versinkt das Bild mit Schauern
Einsam in sich selbst zurück.

Klage

1809

O könnt' ich mich niederlegen
Weit in den tiefsten Wald,
Zu Häupten den guten Degen,
Der noch von den Vätern alt,

Und dürft' von allem nichts spüren
In dieser dummen Zeit,
Was sie da unten hantieren,
Von Gott verlassen, zerstreut;

Von fürstlichen Taten und Werken,
Von alter Ehre und Pracht
Und was die Seele mag stärken,
Verträumend die lange Nacht!

Denn eine Zeit wird kommen,
Da macht der Herr ein End',
Da wird den Falschen genommen
Ihr unechtes Regiment.

Denn wie die Erze vom Hammer,
So wird das lockre Geschlecht
Gehaun sein von Not und Jammer
Zu festem Eisen recht.

Da wird Aurora tagen
Hoch über den Wald hinauf,
Da gibt's was zu singen und schlagen.
Da wacht, ihr Getreuen, auf.

An ...

Wie nach festen Felsenwänden
Muß ich in der Einsamkeit
Stets auf dich die Blicke wenden.
Alle, die in guter Zeit
Bei mir waren, sah ich scheiden
Mit des falschen Glückes Schaum
Du bliebst schweigend mir im Leiden,
Wie ein treuer Tannenbaum,
Ob die Felder lustig blühn,
Ob der Winter zieht heran,
Immer finster, immer grün –
Reich die Hand mir, wackrer Mann.

Nachtfeier

1810

Decket Schlaf die weite Runde,
Muß ich oft am Fenster lauschen,
Wie die Ströme unten rauschen,
Räder sausen kühl im Grunde,
Und mir ist so wohl zur Stunde;
Denn hinab vom Felsenrande
Spür' ich Freiheit, uralt Sehnen,
Fromm zerbrechend alle Bande,
Über Wälder, Strom und Lande
Keck die großen Flügel dehnen.

Was je Großes brach die Schranken,
Seh' ich durch die Stille gehen,
Helden auf den Wolken stehen,
Ernsten Blickes, ohne Wanken,
Und es wollen die Gedanken
Mit den guten Alten hausen,

Sich in ihr Gespräch vermischen,
Das da kommt in Waldesbrausen.
Manchem füllt's die Brust mit Grausen,
Mich soll's laben und erfrischen!

Tag und Regung war entflohen,
Übern See nur kam Geläute
Durch die monderhellte Weite,
Und rings brannten auf den hohen
Alpen still die bleichen Lohen,
Ew'ge Wächter echter Weihe,
Als, erhoben vom Verderben
Und vom Jammer, da die dreie
Einsam traten in das Freie,
Frei zu leben und zu sterben.

Und so wachen heute viele
Einsam über ihrem Kummer;
Unerquickt von falschem Schlummer,
Aus des Wechsels wildem Spiele
Schauend fromm nach Einem Ziele.
Durch die öde, stumme Leere
Fühl' ich mich euch still verbündet;
Ob der Tag das Recht verkehre,
Ewig strahlt der Stern der Ehre,
Kühn in heil'ger Nacht entzündet.

Zorn

1810

Seh' ich im verfallnen, dunkeln
Haus die alten Waffen hangen,
Zornig aus dem Roste funkeln,
Wenn der Morgen aufgegangen,

Und den letzten Klang verflogen,
Wo im wilden Zug der Wetter,
Aufs gekreuzte Schwert gebogen,
Einst gehaust des Landes Retter;

Und ein neu Geschlecht von Zwergen
Schwindelnd um die Felsen klettern,
Frech, wenn's sonnig auf den Bergen,
Feige krümmend sich in Wettern,

Ihres Heilands Blut und Tränen
Spottend noch einmal verkaufen,
Ohne Klage, Wunsch und Sehnen
In der Zeiten Strom ersaufen;

Denk' ich dann, wie Du gestanden
Treu, da niemand treu geblieben:
Möcht' ich, über unsre Schande
Tiefentbrannt in zorn'gem Lieben,

Wurzeln in der Felsen Marke,
Und empor zu Himmels Lichten
Stumm anstrebend, wie die starke
Riesentanne, mich aufrichten.

Symmetrie

1810

O Gegenwart, wie bist du schnelle,
Zukunft, wie bist du morgenhelle,
Vergangenheit so abendrot!
Das Abendrot soll ewig stehen,
Die Morgenhelle frisch drein wehen,
So ist die Gegenwart nicht tot.

Der Tor, der lahmt auf einem Bein,
Das ist gar nicht zu leiden,
Schlagt ihm das andre Bein entzwei,
So hinkt er doch auf beiden!

Heimkehr

1810

Heimwärts kam ich spät gezogen
Nach dem väterlichen Haus,
Die Gedanken weit geflogen
Über Berg und Tal voraus.
„Nur noch hier aus diesem Walde!"
Sprach ich, streichelt' sanft mein Roß,
„Goldnen Haber kriegst du balde,
Ruhn wir aus auf lichtem Schloß."

„Doch warum auf diesen Wegen
Sieht's so still und einsam aus?
Kommt denn keiner mir entgegen,
Bin ich nicht mehr Sohn vom Haus?
Kein' Hoboe hör' ich schallen,
Keine bunte Truppe mehr
Seh' ich froh den Burgpfad wallen –
Damals ging es lust'ger her."

Über die vergold'ten Zinnen
Trat der Monden eben vor.
„Holla ho! Ist niemand drinnen?
Fest verriegelt ist das Tor.
Wer will in der Nacht mich weisen
Von des Vaters Hof und Haus!"
Mit dem Schwert hau' ich die Eisen,
Und das Tor springt rasselnd auf.

Doch was seh' ich! Wüst, verfallen
Zimmer, Hof und Bogen sind,
Einsam meine Tritte hallen,
Durch die Fenster pfeift der Wind.
Alle Ahnenbilder lagen
Glanzlos in den Schutt verwühlt,
Und die Zither drauf, zerschlagen,
Auf der ich als Kind gespielt.

Und ich nahm die alte Zither,
Trat ans Fenster voller Gras,
Wo so ofte hinterm Gitter
Sonst die Mutter bei mir saß:
Gern mit Märlein mich erbaute,
Daß ich still saß, Abendrot,
Strom und Wälder fromm beschaute –
„Mutter, bist du auch schon tot?"

So war ich in' Hof gekommen, –
Was ich da auf einmal sah,
Hat den Atem mir benommen,
Bleibt mir bis zum Tode nah;
Aufrecht saßen meine Ahnen,
Und kein Laut im Hofe ging,
Eingehüllt in ihre Fahnen,
Da im ewig stillen Ring.

Und den Vater unter ihnen
Sah ich sitzen an der Wand,
Streng und steinern seine Mienen,
Doch in tiefster Brust bekannt;
Und in den gefaltnen Händen
Hielt er ernst ein blankes Schwert,
Tät' die Blicke niemals wenden,
Ewig auf den Stahl gekehrt.

Da rief ich aus tiefsten Schmerzen:
„Vater, sprich ein einzig Wort,
Wälz den Fels von deinem Herzen,
Starre nicht so ewig fort!
Was das Schwert mit seinen Scheinen,
Rede, was dein Schauen will;
Denn mir graust durch Mark und Beine,
Wie du so entsetzlich still." –

Morgenleuchten kam geflogen,
Und der Vater ward so bleich,
Adler hoch darüber zogen
Durch das klare Himmelreich,
Und der Väter stiller Orden
Sank zur Ruh' in Ewigkeit,
Steine, wie es lichte worden,
Standen da im Hof zerstreut.

Nur der Degen blieb da droben
Einsam liegen überm Grab;
„Sei denn Hab und Gut zerstoben,
Wenn ich dich, du Schwert, nur hab'!"
Und ich faßt' es. – Leute wühlten
Über'n Berg, hinab, hinauf,
Ob sie für verrückt mich hielten –
Mir ging hell die Sonne auf.

Gebet

1810

Was soll ich, auf Gott nur bauend,
Schlechter sein, als all' die andern,
Die, so wohlbehaglich schauend,
Froh dem eignen Nichts vertrauend,
Die gemeine Straße wandern?

Warum gabst du mir die Güte,
Die Gedanken himmelwärts,
Und ein ritterlich Gemüte,
Das die Treue heilig hüte
In der Zeit treulosem Scherz?

Was hast du mich blank gerüstet,
Wenn mein Volk mich nicht begehrt,
Keinen mehr nach Freiheit lüstet,
Daß mein Herz, betrübt, verwüstet,
Nur dem Grabe zugekehrt? –

Laß die Ketten mich zerschlagen,
Frei zum schönen Gottesstreit
Deine hellen Waffen tragen,
Fröhlich beten, herrlich wagen,
Gib zur Kraft die Freudigkeit!

Mahnung

1810

1

In Wind verfliegen sah ich, was wir klagen,
Erbärmlich Volk um falscher Götzen Thronen,
Wen'ger Gedanken, deutschen Landes Kronen,
Wie Felsen, aus dem Jammer einsam ragen.

Da mocht' ich länger nicht nach euch mehr fragen,
Der Wald empfing, wie rauschend! den Entflohnen,
In Burgen alt, an Stromeskühle wohnen
Wollt' ich auf Bergen bei den alten Sagen.

Da hört' ich Strom und Wald dort so mich tadeln:
„Was willst, Lebend'ger du, hier überm Leben,
Einsam verwildernd in den eignen Tönen?

Es soll im Kampf der rechte Schmerz sich adeln,
Den deutschen Ruhm aus der Verwüstung heben,
Das will der alte Gott von seinen Söhnen!"

2

Wohl mancher, dem die wirbligen Geschichten
Der Zeit das ehrlich deutsche Herz zerschlagen,
Mag, wie Prinz Hamlet, zu sich selber sagen:
Weh! daß zur Welt ich kam, sie einzurichten!

Weich, aufgelegt zu Lust und fröhlichem Dichten,
Möcht' er so gern sich mit der Welt vertragen,
Doch, Rache fordernd, aus den leichten Tagen
Sieht er der Väter Geist sich stets aufrichten.

Ruhlos und tödlich ist die falsche Gabe:
Des Großen Wink im tiefsten Marke spüren,
Gedanken rastlos – ohne Kraft zum Werke.

Entschließ dich, wie du kannst nun, doch das merke:
Wer in der Not nichts mag, als Lauten rühren,
Des Hand dereinst wächst mahnend aus dem Grabe.

Der Tiroler Nachtwache

1810

In stiller Bucht, bei finstrer Nacht,
Schläft tief die Welt im Grunde,
Die Berge rings stehn auf der Wacht,
Der Himmel macht die Runde,
Geht um und um,
Ums Land herum
Mit seinen goldnen Scharen,
Die Frommen zu bewahren.

Kommt nur heran mit eurer List,
Mit Leitern, Strick und Banden,
Der Herr doch noch viel stärker ist,
Macht euren Witz zu Schanden.
Wie wart ihr klug! –
Nun schwindelt Trug
Hinab vom Felsenrande –
Wie seid ihr dumm! o Schande!

Gleichwie die Stämme in dem Wald
Woll'n wir zusammenhalten,
Ein' feste Burg, Trutz der Gewalt,
Verbleiben treu die alten.
Steig, Sonne, schön!
Wirf von den Höhn
Nacht und die mit ihr kamen,
Hinab in Gottes Namen.

An die Tiroler

Im Jahre 1810

Bei Waldesrauschen, kühnem Sturz der Wogen,
Wo Herden einsam läuten an den Klüften,
Habt ihr in eurer Berge heitern Lüften
Der Freiheit Lebensatem eingesogen.

Euch selbst die Retter, seid ihr ausgezogen,
Wie helle Bäche brechen aus den Klüften;
Hinunter schwindelt Tücke nach den Schlüften,
Der Freiheit Burg sind eure Felsenbogen.
Hochherzig Volk, Genosse größrer Zeiten!

Du sinkst nun in der eignen Häuser Brande,
Zum Himmel noch gestreckt die freien Hände.

O Herr! laß diese Lohen wehn, sich breiten
Auffordernd über alle deutschen Lande,
Und wer da fällt, dem schenk' so glorreich Ende!

An die Meisten

1810

Ist denn alles ganz vergebens?
Freiheit, Ruhm und treue Sitte,
Ritterbild des alten Lebens,
Zog im Lied durch eure Mitte
Hohnverlacht als Don Quixote;
Euch deckt Schlaf mit plumper Pfote,
Und die Ehre ist euch Zote.

Ob sich Kampf erneut' vergliche,
Ob sich roh Gebirgsvolk raufe,
Sucht der Klügre Weg' und Schliche,
Wie er nur sein Haus erlaufe.
Ruhet, stützet nur und haltet!
Untersinkt, was ihr gestaltet,
Wenn der Mutterboden spaltet.

Wie so lustig, ihr Poeten,
An den blumenreichen Hagen
In dem Abendgold zu flöten,
Quellen, Nymphen nachzujagen!
Wenn erst mut'ge Schüsse fallen,
Von den schönen Widerhallen
Laßt ihr zart Sonette schallen.

Wohlfeil Ruhm sich zu erringen,
Jeder ängstlich schreibt und treibet;
Keinem möcht' das Herz zerspringen,
Glaubt sich selbst nicht, was er schreibet.

Seid ihr Männer, seid ihr Christen!
Glaubt ihr, Gott zu überlisten,
So in Selbstsucht feig zu nisten?

Einen Wald doch kenn' ich droben,
Rauschend mit den grünen Kronen,
Stämme brüderlich verwoben,
Wo das alte Recht mag wohnen.
Manche auf sein Rauschen merken,
Und ein neu Geschlecht wird stärken
Dieser Wald zu deutschen Werken.

Der Jäger Abschied

Wer hat dich, du schöner Wald,
Aufgebaut so hoch da droben?
Wohl den Meister will ich loben,
So lang noch mein' Stimm' erschallt.
Lebe wohl,
Lebe wohl, du schöner Wald!

Tief die Welt verworren schallt,
Oben einsam Rehe grasen,
Und wir ziehen fort und blasen,
Daß es tausendfach verhallt:
Lebe wohl,
Lebe wohl, du schöner Wald!

Banner, der so kühle wallt!
Unter deinen grünen Wogen
Hast du treu uns auferzogen,
Frommer Sagen Aufenthalt!
Lebe wohl,
Lebe wohl, du schöner Wald!

Was wir still gelobt im Wald,
Wollen's draußen ehrlich halten,
Ewig bleiben treu die Alten:
Deutsch Panier, das rauschend wallt,
Lebe wohl,
Schirm dich Gott, du schöner Wald!

Auf dem Rhein

Kühle auf dem schönen Rheine,
Fuhren wir vereinte Brüder,
Tranken von dem goldnen Weine,
Singend gute deutsche Lieder.
Was uns dort erfüllt die Brust,
Sollen wir halten,
Niemals erkalten,
Und vollbringen treu mit Lust!
Und so wollen wir uns teilen,
Eines Fels verschiedne Quellen,
Bleiben so auf hundert Meilen
Ewig redliche Gesellen!

Trost

Sag an, du helles Bächlein du,
Von Felsen eingeschlossen,
Du rauschst so munter immerzu,
Wo kommst du hergeflossen?

„Dort oben steht des Vaters Haus
Still in den klaren Lüften,
Da ruhn die alten Helden aus
In den kristallnen Klüften.

Ich sah den Morgen freudig stehn
Hoch auf der Felsenschwelle,
Die Adler ziehn und Ströme gehn,
Und sprang hinaus ins Helle."

Sag an, du königlicher Strom,
Was geht mein Herz mir auf,
Seh' ich dich ziehn durch Waldes Dom?
Wohin führt dich dein Lauf?

„Es treibt und rauscht der Eisenquell
Noch fort mir durch die Glieder;
Die Felsenluft, so kühl und hell,
Lockt zu mir alle Brüder."

Zeichen

So Wunderbares hat sich zugetragen:
Was aus uralten Sagen
Mit tief verworrener Gewalt oft sang
Von Liebe, Freiheit, was das Herz erlabe,
Mit heller Waffen Klang
Es richtet sich geharnischt auf vom Grabe,
Und an den alten Heerschild hat's geschlagen,
Daß Schauer jede Brust durchdrang.

Unmut

O Herbst! betrübt verhüllst du
Strom, Wald und Blumenlust,
Erbleichte Flur, wie füllst du
Mit Sehnsucht nun die Brust!

Weit hinter diesen Höhen,
Die hier mich eng umstellt,

Hör' ich eratmend gehen
Den großen Strom der Welt.

In lichtem Glanze wandelt
Der Helden heil'ger Mut,
Es steigt das Land verwandelt
Aus seiner Söhne Blut.

Auch mich füllt' männlich Trauern.
Wie euch, bei Deutschlands Wehn –
Und muß in Sehnsuchtsschauern
Hier ruhmlos untergehn!

Entschluß

Gebannt im stillen Kreise sanfter Hügel,
Schlingt sich ein Strom von ewig gleichen Tagen,
Da mag die Brust nicht nach der Ferne fragen,
Und lächelnd senkt die Sehnsucht ihre Flügel.

Viel' andre stehen kühn im Rossesbügel,
Des Lebens höchste Güter zu erjagen,
Und was sie wünschen, müssen sie erst wagen,
Ein strenger Geist regiert des Rosses Zügel. –

Was singt ihr lockend so, ihr stillen Matten,
Du Heimat mit den Regenbogenbrücken,
Ihr heitern Bilder, harmlos bunte Spiele?

Mich faßt der Sturm, wild ringen Licht und Schatten,
Durch Wolkenriß bricht flammendes Entzücken –
Nur zu, mein Roß! Wir finden noch zum Ziele!

Abschiedstafel

So rückt denn in die Runde!
Es schleicht die Zeit im Dunkeln,
Sie soll uns rüstig finden
Und heiter, stark und gut!
Gar viel ist zu vollbringen,
Gar vieles muß mißlingen.
So mag die letzte Stunde
Nachleuchten uns und funkeln!
Wo unsre Pfad' sich winden,
Wir sind in Gottes Hut.

Dem Bruder meines Lebens,
Der, fern, mit mir zusammen,
Sei denn aus Herzensgrunde
Das erste Glas gebracht!
Ich brauch' ihn nicht zu nennen,
Er aber wird mich kennen.
Viel Land trennt uns vergebens,
Ihm soll dies Wort, die Stunde,
Durch alle Adern flammen,
Wie ich an ihn gedacht!

Zu dir nun, heitre Schöne,
Wend' ich mich voll Gedanken.
Wie sie zu dir sich wenden,
Muß ich so fröhlich sein.
So weit Poeten wohnen,
So weit der Wälder Kronen,
So weit kunstreiche Töne
Die heiteren Gedanken
Und Himmelsgrüße senden:
Ist alles mein und dein.

Laß nie die Schmach mich sehen,
Daß auch dein Herz, der Lüge

Des andern Volks zum Raube,
Bereuend feig und hohl,
An Licht und Schmuck mag zagen!
Nicht wahr ist, was sie sagen:
Daß Lieb' und Lust vergehen,
Nicht wahr, daß uns betrüge
Der schöne, freud'ge Glaube,
Und also lebe wohl!

Ihr aber, klug' Gesellen,
Die hier mit in dem Kreise,
Wohl quält ihr mich seit Jahren
Mit weisem Rat und Wort. –
Stoßt an, es sei vergessen!
Im Meere, ungemessen,
Sind viele tausend Wellen
Und tausend Schiffe fahren,
Ein jedes seine Reise,
Komm jedes in seinen Port!

Vom Berg hinabgewendet,
Seh' ich die Ströme, Zinnen,
Der Liebsten Schloß darunter –
Nun, Morgenlohe, hülle
In Glorie dein Reich!
Dir, tieflebend'ge Fülle,
Schleudr' ich das Glas hinunter,
Mir schwindeln alle Sinnen,
So wend' ich mich geblendet,
Gott segne dich und euch!

An meinen Bruder 1813

Steig aufwärts, Morgenstunde!
Zerreiß die Nacht, daß ich in meinem Wehe
Den Himmel wiedersehe,

Wo ew'ger Frieden in dem blauen Grunde!
Will Licht die Welt erneuern,
Mag auch der Schmerz in Tränen sich befreien.

Mein lieber Herzensbruder!
Still war der Morgen – *Ein* Schiff trug uns beide,
Wie war die Welt voll Freude!
Du faßtest ritterlich das schwanke Ruder,
Und beide treulich lenkend,
Auf froher Fahrt nur einen Stern bedenkend.

Mich irrte manches Schöne,
Viel reizte mich und viel mußt' ich vermissen.
Von Lust und Schmerz zerrissen,
Was so mein Herz hinausgeströmt in Töne:
Es waren Widerspiele
Von deines Busens ewigem Gefühle.

Da ward die Welt so trübe,
Rings stiegen Wetter von der Berge Spitzen,
Der Himmel borst in Blitzen,
Daß neugestärkt sich Deutschland draus erhübe. –
Nun ist das Schiff zerschlagen,
Wie soll ich ohne *dich* die Flut ertragen! –

Auf *Einem* Fels geboren,
Verteilen kühlerauschend sich zwei Quellen,
Die eigne Bahn zu schwellen,
Doch wie sie fern einander auch verloren:
Es treffen echte Brüder
Im ew'gen Meere doch zusammen wieder.

So wolle Gott du flehen,
Daß Er mit meinem Blut und Leben schalte,
Die Seele nur erhalte,
Auf daß wir freudig einst uns wiedersehen,
Wenn nimmermehr hienieden:
So dort, wo Heimat, Licht und ew'ger Frieden!

Aufbruch

Silbern' Ströme ziehn herunter,
Blumen schwanken fern und nah,
Ringsum regt sich's bunt und bunter –
Lenz! Bist du schon wieder da?

„Reiter sind's, die blitzend ziehen,
Wie viel glänz'ger Ströme Lauf,
Fahnen, liliengleich, erblühen,
Lerchenwirbel, Trommelwirbel
Wecken rings den Frühling auf."

Horch! Was hör' ich draußen klingen
Wild verlockend wie zur Jagd?
Ach, das Herz möcht' mir zerspringen,
Wie es jauchzt und weint und klagt.

„Und in Waldes grünen Hallen,
Tiefe Schauer in der Brust,
Lassen wir die Hörner schallen,
In das Blau die Stimmen hallen,
So zum Schrecken wie zur Lust."

Wehe! Dunkle Wolken decken
Seh' ich all' die junge Pracht,
Feur'ge Todeszungen strecken
Durch die grimme Wetternacht.

„Wettern gleich blüht Kampfesfülle,
Blitze zieht das gute Schwert,
Mancher wird auf ewig stille –
Herr Gott, es gescheh' Dein Wille!
Blast Trompeten! Frisch mein Pferd!"

Regenbogen seh' ich steigen,
Wie von Tränen sprühn die Au,

Jenen sich erbarmend neigen
Über den verweinten Gau.

„Also über Graus und Wogen
Hat der Vater gnadenreich
Ein Triumphtor still gezogen.
Wer da fällt, zieht durch den Bogen
Heim ins ew'ge Himmelreich."

Tusch

Fängt die Sonne an zu stechen,
Tapfer schießen Gras und Kräuter
Und die Bäume schlagen aus:
Muß des Feinds Gewalt zerbrechen,
Nimmt der Winter schnell Reißaus,
Erd' und Himmel glänzen heiter;
Und wir Musikanten fahren
Lustig auf dem Fluß hinunter,
Trommeln, pfeifen, blasen, geigen,
Und die Hörner klingen munter.

Appell

Ich hört' viel' Dichter klagen
Von alter Ehre rein,
Doch wen'ge mochtens wagen
Und selber schlagen drein.

Mein Herz wollt' mir zerspringen,
Sucht' mir ein ander Ziel,
Denn anders sein und singen,
Das ist ein dummes Spiel.

So stieg ich mit Auroren
Still ins Gebirg hinan,
Ich war wie neugeboren,
So kühle weht's mich an.

Und als ich, Bahn mir schaffend,
Zum Gipfel trat hinauf,
Da blitzten schon von Waffen
Ringsum die Länder auf.

Die Hörner hört' ich laden,
Die Luft war streng und klar –
Ihr neuen Kameraden,
Wie singt ihr wunderbar!

Frisch auf, wir wollen uns schlagen,
So Gott will, übern Rhein
Und weiter im fröhlichen Jagen
Bis nach Paris hinein!

Soldatenlied

Was zieht da für schreckliches Sausen,
Wie Pfeifen durch Sturmes Wehn?
Das wendet das Herz recht vor Grausen,
Als sollte die Welt vergehn.

Das Fußvolk kommt da geschritten,
Die Trommeln wirbeln voran,
Die Fahne in ihrer Mitten
Weht über den grünen Plan,
Sie prangt in schneeweißem Kleide
Als wie eine milde Braut,
Die gibt Dem hohe Freude,
Wen Gott ihr angetraut.
Sie haben sie recht umschlossen,

Dicht Mann an Mann gerückt,
So ziehen die Kriegsgenossen
Streng, schweigend und ungeschmückt,
Wie Gottes dunkler Wille,
Wie ein Gewitter schwer,
Da wird es ringsum so stille,
Der Tod nur blitzt hin und her.

Wie seltsame Klänge schwingen
Sich dort von der Waldeshöh'!
Ja, Hörner sind es, die singen
Wie rasend vor Lust und Weh.

Die jungen Jäger sich zeigen
Dort drüben im grünen Wald,
Bald schimmernd zwischen den Zweigen,
Bald lauernd im Hinterhalt.
Wohl sinkt da in ewiges Schweigen
Manch schlanke Rittergestalt,

Die anderen über ihn steigen,
Hurrah! in dem schönen Wald,
Es funkelt das Blau durch die Bäume –
„Ach, Vater, ich komme bald!"

Trompeten nun hör' ich werben
So hell durch die Frühlingsluft,
Zur Hochzeit oder zum Sterben
So übermächtig es ruft.

Das sind meine lieben Reiter,
Die rufen hinaus zur Schlacht,
Das sind meine lustigen Reiter,
Nun, Liebchen, gute Nacht!
Wie wird es da vorne so heiter,
Wie sprühet der Morgenwind,
In den Sieg, in den Tod und weiter,
Bis daß wir im Himmel sind!

Die ernsthafte Fastnacht 1814

Wohl vor Wittenberg auf den Schanzen
Sind der edlen Werber viel,
Wollen da zur Fastnacht tanzen
Ein gar seltsam Ritterspiel.

Und die Stadt vom Felsen droben
Spiegelt sich im Sonnenschein,
Wie ein Jungfräulein erhoben –
Jeder will ihr Bräut'gam sein.

„Jäger! Laßt die Hörner klingen
Durch den Morgen kalt und blank!
Wohl, sie läßt sich noch bezwingen,
Hört sie alten deutschen Klang."

Drauf sie einen Reiter schnelle
Senden, der so fröhlich schaut,
Der bläst seinen Gruß so helle,
Wirbt da um die stolze Braut.

„Sieh, wir werben lang verstohlen
Schon um dich in Not und Tod,
Komm! Sonst wollen wir dich holen,
Wann der Mond scheint blutig rot!"

Bleich schon fallen Abendlichter –
Und der Reiter bläst nur zu,
Nacht schon webt sich dicht und dichter –
Doch das Tor bleibt immer zu.

Nun so spielt denn, Musikanten,
Blast zum Tanz aus frischer Brust!
Herz und Sinne mir entbrannten,
O du schöne, wilde Lust!

Wer hat je so'n Saal gesehen?
Strom und Wälder spielen auf,
Sterne auf und nieder gehen,
Stecken hoch die Lampen auf.

Ja der Herr leucht't selbst zum Tanze,
Frisch denn, Kameraden mein!
Funkelnd schön im Mondesglanze
Strenges Lieb, mußt unser sein! –

Und es kam der Morgen heiter,
Mancher Tänzer lag da tot,
Und *Viktoria* blies der Reiter
Von dem Wall ins Morgenrot.

Schlesier wohl zu Ruhm und Preise
Haben sich dies Lieb gewonnen,
Und ein Schlesier diese Weise
Recht aus Herzenslust ersonnen.

Auf der Feldwacht

Mein Gewehr im Arme steh' ich
Hier verloren auf der Wacht,
Still nach jener Gegend seh' ich,
Hab' so oft dahin gedacht!

Fernher Abendglocken klingen
Durch die schöne Einsamkeit;
So wenn wir zusammen gingen,
Hört' ich's oft in alter Zeit.

Wolken da wie Türme prangen,
Als säh' ich im Duft mein Wien,

Und die Donau hell ergangen
Zwischen Burgen durch das Grün.

Doch wie fern sind Strom und Türme!
Wer da wohnt, denkt mein noch kaum,
Herbstlich rauschen schon die Stürme,
Und ich stehe wie im Traum.

Waffenstillstand der Nacht

Windsgleich kommt der wilde Krieg geritten,
Durch das Grün der Tod ihm nachgeschritten,
Manch Gespenst steht sinnend auf dem Feld,
Und der Sommer schüttelt sich vor Grausen,
Läßt die Blätter, schließt die grünen Klausen,
Ab sich wendend von der blut'gen Welt.

Prächtig war die Nacht nun aufgegangen,
Hatte alle mütterlich umfangen,
Freund und Feind mit leisem Friedenskuß,
Und, als wollt' der Herr vom Himmel steigen,
Hört' ich wieder durch das tiefe Schweigen
Rings der Wälder feierlichen Gruß.

In C. S. ... Stammbuch

Dezember 1814

In verhängnisschweren Stunden,
Streitend für das Vaterland,
Haben wir uns brüderlich gefunden,
In der Menge still erkannt.

Sieh! Es ruhet nun der Degen
Und die hohe Brandung fällt,

Sich verlaufend auf den alten Wegen,
Und langweilig wird die Welt.

Doch der Ernst der heil'gen Stunden
Waltet fort in mancher Brust,
Und was sich wahrhaftig hat verbunden,
Bleibt gesellt in Not und Lust.

Unsichtbar geschwungne Brücken
Halten Lieb' und Lieb' vereint,
Und in allen hellen Lebensblicken
Grüß' ich fern den lieben Freund.

Und so mag der Herr dich segnen!
Frische Fahrt durchs Leben wild,
Gleichen Sinn und freudiges Begegnen,
Wo es immer Hohes gilt!

Der Friedensbote

Schlaf ein, mein Liebchen, schlaf ein,
Leis durch die Blumen am Gitter
Säuselt des Laubes Gezitter,
Rauschen die Quellen herein;
Gesenkt auf den schneeweißen Arm,
Schlaf ein, mein Liebchen, schlaf ein,
Wie atmest du lieblich und warm!

Aus dem Kriege kommen wir heim;
In stürmischer Nacht und Regen,
Wenn ich auf der Lauer gelegen,
Wie dachte ich dorten dein!
Gott stand in der Not uns bei,
Nun droben bei Mondenschein
Schlaf ruhig, das Land ist ja frei!

An meinen Bruder

1815

Was Großes sich begeben,
Der Kön'ge Herrlichkeit,
Du sahst's mit freudigem Beben,
Dir war's vergönnt, zu leben
In dieser Wunderzeit.

Und über diese Wogen
Kam hoch ein himmlisch Bild
Durchs stille Blau gezogen,
Traf mit dem Zauberbogen
Dein Herz so fest und mild.

O wunderbares Grauen,
Zur selben Stund' den Herrn
Im Wetterleuchten schauen,
Und über den stummen Gauen
Schuldloser Liebe Stern!

Und hat nun ausgerungen
Mein Deutschland siegeswund:
Was damals Lieb' gesungen,
Was Schwerter dir geklungen,
Klingt fort im Herzensgrund.

Laß bilden die Gewalten!
Was davon himmlisch war,
Kann nimmermehr veralten,
Wird in der Brust gestalten
Sich manches stille Jahr.

Die Fesseln müssen springen,
Ja, endlich macht sich's frei,

Und Großes wird gelingen
Durch Taten oder Singen,
Vor Gott ist's einerlei.

An Philipp

(Nach einer Wiener Redoutenmelodie)

Kennst du noch den Zaubersaal,
Wo süß Melodien wehen,
Zwischen Sternen ohne Zahl
Frauen auf und nieder gehen?

Kennst du noch den Strom von Tönen,
Der sich durch die bunten Reihen schlang,
Von noch unbekannten Schönen
Und von fernen, blauen Bergen sang?

Sieh! Die lichte Pracht erneut
Fröhlich sich in allen Jahren,
Doch die Brüder sind zerstreut,
Die dort froh beisammen waren.

Und der Blick wird irre schweifen,
Einsam stehst du nun in Pracht und Scherz,
Und die alten Töne greifen
Dir mit tausend Schmerzen an das Herz.

Uhren schlagen durch die Nacht,
Drein verschlafne Geigen streichen,
Aus dem Saale, überwacht,
Sich die letzten Paare schleichen.

So ist unser Fest vergangen,
Und die lust'gen Kerzen löschen aus,
Doch die Sterne draußen prangen,
Und die führen mich und dich nach Haus.

Hermanns Enkel

Altdeutsch! – Altdeutsch? – Nun, das ist,
Was man so in Büchern liest: –
Kluge Rosse – prächt'ge Decken,
Händel, Kruzifixe, Recken –
O, wie herrlich strahlt dies Leben!
Göttlich! – Doch mit Unterschied.
Es versteht sich, daß man's deute –
's wär' doch gar zu unbequem,
Wenn man alles wörtlich nähm',
Wie's da durcheinander blüht! –
Diese Ritter – gute Leute,
Ehrlich, tapfer, brave Reiter –
Gegen uns doch Bärenhäuter!
Eigentlich sind wir wohl weiter.
Lehnstreu – Klöster – Barbarei –
Davon machen *wir* uns frei. –
Fangen wir *so* an zu sichten:
Fürcht' ich, bleibt es bei Gedichten –
Nein doch! Eines, geht mir bei,
Eines bleibt doch: Dies Vernichten
Aller Modesklaverei! –
Hohe Vaterländerei!
Schnitt'st du los nicht Hermanns Söhne
Von des Halstuchs schnöden Schlingen,
Worin, sonder Kraft und Schöne,
Unsre Väter schmählich hingen?
Gabst du nicht dem Löwen Mähne,
Die ihm frech die Zeit gestohlen?
Statt des wind'gen Fracks Geflatter
Der Litefka Schurz aus Polen,
Statt des Franzen knabenglatter
Schnauze: seinen *Henri quatre?* –
Bruder, ich sag's unverhohlen,
Und auch du wirst's nicht bestreiten:
Große Zeichen großer Zeiten! –

Wahrlich, säh' ich nicht den Kragen
Über'n schwarzen Rock geschlagen,
Schien' mir alles Ironie.
Doch wie sprech' ich da? Ironisch –
Dieses Wort ist nicht teutonisch.
Undeutsch ist die falsche Freude:
Künsteln am wahrhaften Wort!
Ob auch feige Poesie
Sauere Gesichter schneide:
Durch den welschen Lügenwitz
Schreitet stramm der Deutsche fort
Hinter seiner Nasenspitz',
Aller Ehrlichkeiten Sitz,
Biderb immer grade aus.
Alles Welsche wird mir Graus,
Seit ich steck' im deutschen Kleide:
Du auch, Liebchen, wähle gleich
Deine Tracht dir altdeutsch aus!
Wie's auf Bildern noch zu schauen:
Wedel von dem Schweif der Pfauen,
Dann von Spitzen, blumenreich,
Wie' ne mittelmäß'ge Scheibe,
Eine steife Halsrotunde!
's ist so über'm schlanken Leibe
Wie ein Regenschirm gespannt,
Oben drauf dann statt dem Knopf
Schwebt der holde Frauenkopf,
In das Blütenmeer von Kragen,
Ariadnen gleich, verschlagen. –
O, und ein moral'scher Kragen!
Denn wer ist da so gewandt,
Flüsternd was ins Ohr zu sagen,
Was nicht gleich die andern wissen:
Und – unmöglich ist das Küssen!

*Der Liedsprecher**

1

Und wo ein tüchtig Leben,
Und wo ein Ehrenhaus,
Da geht der Sänger eben
Gern gastlich ein und aus.

Der freudige Geselle
Grüßt Pfaff und Rittersmann,
Und frische Morgenhelle
Weht all im Liede an.

Und kühn im Rossesbügel
Der Ritter waldwärts zieht,
Und das Gebet nimmt Flügel
Und überfliegt das Lied.

Denn ob's mit Schwert, mit Liedern
Sich Bahn zum Himmel schafft,
's ist *eine* Schar von Brüdern
Und *eine* Liebeskraft.

Wo die vereint, da ranken
Sich willig Stein und Erz,
Da pfeilern die Gedanken
Sich freudig himmelwärts.

Die haben diese Bogen
Kühn über'n wilden Strom
Empörter Zeit gezogen
Zum wunderbaren Dom.

---

* Das vorstehende Lied wurde am 20ten Juni 1822 während der Tafel, welche des damaligen Kronprinzen von Preußen, jetzigen Königs, Majestät, in dem großen Rempter des Marienburger Ritterschlosses gab, von einem Freunde des Verfassers in dem Kostüm der alten Liedsprecher gesungen.

Die Burgen sahn wir fallen,
Die Adler zogen aus,
Wehklagend durch die Hallen
Gehn Winde ein und aus.

Doch droben auf der Zinne
Steht noch der Heldengeist,
Der – was die Zeit beginne –
Still nach dem Kreuze weist.

Es wechseln viel' Geschlechter
Und sinken in die Nacht –
Steh fest, du treuer Wächter,
Und nimm dein Land in acht!

Schon hat zum Kreuzeslichte
Dein Volk sich ernst gewandt,
Im Sturm der Weltgerichte
Tief schaudernd dich erkannt.

Nun hebt sich wieder fröhlich
Dein Haus im Morgenschein,
Die Jungfrau minneselig
Schaut weit ins Land hinein.

Gesänge hör' ich schallen,
Durchs Grün geschmückte Gäst'
Wallfahrten nach den Hallen –
Wem gilt das frohe Fest?

Der Königssohn, ihr Preußen,
Weilt auf dem Ritterschloß,
Das ist nach Adlers Weisen,
Daß er der Höh' Genoß.

Das ist des Königs Walten,
Was herrlich, groß und recht,

Im Wechsel zu erhalten
Dem kommenden Geschlecht.

Er hob die Heldenmale
Zu neuer Herrlichkeit,
Damit das Volk im Tale
Gedenk' der großen Zeit.

Das ewig Alt' und Neue,
Das mit den Zeiten ringt,
Das, Fürst, ist's, was das treue
Herz deines Volks durchdringt.

Wo das noch ehrlich waltet,
Da ist zu Gottes Ruhm
Die Kreuzesfahn' entfaltet,
Und rechtes Rittertum.

O, reicht dem Liedersprecher,
Bevor er scheiden muß,
Den hochgefüllten Becher
Zu seinem besten Gruß!

Doch einzeln nicht verhallen
Darf, was ich jetzt gedacht.
Was jeder meint, von *allen*,
Sei's freudig auch gebracht!

All' ritterliche Geister
Umringen fest den Thron,
Und auf zum höchsten Meister
Dringt treuer Liebe Ton:

Dem ritterlichen *König*
Heil, und dem *Königssohn!*

2

(Als die Kaiserin von Rußland
das Schloß Marienburg besuchte)

Will Lust die Tor' erschließen,
Da bleib' ich draußen nicht,
Das Hohe zu begrüßen,
Das ist des Sängers Pflicht.

Das ist die alte Halle,
Hier sang ich manches Mal,
Die hohen Ritter alle
Rings um mich her im Saal.

Und von dem Heldenstreiten
Erklang manch kühnes Lied,
Das noch in nächt'gen Zeiten
Den stillen Bau durchzieht.

Doch farbenlos vergrauen
Ohn' Blüte Fels und Au –
Es fehlt' der Schmuck der Frauen
Dem hochgewalt'gen Bau.

Die Stärke regt das Wilde,
Und nur, der Kraft gesellt,
Die königliche Milde
Bezwingt die starre Welt. –

Welch' Glanz hat mich umflogen
Und füllt das ganze Haus,
Als pfeilerten die Bogen
Ins Himmelreich hinaus!

Und was der Stein will sagen,
Der Mensch in tiefster Brust,

In Klängen anzuschlagen,
Das ist des Sängers Lust:

O du – gleichbar der Hohen,
Die dieses Haus bewacht
Und Morgenrotes Lohen
Im Norden angefacht –

Was Großes hier ersonnen,
All Segen, der hier weilt,
All Wohl, das hier begonnen,
Dir, hohe Frau, zum Heil!

Und so nun will ich neigen
Mich vor der Majestät –
Dann laßt mich gehn und schweigen,
Bis ihr *Sie* wiederseht.

Der neue Rattenfänger

Juchheisa! Und ich führ' den Zug
Hopp über Feld und Graben.
Des alten Plunders ist genug,
Wir wollen neuen haben.

Was! Wir gering? Ihr vornehm, reich?
Planierend schwirrt die Schere,
Seid Lumps wie wir, so sind wir gleich,
Hübsch breit wird die Misere!

Das alte Lied, das spiel' ich neu,
Da tanzen alle Leute,
Das ist die Vaterländerei,
O Herr, mach uns gescheute! –

Der brave Schiffer

Der Sturm wollt' uns zerschmettern,
Was morsch war, lag zerschellt,
Es schrieb mit feur'gen Lettern
Der Herr, und sprach in Wettern
Zu der erschrocknen Welt.

Durch wilder Wogen Spritzen
Vorüber manchem Riff,
Wo auf Korallenspitzen
Die finstern Nornen sitzen,
Flog da das Preußenschiff.

Das war von echtem Kerne;
Gedankenvoll die Wacht
Schaut durch die wüste Ferne
Zum königlichen Sterne,
Der leuchtet aus der Nacht.

Und ob die Nebel decken,
Was groß und heilig war,
Lenkten da aus den Schrecken
Gewaltig die treuen Recken –
Du mitten in dieser Schar.

Da sah man wohl den schlanken
Wald kühner Masten sich
Zum Himmel pfeilernd ranken!
Du lehntest voll Gedanken
Auf deine Harfe dich.

Bald mächtiger, bald leise,
Mit wunderbarem Klang,
Zogst du Gesangeskreise,
Daß eine tiefe Weise
Das wilde Meer bezwang.

Und Sturm und Nacht verzogen,
Schon blitzt' es hier und da,
Das Land stieg aus den Wogen,
Und unter dem Friedensbogen
Die alte Viktoria. –

Fahr wohl! Wie Adlerschwingen
Wird in der Zeiten Schwung
Dein Ringen und dein Singen
Durch deutsche Herzen klingen,
So bleibst du ewig jung!

Ablösung

Wir saßen gelagert im Grünen,
So traulich und lustig gesellt,
Die Lichter des Frühlings schienen
Holdspielend durchs grüne Gezelt.

Im Frühlingsglanz still auf und nieder
Ergingen der Frauen sich viel',
Und liebliche Augen und Lieder,
Sie hielten ein herzliches Spiel.

Und unten von Tälern und Flüssen
Ein schallendes, wirrendes Reich –
O freudiges, erstes Begrüßen
Von Leben und Lieben zugleich!

Verlassen nun stehen die Räume,
Es schauen und rauschen allein
Die groß gewordenen Bäume
So ernst in die Stille herein.

Von allen, die dort sonst gesessen,
Es sehnet sich niemand hierher,

Sie haben den Frühling vergessen,
Kennt keiner den anderen mehr.

Und wie ich so sinn', da erwachen
Die alten Lieder in mir!
Da hör' ich auf einmal ein Lachen
Und Schallen im grünen Revier.

Und fröhliche Lieder erklangen
Aus Herzensgrunde so recht,
Und unter den Bäumen ergangen
Erblick' ich ein ander Geschlecht.

Geöffnet bleibt ewig zum Feste
Des Frühlings lustiges Haus,
Es schwärmen so wechselnd die Gäste
Da immer herein und heraus.

Die vorigen Lieder verhallen,
Wir sinken verblühend hinab,
Und neue Gesänge erschallen
Hoch über dem blühenden Grab.

An die Lützowschen Jäger

Wunderliche Spießgesellen,
Denkt ihr noch an mich,
Wie wir an der Elbe Wellen
Lagen brüderlich?

Wie wir in des Spreewalds Hallen,
Schauer in der Brust,
Hell die Hörner ließen schallen
So zu Schreck wie Lust?

Mancher mußte da hinunter
Unter den Rasen grün,
Und der Krieg und Frühling munter
Gingen über ihn.

Wo wir ruhen, wo wir wohnen:
Jener Waldeshort
Rauscht mit seinen grünen Kronen
Durch mein Leben fort.

Bei Halle

Da steht eine Burg überm Tale
Und schaut in den Strom hinein,
Das ist die fröhliche Saale,
Das ist der Gibichenstein.

Da hab' ich so oft gestanden,
Es blühten Täler und Höh'n,
Und seitdem in allen Landen
Sah ich nimmer die Welt so schön!

Durchs Grün da Gesänge schallten,
Von Rossen, zu Lust und Streit,
Schauten viel' schlanke Gestalten,
Gleichwie in der Ritterzeit.

Wir waren die fahrenden Ritter,
Eine Burg war noch jedes Haus,
Es schaute durchs Blumengitter
Manch schönes Fräulein heraus.

Das Fräulein ist alt geworden,
Und unter Philistern umher
Zerstreut ist der Ritterorden,
Kennt keiner den andern mehr.

Auf dem verfallenen Schlosse,
Wie der Burggeist, halb im Traum,
Steh' ich jetzt ohne Genossen
Und kenne die Gegend kaum.

Und Lieder und Lust und Schmerzen,
Wie liegen sie nun so weit –
O Jugend, wie tut im Herzen
Mir deine Schönheit so leid.

Wechsel

Es fällt nichts vor, mir fällt nichts ein,
Ich glaub', die Welt steht still,
Die Zeit tritt auf so leis und fein,
Man weiß nicht, was sie will.

Auf einmal rührt sich's dort und hier –
Was das bedeuten mag?
Es ist, als hört'st du über dir
Einen frischen Flügelschlag.

Rasch steigen dunkle Wetter auf,
Schon blitzt's und rauscht die Rund,
Der lust'ge Sturmwind fliegt vorauf –
Da atm' ich aus Herzensgrund.

Abschied

Laß, Leben, nicht so wild die Locken wehen!
Es will so rascher Ritt mir nicht mehr glücken,
Hoch überm Land von diamantnen Brücken:
Mir schwindelt, in den Glanz hinabzusehen.

„Vom Rosse spielend meine Blicke gehen
Nach jüngern Augen, die mein Herz berücken,
Horch, wie der Frühling aufjauchzt vor Entzücken,
Kannst du nicht mit hinab, lass' ich dich stehen."

Kaum noch herzinnig mein, wend'st du dich wieder,
Ist das der Lohn für deine treusten Söhne?
Dein trunkner Blick, fast möcht' er mich erschrecken.

„Wer sagt' dir, daß ich treu, weil ich so schöne?
Leb wohl, und streckst du müde einst die Glieder,
Will ich mit Blumen dir den Rasen decken."

Vorbei

Das ist der alte Baum nicht mehr,
Der damals hier gestanden,
Auf dem ich gesessen im Blütenmeer.
Über den sonnigen Landen.

Das ist der Wald nicht mehr, der sacht
Vom Berge rauschte nieder,
Wenn ich vom Liebchen ritt bei Nacht,
Das Herz voll neuer Lieder.

Das ist nicht mehr das tiefe Tal
Mit den grasenden Rehen,
In das wir nachts viel tausendmal
Zusammen hinausgesehen. –

Es ist der Baum noch, Tal und Wald,
Die Welt ist jung geblieben,
Du aber wurdest seitdem alt,
Vorbei ist das schöne Lieben.

Weltlauf

Was du gestern frisch gesungen,
Ist doch heute schon verklungen,
Und beim letzten Klange schreit
Alle Welt nach Neuigkeit.

War ein Held, der legt' verwegen
Einstmals seinen blut'gen Degen
Als wie Gottes schwere Hand
Über das erschrockne Land.

Mußt's doch blühn und rauschen lassen,
Und den toten Löwen fassen
Knaben nun nach Jungenart
Ungestraft an Mähn' und Bart.

So viel' Gipfel als da funkeln,
Sahn wir abendlich verdunkeln,
Und es hat die alte Nacht
Alles wieder gleich gemacht.

Wie im Turm der Uhr Gewichte
Rücket fort die Weltgeschichte,
Und der Zeiger schweigend kreist,
Keiner rät, wohin er weist.

Aber wenn die ehrnen Zungen
Nun zum letztenmal erklungen,
Auf den Turm der Herr sich stellt,
Um zu richten diese Welt.

Und der Herr hat nichts vergessen,
Was geschehen, wird er messen
Nach dem Maß der Ewigkeit –
O wie klein ist doch die Zeit!

IV. Frühling und Liebe

An die Freunde

Der Jugend Glanz, der Sehnsucht irre Weisen,
Die tausend Ströme durch das duft'ge Land,
Es zieht uns all zu seinen Zauberkreisen. –
Wem Gottesdienst in tiefster Brust entbrannt,
Der sieht mit Wehmut ein unendlich Reisen
Zu ferner Heimat, die er fromm erkannt:
Und was sich *spielend* wob als ird'sche Blume,
Wölbt still den Kelch zum *ernsten* Heiligtume.

So schauet denn das buntbewegte Leben
Ringsum von meines Gartens heitrer Zinn'.
Daß hoch die Bilder, die noch dämmernd schweben –
Wo Morgenglanz geblendet meinen Sinn –
An eurem Blick erwachsen und sich heben.
Verwüstend rauscht die Zeit darüber hin;
In euren treuen Herzen neu geboren,
Sind sie im wilden Strome unverloren.

Anklänge

1

Liebe, wunderschönes Leben,
Willst du wieder mich verführen,
Soll ich wieder Abschied geben
Fleißig ruhigem Studieren?

Offen stehen Fenster, Türen,
Draußen Frühlingsboten schweben,
Lerchen schwirrend sich erheben,
Echo will im Wald sich rühren.

Wohl, da hilft kein Widerstreben,
Tief im Herzen muß ich's spüren:
Liebe, wunderschönes Leben,
Wieder wirst du mich verführen!

2

Hoch über stillen Höhen
Stand in dem Wald ein Haus,
So einsam war's zu sehen
Dort übern Wald hinaus.

Ein Mädchen saß darinnen
Bei stiller Abendzeit,
Tät seidne Fäden spinnen
Zu ihrem Hochzeitskleid.

3

Jagdlied

Durch schwankende Wipfel
Schießt güldner Strahl,
Tief unter den Gipfeln
Das neblige Tal.

Fern hallt es am Schlosse,
Das Waldhorn ruft,
Es wiehern die Rosse
In die Luft, in die Luft!

Bald Länder und Seen
Durch Wolkenzug
Tief schimmernd zu sehen
In schwindelndem Flug,
Bald Dunkel wieder
Hüllt Reiter und Roß,
O Lieb', o Liebe
So laß mich los! –

Immer weiter und weiter
Die Klänge ziehn,
Durch Wälder und Heiden
Wohin, ach wohin?
Erquickliche Frische!
Süßschaurige Lust!
Hoch flattern die Büsche,
Frei schlägt die Brust.

Das Zaubernetz

Fraue, in den blauen Tagen
Hast ein Netz zu ausgehangen,
Zart gewebt aus seidnen Haaren,
Süßen Worten, weißen Armen.

Und die blauen Augen sprachen,
Da ich waldwärts wollte jagen:
„Zieh mir, Schöner, nicht von dannen!"
Ach, da war ich dein Gefangner!

Hörst du nun den Frühling laden? –
Jägers Waldhorn geht im Walde,

Lockend grüßen bunte Flaggen,
Nach dem Sänger alle fragen.

Ach, von euch, ihr Frühlingsfahnen,
Kann ich, wie von dir, nicht lassen!
Reisen in den blauen Tagen
Muß der Sänger mit dem Klange.

Flügel hat, den du gefangen –
Alle Schlingen müssen lassen
Und er wird dir weggetragen,
Wenn die ersten Lerchen sangen.

Liebst du, treu dem alten Sange
Wie dem Sänger, mich wahrhaftig:
Laß dein Schloß, den schönen Garten,
Führ' dich heim in Waldesprachten!

Auf dem Zelter sollst du prangen,
Um die schönen Glieder schlanke
Seide, himmelblau, gespannet,
Als ein süßgeschmückter Knabe.

Und der Jäger, sieht uns fahren,
Und er läßt das Wild, das Jagen,
Will nun ewig mit uns wandern
Mit dem frischen Hörnerklange.

Wer von uns verführt den andern,
Ob es deine Augen taten,
Meine Laut', des Jägers Blasen?
Ach, wir können's nicht erraten;

Aber um uns drei zusammen
Wird der Lenz im grünen Walde
Wohl ein Zaubernetze schlagen,
Dem noch keiner je entgangen.

Der Schalk

Läuten kaum die Maienglocken
Leise durch den lauen Wind,
Hebt ein Knabe froh erschrocken
Aus dem Grase sich geschwind,
Schüttelt in den Blütenflocken
Seine feinen blonden Locken,
Schelmisch sinnend wie ein Kind.

Und nun wehen Lerchenlieder,
Und es schlägt die Nachtigall,
Rauschend von den Bergen nieder
Kommt der kühle Wasserfall,
Rings im Walde bunt Gefieder: –
Frühling, Frühling ist es wieder
Und ein Jauchzen überall.

Und den Knaben hört man schwirren,
Goldne Fäden zart und lind
Durch die Lüfte künstlich wirren –
Und ein süßer Krieg beginnt:
Suchen, Fliehen, schmachtend Irren,
Bis sich alle hold verwirren. –
O beglücktes Labyrinth!

Frühlingsgruß

Es steht ein Berg in Feuer,
In feurigem Morgenbrand,
Und auf des Berges Spitze
Ein Tannbaum überm Land.

Und auf dem höchsten Wipfel
Steh' ich und schau' vom Baum,
O Welt, du schöne Welt, du,
Man sieht dich vor Blüten kaum!

Abendlandschaft

Der Hirt bläst seine Weise,
Von fern ein Schuß noch fällt,
Die Wälder rauschen leise
Und Ströme tief im Feld.

Nur hinter jenem Hügel
Noch spielt der Abendschein –
O hätt' ich, hätt' ich Flügel,
Zu fliegen da hinein!

Elfe

Bleib bei uns! Wir haben den Tanzplan im Tal
Bedeckt mit Mondesglanze,
Johanniswürmchen erleuchten den Saal,
Die Heimchen spielen zum Tanze.

Die Freude, das schöne leichtgläubige Kind,
Es wiegt sich in Abendwinden:
Wo Silber auf Zweigen und Büschen rinnt,
Da wirst du die Schönste finden!

Frühlingsmarsch

Hoch über euren Sorgen
Sah ich vom Berg ins Land
Voll tausend guter Morgen,
Die Welt in Blüten stand.

Was zagt ihr träg und blöde?
Was schön ist, wird doch dein!
Die Welt tut nur so spröde
Und will erobert sein.

Laßt die Trompeten laden,
Durchs Land die Trommeln gehn,
Es wimmeln Kameraden,
Wo rechte Banner wehn.

Wir ziehn durch die Provinzen,
Da funkelt manches Schloß,
Schön Lieb', hol dich vom Zwinger
Und schwing' dich mit aufs Roß!

Und wenn das Blühen endet:
Noch taumelnd sprengen wir,
Vom Abendrot geblendet,
Ins letzte Nachtquartier.

Die Lerche

Ich kann hier nicht singen,
Aus dieser Mauern dunklen Ringen
Muß ich mich schwingen
Vor Lust und tiefem Weh,
O Freude, in klarer Höh'
Zu sinken und sich zu heben,
In Gesang
Über die grüne Erde dahin zu schweben,
Wie unten die licht' und dunkeln Streifen
Wechselnd im Fluge vorüberschweifen,
Aus der Tiefe ein Wirren und Rauschen und Hämmern,
Die Erde aufschimmernd im Frühlingsdämmern,
Wie ist die Welt so voller Klang!
Herz, was bist du bang?
Mußt aufwärts dringen!
Die Sonne tritt hervor,
Wie glänzen mir Brust und Schwingen,
Wie still und weit ist's droben am Himmelstor!

Nachtigall

Nach den schönen Frühlingstagen,
Wenn die blauen Lüfte wehen,
Wünsche mit dem Flügel schlagen
Und im Grünen Amor zielt,
Bleibt ein Jauchzen auf den Höhen;
Und ein Wetterleuchten spielt
Aus der Ferne durch die Bäume
Wunderbar die ganze Nacht,
Daß die Nachtigall erwacht,
Von den irren Widerscheinen,
Und durch alle sel'ge Gründe
In der Einsamkeit verkünde,
Was sie alle, alle meinen:
Dieses Rauschen in den Bäumen
Und der Mensch in dunklen Träumen.

Adler

Steig nur, Sonne,
Auf die Höh'n!
Schauer wehn,
Und die Erde bebt vor Wonne.

Kühn nach oben
Greift aus Nacht
Waldespracht,
Noch von Träumen kühl durchwoben.

Und vom hohen
Felsaltar
Stürzt der Aar
Und versinkt in Morgenlohen.

Frischer Morgen!
Frisches Herz,
Himmelwärts!
Laß den Schlaf nun, laß die Sorgen!

Durcheinander

Spatzen schrein und Nachtigallen,
Nelke glüht und Distel sticht,
Rose schön durch Nesseln bricht,
Besser noch hat mir gefallen
Liebchens spielendes Augenlicht;
Aber fehlte auch nur Eins von allen,
's wär eben der närrische Frühling nicht.

Gleichheit

Es ist kein Blümlein nicht so klein,
Die Sonne wird's erwarmen,
Scheint in das Fenster mild herein
Dem König wie dem Armen,
Hüllt alles ein in Sonnenschein
Mit göttlichem Erbarmen.

Gedenk

Es ist kein Vöglein so gemein,
Es spürt geheime Schauer,
Wenn draußen streift der Sonnenschein
Vergoldend seinen Bauer.

Und du hast es vergessen fast
In deines Kerkers Spangen,
O Menschlein, daß du Flügel hast
Und daß du hier gefangen.

Die Sperlinge

Altes Haus mit deinen Löchern,
Geiz'ger Bauer, nun ade!
Sonne scheint, von allen Dächern
Tröpfelt lustig schon der Schnee,
Draußen auf dem Zaune munter
Wetzen unsre Schnäbel wir,
Durch die Hecken rauf und runter,
In dem Baume vor der Tür
Tummeln wir in hellen Haufen
Uns mit großem Kriegsgeschrei,
Um die Liebste uns zu raufen,
Denn der Winter ist vorbei!

Schneeglöckchen

's war doch wie ein leises Singen
In dem Garten heute nacht,
Wie wenn laue Lüfte gingen:
„Süße Glöcklein, nun erwacht,
Denn die warme Zeit wir bringen,
Eh's noch jemand hat gedacht." –
's war kein Singen, 's war ein Küssen,
Rührt' die stillen Glöcklein sacht,
Daß sie alle tönen müssen
Von der künft'gen bunten Pracht.
Ach, sie konnten's nicht erwarten,
Aber weiß vom letzten Schnee
War noch immer Feld und Garten,
Und sie sanken um vor Weh.
So schon manche Dichter streckten
Sangesmüde sich hinab,
Und der Frühling, den sie weckten,
Rauschet über ihrem Grab.

Spaziergang

Ochse, wie bist du so stattlich, bedachtsam, fleißig und nützlich!
Wahrlich, ich brauche dich sehr – aber du bist doch ein Ochs!

Ho da! Kartoffeln und ihr, ökonomische Knollengewächse,
Schreiten kaum kann man; gemach! macht euch nicht gar zu sehr breit!

Grüß dich, Klatschrose und Gänseblum', Butterblum', ländliches Völkchen,
Schmucklos und ohne Geruch, unschuldig, – weiter sonst nichts? –

Nelke, du reizendes Kind, wie hast du so gar nichts Bescheidnes!
Jauchzende Farben voll Lust flammst du ins traurige Grün,

Tief von den eigenen Düften du selber lustig berauschet,
Spiele denn, brenne, von dir lass' ich berauschen mich gern!

Mädchenseele

Gar oft schon fühlt' ich's tief, des Mädchens Seele
Wird nicht sich selbst, dem Liebsten nur geboren.
Da irrt sie nun verstoßen und verloren,
Schickt heimlich Blicke schön als Boten aus,
Daß sie auf Erden suchen ihr ein Haus.
Sie schlummert in der Schwüle, leicht bedeckt,
Lächelt im Schlafe, atmet warm und leise,
Doch die Gedanken sind fern auf der Reise,

Und auf den Wangen flattert träum'risch Feuer,
Hebt buhlend oft der Wind den zarten Schleier.
Der Mann, der da zum ersten Mal sie weckt,
Zuerst hinunterlangt in diese Stille,
Dem fällt sie um den Hals vor Freude bang
Und läßt ihn nicht mehr all ihr Lebelang.

Steckbrief

Grüß euch aus Herzensgrund:
Zwei Augen hell und rein,
Zwei Röslein auf dem Mund,
Kleid blank aus Sonnenschein!

Nachtigall klagt und weint,
Wollüstig rauscht der Hain,
Alles die Liebste meint:
Wo weilt sie so allein?

Weil's draußen finster war,
Sah ich viel hellern Schein,
Jetzt ist es licht und klar,
Ich muß im Dunkeln sein.

Sonne nicht steigen mag,
Sieht so verschlafen drein,
Wünschet den ganzen Tag,
Daß wieder Nacht möcht' sein.

Liebe geht durch die Luft,
Holt fern die Liebste ein;
Fort über Berg und Kluft!
Und Sie wird doch noch mein!

Morgenständchen

In den Wipfeln frische Lüfte,
Fern melod'scher Quellen Fall,
Durch die Einsamkeit der Klüfte
Waldeslaut und Vogelschall,
Scheuer Träume Spielgenossen,
Steigen all' beim Morgenschein
Auf des Weinlaubs schwanken Sprossen
Dir ins Fenster aus und ein.
Und wir nahn noch halb in Träumen,
Und wir tun in Klängen kund,
Was da draußen in den Bäumen
Singt der weite Frühlingsgrund.
Regt der Tag erst laut die Schwingen:
Sind wir alle wieder weit –
Aber tief im Herzen klingen
Lange nach noch Lust und Leid.

Aussicht

Komm zum Garten denn, du Holde!
In den warmen, schönen Tagen
Sollst du Blumenkränze tragen,
Und vom kühl kristallnen Golde
Mit den frischen, roten Lippen,
Eh' ich trinke, lächelnd nippen.
Ohne Maß dann, ohne Richter,
Küssend, trinkend singt der Dichter
Lieder, die von selbst entschweben:
Wunderschön ist doch das Leben!

Abendständchen

Schlafe, Liebchen, weil's auf Erden
Nun so still und seltsam wird!

Oben gehn die goldnen Herden,
Für uns alle wacht der Hirt.

In der Ferne ziehn Gewitter;
Einsam auf dem Schifflein schwank,
Greif' ich draußen in die Zither,
Weil mir gar so schwül und bang.

Schlingend sich an Bäum' und Zweigen,
In dein stilles Kämmerlein
Wie auf goldnen Leitern steigen
Diese Töne aus und ein.

Und ein wunderschöner Knabe
Schifft hoch über Tal und Kluft,
Rührt mit seinem goldnen Stabe
Säuselnd in der lauen Luft.

Und in wunderbaren Weisen
Singt er ein uraltes Lied,
Das in linden Zauberkreisen
Hinter seinem Schifflein zieht.

Ach, den süßen Klang verführet
Weit der buhlerische Wind,
Und durch Schloß und Wand ihn spüret
Träumend jedes schöne Kind.

Nacht

1

Die Vöglein, die so fröhlich sangen,
Der Blumen bunte Pracht,
's ist alles unter nun gegangen,
Nur das Verlangen
Der Liebe wacht.

2

Tritt nicht hinaus jetzt vor die Tür,
Die Nacht hat eignen Sang,
Das Waldhorn ruft, als rief's nach dir,
Betrüglich ist der irre Klang,
Endlos der Wälder Labyrinth –
Behüt' dich Gott, du schönes Kind!

3

Über'm Lande die Sterne
Machen die Runde bei Nacht,
Mein Schatz ist in der Ferne,
Liegt am Feuer auf der Wacht.

Über's Feld bellen Hunde;
Wenn der Mondschein erblich,
Rauscht der Wald auf im Grunde:
Reiter, jetzt hüte dich!

4

Hörst du die Gründe rufen
In Träumen halb verwacht?
O, von des Schlosses Stufen
Steig nieder in die Nacht! –

Die Nachtigallen schlagen,
Der Garten rauschet sacht,
Es will dir Wunder sagen
Die wunderbare Nacht.

Wahl

Der Tanz, der ist zerstoben,
Die Musik ist verhallt,
Nun kreisen Sterne droben,
Zum Reigen singt der Wald.

Sind alle fortgezogen,
Wie ist's nun leer und tot!
Du rufst vom Fensterbogen:
„Wann kommt das Morgenrot!"

Mein Herz möcht' mir zerspringen,
Darum so wein' ich nicht,
Darum so muß ich singen,
Bis daß der Tag anbricht.

Eh' es beginnt zu tagen:
Der Strom geht still und breit,
Die Nachtigallen schlagen,
Mein Herz wird mir so weit!

Du trägst so rote Rosen,
Du schaust so freudenreich,
Du kannst so fröhlich kosen,
Was stehst du still und bleich?

Und laß sie gehn und treiben
Und wieder nüchtern sein,
Ich will wohl bei dir bleiben!
Ich will dein Liebster sein!

Die Stille

Es weiß und rät es doch keiner,
Wie mir so wohl ist, so wohl!

Ach, wüßt' es nur Einer, nur Einer,
Kein Mensch es sonst wissen sollt'!

So still ist's nicht draußen im Schnee,
So stumm und verschwiegen sind
Die Sterne nicht in der Höhe,
Als meine Gedanken sind.

Ich wünscht' es wäre schon Morgen,
Da fliegen zwei Lerchen auf,
Die überfliegen einander,
Mein Herze folgt ihrem Lauf.

Ich wünscht', ich wäre ein Vöglein
Und zöge über das Meer,
Wohl über das Meer und weiter,
Bis daß ich im Himmel wär'!

Frühlingsnetz

Im hohen Gras der Knabe schlief,
Da hört' er's unten singen,
Es war, als ob die Liebste rief,
Das Herz wollt' ihm zerspringen.

Und über ihm ein Netze wirrt
Der Blumen leises Schwanken,
Durch das die Seele schmachtend irrt
In lieblichen Gedanken.

So süße Zauberei ist los,
Und wunderbare Lieder
Gehn durch der Erde Frühlingsschoß,
Die lassen ihn nicht wieder.

Das Mädchen

Stand ein Mädchen an dem Fenster,
Da es draußen Morgen war,
Kämmte sich die langen Haare,
Wusch sich ihre Äuglein klar.

Sangen Vöglein aller Arten,
Sonnenschein spielt' vor dem Haus,
Draußen überm schönen Garten
Flogen Wolken weit hinaus.

Und sie dehnt' sich in den Morgen,
Als ob sie noch schläfrig sei,
Ach, sie war so voller Sorgen,
Flocht ihr Haar und sang dabei:

Wie ein Vöglein hell und reine,
Ziehet draußen muntre Lieb,
Lockt hinaus zum Sonnenscheine,
Ach, wer da zu Hause blieb'!

Die Studenten

Die Jäger ziehn in grünen Wald
Und Reiter blitzend übers Feld,
Studenten durch die ganze Welt,
So weit der blaue Himmel wallt.

Der Frühling ist der Freudensaal,
Viel' tausend Vöglein spielen auf,
Da schallt's im Wald bergab, bergauf:
Grüß dich, mein Schatz, viel tausendmal!

Viel rüstige Bursche ritterlich,
Die fahren hier in Stromes Mitt',

Wie wilde sie auch stellen sich,
Trau mir, mein Kind, und fürcht' dich nicht!

Querüber übers Wasser glatt
Laß werben deine Äugelein,
Und der dir wohlgefallen hat,
Der soll dein lieber Buhle sein.

Durch Nacht und Nebel schleich' ich sacht,
Kein Lichtlein brennt, kalt weht der Wind,
Riegl auf, riegl auf bei stiller Nacht,
Weil wir so jung beisammen sind!

Ade nun, Kind, und nicht geweint!
Schon gehen Stimmen da und dort,
Hoch übern Wald Aurora scheint,
Und die Studenten reisen fort.

Der Gärtner

Wohin ich geh' und schaue,
In Feld und Wald und Tal,
Vom Berg hinab in die Aue:
Viel schöne, hohe Fraue,
Grüß ich dich tausendmal.

In meinem Garten find ich
Viel' Blumen, schön und fein,
Viel' Kränze wohl draus wind' ich
Und tausend Gedanken bind' ich
Und Grüße mit darein.

Ihr darf ich keinen reichen,
Sie ist zu hoch und schön,
Die müssen alle verbleichen,

Die Liebe nur ohnegleichen
Bleibt ewig im Herzen stehn.

Ich schein' wohl froher Dinge
Und schaffe auf und ab,
Und, ob das Herz zerspringe,
Ich grabe fort und singe
Und grab mir bald mein Grab.

Jägerkatechismus

Was wollt ihr in dem Walde haben,
Mag sich die arme Menschenbrust
Am Waldesgruße nicht erlaben,
Am Morgenrot und grüner Lust?

Was tragt ihr Hörner an der Seite,
Wenn ihr des Hornes Sinn vergaßt,
Wenn's euch nicht selbst lockt in die Weite,
Wie ihr vom Berg früh morgens blast?

Ihr werd't doch nicht die Lust erjagen,
Ihr mögt durch alle Wälder gehn;
Nur müde Füß' und leere Magen –
Mir möcht' die Jägerei vergehn!

O nehmet doch die Schneiderelle,
Guckt in der Küche in den Topf!
Sonntags dann auf des Hauses Schwelle,
Krau' euch die Ehfrau auf dem Kopf!

Die Tierlein selber: Hirsch' und Rehen,
Was lustig haust im grünen Haus,
Sie fliehen auf ihre freien Höhen
Und lachen arme Wichte aus.

Doch kommt ein Jäger, wohlgeboren,
Das Horn irrt, er blitzt rosenrot,
Da ist das Hirschlein wohl verloren,
Stellt selber sich zum lust'gen Tod.

Vor allem aber die Verliebten,
Die lad' ich ein zur Jägerlust,
Nur nicht die weinerlich Betrübten;
Die recht von frisch' und starker Brust.

Mein Schatz ist Königin im Walde,
Ich stoß' ins Horn, ins Jägerhorn!
Sie hört mich fern und naht wohl balde,
Und was ich blas, ist nicht verlorn! –

Der Kadett

Meine Liebste, die ist von allen
Grade die Schönste nicht,
Doch hat mir eben gefallen
Ihr spielendes Augenlicht.

Da kann ich von Glücke sagen,
Denn wär' sie die Schönste just,
Müßt' ich mit allen mich schlagen
Um die Eine nach Herzenslust.

Übermut

Ein' Gems auf dem Stein,
Ein' Vogel im Flug,
Ein Mädel, das klug,
Kein Bursch holt die ein.

Der Polack

Und komm' ich, komm' ich ohne Pelz,
Mein' Liebste fragt mich aus:
Wo hast du lassen deinen Pelz?
Und macht sich doch nichts draus.

Da drüben ist gut Schnaps und Bier,
Der Wirt bläst Klarinett,
Da stritten wir, drei oder vier,
Wer's schönste Liebchen hätt'.

Ich aber trank aus deinem Schuh
Ließ meinen Pelz im Haus
Und eine Handvoll Haar' dazu,
Ich mach mir gar nichts draus.

Der Jäger

Was Segeln der Wünsche durch luftige Höh'!
Was bildendes Träumen im blühenden Klee!
Was Hoffen und Bangen, was Schmachten, was Weh!

Und rauscht nicht die Erde in Blüten und Duft?
Und schreitet nicht Hörnerklang kühn durch die Luft?
Und stürzet nicht jauchzend der Quell von der Kluft?

Drum jage du frisch auch dein flüchtiges Reh
Durch Wälder und Felder, durch Täler und See,
Bis es dir ermüdet im Arme vergeh'!

Der Landreiter

Ich ging bei Nacht einst über Land,
Ein Bürschlein traf ich draußen,
Das hat 'nen Stutzen in der Hand
Und zielt auf mich voll Grausen.
Ich renne, da ich mich erbos',
Auf ihn in vollem Rasen,
Da drückt das kecke Bürschlein los
Und ich stürzt' auf die Nasen.
Er aber lacht mir ins Gesicht,
Daß er mich angeschossen,
Kupido war der kleine Wicht –
Das hat mich sehr verdrossen.

Der Bote

Am Himmelsgrund schießen
So lustig die Stern',
Dein Schatz läßt dich grüßen
Aus weiter, weiter Fern'!

Hat eine Zither gehangen
An der Tür unbeacht',
Der Wind ist gegangen
Durch die Saiten bei Nacht.

Schwang sich auf dann vom Gitter
Über die Berge, über'n Wald –
Mein Herz ist die Zither,
Gibt ein'n fröhlichen Schall.

Die Jäger

Wir waren ganz herunter,
Da sprach Diana ein,

Die blickt' so licht und munter,
Nun geht's zum Wald hinein!

Im Dunkeln Äuglein funkeln,
Kupido schleichet leis,
Die Bäume heimlich munkeln –
Ich weiß wohl, was ich weiß!

Der Winzer

Es hat die Nacht geregnet,
Es zog noch grau ins Tal,
Und ruhten still gesegnet
Die Felder überall;
Von Lüften kaum gefächelt,
Durchs ungewisse Blau
Die Sonne verschlafen lächelt'
Wie eine wunderschöne Frau.

Nun sah ich auch sich heben
Aus Nebeln unser Haus,
Du dehntest zwischen den Reben
Dich von der Schwelle hinaus,
Da funkelt' auf einmal vor Wonne
Der Strom und Wald und Au –
Du bist mein Morgen, meine Sonne,
Meine liebe, verschlafene Frau!

Der Poet

Bin ich fern Ihr: schau' ich nieder
Träumend in die Täler hier,
Ach, ersinn' ich tausend Lieder,
Singt mein ganzes Herz von Ihr.
Doch was hilft die Gunst der Musen,

Daß die Welt mich Dichter nennt?
Keiner fragt, wie mir im Busen
Sorge tief und Sehnsucht brennt.

Ja, darf ich bei Liebchen weilen:
Fühl' ich froh der Stunden Schwall
Wohl melodischer enteilen
Als der schönste Silbenfall,
Will ich singen, Lippen neigen
Sich auf mich und leiden's nicht,
Und wie gerne mag ich schweigen,
Wird mein Leben zum Gedicht!

Die Kleine

Zwischen Bergen, liebe Mutter,
Weit den Wald entlang,
Reiten da drei junge Jäger
Auf drei Rößlein blank,
lieb' Mutter,
Auf drei Rößlein blank.

Ihr könnt fröhlich sein, lieb' Mutter,
Wird es draußen still:
Kommt der Vater heim vom Walde,
Küßt Euch, wie er will,
lieb' Mutter,
Küßt Euch, wie er will.

Und *ich* werfe mich im Bettchen
Nachts ohn' Unterlaß,
Kehr' mich links und kehr' mich rechts hin,
Nirgends hab' ich was,
lieb' Mutter,
Nirgends hab' ich was.

Bin ich eine Frau erst einmal,
In der Nacht dann still
Wend' ich mich nach allen Seiten,
Küß, so viel ich will,
lieb' Mutter,
Küß, so viel ich will.

Die Stolze

Sie steckt mit der Abendröte
In Flammen rings das Land,
Und hat samt Manschetten und Flöte
Den verliebten Tag verbrannt.

Und als nun verglommen die Gründe,
Sie stieg auf die stillen Höh'n,
Wie war da rings um die Schlünde
Die Welt so groß und schön!

Waldkönig zog durch die Wälder
Und stieß ins Horn vor Lust,
Da klang über die stillen Felder,
Wovon der Tag nichts gewußt. –

Und wer mich wollt' erwerben,
Ein Jäger müßt's sein zu Roß,
Und müßt' auf Leben und Sterben
Entführen mich auf sein Schloß!

Der Freiwerber

Frühmorgens durch die Winde kühl
Zwei Ritter hergeritten sind,
Im Garten klingt ihr Saitenspiel,
Wach' auf, wach' auf, mein schönes Kind!

Ringsum viel' Schlösser schimmernd stehn,
So silbern geht der Ströme Lauf,
Hoch, weit rings Lerchenlieder wehn,
Schließ Fenster, Herz und Äuglein auf!

So wie du bist, verschlafen heiß,
Laß allen Putz und Zier zu Haus,
Tritt nur herfür im Hemdlein weiß,
Siehst so gar schön verliebet aus.

Ich hab' einen Fremden wohl bei mir,
Der lauert unten auf der Wacht,
Der bittet schön dich um Quartier,
Verschlafnes Kind, nimm dich in acht!

Jäger und Jägerin

Sie

Wär' ich ein muntres Hirschlein schlank,
Wollt' ich im grünen Walde gehn,
Spazieren gehn bei Hörnerklang,
Nach meinem Liebsten mich umsehn.

Er

Nach meiner Liebsten mich umsehn
Tu' ich wohl, zieh' ich früh von hier,
Doch Sie mag niemals zu mir gehn,
Im dunkelgrünen Waldrevier.

Sie

Im dunkelgrünen Waldrevier
Da blitzt der Liebste rosenrot,

Gefällt so sehr dem armen Tier,
Das Hirschlein wünscht, es läge tot.

Er

Und wär' das schöne Hirschlein tot,
So möcht' ich jagen länger nicht;
Scheint über'n Wald der Morgen rot:
Hüt', schönes Hirschlein, hüte dich!

Sie

Hüt', schönes Hirschlein, hüte dich!
Spricht's Hirschlein selbst in seinem Sinn:
Wie soll ich, soll ich hüten mich,
Wenn ich so sehr verliebet bin?

Er

Weil ich so sehr verliebet bin,
Wollt' ich das Hirschlein, schön und wild,
Aufsuchen tief im Walde drin
Und streicheln, bis es stille hielt.

Sie

Ja, streicheln, bis es stille hielt,
Falsch locken so in Stall und Haus!
Zum Wald springt's Hirschlein frei und wild
Und lacht verliebte Narren aus.

Der Tanzmeister

Wohlgerüstet war ich kommen;
Siegsgewiß, doch wie zum Scherz
Hat ein Blick mein Herz genommen –
Wer kann kämpfen ohne Herz?

So vom Augenblick – geschlagen,
Kniet' ich Armer vor ihr hin,
Hatt' kein Herz nun, ihr zu sagen,
Daß ich ihr Entherzter bin.

Die Braut

Wann die Bäume blühn und sprossen
Und die Lerche kehrt zurück,
Denkt die Seele der Genossen,
Fühlet fern' und nahes Glück.

Selig Weinen sel'ger Herzen!
Wenn das Herz nichts weiter will,
Nicht von Lust erfüllt, noch Schmerzen,
Aber fröhlich ist und still.

Frischer sich die Hügel kränzen,
Heitrer lacht das weite Blau,
Alle Blumen schöner glänzen
Durch des Auges süßen Tau.

Und soll denn das Lieben leiden,
Und, wer leidet, krank auch sein,
Ach, so will ich keine Freuden,
Und mag nicht gesund mehr sein!

Die Geniale

Lustig auf den Kopf, mein Liebchen,
Stell dich, in die Luft die Bein'!
Heisa! Ich will sein dein Bübchen,
Heute nacht soll Hochzeit sein!

Wenn du Shakespeare kannst vertragen,
O du liebe Unschuld du!
Wirst du mich wohl auch ertragen
Und noch jedermann dazu. –

Der verzweifelte Liebhaber

Studieren will nichts bringen,
Mein Rock hält keinen Stich,
Meine Zither will nicht klingen,
Mein Schatz, der mag mich nicht.

Ich wollt', im Grünen spazierte
Die allerschönste Frau,
Ich wär' ein Drach' und führte
Sie mit mir fort durchs Blau.

Ich wollt', ich jagt' gerüstet
Und legt' die Lanze aus,
Und jagte all Philister
Zur schönen Welt hinaus.

Ich wollt, ich säß jetzunder
Im Himmel still und weit,
Und früg' nach all dem Plunder
Nichts vor Zufriedenheit.

Der Glückliche

Ich hab' ein Liebchen lieb recht von Herzen,
Hellfrische Augen hat's wie zwei Kerzen,
Und wo sie spielend streifen das Feld,
Ach, wie so lustig glänzet die Welt!

Wie in der Waldnacht zwischen den Schlüften
Plötzlich die Täler sonnig sich klüften,
Funkeln die Ströme, rauscht himmelwärts
Blühende Wildnis – so ist mein Herz!

Wie vom Gebirge ins Meer zu schauen,
Wie wenn der Seefalk, hangend im Blauen,
Zuruft der dämmernden Erd', wo sie blieb? –
So unermeßlich ist rechte Lieb!

Der Nachtvogel

Liegt der Tag rings auf der Lauer,
Blickt so schlau auf Lust und Trauer:
Kann ich kaum mich selbst verstehen.
Laß die Lauscher schlafen gehen!
Nur ein Stündchen unbewacht
Laß in der verschwiegnen Nacht
Mich in deine Augen sehen
Wie in stillen Mondenschein.
In dem Park an der Rotunde,
Wenn es dunkelt, harr' ich dein.
Still und fromm ja will ich sein.
Liebste, ach nur Eine Stunde! –
Sieh mir nicht so böse drein!
Willst du nie dein Schweigen brechen,
Ewig stumm wie Blumen sein:
O so laß mich das Versprechen
Pflücken dir vom stillen Munde:
Liebste, ach nur Eine Stunde!
In dem Park, an der Rotunde,
Wenn es dunkelt, harr' ich dein.

Coda

Und kann ich nicht sein
Mit dir zu zwein,
So will ich, allein,
Der Schwermut mich weihn!

Die Nachtblume

Nacht ist wie ein stilles Meer,
Lust und Leid und Liebesklagen
Kommen so verworren her
In dem linden Wellenschlagen.

Wünsche wie die Wolken sind,
Schiffen durch die stillen Räume,
Wer erkennt im lauen Wind,
Ob's Gedanken oder Träume? –

Schließ' ich nun auch Herz und Mund,
Die so gern den Sternen klagen:
Leise doch im Herzensgrund
Bleibt das linde Wellenschlagen.

Der Dichter

Nichts auf Erden nenn' ich mein
Als die Lieder meiner Laute,
Doch nenn' den, der freud'ger schaute
In die schöne Welt hinein!
Alles Lebens tiefste Schöne
Tun geheimnisvoll ja Töne
Nur dem frommen Sänger kund,
Und *die* Freude sagt kein Mund,
Die Gott wunderbar gelegt

In des Dichters Herzensgrund.
Wenn die Welt, so wild bewegt,
Ängstlich schaut nach ihren Rettern:
Über aller Nebel Wogen
Wölbt Er kühn den Friedensbogen,
Und, wie nach verzognen Wettern,
Rauscht die Erde wieder mild,
Alle Knospen Blüten treiben,
Und der Frühling ist sein Haus,
Und *der* Frühling geht nie aus. –
O du lieblich Frauenbild!
Willst du bei dem Sänger bleiben? –
Blumen bind't ein streng Geschick:
Wenn die tausend Stimmen singen,
Alle Schmerzen, alles Glück
Ewig lautlos zu verschweigen.
Doch bei kühlem Mondenblick
Regt ihr stiller Geist die Schwingen,
Möcht' dem duftgen Kelch entsteigen.
Sieh, schon ist die Sonn' gesunken
Aus der dunkelblauen Schwüle,
Und zerspringt in tausend Funken
An den Felsen rings und Bäumen,
Bis sie alle selig träumen.
Mit den Sternen in der Kühle
Blühn da Wünsche, steigen Lieder
Aus des Herzens Himmelsgrund,
Und ich fühle alles wieder:
Alte Freuden, junges Wagen!
Ach! so viel möcht' ich dir sagen,
Sagen recht aus Herzensgrund,
In dem Rauschen, in dem Wehen,
Möcht' ich fröhlich mit dir gehen,
Plaudern in der lauen Nacht,
Bis der Morgenstern erwacht! –

An eine Tänzerin

Kastagnetten lustig schwingen
Seh' ich dich, du zierlich Kind!
Mit der Locken schwarzen Ringen
Spielt der sommerlaue Wind.
Künstlich regst du schöne Glieder,
Glühendwild,
Zärtlichmild
Tauchest in Musik du nieder
Und die Woge hebt dich wieder.

Warum sind so blaß die Wangen,
Dunkelfeucht der Augen Glanz,
Und ein heimliches Verlangen
Schimmert glühend durch den Tanz?
Schalkhaft lockend schaust du nieder,
Liebesnacht
Süß erwacht,
Wollüstig erklingen Lieder –
Schlag nicht so die Augen nieder!

Wecke nicht die Zauberlieder
In der dunklen Tiefe Schoß,
Selbst verzaubert sinkst du nieder,
Und sie lassen dich nicht los.
Tödlich schlingt sich um die Glieder
Sündlich Glühn,
Und verblühn
Müssen Schönheit, Tanz und Lieder,
Ach, ich kenne dich nicht wieder!

Klage

Ich hab' manch Lied geschrieben,
Die Seele war voll Lust,

Von treuem Tun und Lieben,
Das beste, was ich wußt'.

Was mir das Herz bewogen,
Das sagte treu mein Mund,
Und das ist nicht erlogen,
Was kommt aus Herzensgrund.

Liebchen wußt's nicht zu deuten
Und lacht' mir ins Gesicht,
Dreht' sich zu andern Leuten
Und achtet's weiter nicht.

Und spielt' mit manchem Tropfe,
Weil ich so tief betrübt.
Mir ist so dumm im Kopfe,
Als wär' ich nicht verliebt.

Ach Gott, wem soll ich trauen?
Will Sie mich nicht verstehn,
Tun all so fremde schauen,
Und alles muß vergehn.

Und alles irrt zerstreuet –
Sie ist so schön und rot –
Ich hab' nichts, was mich freuet,
Wär' ich viel lieber tot!

Trauriger Winter

Nun ziehen Nebel, falbe Blätter fallen,
Öd' alle Stellen, die uns oft entzücket!
Und noch einmal tief' Rührung uns beglücket,
Wie aus der Flucht die Abschiedslieder schallen.

Wohl manchem blüht aus solchem Tod Gefallen:
Daß er, nun eng ans blühnde Herz gedrücket,
Von roten Lippen holdre Sträuße pflücket,
Als Lenz je beut mit Wäldern, Wiesen allen.

Mir sagte niemals ihrer Augen Bläue:
„Ruh auch aus! Willst du ewig sinnen?"
Und einsam sah ich so den Sommer fahren.

So will ich tief des Lenzes Blüte wahren,
Und mit Erinnern zaubrisch mich umspinnen,
Bis ich nach langem Traum erwach' im Maie.

Trauriger Frühling

Mir ist's im Kopf so wüste,
Die Zeit wird mir so lang,
Wie auch der Lenz mich grüßte
Mit Glanz und frischem Klang,
Das Herz bleibt mir so wüste,
Mir ist so sterbensbang.

Viel' Vöglein lockend sangen
Im blühenden Revier,
Ich hatt' mir eins gefangen,
Jetzt ist es weit von mir,
Viel' Vöglein draußen sangen,
Ach, hätt' ich meins nur hier!

Begegnung

Ich wandert' in der Frühlingszeit,
Fern auf den Bergen gingen
Mit Geigenspiel und Singen
Viel' lust'ge Hochzeitsleut,

Das war ein Jauchzen und Klingen!
Es blühte rings in Tal und Höh'n,
Ich konnt' vor Lust nicht weitergehn.

Am Dorfe dann auf grüner Au
Begannen sie den Reigen,
Und durch den Schall der Geigen
Lacht' laut die junge Frau,
Ihr Stimmlein klang so eigen,
Ich wußte nicht, wie mir geschehn –
Da wandt' sie sich in wildem Drehn.

Es war mein Lieb! 's ist lange her,
Sie blickt' so ohne Scheue,
Verloren ist die Treue,
Sie kannte mich nicht mehr –
Da jauchzt' und geigt's aufs neue,
Ich aber wandt' mich fort ins Feld,
Nun wandr' ich bis ans End der Welt!

Der Kranke

Vögelein munter
Singen so schön,
Laßt mich hinunter
Spazieren gehn!

„Nacht ist's ja draußen;
's war nur der Sturm,
Den du hörst sausen
Droben vom Turm."

Liebchen im Garten
Seh' ich dort stehn,
Lang' mußt' sie warten,
O laßt mich gehn.

„Still nur, der blasse
Tod ist's, der sacht
Dort durch die Gasse
Schleicht in der Nacht."

Wie mir ergraute,
Bleiches Gesicht!
Gebt mir die Laute,
Mir wird so licht!

„Was willst du singen
In tiefster Not?
Lenz, Lust vergingen,
Liebchen ist tot!" –

Laßt mich, Gespenster!
Lied, riegl auf die Gruft!
Öffnet die Fenster,
Luft, frische freie Luft!

Im Herbst

Der Wald wird falb, die Blätter fallen,
Wie öd' und still der Raum!
Die Bächlein nur gehn durch die Buchenhallen
Lind rauschend wie im Traum,
Und Abendglocken schallen
Fern von des Waldes Saum.

Was wollt ihr mich so wild verlocken
In dieser Einsamkeit?
Wie in der Heimat klingen diese Glocken
Aus stiller Kinderzeit –
Ich wende mich erschrocken,
Ach, was mich liebt, ist weit!

So brecht hervor nur, alte Lieder,
Und brecht das Herz mir ab!
Noch einmal grüß' ich aus der Ferne wieder,
Was ich nur Liebes hab,
Mich aber zieht es nieder
Vor Wehmut wie ins Grab.

Die Hochzeitsänger

Fernher ziehn wir durch die Gassen,
Stehn im Regen und im Wind,
Wohl von aller Welt verlassen
Arme Musikanten sind.
Aus den Fenstern Geigen klingen,
Schleift und dreht sich's bunt und laut,
Und wir Musikanten singen
Draußen da der reichen Braut.

Wollt' sie doch keinen andern haben,
Ging mit mir durch Wald und Feld,
Prächtig in den blauen Tagen
Schien die Sonne auf die Welt,
Heisa: lustig Drehn und Ringen,
Jeder hält sein Liebchen warm,
Und wir Musikanten singen
Lustig so, daß Gott erbarm'.

Lachend reicht man uns die Neigen,
Auf ihr Wohlsein trinken wir;
Wollt' sie sich am Fenster zeigen,
's wäre doch recht fein von ihr.
Und wir fiedeln und wir singen
Manche schöne Melodei,
Daß die besten Saiten springen,
's war, als spräng' mir's Herz entzwei.

Jetzt ist Schmaus und Tanz zerstoben,
Immer stiller wird's im Haus,
Und die Mägde putzen oben
Alle lust'gen Kerzen aus.
Doch wir blasen recht mit Rasen
Jeder in sein Instrument,
Möcht' in meinem Grimm ausblasen
Alle Stern' am Firmament!

Und am Hause seine Runde
Tritt der Wächter gähnend an,
Rufet aus die Schlafensstunde,
Und sieht ganz erbost uns an.
Doch nach ihrem Kabinette
Schwing' ich noch mein Tamburin,
Fahr wohl in dein Himmelbette,
Weil wir müssen weiter ziehn!

Der letzte Gruß

Ich kam vom Walde hernieder,
Da stand noch das alte Haus,
Mein Liebchen, sie schaute wieder
Wie sonst zum Fenster hinaus.

Sie hat einen andern genommen,
Ich war draußen in Schlacht und Sieg.
Nun ist alles anders gekommen,
Ich wollt', 's wär' wieder erst Krieg.

Am Wege dort spielte ihr Kindlein,
Das glich ihr recht auf ein Haar,
Ich küßt's auf sein rotes Mündlein:
„Gott segne dich immerdar!“

Sie aber schaute erschrocken
Noch lange Zeit nach mir hin,
Und schüttelte sinnend die Locken
Und wußte nicht, wer ich bin. –

Da droben hoch stand ich am Baume,
Da rauschten die Wälder so sacht,
Mein Waldhorn, das klang wie im Traume
Hinüber die ganze Nacht.

Und als die Vögelein sangen
Frühmorgens, sie weinte so sehr,
Ich aber war weit schon gegangen,
Nun sieht sie mich nimmermehr!

Bei einer Linde

Seh' ich dich wieder, du geliebter Baum,
In dessen junge Triebe
Ich einst in jenes Frühlings schönstem Traum
Den Namen schnitt von meiner ersten Liebe?

Wie anders ist seitdem der Äste Bug,
Verwachsen und verschwunden
Im härt'ren Stamm der vielgeliebte Zug,
Wie ihre Liebe und die schönen Stunden!

Auch ich seitdem wuchs stille fort, wie du,
Und nichts an mir wollt' weilen,
Doch *meine* Wunde wuchs – und wuchs nicht zu,
Und wird wohl niemals mehr hienieden heilen.

Vom Berge

Da unten wohnte sonst mein Lieb,
Die ist jetzt schon begraben,
Der Baum noch vor der Türe blieb,
Wo wir gesessen haben.

Stets muß ich nach dem Hause sehn,
Und seh' doch nichts vor Weinen,
Und wollt' ich auch hinunter gehn,
Ich stürb' dort so alleine!

Verlorene Liebe

Lieder schweigen jetzt und Klagen,
Nun will ich erst fröhlich sein,
All mein Leid will ich zerschlagen
Und Erinnern – gebt mir Wein!
Wie er mir verlockend spiegelt
Sterne und der Erde Lust,
Stillgeschäftig dann entriegelt
All' die Teufel in der Brust,
Erst der Knecht und dann der Meister,
Bricht er durch die Nacht herein,
Wildester der Lügengeister,
Ring' mit mir, ich lache dein!
Und den Becher voll Entsetzen
Werf' ich in des Stromes Grund,
Daß sich nimmer dran soll letzen
Wer noch fröhlich und gesund!

Lauten hör' ich ferne klingen,
Lust'ge Bursche ziehn vom Schmaus,
Ständchen sie den Liebsten bringen,
Und das lockt mich mit hinaus.
Mädchen hinterm blühnden Baume

Winkt und macht das Fenster auf,
Und ich steige wie im Traume
Durch das kleine Haus hinauf.
Schüttle nur die dunklen Locken
Aus dem schönen Angesicht!
Sieh, ich stehe ganz erschrocken:
Das sind *ihre* Augen licht,

Locken hatte sie wie deine,
Bleiche Wangen, Lippen rot –
Ach, du bist ja doch nicht meine,
Und *mein* Lieb ist lange tot!
Hättest du nur nicht gesprochen
Und so frech geblickt nach mir,
Das hat ganz den Traum zerbrochen
Und nun grauet mir vor dir.
Da nimm Geld, kauf' Putz und Flimmern,
Fort und lache nicht so wild!
O ich möchte dich zertrümmern,
Schönes, lügenhaftes Bild!

Spät von dem verlornen Kinde
Kam ich durch die Nacht daher,
Fahnen drehten sich im Winde,
Alle Gassen waren leer.
Oben lag noch meine Laute
Und mein Fenster stand noch auf,
Aus dem stillen Grunde graute
Wunderbar die Stadt herauf.
Draußen aber blitzt's vom weiten,
Alter Zeiten ich gedacht',
Schaudernd reiß' ich in den Saiten
Und ich sing' die halbe Nacht.
Die verschlafnen Nachbarn sprechen,
Daß ich nächtlich trunken sei –
O du mein Gott! und mir brechen
Herz und Saitenspiel entzwei!

Das Ständchen

Auf die Dächer zwischen blassen
Wolken scheint der Mond herfür,
Ein Student dort auf der Gassen
Singt vor seiner Liebsten Tür.

Und die Brunnen rauschen wieder
Durch die stille Einsamkeit,
Und der Wald vom Berge nieder,
Wie in alter, schöner Zeit.

So in meinen jungen Tagen
Hab' ich manche Sommernacht
Auch die Laute hier geschlagen
Und manch lust'ges Lied erdacht.

Aber von der stillen Schwelle
Trugen sie mein Lieb zur Ruh –
Und du, fröhlicher Geselle,
Singe, sing nur immer zu!

Klang um Klang

1

Es ist ein Klang gekommen
Herüber durch die Luft,
Der Wind hat's gebracht und genommen,
Ich weiß nicht, wer mich ruft.
Es schallt der Grund von Hufen,
In der Ferne fiel ein Schuß –
Das sind die Jäger, die rufen,
Daß ich hinunter muß!

2

Das sind nicht die Jäger – im Grunde
Gehn Stimmen hin und her.
Hüt' dich zu dieser Stunde,
Mein Herz ist mir so schwer!
Wer dich lieb hat, macht die Runde,
Steig nieder und frag nicht, wer!
Ich führ' dich aus diesem Grunde –
Dann siehst du mich nimmermehr.

3

Ich weiß einen großen Garten,
Wo die wilden Blumen stehn,
Die Engel frühmorgens sein warten,
Wenn alles noch still auf den Höh'n.
Manch zackiges Schloß steht darinne,
Die Rehe grasen ums Haus,
Da sieht man weit von der Zinne,
Weit über die Länder hinaus.

Neue Liebe

Herz, mein Herz, warum so fröhlich,
So voll Unruh' und zerstreut,
Als käm' über Berge selig
Schon die schöne Frühlingszeit?

Weil ein liebes Mädchen wieder
Herzlich an dein Herz sich drückt,
Schaust du fröhlich auf und nieder,
Erd' und Himmel dich erquickt.

Und ich hab' die Fenster offen,
Neu zieh' in die Welt hinein
Altes Bangen, altes Hoffen!
Frühling, Frühling soll es sein!

Still kann ich hier nicht mehr bleiben,
Durch die Brust ein Singen irrt,
Doch zu licht ist's mir zum Schreiben,
Und ich bin so froh verwirrt.

Also schlendr' ich durch die Gassen,
Menschen gehen her und hin,
Weiß nicht, was ich tu und lasse,
Nur, daß ich so glücklich bin.

Frühlingsnacht

Über'n Garten durch die Lüfte
Hört' ich Wandervögel ziehn,
Das bedeutet Frühlingsdüfte,
Unten fängt's schon an zu blühn.

Jauchzen möcht' ich, möchte weinen,
Ist mir's doch, als könnt's nicht sein!
Alte Wunder wieder scheinen
Mit dem Mondesglanz herein.

Und der Mond, die Sterne sagen's,
Und in Träumen rauscht's der Hain,
Und die Nachtigallen schlagen's:
Sie ist Deine, sie ist dein!

Frau Venus

Was weckst du, Frühling, mich von neuem wieder?
Daß all' die alten Wünsche auferstehen,
Geht übers Land ein wunderbares Wehen;
Das schauert mir so lieblich durch die Glieder.

Die schöne Mutter grüßen tausend Lieder,
Die, wieder jung, im Brautkranz süß zu sehen;
Der Wald will sprechen, rauschend Ströme gehen,
Najaden tauchen singend auf und nieder.

Die Rose seh' ich gehn aus grüner Klause
Und, wie so buhlerisch die Lüfte fächeln,
Errötend in die laue Flut sich dehnen.

So mich auch ruft ihr aus dem stillen Hause –
Und schmerzlich nun muß ich im Frühling lächeln,
Versinkend zwischen Duft und Klang vor Sehnen.

Erwartung

O schöne, bunte Vögel,
Wie singt ihr gar so hell!
O Wolken, luft'ge Segel,
Wohin so schnell, so schnell?

Ihr alle, ach, gemeinsam
Fliegt zu der Liebsten hin,
Sagt ihr, wie ich hier einsam
Und voller Sorgen bin.

Im Walde steh' und laur' ich,
Verhallt ist jeder Laut,
Die Wipfel nur wehn schaurig,
O komm, du süße Braut!

Schon sinkt die dunkelfeuchte
Nacht rings auf Wald und Feld,
Des Mondes hohe Leuchte
Tritt in die stille Welt.

Wie schauert nun im Grunde
Der tiefsten Seele mich!
Wie öde ist die Runde
Und einsam ohne dich!

Was rauscht? – *Sie* naht von ferne! –
Nun, Wald, rausch', von den Höh'n,
Nun laß Mond, Nacht und Sterne
Nur auf und unter gehn!

Leid und Lust

Euch Wolken beneid' ich
In blauer Luft,
Wie schwingt ihr euch freudig
Über Berg und Kluft!

Mein Liebchen wohl seht ihr
Im Garten gehn,
Am Springbrunnen steht sie
So morgenschön.

Und wäscht an der Quelle
Ihr goldenes Haar,
Die Äuglein helle,
Und blickt so klar.

Und Busen und Wangen
Dürft ihr da sehn. –
Ich brenn' vor Verlangen,
Und muß hier stehn!

Euch Wolken bedaur' ich
Bei stiller Nacht;
Die Erde bebt schaurig,
Der Mond erwacht:

Da führt mich ein Bübchen
Mit Flügelein fein,
Durchs Dunkel zum Liebchen,
Sie läßt mich ein.

Wohl schaut ihr die Sterne
Weit ohne Zahl,
Doch bleiben sie ferne
Euch allzumal.

Mir leuchten zwei Sterne
Mit süßem Strahl,
Die küss' ich so gerne
Viel tausendmal.

Euch grüßt mit Gefunkel
Der Wasserfall,
Und tief aus dem Dunkel
Die Nachtigall.

Doch süßer es grüßet
Als Wellentanz,
Wenn Liebchen hold flüstert:
„Dein bin ich ganz."

So segelt denn traurig
In öder Pracht!
Euch Wolken bedaur' ich
Bei süßer Nacht.

Trennung

1

Denkst du noch jenes Abends, still vor Sehnen,
Wo wir zum letztenmal im Park beisammen?
Kühl standen rings des Abendrotes Flammen,
Ich scherzte wild – du lächeltest durch Tränen.
So spielt der Wahnsinn lieblich mit den Schmerzen
An jäher Schlüfte Rand, die nach ihm trachten;
Er mag der lauernden Gefahr nicht achten;
Er hat den Tod ja schon im öden Herzen.

Ob du die Mutter auch belogst, betrübtest,
Was andre Leute drüber deuten, sagen –
Sonst scheu – heut' mocht'st du nicht nach allem fragen,
Mir einzig zeigen nur, wie du mich liebtest.
Und aus dem Hause heimlich so entwichen,
Gabst du ins Feld mir schweigend das Geleite,
Vor uns das Tal, das hoffnungsreiche, weite,
Und hinter uns kam grau die Nacht geschlichen.

Du gehst nun fort, sprachst du, ich bleib' alleine;
Ach! dürft' ich alles lassen, still und heiter
Mit dir so ziehn hinab und immer weiter –
Ich sah dich an – es spielten bleiche Scheine
So wunderbar um Locken dir und Glieder;
So ruhig, fremd warst du mir nie erschienen,
Es war, als sagten die versteinten Mienen,
Was du verschwiegst: Wir sehn uns niemals wieder!

2

Schon wird es draußen licht auf Berg und Talen;
Aurora, stille Braut, ihr schönen Strahlen,
Die farb'gen Rauch aus Fluß und Wäldern saugen.
Euch grüßen neu die halbverschlafnen Augen.

Verrät'risch, sagt man, sei des Zimmers Schwüle,
Wo nachts ein Mädchen träumte vom Geliebten:
So komm herein, du rote, frische Kühle,
Fliegt in die blaue Luft, ihr schönen Träume!

Ein furchtsam Kind, im stillen Haus erzogen,
Konnt' ich am Abendrot die Blicke weiden,
Tiefatmend in die laue Luft vor Freuden.
Er hat um diese Stille mich betrogen.
Mit stolzen Augen, fremden schönen Worten
Lockt er die Wünsche aus dem stillen Hafen,
Wo sie bei Sternenglanze selig schlafen,
Hinaus ins unbekannte Reich der Wogen;
Da kommen Winde buhlend angeflogen,
Die zarte Hand zwingt nicht die wilden Wellen,
Du mußt, wohin die vollen Segel schwellen.

Da zog er heimlich fort. – Seit jenem Morgen
Da hatt' ich Not, hatt' heimlich was zu sorgen.
Wenn nächtlich unten lag die stille Runde,
Einförmig Rauschen herkam von den Wäldern,
Pfeifend der Wind strich durch die öden Felder
Und hin und her in Dörfern bellten Hunde,
Ach! wenn kein glücklich Herz auf Erden wacht,
Begrüßten die verweinten Augen manche Nacht!

Wie oft, wenn wir im Garten ruhig waren,
Sagte mein Bruder mir vor vielen Jahren:
„Dem schönen Lenz gleicht recht die erste Liebe.
Wann draußen neu geschmückt die Frühlingsbühne,
Die Reiter blitzend unten ziehn durchs Grüne,
In blauer Luft die Lerchen lustig schwirren,
Läßt sie sich weit ins Land hinaus verführen,
Fragt nicht, wohin, und mag sich gern verirren,
Den Stimmen folgend, die sie wirrend führen.
Da wendet auf den Feldern sich der Wind,
Die Vögel hoch durch Nebel ziehn nach Haus;

Es wird so still, das schöne Fest ist aus.
Gar weit die Heimat liegt, das schöne Kind
Find't nicht nach Hause mehr, nicht weiter fort –
Hüt' dich, such früh dir einen sichern Port!"

Glück

Wie jauchzt meine Seele
Und singet in sich!
Kaum, daß ich's verhehle,
So glücklich bin ich.

Rings Menschen sich drehen
Und sprechen gescheut,
Ich kann nichts verstehen,
So fröhlich zerstreut. –

Zu eng wird das Zimmer,
Wie glänzet das Feld,
Die Täler voll Schimmer,
Weit herrlich die Welt!

Gepreßt bricht die Freude
Durch Riegel und Schloß,
Fort über die Heide!
Ach, hätt' ich ein Roß! –

Und frag' ich und sinn' ich,
Wie *so* mir geschehn?: –
Mein Liebchen herzinnig,
Das soll ich heut sehn!

Die Schärpe

Mein Schatz, das ist ein kluges Kind,
Die spricht: „Willst du nicht fechten:
Wir zwei geschiedne Leute sind;
Erschlagen dich die Schlechten:
Auch keins von beiden dran gewinnt."
Mein Schatz, das ist ein kluges Kind,
Für die will ich *leben* und fechten!

Abschied und Wiedersehen

1

In süßen Spielen unter nun gegangen
Sind Liebchens Augen, und sie atmet linde,
Stillauschend sitz ich bei dem holden Kinde,
Die Locken streichelnd ihr von Stirn und Wangen.

Ach! Lust und Mond und Sterne sind vergangen,
Am Fenster mahnen schon die Morgenwinde:
Daß ich vom Nacken leis die Arme winde,
Die noch im Schlummer lieblich mich umfangen.

O öffne nicht der Augen süße Strahle!
Nur einen Kuß noch – und zum letzten Male
Geh' ich von dir durchs stille Schloß hernieder.

Streng greift der eis'ge Morgen an die Glieder,
Wie ist die Welt so klar und kalt und helle –
Tiefschauernd tret' ich von der lieben Schwelle.

2

Ein zart Geheimnis webt in stillen Räumen,
Die Erde löst die diamantnen Schleifen,

Und nach des Himmels süßen Strahlen greifen
Die Blumen, die der Mutter Kleid besäumen.

Da rauscht's lebendig draußen in den Bäumen,
Aus Osten langen purpurrote Streifen,
Hoch Lerchenlieder durch das Zwielicht schweifen –
Du hebst das blühnde Köpfchen hold aus Träumen.

Was sind's für Klänge, die ans Fenster flogen?
So altbekannt verlocken diese Lieder,
Ein Sänger steht im schwanken Dämmerschein.

Wach auf! Dein Liebster ist fernher gezogen,
Und Frühling ist's auf Tal und Bergen wieder,
Wach auf, wach auf, nun bist du ewig mein!

Die Einsame

1

Wenn morgens das fröhliche Licht bricht ein,
Tret' ich zum offenen Fensterlein,
Draußen gehn lau die Lüft' auf den Auen,
Singen die Lerchen schon hoch im Blauen,
Rauschen am Fenster die Bäume gar munter,
Ziehn die Brüder in den Wald hinunter;
Und bei dem Sange und Hörnerklange
Wird mir immer so bange, bange.

Wüßt' ich nur immer, wo du jetzo bist,
Würd' mir schon wohler auf kurze Frist.
Könntest du mich nur über die Berge sehen
Dein gedenkend im Garten gehen:
Dort rauschen die Brunnen jetzt alle so eigen,
Die Blumen vor Trauern im Winde sich neigen.

Ach! von den Vöglein über die Tale
Sei mir gegrüßt viel tausend Male!

Du sagtest gar oft: Wie süß und rein
Sind deine blauen Äugelein!
Jetzo müssen sie immerfort weinen,
Da sie nicht finden mehr, was sie meinen;
Wird auch der rote Mund erblassen,
Seit du mich, süßer Buhle, verlassen.
Eh' du wohl denkst, kann das Blatt sich wenden,
Geht alles gar bald zu seinem Ende.

2

Die Welt ruht still im Hafen,
Mein Liebchen, gute Nacht!
Wann Wald und Berge schlafen,
Treu' Liebe einsam wacht.

Ich bin so wach und lustig.
Die Seele ist so licht,
Und eh' ich liebt', da wußt' ich
Von solcher Freude nicht.

Ich fühl' mich so befreiet
Von eitlem Trieb und Streit,
Nichts mehr das Herz zerstreuet
In seiner Fröhlichkeit.

Mir ist, als müßt' ich singen
So recht aus tiefster Lust
Von wunderbaren Dingen,
Was niemand sonst bewußt.

O könnt' ich alles sagen!
O wär' ich recht geschickt!
So muß ich still ertragen,
Was mich so hoch beglückt.

3

Wär's dunkel, ich läg' im Walde,
Im Walde rauscht's so sacht,
Mit ihrem Sternenmantel
Bedecket mich da die Nacht,
Da kommen die Bächlein gegangen:
Ob ich schon schlafen tu?
Ich schlaf nicht, ich hör noch lange
Den Nachtigallen zu,
Wenn die Wipfel über mir schwanken,
Es klinget die ganze Nacht,
Das sind im Herzen die Gedanken,
Die singen, wenn niemand wacht.

4

Im beschränkten Kreis der Hügel,
Auf des stillen Weihers Spiegel
Scheue, fromme Silberschwäne –
Fassend in des Rosses Mähne
Mit dem Liebsten kühn im Bügel –
Blöde Bande – mut'ge Flügel
Sind getrennter Lieb' Gedanken!

An die Entfernte

1

Denk' ich, du Stille, an dein ruhig Walten,
An jenes letzten Abends rote Kühle,
Wo ich die teure Hand noch durfte halten:
Steh' ich oft sinnend stille im Gewühle,
Und, wie den Schweizer heim'sche Alphornslieder
Auf fremden Bergen, fern den Freunden allen,

Oft unverhofft befallen,
Kommt tiefe Sehnsucht plötzlich auf mich nieder.

Ich hab' es oft in deiner Brust gelesen:
Nie hast du recht mich in mir selbst gefunden,
Fremd blieb, zu keck und treibend dir mein Wesen,
Und so bin ich im Strome dir verschwunden.
O nenn' drum nicht die schöne Jugend wilde,
Die mit dem Leben und mit seinen Schmerzen
Mag unbekümmert scherzen,
Weil sie die Brust reich fühlt und ernst und milde!

Getrennt ist längst schon unsres Lebens Reise,
Es trieb mein Herz durch licht' und dunkle Stunden.
Dem festern Blick erweitern sich die Kreise,
In Duft ist jenes erste Reich verschwunden –
Doch, wie die Pfade einsam sich verwildern,
Was ich seitdem, von Lust und Leid bezwungen
Geliebt, geirrt, gesungen:
Ich knie' vor dir in all' den tausend Bildern.

2

Als noch Lieb' mit mir im Bunde,
Hatt' ich Ruhe keine Stunde;
Wenn im Schloß noch alle schliefen,
War's, als ob süß' Stimmen riefen,
Tönend bis zum Herzensgrunde:
„Auf! schon goldne Strahlen dringen,
Heiter funkeln Wald und Garten,
Neu erquickt die Vögel singen,
Läßt du so dein Liebchen warten?"
Und vom Lager mußt' ich springen.

Doch kein Licht noch sah ich grauen,
Draußen durch die nächtlich lauen
Räume nur die Wolken flogen,

Daß die Seele, mitgezogen,
Gern versank im tiefen Schauen –
Unten dann die weite Runde,
Schlösser glänzend fern erhoben,
Nachtigallen aus dem Grunde,
Alles wie im Traum verwoben,
Mit einander still im Bunde.

Wach blieb ich am Fenster stehen,
Kühler schon die Lüfte wehen,
Rot schon rings des Himmels Säume,
Regten frischer sich die Bäume,
Stimmen hört' ich fernab gehen:
Und durch Türen, öde Bogen,
Zürnend, daß die Riegel klungen,
Bin ich heimlich ausgezogen,
Bin befreit aufs Roß geschwungen,
Morgenwinde mich umflogen.

Läßt der Morgen von den Höhen
Weit die roten Fahnen wehen,
Widerhall in allen Lüften,
Losgerissen aus den Klüften
Silberner die Ströme gehen:
Spürt der Mann die frischen Geister,
Draußen auf dem Feld, zu Pferde,
Alle Ängste keck zerreißt er,
Dampfend unter ihm die Erde,
Fühlt er hier sich Herr und Meister.

Und so öffnet' ich die schwüle
Brust aufatmend in der Kühle!
Locken fort aus Stirn und Wange,
Daß der Strom mich ganz umfange,
Frei das blaue Meer umspüle,
Mit den Wolken, eilig fliehend,
Mit der Ströme lichtem Grüßen

Die Gedanken fröhlich ziehend,
Weit voraus vor Wolken, Flüssen –
Ach! ich fühlte, daß ich blühend!

Und im schönen Garten droben,
Wie aus Träumen erst gehoben,
Sah ich still mein Mädchen stehen,
Über Fluß und Wälder gehen
Von der heitern Warte oben
Ihre Augen licht und helle,
Wann der Liebste kommen werde. –
Ja! da kam die Sonne schnelle,
Und weit um die ganze Erde
War es morgenschön und helle!

Das Flügelroß

Ich hab' nicht viel hienieden,
Ich hab' nicht Geld noch Gut;
Was vielen nicht beschieden,
Ist mein: – der frische Mut.

Was andre mag ergötzen,
Das kümmert wenig mich,
Sie leben in den Schätzen,
In Freuden lebe ich.

Ich hab' ein Roß mit Flügeln,
Getreu in Lust und Not,
Das wiehernd spannt die Flügel
Bei jedem Morgenrot.

Mein Liebchen! wie so öde
Wird's oft in Stadt und Schloß,
Frisch auf und sei nicht blöde,
Besteig mit mir mein Roß!

Wir segeln durch die Räume,
Ich zeig' dir Meer und Land,
Wie wunderbare Träume
Tief unten ausgespannt.

Hellblinkend zu den Füßen
Unzähl'ger Ströme Lauf –
Es steigt ein Frühlingsgrüßen
Verhallend zu uns auf.

Und bunt und immer wilder
In Liebe, Haß und Lust
Verwirren sich die Bilder –
Was schwindelt dir die Brust?

So fröhlich tief im Herzen,
Zieh' ich all' himmelwärts,
Es kommen selbst die Schmerzen
Melodisch an das Herz.

Der Sänger zwingt mit Klängen,
Was störrig, dumpf und wild,
Es spiegelt in Gesängen
Die Welt sich göttlich mild.

Und unten nun verbrauset
Des breiten Lebens Strom,
Der Adler einsam hauset
Im stillen Himmelsdom. –

Und sehn wir dann den Abend
Verhallen und verblühn,
Im Meere, kühle labend,
Die heil'gen Sterne glühn:

So lenken wir hernieder
Zu Waldes grünem Haus,

Und ruhn vom Schwung der Lieder
Auf blühndem Moose aus.

O sterndurchwebtes Düstern,
O heimlich stiller Grund!
O süßes Liebesflüstern
So innig Mund an Mund!

Die Nachtigallen locken,
Mein Liebchen atmet lind,
Mit Schleier zart und Locken
Spielt buhlerisch der Wind.

Und schlaf denn bis zum Morgen
So sanft gelehnt an mich!
Süß sind der Liebe Sorgen,
Dein Liebster wacht für dich.

Ich halt' die blühnden Glieder,
Vor süßen Schauern bang.
Ich laß dich ja nicht wieder
Mein ganzes Leben lang! –

Aurora will sich heben,
Du schlägst die Augen auf,
O wonniges Erbeben,
O schöner Lebenslauf!

Glückwunsch

Brech' der lustige Sonnenschein
Mit der Tür euch ins Haus hinein,
Daß alle Stuben so frühlingshelle;
Ein Engel auf des Hauses Schwelle
Mit seinem Glanze säume

Hof, Garten, Feld und Bäume,
Und geht die Sonne abends aus,
Führ' er die Müden mild nach Haus!

Der junge Ehemann

Hier unter dieser Linde
Saß ich viel tausendmal
Und schaut' nach meinem Kinde
Hinunter in das Tal,
Bis daß die Sterne standen
Hell über ihrem Haus,
Und weit in den stillen Landen
Alle Lichter löschten aus.

Jetzt neben meinem Liebchen
Sitz' ich im Schatten kühl,
Sie wiegt ein muntres Bübchen,
Die Täler schimmern schwül,
Und unten im leisen Winde
Regt sich das Kornfeld kaum,
Und über uns säuselt die Linde –
Es ist mir noch wie ein Traum.

Im Abendrot

Wir sind durch Not und Freude
Gegangen Hand in Hand,
Vom Wandern ruhn wir beide
Nun überm stillen Land.

Rings sich die Täler neigen,
Es dunkelt schon die Luft,
Zwei Lerchen nur noch steigen
Nachträumend in den Duft.

Tritt her, und laß sie schwirren,
Bald ist es Schlafenszeit,
Daß wir uns nicht verirren
In dieser Einsamkeit.

O weiter, stiller Friede!
So tief im Abendrot
Wie sind wir wandermüde –
Ist das etwa der Tod?

Nachklänge

1

Lust'ge Vögel in dem Wald,
Singt, so lang es grün,
Ach, wer weiß, wie bald, wie bald
Alles muß verblühn!

Sah ich's doch vom Berge einst
Glänzen überall,
Wußte kaum, warum du weinst,
Fromme Nachtigall.

Und kaum ging ich über Land,
Frisch durch Lust und Not
Wandelt' alles, und ich stand
Müd' im Abendrot.

Und die Lüfte wehen kalt
Übers falbe Grün,
Vöglein, euer Abschied hallt –
Könnt' ich mit euch ziehn!

2

O Herbst, in linden Tagen
Wie hast du rings dein Reich
Phantastisch aufgeschlagen,
So bunt und doch so bleich!

Wie öde, ohne Brüder,
Mein Tal so weit und breit,
Ich kenne dich kaum wieder
In dieser Einsamkeit.

So wunderbare Weise
Singt nun dein bleicher Mund,
Es ist, als öffnet' leise
Sich unter mir der Grund.

Und ich ruht' überwoben,
Du sängest immerzu,
Die Linde schüttelt' oben
Ihr Laub und deckt' mich zu.

3

Schon kehren die Vögel wieder ein,
Es schallen die alten Lieder,
Ach, die fröhliche Jugend mein
Kommt sie wohl auch noch wieder? –

Ich weiß nicht, was ich so töricht bin!
Wolken im Herbstwind jagen,
Die Vögel ziehn über die Wälder hin,
Das klang wie in Frühlingstagen.

Dort auf dem Berge da steht ein Baum,
Drin jubeln die Wandergäste,
Er aber, müde, rührt wie im Traum
Noch einmal Wipfel und Äste.

4

Mir träumt', ich ruhte wieder
Vor meines Vaters Haus
Und schaute fröhlich nieder
Ins alte Tal hinaus,
Die Luft mit lindem Spielen
Ging durch das Frühlingslaub,
Und Blütenflocken fielen
Mir über Brust und Haupt.

Als ich erwacht, da schimmert
Der Mond vom Waldesrand,
Im falben Scheine flimmert
Um mich ein fremdes Land,
Und wie ich ringsher sehe:
Die Flocken waren Eis,
Die Gegend war vom Schnee,
Mein Haar vom Alter weiß.

5

Es schauert der Wald vor Lust,
Die Sterne nun versanken,
Und wandeln durch die Brust
Als himmlische Gedanken.

6

An meinen Bruder

Gedenkst du noch des Gartens
Und Schlosses überm Wald,
Des träumenden Erwartens:
Ob's denn nicht Frühling bald?

Der Spielmann war gekommen,
Der jeden Lenz singt aus,

Er hat uns mitgenommen
Ins blühnde Land hinaus.

Wie sind wir doch im Wandern
Seitdem so weit zerstreut!
Frägt einer nach dem andern,
Doch niemand gibt Bescheid.

Nun steht das Schloß versunken
Im Abendrote tief,
Als ob dort traumestrunken
Der alte Spielmann schlief'.

Gestorben sind die Lieben,
Das ist schon lange her,
Die wen'gen, die geblieben,
Sie kennen uns nicht mehr.

Und fremde Leute gehen
Im Garten vor dem Haus –
Doch übern Garten sehen
Nach *uns* die Wipfel aus.

Doch rauscht der Wald im Grunde
Fort durch die Einsamkeit
Und gibt noch immer Kunde
Von unsrer Jugendzeit.

Bald mächt'ger und bald leise
In jeder guten Stund'
Geht diese Waldesweise
Mir durch der Seele Grund.

Und stamml' ich auch nur bange,
Ich sing' es, weil ich muß,
Du hörst doch in dem Klange
Den alten Heimatsgruß.

V. Totenopfer

Gewalt'ges Morgenrot,
Weit, unermeßlich – du verzehrst die Erde!
Und in dem Schweigen nur der Flug der Seelen,
Die säuselnd heimziehn durch die stille Luft. –

Wehmut

Ich irr' in Tal und Hainen
Bei kühler Abendstund,
Ach, weinen möcht' ich, weinen
So recht aus Herzensgrund.

Und alter Zeiten Grüßen
Kam da, im Tal erwacht,
Gleichwie von fernen Flüssen
Das Rauschen durch die Nacht.

Die Sonne ging hinunter,
Da säuselt' kaum die Welt,
Ich blieb noch lange munter
Allein im stillen Feld.

Sonette

1

Es qualmt' der eitle Markt in Staub und Schwüle,
So klanglos öde wallend auf und nieder,
Wie dacht' ich da an meine Berge wieder,
An frischen Sang, Felsquell und Waldeskühle!

Doch steht ein Turm dort über dem Gewühle,
Der andre Zeiten sah und beßre Brüder,
Das Kreuz treu halten seine Riesenglieder,
Wie auch der Menschlein Flut den Fels umspüle.

Das war mein Hafen in der weiten Wüste,
Oft kniet' ich betend in des Domes Mitte,
Dort hab' ich dich, mein liebes Kind, gefunden;

Ein Himmelbote wohl, der so mich grüßte:
„Verzweifle nicht! Die Schönheit und die Sitte
Sie sind noch von der Erde nicht verschwunden."

2

Ein alt' Gemach voll sinnger Seltsamkeiten,
Still Blumen aufgestellt am Fensterbogen,
Gebirg' und Länder draußen blau gezogen,
Wo Ströme gehn und Ritter ferne reiten.

Ein Mädchen, schlicht und fromm wie jene Zeiten,
Das, von den Abendscheinen angeflogen,
Versenkt in solcher Stille tiefe Wogen –
Das mocht auf Bildern oft das Herz mir weiten.

Und nun wollt' wirklich sich das Bild bewegen,
Das Mädchen atmet' auf, reicht aus dem Schweigen
Die Hand mir, daß sie ewig meine bliebe.

Da sah ich draußen auch das Land sich regen,
Die Wälder rauschen und Aurora steigen –
Die alten Zeiten all' weckt mir die Liebe.

3

Wenn zwei geschieden sind von Herz und Munde,
Da ziehn Gedanken über Berg' und Schlüfte
Wie Tauben säuselnd durch die blauen Lüfte,
Und tragen hin und wieder süße Kunde.

Ich schweif' umsonst, so weit der Erde Runde,
Und stieg ich hoch auch über alle Klüfte,
Dein Haus ist höher noch als diese Lüfte,
Da reicht kein Laut hin, noch zurück zum Grunde.

Ja, seit du tot – mit seinen blüh'nden Borden
Wich ringsumher das Leben mir zurücke,
Ein weites Meer, wo keine Bahn zu finden.

Doch ist dein Bild zum Sterne mir geworden,
Der nach der Heimat weist mit stillem Blicke,
Daß fromm der Schiffer streite mit den Winden.

Treue

Wie dem Wanderer in Träumen,
Daß er still im Schlafe weint,
Zwischen goldnen Wolkensäumen
Seine Heimat wohl erscheint:

So durch dieses Frühlings Blühen
Über Berg' und Täler tief,
Sah ich oft dein Bild noch ziehen,
Als ob's mich von hinnen rief';

Und mit wunderbaren Wellen
Wie im Traume, halbbewußt,
Gehen ew'ge Liederquellen
Mir verwirrend durch die Brust.

Gute Nacht

Die Höh'n und Wälder schon steigen
Immer tiefer ins Abendgold,
Ein Vöglein frägt in den Zweigen:
Ob es Liebchen grüßen sollt'?

O Vöglein, du hast dich betrogen,
Sie wohnet nicht mehr im Tal,
Schwing' auf dich zum Himmelsbogen,
Grüß' sie droben zum letztenmal!

Am Strom

Der Fluß glitt einsam hin und rauschte,
Wie sonst, noch immer, immerfort,
Ich stand am Strand gelehnt und lauschte,
Ach, was ich liebt', war lange fort!

Kein Laut, kein Windeshauch, kein Singen
Ging durch den weiten Mittag schwül,
Verträumt die stillen Weiden hingen
Hinab bis in die Wellen kühl.

Die waren alle wie Sirenen
Mit feuchtem, langem, grünem Haar,
Und von der alten Zeit voll Sehnen
Sie sangen leis und wunderbar.
Sing, Weide, singe, grüne Weide!
Wie Stimmen aus der Liebsten Grab
Zieht mich dein heimlich Lied voll Leide
Zum Strom von Wehmut mit hinab.

Nachruf an meinen Bruder

Ach, daß auch wir schliefen!
Die blühenden Tiefen,
Die Ströme, die Auen
So heimlich aufschauen,
Als ob sie all' riefen:
„Dein Bruder ist tot!
Unter Rosen rot
Ach, daß wir auch schliefen!"

„Hast doch keine Schwingen,
Durch Wolken zu dringen!
Mußt immerfort schauen
Die Ströme, die Auen –
Die werden dir singen
Von *ihm* Tag und Nacht,
Mit Wahnsinnesmacht
Die Seele umschlingen."

So singt, wie Sirenen,
Von hellblauen, schönen

Vergangenen Zeiten,
Der Abend vom weiten
Versinkt dann in Tönen,
Erst Busen, dann Mund.
Im blühenden Grund.
O schweiget Sirenen!

O wecket nicht wieder!
Denn zaub'rische Lieder
Gebunden hier träumen
Auf Feldern und Bäumen,
Und ziehen mich nieder
So müde vor Weh –
Zu tiefstillem See –
O weckt nicht die Lieder!

Du kanntest die Wellen
Des Sees, sie schwellen
In magischen Ringen.
Ein wehmütig Singen
Tief unter den Quellen
Im Schlummer dort hält
Verzaubert die Welt.
Wohl kennst du die Wellen.

Kühl wird's auf den Gängen,
Vor alten Gesängen
Möcht's Herz mir zerspringen.
So will ich denn singen!
Schmerz fliegt ja auf Klängen
Zu himmlischer Lust,
Und still wird die Brust
Auf kühl grünen Gängen.

Laß fahren die Träume!
Der Mond scheint durch Bäume,
Die Wälder nur rauschen,

Die Täler still lauschen,
Wie einsam die Räume!
Ach, niemand ist mein!
Herz, wie so allein!
Laß fahren die Träume!

Der Herr wird dich führen.
Tief kann ich ja spüren
Der Sterne still Walten.
Der Erde Gestalten
Kaum hörbar sich rühren.
Durch Nacht und durch Graus
Gen Morgen, nach Haus –
Ja. Gott wird mich führen.

Auf meines Kindes Tod

1

Das Kindlein spielt' draußen im Frühlingsschein,
Und freut' sich und hatte so viel zu sehen,
Wie die Felder schimmern und die Ströme gehen –
Da sah der Abend durch die Bäume herein,
Der alle die schönen Bilder verwirrt,
Und wie es nun ringsum so stille wird,
Beginnt aus den Tälern ein heimlich Singen,
Als wollt's mit Wehmut die Welt umschlingen,
Die Farben vergehn und die Erde wird blaß.
Voll Staunen fragt's Kindlein: Ach, was ist das?
Und legt sich träumend ins säuselnde Gras;
Da rühren die Blumen ihm kühle ans Herz
Und lächelnd fühlt es so süßen Schmerz,
Und die Erde, die Mutter, so schön und bleich,
Küßt das Kindlein und läßt's nicht los,
Zieht es herzinnig in ihren Schoß

Und bettet es drunten gar warm und weich,
Still unter Blumen und Moos. –

„Und was weint ihr, Vater und Mutter, um mich?
In einem viel schöneren Garten bin ich,
Der ist so groß und weit und wunderbar,
Viel Blumen stehn dort von Golde klar,
Und schöne Kindlein mit Flügeln schwingen
Auf und nieder sich drauf und singen. –
Die kenn' ich gar wohl aus der Frühlingszeit,
Wie sie zogen über Berge und Täler weit
Und mancher mich da aus dem Himmelblau rief,
Wenn ich drunten im Garten schlief. –
Und mitten zwischen den Blumen und Scheinen
Steht die schönste von allen Frauen,
Ein glänzend Kindlein an ihrer Brust. –
Ich kann nicht sprechen und auch nicht weinen,
Nur singen immer und wieder dann schauen
Still vor großer, seliger Lust."

2

Als ich nun zum ersten Male
Wieder durch den Garten ging,
Busch und Bächlein in dem Tale
Lustig an zu plaudern fing.

Blumen halbverstohlen blickten
Neckend aus dem Gras heraus,
Bunte Schmetterlinge schickten
Sie sogleich auf Kundschaft aus.

Auch der Kuckuck in den Zweigen
Fand sich bald zum Spielen ein,
Endlich brach der Baum das Schweigen:
„Warum kommst du heut allein?"

Da ich aber schwieg, da rührt' er
Wunderbar sein dunkles Haupt,
Und ein Flüstern konnt' ich spüren
Zwischen Vöglein, Blüt' und Laub.

Tränen in dem Grase hingen,
Durch die abendstille Rund
Klagend nun die Quellen gingen,
Und ich weint' aus Herzensgrund.

3

Was ist mir denn so wehe?
Es liegt ja wie im Traum
Der Grund schon, wo ich stehe,
Die Wälder säuseln kaum

Noch von der dunklen Höhe.
Es komme wie es will,
Was ist mir denn so wehe –
Wie bald wird alles still.

4

Das ist's, was mich ganz verstöret:
Daß die Nacht nicht Ruhe hält,
Wenn zu atmen aufgehöret
Lange schon die müde Welt.

Daß die Glocken, die da schlagen,
Und im Wald der leise Wind
Jede Nacht von neuem klagen
Um mein liebes, süßes Kind.

Daß mein Herz nicht konnte brechen
Bei dem letzten Todeskuß,
Daß ich wie im Wahnsinn sprechen
Nun in irren Liedern muß.

5

Freuden wollt' ich dir bereiten,
Zwischen Kämpfen, Lust und Schmerz
Wollt' ich treulich dich geleiten
Durch das Leben himmelwärts.

Doch du hast's allein gefunden,
Wo kein Vater führen kann,
Durch die ernste, dunkle Stunde
Gingst du schuldlos mir voran.

Wie das Säuseln leiser Schwingen
Draußen über Tal und Kluft
Ging zur selben Stund' ein Singen
Ferne durch die stille Luft.

Und so fröhlich glänzt' der Morgen,
's war als ob das Singen sprach:
Jetzo lasset alle Sorgen,
Liebt ihr mich, so folgt mir nach!

6

Ich führt' dich oft spazieren
In Wintereinsamkeit,
Kein Laut ließ sich da spüren,
Du schöne, stille Zeit!

Lenz ist's nun, Lerchen singen
Im Blauen über mir,
Ich weine still – sie bringen
Mir einen Gruß von dir.

7

Die Welt treibt fort ihr Wesen,
Die Leute kommen und gehn,

Als wärst du nie gewesen,
Als wäre nichts geschehn.

Wie sehn' ich mich aufs neue
Hinaus in Wald und Flur!
Ob ich mich gräm', mich freue,
Du bleibst mir treu, Natur.

Da klagt vor tiefem Sehnen
Schluchzend die Nachtigall,
Es schimmern rings von Tränen
Die Blumen überall.

Und über alle Gipfel
Und Blütentäler zieht
Durch stillen Waldes Wipfel
Ein heimlich Klagelied.

Da spür' ich's recht im Herzen,
Daß du's, Herr, draußen bist –
Du weißt's, wie mir von Schmerzen
Mein Herz zerrissen ist!

8

Von fern die Uhren schlagen,
Es ist schon tiefe Nacht,
Die Lampe brennt so düster,
Dein Bettlein ist gemacht.

Die Winde nur noch gehen
Wehklagend um das Haus,
Wir sitzen einsam drinne
Und lauschen oft hinaus.

Es ist, als müßtest leise
Du klopfen an die Tür,

Du hätt'st dich nur verirret,
Und kämst nun müd zurück.

Wir armen, armen Toren!
Wir irren ja im Graus
Des Dunkels noch verloren –
Du fand'st dich längst nach Haus.

9

Dort ist so tiefer Schatten,
Du schläfst in guter Ruh,
Es deckt mit grünen Matten
Der liebe Gott dich zu.

Die alten Weiden neigen
Sich auf dein Bett herein,
Die Vöglein in den Zweigen
Sie singen treu dich ein.

Und wie in goldnen Träumen
Geht linder Frühlingswind
Rings in den stillen Bäumen –
Schlaf wohl mein süßes Kind!

10

Mein liebes Kind, Ade!
Ich konnt' Ade nicht sagen,
Als sie dich fortgetragen,
Vor tiefem, tiefem Weh.

Jetzt auf lichtgrünem Plan
Stehst du im Myrtenkranze,
Und lächelst aus dem Glanze
Mich still voll Mitleid an.

Und Jahre nahn und gehn,
Wie bald bin ich verstoben –
O bitt' für mich da droben,
Daß wir uns wiedersehn!

An einen Offizier, der als Bräutigam starb

Frisch flogst du durch die Felder
Und faßtest ihre Hand,
Ringsum der Kreis der Wälder
In Morgenflammen stand.

O falsches Rot! Verblühen
Mußt' dieses Blütenmeer,
Wer dachte, daß dies Glühen
Das Abendrot schon wär'!

Nun dunkeln schon die Fernen,
Du wirst so still und bleich,
Wie ist da weit von Sternen
Der Himmelsgrund so reich!

Trompeten hört' ich laden
Fern durch sie stille Luft,
Als zögen Kameraden –
Der alte Feldherr ruft.

Es sinken schon die Brücken,
Heut dir und morgen mir.
Du mußt hinüberrücken,
Kam'rad, mach uns Quartier!

Treu' Lieb ist unverloren,
Empfängst – wie bald ist's hin! –
Einst an den Himmelstoren
Die müde Pilgerin.

Angedenken

Berg und Täler wieder fingen
Ringsumher zu blühen an,
Aus dem Walde hört' ich singen
Einen lust'gen Jägersmann.

Und die Tränen drangen leise:
So einst blüht' es weit und breit,
Als mein Lieb dieselbe Weise
Mich gelehrt vor langer Zeit.

Ach, ein solches Angedenken,
's ist nur eitel Klang und Luft,
Und kann schimmernd doch versenken
Rings in Tränen Tal und Kluft!

In der Fremde

Aus der Heimat hinter den Blitzen rot
Da kommen die Wolken her,
Aber Vater und Mutter sind lange tot,
Es kennt mich dort keiner mehr.
Wie bald, wie bald kommt die stille Zeit,
Da ruhe ich auch, und über mir
Rauschet die schöne Waldeinsamkeit.
Und keiner mehr kennt mich auch hier.

Vesper

Die Abendglocken klangen
Schon durch das stille Tal,
Da saßen wir zusammen
Da droben wohl hundertmal.

Und unten war's so stille
Im Lande weit und breit,
Nur über uns die Linde
Rauscht' durch die Einsamkeit.

Was gehn die Glocken heute,
Als ob ich weinen müßt'?
Die Glocken, die bedeuten,
Daß meine Lieb' gestorben ist!

Ich wollt', ich läg' begraben
Und über mir rauschte weit
Die Linde jeden Abend
Von der alten, schönen Zeit!

Die Nachtigallen

Möcht' wissen, was sie schlagen
So schön bei der Nacht,
's ist in der Welt ja doch niemand,
Der mit ihnen wacht.

Und die Wolken, die reisen,
Und das Land ist so blaß,
Und die Nacht wandert leise
Durch den Wald übers Gras.

Nacht, Wolken, wohin sie gehen,
Ich weiß es recht gut,
Liegt ein Grund hinter den Höhen,
Wo meine Liebste jetzt ruht.

Zieht der Einsiedel sein Glöcklein,
Sie höret es nicht,
Es fallen ihr die Löcklein
Übers ganze Gesicht.

Und daß sie niemand erschrecket,
Der liebe Gott hat sie hier
Ganz mit Mondschein bedecket,
Da träumt sie von mir.

Nachruf

Du liebe, treue Laute,
Wie manche Sommernacht,
Bis daß der Morgen graute,
Hab' ich mit dir durchwacht!

Die Täler wieder nachten,
Kaum spielt noch Abendrot,
Doch die sonst mit uns wachten,
Die liegen lange tot.

Was wollen wir nun singen
Hier in der Einsamkeit,
Wenn alle von uns gingen,
Die unser Lied erfreut?

Wir wollen dennoch singen!
So still ist's auf der Welt;
Wer weiß, die Lieder dringen
Vielleicht zum Sternenzelt.

Wer weiß, die da gestorben,
Sie hören droben mich,
Und öffnen leis die Pforten
Und nehmen uns zu sich.

VI. Geistliche Gedichte

Andre haben andre Schwingen,
Aber wir, mein fröhlich Herz,
Wollen grad' hinaus uns singen,
Aus dem Frühling himmelwärts!

Götterdämmerung

1

Was klingt mir so heiter
Durch Busen und Sinn?
Zu Wolken und weiter,
Wo trägt es mich hin?

Wie auf Bergen hoch bin ich
So einsam gestellt
Und grüße herzinnig,
Was schön auf der Welt.

Ja, Bacchus, dich seh' ich,
Wie göttlich bist du!
Dein Glühen versteh' ich,
Die träumende Ruh'.

O rosenbekränztes
Jünglingsbild,
Dein Auge, wie glänzt es,
Die Flammen so mild!

Ist's Liebe, ist's Andacht,
Was so dich beglückt?
Rings Frühling dich anlacht,
Du sinnest entzückt. –

Frau Venus, du Frohe,
So klingend und weich,
In Morgenrots Lohe
Erblick' ich dein Reich

Auf sonnigen Hügeln
Wie ein' Zauberring. –
Zart Bübchen mit Flügeln
Bedienen dich flink.

Durchsäuseln die Räume
Und laden, was fein,
Als goldene Träume
Zur Königin ein.

Und Ritter und Frauen
Im grünen Revier
Durchschwärmen die Auen
Wie Blumen zur Zier.

Und jeglicher hegt sich
Sein Liebchen im Arm,
So wirrt und bewegt sich
Der selige Schwarm. –

Die Klänge verrinnen,
Es bleichet das Grün,
Die Frauen stehn sinnend,
Die Ritter schaun kühn.

Und himmlisches Sehnen
Geht singend durchs Blau,
Da schimmert von Tränen
Rings Garten und Au. –

Und mitten im Feste
Erblick' ich, wie mild!
Den stillsten der Gäste. –
Woher, einsam Bild?

Mit blühendem Mohne,
Der träumerisch glänzt,
Und mit Lilienkrone
Erscheint er bekränzt.

Sein Mund schwillt zum Küssen
So lieblich und bleich,

Als brächt' er ein Grüßen
Aus himmlischem Reich.

Eine Fackel wohl trägt er,
Die wunderbar prangt.
„Wo ist einer", frägt er,
„Den heimwärts verlangt?"

Und manchmal da drehet
Die Fackel er um –
Tiefschauernd vergehet
Die Welt und wird stumm.

Und was hier versunken
Als Blumen zum Spiel,
Siehst oben du funkeln
Als Sterne nun kühl. –

O Jüngling vom Himmel,
Wie bist du so schön!
Ich laß' das Gewimmel,
Mit dir will ich gehn!

Was will ich noch hoffen?
Hinauf, ach hinauf!
Der Himmel ist offen,
Nimm, Vater, mich auf!

2

Von kühnen Wunderbildern
Ein großer Trümmerhauf',
In reizendem Verwildern
Ein blühnder Garten drauf;

Versunknes Reich zu Füßen,
Vom Himmel fern und nah,

Aus anderm Reich ein Grüßen –
Das ist Italia!

Wenn Frühlingslüfte wehen
Hold übern grauen Plan,
Ein leises Auferstehen
Hebt in den Tälern an.

Da will sich's unten rühren
Im stillen Göttergrab,
Der Mensch kann's schauernd spüren
Tief in die Brust hinab.

Verwirrend in den Bäumen
Gehn Stimmen hin und her,
Ein sehnsuchtsvolles Träumen
Weht übers blaue Meer.

Und unterm duft'gen Schleier,
Sooft der Lenz erwacht,
Webt in geheimer Feier
Die alte Zaubermacht.

Frau Venus hört das Locken,
Der Vögel heitern Chor,
Und richtet froh erschrocken
Aus Blumen sich empor.

Sie sucht die alten Stellen,
Das luft'ge Säulenhaus,
Schaut lächelnd in die Wellen
Der Frühlingsluft hinaus.

Doch öd' sind nun die Stellen,
Stumm liegt ihr Säulenhaus,
Gras wächst da auf den Schwellen,
Der Wind zieht ein und aus.

Wo sind nun die Gespielen?
Diana schläft im Wald,
Neptunus ruht im kühlen
Meerschloß, das einsam hallt.

Zuweilen nur Sirenen
Noch tauchen aus dem Grund,
Und tun in irren Tönen
Die tiefe Wehmut kund. –

Sie selbst muß sinnend stehen
So bleich im Frühlingsschein,
Die Augen untergehen,
Der schöne Leib wird Stein. –

Denn über Land und Wogen
Erscheint, so still und mild,
Hoch auf dem Regenbogen
Ein andres Frauenbild.

Ein Kindlein in den Armen
Die Wunderbare hält,
Und himmlisches Erbarmen
Durchdringt die ganze Welt.

Da in den lichten Räumen
Erwacht das Menschenkind,
Und schüttelt böses Träumen
Von seinem Haupt geschwind.

Und, wie die Lerche singend,
Aus schwülen Zaubers Kluft
Erhebt die Seele ringend
Sich in die Morgenluft.

Mariä Sehnsucht

Es ging Maria in den Morgen hinein,
Tat die Erd' einen lichten Liebesschein,
Und über den fröhlichen, grünen Höh'n
Sah sie den bläulichen Himmel stehn.
„Ach, hätt' ich ein Brautkleid von Himmelsschein,
Zwei goldene Flüglein – wie flög' ich hinein!" –

Es ging Maria in stiller Nacht,
Die Erde schlief, der Himmel wacht',
Und durch's Herze, wie sie ging und sann und dacht',
Zogen die Sterne mit goldener Pracht.
„Ach, hätt' ich das Brautkleid von Himmelsschein,
Und goldene Sterne gewoben drein!"

Es ging Maria im Garten allein,
Da sangen so lockend bunt' Vögelein,
Und Rosen sah sie im Grünen stehn,
Viel' rote und weiße so wunderschön.
„Ach hätt' ich ein Knäblein, so weiß und rot,
Wie wollt ich's lieb haben bis in den Tod!"

Nun ist wohl das Brautkleid gewoben gar,
Und goldene Sterne im dunkelen Haar,
Und im Arme die Jungfrau das Knäblein hält,
Hoch über der dunkelerbrausenden Welt,
Und vom Kindlein gehet ein Glänzen aus,
Das ruft uns nur ewig: nach Haus, nach Haus!

Jugendandacht

1

Daß des verlornen Himmels es gedächte,
Schlugen ans Herz des Frühlings linde Wellen,

Wie ew'ger Wonnen schüchternes Vermuten,
Geheimer Glanz der lauen Sommernächte,
Du grüner Wald, verführend Lied der Quellen,
Des Morgens Pracht, stillblühnde Abendgluten,
Ihr fragt: wo Schmerz und Lust so lange ruhten,
Die süß das Herz verdunkeln und es hellen?
Wie tut ihr zaubrisch auf die alten Wunden,
Daß losgebunden in das Licht sie bluten!
O sel'ge Zeit entfloßner Himmelsbläue,
Der ersten Andacht solch inbrünst'ger Liebe,
Die ewig wollte knien vor der Einen!
Demütig in der Glorie des Maien
Hob sie den Schleier oft, daß offen bliebe
Der Augen Himmel, in das Land zu scheinen.
Und stand ich still, und mußt' ich herzlich weinen;
In ihrem Blick gereinigt alle Triebe:
Da war nur Wonne, was ich mußte klagen,
Im Angesicht der Stillen, Ewigreinen
Kein Schmerz, als solcher Liebe Lieb' ertragen!

2

Wie in einer Blume himmelblauen
Grund, wo schlummernd träumen stille Regenbogen,
Ist mein Leben ein unendlich Schauen,
Klar durchs ganze Herz Ein süßes Bild gezogen.

Stille saß ich, sah die Jahre fliegen,
Bin im Innersten Dein treues Kind geblieben;
Aus dem duft'gen Kelche aufgestiegen,
Ach, wann lohnst Du endlich auch mein treues Lieben!

3

Was wollen mir vertraun die blauen Weiten,
Des Landes Glanz, die Wirrung süßer Lieder,

Mir ist so wohl, so bang! Seid ihr es wieder
Der frommen Kindheit stille Blumenzeiten?

Wohl weiß ich's – dieser Farben heimlich Spreiten
Deckt einer Jungfrau strahlend reine Glieder;
Es wogt der große Schleier auf und nieder,
Sie schlummert drunten fort seit Ewigkeiten.

Mir ist in solchen linden, blauen Tagen,
Als müßten alle Farben auferstehen,
Aus blauer Fern' sie endlich zu mir gehen.

So wart' ich still, schau in den Frühling milde,
Das ganze Herz weint nach dem süßen Bilde,
Vor Freud', vor Schmerz? – Ich weiß es nicht zu sagen.

4

Viel' Lenze waren lange schon vergangen,
Vorüber zogen wunderbare Lieder,
Die Sterne gingen ewig auf und nieder,
Die selbst vor großer Sehnsucht golden klangen.

Und wie so tausend Stimmen ferne sangen,
Als riefen mich von hinnen sel'ge Brüder,
Fühlt' ich die alten Schmerzen immer wieder,
Seit Deine Blicke, Jungfrau, mich bezwangen.

Da war's, als ob sich still Dein Auge hübe,
Langst sehnsuchtsvoll nach mir mit offnen Armen,
Fühlst selbst den Schmerz, den Du mir süß gegeben. –

Umfangen fühl' ich innigst mich erwarmen,
Berührt mit goldnen Strahlen mich das Leben,
Ach! daß ich ewig Dir am Herzen bliebe!

5

Wann Lenzesstrahlen golden niederrinnen,
Sieht man die Scharen losgebunden ziehen,
Im Waldrevier, dem neu der Schmuck geliehen,
Die lust'ge Jagd nach Lieb' und Scherz beginnen.

Den Sänger will der Frühling gar umspinnen,
Er, der Geliebteste, darf nicht entfliehen,
Fühlt rings ein Lied durch alle Farben ziehen,
Das ihn so lockend nimmer läßt von hinnen.

Gefangen so, sitzt er viel' sel'ge Jahre;
Des Einsamen spottet des Pöbels Scherzen,
Der aller Glorie möchte Lieb' entkleiden.

Doch er grüßt fröhlich alle, wie sie fahren,
Und mutig sagt er zu den süßen Schmerzen:
„Gern sterb' ich bald, wollt ihr von mir je scheiden!"

6

Wann frisch die buntgewirkten Schleier wallen,
Weit in das Land die Lerchen mich verführen,
Da kann ich's tief im Herzen wieder spüren,
Wie mich die Eine liebt und ruft vor allen.

Wenn Nachtigalln aus grünen Hallen schallen,
Wen möchten nicht die tiefen Töne rühren;
Wen nicht das süße Herzeleid verführen,
Im Liebesschlagen tot vom Baum zu fallen? –

So sag' auch ich bei jedem Frühlingsglanze:
Du süße Laute! laß uns beide sterben,
Beklagt vom Widerhallen zarter Töne,

Kann unser Lied auch nie den Lohn erwerben,
Daß hier mit eignem, frischem Blumenkranze
Uns endlich kröne nun die Wunderschöne! –

7

Der Schäfer spricht, wenn er frühmorgens weidet:
„Dort drüben wohnt sie hinter Berg' und Flüssen!"
Doch seine Wunden deckt sie gern mit Küssen,
Wann lauschend Licht am stillen Abend scheidet.

Ob neu der Morgenschmuck die Erde kleidet,
Ob Nachtigallen Nacht und Stern' begrüßen,
Stets fern und nah bleibt *meine* Lieb' der Süßen,
Die in dem Lenz mich ewig sucht und meidet. –

Doch hör' ich wunderbare Stimmen sprechen:
„Die Perlen, die du treu geweint im Schmerze,
Sie wird sie sorglich all zusammenbinden,

Mit eigner Kette so dich süß umwinden,
Hinaufziehn dich an Mund und blühend Herze –
Was Himmel schloß, mag nicht der Himmel brechen."

8

Wenn du am Felsenhange standst alleine,
Unten im Walde Vögel seltsam sangen
Und Hörner aus der Ferne irrend klangen,
Als ob die Heimat drüben nach dir weine,

War's niemals da, als rief die Eine, Deine?
Lockt' dich kein Weh, kein brünstiges Verlangen
Nach andrer Zeit, die lange schon vergangen,
Auf ewig einzugehn in grüne Scheine?

Gebirge dunkelblau steigt aus der Ferne,
Und von den Gipfeln führt des Bundes Bogen
Als Brücke weit in unbekannte Lande.

Geheimnisvoll gehn oben goldne Sterne,
Unten erbraust viel Land im dunklen Wogen –
Was zögerst du am unbekannten Rande?

9

Es wendet zürnend sich von mir die Eine,
Versenkt die Ferne mit den Wunderlichtern.
Es stockt der Tanz – ich stehe plötzlich nüchtern,
Musik läßt treulos mich so ganz alleine.

Da spricht der Abgrund dunkel: Bist nun meine;
Zieht mich hinab an bleiernen Gewichtern,
Sieht stumm mich an aus steinernen Gesichtern,
Das Herz wird selber zum kristallnen Steine.

Dann ist's, als ob es dürstend Schmerzen sauge
Aus lang vergeßner Zeit Erinnerungen,
Und kann sich rühren nicht, von Frost bezwungen.

Versteinert schweigen muß der Wehmut Welle,
Wie willig auch, schmölz' ihn ein wärmend Auge,
Kristall zerfließen wollt' als Tränenquelle.

10

Durch's Leben schleichen feindlich fremde Stunden,
Wo Ängsten aus der Brust hinunterlauschen,
Verworrne Worte mit dem Abgrund tauschen,
Drin bodenlose Nacht nur ward erfunden.

Wohl in des Dichters Seele stumm verbunden
Mit Mächten, die am Volk vorüberrauschen;

Sehnsucht muß wachsen an der Tiefe Rauschen
Nach hellem Licht und nach des Himmels Kunden.

O Herr! du kennst allein den treuen Willen,
Befrei ihn von der Kerkerluft des Bösen,
Laß nicht die eigne Brust mich feig zerschlagen!

Und wie ich schreibe hier, den Schmerz zu stillen,
Fühl' ich den Engel schon die Riegel lösen,
Und kann vor Glanze nicht mehr weiter klagen.

Der Fromme

Es saß ein Kind gebunden und gefangen,
Wo vor der Menschen eitlem Tun und Schallen
Der Vorzeit Wunderlaute trüb verhallen;
Der alten Heimat dacht' es soll Verlangen.

Da sieht es draußen Ströme, hell ergangen,
Durch zaubrisch Land viel' Pilger, Sänger wallen,
Kühl rauscht der Wald, die lust'gen Hörner schallen,
Aurora scheint, so weit die Blicke langen. –

O laß die Sehnsucht ganz dein Herz durchdringen!
So legt sich blühend um die Welt dein Trauern
Und himmlisch wird dein Schmerz und deine Sorgen.

Ein frisch Gemüt mag wohl die Welt bezwingen,
Ein recht Gebet bricht Banden bald und Mauern:
Und frei springst du hinunter in den Morgen.

Willkommen, Liebchen, denn am Meeresstrande!
Wie rauschen lockend da ans Herz die Wellen
Und tiefe Sehnsucht will die Seele schwellen,
Wenn andre träge schlafen auf dem Lande.

So walte Gott! – Ich lös' des Schiffleins Bande,
Wegweiser sind die Stern', die ewig hellen,
Viel' Segel fahren da und frisch' Gesellen
Begrüßen uns von ihrer Schiffe Rande.

Wir sitzen still, gleich Schwänen zieht das Segel,
Ich schau' in deiner Augen lichte Sterne,
Du schweigst und schauerst heimlich oft zusammen.

Blick auf! Schon schweifen Paradiesvögel,
Schon wehen Wunderklänge aus der Ferne,
Der Garten Gottes steigt aus Morgenflammen.

Lieder

1

Frisch eilt der helle Strom hinunter,
Drauf ziehn viel' bunte Schifflein munter,
Und Strom und Schiff und bunte Scheine,
Sie fragen alle: was ich weine?
Mir ist so wohl, mir ist so weh,
Wie ich den Frühling fahren seh'.

Viel' Lenze sitz' ich schon da oben,
Ein Regenbogen steht im Land erhoben
Und durch die Täler, Wiesen, Wogen
Still, wie ein fernes Lied, gezogen.
Schifft immerfort Dein himmlisch! Bild –
Doch Strom und Schiff nie stille hielt.

2

Denk' ich Dein, muß bald verwehen
Alle Trübnis weit und breit,

Und die frischen Blicke gehen
Wie in einen Garten weit.

Wunderbare Vögel wieder
Weiden dort auf grüner Au,
Einsam' Engel, alte Lieder
Ziehen durch den Himmel blau.

Wolken, Ströme, Schiffe, alle
Segeln in die Pracht hinein –
Keines kehrt zurück von allen,
Und ich stehe so allein.

An den heiligen Joseph

Wenn trübe Schleier alles grau umweben,
Zur bleichen Ferne wird das ganze Leben,
Will Heimat oft sich tröstend zeigen;
Aus Morgenrot die goldnen Höhen steigen,
Und aus dem stillen, wundervollen Duft
Eine wohlbekannte Stimm' hinüberruft.

Du warst ja auch einmal hier unten,
Hast ew'ger Treue Schmerz empfunden;
Längst war Maria fortgezogen,
Wie einsam rauschten rings die dunklen Wogen!
Da breitet' oben sie die Arme aus:
Komm, treuer Pilger, endlich auch nach Haus!

Seitdem ist wohl viel anders worden,
Treu' Lieb auf Erden ist ausgestorben.
Wem könnt ich's außer Dir wohl klagen,
Wie oft in kummervollen Tagen
Mein ganzes Herz hier hofft und bangt,
Und nach der Heimat immer fort verlangt!

Kirchenlied

O Maria, meine Liebe!
Denk ' ich recht im Herzen Dein:
Schwindet alles Schwer' und Trübe,
Und, wie heller Morgenschein,
Dringt's durch Lust und ird'schen Schmerz
Leuchtend mir durchs ganze Herz.

Auf des ew'gen Bundes Bogen,
Ernst von Glorien umblüht,
Stehst du über Land und Wogen;
Und ein himmlisch Sehnen zieht
Alles Leben himmelwärts
An das große Mutterherz.

Wo Verlaßne einsam weinen,
Sorgenvoll in stiller Nacht,
Den' vor allen läßt Du scheinen
Deiner Liebe milde Pracht,
Daß ein tröstend Himmelslicht
In die dunklen Herzen bricht.

Aber wütet wildverkehrter
Sünder frevelhafte Lust:
Da durchschneiden neue Schwerter
Dir die treue Mutterbrust;
Und voll Schmerzen flehst Du doch:
Herr! Vergib, o schone noch!

Deinen Jesus in den Armen,
Über'n Strom der Zeit gestellt,
Als das himmlische Erbarmen
Hütest Du getreu die Welt,
Daß im Sturm, der trübe weht,
Dir kein Kind verloren geht.

Wenn die Menschen mich verlassen
In der letzten stillen Stund',
Laß mich fest das Kreuz umfassen.
Aus dem dunklen Erdengrund
Leite liebreich mich hinaus,
Mutter, in des Vaters Haus!

Morgengebet

O wunderbares, tiefes Schweigen,
Wie einsam ist's noch auf der Welt!
Die Wälder nur sich leise neigen,
Als ging' der Herr durchs stille Feld.

Ich fühl' mich recht wie neu geschaffen,
Wo ist die Sorge nun und Not?
Was mich noch gestern wollt' erschlaffen,
Ich schäm' mich des im Morgenrot.

Die Welt mit ihrem Gram und Glücke
Will ich, ein Pilger, frohbereit
Betreten nur wie eine Brücke
Zu Dir, Herr, übern Strom der Zeit.

Und buhlt mein Lied, auf Weltgunst lauernd,
Um schnöden Sold der Eitelkeit:
Zerschlag' mein Saitenspiel, und schauernd
Schweig' ich vor Dir in Ewigkeit.

Mittag

Vergeht mir der Himmel
Vor Staube schier,
Herr, im Getümmel
Zeig' Dein Panier!

Wie schwank' ich sündlich,
Läßt Du von mir;
Unüberwindlich
Bin ich mit Dir!

Abend

Gestürzt sind die goldnen Brücken
Und unten und oben so still!
Es will mir nichts mehr glücken,
Ich weiß nicht mehr, was ich will.

Von üppig blühenden Schmerzen
Rauscht eine Wildnis im Grund,
Da spielt wie in wahnsinnigen Scherzen
Das Herz an dem schwindligen Schlund. –

Die Felsen möchte ich packen
Vor Zorn und Wehe und Lust,
Und unter den brechenden Zacken
Begraben die wilde Brust.

Da kommt der Frühling gegangen,
Wie ein Spielmann aus alter Zeit,
Und singt von uraltem Verlangen
So treu durch die Einsamkeit.

Und über mir Lerchenlieder
Und unter mir Blumen bunt,
So werf' ich im Grase mich nieder
Und weine aus Herzensgrund.

Da fühl' ich ein tiefes Entzücken,
Nun weiß ich wohl, was ich will,
Es bauen sich andere Brücken,
Das Herz wird auf einmal still.

Der Abend streut rosige Flocken,
Verhüllet die Erde nun ganz,
Und durch des Schlummernden Locken
Ziehn Sterne den heiligen Kranz.

Nachtgruß

Weil jetzo alles stille ist
Und alle Menschen schlafen,
Mein' Seel' das ew'ge Licht begrüßt,
Ruht wie ein Schiff im Hafen.

Der falsche Fleiß, die Eitelkeit,
Was keinen mag erlaben,
Darin der Tag das Herz zerstreut,
Liegt alles tief begraben.

Ein andrer König wunderreich
Mit königlichen Sinnen,
Zieht herrlich ein im stillen Reich,
Besteigt die ew'gen Zinnen.

Morgenlied

Kein Stimmlein noch schallt von allen
In frühester Morgenstund',
Wie still ist's noch in den Hallen
Durch den weiten Waldesgrund.

Ich stehe hoch überm Tale
Stille vor großer Lust,
Und schau' nach dem ersten Strahle,
Kühl schauernd in tiefster Brust.

Wie sieht da zu dieser Stunde
So anders das Land herauf,
Nichts hör' ich da in der Runde
Als von fern der Ströme Lauf.

Und ehe sich alle erhoben
Des Tages Freuden und Weh,
Will ich, Herr Gott, Dich loben
Hier einsam in stiller Höh'. –

Nun rauschen schon stärker die Wälder,
Morgenlicht funkelt herauf,
Die Lerche singt über den Feldern,
Schöne Erde, nun wache auf!

In der Nacht

Das Leben draußen ist verrauschet,
Die Lichter löschen aus,
Schauernd mein Herz am Fenster lauschet
Still in die Nacht hinaus.

Da nun der laute Tag zerronnen
Mit seiner Not und bunten Lust,
Was hast du in dem Spiel gewonnen,
Was blieb der müden Brust? –

Der Mond ist trostreich aufgegangen,
Da unterging die Welt,
Der Sterne heil'ge Bilder prangen
So einsam hoch gestellt!

O Herr! auf dunkelschwankem Meere
Fahr' ich im schwachen Boot,
Treufolgend Deinem goldenen Heere
Zum ew'gen Morgenrot.

Werktag

Wir wandern nun schon viel' hundert Jahr',
Und kommen doch nicht zur Stelle –
Der Strom wohl rauscht an die tausend gar,
Und kommt doch nicht zur Quelle.

Sonntag

Weit in das Land die Ström' ihr Silber führen,
Fern blau Gebirge duftig hingezogen,
Die Sonne scheint, die Bäume sanft sich rühren,
Und Glockenklang kommt auf den linden Wogen;
Hoch in den Lüften Lerchen jubilieren,
Und, so weit klar sich wölbt des Himmels Bogen,
Von Arbeit ruht der Mensch rings in die Runde,
Atmet zum Herren auf aus Herzensgrunde.

Frühling

Und wenn die Lerche hell anstimmt
Und Frühling rings bricht an:
Da schauert tief und Flügel nimmt,
Wer irgend fliegen kann.

Die Erde grüßt er hochbeglückt,
Die, eine junge Braut,
Mit Blumen wild und bunt geschmückt,
Tief in das Herz ihm schaut.

Den Himmel dann, das blaue Meer
Der Sehnsucht, grüßt er treu,
Da stammen Lied und Sänger her
Und spüren's immer neu.

Die dunkeln Gründe säuseln kaum,
Sie schaun so fremd herauf.
Tiefschauernd fühlt er, 's war ein Traum –
Und wacht im Himmel auf.

Herbst

Es ist nun der Herbst gekommen,
Hat das schöne Sommerkleid
Von den Feldern weggenommen
Und die Blätter ausgestreut,
Vor dem bösen Winterwinde
Deckt er warm und sachte zu
Mit dem bunten Laub die Gründe,
Die schon müde gehn zur Ruh.

Durch die Felder sieht man fahren
Eine wunderschöne Frau,
Und von ihren langen Haaren
Goldne Fäden auf der Au
Spinnet sie und singt im Gehen:
Eia, meine Blümelein,
Nicht nach andern immer sehen,
Eia, schlafet, schlafet ein.

Und die Vöglein hoch in Lüften
Über blaue Berg' und Seen
Ziehn zur Ferne nach den Klüften,
Wo die hohen Zedern stehn,
Wo mit ihren goldnen Schwingen
Auf des Benedeiten Gruft
Engel Hosianna singen
Nächtens durch die stille Luft.

Winter

Wie von Nacht verhangen,
Wußt' nicht, was ich will,
Schon so lange, lange
War ich totenstill.

Liegt die Welt voll Schmerzen,
Will's auch draußen schnei'n:
Wache auf, mein Herze,
Frühling muß es sein!

Was mich frech wollt' fassen,
's ist nur Wogenschaum,
Falsche Ehr', Not, Hassen,
Welt, ich spür' dich kaum.

Breite nur die Flügel
Wieder, schönes Roß,
Frei laß ich die Zügel,
So brich durch, Genoß!

Und hat ausgeklungen
Liebes-Lust und Leid,
Um die wir gerungen
In der schönsten Zeit;

Nun so trag' mich weiter,
Wo das Wünschen aus –
Wie wird mir so heiter,
Roß, bring mich nach Haus!

Der Schiffer

Die Lüfte linde fächeln,
Aus stillen Meeres Schaum
Sirenen tauchend lächeln,
Der Schiffer liegt im Traum.

Da faßt der Sturm die Wellen,
Durchwühlt die Einsamkeit:
Wacht auf, ihr Traumgesellen,
Nun ist's nicht Schlafenszeit! –

In jenen stillen Tagen
Wie war ich stolz und klug,
In sichern Glücks Behagen
Mir selber gut genug.

Du hast das Glück zerschlagen;
Nimm wieder, was Du gabst,
Ich schweig' und will nicht klagen,
Jetzt weiß ich, wie Du labst.

Das sind die mächt'gen Stürme,
Die wecken, was da ruht,
Es sinken Land und Türme
Allmählich in die Flut.

Kein Meerweib will sich zeigen,
Kein Laut mehr langt zu mir,
Und in dem weiten Schweigen
Steh' ich allein mit Dir.

O führe an den Riffen
Allmächtig Deine Hand,
Wohin wir alle schiffen,
Uns zu dem Heimatsstrand!

Der Soldat

Und wenn es einst dunkelt,
Der Erd' bin ich satt,
Durchs Abendrot funkelt
Eine prächt'ge Stadt:

Von den goldenen Türmen
Singet der Chor,
Wir aber stürmen
Das himmlische Tor.

Der Wächter

Nächtlich macht der Herr die Rund',
Sucht die Seinen unverdrossen,
Aber überall verschlossen
Trifft er Tür und Herzensgrund,
Und er wendet sich voll Trauer:
Niemand ist, der mit mir wacht. –
Nur der Wald vernimmt's mit Schauer,
Rauschet fromm die ganze Nacht.

Waldwärts durch die Einsamkeit
Hört' ich über Tal und Klüften
Glocken in den stillen Lüften,
Wie aus fernem Morgen weit –
An die Tore will ich schlagen,
An Palast und Hütten: Auf!
Flammend schon die Gipfel ragen,
Wachet auf, wacht auf, wacht auf!

Gottes Segen

Das Kind ruht aus vom Spielen,
Am Fenster rauscht die Nacht,

Die Engel Gott's im Kühlen
Getreulich halten Wacht.

Am Bettlein still sie stehen,
Der Morgen graut noch kaum,
Sie küssen's, eh' sie gehen,
Das Kindlein lacht im Traum.

Der Umkehrende

1

Du sollst mich doch nicht fangen,
Duftschwüle Zaubernacht!
Es stehn mit goldnem Prangen
Die Stern' auf stiller Wacht,
Und machen überm Grunde,
Wo du verirret bist,
Getreu die alte Runde –
Gelobt sei Jesus Christ!

Wie bald in allen Bäumen
Geht nun die Morgenluft,
Sie schütteln sich in Träumen,
Und durch den roten Duft
Eine fromme Lerche steiget,
Wenn alles still noch ist,
Den rechten Weg dir zeiget –
Gelobt sei Jesus Christ!

2

Hier bin ich, Herr! Gegrüßt das Licht,
Das durch die stille Schwüle
Der müden Brust gewaltig bricht
Mit seiner strengen Kühle.

Nun bin ich frei! Ich taumle noch
Und kann mich noch nicht fassen –
O Vater, Du erkennst mich doch,
Und wirst nicht von mir lassen!

3

Was ich wollte, liegt zerschlagen,
Herr, ich lasse ja das Klagen,
Und das Herz ist still.
Nun aber gib auch Kraft, zu tragen,
Was ich *nicht* will!

4

Es wandelt, was wir schauen,
Tag sinkt ins Abendrot,
Die Lust hat eignes Grauen,
Und alles hat den Tod.

Ins Leben schleicht das Leiden
Sich heimlich wie ein Dieb,
Wir alle müssen scheiden
Von allem, was uns lieb.

Was gäb' es doch auf Erden,
Wer hielt' den Jammer aus,
Wer möcht' geboren werden,
Hielt'st Du nicht droben Haus!

Du bist's, der, was wir bauen,
Mild über uns zerbricht,
Daß wir den Himmel schauen –
Darum so klag' ich nicht.

5

Waldeinsamkeit!
Du grünes Revier,
Wie liegt so weit
Die Welt von hier!
Schlaf' nur, wie bald
Kommt der Abend schön,
Durch den stillen Wald
Die Quellen gehn,
Die Mutter Gottes wacht,
Mit ihrem Sternenkleid
Bedeckt sie dich sacht
In der Waldeinsamkeit,
Gute Nacht, gute Nacht! –

Der Kranke

Soll ich dich denn nun verlassen,
Erde, heitres Vaterhaus?
Herzlich Lieben, mutig Hassen,
Ist denn alles, alles aus?

Vor dem Fenster durch die Linden
Spielt es wie ein linder Gruß,
Lüfte, wollt ihr mir verkünden,
Daß ich bald hinunter muß? –

Liebe, ferne, blaue Hügel,
Stiller Fluß im Talesgrün,
Ach, wie oft wünscht' ich mir Flügel,
Über euch hinweg zu ziehn!

Da sich jetzt die Flügel dehnen,
Schaur' ich in mich selbst zurück,
Und ein unbeschreiblich Sehnen
Zieht mich zu der Welt zurück.

Sterbeglocken

Nun legen sich die Wogen,
Und die Gewitter schwül'
Sind all' hinabgezogen,
Mir wird das Herz so kühl.

Die Täler alle dunkeln,
Ist denn das Morgenzeit?
Wie schön die Gipfel funkeln,
Und Glocken hör' ich weit.

So hell noch niemals klangen
Sie übern Waldessaum –
Wo war ich denn so lange?
Das war ein schwerer Traum.

Der Pilger

1

Man setzt uns auf die Schwelle,
Wir wissen nicht, woher?
Da glüht der Morgen helle,
Hinaus verlangt uns sehr.
Der Erde Klang und Bilder,
Tiefblaue Frühlingslust,
Verlockend wild und wilder,
Bewegen da die Brust.
Bald wird es rings so schwüle,
Die Welt eratmet kaum,
Berg', Schloß und Wälder kühle
Stehn lautlos wie im Traum,
Und ein geheimes Grausen

Beschleichet unsern Sinn:
Wir sehnen uns nach Hause
Und wissen nicht, wohin?

2

Dein Wille, Herr, geschehe!
Verdunkelt schweigt das Land,
Im Zug der Wetter sehe
Ich schauernd Deine Hand.
O mit uns Sündern gehe
Erbarmend ins Gericht!
Ich beug' im tiefsten Wehe
Zum Staub mein Angesicht,
Dein Wille, Herr, geschehe!

3

Schlag mit den flamm'gen Flügeln!
Wenn Blitz aus Blitz sich reißt:
Steht wie in Rossesbügeln
So ritterlich mein Geist.

Waldesrauschen, Wetterblicken
Macht recht die Seele los,
Da grüßt sie mit Entzücken,
Was wahrhaft, ernst und groß.

Es schiffen die Gedanken
Fern wie auf weitem Meer,
Wie auch die Wogen schwanken:
Die Segel schwellen mehr.

Herr Gott, es wacht Dein Wille,
Ob Tag und Lust verwehn,
Mein Herz wird mir so stille
Und wird nicht untergehn.

4

So laß herein nun brechen
Die Brandung, wie sie will,
Du darfst ein Wort nur sprechen,
So wird der Abgrund still;
Und bricht die letzte Brücke,
Zu Dir, der treulich steht,
Hebt über Not und Glücke
Mich einsam das Gebet.

5

Wie ein todeswunder Streiter,
Der den Weg verloren hat,
Schwank' ich nun und kann nicht weiter,
Von dem Leben sterbensmatt.
Nacht schon decket alle Müden
Und so still ist's um mich her,
Herr, auch mir gib endlich Frieden,
Denn ich wünsch' und hoff' nichts mehr.

6

Wie oft wollt' mich die Welt ermüden,
Ich beugt' aufs Schwert mein Angesicht
Und bat Dich frevelhaft um Frieden –
Du wußtest's besser, gabst ihn nicht.

Ich sah in Nacht das Land vergehen,
In Blitzen Du die Wetter brachst,
Da konnt' ich schauernd erst verstehen,
Was Du zu mir Erschrocknem sprachst:

„Meine Lieder sind nicht deine Lieder,
Leg ab den falschen Schmuck der Zeit
Und nimm das Kreuz, dann komme wieder
In deines Herzens Einsamkeit."

Und alle Bilder ferne treten,
Und tief noch rauschet kaum die Rund' –
Wie geht ein wunderbares Beten
Mir leuchtend durch der Seele Grund!

Der Pilot

Glaube stehet still erhoben
Überm nächt'gen Wellenklang,
Lieset in den Sternen droben
Fromm des Schiffleins sichern Gang.

Liebe schwellet sanft die Segel,
Dämmernd zwischen Tag und Nacht
Schweifen Paradiesesvögel,
Ob der Morgen bald erwacht?

Morgen will sich kühn entzünden,
Nun wird's mir auf einmal kund:
Hoffnung wird die Heimat finden
Und den stillen Ankergrund.

Der Einsiedler

Komm, Trost der Welt, du stille Nacht!
Wie steigst du von den Bergen sacht,
Die Lüfte alle schlafen,
Ein Schiffer nur noch, wandermüd',
Singt übers Meer sein Abendlied
Zu Gottes Lob im Hafen.

Die Jahre wie die Wolken gehn
Und lassen mich hier einsam stehn,
Die Welt hat mich vergessen,
Da trat'st du wunderbar zu mir,

Wenn ich beim Waldesrauschen hier
Gedankenvoll gesessen.

O Trost der Welt, du stille Nacht!
Der Tag hat mich so müd' gemacht,
Das weite Meer schon dunkelt,
Laß ausruhn mich von Lust und Not,
Bis daß das ew'ge Morgenrot
Den stillen Wald durchfunkelt.

Der Sänger

1

Siehst du die Wälder glühen,
Die Ströme flammend sprühen,
Die Welt in Abendgluten
Wie träumerische Fluten,
Wo blüh'nde Inseln trunken
Sich spiegeln in dem Duft? –
Es weht und rauscht und ruft:
O komm, eh wir versunken!

Eh' noch die Sonn' versunken:
Gehn durch die goldnen Funken
Still Engel in den Talen,
Das gibt so leuchtend Strahlen
In Blumen rings und Zweigen. –
Wie frommer Widerhall
Weht noch der Glocken Schall,
Wenn längst die Täler schweigen.

Leis wächst durchs dunkle Schweigen
Ein Flüstern rings und Neigen
Wie ein geheimes Singen,
In immer weitern Ringen

Zieht's alle, die da lauschen,
In seine duft'ge Rund',
Wo kühl im stillen Grund
Die Wasserkünste rauschen.

Wie Wald und Strom im Rauschen
Verlockend Worte tauschen!
Was ist's, daß ich ergrause? –
Führt doch aus stillem Hause
Der Hirt die goldne Herde,
Und hütet treu und wacht,
So lieblich weht die Nacht,
Lind säuselt kaum die Erde.

2

Und zu den Felsengängen
Der nächt'ge Sänger flieht,
Denn wie mit Wahnsinns Klängen
Treibt ihn sein eignes Lied.

Bei leuchtenden Gewittern
Schreckt ihn das stille Land,
Ein wunderbar Erschüttern
Hat ihm das Herz gewandt.

Bereuend sinkt sein Auge –
Da blickt durch Nacht und Schmerz
Ein unsichtbares Auge –
Ihm klar ins tiefste Herz.

Sein Saitenspiel zur Stunde
Wirft er in tiefsten Schlund,
Und weint aus Herzensgrunde,
Und ewig schweigt sein Mund.

Morgendämmerung

Es ist ein still Erwarten in den Bäumen,
Die Nachtigallen in den Büschen schlagen
In irren Klagen, können's doch nicht sagen,
Die Schmerzen all' und Wonne, halb in Träumen.

Die Lerche auch will nicht die Zeit versäumen,
Da solches Schallen bringt die Luft getragen,
Schwingt sich vom Tal, eh's noch beginnt zu tagen,
Im ersten Strahl die Flügel sich zu säumen.

Ich aber stand schon lange in dem Garten
Und bin ins stille Feld hinausgegangen,
Wo leis die Ähren an zu wogen fingen.

O fromme Vöglein, ihr und ich, wir warten
Aufs frohe Licht, da ist uns vor Verlangen
Bei stiller Nacht erwacht so sehnend Singen.

Das Gebet

Wen hat nicht einmal Angst befallen,
Wenn Trübnis ihn gefangen hält,
Als müßt' er ewig rastlos wallen
Nach einer wunderbaren Welt?
All' Freunde sind lang fortgezogen,
Der Frühling weint in einem fort,
Eine Brücke ist der Regenbogen
Zum friedlich sichern Heimatsport.

Hinauszuschlagen in die Töne,
Lockt dich Natur mit wilder Lust,
Zieht Minne, holde Frauenschöne
Zum Abgrund süß die sel'ge Brust;
Den Tod siehst du verhüllet gehen

Durch Lieb' und Leben himmelwärts,
Ein einzig' Wunder nur bleibt stehen
Einsam über dem öden Schmerz. –

Du seltner Pilger, laß dich warnen!
Aus ird'scher Lust und Zauberei,
Die freud- und leidvoll dich umgarnen,
Strecke zu Gott die Arme frei!
Nichts mehr mußt du hienieden haben,
Himmlisch betrübt, verlassen, arm,
Ein treues Kind, dem Vater klagen
Die ird'sche Lust, den ird'schen Harm.

Es breitet diese einz'ge Stunde
Sich übers ganze Leben still,
Legt blühend sich um deine Wunde,
Die niemals wieder heilen will.
Treu bleibt der Himmel stets dem Treuen,
Zur Erd' das Ird'sche niedergeht,
Zum Himmel über Zaubereien
Geht ewig siegreich das Gebet.

Sonntag

Die Nacht war kaum verblühet,
Nur eine Lerche sang
Die stille Luft entlang.
Wen grüßt sie schon so frühe?

Und draußen in dem Garten
Die Bäume übers Haus
Sahn weit ins Land hinaus,
Als ob sie wen erwarten.

In festlichen Gewanden
Wie eine Kinderschar,

Tauperlen in dem Haar,
Die Blumen alle standen.

Ich dacht': ihr kleinen Bräute,
Was schmückt ihr euch so sehr? –
Da blickt' die eine her:
„Still, still, 's ist Sonntag heute."

„Schon klingen Morgenglocken,
Der liebe Gott nun bald
Geht durch den stillen Wald."
Da kniet' ich froherschrocken.

Nachtgebet

Es rauschte leise in den Bäumen,
Ich hörte nur der Ströme Lauf,
Und Berg' und Gründe, wie aus Träumen,
Sie sahn so fremd zu mir herauf.

Drin aber in der stillen Halle
Ruht' Sang und Plaudern müde aus,
Es schliefen meine Lieben alle,
Kaum wieder kannt' ich nun mein Haus.

Mir war's, als lägen sie zur Stunde
Gestorben, bleich im Mondenschein,
Und schauernd in der weiten Runde
Fühlt' ich auf einmal mich allein.

So blickt in Meeres öden Reichen
Ein Schiffer einsam himmelan –
O Herr, wenn einst die Ufer weichen,
Sei gnädig Du dem Steuermann!

Ostern

Vom Münster Trauerglocken klingen,
Vom Tal ein Jauchzen schallt herauf.
Zur Ruh' sie dort dem Toten singen,
Die Lerchen jubeln: Wache auf!
Mit Erde sie ihn still bedecken,
Das Grün aus allen Gräbern bricht,
Die Ströme hell durchs Land sich strecken,
Der Wald ernst wie in Träumen spricht,
Und bei den Klängen, Jauchzen, Trauern,
Soweit ins Land man schauen mag,
Es ist ein tiefes Frühlingsschauern
Als wie ein Auferstehungstag.

Weihnachten

Markt und Straßen stehn verlassen,
Still erleuchtet jedes Haus,
Sinnend geh' ich durch die Gassen,
Alles sieht so festlich aus.

An den Fenstern haben Frauen
Buntes Spielzeug fromm geschmückt,
Tausend Kindlein stehn und schauen,
Sind so wunderstill beglückt.

Und ich wandre aus den Mauern
Bis hinaus ins freie Feld,
Hehres Glänzen, heil'ges Schauern!
Wie so weit und still die Welt!

Sterne hoch die Kreise schlingen,
Aus des Schnees Einsamkeit
Steigt's wie wunderbares Singen –
O du gnadenreiche Zeit!

Abschied

Abendlich schon rauscht der Wald
Aus den tiefen Gründen,
Droben wird der Herr nun bald
An die Sterne zünden,
Wie so stille in den Schlünden,
Abendlich nur rauscht der Wald.

Alles geht zu seiner Ruh',
Wald und Welt versausen,
Schauernd hört der Wandrer zu,
Sehnt sich recht nach Hause,
Hier in Waldes grüner Klause
Herz, geh endlich auch zur Ruh'!

Mondnacht

Es war, als hätt' der Himmel
Die Erde still geküßt,
Daß sie im Blütenschimmer
Von ihm nun träumen müßt'.

Die Luft ging durch die Felder,
Die Ähren wogten sacht,
Es rauschten leis die Wälder,
So sternklar war die Nacht.

Und meine Seele spannte
Weit ihre Flügel aus,
Flog durch die stillen Lande,
Als flöge sie nach Haus.

Glück auf

Gar viel hab' ich versucht, gekämpft, ertragen;
Das ist der tiefen Sehnsucht Lebenslauf,
Daß brünstig sie an jeden Fels muß schlagen,
Ob sich des Lichtes Gnadentür tät' auf,
Wie ein verschütt'ter Bergmann in den Klüften
Heraus sich hauet zu den heitern Lüften.

Auch ich gelang' einst zu dem stillen Gipfel,
Vor dem mich schaudert in geheimer Lust.
Tief unten rauschen da des Lebens Wipfel
Noch einmal dunkelrührend an die Brust,
Dann wird es unten still im weiten Grunde
Und oben leuchtet streng des Himmels Runde.

Wie klein wird sein da, was mich hat gehalten,
Wie wenig, was ich Irrender vollbracht,
Doch was den Felsen gläubig hat gespalten:
Die Sehnsucht treu steigt mit mir aus der Nacht.
Und legt mir an die wunderbaren Schwingen,
Die durch die Stille mich nach Hause bringen.

Nachtlied

Vergangen ist der lichte Tag,
Von ferne kommt der Glocken Schlag;
So reist die Zeit die ganze Nacht,
Nimmt manchen mit, der's nicht gedacht.

Wo ist nun hin die bunte Lust,
Des Freundes Trost und treue Brust,
Des Weibes süßer Augenschein?
Will keiner mit mir munter sein?

Da's nun so stille auf der Welt,
Ziehn Wolken einsam übers Feld,

Und Feld und Baum besprechen sich, –
O Menschenkind! was schauert dich?

Wie weit die falsche Welt auch sei,
Bleibt mir doch einer nur getreu,
Der mit mir weint, der mit mir wacht,
Wenn ich nur recht an ihn gedacht.

Frisch auf denn, liebe Nachtigall,
Du Wasserfall mit hellem Schall!
Gott loben wollen wir vereint,
Bis daß der lichte Morgen scheint!

Stimmen der Nacht

1

Weit tiefe, bleiche, stille Felder –
O wie mich das freut,
Über alle, alle Täler, Wälder
Die prächtige Einsamkeit!

Aus der Stadt nur schlagen die Glocken
Über die Wipfel herein,
Ein Reh hebt den Kopf erschrocken
Und schlummert gleich wieder ein.

Der Wald aber rühret die Wipfel
Im Schlaf von der Felsenwand,
Denn der Herr geht über die Gipfel
Und segnet das stille Land.

2

Nächtlich wandern alle Flüsse,
Und der Himmel, Stern auf Stern,

Sendet so viel tausend Grüße,
Daß die Wälder nah und fern
Schauernd rauschen in den Gründen;
Nur der Mensch, dem Tod geweiht,
Träumet fort von seinen Sünden
In der stillen Gnadenzeit.

Herbstweh

1

So still in den Feldern allen,
Der Garten ist lange verblüht,
Man hört nur flüsternd die Blätter fallen,
Die Erde schläfert – ich bin so müd'.

2

Es schüttelt die welken Blätter der Wald,
Mich friert, ich bin schon alt,
Bald kommt der Winter und fällt der Schnee,
Bedeckt den Garten und mich und alles, alles Weh.

Winternacht

Verschneit liegt rings die ganze Welt,
Ich hab' nichts, was mich freuet,
Verlassen steht der Baum im Feld,
Hat längst sein Laub verstreuet.

Der Wind nur geht bei stiller Nacht
Und rüttelt an dem Baume,
Da rührt er seinen Wipfel sacht
Und redet wie im Traume.

Er träumt von künft'ger Frühlingszeit,
Von Grün und Quellenrauschen,
Wo er im neuen Blütenkleid
Zu Gottes Lob wird rauschen.

Trost

Der jagt dahin, daß die Rosse schnaufen,
Der muß im Staub daneben laufen;
Aber die Nacht holt beide ein,
Setzt jenen im Traume neben die Rosse
Und den andern in seine Karosse. –
Wer fährt nun fröhlicher? Der da wacht,
Oder der blinde Passagier bei Nacht?

Dank

Mein Gott, Dir sag' ich Dank,
Daß Du die Jugend mir bis über alle Wipfel
In Morgenrot getaucht und Klang,
Und auf des Lebens Gipfel,
Bevor der Tag geendet,
Vom Herzen unbewacht
Den falschen Glanz gewendet,
Daß ich nicht taumle ruhmgeblendet,
Da nun herein die Nacht
Dunkelt in ernster Pracht.

Kurze Fahrt

Posthorn, wie so keck und fröhlich
Brachst du einst den Morgen an,
Vor mir lag's so frühlingsselig,
Daß ich still auf Lieder sann.

Dunkel rauscht es schon im Walde,
Wie so abendkühl wird's hier,
Schwager, stoß ins Horn – wie balde
Sind auch wir im Nachtquartier!

Schifferspruch

Wenn die Wogen unten toben,
Menschenwitz zu Schanden wird,
Weist mit feur'gen Zügen droben
Heimwärts dich der Wogen Hirt.
Sollst nach keinem andern fragen,
Nicht zurückschaun nach dem Land,
Faß das Steuer, laß das Zagen!
Aufgerollt hat Gottes Hand
Diese Wogen zum Befahren
Und die Sterne, dich zu wahren.

So oder so

Die handeln und die dichten,
Das ist der Lebenslauf,
Der eine macht Geschichten,
Der andre schreibt sie auf,
Und der will beide richten;
So schreibt und treibt sich's fort,
Der Herr wird alles schlichten,
Verloren ist kein Wort.

Walt' Gott!

Gestern stürmt's noch, und am Morgen
Blühet schon das ganze Land –
Will auch nicht für morgen sorgen,
Alles steht in Gottes Hand.

Putz dich nur in Gold und Seiden:
In dem Felde über Nacht
Engel Gott's die Lilien kleiden,
Schöner als du's je gedacht.

Sonn' dich auf des Lebens Gipfeln:
Über deinem stolzen Haus
Singt der Vogel in den Wipfeln,
Schwingt sich über dich hinaus!

Vögel nicht, noch Blumen sorgen,
hat doch jedes sein Gewand –
Wie so fröhlich rauscht der Morgen!
Alles steht in Gottes Hand.

Schiffergruß

Stolzes Schiff mit seidnen Schwingen,
Fährst mein Boot zu Grunde schier,
Sang von Bord und Lauten klingen,
O du fröhlicher Schiffsherr dir;
Ich muß selbst mein Lied mir singen,
Nur der Sturmwind singt mit mir.

Stolzes Schiff, wenn deine Feuer
Nachts verlöscht: beim falben Licht
Steht ein Fremder an dem Steuer,
Mit den Winden laut er spricht,
Und die Wogen rauschen scheuer –
Trau' dem finstern Bootsmann nicht!

Gleiche Winde, gleiche Wellen,
Reiches Schiff und armes Boot
Nach demselben Strande schwellen,
Deine Hoffart, meine Not
Wird an *einem* Riff zerschellen,
Denn der Bootsmann ist der Tod.

Todeslust

Bevor er in die blaue Flut gesunken,
Träumt noch der Schwan und singet todestrunken;
Die sommermüde Erde im Verblühen
Läßt all ihr Feuer in den Trauben glühen;
Die Sonne, Funken sprühend, im Versinken,
Gibt noch einmal der Erde Glut zu trinken,
Bis, Stern auf Stern, die Trunkne zu umfangen,
Die wunderbare Nacht ist aufgegangen.

Warnung

Aus ist dein Urlaub und die Laut' zerschlagen,
Nachts aus der stillen Stadt nun mußt du gehen,
Die Wetterfahnen nur im Wind sich drehen,
Dein Tritt verhallt, mag niemand nach dir fragen.

Doch draußen waldwärts, wo du herstammst, ragen
Die Zinnen noch der goldnen Burg, es gehen
Die Wachen schildernd auf dem Wall, das Wehen
Der Nacht bringt ihren Ruf ins Land getragen.

Der Engel dort mit seinem Flammendegen
Steht blankgerüstet noch, das Tor zu hüten,
Und wird dich mit den ernsten Blicken messen,

Die manches Herze schon zu Asche glühten.
Hast du Parol' und Feldgeschrei vergessen:
Weh! wo nun willst dein müdes Haupt hinlegen?

Die heilige Mutter

Es ist ein Meer, von Schiffen irr' durchflogen,
Die steuern rastlos nach den falschen Landen,

Die alle suchen und wo alle stranden
Auf schwanker Flut, die jeden noch betrogen.

Es ist im wüsten Meer ein Felsenbogen,
An dem die sturmgepeitschten Wellen branden
Und aller Zorn der Tiefe wird zu Schanden,
Die nach dem Himmel zielt mit trüben Wogen.

Und auf dem Fels die mildeste der Frauen
Zählt ihre Kinder und der Schiffe Trümmer,
Stillbetend, daß sich rings die Stürme legen.

Das sind die treuen Augen, himmelblauen –
Mein Schiff versenk' ich hinter mir auf immer,
Hier bin ich, Mutter, gib mir Deinen Segen!

Mahnung

Genug gemeistert nun die Weltgeschichte!
Die Sterne, die durch alle Zeiten tagen,
Ihr wolltet sie mit frecher Hand zerschlagen
Und jeder leuchten mit dem eignen Lichte.

Doch unaufhaltsam rucken die Gewichte,
Von selbst die Glocken von den Türmen schlagen,
Der alte Zeiger, ohne euch zu fragen,
Weist flammend auf die Stunde der Gerichte.

O stille Schauer, wunderbares Schweigen,
Wenn heimlichflüsternd sich die Wälder neigen,
Die Täler alle geisterbleich versanken,

Und in Gewittern von den Bergesspitzen
Der Herr die Weltgeschichte schreibt mit Blitzen –
Denn seine sind nicht euere Gedanken.

Wacht auf!

Es ist kein Kirchlein zwischen Felsenbogen
So tief versteckt: wie in den alten Sagen
Hat nächtens drin die Glocke angeschlagen,
Weiß keiner, wer die Glocken hat gezogen.

Erwache, Steuermann! hoch gehn die Wogen;
Ihr Hirten auf, die Herden nach euch fragen;
Ihr Wächter sollt an Schloß und Hütten schlagen,
Wacht auf, wacht auf, bevor der Klang verflogen!

Denn Heerschau halten will in deutschen Gauen
Der Herr und zählen, die ihm treu geblieben,
Eh' er den Engel mit dem Schwerte sendet.

Schon bricht's so dunkelrot durchs Morgengrauen,
Ob's Blut bedeutet oder feur'ges Lieben,
Es steht in Gottes Hand, die niemand wendet.

Im Alter

Wie wird nun alles so stille wieder!
So war mir's oft in der Kinderzeit,
Die Bäche gehen rauschend nieder
Durch die dämmernde Einsamkeit,
Kaum noch hört man einen Hirten singen,
Aus allen Dörfern, Schluchten weit,
Die Abendglocken herüberklingen,
Versunken nun mit Lust und Leid
Die Täler, die noch einmal blitzen,
Nur hinter dem stillen Walde weit
Noch Abendröte an den Bergesspitzen,
Wie Morgenrot der Ewigkeit.

Memento mori

Schnapp Austern, Dukaten,
Mußt dennoch sterben!
Dann tafeln die Maden
Und lachen die Erben.

Die Flucht der heiligen Familie

Länger fallen schon die Schatten,
Durch die kühle Abendluft,
Waldwärts über stille Matten
Schreitet Joseph von der Kluft,
Führt den Esel treu am Zügel;
Linde Lüfte fächeln kaum,
's sind der Engel leise Flügel,
Die das Kindlein sieht im Traum,
Und Maria schauet nieder
Auf das Kind voll Lust und Leid,
Singt im Herzen Wiegenlieder
In der stillen Einsamkeit.
Die Johanniswürmchen kreisen
Emsig leuchtend übern Weg,
Wollen der Mutter Gottes weisen
Durch die Wildnis jeden Steg.
Und durchs Gras geht süßes Schaudern,
Streift es ihres Mantels Saum;
Bächlein auch läßt jetzt sein Plaudern
Und die Wälder flüstern kaum,
Daß sie nicht die Flucht verraten.
Und das Kindlein hob die Hand,
Da sie ihm so Liebes taten,
Segnete das stille Land,
Daß die Erd' mit Blumen, Bäumen
Fernerhin in Ewigkeit
Nächtlich muß vom Himmel träumen –
O gebenedeite Zeit!

Marienlied

Wenn ins Land die Wetter hängen
Und der Mensch erschrocken steht,
Wendet, wie mit Glockenklängen,
Die Gewitter Dein Gebet,
Und wo aus den grauen Wogen
Weinend auftaucht das Gefild,
Segnest Du's vom Regenbogen –
Mutter, ach, wie bist Du mild!

Wenn's einst dunkelt auf den Gipfeln
Und der kühle Abend sacht
Niederrauschet in den Wipfeln:
O Maria, heil'ge Nacht!
Laß mich nimmer wie die andern,
Decke zu der letzten Ruh'
Mütterlich den müden Wandrer
Mit dem Sternenmantel zu.

Durch!

Ein Adler saß am Felsenbogen,
Den lockt' der Sturm weit übers Meer,
Da hatt' er droben sich verflogen,
Er fand sein Felsennest nicht mehr,
Tief unten sah er kaum noch liegen
Verdämmernd Wald und Land und Meer,
Mußt' höher, immer höher fliegen,
Ob nicht der Himmel offen wär'.

VII. Romanzen

Aus schweren Träumen
Fuhr ich oft auf und sah durch Tannenwipfel
Den Mond ziehn übern stillen Grund und sang
Vor Bangigkeit und schlummert' wieder ein. –

Ja, Menschenstimme, hell aus frommer Brust!
Du bist doch die gewaltigste, und triffst
Den rechten Grundton, der verworren anklingt
In all den tausend Stimmen der Natur! –

Die Zauberin im Walde

„Schon, vor vielen, vielen Jahren
Saß ich drüben an dem Ufer,
Sah manch Schiff vorüber fahren
Weit hinein ins Waldesdunkel."

„Denn ein Vogel jeden Frühling
An dem grünen Waldessaume
Sang mit wunderbarem Schalle,
Wie ein Waldhorn klang's im Traume."

„Und gar seltsam hohe Blumen
Standen an dem Rand der Schlünde,
Sprach der Strom so dunkle Worte,
's war, als ob ich sie verstünde."

„Und wie ich so sinnend atme
Stromeskühl' und Waldesdüfte,
Und ein wundersam Gelüsten
Mich hinabzog nach den Klüften:"

„Sah ich auf kristallnem Nachen,
Tief im Herzensgrund erschrocken,
Eine wunderschöne Fraue,
Ganz umwallt von goldnen Locken."

„Und von ihrem Hals behende
Tät' sie lösen eine Kette,
Reicht' mit ihren weißen Händen
Mir die allerschönste Perle."

„Nur ein Wort von fremdem Klange
Sprach sie da mit rotem Munde,
Doch im Herzen ewig stehen
Wird des Worts geheime Kunde."

„Seitdem saß ich wie gebannt dort,
Und wenn neu der Lenz erwachte,
Immer von dem Halsgeschmeide
Eine Perle sie mir brachte."

„Ich barg all' im Waldesgrunde,
Und aus jeder Perl' der Fraue
Sproßte eine Blum' zur Stunde,
Wie ihr Auge anzuschauen."

„Und so bin ich aufgewachsen,
Tät' der Blumen treulich warten,
Schlummert' oft und träumte golden
In dem schwülen Waldesgarten."

„Fortgespült ist nun der Garten
Und die Blumen all' verschwunden,
Und die Gegend, wo sie standen,
Hab' ich nimmermehr gefunden."

„In der Fern' liegt jetzt mein Leben,
Breitend sich wie junge Träume,
Schimmert stets so seltsam lockend
Durch die alten, dunklen Bäume."

„Jetzt erst weiß ich, was der Vogel
Ewig ruft so bange, bange,
Unbekannt zieht ew'ge Treue
Mich hinunter zu dem Sange."

„Wie die Wälder kühle rauschen,
Zwischendurch das alte Rufen,
Wo bin ich so lang gewesen? –
O ich muß hinab zur Ruhe!"

Und es stieg vom Schloß hinunter
Schnell der süße Florimunde,

Weit hinab und immer weiter
Zu dem dunkelgrünen Grunde.

Hört' die Ströme stärker rauschen,
Sah in Nacht des Vaters Burge
Stillerleuchtet ferne stehen,
Alles Leben weit versunken.

Und der Vater schaut' vom Berge,
Schaut' zum dunklen Grunde immer,
Regte sich der Wald so grausig,
Doch den Sohn erblickt' er nimmer.

Und es kam der Winter balde,
Und viel' Lenze kehrten wieder,
Doch der Vogel in dem Walde
Sang nie mehr die Wunderlieder.

Und das Waldhorn war verklungen
Und die Zauberin verschwunden,
Wollte keinen andern haben
Nach dem süßen Florimunde. –

Die Riesen

Hoch über blauen Bergen
Da steht ein schönes Schloß,
Das hütet von Gezwergen
Ein wunderlicher Troß.

Da ist ein Lautenschlagen
Und Singen insgemein,
Die Lüfte es vertragen
Weit in das Land hinein.

Und wenn die Länder schweigen,
Funkelnd im Abendtau,

Soll manchmal dort sich zeigen
Eine wunderschöne Frau.

Da schworen alle Riesen,
Zu holen sie als Braut,
Mit Leitern da und Spießen
Sie stapften gleich durchs Kraut.

Da krachte manche Leiter,
Sie wunderten sich sehr:
Die Wildnis wuchs, je weiter
Je höher rings umher.

Sie waren recht bei Stimme
Und zankten um ihren Schatz,
Und fluchten in großem Grimme,
Und fanden nicht den Platz.

Und bei dem Lärm sie stunden
In Wolken bis an die Knie,
Das Schloß, das war verschwunden,
Und wußten gar nicht wie. –

Aber wie ein Regenbogen
Glänzt's drüben durch die Luft,
Sie hatt' indes gezogen
Neue Gärten in den Duft.

Der Götter Irrfahrt

(Nach einer Volkssage der Tongainseln)

1

Unten endlos nichts als Wasser,
Droben Himmel still und weit,

Nur das Götterland, das blasse,
Lag in Meereseinsamkeit,
Wo auf farbenlosen Matten
Gipfel wie in Träumen stehn,
Und Gestalten ohne Schatten
Ewig lautlos sich ergehn.

Zwischen grauen Wolkenschweifen,
Die verschlafen Berg und Flut
Mit den langen Schleiern streifen,
Hoch der Göttervater ruht.
Heut zu fischen ihn gelüstet,
Und vom zack'gen Felsenhang
In des Meeres grüne Wüste
Senket er die Schnur zum Fang.

Sinnend sitzt er, und es flattern
Bart und Haar im Sturme weit,
Und die Zeit wird ihm so lange
In der stillen Einsamkeit.
Da fühlt er die Angel zucken:
„Ei, da ist ein schwerer Fisch!"
Freudig fängt er an zu rucken,
Stemmt sich, zieht und windet frisch.

Sieh, da hebt er Felsenspitzen
Langsam aus der Wasser Grund,
Und erschrocken aus den Ritzen
Schießen schupp'ge Schlangen bunt;
Ringelnd' Ungetüm' der Tiefen,
Die im öden Wogenhaus
In der grünen Dämmrung schliefen,
Stürzen sich ins Meer hinaus.

Doch der Vater hebt aufs neue,
Und Gebirge, Tal und Strand
Taucht allmählich auf ins Freie,

Und es grünt das junge Land,
Irrend farb'ge Lichter schweifen
Und von Blumen glänzt die Flur,
Wo des Vaters Blick' sie streifen –
Da zerreißt die Angelschnur.

Wie 'ne liebliche Sirene
Halb nun überm Wellenglanz,
Staunend ob der eignen Schöne,
Schwebt es mit dem Blütenkranz,
Bei der Lüfte lindem Fächeln
Sich im Meer, das rosig brennt,
Spiegelnd mit verschämtem Lächeln –
Erde sie der Vater nennt.

2

Staunend auf den Göttersitzen
Die Unsterblichen nun stehn,
Sehn den Morgen drüben blitzen,
Fühlen Duft herüberwehn,
Und so süßes Weh sie spüren,
Lösen leis ihr Schiff vom Strand,
Und die Lüfte sie verführen
Fern durchs Meer zum jungen Land.

O wie da die Quellen sprangen
In die tiefe Blütenpracht
Und Lianen dort sich schlangen
Glühend durch die Waldesnacht!
Und die Wandrer trunken lauschen,
Wo die Wasserfälle gehn,
Bis sie in dem Frühlingsrauschen
Plötzlich all' erschrocken stehn:

Denn sie sehn zum ersten Male
Nun die Sonne niedergehn

Und verwundert Berg' und Tale
Tief im Abendrote stehn,
Und der schönste Gott von allen
Sank erbleichend in den Duft,
Denn dem Tode ist verfallen,
Wer geatmet ird'sche Luft.

Die Genossen faßt ein Grauen,
Und sie fahren weit ins Meer,
Nach des Vaters Haus sie schauen,
Doch sie finden's nimmermehr.
Mußten aus den Wogenwüsten
Ihrer Schiffe Schnäbel drehn
Wieder nach des Eilands Küsten,
Ach, das war so falsch und schön!

Und für immer da verschlagen
Blieben sie im fremden Land,
Hörten nachts des Vaters Klagen
Oft noch fern vom Götterstrand. –
Und nun Kindeskinder müssen
Nach der Heimat sehn ins Meer,
Und es kommt im Wind ein Grüßen,
Und sie wissen nicht woher.

Die Brautfahrt

Durch des Meeresschlosses Hallen
Auf bespültem Felsenhang
Weht der Hörner festlich Schallen;
Froher Hochzeitsgäste Drang,
Bei der Kerzen Zauberglanze,
Wogt im buntverschlungnen Tanze.

Aber an des Fensters Bogen,
Ferne von der lauten Pracht,

Schaut der Bräut'gam in die Wogen
Draußen in der finstern Nacht,
Und die trunknen Blicke schreiten
Furchtlos durch die öden Weiten.

„Lieblich", sprach der wilde Ritter
Zu der zarten, schönen Braut,
„Lieblich girrt die sanfte Zither –
Sturm ist *meiner* Seele Laut,
Und der Wogen dumpfes Brausen
Hebt das Herz in kühnem Grausen.

Ich kann hier nicht müßig lauern,
Treiben auf dem flachen Sand,
Dieser Kreis von Felsenmauern
Hält mein Leben nicht umspannt;
Schön're Länder blühen ferne,
Das verkünden mir die Sterne.

Du mußt glauben, du mußt wagen,
Und, den Argonauten gleich,
Wird die Woge fromm dich tragen
In das wunderbare Reich;
Mutig streitend mit den Winden,
Muß ich meine Heimat finden!

Siehst du, heißer Sehnsucht Flügel,
Weiße Segel dort gespannt?
Hörst du tief die feuchten Hügel
Schlagen an die Felsenwand?
Das ist Sang zum Hochzeitsreigen –
Willst du mit mir niedersteigen?

Kannst du rechte Liebe fassen,
Nun so frage, zaudre nicht!
Schloß und Garten mußt du lassen
Und der Eltern Angesicht –

Auf der Flut mit mir alleine,
Da erst, Liebchen, bist du meine!"

Schweigend sieht ihn an die milde
Braut mit schauerlicher Lust,
Sinkt dem kühnen Ritterbilde
Trunken an die stolze Brust:
„Dir hab ich mein Los ergeben,
Schalte nun mit meinem Leben."

Und er trägt die süße Beute
Jubelnd aus dem Schloß aufs Schiff,
Drunten harren seine Leute,
Stoßen froh vom Felsenriff;
Und die Hörner leis verhallen,
Einsam rings die Wogen schallen.

Wie die Sterne matter blinken
In die morgenrote Flut,
Sieht sie fern die Berge sinken,
Flammend steigt die hehre Glut,
Überm Spiegel trunkner Wellen
Rauschender die Segel schwellen.

Monde steigen und sich neigen,
Lieblich weht schon fremde Luft,
Da sehn sie ein Eiland steigen
Feenhaft aus blauem Duft,
Wie ein farb'ger Blumenstreifen –
Meerwärts fremde Vögel schweifen.

Alle faßt ein freud'ges Beben –
Aber dunkler rauscht das Meer,
Schwarze Wetter schwer sich heben,
Stille wird es rings umher,
Und nur freudiger und treuer
Steht der Ritter an dem Steuer.

Und nun flattern wilde Blitze,
Sturm rast um den Felsenriff,
Und von grimmer Wogen Spitze
Stürzt geborsten sich das Schiff.
Schwankend auf des Mastes Splitter,
Schlingt die Braut sich um den Ritter.

Und die Müde in den Armen,
Springt er abwärts, sinkt und ringt,
Hält den Leib, den blühend warmen,
Bis er alle Wogen zwingt,
Und am Blumenstrand gerettet,
Auf das Gras sein Liebstes bettet.

„Wache auf, wach' auf, du Schöne!
Liebesheimat ringsum lacht,
Zaubrisch ringen Duft und Töne,
Wunderbarer Blumen Pracht
Funkelnd rings im Morgengolde –
Schau um dich! Wach' auf, du Holde!"

Aber frei von Lust und Kummer
Ruht die liebliche Gestalt,
Lächelnd noch im längsten Schlummer,
Und das Herz ist still und kalt,
Still der Himmel, still im Meere,
Schimmernd rings des Taues Zähre.

Und er sinkt zu ihr vor Schmerzen,
Einsam in dem fremden Tal,
Tränen aus dem wilden Herzen
Brechen da zum ersten Mal,
Und vor diesem Todesbilde
Wird die ganze Seele milde.

Von der langen Täuschung trennt er
Schauernd sich – der Stolz entweicht,

Andre Heimat nun erkennt er,
Die kein Segel hier erreicht,
Und an echten Schmerzen ranken
Himmelwärts sich die Gedanken.

Scharrt die Tote ein in Stille,
Pflanzt ein Kreuz hoch auf ihr Grab,
Wirft von sich die seidne Hülle,
Leget Schwert und Mantel ab,
Kleidet sich in rauhe Felle,
Haut in Fels sich die Kapelle.

Überm Rauschen dunkler Wogen
In der wilden Einsamkeit,
Hausend auf dem Felsenbogen,
Ringt er fromm mit seinem Leid,
Hat, da manches Jahr entschwunden,
Heimat, Braut und Ruh' gefunden. –

Viele Schiffe drunten gehen
An dem schönen Inselland,
Sehen hoch das Kreuz noch stehen,
Warnend von der Felsenwand;
Und des strengen Büßers Kunde
Gehet fromm von Mund zu Munde.

Vom heiligen Eremiten Wilhelm

Von Jerusalem die Warten
Lagen schon in rotem Duft,
Stand der Patriarch im Garten,
Glockenklang ging durch die Luft.

Kommt ein Pilger da gezogen,
Tritt zu ihm im Abendrot,

Bleich, von strupp'gem Haar umflogen,
Bettelt um ein Stücklein Brot.

Kommst aus Frankreich, frommer Pilger,
Hör' der Heimat Laut so gern!
Kennst du dort den Grafen Wilhelm,
Meinen vor'gen Landesherrn?

„Kenn' ihn wohl, er hat geschrieben
Feur'ge Schrift mit blut'ger Hand,
Hat aus Frankreich dich vertrieben,
Und dein Kloster liegt verbrannt."

Gott im Himmel, sollt' dich kennen,
Wie du so den Blick gewandt,
Bist Graf Wilhelm der Ardennen –
„Also ward ich sonst genannt."

O mein lieber Herr, am Grabe
Stehen beid' als Sünder wir –
Haus und Garten, was ich habe,
Nehmt es hin und rastet hier!

„Bet' für mich, ich darf nicht rasten,
Denn ohn' Rasten geht die Zeit,
Hart mit Geißeln, Wachen, Fasten
Lieg' ich mit der Höll' in Streit.

Kron' und Land ließ ich den Erben,
Muß mit stürmender Gewalt
Mir ein andres Reich erwerben." –
Und so schritt er fort zum Wald.

Der Kühne

Und wo noch kein Wandrer gegangen,
Hoch über Jäger und Roß

Die Felsen im Abendrot hangen
Als wie ein Wolkenschloß.

Dort zwischen den Zinnen und Spitzen
Von wilden Nelken umblüht,
Die schönen Waldfrauen sitzen
Und singen im Wind ihr Lied.

Der Jäger schaut nach dem Schlosse:
Die droben das ist mein Lieb! –
Er sprang vom scheuenden Rosse,
Weiß keiner, wo er blieb.

Der Wachtturm

Ich sah im Mondschein liegen
Die Felsen und das Meer,
Ich sah ein Schifflein fliegen
Still durch die Nacht daher.

Ein Ritter saß am Steuer,
Ein Fräulein stand am Bord,
Im Winde weht' ihr Schleier,
Die sprachen kein einzig Wort.

Ich sah verfallen grauen
Das hohe Königshaus,
Den König stehn und schauen
Vom Turm ins Meer hinaus.

Und als das Schiff verschwunden,
Er warf seine Krone nach,
Und aus dem tiefen Grunde
Das Meer wehklagend brach.

Das war der kühne Buhle,
Der ihm sein Kind geraubt,
Der König, der verfluchet
Der eignen Tochter Haupt.

Da hat das Meer mit Toben
Verschlungen Ritter und Maid,
Der König starb da droben
In seiner Einsamkeit.

Nun jede Nacht vor Sturme
Das Schiff vorüberzieht,
Der König von dem Turme
Nach seinem Kinde sieht.

Nachtwanderer

Er reitet nachts auf einem braunen Roß,
Er reitet vorüber an manchem Schloß:
Schlaf droben, mein Kind, bis der Tag erscheint,
Die finstre Nacht ist des Menschen Feind!

Er reitet vorüber an einem Teich,
Da stehet ein schönes Mädchen bleich
Und singt, ihr Hemdlein flattert im Wind:
Vorüber, vorüber, mir graut vor dem Kind!

Er reitet vorüber an einem Fluß,
Da ruft ihm der Wassermann seinen Gruß,
Taucht wieder unter dann mit Gesaus',
Und stille wird's über dem kühlen Haus.

Wann Tag und Nacht in verworrenem Streit,
Schon Hähne krähen in Dörfern weit,
Da schauert sein Roß und wühlet hinab,
Scharret ihm schnaubend sein eigenes Grab.

Der Knabe

Es war ein zartes Vögelein,
Das saß in Lieb' gefangen,
Ein Knabe hegt' und pflegt' sich's fein
Wohl hinter goldnen Stangen.

Und draußen hört's auf grünem Plan
Verschiedner Vögel Weisen,
Sah Tag und Nacht den Knaben an,
Mocht' nicht mit ihnen reisen.

Und als der Frühling weit und breit
Von neuem schien und schwärmte,
Da tat dem Knaben 's Vöglein leid,
Daß es kein Strahl erwärmte.

Da nahm er aus dem stillen Haus
Das Vöglein fromm und treue,
Und schweift' mit ihm durchs Feld hinaus
Ins himmelblaue Freie.

Er setzt' es vor sich auf die Hand,
Da wend't und putzt sich's feine,
In bunten Farben spielt' und brannt'
Sein Kleid im Sonnenscheine.

Doch aus dem Wald ein Singen rief,
Bunt' Vöglein ziehn und reisen,
Das lockt so hell, das lockt so tief
In wundersüßen Weisen.

Das Vöglein frisch die Flügel rührt –
Es ruft: Kommst du nicht balde? –
Das hat das Vögelein verführt,
Fort flog's zum grünen Walde –

Nun muß der Knabe einsam gehn,
Klagt über Tal und Hügel:
„Süß' Lieb', süß' Lieb', wie bist du schön:
Ach, hätt'st du keine Flügel!" –

Die Nonne und der Ritter

Da die Welt zur Ruh' gegangen,
Wacht mit Sternen mein Verlangen;
In der Kühle muß ich lauschen,
Wie die Wellen unten rauschen.

„Fernher mich die Wellen tragen,
Die ans Land so traurig schlagen.
Unter deines Fensters Gitter,
Fraue, kennst du noch den Ritter?"

Ist's doch, als ob seltsam' Stimmen
Durch die lauen Lüfte schwimmen;
Wieder hat's der Wind genommen –
Ach, mein Herz ist so beklommen!

„Drüben liegt dein Schloß verfallen,
Klagend in den öden Hallen
Aus dem Grund der Wald mich grüßte –
's war, als ob ich sterben müßte."

Alte Klänge blühend schreiten!
Wie aus lang versunknen Zeiten
Will mich Wehmut noch bescheinen,
Und ich möcht' von Herzen weinen.

„Überm Walde blitzt's vom weiten,
Wo um Christi Grab sie streiten;
Dorthin will mein Schiff sich wenden,
Da wird alles, alles enden!"

Geht ein Schiff, ein Mann stand drinne –
Falsche Nacht, verwirrst die Sinne,
Welt Ade! Gott woll' bewahren,
Die noch irr' im Dunkeln fahren.

Der stille Grund

Der Mondenschein verwirret
Die Täler weit und breit,
Die Bächlein, wie verirret,
Gehn durch die Einsamkeit.

Da drüben sah ich stehen
Den Wald auf steiler Höh',
Die finstern Tannen sehen
In einen tiefen See.

Ein' Kahn wohl sah ich ragen,
Doch niemand, der es lenkt,
Das Ruder war zerschlagen,
Das Schifflein halb versenkt.

Eine Nixe auf dem Steine
Flocht dort ihr goldnes Haar,
Sie meint', sie wär' alleine,
Und sang so wunderbar.

Sie sang und sang, in den Bäumen
Und Quellen rauscht' es sacht,
Und flüsterte wie in Träumen
Die mondbeglänzte Nacht.

Ich aber stand erschrocken,
Denn über Wald und Kluft
Klangen die Morgenglocken
Schon ferne durch die Luft.

Und hätt' ich nicht vernommen
Den Klang zu guter Stund,
Wär' nimmermehr gekommen
Aus diesem stillen Grund.

Der Kämpe

Nach drei Jahren kam gefahren
Einsam auf dem Rhein ein Schiff,
Drin gebunden und voll Wunden
Lag ein Rittersmann und rief:

„Still den Garten schön tust warten,
Bleibst am Fenster ofte stehn,
Ruhig scheinst du, heimlich weinst du,
Wie die Schiffe unten gehn."

„Was vertraust du, warum baust du
Auf der Männer wilde Brust,
Die das Blut ziert und der Streit rührt
Und die schöne Todeslust!"

Oben spinnend, saß sie sinnend –
Schwanden Schiff und Tageslicht,
Was er sunge, war verklungen,
Sie erkannt' den Liebsten nicht.

Waldmädchen

Bin ein Feuer hell, das lodert
Von dem grünen Felsenkranz,
Seewind ist mein Buhl' und fodert
Mich zum lust'gen Wirbeltanz,
Kommt und wechselt unbeständig.
Steigend wild,

Neigend mild,
Meine schlanken Lohen wend' ich:
Komm nicht nah mir, ich verbrenn' dich!

Wo die wilden Bäche rauschen
Und die hohen Palmen stehn,
Wenn die Jäger heimlich lauschen,
Viele Rehe einsam gehn.
Bin ein Reh, flieg' durch die Trümmer,
Über die Höh',
Wo im Schnee
Still die letzten Gipfel schimmern,
Folg' mir nicht, erjagst mich nimmer!

Bin ein Vöglein in den Lüften,
Schwing' mich übers blaue Meer,
Durch die Wolken von den Klüften
Fliegt kein Pfeil mehr bis hieher,
Und die Au'n und Felsenbogen,
Waldeseinsamkeit
Weit, wie weit,
Sind versunken in die Wogen –
Ach, ich habe mich verflogen!

Der Unbekannte

Vom Dorfe schon die Abendglocken klangen,
Die müden Vöglein gingen auch zur Ruh',
Nur auf den Wiesen noch die Heimchen sangen
Und von den Bergen rauscht' der Wald dazu;
Da kam ein Wandrer durch die Ährenwogen,
Aus fernen Landen schien er hergezogen.

Vor seinem Hause, unter blühnden Lauben
Lud ihn ein Mann zum fröhl'chen Rasten ein,
Die junge Frau bracht' Wein und Brot und Trauben,

Setzt' dann, umspielt vom letzten Abendschein,
Sich neben ihn und blickt halb scheu, halb lose,
Ein lockig Knäblein lächelnd auf dem Schoße.

Ihr dünkt', er wär' schon einst im Dorf gewesen,
Und doch so fremd und seltsam war die Tracht,
In seinen Mienen feur'ge Schrift zu lesen
Gleich Wetterleuchten fern bei stiller Nacht,
Und traf sein Auge sie, wollt' ihr fast grauen,
Denn's war, wie in den Himmelsgrund zu schauen.

Und wie sich kühler nun die Schatten breiten:
Vom Berg Vesuv, der über Trümmern raucht,
Vom blauen Meer, wo Schwäne singend gleiten,
Kristallnen Inseln, blühend draus getaucht,
Und Glocken, die im Meeresgrunde schlagen,
Wußt' wunderbar der schöne Gast zu sagen.

„Hast viel erfahren, willst du ewig wandern?"
Sprach drauf sein Wirt mit herzlichem Vertraun,
„Hier kannst du froh genießen wie die andern,
Am eignen Herd dein kleines Gärtchen baun,
Des Nachbars Töchter haben reiche Truhen,
Ruh endlich aus, brauchst nicht allein zu ruhen."

Da stand der Wandrer auf, es blühten Sterne
Schon aus dem Dunkel überm stillen Land,
„Gesegn' euch Gott! Mein Heimatland liegt ferne. –"
Und als er von den beiden sich gewandt,
Kam himmlisch Klingen von der Waldeswiese. –
So sternklar war noch keine Nacht wie diese.

Der stille Freier

Mond, der Hirt, lenkt seine Herde
Einsam überm Wald herauf,

Unten auf der stillen Erde
Wacht verschwiegne Liebe auf.

Fern vom Schlosse Glocken schlagen,
Übern Wald her von der Höh'
Bringt der Wind den Schall getragen,
Und erschrocken lauscht das Reh.

Nächtlich um dieselbe Stunde
Hallet Hufschlag, schnaubt ein Roß,
Macht ein Ritter seine Runde
Schweigend um der Liebsten Schloß.

Wenn die Morgensterne blinken,
Totenbleich der Hirte wird,
Und sie müssen all' versinken:
Reiter, Herde und der Hirt.

Waldgespräch

Es ist schon spät, es wird schon kalt,
Was reit'st du einsam durch den Wald?
Der Wald ist lang, du bist allein,
Du schöne Braut! Ich führ' dich heim!

„Groß ist der Männer Trug und List,
Vor Schmerz mein Herz gebrochen ist,
Wohl irrt das Waldhorn her und hin,
O flieh! Du weißt nicht, wer ich bin."

So reich geschmückt ist Roß und Weib,
So wunderschön der junge Leib,
Jetzt kenn' ich dich – Gott steh' mir bei!
Du bist die Hexe Lorelei.

„Du kennst mich wohl – von hohem Stein
Schaut still mein Schloß tief in den Rhein.
Es ist schon spät, es wird schon kalt,
Kommst nimmermehr aus diesem Wald!“

Die Saale

Doch manchmal in Sommertagen
Durch die schwüle Einsamkeit
Hört man mittags die Turmuhr schlagen,
Wie aus einer fremden Zeit.

Und ein Schiffer zu dieser Stunde
Sah einst eine schöne Frau
Vom Erker schaun zum Grunde –
Er ruderte schneller vor Graun.

Sie schüttelt’ die dunklen Locken
Aus ihrem Angesicht:
„Was ruderst du so erschrocken?
Behüt’ dich Gott, dich mein’ ich nicht!“

Sie zog ein Ringlein vom Finger,
Warf’s tief in die Saale hinein:
„Und der mir es wiederbringet,
Der soll mein Liebster sein!“

Der alte Garten

Kaiserkron’ und Päonien rot,
Die müssen verzaubert sein,
Denn Vater und Mutter sind lange tot,
Was blühn sie hier so allein!

Der Springbrunn plaudert noch immerfort
Von der alten schönen Zeit,
Eine Frau sitzt eingeschlafen dort,
Ihre Locken bedecken ihr Kleid.

Sie hat eine Laute in der Hand,
Als ob sie im Schlafe spricht,
Mir ist, als hätt' ich sie sonst gekannt –
Still, geh vorbei und weck' sie nicht!

Und wenn es dunkelt das Tal entlang,
Streift sie die Saiten sacht,
Da gibt's einen wunderbaren Klang
Durch den Garten die ganze Nacht.

Verloren

Still bei Nacht fährt manches Schiff,
Meerfei kämmt ihr Haar am Riff,
Hebt von Inseln an zu singen,
Die im Meer dort untergingen.

Wann die Morgenwinde wehn,
Ist nicht Riff noch Fei zu sehn,
Und das Schifflein ist versunken,
Und der Schiffer ist ertrunken.

Der Schnee

Wann der kalte Schnee zergangen,
Stehst du draußen in der Tür,
Kommt ein Knabe schön gegangen,
Stellt sich freundlich da zu dir,

Lobet deine frischen Wangen,
Dunkle Locken, Augen licht,
Wann der kalte Schnee zergangen,
Glaub' dem falschen Herzen nicht!

Wann die lauen Lüfte wehen,
Scheint die Sonne lieblich warm:
Wirst du wohl spazieren gehen,
Und er führet dich am Arm,
Tränen dir im Auge stehen,
Denn so schön klingt, was er spricht,
Wann die lauen Lüfte wehen,
Glaub' dem falschen Herzen nicht!

Wann die Lerchen wieder schwirren,
Trittst du draußen vor das Haus,
Doch er mag nicht mit dir irren,
Zog weit in das Land hinaus;
Die Gedanken sich verwirren,
Wie du siehst den Morgen rot, –
Wann die Lerchen wieder schwirren,
Armes Kind, ach wärst du tot!

Die weinende Braut

Du warst so herrlich anzuschauen,
So kühn und wild und doch so lieb,
Dir mußt' ich Leib und Seel' vertrauen,
Ich mocht' nichts mehr, das meine blieb!
Da hast du, Falscher, mich verlassen
Und Blumen, Lust und Frühlingsschein,
Die ganze Welt sah ich erblassen,
Ach Gott, wie bin ich nun allein!

Wohl jahrlang sah ich von den Höhen
Und grüßte dich viel tausendmal,

Und unten sah ich viele gehen,
Doch du erschienst nicht in dem Tal.
Und mancher Lenz mit bunten Scherzen
Kam und verflog im lust'gen Lauf,
Doch ach, in dem betrognen Herzen
Geht niemals mehr der Frühling auf.

Ein Kränzlein trag' ich nun im Haare,
In reichen Kleidern schön geschmückt,
Führt mich ein andrer zum Altare,
Die Eltern sind so hoch beglückt.
Und fröhlich kann ich mich wohl zeigen,
Die Sonne hell wie damals scheint,
Und vor dem Jauchzen und dem Geigen
Hört keiner, wie die Braut still weint.

Die Frühlingslieder neu beginnen –
Du kehrst nach manchem Jahr zurück,
Und stehest still, dich zu besinnen,
Wie auf ein längstvergangnes Glück.
Doch wüst verwachsen liegt der Garten,
Das Haus steht lange still und leer,
Kein Lieb will dein am Fenster warten,
Und dich und mich kennt niemand mehr.

Doch eine Lerche siehst du steigen
Vom Tal zum blauen Himmelsport,
Ein Bächlein rauschet da so eigen,
Als weinte es in einem fort.
Dort haben sie mich hingetragen,
Bedeckten mir mit Stein den Mund –
Nun kann ich dir nicht einmal sagen,
Wie ich dich liebt' aus Herzensgrund.

Das zerbrochene Ringlein

In einem kühlen Grunde
Da geht ein Mühlenrad,

Mein' Liebste ist verschwunden,
Die dort gewohnet hat.

Sie hat mir Treu' versprochen,
Gab mir ein'n Ring dabei,
Sie hat die Treu' gebrochen,
Mein Ringlein sprang entzwei.

Ich möcht' als Spielmann reisen
Weit in die Welt hinaus,
Und singen meine Weisen,
Und gehn von Haus zu Haus.

Ich möcht' als Reiter fliegen
Wohl in die blut'ge Schlacht,
Um stille Feuer liegen
Im Feld bei dunkler Nacht.

Hör' ich das Mühlrad gehen:
Ich weiß nicht, was ich will –
Ich möcht' am liebsten sterben,
Da wär's auf einmal still!

Der Gefangene

In goldner Morgenstunde,
Weil alles freudig stand,
Da ritt im heitern Grunde
Ein Ritter über Land.

Rings sangen auf das beste
Die Vöglein mannigfalt,
Es schüttelte die Äste
Vor Lust der grüne Wald.

Den Nacken, stolz gebogen,
Klopft er dem Rösselein –
So ist er hingezogen
Tief in den Wald hinein.

Sein Roß hat er getrieben,
Ihn trieb der frische Mut:
„Ist alles fern geblieben,
So ist mir wohl und gut!"

Mit Freuden mußt' er sehen
Im Wald ein' grüne Au,
Wo Brünnlein kühle gehen,
Von Blumen rot und blau.

Vom Roß ist er gesprungen,
Legt' sich zum kühlen Bach,
Die Wellen lieblich klungen,
Das ganze Herz zog nach.

So grüne war der Rasen,
Es rauschte Bach und Baum,
Sein Roß tät' stille grasen,
Und alles wie ein Traum.

Die Wolken sah er gehen,
Die schifften immer zu,
Er konnt' nicht widerstehen, –
Die Augen sanken ihm zu.

Nun hört' er Stimmen rinnen,
Als wie der Liebsten Gruß,
Er konnt' sich nicht besinnen –
Bis ihn erweckt' ein Kuß.

Wie prächtig glänzt' die Aue!
Wie Gold der Quell nun floß,

Und einer süßen Fraue
Lag er im weichen Schoß.

„Herr Ritter! Wollt Ihr wohnen
Bei mir im grünen Haus:
Aus allen Blumenkronen
Wind' ich Euch einen Strauß!

Der Wald ringsum wird wachen,
Wie wir beisammen sein,
Der Kuckuck schelmisch lachen,
Und alles fröhlich sein."

Es bog ihr Angesichte
Auf ihn, den süßen Leib,
Schaut' mit den Augen lichte
Das wunderschöne Weib.

Sie nahm sein'n Helm herunter,
Löst' Krause ihm und Bund,
Spielt' mit den Locken munter,
Küßt' ihm den roten Mund.

Und spielt' viel süße Spiele
Wohl in geheimer Lust,
Es flog so kühl und schwüle
Ihm um die offne Brust.

Um ihn nun tät' sie schlagen
Die Arme weich und bloß,
Er konnte nichts mehr sagen,
Sie ließ ihn nicht mehr los.

Und diese Au zur Stunde
Ward ein kristallnes Schloß,
Der Bach ein Strom, gewunden
Ringsum, gewaltig floß.

Auf diesem Strome gingen
Viel' Schiffe wohl vorbei,
Es konnt ihn keines bringen
Aus böser Zauberei.

Der traurige Jäger

Zur ew'gen Ruh' sie sangen
Die schöne Müllerin,
Die Sterbeglocken klangen
Noch übern Waldgrund hin.

Da steht ein Fels so kühle,
Wo keine Wandrer gehn,
Noch einmal nach der Mühle
Wollt' dort der Jäger sehn.

Die Wälder rauschten leise,
Sein Jagen war vorbei,
Der blies so irre Weise,
Als müßt' das Herz entzwei.

Und still dann in der Runde
Ward's über Tal und Höh'n,
Man hat seit dieser Stunde
Ihn nimmermehr gesehn.

Der Bräutigam

Von allen Bergen nieder
So fröhlich Grüßen schallt –
Das ist der Frühling wieder,
Der ruft zum grünen Wald!

Ein Liedchen ist erklungen
Herauf zum stillen Schloß –
Dein Liebster hat's gesungen,
Der hebt dich auf sein Roß.

Wir reiten so geschwinde,
Von allen Menschen weit. –
Da rauscht die Luft so linde
In Waldeseinsamkeit.

Wohin? Im Mondenschimmer
So bleich der Wald schon steht. –
Leis' rauscht die Nacht – frag' nimmer,
Wo Lieb' zu Ende geht!

Die falsche Schwester

Meine Schwester, die spielt' an der Linde –
Stille Zeit, wie so weit, so weit!
Da spielten so schöne Kinder
Mit ihr in der Einsamkeit.

Von ihren Locken verhangen
Schlief sie und lachte im Traum,
Und die schönen Kinder sangen
Die ganze Nacht unterm Baum.

Die ganze Nacht hat gelogen,
Sie hat mich so falsch gegrüßt,
Die Engel sind fortgeflogen,
Und Haus und Garten stehn wüst.

Es zittert die alte Linde
Und klaget der Wind so schwer,
Das macht, das macht die Sünde –
Ich wollt', ich läg' im Meer!

Die Sonne ist untergegangen
Und der Mond im tiefen Meer,
Es dunkelt schon über dem Lande,
Gute Nacht! Seh' dich nimmermehr!

Der Reitersmann

Hoch über den stillen Höhen
Stand in dem Wald ein Haus,
Dort war's so einsam zu sehen
Weit übern Wald hinaus.

Drin saß ein Mädchen am Rocken
Den ganzen Abend lang,
Der wurden die Augen nicht trocken,
Sie spann und sann und sang:

„Mein Liebster, der war ein Reiter,
Dem schwur ich Treu' bis in Tod,
Der zog über Land und weiter,
Zu Krieges Lust und Not.

Und als ein Jahr war vergangen,
Und wieder blühte das Land,
Da stand ich voller Verlangen
Hoch an des Waldes Rand.

Und zwischen den Bergesbogen,
Wohl über den grünen Plan
Kam mancher Reiter gezogen,
Der meine kam nicht mit an.

Und zwischen den Bergesbogen,
Wohl über den grünen Plan,
Ein Jägersmann kam geflogen,
Der sah mich so mutig an.

So lieblich die Sonne schiene,
Das Waldhorn scholl weit und breit,
Da führt' er mich in das Grüne,
Das war eine schöne Zeit! –

Der hat so lieblich gelogen
Mich aus der Treue heraus,
Der Falsche hat mich betrogen,
Zog weit in die Welt hinaus."

Sie konnte nicht weiter singen,
Vor bitterem Schmerz und Leid,
Die Augen ihr übergingen
In ihrer Einsamkeit.

Die Muhme, die saß beim Feuer
Und wärmte sich am Kamin,
Es flackert' und sprüht' das Feuer,
Hell über die Stube es schien.

Sie sprach: „Ein Kränzlein in Haaren,
Das stünde dir heut gar schön,
Willst draußen auf dem See nicht fahren?
Hohe Blumen am Ufer dort stehn."

Ich kann nicht holen die Blumen,
Im Hemdlein weiß am Teich
Ein Mädchen hütet die Blumen,
Die sieht so totenbleich.

„Und hoch auf des Sees Weite,
Wenn alles finster und still,
Da rudern zwei stille Leute, –
Der eine dich haben will."

Sie schauen wie alte Bekannte,
Still, ewig stille sie sind,

Doch einmal der eine sich wandte,
Da faßt' mich ein eiskalter Wind. –

Mir ist zu wehe zum Weinen –
Die Uhr so gleichförmig pickt,
Das Rädlein, das schnurrt so in einem,
Mir ist, als wär' ich verrückt. –

Ach Gott! wann wird sich doch röten
Die fröhliche Morgenstund!
Ich möchte hinausgehn und beten,
Und beten aus Herzensgrund!

So bleich schon werden die Sterne,
Es rührt sich stärker der Wald,
Schon krähen die Hähne von ferne,
Mich friert, es wird so kalt!

Ach, Muhme! was ist Euch geschehen?
Die Nase wird Euch so lang,
Die Augen sich seltsam verdrehen –
Wie wird mir vor Euch so bang! –

Und wie sie so grauenvoll klagte,
Klopft's draußen ans Fensterlein,
Ein Mann aus der Finsternis ragte,
Schaut' still in die Stube herein.

Die Haare wild umgehangen,
Von blutigen Tropfen naß,
Zwei blutige Streifen sich schlangen,
Wie Kränzlein, ums Antlitz blaß.

Er grüßt' sie so fürchterlich heiter,
Seine Braut wohl heißet er sie,
Da kannt' sie mit Schaudern den Reiter,
Fällt nieder auf ihre Knie.

Er zielt' mit dem Rohre durchs Gitter
Auf die schneeweiße Brust hin;
„Ach, wie ist das Sterben so bitter,
Erbarm' dich, weil ich so jung noch bin!" –

Stumm blieb sein steinerner Wille,
Es blitzte so rosenrot,
Da wurd' es auf einmal stille
Im Walde und Haus und Hof. –

Frühmorgens da lag so schaurig
Verfallen im Walde das Haus,
Ein Waldvöglein sang so traurig,
Flog fort über den See hinaus.

Das kalte Liebchen

Er. Laß mich ein, mein süßes Schätzchen!
Sie. Finster ist mein Kämmerlein.
Er. Ach, ich finde doch ein Plätzchen.
Sie. Und mein Bett ist eng und klein.

Er. Fern komm' ich vom weichen Pfühle.
Sie. Ach, mein Lager ist von Stein.
Er. Draußen ist die Nacht so kühle.
Sie. Hier wirds' noch viel kühler sein.

Er. Sieh! die Sterne schon erblassen.
Sie. Schwerer Schlummer fällt mich an. –
Er. Nun, so will ich schnell dich fassen!
Sie. Rühr mich nicht so glühend an.

Er. Fieberschauer mich durchbeben.
Sie. Wahnsinn bringt der Toten Kuß. –
Er. Weh! es bricht mein junges Leben!
Sie. Mit ins Grab hinunter muß.

Die verlorene Braut

Vater und Kind gestorben
Ruhten im Grabe tief,
Die Mutter hatt' erworben
Seitdem ein ander Lieb.

Da droben auf dem Schlosse
Da schallt das Hochzeitsfest,
Da lacht's und wiehern Rosse,
Durchs Grün ziehn bunte Gäst'.

Die Braut schaut' ins Gefilde
Noch einmal vom Altan,
Es sah so ernst und milde
Sie da der Abend an.

Rings waren schon verdunkelt
Die Täler und der Rhein,
In ihrem Brautschmuck funkelt
Nur noch der Abendschein.

Sie hörte Glocken gehen
Im weiten, tiefen Tal,
Es bracht' der Lüfte Wehen
Fern übern Wald den Schall.

Sie dacht': „O falscher Abend!
Wen das bedeuten mag?
Wen läuten sie zu Grabe
An meinem Hochzeitstag?"

Sie hört' im Garten rauschen
Die Brunnen immerdar,
Und durch der Wälder Rauschen
Ein Singen wunderbar.

Sie sprach: „Wie wirres Klingen
Kommt durch die Einsamkeit,
Das Lied wohl hört' ich singen
In alter, schöner Zeit.“

Es klang, als wollt' sie's rufen
Und grüßen tausendmal –
So stieg sie von den Stufen,
So kühle rauscht' das Tal.

So zwischen Weingehängen
Stieg sinnend sie ins Land
Hinunter zu den Klängen,
Bis sie im Walde stand.

Dort ging sie, wie in Träumen,
Im weiten, stillen Rund,
Das Lied klang in den Bäumen,
Von Quellen rauscht' der Grund. –

Derweil von Mund zu Munde
Durchs Haus, erst heimlich sacht,
Und lauter geht die Kunde
Die Braut irrt in der Nacht!

Der Bräut'gam tät' erbleichen,
Er hört im Tal das Lied,
Ein dunkelrotes Zeichen
Ihm von der Stirne glüht.

Und Tanz und Jubel enden,
Er und die Gäst' im Saal,
Windlichter in den Händen,
Sich stürzen in das Tal.

Da schweifen rote Scheine,
Schall nun und Rosseshuf,

Es hallten die Gesteine
Rings von verworrnem Ruf.

Doch einsam irrt die Fraue
Im Walde schön und bleich,
Die Nacht hat tiefes Grauen,
Das ist von Sternen so reich.

Und als sie war gelanget
Zum allerstillsten Grund,
Ein Kind am Felsenhange
Dort freundlich lächelnd stund.

Das trug in seinen Locken
Einen weißen Rosenkranz,
Sie schaut' es an erschrocken
Beim irren Mondesglanz.

„Solch' Augen hat das meine,
Ach meines bist du nicht,
Das ruht ja unterm Steine,
Den niemand mehr zerbricht.

Ich weiß nicht, was mir grauset,
Blick nicht so fremd auf mich!
Ich wollt', ich wär zu Hause." –
„Nach Hause führ' ich dich."

Sie gehn nun miteinander,
So trübe weht der Wind,
Die Fraue sprach im Wandern:
„Ich weiß nicht, wo wir sind.

Wen tragen sie beim Scheine
Der Fackeln durch die Schluft?
O Gott, der stürzt' vom Steine
Sich tot in dieser Kluft!"

Das Kind sagt: „Den sie tragen,
Dein Bräut'gam heute war,
Er hat meinen Vater erschlagen,
's ist diese Stund' ein Jahr.

Wir alle müssen's büßen,
Bald wird es besser sein,
Der Vater läßt dich grüßen,
Mein liebes Mütterlein."

Ihr schauert's durch die Glieder:
„Du bist mein totes Kind!
Wie funkeln die Sterne nieder,
Jetzt weiß ich, wo wir sind." –

Da löst' sie Kranz und Spangen,
Und über ihr Angesicht
Perlen und Tränen rannen,
Man unterschied sie nicht.

Und über die Schultern nieder
Rollten die Locken sacht,
Verdunkelnd Augen und Glieder,
Wie eine prächtige Nacht.

Ums Kind den Arm geschlagen,
Sank sie ins Gras hinein –
Dort hatten sie erschlagen
Den Vater im Gestein.

Die Hochzeitsgäste riefen
Im Walde auf und ab,
Die Gründe alle schliefen,
Nur Echo Antwort gab.

Und als sich leis erhoben
Der erste Morgenduft,
Hörten die Hirten droben
Ein Singen in stiller Luft.

Parole

Sie stand wohl am Fensterbogen
Und flocht sich traurig ihr Haar,
Der Jäger war fortgezogen,
Der Jäger ihr Liebster war.

Und als der Frühling gekommen,
Die Welt war von Blüten verschneit,
Da hat sie ein Herz sich genommen
Und ging in die grüne Heid'.

Sie legt das Ohr an den Rasen,
Hört ferner Hufe Klang –
Das sind die Rehe, die grasen
Am schattigen Bergeshang.

Und abends die Wälder rauschen,
Von fern nur fällt noch ein Schuß,
Da steht sie stille, zu lauschen:
„Das war meines Liebsten Gruß!"

Da sprangen vom Fels die Quellen,
Da flogen die Vöglein ins Tal.
„Und wo ihr ihn trefft, ihr Gesellen,
Grüßt mir ihn tausendmal!"

Zauberblick

Die Burg, die liegt verfallen
In schöner Einsamkeit,
Dort saß ich vor den Hallen
Bei stiller Mittagszeit.

Es ruhten in der Kühle
Die Rehe auf dem Wall

Und tief in blauer Schwüle
Die sonn'gen Täler all'.

Tief unten hört' ich Glocken
In weiter Ferne gehn,
Ich aber mußt' erschrocken
Zum alten Erker sehn.

Denn in dem Fensterbogen
Ein' schöne Fraue stand,
Als hütete sie droben
Die Wälder und das Land.

Ihr Haar, wie'n goldner Mantel,
War tief herabgerollt;
Auf einmal sie sich wandte,
Als ob sie sprechen wollt'.

Und als ich schauernd lauschte –
Da war ich aufgewacht,
Und unter mir schon rauschte
So wunderbar die Nacht.

Träumt' ich im Mondesschimmer?
Ich weiß nicht, was mir graut,
Doch das vergess' ich nimmer,
Wie sie mich angeschaut!

Der verirrte Jäger

„Ich hab' gesehn ein Hirschlein schlank
Im Waldesgrunde stehn,
Nun ist mir draußen weh und bang,
Muß ewig nach ihm gehn.

Frischauf, ihr Waldgesellen mein!
Ins Horn, ins Horn frischauf!
Das lockt so hell, das lockt so fein,
Aurora tut sich auf!“

Das Hirschlein führt den Jägersmann
In grüner Waldesnacht,
Talunter schwindelnd und bergan,
Zu nie gesehner Pracht.

„Wie rauscht schon abendlich der Wald,
Die Brust mir schaurig schwellt!
Die Freunde fern, der Wind so kalt,
So tief und weit die Welt!“

Es lockt so tief, es lockt so fein
Durchs dunkelgrüne Haus,
Der Jäger irrt und irrt allein,
Find’t nimmermehr heraus.

Die späte Hochzeit

Der Mond ging unter – jetzt ist’s Zeit. –
Der Bräut’gam steigt vom Roß,
Er hat so lange schon gefreit –
Da tut sich auf das Schloß,
Und in der Halle sitzt die Braut
Auf diamantnem Sitz,
Von ihrem Schmuck tut’s durch den Bau
Ein’n langen roten Blitz. –

Blass’ Knaben warten schweigend auf,
Still’ Gäste stehn herum,
Da richt’t die Braut sich langsam auf,
So hoch und bleich und stumm.
Sie schlägt zurück ihr Goldgewand,

Da schauert ihn vor Lust,
Sie langt mit kalter, weißer Hand
Das Herz ihm aus der Brust.

Die stille Gemeinde

Von Bretagnes Hügeln, die das Meer
Bühend hell umsäumen,
Schaute ein Kirchlein trostreich her
Zwischen uralten Bäumen.

Das Kornfeld und die Wälder weit
Rauschten im Sonntagsglanze,
Doch keine Glocken klangen heut
Vom grünen Felsenkranze.

Denn auf des Kirchhofs schatt'gem Grund
Die Jakobiner saßen,
Ihre Pferde alle Blumen bunt
Von den Grabeshügeln fraßen.

Sie hatten am Kreuz auf stiller Höh'
Feldflasch' und Säbel hangen,
Derweil sie, statt des Kyrie,
Die Marseillaise sangen.

Ihr Hauptmann aber lehnt' am Baum,
Todmüde von schweren Wunden,
Und schaute wie im Fiebertraum
Nach dem tiefschwülen Grunde.

Er sprach verwirrt: „Da drüben stand
Des Vaters Schloß am Weiher,
Ich selbst steckt's an; das war ein Brand,
Der Freiheit Freudenfeuer!"

„Ich seh' ihn noch: Wie durch den Sturm
Zwischen den feur'gen Zungen
Mein stolzer Vater da vom Turm
Sein Banner hat geschwungen."

„Und als es war entlaubt vom Brand,
Die Fahn' im Wind zerflogen:
Den Schaft als Kreuz nun in der Hand
Teilt' er die Flammenwogen."

„Er sah so wunderbar auf mich,
Ich konnt' ihn nicht ermorden –
Da sank die Burg, er wandte sich
Und ist ein Pfaff' geworden."

„Seitdem hör' ich in Träumen schwer
Von ferne Glocken gehen
Und seh' in totem Feuermeer
Ein Kreuz allnächtlich stehen."

„Es sollen keine Glocken gehn,
Die Nächte zu verstören,
Kein Kreuz soll mehr auf Erden stehn,
Um Narren zu betören!"

„Und dieses Kirchlein hier bewacht,
Sie sollen nicht Messe singen,
Wir reißen's nieder über Nacht,
Licht sei, wohin wir dringen!"

Und als die Nacht schritt leis daher,
Der Hauptmann stand am Strande,
So still im Wald, so still das Meer,
Nur die Wachen riefen im Lande.

Im Wind die Glock' von selbst anschlug,
Da wollt' ein Hauch sich heben,

Wie unsichtbarer Engel Flug,
Die übers Wasser schweben.

Nun sieht er auch im Meere fern
Ein Lichtlein hell entglommen;
Er dacht', wie ist der schöne Stern
Dort in die Flut gekommen?

Am Ufer aber durch die Nacht
In allen Felsenspalten
Regt sich's und schlüpft es leis und sacht,
Viel' dunkle, schwanke Gestalten.

Nur manchmal von den Buchten her
Schallt Ruderschlag von weitem,
Auf Barken lautlos in das Meer
Sie nach dem Stern hin gleiten.

Der wächst und breitet sich im Nahn
Und streift mit Glanz die Wellen,
Es ist ein kleiner Fischerkahn,
Den Fackeln mild erhellen.

Und einsam auf des Schiffleins Rand
Ein Greis kommt hergezogen
In wunderbarem Meßgewand
Als wie der Hirt der Wogen.

Die Barken eine weite Rund'
Dort um den Hirten machen,
Der laut nun überm Meeresgrund
Den Segen spricht im Nachen.

Da schwieg der Wind und rauscht' das Meer
So wunderbare Weise,
Und auf den Knien lag ringsher
Die stille Gemeinde im Kreise.

Und als er das Kreuz hob in die Luft,
Hoch zwischen die Fackeln trat er –
Den Hauptmann schauert im Herzensgrund,
Es war sein alter Vater.

Da taumelt' er und sank ins Gras
Betend im stillen Grunde,
Und wie Felsenquellen im Frühling brach
Sein Herzblut aus allen Wunden.

Und als die Gesellen kommen zum Strand,
Einen toten Mann sie finden –
Voll Grau'n sie sprengen fort durchs Land,
Als jagt' sie der Tod in den Winden.

Die stürzten sich in den Krieg so weit,
Sie sind verweht und zerstoben,
Das Kirchlein aber steht noch heut
Unter den Linden droben.

Die deutsche Jungfrau

Es stand ein Fräulein auf dem Schloß,
Erschlagen war im Streit ihr Roß,
Schnob wie ein See die finstre Nacht,
Wollt' überschrein die wilde Schlacht.

Im Tal die Brüder lagen tot,
Es brannt' die Burg so blutigrot,
In Lohen stand sie auf der Wand,
Hielt hoch die Fahne in der Hand.

Da kam ein röm'scher Rittersmann,
Der ritt keck an die Burg hinan,
Es blitzt' sein Helm gar mannigfach,
Der schöne Ritter also sprach:

„Jungfrau, komm in die Arme mein!
Sollst deines Siegers Herrin sein.
Will baun dir einen Palast schön,
In prächt'gen Kleidern sollst du gehn.

Es tun dein' Augen mir Gewalt,
Kann nicht mehr fort aus diesem Wald,
Aus wilder Flammen Spiel und Graus
Trag' ich mir meine Braut nach Haus!"

Der Ritter ließ sein weißes Roß,
Stieg durch den Brand hinauf ins Schloß,
Viel Knecht' ihm waren da zur Hand,
Zu holen das Fräulein von der Wand.

Das Fräulein stieß die Knecht' hinab,
Den Liebsten auch ins heiße Grab,
Sie selber dann in die Flamme sprang.
Über ihnen die Burg zusammen sank.

Die wunderliche Prinzessin

Weit in einem Walde droben
Zwischen hoher Felsen Zinnen,
Steht ein altes Schloß erhoben,
Wohnet eine Zaub'rin drinnen.
Von dem Schloß, der Zaub'rin Schöne
Gehen wunderbare Sagen,
Lockend schweifen fremde Töne
Plötzlich her oft aus dem Walde.
Wem sie recht das Herz getroffen,
Der muß nach dem Walde gehen,
Ewig diesen Klängen folgend,
Und wird nimmermehr gesehen.
Tief in wundersamer Grüne
Steht das Schloß, schon halb verfallen,

Hell die goldnen Zinnen glühen,
Einsam sind die weiten Hallen.
Auf des Hofes stein'gem Rasen
Sitzen von der Tafelrunde
All' die Helden dort gelagert,
Überdeckt mit Staub und Wunden.
Heinrich liegt auf seinem Löwen,
Gottfried auch, Siegfried der Scharfe,
König Alfred, eingeschlafen
Über seiner goldnen Harfe.
Don Quixote hoch auf der Mauer
Sinnend tief in nächt'ger Stunde,
Steht gerüstet auf der Lauer
Und bewacht die heil'ge Runde.
Unter fremdes Volk verschlagen,
Arm und ausgehöhnt, verraten,
Hat er treu sich durchgeschlagen,
Eingedenk der Heldentaten
Und der großen, alten Zeiten,
Bis er, ganz von Wahnsinn trunken,
Endlich so nach langem Streiten
Seine Brüder hat gefunden.

Einen wunderbaren Hofstaat
Die Prinzessin dorten führet,
Hat ein'n wunderlichen Alten,
Der das ganze Haus regieret.
Einen Mantel trägt der Alte,
Schillernd bunt in allen Farben
Mit unzähligen Zierraten,
Spielzeug hat er in den Falten.
Scheint der Monden helle draußen
Wolken fliegen überm Grunde:
Fängt er draußen an zu hausen,
Kramt sein Spielzeug aus zur Stunde.
Und das Spielzeug um den Alten
Rührt sich bald beim Mondenscheine,

Zupfet ihn beim langen Barte,
Schlingt um ihn die bunten Kreise,
Auch die Blümlein nach ihm langen,
Möchten doch sich sittsam zeigen,
Ziehn verstohlen ihn beim Mantel,
Lachen dann in sich gar heimlich.
Und ringsum die ganze Runde
Zieht Gesichter ihm und rauschet,
Unterhält aus dunklem Grunde
Sich mit ihm als wie im Traume.
Und er spricht und sinnt und sinnet,
Bunt verwirrend alle Zeiten,
Weinet bitterlich und lachet,
Seine Seele ist so heiter.

Bei ihm sitzt dann die Prinzessin,
Spielt mit seinen Seltsamkeiten,
Immer neue Wunder blinkend
Muß er aus dem Mantel breiten.
Und der wunderliche Alte
Hielt sie sich bei seinen Bildern
Neidisch immerfort gefangen,
Weit von aller Welt geschieden.
Aber der Prinzessin wurde
Mitten in dem Spiele bange
Unter diesen Zauberblumen,
Zwischen dieser Quellen Rauschen.
Frisches Morgenrot im Herzen
Und voll freudiger Gedanken,
Sind die Augen wie zwei Kerzen,
Schön, die Welt dran zu entflammen.
Und die wunderschöne Erde,
Wie Aurora sie berühret,
Will mit ird'scher Lust und Schmerzen
Ewig neu sie stets verführen.
Denn aus dem bewegten Leben
Spüret sie ein Hochzeitsgrüßen,

Mitten zwischen ihren Spielen
Muß sie sich bezwungen fühlen.

Und es hebt die ewig Schöne,
Da der Morgen herrlich schiene,
In den Augen große Tränen,
Hell die jugendlichen Glieder.
„Wie so anders war es damals,
Da mich, bräutlich Ausgeschmückte,
Aus dem heimatlichen Garten
Hier herab der Vater schickte!
Wie die Erde frisch und jung noch,
Von Gesängen rings erklingend,
Schauernd in Erinnerungen,
Helle in das Herz mir blickte,
Daß ich, schamhaft mich verhüllend,
Meinen Ring, vom Glanz geblendet,
Schleudert' in die prächt'ge Fülle,
Als die ew'ge Braut der Erde.
Wo ist nun die Pracht geblieben,
Treuer Ernst im rüst'gen Treiben,
Rechtes Tun und rechtes Lieben
Und die Schönheit und die Freude?
Ach! ringsum die Helden alle,
Die sonst schön und helle schauten,
Um mich in den lichten Tagen
Durch die Welt sich fröhlich hauten,
Strecken steinern nun die Glieder,
Eingehüllt in ihre Fahnen,
Sind seitdem so alt geworden,
Nur *ich* bin so jung wie damals. –
Von der Welt kann ich nicht lassen,
Liebeln nicht von fern mit Reden,
Muß im Arm lebendig fassen! –
Laß mich lieben, laß mich leben!"

Nun verliebt die Augen gehen
Über ihres Gartens Mauer,

War so einsam dort zu sehen
Schimmernd Land und Ström' und Auen.
Und wo ihre Augen gingen:
Quellen aus der Grüne sprangen,
Berg und Wald verzaubert standen,
Tausend Vögel schwirrend sangen.
Golden blitzt es überm Grunde,
Seltne Farben irrend schweifen,
Wie zu lang entbehrtem Feste
Will die Erde sich bereiten.
Und nun kamen angezogen
Freier bald von allen Seiten,
Federn bunt im Winde flogen,
Jäger schmuck im Walde reiten.
Hörner munter drein erschallen
Auf und unter durch das Grüne,
Pilger fromm dazwischen wallen,
Die das Heimatsfieber spüren.
Auf vielsonn'gen Wiesen flöten
Schäfer bei schneeflock'gen Schafen,
Ritter in der Abendröte
Knien auf des Berges Hange,
Und die Nächte von Guitarren
Und Gesängen weich erschallen,
Daß der wunderliche Alte
Wie verrückt beginnt zu tanzen.
Die Prinzessin schmückt mit Kränzen
Wieder sich die schönen Haare,
Und die vollen Kränze glänzen
Und sie blickt verlangend nieder.

Doch die alten Helden alle,
Draußen vor der Burg gelagert,
Saßen dort im Morgenglanze,
Die das schöne Kind bewachten.
An das Tor die Freier kamen
Nun gesprengt, gehüpft, gelaufen,

Ritter, Jäger, Provençalen,
Bunte, helle, lichte Haufen.
Und vor allen junge Recken
Stolzen Blicks den Berg berannten,
Die die alten Helden weckten,
Sie vertraulich Brüder nannten.
Doch wie diese uralt blicken,
An die Eisenbrust geschlossen,
Brüderlich die Jungen drücken,
Fallen die erdrückt zu Boden.
Andre lagern sich zum Alten,
Graust ihn'n gleich bei seinen Mienen,
Ordnen sein verworrnes Walten,
Daß es jedem wohlgefiele;
Doch sie fühlen schauernd balde,
Daß sie ihn nicht können zwingen,
Selbst zu Spielzeug sind verwandelt,
Und der Alte spielt mit ihnen.
Und sie müssen töricht tanzen,
Manche mit der Kron' geschmücket
Und im purpurnen Talare
Feierlich den Reigen führen.
Andre schweben lispelnd lose,
Andre müssen männlich lärmen,
Rittern reißen aus die Rosse,
Und die schreien gar erbärmlich.
Bis sie endlich alle müde
Wieder kommen zu Verstande,
Mit der ganzen Welt im Frieden,
Legen ab die Maskerade.
„Jäger sind wir nicht, noch Ritter",
Hört man sie von fern noch summen,
„Spiel nur war das – wir sind Dichter!"
So vertost der ganze Plunder,
Nüchtern liegt die Welt wie ehe,
Und die Zaub'rin bei dem Alten
Spielt' die vor'gen Spiele wieder
Einsam wohl noch lange Jahre.

Meeresstille

Ich seh' von des Schiffes Rande
Tief in die Flut hinein:
Gebirge und grüne Lande
Und Trümmer im falben Schein
Und zackige Türme im Grunde,
Wie ich's oft im Traum mir gedacht,
Wie dämmert alles da unten
Als wie eine prächtige Nacht.

Seekönig auf seiner Warte
Sitzt in der Dämm'rung tief,
Als ob er mit langem Barte
Über seiner Harfe schlief';
Da kommen und gehen die Schiffe
Darüber, er merkt es kaum,
Von seinem Korallenriffe
Grüßt er sie wie im Traum.

Der zaubrische Spielmann

Nächtlich in dem stillen Grunde,
Wenn das Abendrot versank,
Um das Waldschloß in die Runde
Ging ein lieblicher Gesang.

Fremde waren diese Weisen
Und der Sänger unbekannt,
Aber, wie in Zauberkreisen,
Hielt er jede Brust gebannt.

Hinter blüh'nden Mandelbäumen
Auf dem Schloß das Fräulein lauscht –
Drunten alle Blumen träumen,
Wollüstig der Garten rauscht.

Und die Wellen buhlend klingen,
Ringend in geheimer Lust
Kommt das wunderbare Singen
An die süß verträumte Brust.

„Warum weckst du das Verlangen,
Das ich kaum zur Ruh' gebracht?
Siehst du hoch die Lilien prangen?
Böser Sänger, gute Nacht!

Sieh, die Blumen stehn voll Tränen,
Einsam die Viole wacht,
Als wollt' sie sich schmachtend dehnen
In die warme Sommernacht.

Wohl von süßem, rotem Munde
Kommt so holden Sanges Macht –
Bleibst du ewig dort im Grunde,
Unerkannt in stiller Nacht?

Ach, im Wind verfliegt meine Grüßen!
Einmal, eh' der Tag erwacht,
Möcht' ich deinen Mund nur küssen,
Sterbend so in süßer Nacht!

Nachtigall, verliebte, kluge
Nicht so schmeichelnd durch die Nacht! –
Ach! ich weiß nicht, was ich sage,
Krank bin ich und überwacht."

Also sprach sie, und die Lieder
Lockten stärker aus dem Tal,
Rings durchs ganze Tal hallt's wider
Von der Liebe Lust und Qual.

Und sie konnt' nicht widerstehen,
Enge ward ihr das Gemach,

Aus dem Schlosse mußt' sie gehen
Diesem Zauberstrome nach.

Einsam steigt sie von den Stufen,
Ach! so schwüle weht der Wind:
Draußen süß die Stimmen rufen
Immerfort das schöne Kind.

Alle Blumen trunken lauschen,
Von den Klängen hold durchirrt,
Lieblicher die Brunnen rauschen,
Und sie eilet süß verwirrt. –

Wohl am Himmel auf und nieder
Trieb der Hirt die goldne Schar,
Die Verliebte kehrt nicht wieder,
Leer nun Schloß und Garten war.

Und der Sänger seit der Stunde
Nicht mehr weiter singen will,
Rings im heimlich kühlen Grunde
War's vor Liebe selig still.

Das kranke Kind

Die Gegend lag so helle,
Die Sonne schien so warm,
Es sonnt sich auf der Schwelle
Ein Kindlein krank und arm.

Geputzt zum Sonntag heute
Ziehn sie das Tal entlang,
Das Kind grüßt alle Leute,
Doch niemand sagt ihm Dank.

Viel' Kinder jauchzen ferne,
So schön ist's auf der Welt!
Ging' auch spazieren gerne,
Doch müde stürzt's im Feld.

„Ach, Vater, liebe Mutter,
Helft mir in meiner Not! –“
Du armes Kind! die ruhen
Ja unterm Grase tot.

Und so im Gras alleine
Das kranke Kindlein blieb,
Frug keiner, was es weine,
Hat jeder seins nur lieb.

Die Abendglocken klangen
Schon durch die stille Welt,
Die Engel Gottes sangen
Und gingen übers Feld.

Und als die Nacht gekommen
Und alles das Kind verließ,
Sie haben's mitgenommen,
Nun spielt's im Paradies.

Der Schatzgräber

Wenn alle Wälder schliefen,
Er an zu graben hub,
Rastlos in Berges Tiefen
Nach einem Schatz er grub.

Die Engel Gottes sangen
Derweil in stiller Nacht,
Wie rote Augen drangen
Metalle aus dem Schacht.

„Und wirst doch mein!“ und grimmer
Wühlt er und wühlt hinab,
Da stürzen Steine und Trümmer
Über dem Narren herab.

Hohnlachen wild erschallte
Aus der verfallnen Kluft,
Der Engelsang verhallte
Wehmütig in der Luft.

Die Räuberbrüder

„Vorüber ist der blut'ge Strauß,
Hier ist's so still, nun ruh dich aus.“

„Vom Tal herüber kommt die Luft;
Horch, hörst du nichts? Die Mutter ruft.“

„Die Mutter ist ja lange tot,
Eine Glocke klingt durchs Morgenrot,“

„Lieb' Mutter, hab' nicht solches Leid,
Mein wildes Leben mich gereut. –“

„Was sinkst du auf die Knie ins Gras?
Deine Augen dunkeln, du wirst so blaß.“ –

Es war von Blut der Grund so rot,
Der Räuber lag im Grase tot.

Da küßt der Bruder den bleichen Mund:
„Dich liebt' ich recht aus Herzensgrund.“

Vom Fels dann schoß er noch einmal
Und warf die Büchse tief ins Tal.

Drauf schritt er durch den Wald zur Stadt:
„Ihr Herrn, ich bin des Lebens satt."

„Hie ist mein Haupt, nun richtet bald,
Zum Bruder legt mich in den Wald."

Sonst

Es glänzt der Tulpenflor, durchschnitten von Alleen,
Wo zwischen Taxus still die weißen Statuen stehen,
Mit goldnen Kugeln spielt die Wasserkunst im Becken,
Im Laube lauert Sphinx, anmutig zu erschrecken.

Die schöne Chloe heut spazieret in dem Garten,
Zur Seit' ein Kavalier, ihr höflich aufzuwarten,
Und hinter ihnen leis Kupido kommt gezogen,
Bald duckend sich im Grün, bald zielend mit dem Bogen.

Es neigt der Kavalier sich in galantem Kosen,
Mit ihrem Fächer schlägt sie manchmal nach dem Losen,
Es rauscht der taftne Rock, es blitzen seine Schnallen,
Dazwischen hört man oft ein art'ges Lachen schallen.

Jetzt aber hebt vom Schloß, da sich's im West will röten,
Die Spieluhr schmachtend an, ein Menuett zu flöten,
Die Laube ist so still, er wirft sein Tuch zur Erde
Und stürzet auf ein Knie mit zärtlicher Gebärde.

„Wie wird mir, ach, ach, ach, es fängt schon an zu dunkeln –"
„So angenehmer nur seh' ich zwei Sterne funkeln –"
„Verwegner Kavalier!" – „Ha, Chloe, darf ich hoffen? –"
Da schießt Kupido los und hat sie gut getroffen.

Der Kehraus

Es fiedeln die Geigen,
Da tritt in den Reigen
Ein seltsamer Gast,
Kennt keiner den Dürren,
Galant aus dem Schwirren
Die Braut er sich faßt.

Hebt an, sich zu schwenken
In allen Gelenken.
Das Fräulein im Kranz:
„Euch knacken die Beine –"
„Bald rasseln auch deine,
Frisch auf spielt zum Tanz!"

Die Spröde hinterm Fächer,
Der Zecher vom Becher,
Der Dichter so lind,
Muß auch mit zum Tanze,
Daß die Lorbeern vom Kranze
Fliegen im Wind.

So schnurret der Reigen
Zum Saal 'raus ins Schweigen
Der prächtigen Nacht,
Die Klänge verwehen,
Die Hähne schon krähen,
Da verstieben sie sacht. –

So ging's schon vor Zeiten
Und geht es noch heute,
Und hörest du hell
Aufspielen zum Reigen,
Wer weiß, wem sie geigen –
Hüt' dich, Gesell!

Der armen Schönheit Lebenslauf

Die arme Schönheit irrt auf Erden,
So lieblich Wetter draußen ist,
Möcht' gern recht viel gesehen werden,
Weil jeder sie so freundlich grüßt.

Und wer die arme Schönheit schauet,
Sich wie auf großes Glück besinnt,
Die Seele fühlt sich recht erbauet,
Wie wenn der Frühling neu beginnt.

Da sieht sie viele schöne Knaben,
Die reiten unten durch den Wind,
Möcht' manchen gern im Arme haben,
Hüt' dich, hüt' dich, du armes Kind!

Da ziehn manch' redliche Gesellen,
Die sagen: Hast nicht Geld, noch Haus,
Wir fürchten deine Augen helle,
Wir haben nichts zum Hochzeitsschmaus.

Von andern tut sie sich wegdrehen,
Weil keiner ihr so wohl gefällt,
Die müssen traurig weitergehen,
Und zögen gern ans End' der Welt.

Da sagt sie: Was hilft mir mein Sehen,
Ich wünscht', ich wäre lieber blind,
Da alle furchtsam von mir gehen,
Weil gar so schön mein' Augen sind. –

Nun sitzt sie hoch auf lichtem Schlosse,
In schöne Kleider putzt sie sich,
Die Fenster glühn, sie winkt vom Schlosse,
Die Sonne sinkt, das blendet dich.

Die Augen, die so furchtsam waren,
Die haben jetzt so freien Lauf,
Fort ist das Kränzlein aus den Haaren,
Und hohe Federn stehn darauf.

Das Kränzlein ist herausgerissen,
Ganz ohne Scheu sie mich anlacht;
Geh du vorbei: sie wird dich grüßen,
Winkt dir zu einer schönen Nacht. –

Da sieht sie die Gesellen wieder,
Die fahren unten auf dem Fluß,
Es singen laut die lust'gen Brüder,
So furchtbar schallt des Einen Gruß:

„Was bist du für 'ne schöne Leiche!
So wüste ist mir meine Brust,
Wie bist du nun so arm, du Reiche,
Ich hab' an dir nicht weiter Lust!"

Der Wilde hat ihr so gefallen,
Laut schrie sie auf bei seinem Gruß,
Vom Schloß möcht' sie herunter fallen,
Und unten ruhn im kühlen Fluß. –

Sie blieb nicht länger mehr da oben,
Weil alles anders worden war,
Vor Schmerz ist ihr das Herz erhoben,
Da ward's so kalt, doch himmlisch klar.

Da legt sie ab die goldnen Spangen,
Den falschen Putz und Ziererei,
Aus dem verstockten Herzen drangen
Die alten Tränen wieder frei.

Kein Stern wollt' nicht die Nacht erhellen,
Da mußte die Verliebte gehn,

Wie rauscht der Fluß! Die Hunde bellen,
Die Fenster fern erleuchtet stehn.

Nun bist du frei von deinen Sünden,
Die Lieb' zog triumphierend ein,
Du wirst noch hohe Gnade finden,
Die Seele geht in Hafen ein.

Der Liebste war ein Jäger worden,
Der Morgen schien so rosenrot,
Da blies er lustig auf dem Horne,
Blies immerfort in seiner Not.

Die Hochzeitsnacht

Nachts durch die stille Runde
Rauschte des Rheines Lauf,
Ein Schifflein zog im Grunde,
Ein Ritter stand darauf.

Die Blicke irre schweifen
Von seines Schiffes Rand,
Ein blutigroter Streifen
Sich um das Haupt ihm wand.

Der sprach: „Da oben stehet
Ein Schlößlein überm Rhein,
Die an dem Fenster stehet:
Das ist die Liebste mein.

Sie hat mir Treu' versprochen,
Bis ich gekommen sei,
Sie hat die Treu' gebrochen,
Und alles ist vorbei."

Viel' Hochzeitleute drehen
Sich oben laut und bunt,

Sie bleibet einsam stehen,
Und lauschet in den Grund.

Und wie sie tanzen munter
Und Schiff und Schiffer schwand,
Stieg sie vom Schloß herunter,
Bis sie im Garten stand.

Die Spielleut' musizierten,
Sie sann gar mancherlei,
Die Töne sie so rührten,
Als müßt' das Herz entzwei.

Da trat ihr Bräut'gam süße
Zu ihr aus stiller Nacht,
So freundlich er sie grüßte,
Daß ihr das Herze lacht.

Er sprach: „Was willst du weinen,
Weil alle fröhlich sein?
Die Stern' so helle scheinen,
So lustig geht der Rhein.

Das Kränzlein in den Haaren
Steht dir so wunderfein,
Wir wollen etwas fahren
Hinunter auf dem Rhein."

Zum Kahn folgt' sie behende,
Setzt' sich ganz vorne hin,
Er setzt' sich an das Ende
Und ließ das Schifflein ziehn.

Sie sprach: „Die Töne kommen
Verworren durch den Wind,
Die Fenster sind verglommen,
Wir fahren so geschwind.

Was sind das für so lange
Gebirge weit und breit?
Mir wird auf einmal bange
In dieser Einsamkeit!

Und fremde Leute stehen
Auf mancher Felsenwand,
Und stehen still und sehen
So schwindlig übern Rand." –

Der Bräut'gam schien so traurig
Und sprach kein einzig Wort,
Schaut' in die Wellen schaurig
Und rudert' immerfort.

Sie sprach: „Schon seh' ich Streifen
So rot im Morgen stehn,
Und Stimmen hör' ich schweifen,
Vom Ufer Hähne krähn.

Du siehst so still und wilde,
So bleich wird dein Gesicht,
Mir graut vor deinem Bilde –
Du bist mein Bräut'gam nicht!" –

Da stand er auf – das Sausen
Hielt an in Flut und Wald –
Es rührt mit Lust und Grausen
Das Herz ihr die Gestalt.

Und wie mit steiner'n Armen
Hob er sie auf voll Lust,
Drückt ihren schönen, warmen
Leib an die eis'ge Brust. –

Licht wurden Wald und Höhen,
Der Morgen schien blutrot,
Das Schifflein sah man gehen,
Die schöne Braut drin tot.

Von Engeln und von Bengeln

Im Frühling auf grünem Hügel
Da saßen viel' Engelein,
Die putzten sich ihre Flügel
Und spielten im Sonnenschein.

Da kamen Störche gezogen,
Und jeder sich eines nahm,
Und ist damit fortgeflogen,
Bis daß er zu Menschen kam.

Und wo er anklopft' bescheiden
Der kluge Adebar,
Da war das Haus voller Freuden –
So geht es noch alle Jahr.

Die Engel weinten und lachten
Und wußten nicht, wie ihn'n geschehn. –
Die einen doch bald sich bedachten,
Und meinten: Das wird wohl gehn!

Die machten bald wichtige Mienen
Und wurden erstaunlich klug,
Die Flügel gar unnütz ihn'n schienen,
Sie schämten sich deren genug.

Und mit dem Flügelkleide
Sie ließen den Flügelschnack,
Das war keine kleine Freude:
Nun stattlich in Hosen und Frack!

So wurden sie immer gescheuter
Und applizierten sich recht –
Das wurden ansehnliche Leute,
Befanden sich gar nicht schlecht.

Den andern war's, wenn die Aue
Noch dämmert' im Frühlingsschein,
Als zöge ein Engel durch's Blaue
Und rief' die Gesellen sein.

Die suchten den alten Hügel,
Der lag so hoch und weit –
Und dehnten sehnsüchtig die Flügel
Mit jeder Frühlingszeit.

Die Flügeldecken zersprangen,
Weit, morgenschön strahlt' die Welt,
Und übers Grün sie sich schwangen
Bis an das Himmelszelt.

Das fanden sie droben verschlossen,
Versäumten unten die Zeit –
So irrten die kühnen Genossen,
Verlassen in Lust und Leid. –

Und als es nun kam zum Sterben,
Gott Vater zur Erden trat,
Seine Kinder wieder zu werben,
Die der Storch vertragen hat.

Die einen konnten nicht fliegen,
So wohlleibig, träg und schwer,
Die mußt' Er da lassen liegen,
Das tat ihm leid so sehr.

Die andern streckten die Schwingen
In den Morgenglanz hinaus,
Und hörten die Engel singen,
Und flogen jauchzend nach Haus!

Valet

Ade nun, liebe Lieder,
Ade, du schöner Sang!
Nun sing' ich wohl nicht wieder
Vielleicht mein Leben lang.

Einst blüht' von Gottes Odem
Die Welt so wunderreich,
Da in den grünen Boden
Senkt' ich als Reiser euch.

Jetzt eure Wipfel schwanken
So kühle über mir,
Ich stehe in Gedanken
Gleichwie im Walde hier.

Da muß ich oft noch lauschen
In meiner Einsamkeit,
Und denk' bei eurem Rauschen
Der schönen Jugendzeit.

VIII. Aus dem Spanischen

Vom Strande

Ich rufe vom Ufer
Verlorenes Glück,
Die Ruder nur schallen
Zum Strande zurück.

Vom Strande, lieb' Mutter,
Wo der Wellenschlag geht,
Da fahren die Schiffe,
Mein Liebster drauf steht.
Je mehr ich sie rufe,
Je schneller ihr Lauf,
Wenn ein Hauch sie entführet,
Wer hielte sie auf?
Der Hauch meiner Klagen
Die Segel nur schwellt,
Je mehr mein Verlangen
Zurücke sie hält!
Verhielt' ich die Klagen:
Es löst' sie der Schmerz,
Und Klagen und Schweigen
Zersprengt mir das Herz.

Ich rufe vom Ufer
Verlorenes Glück,
Die Ruder nur schallen
Zum Strande zurück.

So flüchtige Schlösser,
Wer könnt' ihn'n vertraun,
Und Liebe, die bliebe,
Mit Freuden drauf baun?
Wie Vögel im Fluge,
Wo ruhen sie aus?
So eilige Wandrer
Sie finden kein Haus,

Zertrümmern der Wogen
Grünen Kristall,
Und was sie berühren
Verwandelt sich all',
Es wandeln die Wellen
Und wandelt der Wind –
Meine Schmerzen im Herzen
Beständig nur sind.

Ich rufe vom Ufer
Verlorenes Glück,
Die Ruder nur schallen
Zum Strande zurück.

Die Musikantin

Schwirrend Tamburin, dich schwing' ich,
Doch mein Herz ist weit von hier.

Tamburin, ach könnt'st du's wissen,
Wie mein Herz von Schmerz zerrissen,
Deine Klänge würden müssen
Weinen um mein Leid mit mir.

Weil das Herz mir will zerspringen,
Lass' ich hell die Schellen klingen,
Die Gedanken zu versingen
Aus des Herzens Grunde mir.

Schöne Herren, tief im Herzen
Fühl' ich immer neu die Schmerzen,
Wie ein Angstruf ist mein Scherzen,
Denn mein Herz ist weit von hier.

Turteltaube und Nachtigall

Bächlein, das so kühle rauschet,
Tröstet alle Vögelein,
Nur das Turteltäubchen trauert,
Weil's verwitwet und allein.

Nachtigallenmännchen draußen
Schmettert so verlockend drein:
Mir vertraue, süße Fraue,
Will dein Lieb, dein Liebster sein!

„Böser, laß die falschen Lieder!
Ruh' auf keinem Zweig, der blüht,
Laß auf keiner Au mich nieder,
Die von schönen Blumen glüht."

„Wo ich finde eine Quelle
Helle in dem grünen Haus,
Mit dem Schnabel erst die Welle
Trüb' ich, eh' ich trink daraus."

„Einsam soll man mich begraben,
Laß mich trauernd hier allein,
Will nicht Trost, nicht Lust mehr haben,
Nicht dein Weib, noch Liebchen sein!"

Graf Arnold und der Schiffer

Wem begegnet' je solch Wunder,
Als Graf Arnold ist geschehn,
Da er am Johannismorgen
Wollt' am Meere jagen gehn?

Auf dem Meer ein Schifflein fahren
Sah er, als ob's landen wollt',

Seiden seine Segel waren
Und das Tauwerk war von Gold.

Fing der Schiffer da zu singen,
Wunderbar zu singen an.
Daß die Wogen leiser gingen,
Wind hielt seinen Atem an;

Daß die Fische lauschend stiegen
Tief aus ihrem kühlen Haus,
Und die Vögel, die da fliegen,
Auf dem Maste ruhten aus:

„Durch die Einsamkeit der Wogen,
Schifflein, lenk' dich Gottes Hand
An Gibraltars Felsenbogen,
An dem tück'schen Mohrenstrand."

„Flandern gürten sand'ge Banken,
Bei Leon da steht ein Riff,
Wo schon viele Schiffe sanken,
Hüt' dich Gott, mein schönes Schiff!"

„Schiffer!" rief der Graf am Strande,
„Schiffer, lehre mich dein Lied!" –
Doch der Schiffer lenkt' vom Lande:
„Lehr's nur den, der mit mir zieht."

Der Hochzeitstanz

Wie so zierlich in dem Saale
Führt die Braut den Hochzeitsreih'n,
Wie so mutig schaut Graf Martin
In die freud'gen Klänge drein!

Und sie im Vorüberschweifen
Flüstert: „Graf, was sinnet Ihr?

Sagt mir, schaut Ihr nach dem Tanze,
Oder blicket Ihr nach mir?"

„Hab' schon manchen Tanz gesehen,
Und das war's nicht, was ich sann,
Eure Schönheit mich verblendet,
Eure Augen tun mir's an."

„Wenn so schöne meine Augen,
Führt mich hier vom Tanze heim,
Alt und grau ist schon mein Bräut'gam
Und er holt uns nimmer ein."

Blanka

„Blanker seid Ihr, meine Herrin,
Blanker als der Sonne Strahl!
Einmal sorglos möcht ich schlafen
Ohne Waffen diese Nacht,
Denn wohl sieben lange Jahre
Legt' ich nicht die Rüstung ab,
Dunkler schon als ruß'ge Kohlen
Ist mein junger Leib vom Stahl."

Ruhet diese Nacht nur, Ritter,
Schlaft entwaffnet ohne Arg,
Denn der Graf ist fern im Walde,
Jagend über Berg und Tal.
Wollt', der Sturm zerriss' die Hunde
Und der Adler ihm den Falk,
Und die Berg', im Grunde wankend,
Stürzten ihn vom Fels herab!

Drauf, heimkehrend aus dem Walde,
Trat ins Zimmer ihr Gemahl:
„Was hier einsam sinnt Ihr, Dame?

Euer Stamm ist voll Verrat." –
„Herr, ich kämme meine Locken,
Kämme sie mit großem Gram,
Weil Ihr so allein mich lasset,
Draußen schweifend auf der Jagd." –
„Diese Worte, schöne Blanka,
Haben einen falschen Klang,
Wessen ist das Roß im Hofe,
Dessen Wiehern dort erschallt?" –

„Meines Vaters Rößlein ist es,
Das er Euch geschickt zur Jagd." –
„Wessen sind die blanken Waffen,
Die ich leuchten sah im Gang?" –
„Herr, 's sind meines Bruders Waffen,
Euch hat er sie heut gesandt." –
„Wessen ist die fremde Lanze,
Die dort herblinkt von der Wand?" –
„Nehmt sie rasch und stoßt mich nieder,
Das verdien' ich, guter Graf!"

Die Jungfrau und der Ritter

Eine Jungfrau wandert' einsam
In dem wunderschönen Frankreich,
Gen Paris sie wollte ziehen,
Wo die Eltern ihrer harrten;
Von den Ihren abgekommen,
Hatt' sie sich verirrt im Walde,
Lehnte sich an eine Eiche,
Andre Wandrer abzuwarten.

Kam ein Ritter da geritten,
Gleichfalls gen Paris er trabte.
„Wenn es Euch beliebt, Herr Ritter,
Nehmt mich mit aus diesem Walde." –

„Herzlich gerne, schöne Herrin!"
Und, ihr höflich aufzuwarten,
Sprang der Ritter von dem Rosse,
Hob hinauf sie, in den Sattel
Drauf sich selber zu ihr schwingend.

Aber als sie so im Walde
Einsam ritten, da begann er
Ihr verliebt den Hof zu machen.
„Hüt' dich, Ritter, sei nicht schändlich,
Ein Todkranker war mein Vater
Und verpestet meine Mutter,
Siech und elend müßt' verschmachten,
Wer mich frevelhaft berührte." –
Und der Ritter schwieg erblassend.
Aber in Paris am Tore
Still in sich die Jungfrau lachte.

„Warum lacht Ihr, schöne Herrin?" –
„Über den feigen Ritter lach' ich,
Der sein Mädchen hat im Freien
Und nichts macht als Redensarten!"

Voller Scham sprach da der Ritter:
„Kehrt' noch einmal um zum Walde,
Habe draußen was vergessen."
Doch die schlaue Jungfrau sagte:
„Nimmer kehr' ich um, und tät' ich's,
Keiner doch wagt's, mir zu nahen,
Denn ich bin die Tochter Frankreichs,
Und der König ist mein Vater,
Und wer meinen Leib berührte,
Müßt's mit seinem Kopf bezahlen."

Herkules' Haus

König Rodrich in Toledo,
Seiner Krone Glanz zu mehren,
Ließ ein groß' Turnier verkünden.
Hell schon die Trompeten schmettern,
Sechzigtausend Ritter kamen,
Die zu kämpfen dort begehrten.
Doch bevor der Kampf begonnen,
Zu ihm die Toleder treten
Bittend, daß er Tor und Riegel
Woll' mit neuem Schloß versehen
An des Herkules Palaste,
Wie's bisher der Brauch gewesen.
Aber in dem alten Hause
Dacht' er, reichen Schatz zu heben,
Ließ die Riegel all' zerbrechen
Und des Tempels Tore sprengen.

Als er eintrat, war's so still drin,
Nur ein Spruch glänzt ihm entgegen:
Weh dir, Rodrich, denn der König,
Der betreten diese Schwelle,
Der gebrochen diese Stille,
Wird Hispanien versengen!"
Seitwärts hinter einem Pfeiler
War ein prächt'ger Schrank zu sehen,
Drinnen lagen fremde Banner
Mit Figuren zum Erschrecken,
Und Araber, hoch zu Rosse,
Funkelnd mit gezückten Schwertern,
Hielten an dem Schrein die Wache,
Lautlos, ohne sich zu regen. –
Rodrich wandt' sich vor Entsetzen,
Wollt' fortan nichts weiter sehen,
Und ein Blitzstrahl zuckt' vom Himmel
Und verbrannt' den Zaubertempel.

Übers Meer wohl sandt' er Kriegsvolk,
Sollten Afrika erwerben,
Wetter stiegen, wo sie fuhren,
Mußten all' im Meer verderben.

Donna Urraca

Schon in Trümmern lag Zamora,
Das der stolze Cid umzingelt,
Auf den Turm da trat Urraca,
Rief von den zerschoßnen Zinnen:
„Übermüt'ger Cid da drunten,
Solltest dich der Zeit erinnern,
Da am Altar von Sankt Jago,
Sie geschlagen dich zum Ritter!
An dem Tage gab mein Vater
Waffen dir zum Angebinde,
Meine Mutter gab dein Roß dir.
Wie so fein die Sporen klingen!
Ich hab' dir sie umgebunden –
Damals schien's, wir schieden nimmer,
Anders wollten's meine Sünden,
Anders wandten's die Geschicke:
Mit Ximene von Lozano
Tauschtest treulos du die Ringe.
Schlecht gezielet, Don Rodrigo!
Höh'res Ziel war dir beschieden,
Kron' und Reich, die ich dir brachte,
Gabst du hin für Silberlinge
Und verlorst die Königstochter,
Um die Magd dir zu gewinnen!"

„Auf mein Volk", rief da der Ritter,
„Auf und wendet euch von hinnen!
Denn ein Pfeil dort durch die Lüfte
Schwirrte von des Turmes Zinnen,

Ohne Eisen war die Spitze,
Hat mir doch das Herz zerrissen,
Und kein Heilkraut gibt's auf Erden,
Muß fortan nun trostlos irren!"

Durandartes Abschied

Durandarte, Durandarte,
Ritterlich in Lust und Streit,
Bitt' dich, laß uns einmal plaudern
Wieder von der alten Zeit.

Denkst du noch der schönen Tage,
Wo du mir dein Herz geweiht,
Und in Sang und Ritterspielen
Vor der Welt um mich gefreit?

Wie viel Mohren warfst du nieder,
Rief ich zum Turniere dich!
Fast kenn' ich dich jetzt nicht wieder,
Sag, warum vergaß'st du mich? –

„Schmeichelnd klingen solche Worte
Und verlockend ist die Huld,
Aber wenn mein Herz sich wandte,
Euer, Dame, ist die Schuld."

„Wohl weiß ich's, für Gaiferos
Waret Ihr in Lieb' entbrannt,
Als ich trostlos und geächtet
Irrte fern im fremden Land."

„Drum, wenn Ihr von Lieb' jetzt redet,
Habt Ihr's weislich nicht bedacht,
Denn um nicht die Schmach zu tragen,
Wend' ich mich in Todesnacht."

Durandartes Tod

O Belerma, o Belerma,
Du geboren mir zum Unheil!
Sieben Jahr' dient' ich dir treulich,
Hab' mir doch kein' Lieb errungen,
Und jetzt, da du mich erhörtest,
Muß ich in der Schlacht verbluten.
Nicht die Todesstimmen fürcht' ich,
Wenn sie auch so früh mich rufen,
Darum nur ist Tod so bitter,
Weil er mir dein Bild verdunkelt.
O mein Vetter Montesinos,
Wenn sich meine Seel' entschwungen,
Bringt mein Herze zu Belerma,
Wollt ihr meinetwegen huld'gen,
Bitten, daß sie mein gedenke,
Der so treu um sie gerungen.
Gebt ihr alle meine Länder,
Die ich freudig einst bezwungen;
Da mein' Lieb' nun untergehet,
Sei all' Gut mit ihr versunken! –
Montesinos, Montesinos,
Heiß brennt diese Lanzenwunde,
Müde schon ist meine Rechte,
Aus viel' Quellen hier verblut' ich,
's wird so kühl nun – ach die Augen,
Die uns ausziehn sahn so mutig,
Sehn uns nimmermehr in Frankreich. –
Drückt noch einmal an die Brust mich,
Vetter, denn ich sprech' verworren
Und vor meinen Augen dunkelt's,
Euch befehl' ich all mein Sorgen
Und vertraue Eurem Schwure,
Denn der Herr, an den Ihr glaubet,
Höret uns in dieser Stunde.

Tot nun ruhet Durandarte
In dem stillen Felsengrunde,
Weinend löst ihm Montesinos
Helm und seiner Rüstung Gurte,
Löst sein Herze für Belerma
Mit dem Dolche aus der Brust ihm
Und begrub ihn unterm Felsen,
Sprach dabei aus Herzensgrunde:
„O mein Vetter Durandarte,
Tapfrer Degen, Herzensbruder,
Was soll ich fortan auf Erden,
Da die Mohren dich erschlugen!"

Donna Alda

In Paris saß Donna Alda,
Rolands Braut, im hohen Saal
Und mit ihr dreihundert Damen,
Ihrer Gespielinnen Schar;
Alle waren gleich beschuhet,
Alle trugen gleich Gewand,
Aßen rund um eine Tafel
Von demselben Brot zumal,
Donna Alda ausgenommen,
Weil sie ihre Herrin war,
Hundert spannen goldne Fäden,
Hundert woben Tepp'che zart,
Hundert aber musizierten,
Sie zu trösten mit Gesang.

Donna Alda war entschlummert
Bei der Instrumente Klang,
Plötzlich fuhr sie auf, lautschreiend,
Daß man's hört' bis in die Stadt.

Zu ihr sprachen da die Jungfraun:
„Wer tat Euch was Schlimmes an? –"

„Einen Traum hatt' ich, ihr Mädchen,
Der mir großen Schrecken gab:
Einsam im Gebirge stand ich,
Durch die Öde flog ein Falk',
Hinterdrein ein junger Adler,
Drängend ihn in wilder Jagd,
So geängstigt stürzt' der Falke
Flüchtend sich in mein Gewand,
Doch der Aar mit seinen Fängen
Hatt' ihn zornig schon umkrallt,
Riß den Falken mir in Stücke,
Streut' die Federn übern Plan."

Drauf zu der erschrocknen Herrin
Eins der Kammerfräulein sprach:
„Diesen Traum will ich Euch deuten:
Euer Bräut'gam ist der Falk',
Der sich übers Meer verflogen,
Eure Schönheit ist der Aar,
Der den wilden Edelfalken
Sich im Flug gefangen hat,
Und das Hochgebirg die Kirche,
Wo man traut Euch am Altar. –"
„Reichlich wohl will ich dir's lohnen,
Liebes Mädchen, sprichst du wahr."

Kam ein Brief am andern Morgen,
Drin mit Blut geschrieben war,
Daß ihr Roland war gefallen
In der Schlacht von Roncesval.

Das Waldfräulein

Falke war im Wald verflogen
Und die Hunde irrten weit,
Jagdmüd' lehnt' an eine Eiche

Sich der Ritter im Gestein,
Eine Jungfrau da erschrocken
In des Wipfels Dunkelheit
Sah er stehen, ihre Locken
Rings umgaben Stamm und Zweig.
„Staune nicht und laß dein Grauen,
Bin ein Königstöchterlein,
Sieben Zauberfrau'n mich haben
Auf der Amme Schoß gefeit,
Daß ich sieben Jahr' muß wohnen
Hier in Waldeseinsamkeit.
Sieben Jahr' sind heut verflossen
Oder morgen um die Zeit,
Bitte dich um Gottes willen,
Führ' mich aus dem Walde heim,
Will als Ehefrau dir dienen,
Oder auch dein Liebchen sein."

„Fräulein, noch bis morgen frühe
Harret in dem Walde mein,
Hab' zu Haus 'ne weise Mutter,
Will erst fragen, was sie meint." –
Sie vom Baum rief: „Weh dem Ritter,
Der die Jungfrau läßt allein!"
Er ritt fort, sie blieb im Walde,
Mutter riet, er sollt' sie frein.
Als er morgens kehrt' zurücke,
War's so stille im Gestein,
Konnt' den Baum nicht wiederfinden,
Aber weit, vom Walde weit
Sah er ziehn ein Fähnlein Reiter,
Führten fort das Waldfräulein;
Und er stürzt zu Boden nieder
In der grünen Einsamkeit:
„Schwer' Gericht verdient der Ritter,
Der verloren solche Maid!
Ich will selbst den Stab mir brechen,

Ich will selbst mein Richter sein,
Abhau'n soll man mir die Rechte
Und mich schleifen durch die Heid'!"

Weh Valencia!

Eingeschlossen war Valencia,
Konnte kaum sich länger wahren,
Weil sich die Almoraviden
Zögernd nicht zum Beistand wandten.
Da dies sah ein alter Maure:
Auf des höchsten Turmes Warte
Stieg er schweigend da, noch einmal
Zu beschauen Stadt und Lande.
Und wie sie herauf so leuchten,
Brach das Herz ihm bei dem Glanze;
Gramvoll mit prophet'schem Munde
Also von dem Turme sprach er:
„O Valencia, o Valencia,
Würd'ge Herrscherin der Lande,
Deine heitre Pracht muß sinken,
So sich Gott nicht dein erbarmet!
Die vier Felsen, drauf du thronest,
Würden, wenn sie könnten, klagen,
Deine festen Mauern seh' ich
Von dem wilden Anlauf wanken,
Deine Türme, die so trostreich
Über Land und Völker ragen,
Werden unaufhaltsam stürzen,
Deine Zinnen, gleich Kristallen,
Ihren Wunderglanz verlöschen,
Und dein mächt'ger Guadalaviar
Wird aus seinen Ufern steigen,
Trüben jeden Bach im Lande.
In den trocknen Wasserkünsten
Funkeln nimmermehr die Strahlen,

Rings in deinen schönen Gärten,
Die fortan verwildernd ranken,
Werden Hirsche einsam grasen,
Alles fröhl'che Grün zernagend,
Keinen Duft mehr haucht die Luft her,
Wo viel' tausend Blumen standen,
Muß das Glühen all verblühen;
Wo jetzt Schiffe kommen, fahren,
Liegt verödet Strand und Hafen,
Und vom weiten Bergeskranze,
Den du mächtig einst beherrschtest,
Schlagen blutrot auf die Flammen,
Daß das Qualmen dich erblindet
Rings von deiner Länder Brande,
Bis, als eine Todeswunde,
Alles Volk hat dich verlassen. –
O Valencia, o Valencia,
Helf' dir Gott in jenen Tagen!
Oft schon hab' ich es verkündet,
Was ich weinend jetzt beklage."

Gedicht-Nachlese

Eine Auswahl

Frühe Gedichte

(Bis zum Erscheinen von „Ahnung und Gegenwart“)

Der Morgen

Sei mir gegrüßt, o Morgensonne!
Erquickend strömt dein Strahlenhaupt
Des neuen Lebens Frohgefühle
Schnell über eine halbe Welt!

Wie einst des Abgrunds schwarze Nächte
Als Licht des Schöpfers Hand entrann,
Erstaunten Welten sich entwälzten,
Und Ordnung aus dem Chaos stieg;

So windet mir aus düstern Nebeln
Auch jetzt sich Flur und Stadt und Dorf,
Bis, ganz entschleiert, meinen Blicken
Das bunt Gemisch entgegenlacht.

Wie rings um mich die Täler dampfen;
Es rauscht der Wald, es tönt der Fluß,
Das Gräschen lispelt – Gott – dein Lob!
Und ich – ich bete, danke nicht?

Einst, ach! wer fasset den Gedanken?
Wenn meines Lebens Genius winkt,
Strahlt so mir der Vollendung Sonne
Des ew’gen Tages Morgenrot.

Da schwindet Nebel – jedes Wölkchen,
Das sonst den Späherblick gehemmt,
Und in das Wonnemeer des Urlichts
Taucht sich der fesselfreie Geist.

O, schwebe bald herab vom Himmel!
Du schöner Tag! – zum Vater mich,
In meine Heimat mich zu schwingen.
Ich zage nicht, o komm, o komm!

Beim Erwachen

An M. H.

Tiefer ins Morgenrot versinken die Sterne alle,
Fern nur aus Träumen dämmert dein Bild noch vorüber,
Und weinender tauch' ich aus seliger Flut. –
Aber im Herzen tief bewahr' ich die lieben Züge,
Trage sie schweigend durch des Tages Gewühle
Bis wieder zur stillen träumenden Nacht. –

An Isidorus Orientalis

Zu den Sonetten an Novalis

Erwartung wob sich grün um alle Herzen
Als wir die blaue Blume sahen glühen,
Das Morgenrot aus langen Nächten blühen –
Da zog Maria ihn zu ihrem Herzen.

Die Treuen schlossen sich in tausend Schmerzen,
Erfüllung betend wolltn sie ewig knieen;
Da sahn sie neuen Glanz die Blumen sprühen,
Ein Kind stieg licht aus ihrem duft'gen Herzen.

Solch Glühen muß der Erde Mark durchdringen,
In Flammen alle Farben jauchzend schwingen,
Ein gotterklungner unermeßner Brand!

Wie ruft es mich! – Reich fester mir die Hand –
Hinunter in den Opfertod zu springen!
Du wirst uns all' dem Vater wiederbringen! –

An I – –

Von trüber Bangnis war ich so befangen,
Da sprach Waldhorn zu mir aus grünen Weiten:
Mir nach! durch unbekannte Lande schreiten!
Rief immer fern und fern – konnt's nie erlangen.

Wo führst mich endlich hin? sprach ich voll Bangen,
Weit Freund' und Welt von diesen Einsamkeiten!
Da klang es fern und nah wie alte Zeiten,
Dich sah ich fröhlich stehn am Bergeshange.

Und unten lag ein weites Land so helle,
War aufgetan die ew'ge Farbenquelle,
Nach Osten sah man fromme Pilger ziehen.

So nimm nur alles, was ich lieb' und habe,
Gern laß' ich ja die Welt und ihre Gabe,
Mit dir nur, Liebster, will ich ewig ziehen!

Antwort

An H. Gf. v. Loeben

Demütig kniet ich vor der Jungfrau Bilde,
Erflehend nur ein einzig Liebeszeichen,
Das nicht in Angst und Pein möcht' von mir weichen.
Sie gab mir – Mut und Andacht milde.

Nun drängt ein Schmerz mich süß und sanft und wilde,
Daß ich mir ihrer Wunder Himmelreichen,
Die weiter als mein ird'sches Leben reichen,
Wie ich sie himmlisch schau', die Schöne bilde.

Mir fehlen Töne noch und Himmelsfrieden;
Dir ward Erfüllung frühe schon beschieden,
Dein Himmel ist, wo zauberte dein Beten.

Hast du den höchsten Wunsch mir nun genommen,
Werd' ich demutsvoll wieder vor dich treten;
Eins sein mit dir, kann nur allein mir frommen.

Der Lenz mit Klang und roten Blumenmunden

Der Lenz mit Klang und roten Blumenmunden,
Holdsel'ge Pracht! wird bleich in Wald und Aue;
Tonlos schweift' ich damals durchs heitre Blaue,
Hatt' nicht das Glühn im Tiefsten noch empfunden.

Da sprach Waldhorn von übersel'gen Stunden,
Und wie ich mutig in die Klänge schaue,
Reit't aus dem Wald die wunderschöne Fraue,
Oh! Niederknien, erst's Aufblühn ewiger Wunden!

Zu weilen, fortzuziehn, schien sie zu zagen,
Verträumt blühten ins Grün der Augen Scheine,
Der Wald schien schnell zu wachsen mit Gefunkel.

Aus meiner Brust quoll ein unendlich Fragen,
Da blitzten noch einmal die Edelsteine,
Und um den Zauber schlug das grüne Dunkel.

Aussichten

Es will der Morgen sich von weitem zeigen,
Das dunkle Meer im Innern still erglühen,
Erwartungsvoll die reinen Segel blühen,
Doch deckt noch all' geheimnisvolles Schweigen.

Wird erst die Sonne auf die Berge steigen,
Gewaltig Licht in alle Lande blühen,
Sieht man ein frei Geschlecht nach Angst und Mühen
In stolzer Demut fromm die Kniee beugen.

Unendlich' Wunderfernen sind gelichtet,
Unzählig' Lieder himmelwärts auflangen,
Daß treue Liebe Gegenlieb erreiche. –

Wer frei geboren, ist schon längst geflüchtet,
Die andern faßt ein unaussprechlich Bangen,
Der Sieger zieht zum alten ew'gen Reiche.

Angedenken

1

Sie band die Augen mir an jenen Bäumen,
„Geh schöner Blinder!" sagt' sie dabei sachte,
Wußt' nicht, wie Wunden süß dies Flüstern brachte,
Und stieß mich in des Spieles wogend' Schäumen.

Nun in der Augen Nacht quoll blühend Träumen,
Der Mienen Huld, wie Zauberblum'n, erwachte,
Da end't das Spiel, ins Aug' Licht wieder lachte,
Doch sehend träumt' ich fort von jenen Träumen.

So stand ich unter holden Farbenbogen,
Und wie mein ganzes Leben schwellend blühte,
Dankt' ich dem Lenz für solch zaubrisch Verschönen.

Noch blüht der Lenz, doch sie ist fortgezogen,
Nun weiß ich, daß nur sie den Lenz beglühte,
Und einsam traur' ich in den Strahlen, Tönen.

2

Wie wenn aus Tänzen, die sich lockend drehten,
Von müder Augen süßen Himmelsträumen,
Daß nun Gewährung nicht wollt' länger säumen,
Verratend die schamhaften Schleier wehten,

Ein einz'ger in die Nacht hinausgetreten,
Schauend wie draußen Land und Seen träumen,
Die Töne noch verklingen in den Bäumen,
Ans Herz nun schwellend tritt einsames Beten:

Also, seit du erhörend mich verlassen,
Grüßt mich Musik und Glänzen nur von ferne,
Wie Tauben, Botschaft bring'nd durch blaue Lüfte.

Nacht legt sich um die Augen hold, die nassen,
Als Blume sprieß' ich in die Klänge, Sterne,
Der goldnen Ferne hauchend alle Düfte.

Der Schiffer

Du schönste Wunderblume süßer Frauen!
Ein Meer bist du, wo Flut und Himmel laden,
Fröhlich zu binden von des Grüns Gestaden
Der Wünsche blühnde Segel voll Vertrauen.

So schiffend nun auf stillerblühten Auen,
In Lockennacht, wo Blicke zaub'risch laden,
Des Munds Koralln in weißem Glanze baden,
Wen füllt' mit süßem Schauer nicht solch' Schauen!

Viel hab ich von Sirenen sagen hören,
Stimmen, die aus dem Abgrund lockend schallen
Und Schiff und Schiffer ziehn zum kühlen Tode.

Ich muß dem Zauber ew'ge Treue schwören,
Und Ruder, Segel laß' ich gerne fallen,
Denn schönres Leben blüht aus solchem Tode.

Die Wunderblume

Es war die Nacht so wunderbar, so schwüle,
Weit ab wohl lagen dunkle Länder viele,
Die Ströme hört' ich ferne gehen,
Doch, wo ich war, konnt' ich nicht sehen.

Und ferne sah ich aus dem grauen Schweigen
Seltsam verschlungne Wunder dunkel steigen,
Stumm gehen in den Finsternissen –
„Ach! Sind es Berge, sind es Riesen!"

Aus solchen Ängsten wollt' mein Herz verlangen,
Nie fühlt' ich noch so unaussprechlich' Bangen.
„Wann wird der Morgen endlich röten?
Oh! Jesus hilf aus tiefsten Nöten!"

Und wie ich rief, sah ich fern Funken sprühen,
Ein Wunderglänzen aus der Nacht erblühen,
Und eine Blume drin erhoben,
Aus milden Flammen bunt gewoben.

Und wundersüße Scheine sandten
Die Blätter bald nach allen Strömen, Landen,
Rings wurd' es weit und immer weiter,
Der Himmel blau, die Erde heiter.

Wie weit liegt alle Bangigkeit dahinten!
Es wollen brünstig mich die Scheine zünden,
Frisch bluten alle Liebes-Wunden,
Verbrennt mich nur! – Bin euch ja längst verbunden!

Frühlingsandacht

In Lust und Scherzen drehn sich leichte Tage,
Von weißen Armen ruhet Lieb' umwunden,

Der Sänger schweift allein in Waldesgrunde,
Nur Waldhorns-Klang will, was er sucht, ihm sagen.

Es bringt der Lenz so glänzend' Spiel getragen,
Durchs farb'ge Land die Ströme hell gewunden,
All' bunte Schifflein wieder losgebunden!
So zieh' doch fröhlich mit! – Wer wollt' noch zagen?

Doch daß im bunten, lichten Tanz des Maien
Der Einz'ge nur allein nicht länger weine,
Sieht er als Blume sich den Lenz erschließen;

Und aus dem duft'gen Kelch im Glorienscheine
Neigt sich die ew'ge Jungfrau, hebt den Treuen
An ihre Mutterbrust mit tausend Küssen.

Sonett

Rasch sprengt der Ritter an ertosnden Flüssen
Funkelnd durch Waldes dunkelgrüne Dichten,
Die schlanker sich in Himmelsglanz aufrichten,
Den König kühlerauschend zu begrüßen.

Viel' schöne Augen werden weinen müssen,
Daß er Visier und Locken nie will lichten,
Daß zu dem Hohen sie sich mußten richten,
In süßen Himmelsqualen gerne büßen.

Schön ist's, von ird'schen Banden losgebunden,
In grüner Nacht, in dunkler Wetter Blicke
Einsam den Sinn zu weidn, den wilden, reinen;

Doch Schönres wird auf Erden nicht gefunden,
Als wenn der Stolze senkt die dunklen Blicke,
Sanft niederknieend vor der Ersten, Einen.

Wohl kann ich, wie die andern, tun und lassen ...

Wohl kann ich, wie die andern, tun und lassen,
Auf kurze Frist von ird'schem Wahn befangen,
Mitspielen ohne Klage und Verlangen,
Manch' Mädchen will mich nicht vom Herzen lassen.

Die Erde seh' ich schauernd süß erblassen,
Den Himmel überschwenglich aufgegangen,
Da faßt micht alte Liebe, altes Bangen,
Weiß nicht, soll ich das Kreuz, die Fahne fassen.

Es stürzt der Bach, hoch brausen Waldeswipfel,
Durch fliehnde Wolken Waldhornsklang geflogen,
Und wenn der Blitz die grimme Nacht durchzücket,

Sehn fern die Furchtsamen auf steilem Gipfel
Den Fremdling knien, auf das Schwert gebogen,
Das zornigleuchtend aus dem Dunkel blicket.

Madrigal

O Strom auf morgenroten Matten!
Rubin, smaragden deine Wellen,
Dann in des blauen Mittags schwülen Schatten,
Rauschend in Abendglanz versunken,
Bis du der Nächte Licht getrunken,
So muß mein Leben rastlos quellen,
Sich selber lauschend oft das sel'ge Herze,
O Liebe süß, o Lust im Schmerze!

Bin ich denn nicht auch ein Kind gewesen?

Bin ich denn nicht auch ein Kind gewesen? –
Spielte goldne, goldne Stunden,
Unbekannt noch mit dem Bösen,
Furchtlos an dem finstern Schlunde.

Spielt' so lang am Felsenrande,
Sah viel Ströme unten fließen,
Fromme Pilger ziehn im Lande –
Doch mich wollte niemand grüßen.

Denn so wild war schon mein Spielen,
Und es zuckten furchtbar munter
Schon im Aug die Flammen kühle,
Die zum Abgrund langten runter.

Und ich wandt' mich ohne Klage
Stumm im Zorne von dem Lichte,
Hinter mir die bunten Tage,
Wob die Nacht sich um mich dichte.

Und durchs Dunkel flogen Blitze
Und die flamm'nden Fahnen wehten;
Keck sucht' ich des Felsen Spitze
Mußte fluchen nur, statt beten.

Minnelied

Blaue Augen, blaue Augen!
Ach! wie gebt ihr süße Peine!
Aus dem schönen Wald unzählig
Stimmen zielen, grüne Scheine,
Und ich laß' mich gern verführen;
Locken Schmerzen so von weitem.
Draußen auf der Waldeswiese

Laß' ich wohl mein Rößlein weiden,
Sinnend steh' ich lang' daneben,
Grüßt mich wie aus fremden Zeiten
Waldesrauschen, Lied der Bronnen,
Ewigblühend grünes Schweigen,
Aus der tiefsten Brust Erinnern
Lang' verlorner Himmelsweide –
Und ich muß dann immer klagen,
Ewig fragen: wo Sie weile?
Und das Waldhorn will mir's sagen,
Ewig will das Herz nur weinen:
Süße Peine! blaue Augen! –
Ewig stehst du in der Weite,
Blühend in den blauen Tagen.
Wolken durch den Himmel eilen,
Liebesblick kömmt oft geschossen,
Dann erglänzen Kelch' und Zweige;
Doch die Klarheit schließt sich wieder,
Und ich stehe so alleine,
Und ich kann mich gar nicht retten
Vor den Freuden, vor den Leiden,
Und ich kniee und ich bete:
Schöne Fraue, liebe, reine!
Blaue Augen, blaue Augen,
Ach! wie gebt ihr süße Peine!

Sehnsucht

Selig, wer zur Kunst erlesen,
Ruhig in getreuer Lust,
Hoher Dinge seltsam' Wesen,
Selber froh erschreckt, mag lesen
In der Wunder vollen Brust!

Wie die Rosse mutig scharrten!
Ach, die Freunde sind voraus!

Draußen blüht der schöne Garten,
Draußen Wald und Liebchen warten,
Und ich kann nicht, kann nicht raus!

Bleib ich ewig fern vom Glücke? –
Wen die Treue ganz durchdrang,
Einmal trafen Liebesblicke,
Ach! er kann nicht mehr zurücke,
Und ich kniee Lebens lang.

Lodert, lodert heil'ge Kerzen!
Bleibet unerhört mein Flehn:
Will ich in den Freuden, Schmerzen,
Mit dem unentweihten Herzen
Treu und heilig untergehn.

Minnelied

Über blaue Berge fröhlich
Kam der bunte Schein geflossen,
In den Schimmer rief ich selig:
„Freu' dich nur, jetzt wird's vollendet!"
Doch der Frühling ist vergangen,
Was ich innigst hofft' und strebte,
Blieb ein unbestimmt Verlangen.

Und nach langem trübem Schweigen
Kamen goldne Tage wieder,
Blaue Berge, alte Zeiten,
Blumen, Sterne, Ström' und Lieder
Woben wunderbar ein Netze,
Schüchtern schlang sich's um die Glieder,
Zog so innig fest und fester
Mich ans Herz der Erde nieder,
Und in diesem Netz die Blüte
Ward zum himmlischen Gefieder.

Der arme Blondel

Wie sie in den Blumentagen,
Über mir mit rotem Munde,
Daß die Locken mich umwunden,
Mich verführt aus Herzensgrunde,
Wollt' es immer, konnt's nie sagen!
Fortgezohn ist nun die Eine,
Weine, armer (Blondel), weine!

Nachts die Berge stille stehen,
Ferne Schlösser, Strom und Bäume
Sahn mich seltsam an, wie Träume.
Drüber Wolken schnelle gehen,
Fest am Herzen steht die Eine,
Weine, armer (Blondel), weine!

Herbstliedchen

Flog Waldvögelein über den See,
Lieb' grüne Zeit, lieb' grüne Zeit!
Es zogen die Wolken: Ade! Ade!
Wir fliegen mitsammen gar weit, gar weit!

Es schaut Feinsliebchen vom hohen Saal,
Fern zohe der Ritter im grünen Tal;
Waldvöglein sang immerfort: Ade! –
Das tat Feinsliebchen im Herzen so weh.

Es wächst und strömt in ewigen Gedichten

Es wächst und strömt in ewigen Gedichten
Jauchzend im Innersten das freie Leben;
Des Tempels strahlnde Säulen klingend beben,
Unübersehbar will sich's himmlisch lichten.

Den heil'gen Kampf sie irdisch möchten schlichten,
Er spült sonst mit sich fort ihr schwankes Leben;
Die Arme wollen sie nicht gläubig heben,
Zur Nacht kein Herz, nicht Lieb' sich aufzurichten.

Es bäumt das Roß in zorn'gem Mut sich raffend,
Durch eure Netze funkeln Schwert und Lanze,
Bricht Liebesblick aus tiefer, ew'ger Bläue.

Und wie ihr stehet, euch verwundernd, gaffend,
Glüht ferne Helm und Speer im Morgenglanze,
Und über die Berge sprenget froh der Freie.

Gebet

Gott, inbrünstig möcht' ich beten,
Doch der Erde Bilder treten
Immer zwischen dich und mich,
Und die Seele muß mit Grauen
Wie in einen Abgrund schauen,
Strenger Gott, ich fürchte dich!

Ach, so brich auch meine Ketten!
Alle Menschen zu erretten,
Gingst du ja in bittern Tod.
Irrend an der Hölle Toren,
Ach, wie bald bin ich verloren,
Hilfst du nicht in meiner Not!

Sei stark, getreues Herze ...

Sei stark, getreues Herze!
Laß' ab von Angst und Schmerze!
Steh' auf und geh' mit mir,
Viel' Freude zeig' ich dir.

Die Lerchen jubilieren,
Und fröhlich musizieren
Aus grünem frischem Wald,
Von Stimmlein mannigfalt.

Geschmückt mit Edelsteinen
Die Erd' in bunten Scheinen
Als junge fromme Braut
Dir froh ins Herze schaut.

Im Garten zu spazieren
Die Blumen mich verführen,
Die Augen aus dem Grün,
Das Quellen und das Blühn.

Maria, schöne Rose!
Wie stünd' ich freudenlose,
Hätt' ich nicht dich ersehn
Vor allen Blumen schön.

Nun laß' den Sommer gehen,
Laß' kommen Wind' und Schneen;
Bleibt diese Rose mein,
Wie könnt' ich traurig sein?

An ...

Was lebte, rollt' zum Himmel aus dem Tale,
Des Ritters Mut, Gesanges feur'ge Zungen,
Und aus den Felsen Münster kühn geschwungen,
Das Kreuz erhebend hoch im Morgenstrahle.

Versunken sind die alten Wundermale,
Nur eine Waldkapelle unbezwungen,
Blieb einsam stehen über Niederungen,
Die läutet fort und fort hinab zum Tale.

Was frägt die Menge, ob's der Wind verwehe –
Nur ein'ge trifft der Laut, die stehn erschrocken,
Und mahnend lockt's wie Heimweh sie zur Höhe.

Ein heitrer Greis zieht oben still die Glocken,
Reicht fest die Hand und führt aus der Verheerung
Durchs alte Tor die Treuen zur Verklärung.

Kaiser Alberts I. Tod

Romanze

Lebewohl noch schnell zu sagen,
Da der Tag zu graun begann,
Noch ein Mal der Kaiser trate
In den dunkelhellen Saal.

Sein Gemahl er dorten fande,
Die in bittern Tränen saß,
Halb bescheint vom Morgenstrahle,
Nahm sie herzlich noch in Arm.

„Zieh nur heute nicht von dannen,
Denn so blutrot ist der Tag
Überm Walde aufgegangen;
Dunkel geht noch durch das Tal.

Mutig steht empor der Jammer,
Wenn die Feldschlacht draußen ras't,
Heut' so grausig stummes Bangen
Aus der Nacht ans Herze langt."

„Also will das Herz verzagen,
Ist es draußen finstre Nacht;
Fraget fröhlich, was es bange,
Wenn der Morgen golden wacht.

Fern schon wehen meine Fahnen,
Aus dem Tal ruft Hörnerklang,
Deine Lieb' wird Gott bewachen,
Durch das Dunkel ficht der Mann."

Und es legte Helm und Panzer
Schnell nun Kaiser Albert an,
Stiege freudig auf den Rappen,
Funkelnd hoch im Morgenglanz.

Von dem Schloß, von der Altane
Weint sie lang hinaus ins Land,
Grüßt die Ziehnden in dem Tale
Noch vieltausend, tausendmal. –

Wie sie nun hinunter kamen
Tiefer in den dunkeln Wald,
Traten aus dem Wald Gedanken
Seltsam Kaiser Albert an.

Jetzo erst so ganz empfand er,
Jener Worte tiefe Kraft,
Diese Treu', das holde Bangen,
Ihres süßen Leibs Gestalt.

Und die Tränen linde drangen,
Und so gar betrübt er sang.
Da die Vögel lustig sangen,
Schloß und Berg versunken war:

„Wie so wunderschön die Matte!
Ist's doch, als ob Wald und Bach
Mir hier Seltsams wollten sagen,
Alles doch so unbekannt!

Mögen weiterziehn die andern,
Freudig grüßt von fern ihr Klang,

Ich will hier ein wenig rasten,
Denn so schwül wird dieser Tag."

Kaiser Albert! Kaiser Albert!
Bleib' zu dieser Stunde wach!
Stimmen gehen in dem Walde,
Näher schleichet schon Verrat. –

„Schönes Schloß, viel heitre Tage!
Schlummernd Rauschen, Vogelsang –
Wolken, über mir gegangen –
Schöner, grüner Wiesenplan." –

Und dort hat ihn überfallen
Böser Ritter dunkle Schar,
Herzog Johann war's von Schwaben,
Der sein eigner Neffe war.

Ferne wohl die Hörner klangen,
Irrend durch die Waldesnacht,
Euer Herre ist erschlagen
Auf dem grünen Wiesenplan.

Zum Abschied

An L. 1812

Wenn vom Gebirg' der Quell kommt hell geschossen,
Die Lerchen schwirrend sich ins Blaue schwingen,
Da fühlt die Seele in dem Rauschen, Singen,
Bald sei des Frühlings Wunderpracht erschlossen.

So schauend auch in deiner Brust das Sprossen,
Verborgner Quellen Gang und sehnend Ringen,
Jauchz' ich dir zu: es wird die Knospe springen,
Die deine Blüte neidisch hielt umschlossen.

Wer möchte nicht die weite Öde rühren,
Der ew'ge Winter auf den deutschen Auen,
Die lang' in dumpfer Trägheit ruhmlos ruhten?

Nur wen'ge will des Himmels Licht berühren,
Die mögen fromm den Frühling Gottes schauen,
Sich selig tauchen in die farb'gen Fluten.

Reiselied

So ruhig geh' ich meinen Pfad,
So still ist mir zumut;
Es dünkt mir jeder Weg gerad'
Und jedes Wetter gut.

Wohin mein Weg mich führen mag,
Der Himmel ist mein Dach,
Die Sonne kommt mit jedem Tag,
Die Sterne halten Wach.

Und komm' ich spät und komm' ich früh
Ans Ziel, das mir gestellt:
Verlieren kann ich mich doch nie,
O Gott, aus deiner Welt!

An Luise
1816

Ich wollt' in Liedern oft dich preisen,
Die wunderstille Güte,
Wie du ein halbverwildertes Gemüte
Dir liebend hegst und heilst auf tausend süße Weisen,
Des Mannes Unruh' und verworrnem Leben
Durch Tränen lächelnd bis zum Tod ergeben.

Doch wie den Blick ich dichtend wende,
So schön in stillem Harme
Sitzt'st du vor mir, das Kindlein auf dem Arme,
Im blauen Auge Treu' und Frieden ohne Ende,
Und alles laß' ich, wenn ich dich so schaue –
Ach, wen Gott lieb hat, gab er solche Fraue!

An die Freunde

Es löste Gott das lang' verhaltne Brausen
Der Ströme rings – und unser ist der Rhein!
Auf freien Bergen darf der Deutsche hausen
Und seine Wälder nennt er wieder sein.
So brach gewaltig und mit kühnem Grausen
Ein mächt'ger Frühling in die Welt herein,
Und alle sah man ringen, fechten, streben –
O Heldenlust, in solchem Lenz zu leben!

Nun ist der Friede wieder wohl gekommen,
Gesühnt ist manche Sünde vor'ger Zeit. –
Doch wird der Kampf nicht von der Welt genommen,
Solang' der Mensch sich ernstrem Streben weiht.

Es hat der Krieg den Funken kühn entglommen,
Das Schlechte stürzt er um im blut'gen Streit:
Das *Beßre* überm Schutte aufzuführen,
Muß sich nun Geister-Kampf lebendig rühren.

Nennt mir die Palme eures hohen Strebens!
Wollüst'ge Rast ist nicht des Lebens wert.
Nach Ruh' sehnt sich die Menschenbrust vergebens,
Erkämpft will sein, was hoher Sinn begehrt,
Ein Krieger bleibt der größre Mann zeitlebens,
Er kämpf' mit Rede, Büchern oder Schwert,
Und rechter Friede wird nur dort geschlossen,
Wo jedem Streiter seine Palmen sprossen.

Wüst ras't der Krieg, Land, Herzen, Städte brennen,
Der Tag, er kommt und scheidet blutigrot;
Doch spannt der Friede ab die tapfern Sennen,
Dann hüte dich, mein Volk, vor größrer Not!
Denn tiefres Wehe weiß ich noch zu nennen:
Erschlaffter Hochmut ist der Völker Tod.
Umsonst geflossen ist das Blut im Kriege,
Sind wir unwürdig selbst der hohen Siege.

So laßt uns unser Deutschland denn umstellen
Bewachend brüderlich in treuer Hut,
Mit Lehren, Rat und Sang die Herzen schwellen,
Daß es bewahre rein die heil'ge Glut,
Den Ernst, den es erkämpft in Bluteswellen,
Der Ehre Hort, Eintracht und festen Mut.
Friede dem Herd und ew'ger Krieg dem Bösen!
So mag uns Gott von aller Schmach erlösen.

Herbstklage

Herbstnebel ziehn über den Weiher,
Das ist recht des Todes Bild!
Und tagelang sinnet der Reiher
Am Ufer dort einsam wild.

Mein Liebchen das hat mich verlassen,
Die Freunde sind alle weit,
Und Garten und Wälder erblassen,
Und singen von tiefem Leid.

Verschneit liegt bald alles danieder,
Wir selber wir werden alt,
Und kennen einander nicht wieder,
Verkümmert, zerstreut und kalt.

Zum Wald denn! da raset lautschallend
Das Horn durch des Windes Schrei'n,
Da krachen die Wipfel und fallen
Zum Abgrund Strom, Baum und Stein.

Und Schneewolken jagt's übern Weiher,
Die Windsbraut singt ihren Gruß,
Rasch stürzt in den Sturm sich der Reiher –
Ach, daß ich hier stehen muß!

Memento

Solange Recht regiert und schöne Sitte,
Du schlicht und gläubig gehst in sichrer Mitte,
Da trittst du siegreich zwischen Molch und Drachen,
Und wo du ruhst, da wird ein Engel wachen.
Doch wenn die Kräft, die wir *„Uns selber"* nennen
Die wir mit Schaudern raten und nicht kennen,
Gebundne Bestien, wie geklemmt in Mauern,

Die nach der alten Freiheit dunkel lauern –
Wenn die rebellisch sich von dir lossagen,
Gewohnheit, Glauben, Sitt' und Recht zerschlagen,
Und stürmend sich zum Elemente wenden:
Mußt Gott du werden oder teuflisch enden.

Durch!

Laß' dich die Welt nicht fangen,
Brich durch, mein freudig Herz,
Ein ernsteres Verlangen
Erheb' dich himmelwärts!

Greif' in die goldnen Saiten,
Da spürst du, daß du frei,
Es hellen sich die Zeiten,
Aurora scheinet neu.

Es mag, will alles brechen,
Die gotterfüllte Brust
Mit Tönen wohl besprechen
Der Menschen Streit und Lust.

Und eine Welt von Bildern
Baut sich da auf so still,
Wenn draußen dumpf verwildern
Die alte Schönheit will.

Sängerglück

Herbstlich alle Fluren rings verwildern,
Und unkenntlich wird die Welt.
Dieses Scheidens Schmerzen sich zu mildern,
Wenn die Zauberei zerfällt,
Sinnt der Dichter, treulich abzuschildern

Den versunknen Glanz der Welt.
Selig' Herze, das in kühnen Bildern
Ewig sich die Schönheit hält!

Frisch auf!

Vergangen ist die finstre Nacht,
Des Bösen Trug und Zaubermacht,
Zur Arbeit weckt der lichte Tag;
Frisch auf, wer Gott noch loben mag!

Jagdlied

Wenn die Bergesbäche schäumen
Und der Mond noch schweigend wacht,
Zwischen Felsen rings und Bäumen
Wie ein Feenland von Träumen
Ruht die wunderbare Nacht.

Da wird bei der Wipfel Wehen
Recht das Herz dem Jäger weit,
Talwärts von den stillen Höhen
Läßt er Hörnerklang ergehen
Durch die schöne Einsamkeit.

Und er weckt die Götter alle,
Von dem Berg Aurora lacht,
Venus folgt dem mut'gen Schalle,
Doch Diana, sie vor allen
Stürzt hervor aus Waldespracht.

Aus der Büchse sprühend Funken!
Immer höher schwillt die Brust!
Wild und Jäger todestrunken
In die grüne Nacht versunken —
O du schöne Jägerlust!

Treue

Wenn schon alle Vögel schweigen
In des Sommers schwülem Drang,
Sieht man, Lerche, dich noch steigen
Himmelwärts mit frischem Klang.

Wenn die Bäume all' verzagen
Und die Farben rings verglühn,
Tannbaum! deine Kronen ragen
Aus der Öde ewiggrün.

Darum halt nur fest die Treue!
Wird die Welt auch alt und bang:
Brich den Frühling an aufs neue,
Wunder tut ein rechter Klang!

Die Heimat
An meinen Bruder

Denkst du des Schlosses noch auf stiller Höh'?
Das Horn ruft nächtlich dort, als ob's dich riefe,
Am Abgrund grast das Reh,
Es rauscht der Wald verwirrend aus der Tiefe –
O stille! wecke nicht! es war, als schliefe
Da drunten unnennbares Weh. –

Kennst du den Garten? – Wenn sich Lenz erneut,
Geht dort ein Fräulein auf den kühlen Gängen
Still durch die Einsamkeit
Und weckt den leisen Strom von Zauberklängen,
Als ob die Bäume und die Blumen sängen,
Von der alten schönen Zeit.

Ihr Wipfel und ihr Brunnen, rauscht nur zu!
Wohin du auch in wilder Flucht magst dringen:

Du findest nirgends Ruh'!
Erreichen wird dich das geheime Singen,
In dieses Sees wunderbaren Ringen
Gehn wir doch unter, ich und du! –

Sprüche

Laß nur die Wetter wogen!
Wohl übers dunkle Land
Zieht einen Regenbogen
Barmherzig Gottes Hand.

Auf dieser schönen Brücke,
Wenn alles wüst und bleich,
Gehn über Not und Glücke
Wir in das Himmelreich.

*

Hast du doch Flügel eben
Und das gewalt'ge Wort;
Halt' hoch dich über dem Leben,
Sonst geht's über dich fort.

*

Gleichwie auf dunklem Grunde
Der Friedensbogen blüht,
So durch die böse Stunde
Versöhnend geht das Lied.

*

Magst du zu dem Alten halten
Oder Altes neu gestalten,
Mein's nur treu und laß Gott walten!

Der Verirrte

Vor dem Schloß in den Bäumen es rauschend weht,
Unter den Fenstern ein Spielmann geht,
Mit irren Tönen verlockend den Sinn –
Der Spielmann aber ich selber bin.

Vorüber jag' ich an manchem Schloß,
Die Locken zerwühlet, verwildert das Roß,
Du frommes Kindlein im stillen Haus,
Schau' nicht nach mir zum Fenster hinaus.

Von Lüsten und Reue zerrissen die Brust,
Wie rasend in verzweifelter Lust,
Brech' ich im Fluge mir Blumen zum Strauß,
Wird doch kein fröhlicher Kranz nicht daraus!

Wird aus dem Schrei doch nimmer Gesang,
Herz, o mein Herz, bist ein irrer Klang,
Den der Sturm in alle Lüfte verweht –
Lebt wohl, und fragt nicht, wohin es geht!

Morgendämmerung

Gedenk' ich noch der Frühlingsnächte
Vor manchem, manchem Jahr,
Wie wir zusammen im Garten standen
Und unten über den Landen
Alles so still noch war.

Wie wir standen in Gedanken,
Bis eine Morgenglocke erwacht' –
Das ist alles lange vergangen;
Aber die Glocken, die da klangen,
Hör' ich noch oft bei Nacht.

In der Nacht

Wie rauscht so sacht
Durch alle Wipfel
Die stille Nacht,
Hat Tal und Gipfel
Zur Ruh' gebracht.
Nur der Mensch in Träumen
Sinnt fort, was er bei Tag gedacht,
Weiß nichts von dem Lied in den Bäumen
Und von des Himmels Pracht,
Der in den stillen Räumen
Über allen wacht.

Entschluß

Noch schien der Lenz nicht gekommen,
Es lag noch so stumm die Welt,
Da hab' den Stab ich genommen,
Zu pilgern ins weite Feld.

Und will auch kein' Lerch' sich schwingen,
Du breite die Flügel, mein Herz,
Laß' hell und fröhlich uns singen
Zum Himmel aus allem Schmerz!

Da schauen im Tale erschrocken
Die Wandrer rings in die Luft,
Mein Liebchen schüttelt die Locken,
Sie weiß es wohl, wer sie ruft.

Und wie sie noch stehn und lauschen,
Da blitzt es schon fern und nah,
All' Wälder und Quellen rauschen,
Und Frühling ist wieder da!

Spruch

Drüben von dem sel'gen Lande
Kommt ein seltsam Grüßen her,
Warum zagst du noch am Strande?
Graut dir, weil im falschen Meer
Draußen auf verlornem Schiffe
Mancher frische Segler sinkt,
Und von halbversunknem Riffe
Meerfei nachts verwirrend singt?
Wagst du's nicht draufhin zu stranden,
Nimmer wirst du drüben landen!

Die Mahnung

1837

O heil'ge Stadt, dein Hirte ist gefangen,
Die halbe Welt steht jubelnd auf der Lauer,
Doch andre sinnen ernst in stummer Trauer,
Er mitten drin, von greisem Haar umhangen.

Da, als die Nacht und Trübsal näher drangen,
Ging durch die Seele ihm ein ahnend' Schauer,
Ein recht' Gebet hebt über Schloß und Mauer –
Still segnet er das Land, das ihn gefangen.

Und wie er segnet, klang's vom hohen Dome,
Die Glocken fingen an von selbst zu schlagen,
Und weithin drang ihr Ruf vom deutschen Strome.

Die Nacht entfloh, der Morgen strahlte nieder,
Und betend sah man in des Frührots Tagen
Sich alle sammeln um den Herren wieder.

An Görres

(Oktober 1839)

Wo einer noch Christi Fahne hält
Hoch über dem Erden-Plunder,
Für einen Narren hält ihn die Welt
Für ein gar fabelhaft Wunder.

Der Alte vom Berge nachts umgeht
Und zieht die alten Glocken,
Im Schlaf auf die andre Seite dreht
Die Welt sich, fast erschrocken.

Einsiedler du, ja stürme nur fort!
Manch Wandrer noch irrt verspätet,
Spürt in dem Klange des Herren Wort
Und sinkt in die Knie und betet.

Nachts

Ich stehe in Waldesschatten
Wie an des Lebens Rand,
Die Länder wie dämmernde Matten,
Der Strom wie ein silbern' Band.

Von fern nur schlagen die Glocken
Über die Wälder herein,
Ein Reh hebt den Kopf erschrocken
und schlummert gleich wieder ein.

Der Wald aber rühret die Wipfel
Im Traum von der Felsenwand.
Denn der Herr geht über die Gipfel
Und segnet das stille Land.

Moderne Ritterschaft

O große heldenmüt'ge Zeit
In diesen Friedenstagen,
Wo man in ganzer Christenheit
Nichts tut, als Ritter schlagen!

Wer so viel Adlerbrut nur treibt
Aus ihren Felsenhängen,
Daß drunten jedem Wichte bleibt
Im Knopfloch einer hängen!

O wunderbare Ritterschaft!
Wie würdet ihr wohl zagen,
Wenn Sankt Georgens Lanzenschaft
Zu Rittern euch wollt' schlagen?

Auf offener See

Ade, du Küste mit den falschen Sorgen,
Furcht, Glück und Not, sinkt unter in das Meer!
Nun bin ich frei, jetzt bin ich erst geborgen,
Kein eitles Hoffen langet bis hierher.
Wie still, wohin ich auch die Blicke wende,
Wie tief und hoch und ringsum ohne Ende!

Gestirne, Wolken gehen auf und unter
Und spiegeln sich im stillen Ozean,
Hoch Himmel über mir und Himmel drunter,
Inmitten wie so klein mein schwacher Kahn!
Walt' Gott, ihm hab' ich alles übergeben,
Nun komm' nur, Sturm, ich fürcht' nicht Tod noch Leben!

Nachtgruß

Den 16. Mai 1840

Nachts an der Küste wir vorüberfahren,
Von der ich schied vor vielen, vielen Jahren.
Der Liebsten Haus steht dort, wo wir so fröhlich waren;
Gott wolle sie bewahren!

Die Nachtigallen noch im Garten schlagen,
Wie damals in den schönen, stillen Tagen.
Was mögen sie jetzt klagen?
Nach mir will niemand fragen.

Sonst wenn die Linden in der Blüte waren,
Sahst du dort aus nach mir vor vielen, vielen Jahren –
Ich muß vorüberfahren, Gott wolle dich bewahren!

Das Lied

Halb Worte sind's, halb Melodie,
Was mir durchs Herze zieht,
Weiß nicht, woher, wozu und wie,
Mit *einem* Wort: ein Lied.

Der Lenz rauscht durch die offne Tür,
Da hab' ich's mir erdacht,
Ich sel'ger Mensch kann nichts dafür,
Gott hat's so schön gemacht!

Die blaue Blume

Ich suche die blaue Blume,
Ich suche und finde sie nie,
Mir träumt, daß in der Blume
Mein gutes Glück mir blüh'.

Ich wandre mit meiner Harfe
Durch Länder, Städt und Aun,
Ob nirgends in der Runde
Die blaue Blume zu schaun.

Ich wandre schon seit lange
Hab' lang' gehofft, vertraut,
Doch ach, noch nirgends hab' ich
Die blaue Blum geschaut.

Späte Gedichte

(Entstanden nach dem Erscheinen der Ausgabe von 1841)

Fata Morgana

Du Pilger im Wüstensande,
Ich spiegle Wälder und Kluft,
Der Heimat blühende Lande
Dir wunderbar in der Luft.

Wer hielte in dieser Wüste
Das einsame Wandern aus,
Wenn ich ihn barmherzig nicht grüßte
Mit Frühlingsdüften von Haus?

Und ob's auch wieder verflogen
In Luft – und schien doch so nah –
Nur frisch durch die sengenden Wogen.
Wer weiß, wie bald bist du da!

In Danzig

1842

Dunkle Giebel, hohe Fenster,
Türme tief aus Nebeln sehn,
Bleiche Statuen wie Gespenster
Lautlos an den Türen stehn.

Träumerisch der Mond drauf scheinet,
Dem die Stadt gar wohl gefällt,
Als läg' zauberhaft versteinet
Drunten eine Märchenwelt.

Ringsher durch das tiefe Lauschen,
Über alle Häuser weit,
Nur des Meeres fernes Rauschen –
Wunderbare Einsamkeit!

Und der Türmer wie vor Jahren
Singet ein uraltes Lied:
Wolle Gott den Schiffer wahren,
Der bei Nacht vorüberzieht!

Lied des Armen

Stände noch das Feld im Flore
Wie in warmer Sommerzeit,
Ging' ich aus dem dunklen Tore
In die Waldeseinsamkeit.

Legt' im tiefsten Wald mich nieder,
Wo der Vöglein Nachtquartier,
Und es sängen ihre Lieder
Nachtigallen über mir.

Doch verschneiet Markt und Gassen
Nun der böse Winter hat,
Und ich wandre arm, verlassen
Durch die stille fremde Stadt.

Späte Gäste gleich Gespenstern
Schlüpfen da und dort ins Haus,
Und der Nachtwind an den Fenstern
Löscht die letzten Lampen aus.

Nur aus einem noch sprüht Glänzen
Weithin in den bleichen Schnee,
Spielen auf da drinn' zu Tänzen,
Klingt hier draußen fast wie Weh.

Und im mitternächt'gen Sturme,
Der am Himmel brausend zieht,
Singt das Glockenspiel vom Turme
Über mir ein frommes Lied.

An dem Kirchhof die Kapelle
Ladet mich zur müden Ruh,
Lege stumm mich auf die Schwelle,
Und die Nacht, sie deckt mich zu.

Wolle Gott die Stadt bewahren,
Mild behüten Hof und Haus –
Die da tanzen, die da fahren,
Hier doch ruhen alle aus!

Der brave Schiffer

Meinem Theodor von Schön

1842

Solang an Preußens grünem Strand
Die Meereswogen schlagen,
Wird Kindeskind im ganzen Land
Vom braven Schiffer sagen.

In wilden Wettern trieb das Schiff,
Die wollten es begraben,
Da sprach er stolz zu Sturm und Riff:
Ihr sollt es nimmer wagen!

Kühn um der Nornen Felsenwand,
Durch Meeresungeheuer,
Weil er das hohe Wort verstand,
Lenkt' mächtig er das Steuer.

Und als die Brandung sich verlief,
Die Waffen müde sanken,
Gerettet hatte aus der Tief'
Den Hort er der Gedanken.

Und ob auch Stern auf Stern versank
Und schlaff die Segel hingen,
Der Teufel, nicht das Schiff, ertrank,
Gedanken sind ja Schwingen.

So zwischen Schrecken, träger Ruh
Und Sandbank des Gemeinen
Dem ritterlichen König zu
Führt' er getreu die Seinen.

Jetzt überm Lande auf der Wacht
Steht rastend er im Hafen:
Die See geht hoch, gebt acht, gebt acht,
Ihr Schiffer sollt nicht schlafen!

Ja, und so oft wir wogenwärts
Noch frische Fahrten wagen,
Soll hell an jedes Preußenherz
Des Schiffers Mahnung schlagen!

Den Dichtern Wiens

1846

Lerche, wo sie's grünen sieht,
Lenkt sie hin von ferne –
Wo ein Liederfrühling blüht,
Weilt der Dichter gerne.

Segnet dankbar Stadt und Tal,
Die ihn traut empfangen,

Grüßt die Sänger allzumal,
Die so liebreich sangen.

Und senkt alternd sich sein Schwung,
Nimmer mag's ihn schmerzen,
Bleibt doch Dichtung ewig jung
In den deutschen Herzen!

Drum laßt aus der Seele Grund
Seinen Gruß euch klingen:
Heil dem schönen Sängerbund,
Heil dem wackern Ringen!

Dichterweisheit

Scherz im Ernst und Ernst im Scherz,
Lachst der Grillen, die sie fingen,
Daß des Lebens Kampf und Schmerz
Selber heiter muß erklingen.
Alter Dichter, junges Herz,
Sollst noch lang' auf Erden singen
Und dereinst dich himmelwärts
Jubelnd, wie die Lerche, schwingen.

Zum Abschied an J. und R.

Wien 1847

Selig, wo sich zwei gesellt
In den schlimmen Tagen
Und ihr häuslich frommes Zelt
Schirmend aufgeschlagen!

Vor dem Zelte halten Wacht
Engel Stund' um Stunde,

Drüber machen bei der Nacht
Sterne ihre Runde.

Und wenn's draußen wirbelnd schwirrt,
Flüchten hin viel' Gäste,
Wahrlich, jede Einkehr wird
So zum heitern Feste.

Einer scheidet, zögernd noch,
Denn er blieb' so gerne!
Mit dem Herzen bleibt er, doch –
Denk't sein' in der Ferne!

Die Altliberalen

1848

Die wilden Wasser, sagt man, hat entbunden
Ein Lehrling einst, vorwitzig und vermessen,
Doch hinterdrein den Zauberspruch vergessen,
Der streng die Elemente hält gebunden.

Ein tödlich Pulver, sagt man, zu erkunden,
Hat einst ein Mönch sich überklug vermessen,
Und als er eben recht darauf versessen,
Im Zauberdampf den eignen Tod gefunden.

So habt den Zeitgeist ihr gebraut, gemodelt,
Und wie so lustig dann der Brei gebrodelt,
Ihm eure Zaubersprüche zugejodelt.

Und da's nun gärt und schwillt und quillt –
was Wunder,
Wenn platzend dieser Hexentopf jetzunder
Euch in die Lüfte sprengt mit allem Plunder!

Ihr habt es ja nicht anders haben wollen

1848

Es fährt die Welt mit Dampf, die Meister grollen
Dem treuen Roß ob seinem trägen Schritte,
Und stacheln es, daß es den Zaum nicht litte,
Und stachelten, bis ihm der Kamm geschwollen.

O wunderschön, ein edles Roß im vollen
Kühnfreien Lauf durch grüner Wälder Mitte!
Lichtfunken sprühen hinter jedem Tritte,
Die Mähne flattert und die Augen rollen!

Was ruft ihr nun so ängstlich? Euren Winken
Hat es zum Ritt sich wieder stellen sollen?
Zu spät! Das Roß riß plötzlich aus zur Linken.

Ihr müßt zur Rechten hinterdrein jetzt hinken,
Da ist es nicht mehr Zeit, vornehm zu schmollen,
Ihr habt es ja nicht anders haben wollen!

Kein Pardon

1848

Hervor jetzt hinter euren rost'gen Gittern,
Heraus, ihr Schriftgelehrten, Hochmutstollen!
An euch ist der Posaunenruf erschollen,
Vor dem die Schlechten und Gerechten zittern.

Denn Deutschland dunkelt tief in Ungewittern,
Wo alle Quellen, Bäche, zorngeschwollen
Als Ströme donnernd von den Höhen rollen,
Und Blitze, was der Sturm verschont, zersplittern.

Die Ströme werden immer rückwärts stauen,
Die Blitze werden zielen nach den Kronen,
Die Stürme rastlos fegen durch die Gauen,

All' Türme brechend, wo die Stolzen wohnen,
Bis all' erkannt demütig in dem Grauen
Den *einen* König über allen Thronen.

Wer rettet?

1848

Es ist den frischen hellen Quellen eigen,
Was alt und faul, beherzt zu unterwühlen
Und Wasserkünste unversehns und Mühlen
Wild zu zerreißen, wenn die Fluten steigen.

Es liebt das Feuer frei emporzusteigen,
Verzehrend, die mit seinen Lohen spielen,
Es liebt der Sturm, was leicht, hinwegzuspülen,
Und bricht, was sich hochmütig nicht will neigen.

Sahn wir den Herren nun in diesen Tagen
Ernstrichtend durch das deutsche Land geschritten,
Und Wogenrauschen hinter seinen Tritten,

Und Flammen aus dem schwanken Boden schlagen,
Empor sich ringelnd in des Sturmes Armen:
Wer rettet uns noch da, als Sein Erbarmen?

Das Schiff der Kirche

1848

Die alten Türme sah man längst schon wanken,
Was unsre Väter fromm gebaut, errungen,

Thron, Burg, Altar, es hat sie all' verschlungen
Ein wilder Strom entfesselter Gedanken;

Der wühlt sich breit und breiter ohne Schranken,
Ein Meer, wo zornigbäumend aufgeschwungen
Die trüben Fluten Fels und Fels bezwungen,
Und alle Rettungsufer rings versanken.

Deutschlands künftiger Retter

Kein Zauberwort kann mehr den Ausspruch mildern,
Das sündengraue Alte ist gerichtet,
Da Gott nun selbst die Weltgeschichte dichtet
Und auf den Höhen zürnend Engel schildern.

Die Babel bricht mit ihren Götzenbildern
Ein junger Held, der mit dem Schwerte schlichtet,
Daß Stein auf Stein, ein Trümmerhauf, geschichtet,
Die Welt vergeht in schauerndem Verwildern.

Doch *eins,* das hastig alle übersehen,
Das Kreuz, bleibt auf den Trümmern einsam stehen,
Da sinkt ins Knie der Held, ein arbeitsmüder,

Doch drüberhin gewölbt ein Friedensbogen,
Wohin nicht reichen die empörten Wogen,
Und unter ihm ein Schiff dahingezogen,

Das weiß nichts von der Wasser wüstem Branden,
Das macht der Stürme Wirbeltanz zuschanden –
O Herr, da laß' uns alle selig landen!

Der Freiheit Wiederkehr

1849

1

Um mich wogt es wie ein Meer,
Fast wie in vergangnen Tagen,
Da die Wälder ringsumher
Rauschten von uralten Sagen.

Dort blitzt's auf, das ist der Rhein,
Wo sich zwischen Rebenhügeln
Bei dem glühen Morgenschein
Burgen in den Fluten spiegeln.

Sei gegrüßt, du schöner Strom!
Brüderlich wob seine Äste
Damals deiner Wälder Dom
Dir zum Schutz und Trutz als Veste,

Als der Römer-Adler flog
Und ich flüchtet' vor dem Volke,
Das den Erdkreis überzog,
Eine Zornes-Wetterwolke;

Das einst kühn nach Heldenart,
Mit dem Schwert die Welt gemessen,
Doch geworden stolz und hart,
Seit es meiner hatt' vergessen.

Hinter mir in Schmach und Tod
Sah ich da die Länder dunkeln,
Vor mir frisches Morgenrot
Rings von deinen Bergen funkeln.

Freudig zog ich zu dir hin,
Bracht' dir aller Länder Kronen,

Bis auch du in blödem Sinn
Mir's nicht länger mochtest lohnen.

Jetzt nach langer banger Fahrt,
Hab' ich wieder dich gefunden,
Und es grüßt nach Landesart,
Mich die ganze weite Runde.

Feuerzeichen steigen auf,
Von den Gipfeln ringsum schallt es,
Und zum Willkomm mir herauf
Rauscht der Rhein und widerhallt es.

Und von Berg zu Bergeswand,
Weit hinab durch alle Gaue
Segn' ich dich, du deutsches Land,
Dem ich wieder mich vertraue.

2

Geht ein Klingen in den Lüften,
Aus der Tiefe rauscht der Fluß,
Quellen stürzen von den Klüften,
Bringen ihr der Höhen Gruß.

Denn es naht in Morgenblitzen
Eine hohe Frau zu Roß,
Als wär' mit den Felsenspitzen
Das Gebirge dort ihr Schloß.

Und die grauen Schatten senken,
Wie sie durch die Dämm'rung bricht,
Und die Kreaturen trinken
Durstend alle wieder Licht.

Ja, sie ist's, die wir da schauen,
Unsre Königin im Tal,
Holde Freiheit, schöne Frauen,
Grüß' dich Gott vieltausendmal!

Der Freiheit Klage

1849

Weh' du Land, das keck mich bannte,
Und da ich zu dir mich wandte,
Mich blödsinnig nicht erkannte;

Wo aus Trümmern nun die blassen
Geister stieren: Stolz und Hassen,
Brüder sich ingrimmig fassen.

Habt ihr euch von dem gewendet,
Der barmherzig mich gesendet,
Wird in Schmach die Ehr' geendet.

Wer will meinen Banner schwingen,
Muß erst mit dem Teufel ringen,
Der ihn selber hält in Schlingen.

Wer so kühn, um mich zu werben,
Zage nicht, für mich zu sterben,
Um das Himmelreich zu erben,

Lieble nicht, nach andern lugend,
Denn ich bin des Herzens Jugend
Und der Völker strenge Tugend.

Bin die Lebensluft der Höhen,
Wo der Atem mag vergehen
Allen, die zur Tiefe sehen,

Flamme, schlank emporgelodert,
Die in Zornesmut, was modert,
Sengend zu Gerichte fodert.

’s war ein mächt’ger Wald da droben,
Treulich Stamm in Stamm verwoben,
Mir zum grünen Dom erhoben.

Weh’, du schönes Land der Eichen!
Bruderzwist schon, den todbleichen,
Seh’ ich mit der Mordaxt schleichen.

Und in künft’gen öden Tagen
Werden nur verworrne Sagen
Um den deutschen Wald noch klagen.

Frühlingsdämmerung

In der stillen Pracht,
In allen frischen Büschen und Bäumen
Flüstert’s wie Träumen
Die ganze Nacht.
Denn über den mondbeglänzten Ländern
Mit langen weißen Gewändern
Ziehen die schlanken
Wolkenfraun wie geheime Gedanken,
Senden von den Felsenwänden
Hinab die behenden
Frühlingsgesellen, die hellen Waldquellen,
Die’s unten bestellen
An die duft’gen Tiefen,
Die gerne noch schliefen.
Nun wiegen und neigen in ahnendem Schweigen
Sich alle so eigen
Mit Ähren und Zweigen,
Erzählen’s den Winden,

Die durch die blühenden Linden
Vorüber den grasenden Rehen
Säuselnd über die Seen gehen,
Daß die Nixen verschlafen auftauchen
Und fragen,
Was sie so lieblich hauchen –
Wer mag es wohl sagen?

Frühlingsahnen

Was gibt's, daß vom Horste
An der zackigen Kluft
Der Adler schon steigt
Und hängt überm Forste
In der stillen Luft,
Wenn alles noch schweigt?

Ich hörte in Träumen
Ein Rauschen gehn,
Sah die Gipfel sich säumen
Von allen Höhn –
Ist's ein Brand, ist's die Sonne?
Ich weiß es nicht,
Aber ein Schauer voll Wonne
Durch die Wälder bricht.

Die gebunden da lauern,
Sprengt' Riegel und Gruft,
Du ahnend' Schauern
Der Felsenkluft,
Unsichtbar ringen in der stillen Luft,
Du träumend' Singen
Im Morgenduft!
Brecht auf, schon ruft
Der webende, blaue
Frühling durchs Tal.

Der Vögel Abschied

Ade, ihr Felsenhallen,
Du schönes Waldrevier,
Die falben Blätter fallen,
Wir ziehen weit von hier.

Träumt fort im stillen Grunde!
Die Berg stehn auf der Wacht,
Die Sterne machen Runde
Die lange Winternacht.

Und ob sie all' verglommen,
Die Täler und die Höhn –
Lenz muß doch wiederkommen
Und alles auferstehn!

Der Auswanderer

Fragment

I

Europa, du falsche Creatur!
Man quält sich ab mit der Cultur,
Spannt vorn die Lokomotive an.
Gleich hängen sie hinten eine andre dran,
Die eine schiebt vorwärts, die andre retour,
So bleibt man stecken mit der ganzen Cultur.
Und Ärger hier, und Ärger da,
Und Prügel! – Vivat Amerika!
Mit den vereinigten Provinzen,
Wo die Einwohner alle Prinzen
Und alle Berge in Gold verhext,
Wo die Cigarre und der Pfeffer wächst!

Und also flog ich dahin wie ein Pfeil,
Über uns Wolken in großer Eil',
England zur Rechten und Frankreich links,
Jetzt in den Ozean grad hinein ging's,
Daß mir der Wind am Hute pfiff.
Ich stand ganz vorne in dem Schiff;
Und als die alte Welt versank,
Nahm ich mein Waldhorn und blies Ade,
Das gab einen prächtigen Klang,
Mir aber tat's doch im Herzen weh'.

II

Auf einmal stößt das Schiff ans Land,
Greift jeder nach seinem Plunder;
Am Land hat man mich gleich erkannt,
Das war ein Lärm, Gott's Wunder!
Da wurden Böller abgebrannt,
Entgegen uns gegangen
Kam ganz Neuyork heraus zum Strand,
Mich würdig zu empfangen.

Der bot mir fürstliches Quartier,
Der bat um meine Sachen,
Man riß sich ordentlich nach mir,
Ich aber mußte lachen.
„Mein' Herberg heißt zum Himmelszelt!"
Mein Ränzel nun geschwungen,
So bin ich in die neue Welt
Vom Schiff hineingesprungen.

Doch kaum hatt' ich zum Umsehn Zeit,
Spricht einer aus dem Haufen:
Mein Schiff das läge schon bereit,
Um eben auszulaufen;
Geheimer Rat und Hofmarschall,
Kurz meine ganze Suite,

Sei schon in freudenreichem Schall
An Bord, und tät sich Güte.

Ich: wie? – Er: Jes! Ich: zuviel Ehr! –
Sehr gütig, und so weiter.
Das half nun alles nimmermehr,
Ein Dampfboot braust! Zur Leiter
Trug man mich auf den Händen fast,
Wie einen Potentaten;
Und stromauf ging's sogleich voll Hast
In die vereinten Staaten.

Auf dem Verdecke aber dort
Sah ich viel' Herrn, die lasen
In langen Blättern immerfort: –
Nichts, als Papier und Nasen!
Zuweilen nur ein Rauschen schallt',
Wenn einer 's Blatt umdrehte.
Da merkt' ich's wohl, und wußt' es bald:
Das sind Geheime Räte!

Nur einer ging stolz her und hin,
Die Hände in den Taschen,
In seinen Rock geknöpft, vom Kinn
Bis unter die Gamaschen.
Aha, dacht ich, der Hofmarschall,
Der scheint, als wollt' er beißen?
Engländer waren's aber all',
Die dorten Yankees heißen.

Ich bracht' gleich ein Gespräch ins Gleis,
Wir sprachen erst ganz gelassen,
Doch konnten wir bei allem Fleiß
Einander nicht recht fassen.
Da fiel mir grad zum Glücke ein,
Was ich gehört schon häufig,
Musik soll eine Weltsprach' sein,
Die überall geläufig.

Sprach nun zur mir der Hofmarschall,
Als wie ein Puter im Zorne,
Gab ich ihm gleich mit sanftem Schall
Antwort auf meinem Horne.
Er blickte martialisch dann
Durch seine gold'ne Brille,
Ich aber blies ihn tapfer an
Mit einem langen Trille-r.

Und das gelang erstaunlich gut,
Je mehr ich blies und lauter,
Je mehr bekam er frischen Mut,
Je aufgeweckter schaut' er.
Und nun ging's immer rascher los,
Ein lebhaft Diskurieren,
Er Wort auf Wort, ich Stoß auf Stoß,
Als wollten wir duellieren.

Geheime Räte alt und jung,
Die nahten auch ganz leise,
Und standen voll Bewunderung
Rings um mich her im Kreise.
Ich aber brech' auf einmal aus,
Und fröhlich Platz mir mache –
Ich hörte mitten durchs Gebraus
Meine Fraumuttersprache!

Da kommt auch einer schon gerannt,
Tritt fast mir auf die Zehen:
„Ei, ei, grüß Gott, Herr Musikant,
Freut mich, Sie wohl zu sehen?"
Ich drauf: Bitt' sehr, ein Musi*kus*!
Wie fuhr da der Geselle
Zu einem ehrfurchtsvollen Gruß
Nach seinem Hut so schnelle.

Der Abend aber unterdes
War schon hereingebrochen,
Und plötzlich Preuße, Schwab' und Hess',
Da wir so deutsch gesprochen,
Kam's Kopf auf Kopf, und Hut und Mütz'
Da aus dem Schiff gekrochen,
Wie Fledermäus aus jedem Ritz,
Weiß nicht, wo all' gestochen.

„Auf Ehre", da der eine rief,
„Das heiß ich einmal blasen!
Der Hut sitzt Euch schon ganz windschief,
Vor kunstreichen Ekstasen."
Ein andrer mich erstaunt besieht:
„Wir möchten gerne wissen,
Ob Sie vielleicht europamüd',
Von Weltschmerz so zerrissen?"

Zerrissen! Ja, das einz'ge Loch,
Der Schalk hat's gleich erspähet,
Und hatt's am Ellenbogen doch
Erst gestern zugenähet!
Auch mein Castor, das leugn' ich nicht,
Hatt' manchen Bug erlitten,
Weil ich so *rasch* mit Zelt und Licht
Und Bildung fortgeschritten.

„O ho", rief ich den Schälken zu,
„Gemach, Ihr Herrn Landsleute,
Es sind wohl meine Reiseschuh
Von gestern nicht und heute.
Bin ich in Linz, im Binger Loch,
Im Kuhstall doch gewesen;
Man kann da meinen Namen noch
Auf allen Bänken lesen.

Durch Polen nahm ich meinen Weg,
Da trägt man noch Weichselzöpfe,
Nach München zur Pinaglypthek,
Und von den Bockbiertöpfen
Bis nach Savoyen immerfort,
Wo das Gebirg' wie Lanzen
Hellfunkelnd in die Wolken bohrt
Und Murmeltiere tanzen.

Und weiter sah ich's Schwarze Meer,
In Wien den Wurstelprater,
Viel Residenzen hin und her
Von manchem Landesvater;
Die Jungfrau von Neu-Orleans,
Mit dem schwarzen Ritter fechten,
Vom großen Schill den Posner dann
Mit seinen Menschenrechten.

Und Dampfschiff, Treckschuit, Eisenbahn
Und Pesth, Triest und Halle ..."
„Halt, halt, Herr Landsmann, haltet an!"
Schrien die voll Staunen alle.
Und alles jubiliert und ruft,
Und ihre Hüte schwingen
Sie wie besessen in die Luft,
Ein Vivat mir zu bringen.

Jetzt erst erkannt' ich bei dem Lärm
Verwundert manch' Bekannten
Von Deutschland her in dem Geschwärm, –
Es waren Komödianten.
Und der Direktor tät alsbald
Als Staberl mich engagieren,
Um bei den Yankees im Urwald
Die Bildung einzuführen.

Die Zeit geht schnell

Lieb Vöglein, vor Blüten
Sieht man dich kaum
Im dämmernd beglühten
Flüsternden Baum;
Wann von blitzenden Funken
Sprühn Täler und Quell,
Singst du frühlingstrunken –
Aber die Zeit geht schnell.

Wie balde muß lassen
Sein' Blätter der Wald,
Die Blumen erblassen,
Die Gegend wird alt,
Erstarrt ist im Eise
Der muntere Quell –
Rüst' die Flügel zur Reise,
Denn die Zeit geht schnell!

An meinem Geburtstage

1850

Sonnenglanz lag überm Garten,
Warm und herrlich aufgetan
Lenz und Licht des Reisleins harrten,
Daß es wuchs zum Himmel an.

Wie die Blätter ringsum glühten
In der frohen Morgenzeit,
Alle Zweige voller Blüten,
Vögel sangen weit und breit!

Mittag kam, die Blätter hingen,
In dem Wipfel säuselt's kaum,

Wetter stiegen auf und gingen,
Stumm erwartend stand der Baum.

Jetzo sinkt die Abendröte,
Blüte fällt, es schweigt der Sang,
Und ich rausch' wie im Gebete
Mit den Zweigen: Gott sei Dank!

Nachtzauber

Hörst du nicht die Quellen gehen
Zwischen Stein und Blumen weit
Nach den stillen Waldesseen,
Wo die Marmorbilder stehen
In der schönen Einsamkeit?
Von den Bergen sacht hernieder,
Weckend die uralten Lieder,
Steigt die wunderbare Nacht,
Und die Gründe glänzen wieder,
Wie du's oft im Traum gedacht.

Kennst die Blume du, entsprossen
In dem mondbeglänzten Grund?
Aus der Knospe, halb erschlossen,
Junge Glieder blühend sprossen,
Weiße Arme, roter Mund,
Und die Nachtigallen schlagen,
Und rings hebt es an zu klagen,
Ach, vor Liebe todeswund,
Von versunknen schönen Tagen –
Komm, o komm zum stillen Grund!

An Jegór von Sivers

Wo sie schwindeln, und vor Bangen
In der zack'gen Gipfel Bann

Andern längst der Mut vergangen,
Geht erst deine Lust recht an,
Und bei wilder Brandung Schäumen
An des Nordlands Felsenhang
Oder unter Palmenbäumen:
Wo du stehst, gibt's frischen Klang.

Wanderdichter, nimm vom Greise,
Da er von dir scheiden muß,
Recht aus Herzensgrund zur Reise
Seinen allerbesten Gruß:
Wo die Pfade kühn sich schlingen
Nach des Lebens lichten Höhn,
Freud'ges Ringen, herzhaft Singen
Und ein bald'ges Wiedersehn!

Der verspätete Wanderer

Wo aber werd' ich sein im künft'gen Lenze?
So frug ich sonst wohl, wenn beim Hüteschwingen
Ins Tal wir ließen unser Lied erklingen,
Denn jeder Wipfel bot mir frische Kränze.

Ich wußte nur, daß rings der Frühling glänze,
Daß nach dem Meer die Ströme leuchtend gingen,
Von fernem Wunderland die Vögel singen:
Da hatt' das Morgenrot noch keine Grenze.

Jetzt aber wird's schon Abend, alle Lieben
Sind wandermüde längst zurückgeblieben,
Die Nachtluft weht durch meine welken Kränze,
Und heimwärts rufen mich die Abendglocken,
Und in der Stille frag' ich froh erschrocken:
Wo aber werd' ich sein im künft'gen Lenze?

Einem Paten zu seinem ersten Geburtstage

1854

Noch singt der Wind, der durch die Bäume
Am Fenster lind vorüberzieht,
Das Meer von fern in deine Träume,
Du Dichterkind, ein Schlummerlied.

Doch wenn dereinst die Segel schwellen:
Glücksel'ge Fahrt durch Ebb' und Flut,
Lenzfrischen Hauch beim Klang der Wellen,
Ein fröhlich' Herz in Gottes Hut!

Und so mag dich von Strand zu Strande
Ein milder Wind hinüberwehn
Einst zum geheimnisvollen Lande,
Wohin wir alle hoffend sehn.

Prinz Rokoko

Prinz Rokoko, hast dir Gassen
Abgezirkelt fein von Bäumen
Und die Bäume scheren lassen,
Daß sie nicht vom Wald mehr träumen.

Wo sonst nur gemein Gefieder
Ließ sein bäurisch' Lied erschallen,
Muß ein Papagei jetzt bieder:
Vivat Prinz Rokoko! lallen.

Quellen, die sich unterfingen,
Durch die Waldesnacht zu tosen,
Läßt du als Fontänen springen
Und mit goldnen Bällen kosen.

Und bei ihrem sanften Rauschen
Geht Damöt bebändert flöten,
Und in Rosenhecken lauschen
Daphnen frommentzückt Damöten.

*

Prinz Rokoko, Prinz Rokoko,
Laß dir raten, sei nicht dumm!
In den Bäumen, wie in Träumen,
Gehen Frühlingslüfte um.

Springbrunn' in dem Marmorbecken
Singt ein wunderbares Lied,
Deine Taxusbäume recken
Sehnend sich aus Reih und Glied.

Daphne mag nicht weiter schweifen
Und Damöt erschrocken schmält,
Könnten beide nicht begreifen,
Was sich da der Wald erzählt.

Laß die Wälder ungeschoren,
Anders rauscht's, als du gedacht.
Sie sind mit dem Lenz verschworen,
Und der Lenz kommt über Nacht.

Verschwiegene Liebe

Über Wipfel und Saaten
In den Glanz hinein –
Wer mag sie erraten,
Wer holte sie ein? –
Gedanken sich wiegen,
Die Nacht ist verschwiegen,
Gedanken sind frei.

Es rät es nur eine,
Wer an sie gedacht
Beim Rauschen der Haine,
Wenn niemand mehr wacht,
Als die Wolken, die fliegen,
Mein' Lieb' ist verschwiegen
Und schön wie die Nacht.

Lied der Pilger

Fromme Vöglein hoch in Lüften
Über blaue Berg' und Seen
Ziehn zur Ferne nach den Klüften,
Wo die hohen Zedern stehn,
Wo mit ihren goldnen Schwingen
Auf des Benedeiten Gruft
Engel Hosianna singen
Nächtlich durch die stille Luft.

Morgenschauer, still Erwarten!
Hören wir doch Stimmen gehn,
Wie aus einem fernen Garten
Heimatsgruß herüberwehn;
Warum sollten wir verzagen?
Aus der Fremde wüst und fern,
Wo wir irrend hier verschlagen,
Führe heim uns, Morgenstern!

Mahnung

Was blieb dir nun nach allen Müh' und Plagen?
So viel der Ehre dir die Welt gespendet,
Es treibt vom stolzen Ziele, kaum geendet,
Nach neuem Ziel dich neues Unbehagen.

Hätt'st du zu Ihm, von dem die Himmel sagen,
Den kleinsten Teil der Liebe nur gewendet,
Die du an eitel Hoffart hast verschwendet,
Du würdest jetzt nicht rettungslos verzagen.

Wohl liebt die Welt, den Günstling zu erheben,
Doch wenn du glaubst, im Siegesschmuck zu prangen,
Sind's Ketten nur, die rasselnd dich umhangen.

Laß, eh's zu spät, von dem verlornen Leben;
Gott wartet deiner noch, in seinen Armen,
Da find'st du, was die Welt nicht kennt, Erbarmen.

Das Alter

Hoch mit den Wolken geht der Vögel Reise,
Die Erde schläfert, kaum noch Astern prangen,
Verstummt die Lieder, die so fröhlich klangen,
Und trüber Winter deckt die weiten Kreise.

Die Wanduhr pickt, im Zimmer singet leise
Waldvöglein noch, so du im Herbst gefangen.
Ein Bilderbuch scheint alles, was vergangen,
Du blätterst drin, geschützt vor Sturm und Eise.

So mild ist oft das Alter mir erschienen:
Wart' nur, bald taut es von den Dächern nieder,
Und über Nacht hat sich die Luft gewendet.

Ans Fenster klopft ein Bot' mit frohen Mienen,
Du trittst erstaunt heraus – und kehrst nicht wieder,
Dann endlich kommt der Lenz, der nimmer endet.

Alphabetisches Verzeichnis der Gedichtanfänge und -überschriften

schallen. Es war eine Prozession weißgekleideter liberaler Mädchen, die sich abquälten, einen gestickten Banner zu dem Feste hinaufzutragen. Der Wind zerarbeitete gar wacker die große Fahne, in deren flatternde Zipfel, so oft sie die Erde streiften, sich Eidechsen und dicke Kröten anhingen. Noch schlimmer schien es weiter unten mehreren anständig gekleideten Männern zu ergehen, die sich vergeblich dem anderen lustigen Gesindel nachzukommen bemühten. Der Professor rieb sich lustig die Hände. „Es geschieht ihnen schon recht", sagte er, „das sind die Doktrinärs, halb des Himmels und halb des Teufels, sie wollen es mit keinem verderben." – Ich konnte nun deutlich vernehmen, wie diese Unglücklichen jede an ihnen vorüberhuschende Gestalt mit weitläufigen Demonstrationen beredt haranguierten. Aber, ehe sie sich's versahen, kehrte ein fliegender Besen sich schnell in der Luft um und schlug ihnen die Hüte vom Kopf, oder ein Bock, den sie eben überzeugt zu haben glaubten, stieß sie plötzlich von der mühsam erklommenen Höhe kopfüber wieder hinab. Noch lange hörte ich sie aus ferner Tiefe kläglich rufen: „Nehmt uns mit, nehmt uns doch mit!" worauf jedesmal ein schadenfrohes Gelächter aus allen Schluften erschallte. Lärm, Gewirre, Drängen, Fluchen und Stoßen nahmen jetzt mit jeder Minute betäubend zu. Von Zeit zu Zeit aber schoß zwischen dem Gestrüpp und Geklüfte eine ungeheure goldflammende Schlange, wie glühende Lava das unermeßliche Getümmel plötzlich beleuchtend, den ganzen Berg hinunter, und ein allgemeines Hurra begrüßte sie vom Gipfel bis in die tiefsten Gründe hinab. Ich glaubte, das gölte unserer Ankunft, und dankte, mit gebührender Höflichkeit mein Haupt entblößend. – „Aber sind Sie toll?" fuhr mich der Professor zornig an, indem er mir den Hut bis über die Augen wieder aufstülpte – „solche servile Gewohnheiten deutschen Knechtsinns!"

Hier stießen wir, etwa in der Mitte des Berges, plötzlich ans Land. Unser Roß wälzte sich sogleich zur Seite und nahm nach dem ermüdenden Fluge, ein Schlammbad. Wir

aber drangen weiter vor. „Halten Sie sich nur an meinen Rockschoß“, rief mir der Professor zu und machte ohne Umstände mit beiden Ellenbogen Platz. Da konnte ich bemerken, in welchem Ansehen der starke Mann hier stand. Von allen Seiten wichen die Wimmelnden, so gut es gehen wollte, ehrerbietig aus, obgleich es mir vorkam, als zwickten sie, so oft er sich wandte, mich hinterrücks heimlich in die Waden.

Unter solchen Gewaltstreichen erreichten wir endlich eine Restauration, die, ziemlich geschmacklos, sich unter einem dreifarbigen Zelte befand, auf welchem ein fuchsroter alter Hahn saß und unaufhörlich krähte. Sieben Pfeifer saßen zur Seiten auf einem Stein und bliesen da Ça ira vom Anfang bis zu Ende und wieder und immer wieder von vorn, so langweilig, als bliesen sie schon auf dem letzten Loche. Auf der Tribüne der Restauration aber stand der Wirt und schrie mitten durch das Geblase mit durchdringender Stimme seine Wunderbüchsen und Likörflaschen aus: Konstitutionswasser, doppelte Freiheit usw. Unten schossen Kinder Purzelbäume und warfen jauchzend ihre roten Mützchen in die Luft, das Volk war wie besessen, sie würgten einander ordentlich, jeder wollte sein Geld zuerst los sein.

Hatt' ich nun aber den Professor schon im „Goldenen Zeitgeist“ bewundert, so mußte ich ihn jetzt fast vergöttern. Stürzte er doch fünf, sechs Flaschen abgezogene Garantie hinunter, ohne sich zu schütteln, und fand zuletzt alle das Zeug noch nicht scharf genug! Auch ich mußte davon kosten, konnte es aber nicht herunterbringen, so widerlich fuselte der Schnaps. „Alles Pariser Fabrikat!“ rief mir der Professor zu. – „Muß auf dem Transport ein wenig gelitten haben“, erwiderte ich bescheiden. – „Kleinigkeit!“ mengte sich der Wirt herein, „man tut etwas gestoßenen Pfeffer daran, die Leute mögen's nicht, wenn es sie nicht in die Zunge beißt.“

Währenddes war der Professor schon mit beiden Füßen in ein Paar dicke Schmierstiefel gefahren; ich mußte eiligst

desgleichen tun. „Wir müssen nun immer weiter hinauf“, sagte er, „wer mit der Zeit fortgehen will, der muß sich vorsehn, da geht’s durch dick und dünn.“ In der Tat begann nun auch von allen Seiten ein allgemeiner Aufbruch, als wenn man kochenden Brei im Kessel umrührte. Bald darauf aber schien der ganze Zug an der Spitze auf einmal wieder in Stocken zu geraten. Es entstand vorn ein Drängen und Wogen, dann ein heftiges Gezänk, das sich nach und nach, wie ein Lauffeuer, nach allen Richtungen hin verbreitete; man konnte zuletzt durch den Lärm nur noch einzelne grobe Stimmen deutlicher unterscheiden, die beinah wie Rebellion klangen. – „Was gibt’s denn?“ schrie der Professor voller Ungeduld. Da kamen mehrere junge Doktoren plötzlich herangestürzt, schreckensbleich und mit allen Zeichen der Verzweifelung, der eine hatte seinen Hut, der andere seinen Rockschoß in dem Getümmel verloren. „Alles aus!“ riefen sie atemlos, „sie wollen hier bei der Schnapsbude bleiben, es geht ein Schrei durchs ganze Volk nach Braten und Likör, sie mögen nichts von Freiheit und Prinzipien mehr wissen, sie wollen durchaus nicht weiter fortschreiten!“ – „So fraternisiert doch mit dem Lumpengesindel“ rief der Professor. – „Zu spät!“ erwiderten jene, „sie sind alle schon betrunken. O unsere Reputation! Was wird die öffentliche Meinung sagen? wir kommen um ein Dezennium zurück!“ – „Nun so soll sie doch!“, donnerte der Professor mit seiner Stentorstimme ganz wütend in das dickste Getümmel hinein, „wollt ihr wohl frei und patriotisch und gebildet sein in des Teufels Namen!“ Hiermit stemmte er mit hinreißender Gewalt seinen breiten Rücken gegen die rebellische Masse; die entlaufenen Doktoren und andere Honoratioren folgten mutig seinem Beispiel, die liberalen Mädchen mit ihrer Fahne wallten singend voran, die sieben Pfeifer spielten auf, und so rückte über liederliche Handwerker und betrunkene alte Weiber hinweg, die noch auf dem Boden keiften, die ganze Konfusion unter dem ungeheuersten Lärm und Gezänke langsam der Höhe zu.

Mir klopfte das Herz, als wir uns endlich der Stelle näherten, wo der berühmte Hexenaltar steht; ich blickte nach allen Seiten, ob nicht bald eine Teufelsklaue aus den Nebeln langte, die, wie Drachenleiber, vor uns den Boden streiften. Auf einmal tat es einen kurzen matten Blitz, als wenn es dem Himmel von der Pfanne gebrannt wäre. Was auch der Professor sagen mag, ich laß es mir nicht ausstreiten, ich sah damals einen Kerl mit Kolophonium und Laterne schnell hinter den einen Felsen huschen. Eh' ich indes noch darüber reiflich nachdenken konnte, erfolgte ein zweiter, ordentlicher Blitz, das Nachtgewölk teilte sich knarrend – und auf dem Hexensteine vor uns, in bläulicher bengalischer Beleuchtung, stand plötzlich ein ziemlich leichtfertig angezogenes Frauenzimmer zierlich auf einem Beine, beide Arme über sich emporgeschwungen, zu ihren beiden Seiten zwei elegant gekleidete junge Männer in Schuh und Strümpfen und Klapphüte unter den Armen, mit den beiden anderen Armen über dem Haupte der Dame in malerischer Stellung einen luftigen Schwibbogen bildend.

In demselben Augenblick lag auch die ganze Schar der Wallfahrer, mit den Angesichtern auf den Boden gestreckt, in tiefster Anbetung versunken. Ich erschrak, als ich fragend um mich her schaute und mich auf einmal als den einzigen Aufrechtstehenden befand in der kuriosen Gemeine. – „Die öffentliche Meinung!" rief da leise eine Stimme hinter mir, und zugleich fühlte ich ein Paar Fäuste so derb in beiden Kniekehlen, daß ich gleichfalls auf meine Knie hinstürzte.

Als ich einigermaßen wieder zur Besinnung gekommen war, stand mein Professor schon vor dem Altare und hielt eine gutgesetzte Rede an die öffentliche Meinung. Er sprach und log wie gedruckt: von ihren außerordentlichen Eigenschaften, dann von den Volkstugenden, von der Preßfreiheit und dem allgemeinen Schrei darnach. Ich aber wußte wohl, was sie geschrien hatten, und wer eigentlich gepreßt worden war.

Die Rede dauerte erstaunlich lange. Die arme öffentliche Meinung konnt' es kaum mehr aushalten, sie stellte sich bald auf dieses, bald auf jenes Bein, das andere vor sich in die Luft streckend, wie eine Gans, die Langeweile hat. Da hatte ich denn Zeit genug, sie mir recht genau zu betrachten. Sie trug ein prächtiges Ballkleid von Schillertaft, der bei der bengalischen Beleuchtung wechselnd in allen Farben spielte, ihre Finger funkelten von Ringen, von der Stirn blitzte ein ungeheures regardez-moi, aber alles, wie mir schien, von böhmischen Steinen. Übrigens war sie etwas kurzer, derber Konstitution, daher stand sie auf dem Kothurn, während dicke Sträuße hoher Pfauenfedern von ihrer turmähnlichen Frisur herabnickten. Ein leises Bärtchen auf der Oberlippe stand ihr gar nicht übel; dabei hatte sie ein gewisses air enragé, ich weiß nicht, ob von Schminke oder von der gezwungenen Stellung, oder ob sie gleichfalls gegen die Nachtluft einen Schnaps genommen hatte.

Währenddes war der Professor allmählich in seiner Redewut fast außer sich geraten. „Triumph! Triumph!" schrie er, ganz rotblau im Gesicht, „das Volk hat sich selbst geistig emanzipiert. Die Augen Europas – was sag' ich Europas! – des Weltbaues, sind in diesem hochwichtigen Augenblick auf uns gerichtet. Ja, wenn man mich hier niederwürfe und knebelte, die Gewalt der Wahrheit würde den Knebel aus dem Munde speien, und gefesselt von dem Boden noch würde himmelwärts ich schreien: Es werde Licht, es weiche die Finsternis, nieder mit der Zensur!"

Ein ungeheures Bravogebrüll donnerte den ganzen Berg hinab und wieder herauf. Einige Stimmen riefen: Da capo!, der Professor, der sich unterdes ein wenig erholt hatte, schickte sich auch unverdrossen an, von neuem loszulegen, und ich glaube in der Tat, er spräche noch heut, wenn die öffentliche Meinung, die sich seit geraumer Zeit schon zu ennuyieren schien, nicht schnell vom Altar herabgesprungen wäre, sein Haupt mit ihrem Fächer berührend, als wollte sie ihn zum Ritter schlagen.

Darauf rauschte sie in ihrem Taftgewande wohlgefällig durch die Reihen ihrer Getreuen. Da entstand aber bald ein außerordentliches Gedränge um sie her. Jeder wollte wenigstens den Saum ihres Kleides küssen, wobei sie denn manchem mit ihren Pantöffelchen unversehens einen derben Tritt versetzte, oder wohl auch ihr Schnupftuch fallen ließ und sich dann totlachen wollte, wenn sie sich darum rissen, um es ihr zu apportieren. Viele junge Autoren umschwärmten sie von allen Seiten und suchten sich durch elegante Konversation und politische Witze bei ihr zu insinuieren, während sie jeden Laut aus dem Munde der Angebeteten eifrig in ihre Etuikalender notierten. Mehrere ernstere Männer dagegen schritten nebenher und lasen ihr mit lauter Stimme die schönsten Paragraphen ihrer neuen Kompendien vor. Sie aber ließ ihre spielenden Augen durch die Scharen ergehen, und hatte gar bald einen Studenten erspäht, der, unablässig nach Freiheit schreiend, sich mit Ziegenhainer und Kanonen in dem Gedränge Bahn machte. Er war auf seinem Stiefelknecht hergeritten, ein junger Bursch von kräftigem Gliederbau, mehr Bart als Gesicht, mehr Stiefel als Mann. Sie winkte ihn heran, hing sich ohne weiteres an seinen Arm und, eh' ich's mich versah, war sie mitten durch das Getümmel im Dunkel der verschwiegenen Nacht mit ihm verschwunden.
Ich schaute dem Paar, ganz erstaunt, noch lange nach, wäre aber dabei um ein Haar umgerannt worden. Denn die andern schienen eben nicht viel aus dem Verschwinden zu machen, vielmehr sah ich sie nun, mit einer mir unerklärlichen Geschäftigkeit, plötzlich in großer Eile hin und her laufen, den Professor mitten unter ihnen, voller Eifer anordnend, rufend und treibend. Einige hatten sich an den Zipfel eines vorüberfliegenden Nebelstreifs gehängt und bogen ihn herunter, andere rollten ein leichtes Gewölk wie einen Vorhang auf, während wieder andere sich wunderlich in eine schwere dicke Wolke hineinarbeiteten, die sich auch wirklich nach und nach in Bäume, Felsen und Häuser zu gestalten anfing. Im Hintergrunde aber schien

sich ein seltsames Wolkengerüst mit Bogen und Galerien langsam aufzubauen, alles grau in grau; dazwischen pfiff ein heftiger Zugwind, daß ich meinen Hut mit beiden Händen auf dem Kopfe festhalten mußte, und die Fackeln warfen wilde rote Streiflichter zwischen die Wolkengebilde, überall ein chaotisches Dehnen und Wogen, als sollte die Welt von neuem erschaffen werden. – Vom Professor erfuhr ich endlich im Fluge, daß man in aller Geschwindigkeit eine Bühne einrichte, um vor den Augen der öffentlichen Meinung sich die Zukunft ein wenig einzuexerzieren.

In der Tat, ich bemerkte nun auch bald, wie jene Galerien sich allmählich mit Zuschauern füllten, aber lauter nur halbkenntliche Gestalten, deren Gliedmaßen allmählich nebelhaft auseinanderzufließen schienen; ich glaube, es war auch ein zukünftiges Publikum, das in der Eile noch nicht ganz fertig geworden war, aber doch schon sehr laut plauderte. Nur die Hauptloge stand noch leer; sie war prächtig ausgeschmückt, über ihr funkelte eine Sonne im Brillantfeuer, deren Gesicht, zu meinem großen Erstaunen, grauenhaft die Augen rollte und bald schmunzelte, bald gähnte. – Endlich erschien die öffentliche Meinung mit bedeutendem Geräusch in der Loge, das ganze Publikum stand auf und verneigte sich ehrerbietig. In demselben Augenblick wurde ein Böller gelöst, und ohne Ouvertüre, Prolog oder andern Übergang ging unten sogleich die Zukunft los.

Zuerst kam ein langer Mann in schlichter bürgerlicher Kleidung plötzlich dahergestürzt, ein Purpurmantel flog von seiner Schulter hinter ihm her, eine Krone saß ihm in der Eile etwas schief auf dem Haupt; dabei die Adlernase, die kleinen blitzenden Augen, die flammenrote Stirn: er war offenbar seines Gewerbes ein Tyrann. Er schritt hastig auf und ab, sich manchmal mit dem Purpurmantel den Schweiß von der Stirn wischend, und studierte in einem dicken Buche über Urrecht und Menschheitswohl, wie ich an den großen goldnen Buchstaben auf dem Rücken

des Buches erkennen konnte. Ein Oberpriester im Talar eines ägyptischen Weisen schritt ihm mit einer brennenden Kerze feierlich voran. Ich hätte beinah laut aufgelacht: es war wahrhaftig niemand anders, als mein Professor! Er hatte nicht geringe Not hier, denn, um immer in gehöriger Distanz voranzubleiben, suchte er, halb rückwärts gewendet, Schnelligkeit und Richtung in den Augen des Tyrannen vorauszulesen, der oft anhielt, oft plötzlich wieder rasch vorschritt und dem Professor unverhofft auf die Fersen trat. Auf einmal blieb der Tyrann mit über der Brust verschränkten Armen, wie in tiefes Nachsinnen versunken, stehen. Dann, nach einer gedankenschweren Pause, rief er plötzlich: „Ja, seid umschlungen, Millionen! Es weiche die Finsternis, nieder mit der Zensur!" – Da klatschte die öffentliche Meinung von neuem, die andern folgten, der Tyrann verneigte sich, die Krone vom Kopfe lüftend, und verschwand mit Würde hinter den Wolkenkulissen.

Jetzt blieb der Professor in seinem Priestertalar allein zurück. Er schien die Exposition des Ganzen machen zu wollen und freute sich in einem salbungsreichen Monologe weitläufig über die gute Applikation des Tyrannen, wie er schon seit geraumer Frist sich auf den Patriotismus lege und es sich recht sauer werden lasse, mit der Zeit fortzuschreiten usw. Während er so deklamierte, traten noch andere und immer mehrere Oberpriester von allen Seiten herzu, jeder von ihnen hatte gleichfalls ein brennendes Licht in der Hand. Sie verneigten sich erst verbindlich einer vor dem andern und drückten dann ihr gerechtes Erstaunen aus, wie sie in Behandlung des Tyrannen und sonst im Fache der Vaterländerei bereits so Großes vollbracht, wobei sie sich wechselseitig auf das vergnüglichste lobten. Das schien aber nicht ernstlich gemeint, denn jeder Lobende wandte sich jedesmal mit einem verächtlichen Achselzucken von dem eben Belobten, und suchte ihm heimlich von seinem tropfenden Lichte einige Kleckse auf den weißen Talar beizubringen, bei welcher Gele-

genheit ich denn bemerkte, daß ihre Kerzen bloße Talglichter waren und einen üblen Dunst verbreiteten.
Zwei von den Oberpriestern schienen besonders ihr vertrauliches Stündchen zu haben. Sie nahmen eine Prise Tabak zusammen und beklagten sich, daß es so langsam ginge in der Welt. Sie würden endlich auch alt und schäbig, und ihre Kerzen brennten sie bald auf die Finger. Das Volk werde es am Ende noch merken, daß sie den Tyrannen nur darum in solchen Edelmut und Resignation brächten, um dann selber auf seinem Throne Platz zu nehmen und kommode zu regieren, wie es ihnen eben konveniere. Jeder von ihnen habe doch unten, der eine sein Schätzchen, die durchaus Königin, der andere einen lüderlichen Vetter, der Minister werden wolle. – Vergeblich hustete der Professor immer lauter und lauter; vergebens schimpfte er halbleise: „Seid ihr betrunken, daß ihr das alles hier vor dem Volke ausplaudert!" – Endlich erscholl ein Schrei des einen plauderhaften Oberpriesters; der Professor hatte dem Unglücklichen insgeheim auf sein bestes Hühnerauge getreten.
Glücklicherweise indes war das ganze Gespräch nicht bis zu den Ohren der öffentlichen Meinung gekommen. Diese hatte schon lange nicht mehr aufgepaßt, sie schwatzte mit ihren Nachbarn, bog sich weit aus der Loge hervor und musterte das Publikum durch ihr Opernglas. Der Schrei des Getretenen erregte endlich ihre Aufmerksamkeit. Sie meinte, sie hätten da unten wieder einen philosophischen Zank, was sie jederzeit gewaltig langweilte. Sie ergriff daher rasch ihre Papagenoflöte, die sie beständig am Halse trug, und fing in ihrer Launenhaftigkeit einen Kontertanz zu blasen an. Umsonst protestierten die erschrockenen Oberpriester, das liege ja gar nicht im Plane des Stückes, es half alles nichts, sie mußten, ohne alles vernünftige Motiv, nach ihrer Pfeife tanzen. Das war wie ein Fackeltanz betrunkener Derwische, die langen Habite flogen, bläuliche Irrlichter, wie sie sprangen, schlugen foppend zwischen ihnen auf dem Boden auf, sie betropf-

ten sich mit den Talglichtern von oben bis unten, daß es eine Schande war, und der Schweiß strömte von ihren Angesichtern, bis sie endlich in verwegenen Luftsprüngen plötzlich nach allen Seiten auseinanderstoben. Mein armer Professor war dabei unversehens in einen Sumpf geraten; ich sprang herbei und half ihm heraus, aber den einen Schmierstiefel mußte er doch drin stecken lassen.

Die hurtige Zukunft inzwischen ging über umgefallene Oberpriester und Schmierstiefel unaufhaltsam ihren Gang weiter fort. Ein Mittelgewölk wurde schnell aufgerollt, und man übersah auf einmal einen weiten Marktplatz voll der lebhaftesten Geschäftigkeit, von den schönsten Palästen umgeben. Aber die Besitzer der letzteren schienen ausgezogen oder verstorben zu sein; wenigstens erblickte man überall nur Tagelöhner und Fabrikarbeiter, die sich selbst ihre Stiefel putzten, ihre Frauen hingen durchlöcherte Wäsche über die marmornen Fensterbrüstungen zum Trocknen aus, mit den offnen Fenstern klappte der Wind, und von Zeit zu Zeit flogen die Scherben einer zerbrochenen Scheibe den Vorüberwandelnden an die Köpfe. Anderes Volk, als hätte man einen Sack voll Lumpen ausgeschüttet, sonnte sich, behaglich über die Marmortreppen der Paläste hingestreckt. Eine prächtige, mit vier Pferden bespannte Staatskarosse rollte über den Platz; mit Erstaunen sah ich am Wagenfenster den nackten Ellenbogen eines Handwerkers, der aus dem zerrissenen Ärmel sich in der Sonne spiegelte. Hinten auf dem Wagentritt aber standen zwei Kavaliere und blickten im Bewußtsein aufgeklärten Edelmuts stolz von der Höhe herab, zu der ihre starken Seelen sich zu erheben gewußt.

Das Patriarchalische dieses rührenden Völkerglücks wurde nur durch einen betäubenden Lärm auf dem Platze selbst unterbrochen. Da gab's ein Heben, Messen, Hämmern und Klappern. Es waren die Oberpriester und andere Gelehrte, sie bauten eine große Regierungsmaschine nach der neuesten Erfindung des Professors, der sich darauf ein Patent erteilen zu lassen im Sinne führte.

Mitten durch dieses Getümmel aber sah man den Tyrannen in Pantoffeln und Schlafrock, als Landesvater unter seinen Kindern, mit einer langen Pfeife auf und nieder wandeln. Krone und Mantel hatte er unterdes an einen Türpfosten an den Nagel gehängt, mit dem Zepter rührte eine rüstige Schneiderfrau im Kessel den Brei für ihre Gesellen um. Er selbst hatte, des Budgets eingedenk, sogar den Gebrauch eines Hutes verschmäht, um ihn nicht durch vieles Grüßen abzunutzen. Überhaupt schien er es in der Popularität schon ziemlich weit gebracht zu haben, nur faßte er es offenbar noch etwas ungeschickt an. So kostete er zum Beispiel unnützerweise von dem Brei im Kessel und verbrannte sich den Mund, ja alle zehn Schritte rief er wiederholt: „Où peut on être mieux, qu'au sein de sa famille!" was die Kerls, die kein Französisch verstanden, für eine jesuitische Zauberformel hielten.
Dazwischen gähnte er dann zuweilen wie eine Hyäne, als wollte er seine Untertanen verschlingen. Da wurde dem Professor, der es bemerkte, ein wenig angst. Er suchte seine Aufmerksamkeit auf die neue Regierungsmaschine zu lenken. Aber der Tyrann konnte sich durchaus nicht darein verstehen, die Pfeife ging ihm aus, sein Verstand stand ihm still dabei. Vergeblich sprachen die Oberpriester, erklärend, von Intelligenz, Garantien, Handels-, Rede-, Gedanken-, Gewerbe-, Preß- und anderer Freiheit. „Ja, wenn ich nur etwas davon hätt'", entgegnete der Tyrann, kaltblütig seine Pfeife ausklopfend. Man sah es ihm an, wie er sich bezwang und abstrapazierte, human zu sein, er sah schon ordentlich angegriffen aus von den Bürgertugenden.
Bis hierher war nun alles ganz vortrefflich gegangen. Aber wie es wohl im Leben geschieht, es gehört oft nur ein kleiner Stein dazu, um in den weisesten Kopf ein Loch zu schlagen. So begab sich's nun auch hier. Der Tyrann, an nichts als an seine Fortschritte denkend, war eben bescheiden zur Seite getreten, um seine Tabakspfeife von neuem zu stopfen, als er plötzlich mit langen Schritten

und allen Symptomen langverhaltener Wut, wie ein leuchtendes Ungewitter, wieder hervorstürzte; seine Stirn glühte aus dem bleichen Gesicht, die Augen funkelten, der Schlafrock rauschte weit im Winde – das Volk hatte ihm seinen Tabaksbeutel gestohlen. Der Professor, als er ihn so daherfliegen sah, erschrak sehr. „Um Gottes willen“, rief er ihm entgegen, „wie wird Ihnen? Woher dieser unverhoffte Rückfall? Sie bringen uns das ganze Stück ins Wackeln!“ – Die öffentliche Meinung pfiff aus Leibeskräften, das gebildete Publikum pochte in gerechtem Unwillen, die Oberpriester langten in der Angst eine Konstitution nach der andern aus den Taschen und warfen sie dem Wüterich zwischen die langen Beine, um ihn zum Stolpern zu bringen. Alles vergebens! Er wollte von Bürgertugend, Popularität und Völkerglück nichts mehr hören und nahm, wie ein Stier, einen entsetzlichen Anlauf, um die ganze Zukunft umzurennen.

Doch die Konfusion sollte noch immer größer werden. Den Faulenzern auf dem Platze, die sich hier eigentlich durch Selbstdenken hatten emanzipieren sollen, war inzwischen auch die Zeit lang geworden. Was haben sie zu tun? Während die andern an der Regierungsmaschine arbeiten, nehmen sie, ganz wider den Plan des Stückes, heimlich Krone und Purpurmantel vom Nagel, holen das Zepter dazu und begeben sich damit ohne weiteres nach der Restauration. Unterwegs kriegen sie Händel untereinander, zerreißen sich und ihre Beute, und lassen sich für die Stücke in der Restauration Schnaps geben. Der Wirt, ein anschlägiger Kopf, wie er diese unerwartete Wendung der Staatsaktion sieht, besinnt sich nicht lange, zapft und läßt laufen, was er hat, leimt und flickt die Stücke schnell wieder zusammen, legt selber Kron' und Mantel an, nimmt das Zepter in die Rechte und führt die freudetrunkene Bande, wie einen Kometenschweif, nach der Bühne zurück.

War nun die Zukunft vorhin schon im Wackeln, so schien sie jetzt ganz und gar in Stücke gehen zu wollen. Derweil die Oberpriester und Schriftgelehrten noch immer beflis-

sen waren, den empörten Tyrannen wieder zu zähmen, ging auf einmal ein Mittelvorhang auf, und man erblickte im Hintergrunde den Thron selbst, auf dem soeben der Wirt aus der Restauration sich breit und vergnüglich zurechtsetzte, wie einer, der mit seiner eigenen Pfiffigkeit wohl zufrieden war. Seine ganze Nation drängte sich, taumelte, lag und hing über Stufen und Lehne des Thrones um ihn her, so daß er gleich zu Anfang von seinem Zepter einen nachdrücklichen Gebrauch machen mußte.

Der Professor und die Seinigen aber standen unten wie angedonnert, sie trauten sich nicht an den unerwarteten Usurpator und feuerten nur aus der Ferne mit wütenden Blicken. Dann traten sie schnell auf die Seite, steckten die Köpfe zusammen und schienen zu konspirieren. Mit Erstaunen glaubte ich dabei einigemal meinen Namen nennen zu hören, und konnte wohl bemerken, daß sie mich öfters bedeutungsvoll ansahen. Mein Gott, dachte ich, nun kommst du am Ende noch selbst mit in das Stück hinein, und ein heimliches Entsetzen rieselte mir durch alle Glieder. Es dauerte auch nicht lange, so kam der Professor auf mich zugeflogen, riß mir meinen Oberrock vom Leibe und zog mir rasch ein prächtiges Hofkleid an, ein anderer rasierte mich, ein dritter steckte mir einen dicken Blumenstrauß vorn ins Knopfloch – ich wußte nicht, wie mir geschah. In der Eile erfuhr ich dann: wie sie der Meinung seien, ich als Uneingeweihter bringe hier alles in solche Unordnung durch meine kritische Gegenwart; auch könnte ich wohl, wenn ich morgen vom Blocksberg käme, unten alles ausplaudern. Umbringen wollten sie mich nicht, weil ich der öffentlichen Meinung ausnehmend gefalle; ich mußte mich daher mit der letzteren sogleich vermählen, um ganz der Ihrige zu werden. – „Aber das ist ja ein Vergnügen zum Tollwerden!" rief ich auf das heftigste erschrocken aus. – „Bah, Kleinigkeit", fiel mir der Professor in die Rede, „wir alle, die Sie hier sehen, sind schon mit ihr verheiratet." – Mich schauerte bei dem Gedanken dieser ungeheuern Schwägerschaft!

Unterdes waren die anderen Gelehrten dennoch mit dem Volke um den Besitz des Thrones handgemein geworden; darüber bekamen die Prinzipien Luft, die sie in die Regierungsmaschine verbaut hatten. Eins nach dem andern streckte neugierig den Kopf hervor, und da es so lustig herging draußen, rüttelten und schüttelten sie und brachen den ganzen Plunder entzwei. Da sah man dort einen dünnen Paragraphen, dort ein schweres Korollarium, hier einen luftigen Heischesatz aus den Trümmern steigen, und kaum fühlten sie sich frei, so lagen sie einander auch schon wieder in den Haaren und stürzten raufend in das dickste Getümmel.
Nun entstand eine allgemeine Schlägerei, da wußte keiner mehr, wer Freund oder Feind war! Dazwischen raste der Sturm, Besen flogen, tiefer unten krähte der rote Hahn wieder, bliesen die sieben Pfeifer, schrie der Wirt, die Bühne suchte die alte Freiheit und rührte und reckte sich in wilde Nebelqualme auseinander, ein entsetzliches übermenschliches Lachen ging durch die Lüfte, der ganze Berg schien auf einmal sich in die Runde zu drehen, erst langsam, dann geschwinder und immer geschwinder – mir vergingen die Gedanken, ich stürzte besinnungslos zu Boden.
Als ich die Augen wieder aufschlug, lag ich ruhig in dem Gasthofe „Zum goldenen Zeitgeist“ im Bett. Die Sonne schien schon hell ins Zimmer, der fatale Kellner stand neben mir und lächelte wieder so ironisch, daß ich mich schämte, nach dem Professor, dem Pegasus und dem Blocksberg zu fragen. Ich griff verwirrt nach meinem Kopf: ich fühlte so etwas von Katzenjammer. Und in der Tat, da ich's jetzt recht betrachte, ich weiß nicht, ob nicht am Ende alles bloß ein Traum war, der mir, wie eine Fata Morgana, die duftigen Küsten jenes volksersehnten Eldorados vorgespiegelt. Dem aber sei nun wie ihm wolle, genug: auch ich war in Arkadien!

Eine Meerfahrt

Es war im Jahre 1540, als das valenzische Schiff „Fortuna" die Linie passierte und nun in den Atlantischen Ozean hinausstach, der damals noch einem fabelhaften Wunderreiche glich, hinter dem Kolumbus kaum erst die blauen Bergesspitzen einer neuen Welt gezogen hatte. Das Schiff hatte eben nicht das beste Aussehen, der Wind pfiff wie zum Spott durch die Löcher in den Segeln, aber die Mannschaft, lumpig, tapfer und allezeit vergnügt, fragte wenig darnach, sie fuhren immerzu und wollten mit Gewalt neue Länder entdecken. Nur der Schiffshauptmann *Alvarez* stand heute nachdenklich an den Mast gelehnt, denn eine rasche Strömung trieb sie unaufhaltsam ins Ungewisse von Amerika ab, wohin er wollte. Von der Spitze des Verdecks aber schaute der fröhliche Don *Antonio* tief aufatmend in das fremde Meer hinaus, ein armer Student aus Salamanka, der von der Schule neugierig mitgefahren war, um die Welt zu sehen. Dabei hatte er heimlich noch die Absicht und Hoffnung, von seinem Oheim Don Diego Kunde zu erhalten, der vor vielen Jahren auf einer Seereise verschollen war, und von dessen Schönheit und Tapferkeit er als Kind so viel erzählen gehört, daß es noch immer wie ein Märchen in seiner Seele nachhallte. – Ein frischer Wind griff unterdes in die geflickten Segel, die künstlich geschnitzte bunte Glücksgöttin am Vorderteil des Schiffes glitt heiter über die Wogen, den wandelbaren Tanzboden Fortunas. Und so segelten die kühnen Gesellen wohlgemut in die unbekannte Ferne hinaus, aus der ihnen seltsame Abenteuer, zackiges Gebirge und stille blühende Inseln wie im Traume allmählich entgegendämmerten. Schon zwei Tage waren sie in derselben Richtung fortgesegelt, ohne ein Land zu erblicken, als sie unerwartet in

den Zauberbann einer Windstille gerieten, die das Schiff fast eine Woche lang mit unsichtbarem Anker festhielt. Das war eine entsetzliche Zeit. Der hagere, gelbe Alvarez saß unbeweglich auf seinem ledernen Armstuhle und warf kurze, scharfe Blicke in alle Winkel, ob ihm nicht jemand guten Grund zu ordentlichem Zorne geben wollte, die Schiffsleute zankten um nichts vor Langeweile, dann wurde oft alles auf einmal wieder so still, daß man die Ratten im unteren Raum schaben hörte. Antonio hielt es endlich nicht länger aus und eilte auf das Verdeck, um nur frische Luft zu schöpfen. Dort hingen die Segel und Taue schlaff an den Masten, ein Matrose mit offener brauner Brust lag auf dem Rücken und sang ein valenzianisches Lied, bis auch er einschlief. Antonio aber blickte in das Meer, es war so klar, daß man bis auf den Grund sehen konnte, das Schiff hing in der Öde wie ein dunkler Raubvogel über den unbekannten Abgründen, ihm schwindelte zum erstenmal vor dem Unternehmen, in das er sich so leicht gestürzt. Da gedachte er der fernen schattigen Heimat, wie er dort als Kind an solchen schönen Sommertagen mit seinen Verwandten oft vor dem hohen Schloß im Garten gesessen, wo sie nach den Segeln fern am Horizonte aussahen, ob nicht Diegos Schiff unter ihnen. Aber die Segel zogen wie stumme Schwäne vorüber, die Wartenden droben wurden alt und starben, und Diego kam nicht wieder, kein Schiffer brachte jemals Kunde von ihm. – Das Angedenken an diese stille Zeit wollte ihm das Herz abdrücken, er lehnte sich an den Bord und sang für sich:

Ich seh' von des Schiffes Rande
Tief in die Flut hinein:
Gebirge und grüne Lande,
Der alte Garten mein,
Die Heimat im Meeresgrunde,
Wie ich's oft im Traum mir gedacht,
Das dämmert alles da drunten
Als wie eine prächtige Nacht.

Die zackigen Türme ragen,
Der Türmer, er grüßt mich nicht,
Die Glocken nur hör' ich schlagen
Vom Schloß durch das Mondenlicht,
Und der Strom und die Wälder rauschen
Verworren vom Grunde her,
Die Wellen vernehmen's und lauschen
So still übers ganze Meer.

Don Diego auf seiner Warte
Sitzet da unten tief,
Als ob er mit langem Barte
Über seiner Harfe schlief.
Da kommen und gehn die Schiffe
Darüber, er merkt es kaum,
Von seinem Korallenriffe
Grüßt er sie wie im Traum.

Und wie er noch so sann, kräuselte auf einmal ein leiser Hauch das Meer immer weiter und tiefer, die Segel schwellten allmählich, das Schiff knarrte und reckte sich wie aus dem Schlaf, und aus allen Luken stiegen plötzlich wilde, gebräunte Gestalten empor, da sie die neue Bewegung spürten, sie wollten sich lieber mit dem ärgsten Sturme herumzausen als länger so lebendig begraben liegen. Auf einmal schrie es „Land!" vom Mastkorbe, „Land, Land!" Antonio kletterte in seinem buntseidenen Wams wie ein Papagei auf der schwankenden Strickleiter den Hauptmast hinan, er wollte das Land zuerst begrüßen. Alvarez eilte nach seiner Karte, da war aber alles leer auf der Stelle, wo sie soeben sich befinden mußten. „Bakkalaureus, Herzensjunge!" schrie er herauf, „schaff mir einen schwarzen Punkt auf die Karte hier, ich mach' dich zum Doktor drin, was siehst du?" – „Ein blauer Berg taucht auf", rief Antonio hinab, „jetzt wieder einer – ich glaub', es sind Wolken, es dehnt sich und steigt im Nebel wie Turmspitzen. – Nein, jetzt unterscheide ich Gipfel, o,

wie das schön ist! und helle Streifen dazwischen in der Abendsonne, unten dunkelt's schon grün, die Gipfel brennen wie Gold." – „Gold?" rief der Hauptmann und hatte sein altes Perspektiv genommen, er zielte und zog es immer länger und länger, er schwor, es sei das reiche Indien, das unbekannte große Südland, das damals alle Abenteurer suchten.

In diesem Augenblicke aber waren plötzlich alle Gesichter erbleichend in die Höhe gerichtet: ein dunkler Geier von riesenhafter Größe hing mit weit ausgespreizten Flügeln gerade über dem Schiff, als könnt' er die Beute von Galgenvögeln nicht erwarten. Bei dem Anblick ging ein Gemurmel, erst leise, dann immer lauter, durch das ganze Schiff, alle hielten es für ein Unglückszeichen. Endlich brach das Schiffsvolk los, sie wollten nicht weiter und drangen ungestüm in den Hauptmann, von dem verhängnisvollen Eiland wieder abzulenken. Da zog Alvarez heftig seinen funkelnden Ring vom Finger, lud ihn schweigend in seine Muskete und schoß nach dem Vogel. Dieser, tödlich getroffen, wie es schien, fuhr pfeilschnell durch die Lüfte, dann sah man ihn taumelnd immer tiefer nach dem Lande hin in der Abendglut verschwinden. „Meld' dem Land, daß sein Herr kommt" – sagte Alvarez nachschauend, auf seine Muskete gestützt – „und wer mir den Ring wiederbringt, soll Statthalter des Reiches sein!" – „Hat sich was, wiederzubringen", brummte einer, „der Ring war nur von böhmischen Steinen!"

Indem aber fing die Luft schon zu dunkeln an, man beschloß daher, den folgenden Tag abzuwarten, bevor man sich der unbekannten Küste näherte. Die Segel wurden eiligst eingezogen, die Anker geworfen und auf Bord und Masten Wachen ausgestellt. Aber keiner konnte schlafen vor Erwartung und Freude, die Matrosen lagen in der warmen Sommernacht plaudernd auf dem Verdecke umher, Alvarez, Antonio und die Offiziere saßen zusammen vorn auf Fortunas Schopfe, unter ihnen schlugen die Wellen leise ans Schiff, während fern am Horizont die Nacht

sich mit Wetterleuchten kühlte. Der vielgereiste Alvarez erzählte vergnügt von seinen früheren Fahrten, von ganz smaragdenen Felsenküsten, an denen er einmal gescheitert, von prächtigen Vögeln, die wie Menschen sängen und die Seeleute tief in die Wälder verlockten, von wilden Prinzessinnen auf goldenen Wagen, die von Pfauen gezogen würden. – „Wer da!" rief da auf einmal eine Wache an, alles sprang rasch hinzu. „Wer da, oder ich schieße!" schrie der Posten von neuem. Da aber alles stille blieb, ließ er langsam seine Muskete wieder sinken und sagte nun aus, es sei ihm schon lange gewesen, als hörte er in der See flüstern, immer näher, bald da, bald dort, dann habe plötzlich die Flut ganz in der Nähe aufgerauscht. Alle lauschten neugierig hinaus, sie konnten aber nichts entdecken, nur einmal war's ihnen selber, als hörten sie Ruderschlag von ferne. – Unterdes aber war der Mond aufgegangen, und sie bemerkten nun, daß sie dem Lande näher waren, als sie geglaubt hatten. Dunkle Wolken flogen wechselnd darüber, der Mond beleuchtete verstohlen ein Stück wunderbares Gebirge mit Zacken und jähen Klüften, immer höher stieg eine Reihe Gipfel hinter der andern empor, der Wind kam vom Lande, sie hörten drüben einen Vogel melancholisch singen und ein tiefes Rauschen dazwischen, sie wußten nicht, ob es die Wälder waren oder die Brandung. So starrten sie lange schweigend in die dunkle Nacht, als auf einmal einer den andern flüsternd anstieß. „Sirenen!" hieß es da plötzlich von Mund zu Munde, „seht da, ein ganzes Nest von Sirenen!" – und in der Ferne glaubten sie wirklich schlanke weibliche Gestalten in der schimmernden Flut spielend auftauchen und wieder verschwinden zu sehen. – „Die erwisch' ich!" rief Alvarez, der sich indes rasch mit Degen, Muskete und Pistolen schon bis an die Zähne bewaffnet hatte und eiligst auf der Schiffsleiter in das kleine Boot hinabstieg. Antonio folgte fast unwillkürlich. – „Gott schütz', der Hauptmann wird verliebt, bindet ihn!" riefen da mehrere Stimmen verworren durcheinander. Alle wollten nun die tolle Ab-

fahrt hindern, da sie aber das Boot festhielten, zerhieb Alvarez zornig mit seinem Schwerte das Tau, und die beiden Abenteurer ruderten allein in den Mondglanz hinaus. Die zurückkehrende Flut trieb sie unmerklich immer weiter dem Lande zu, ein erquickender Duft von unbekannten Kräutern und Blüten wehte ihnen von der Küste entgegen, so fuhren sie dahin. Auf einmal aber bedeckte eine schwere Wolke den Mond, und als er endlich wieder hervortrat, war See und Ufer still und leer, als hätte der fliegende Wolkenschatten alles abgefegt. Betroffen blickten sie umher, da hatten sie zu ihrem Schrecken hinter einer Landzunge nun auch ihr Schiff aus dem Gesicht verloren. Die wachsende Flut riß sie unaufhaltsam nach dem Strande, das Ufer, wie sie so pfeilschnell dahinflogen, wechselte grauenhaft im verwirrenden Mondlicht, auf einsamem Vorsprunge aber saß es wie ein Riese in weiten, grauen Gewändern, der über dem Rauschen des Meeres und der Wälder eingeschlafen. – „Diego!" sagte Antonio halb für sich. – Alvarez aber, in Zorn und Angst, feuerte wütend sein Pistol nach der grauen Gestalt ab. In demselben Augenblick stieß das Boot so hart auf den Grund, daß der weiße Gischt der Brandung hoch über ihnen zusammenschlug. Alvarez schwang sich kühn auf einen Uferfels, den erschrockenen Antonio gewaltsam mit sich emporreißend, hinter ihnen zerschellte das Boot in tausend Trümmer. Aber so zerschlagen und ganz durchnäßt, wie er war, kletterte der Hauptmann eilig weiter hinan, und auf dem ersten Gipfel zog er sogleich seinen Degen, stieß ihn in den Boden und nahm feierlich Besitz von diesem Lande mit allen seinen Buchten, Vorgebirgen und etwa dazugehörigen Inseln. „Amen!" sagte Antonio, sich das Wasser von den Kleidern schüttelnd, „nun aber wollt' ich, wir wären mit Ehren wieder von dieser fürstlichen Höhe hinunter, ich gebe Euch keinen Pfeffersack für Euer ganzes zukünftiges Königreich!" – „Zukünftiges?" erwiderte Alvarez, „das ist mir just das liebste dran! Mit Kron' und Zepter auf dem Throne sitzen, Audienz geben, mit den

Gesandten parlieren: ‚was macht unser Herr Vetter von England' usw. Langweiliges Zeug! Da lob' ich mir einen Regenbogen, zweifelhafte Türme von Städten, die ich noch nicht sehe, blaues Gebirge im Morgenschein, es ist, als rittst du in den Himmel hinein; kommst du erst hin, ist's langweilig. Um ein Liebchen werben ist scharmant; heiraten: wiederum langweilig! Hoffnung ist meine Lust, was ich liebe, muß fern liegen wie das Himmelreich.

Soll Fortuna mir behagen
Will ich über Strom und Feld
Wie ein schlankes Reh sie jagen
Lustig bis ans End' der Welt!"

Eigentlich aber sang er mit seiner heiseren Stimme nur, um sich selber die Grillen zu versingen, denn ihre Lage war übel genug. Zu den Ihrigen wieder zurückzufinden, konnten sie nicht hoffen, ohne sich ihnen durch Signale kundzugeben; Feuer anzünden aber, schießen oder sonstigen Lärm machen, wollten sie nicht, um das wilde Gesindel nicht gegen sich aufzustören, das vielleicht in den umherliegenden Klüften nistete. Da beschlossen sie endlich, einen der höhern Berggipfel zu besteigen, dort wollten sie sich erst umsehen und im schlimmsten Falle den Morgen abwarten. Als sie nun aber in solchen Gedanken immer tiefer in das Gebirge hineingingen, kam ihnen nach und nach alles gar seltsam vor. Der Mondschein beleuchtete wunderlich Wälder, Berge und Klüfte, zuweilen hörten sie Quellen aufrauschen, dann wieder tiefe, weite Täler, wo hohe Blumen und Palmen wie in Träumen standen. Fremde Rehe grasten auf einem einsamen Bergeshange, die reckten scheu die langen, schlanken Hälse empor, dann flogen sie pfeilschnell durch die Nacht, daß es noch weit zwischen den stillen Felswänden donnerte.
Jetzt glaubte Antonio in der Ferne ein Feuer zu bemerken. Alvarez sagte: wo in diesen Ländern eine reiche Goldader

durchs Gebirge ginge, da gäbe es oft solchen Schein in stillen Nächten. Sie verdoppelten daher ihre Schritte, leis und vorsichtig ging es über mondbeglänzte Heiden, das Licht wurde immer breiter und breiter, schon sahen sie den Widerschein jenseits an den Klippen des gegenüberliegenden Berges spielen. Auf einmal standen sie vor einem jähen Abhange und blickten erstaunt in ein tiefes, rings von Felsen eingeschlossenes Tal hinab; kein Pfad schien zwischen den starren Zacken hinabzuführen, die Felswände waren an manchen Stellen wunderbar zerklüftet, aus einer dieser Klüfte drang der trübe Schein hervor, den sie von weitem bemerkt hatten. Zu ihrem Entsetzen sahen sie dort einen wilden Haufen dunkler Männer, Windlichter in den Händen, abgemessen und lautlos im Kreise herumtanzen, während sie manchmal dazwischen bald mit ihren Schilden, bald mit den Fackeln zusammenschlugen, daß die sprühenden Funken sie wie ein Feuerregen umgaben. Inmitten dieses Kreises aber, auf einem Moosbette, lag eine junge, schlanke Frauengestalt, den schönen Leib ganz bedeckt von ihren langen Locken, und Arme, Haupt und Brust mit funkelnden Spangen und wilden Blumen geschmückt, als ob sie schliefe, und so oft die Männer ihre Fackeln schüttelten, konnten sie deutlich das schöne Gesicht der Schlummernden erkennen.
„Es ist Walpurgis heut", flüsterte Alvarez nach einer kleinen Pause, „da sind die geheimen Fenster der Erde erleuchtet, daß man bis ins Zentrum schauen kann." Aber Antonio hörte nicht, er starrte ganz versunken und unverwandt nach dem schönen Weibe hinab. „Vermaledeiter Hexensabbat ist's", sagte der Hauptmann wieder, „Frau Venus ist's! In dieser Nacht alljährlich opfern sie ihr heimlich, *ein* Blick von ihr, wenn sie erwacht, macht wahnsinnig." – Antonio, so verwirrt er von dem Anblick war, ärgerte doch die Unwissenheit des Hauptmanns. „Was wollt Ihr?" entgegnete er leise, „die Frau Venus hat ja niemals auf Erden wirklich gelebt, sie war immer nur so ein Symbolum der heidnischen Liebe, gleichsam ein Luftge-

bild, eine Schimäre. Horatius sagt von ihr: Mater saeva cupidinum." – „Sprecht nicht lateinisch hier, das ist just ihre Muttersprache!" unterbrach ihn Alvarez heftig und riß den Studenten vom Abgrund durch Hecken und Dornen mit sich fort. „Der Teufel", sagte er, als sie schon eine Strecke fortgelaufen waren, „der Teufel – wollt' sagen: der – nun, Ihr wißt schon, man darf ihn heut nicht beim Namen nennen – der hat für jeden seine besondern Finten, unsereins faßt er geradezu beim Schopf, eh' man sich's versieht, euch Gelehrte nimmt er säuberlich zwischen zwei Finger wie eine Prise Tabak."

Unter diesem Diskurs stolperten sie, von Schweiß triefend, im Dunkeln über Stock und Stein, einmal kam's ihnen vor, als flöge eine Mädchengestalt über die Heide, aber der Hauptmann drückte fest die Ohren an. So waren sie in größter Eile, ohne es selbst zu bemerken, nach und nach schon wieder tief ins Tal hinabgekommen, als ihnen plötzlich ein: „Halt, wer da!" entgegenschallte. Da war es ihnen doch nicht anders, als ob sie eine Engelsposaune vom Himmel anbliese! – „He, Landsmann, Kameraden, Hollaho!" schrie Alvarez aus vollem Halse; sie traten aus dem Wald und sahen nun die Schiffsmannschaft auf einer Wiese am Meere um Feldfeuer gelagert, die warfen so lustige Scheine über die Gestalten mit den wilden Bärten, breit aufgekrämpten Hüten und langen Flinten, daß Antonio recht das Herz im Leibe lachte.

Alvarez aber, noch ganz verstört von der verworrenen Nacht, trat sogleich mitten unter die Überraschten und erzählte, wie sie eben aus dem Venusberge kämen und die Frau Venus auf diamantenem Throne gesehen hätten, was sie da erlebt, wollt' er keinem wünschen, denn er müßte gleich toll werden darüber. – „Kerl, warum senkst du die Hellebarde nicht, wenn dein Hauptmann vor dir steht?" fuhr er inzwischen die Schildwache an, die sich neugierig ebenfalls genähert hatte. Der Soldat aber schüttelte den Kopf, als kennte er ihn nicht mehr. Da trat der Schiffsleutnant Sanchez keck aus dem Gedränge hervor, er trug das

Hauptmannszeichen an seinem Hut. Es sei hier alles in guter Ordnung, sagte er zu Alvarez, er habe sie verlassen in der Not und Fremde, auch hätten sie sein Boot zertrümmert gefunden, da habe die Mannschaft nach Seegebrauch einen neuen Anführer gewählt, er sei jetzt der Hauptmann! – „Was", schrie Alvarez, „Hauptmann geworden, wie man einen Handschuh umdreht, wie ein Pilz über Nacht?" – Der schlaue Sanchez aber lächelte sonderbar. „Über Nacht" sagte er, „könnt Ihr etwa im Venusberge wissen, was es an der Zeit ist? Oho, wie lange denkt Ihr denn, daß Ihr fort gewesen, nun?" – Alvarez war ganz betreten, die furchtbare Sage vom Venusberg fiel ihm jetzt erst recht aufs Herz, er traute sich selber nicht mehr. – „Wißt Ihr denn nicht", sagte Sanchez, ihm immer dreister unter das Gesicht tretend, „wißt Ihr nicht, daß mancher als schlanker Jüngling in den Venusberg gegangen und als alter Greis mit grauem Barte zurückgekommen, und meint doch, er sei nur ein Stündlein oder vier zu Biere gewesen, und keiner im Dorfe kannte ihn mehr, und" – Wie er aber dem Alvarez so nahe trat, gab ihm dieser auf einmal eine so derbe Ohrfeige, daß der Hauptmannshut vom Kopfe fiel, denn er hatte sich unterdes rund umgesehen und wohl bemerkt, daß die andern kaum um ein paar Stunden älter geworden, seitdem er sie verlassen. Sanchez griff wütend nach seinem Degen, Alvarez auch, die andern drängten sich wild heran, einige wollten dem alten Hauptmann, andere dem neuen helfen. Da sprang Antonio mitten in den dichtesten Haufen, die Streitenden teilend. „Seid ihr Christen?" rief er, „blickt um euch her, auf was habt ihr eure Sach' gestellt, daß ihr so übermütig seid? Diese alten starren Felsen, die nur mit den Wolken verkehren, fragen nichts nach euch und werden sich eurer nimmermehr erbarmen. Oder baut ihr auf die Nußschale, die da draußen auf den Wellen schwankt? Der Herr allein tut's! Er hat uns mit seinen himmlischen Sternen durch die Einsamkeit der Nächte nach einer fremden Welt herübergeleuchtet und geht nun im stillen Morgengrauen über die

Felsen und Wogen, daß es wie Morgenglocken fern durch die Lüfte klingt, wer weiß, welchen von uns sie abrufen – und anstatt niederzusinken im Gebet, laßt ihr eure blutdürstigen Leidenschaften wie Hunde gegeneinander los, daß wir alle davon zerrissen werden." – „Er hat recht!" sagte Alvarez, seinen Degen in die Scheide stoßend. Sanchez traute dem Alvarez nicht, doch hätte er auffahren mögen vor Ärger, und wußte nicht, an wem er ihn auslassen sollte. „Ihr seid ein tapferer Ritter Rhetorio", sagte er, „habt Ihr noch mehr so schöne Sermone im Halse?" – „Ja, um jeden frechen Narren damit zu Grabe zu sprechen", entgegnete Antonio. – „Oho", rief Sanchez, „so müßt Ihr Feldpater werden, ich will Euch die Tonsur scheren, mein Degen ist just heute haarscharf." Da fuhr Alvarez auf: wer dem Antonio ans Leder wolle, müsse erst durch seinen eignen Koller hindurch. Aber Antonio hatte schon seinen Degen gezogen, trat mit zierlichem Anstande vor und sagte zum Leutnant, daß sie die Sache als Edelleute abmachen wollten. Alvarez und mehrere andere begleiteten nun die beiden weiterhin bis zum Saume des Waldes, die Schwerter wurden geprüft, und der Kampfplatz mit feierlichem Ernst umschritten. Die Palmen streckten ihre langen Blätter und Fächer verwundert über die fremden Gesellen hinaus. Gar bald aber blitzte der Mond in den blanken Waffen, denn Sanchez griff sogleich an und verschwor sich im Fechten, Antonio solle seinen Degen hinunterschlucken bis an den Griff. Der Student aber wußte schöne, gute Hiebe und Finten von der Schule zu Salamanka her, parierte künstlich, maß und stach und versetzte dem Prahlhans, ehe er sich's versah, einen Streich über den rechten Arm, daß ihm der Degen auf die Erde klirrte. Nun faßte Sanchez das Schwert mit der Linken und stürzte in blinder Wut von neuem auf seinen Gegner; er hätte sich selbst Antonios Degenspitze in den Leib gerannt, aber die andern unterliefen ihn schnell und warfen ihn rücklings zu Boden, denn jetzt erst bemerkten sie, daß er schwer betrunken war. In der Hitze des Kampfes hatte er völlig die Be-

sinnung verloren, sie mußten ihn an die Lagerfeuer zurücktragen, wo sie nun seine Wunde verbanden. Da hielt er sich für tot und fing sich selber ein Grablied zu singen an, aber es wollte nicht stimmen, er sah ganz unkenntlich aus, bis er endlich umsank und fest einschlief. – „Das ist gut, er hat die Rebellion mit seinem Blute wieder abgewaschen", sagte Alvarez vergnügt, denn alle waren dem Leutnant gewogen, weil er Not und Lust brüderlich mit seinen Kameraden teilte und in der Gefahr allezeit der erste war.

Unterdes aber hatte die Schiffsmannschaft eilig bunte Zelte aufgeschlagen und plauderte und schmauste vergnügt. Antonio mußte auf viele Gesundheiten fleißig Bescheid tun, sie erklärten ihn alle für einen tüchtigen Kerl. Dazwischen schwirrte eine Zither vom letzten Zelte, der Schiffskoch spielte den Fandango, während einige Soldaten auf dem Rasen dazu tanzten. Von Zeit zu Zeit aber rief Alvarez den Schildwachen zu, auf ihrer Hut zu sein, denn weit in der Nacht hörte man zuweilen ein seltsames Rufen im fernen Gebirge. Nach einer Stunde etwa erwachte der Leutnant plötzlich und sah verwirrt bald seinen Arm an, bald in der fremden Runde umher, aber er verwunderte sich nicht lange, denn dergleichen war ihm oft begegnet. Vom Meere wehte nun schon die Morgenluft erfrischend herüber, ihn schauerte innerlich, da faßte er einen Becher mit Wein und tat einen guten Zug; dann sang er, noch halb im Taumel, und die andern stimmten fröhlich mit ein:

„Ade, mein Schatz, du mocht'st mich nicht,
Ich war dir zu geringe,
Und wenn mein Schiff in Stücken bricht,
Hörst du ein süßes Klingen,
Ein Meerweib singt, die Nacht ist lau,
Die stillen Wolken wandern,
Da denk an mich, 's ist meine Frau,
Nun such dir einen andern.

Ade, ihr Landsknecht', Musketier'!
Wir ziehn auf wildem Rosse,
Das bäumt und überschlägt sich schier
Vor manchem Felsenschlosse,
Lindwürmer links bei Blitzesschein,
Der Wassermann zur Rechten,
Der Haifisch schnappt, die Möwen schrein –
Das ist ein lustig Fechten!

Streckt nur auf eurer Bärenhaut
Daheim die faulen Glieder,
Gott Vater aus dem Fenster schaut,
Schickt seine Sündflut wieder.
Feldwebel, Reiter, Musketier,
Sie müssen all' ersaufen,
Derweil auf der Fortuna wir
Im Paradies einlaufen."

Hier wurden sie auf einmal alle still, denn zwischen den Morgenlichtern über der schönen Einsamkeit erschien plötzlich auf einem Felsen ein hoher Mann, seltsam in weite bunte Gewande gehüllt. Als er in der Ferne das Schiff erblickte, tat er einen durchdringenden Schrei, dann, beide Arme hoch in die Lüfte geschwungen, stürzte er durch das Dickicht herab und warf sich unten auf seine Knie auf den Boden, die Erde inbrünstig küssend. Nach einigen Minuten aber erhob er sich langsam und überschaute verwirrt den Kreis der Reisenden, die sich neugierig um ihn versammelt hatten; es war ein Greis von fast grauenhaftem verwildertem Ansehen. Wie erschraken sie aber, als er sie auf einmal spanisch anredete, wie einer, der die Sprache lange nicht geredet und fast vergessen hatte. „Ihr habt euch", sagte er, „alle sehr verändert in der einen langen Nacht, daß wir uns nicht gesehen." Darauf nannte er mehrere unter ihnen mit fremden Namen und erkundigte sich nach Personen, die ihnen gänzlich unbekannt waren.

Die Umstehenden bemerkten jetzt mit Erstaunen, daß sein Gewand aus europäischen Zeugen bunt zusammengeflickt war; um die Schultern hatte er phantastisch einen köstlichen, halbverblichenen Teppich wie einen Mantel geworfen. Sie fragten ihn, wer er sei und wie er hierher gekommen. Darüber schien der Unbekannte in ein tiefes Nachsinnen zu versinken. „In Valencia", sagte er endlich halb für sich leise und immer leiser sprechend, „in Valencia zwischen den Gärten, die nach dem Meere sich senken, da wohnt ein armes, schönes Mädchen, und wenn es Abend wird, öffnet sie das kleine Fenster und begießt ihre Blumen, da sang ich manche Nacht vor ihrer Tür. Wenn ihr sie wiederseht, sagt ihr – daß ich – sagt ihr" – Hier stockte er, starr vor sich hinsehend, und stand wie im Traume. Alvarez entgegnete: das Mädchen, wenn sie etwa seine Liebste gewesen, müsse nun schon hübsch alt oder längst gestorben sein. – Da sah ihn der Fremde plötzlich mit funkelnden Augen an. „Das lügt Ihr", rief er, „sie ist nicht tot, sie ist nicht alt!" – „Wer lügt?" entgegnete Alvarez ganz hitzig. – „Elender", erwiderte der Alte, „die schläft nur jetzt, bei stiller Nacht erwacht sie oft und spricht mit mir. Ich dürfte nur ein einz'ges Wort ins Ohr ihr sagen, und ihr seid verloren, alle verloren." – „Was will der Prahlhans?" fuhr Alvarez von neuem auf.
Sie wären gewiß hart aneinander geraten, aber der Unbekannte hatte sich schon in die Klüfte zurückgewandt. Vergeblich setzten ihm die Kühnsten nach, er kletterte wie ein Tiger, sie mußten vor den entsetzlichen Abgründen stillstehen; nur einmal noch sahen sie seine Gewänder durch die Wildnis fliegen, dann verschlang ihn die Öde.
„Wunderbar –", sagte Antonio, ihm in Gedanken nachsehend – „es ist, als wäre er in dieser Einsamkeit in seiner Jugend eingeschlummert, den Wechsel der Jahre verschlafend, und spräch' nun irre aus der alten Zeit." – Hier wurden sie von einigen Schiffssoldaten unterbrochen, die währenddes einen Berggipfel erstiegen hatten und nun ihren Kameraden unten unablässig zuriefen und winkten.

Alles kletterte eilfertig hinauf, auch Alvarez und Antonio folgten, und bald hörte man droben ein großes Freudengeschrei und sah Hüte, Degenkoppeln und leere Flaschen durcheinander in die Luft fliegen. Denn von dem vorspringenden Berge sahen sie auf einmal in ein weites gesegnetes Tal wie in einen unermeßlichen Frühling hinein. Blühende Wälder rauschten herauf, unter Kokospalmen standen Hütten auf luftigen Auen, von glitzernden Bächen durchschlängelt, fremde bunte Vögel zogen darüber wie abgewehte Blütenflocken. – „Vivat der Herr Vizekönig Don Alvarez!" rief die Schiffsmannschaft jubelnd und hob den Hauptmann auf ihren Armen hoch empor. Dieser, auf ihren breiten Schultern sich zurechtsetzend, nahm das lange Perspektiv und musterte zufrieden sein Land. Der Student Antonio aber saß doch noch höher zwischen den Blättern einer Palme, wo er mit den jungen Augen weit über Land und Meer sehen konnte. Es war ihm fast wehmütig zumute, als er in der stillen Morgenzeit unten Hähne krähen hörte und einzelne Rauchsäulen aufsteigen sah. Aber die Hähne krähten nicht in den Dörfern, sondern wild im Walde, und der Rauch stieg aus fernen Kratern, zur Warnung, daß sie auf unheimlichem vulkanischem Boden standen.

Plötzlich kam ein Matrose atemlos dahergerannt und erzählte, wie er tiefer im Gebirge auf Eingeborene gestoßen; die wären anfangs scheu und trotzig gewesen, auf seine wiederholten Fragen aber hätten sie ihn endlich an ihren König verwiesen und ihm das Schloß desselben in der Ferne gezeigt. – Er führte die andern sogleich höher zwischen den Klippen hinauf, und sie erblickten nun wirklich gegen Osten hin wunderbare Felsen am Strande, seltsam zerrissen und gezackt gleich Türmen und Zinnen. Unten schien ein Garten wie ein bunter Teppich sich auszubreiten, von dem Felsen aber blitzte es in der Morgensonne, sie wußten nicht, waren es Waffen oder Bäche; der Wind kam von dort her, da hörten sie es zuweilen wie ferne Kriegsmusik durch die Morgenluft herüberklingen.

Einige meinten, man müsse den wilden Landsmann wieder aufsuchen, als Wegweiser und Dolmetsch, aber wer konnte ihn aus dem Labyrinth des Gebirges herausfinden, auch schien es töricht, sich einem Wahnsinnigen zu vertrauen, denn für einen solchen hielten sie alle den wunderlichen Alten. Alvarez beschloß daher, die Verwegensten zu einer bewaffneten feierlichen Gesandtschaft auszuwählen, er selbst wollte sie gleich am folgenden Morgen zu der Residenz des Königs führen, dort hofften sie nähere Auskunft von der Natur und Beschaffenheit des Landes und vielleicht auch über den rätselhaften Spanier zu erhalten.
Das war den abenteuerlichen Gesellen eben recht, sie schwärmten nun in aller Eile wieder den Berg hinab und bald sah man ihr Boot zwischen dem Schiffe und dem Ufer hin und her schweben, um alles Nötige zu der Fahrt herbeizuholen. Auf dem Lande aber wurde das kleine Lager schleunig mit Wällen umgeben, einige fällten Holz zu den Palisaden, andere putzten ihre Flinten, Alvarez stellte die Wachen aus, alles war in freudigem Alarm und Erwartung der Dinge, die da kommen sollten. – Mitten in diesen Vorbereitungen saß Antonio in seinem Zelt und arbeitete mit allem Fleiß eine feierliche Rede aus, die der Hauptmann morgen an dem wilden Hofe halten wollte. Der Abend dunkelte schon wieder, draußen hörte er nur noch die Stimmen und den Klang der Äxte im Wald, seine Rede war ihm zu seiner großen Zufriedenheit geraten, er war lange nicht so vergnügt gewesen.

Die Sonne ging eben auf, das ganze Land schimmerte wie ein stiller Sonntagsmorgen, da hörte man ein Kriegslied von ferne herüberklingen, eine weiße Fahne mit dem kastilianischen Wappen flatterte durch die grüne Landschaft. Don Alvarez war's, der zog schon so früh mit dem Häuflein, das er zu der Ambassade ausgewählt, nach der Richtung ins Blaue hinein, wo sie gestern die Residenz des

Königs erblickt hatten. Die Schalksnarren hatten sich zu dem Zuge auf das allervortrefflichste ausgeputzt. Voran mit der Fahne schritt ein Trupp Soldaten, die Morgensonne vergoldete ihnen lustig die Bärte und flimmerte in ihren Hellebarden, als hätten sich einige Sterne im Morgenrot verspätet. Ihnen folgten mehrere Matrosen, welche auf einer Bahre die für den König bestimmten Geschenke trugen: Pfannen, zerschlagene Kessel, und was sonst die Armut an altem Gerümpel zusammengefegt. Darauf kam Alvarez selbst. Er hatte, um sich bei den Wilden ein vornehmes Ansehen zu geben, den Schiffsesel bestiegen, eine große Allongenperücke aufgesetzt und einen alten weiten Scharlachmantel umgehängt, der ihn und den Esel ganz bedeckte, so daß es aussah, als ritt der lange hagre Mann auf einem Steckenpferde über die grüne Au. Der dicke Schiffskoch aber war als Page ausgeschmückt, der hatte die größte Not, denn der frische Seewind wollte ihm alle Augenblicke das knappe Federbarett vom Kopfe reißen, während der Esel von Zeit zu Zeit gelassen einen Mund voll frischer Kräuter nahm. Antonio ging als Dolmetsch neben Alvarez her, denn er hatte schon zu Hause die indischen Sprachen mit großem Fleiße studiert. Alvarez aber zankte in einem fort mit ihm; er wollte in die Rede, die er soeben memorierte, noch mehr Figuren und Metaphern haben, gleichsam einen gemalten Schnörkel vor jeder Zeile. Dem Antonio aber fiel durchaus nichts mehr ein, denn der steigende Morgen vergoldete rings um sie her die Anfangsbuchstaben einer wunderbaren, unbekannten Schrift, daß er innerlich still wurde vor der Pracht.

Ihre Fahrt ging längs der Küste fort, bald sahen sie das Meer über die Landschaft leuchten, bald waren sie wieder in tiefer Waldeinsamkeit. Der rüstige Sanchez streifte unterdes jägerhaft umher.

Kaum hatte der Zug die Gebirgsschluchten erreicht, als ein Wilder, im Dickicht versteckt, in eine große Seemuschel stieß. Ein zweiter gab Antwort und wieder einer, so lief der Schall plötzlich von Gipfel zu Gipfel über die gan-

ze Insel, daß es tief in den Bergen widerhallte. Bald darauf sahen sie's hier und da im Walde aufblitzen, bewaffnete Haufen mit hellen Speeren und Schilden brachen in der Ferne aus dem Gebirge wie Waldbäche und schienen alle auf einen Punkt der Küste zuzueilen. Antonio klopfte das Herz bei dem unerwarteten Anblick. Sanchez aber schwenkte seinen Hut in der Morgenluft vor Lust. So rückte die Gesandtschaft unerschrocken fort; die Hütten, die sie seitwärts in der Ferne sahen, schienen verlassen, die Gegend wurde immer höher und wilder. Endlich, um eine Bergesecke biegend, erblickten sie plötzlich das Ziel ihrer Wanderschaft: den senkrechten Fels mit seinen wunderlichen Bogen, Zacken und Spitzen, von Bächen zerrissen, die sich durch die Einsamkeit herabstürzten, dazwischen saßen braune Gestalten, so still, als wären sie selber von Stein, man hörte nichts als das Rauschen der Wasser und jenseits die Brandung im Meere. In demselben Augenblick aber tat es einen durchdringenden Metallklang wie auf einen großen Schild, alle die Gestalten auf den Klippen sprangen plötzlich rasselnd mit ihren Speeren auf und rasch zwischen dem Waldesrauschen, den Bächen und Zacken stieg ein junger, hoher, schlanker Mann herab mit goldenen Spangen, den königlichen Federmantel um die Schultern und einen bunten Reiherbusch auf dem Haupt wie ein Goldfasan. Er sprach noch im Herabkommen mit den andern und rief den Spaniern gebieterisch zu. Da aber niemand Antwort gab, blieb er auf seine Lanze gestützt vor ihnen stehen. Alvarez's Perücke schien ihm besonders erstaunlich, er betrachtete sie lange unverwandt, man sah fast nur das Weiße in seinen Augen.

Antonio war ganz konfus, denn zu seinem Schrecken hatte er schon bemerkt, daß er trotz seiner Gelehrsamkeit kein Wort von des Königs Sprache verstand. Der unverzagte Alvarez aber fragte nach nichts, er ließ die Tragbahre mit dem alten Gerümpel dem König vor die Füße setzen, rückte sich auf seinem Esel zurecht und hielt sogleich mit großem Anstande seine wohlverfaßte Anrede, wäh-

rend einige andere hinten feierlich die Zipfel seines Scharlachmantels hielten. Da konnte sich der König endlich nicht länger überwinden, er rührte neugierig mit seinem Speer an Alvarez's Perücke, sie ließ zu seiner Verwunderung und Freude wirklich vom Kopfe des Redners los, und, mit leuchtenden Augen zurückgewandt, wies er sie hoch auf der Lanze seinem Volke. Ein wildes Jauchzen erfüllte die Luft, denn ein großer Haufe brauner Gestalten hatte sich unterdes nachgedrängt, Speer an Speer, daß der ganze Berg wie ein ungeheurer Igel anzusehen war.
Der König hatte unterdes gewinkt, einige Wilde traten mit großen Körben heran, der König griff mit beiden Händen hinein und schüttete auf einmal Platten, Körner und ganze Klumpen Goldes auf seine erstaunten Gäste aus, daß es lustig durcheinanderrollte. Da sah man in dem unverhofften Goldregen plötzlich ein Streiten und Jagen unter den Spaniern, jeder wollte alles haben, und je mehr sie lärmten und zankten, je mehr warf der König aus; ein spöttisches Lächeln zuckte um seinen Mund, daß seine weißen Zähne manchmal hervorblitzten, wie bei einem Tiger. Währenddes aber schwärmten die Eingeborenen von beiden Seiten aus den Schluchten hervor, mit ihren Schilden und Speeren die Raufenden wild umtanzend.
Da war Alvarez der erste, der sich schnell besann. „Ehre über Gold und Gott über alles!“ rief er, seinen Degen ziehend, und stürzte in den dicken Knäuel der Seinigen, um sie mit Gewalt auseinanderzuwirren. „Christen“, schrie er, „wollt ihr euch vom Teufel mit Gold mästen lassen, damit er euch nachher die Hälse umdreht wie Gänsen? Seht ihr nicht, wie er mit seiner Leibgarde den Ring um euch zieht?“ – Aber der Teufel hatte sie schon verblendet; um nichts von ihrem Golde zurückzugeben, entflohen sie einzeln vor dem Hauptmann, sich im Walde verlaufend mit den lächerlich vollgepfropften Taschen. Nur einige alte Soldaten sammelten sich um Alvarez und den Leutnant. Die Eingeborenen stutzten, da sie die bewegliche Burg und die Musketen plötzlich zielend auf sich gerichtet sa-

hen, sie blieben zaudernd stehen. So entkam der Hauptmann mit seinen Getreuen dem furchtbaren Kreise der Wilden, ehe er sich noch völlig hinter ihnen geschlossen hatte.
In der Eile aber hatte auch dieses Häuflein den ersten besten Pfad eingeschlagen und war, ohne es zu bemerken, immer tiefer in den Wald geraten. Der nahm kein Ende, die Sonne brannte auf die nackten Felsen, und als sie sich endlich senkte, hatten sie sich gänzlich verirrt. Jetzt brach die Nacht herein, ein schweres Gewitter, das lange in der Ferne über dem Meere gespielt, zog über das Gebirge; den armen Antonio hatten sie gleich bei Anbruch der Dunkelheit verloren. So stoben sie wie zerstreute Blätter im Sturme durch die schreckliche Nacht, nur die angeschwollenen Bäche rauschten zornig in der Wildnis, dazwischen das blendende Leuchten der Blitze, das Schreien der Wilden und die Signalschüsse der Verirrten aus der Ferne. – „Horcht", sagte Sanchez, „das klingt so hohl unter den Tritten, als ging' ich über mein Grab, und die Wetter breiten sich drüber wie schwarze Bahrtücher, mit feurigen Blumen durchwirkt, das wär' ein schönes Soldatengrab!" – „Schweig", fuhr ihn Alvarez an, „wie kommst du jetzt darauf?" – „Das kommt von dem verdammten Trinken", entgegnete Sanchez, „da werd' ich zuzeiten so melancholisch darnach." Er sang:

„Und wenn es einst dunkelt,
Der Erd' bin ich satt,
Durchs Abendrot funkelt
Eine prächtige Stadt;
Von den goldenen Türmen
Singet der Chor,
Wir aber stürmen
Das himmlische Tor!"

„Was ist das!" rief plötzlich ein Soldat. Sie sahen einen Fremden mit bloßem Schwerte durch die Nacht auf sich

zustürzen, sein Mantel flatterte weit im Winde. – Beim Glanz der Blitze erkannten sie ihren wahnsinnigen Landsmann wieder. – „Hallo!" rief ihm Sanchez freudig entgegen, „hat dich der Lärm und das Schießen aus deinen Felsenritzen herausgelockt, kannst du das Handwerk nicht lassen?" – Der Alte aber, scheu zurückblickend, ergriff hastig die Hand des Leutnants und drängte alle geheimnisvoll und wie in wilder Flucht mit sich fort. „Noch ist es Zeit", sagte er halbleise, „ich rette euch noch, nur rasch, rasch fort, es brennt, seht, wie die blauen Flämmchen hinter mir aus dem Boden schlagen, wo ich trete!" – „Führ uns ordentlich und red nicht so toll in der verrückten Nacht!" entgegnete Alvarez ärgerlich. – Da leuchtete ein Blitz durch des Alten fliegendes Haar. Er blieb stehen und zog die Locken über das Gesicht durch seine weitausgespreizten Finger. „Grau, alles grau geworden in *einer* Nacht", sagte er mit schmerzlichem Erstaunen – „aber es könnte noch alles gut werden", setzte er nach einem Augenblick hinzu, „wenn sie mich nur nicht immer verfolgte." – „Wo? wer?" fragte Sanchez. – „Die grausilberne Schlange", erwiderte der Alte heimlich und riß die Erstaunten wieder mit sich durch das Gestein. Plötzlich aber schrie er laut auf: „Da ist sie wieder!" – Alles wandte sich erschrocken um. – Er meinte den Strom, der, soeben tief unter dem Felsen vorüberschießend, im Wetterleuchten heraufblickte. – Ehe sie sich aber noch besannen, flog der Unglückliche schon durch das Dickicht fort, die Haare stiegen ihm vor Entsetzen zu Berge, so war er ihnen bald in der Dunkelheit zwischen den Klüften verschwunden.

Währenddes irrte Antonio verlassen im Gebirge umher. In der Finsternis war er unversehens von den Seinigen abgekommen. Als er's endlich bemerkte, waren sie schon weit; da hörte er plötzlich wieder Tritte unter sich und eilte darauf zu, bis er mit Schrecken gewahr wurde, daß es Eingeborene waren, die hastig und leise, als hätten sie

einen heimlichen Anschlag, vorüberstreiften, ohne ihn zu sehen. Ihn schauerte, und doch war's ihm eigentlich recht lieb so. Er dachte übers Meer nach Hause, wie nun alle dort ruhig schliefen und nur die Turmuhr über dem mondbeschienenen Hof schlüge und die Bäume dunkel rauschten im Garten. Wie grauenhaft waren ihm da vom Balkon oft die Wolken vorgekommen, die über das stille Schloß gingen, wie Gebirge im Traum. Und jetzt stand er wirklich mitten in dem Wolkengebirge, so rätselhaft sah hier alles aus in dieser wilden Nacht! „Nur zu, blas nur immer zu, blinder Sturm, glühet, ihr Blitze!" rief er aus und schaute recht zufrieden und tapfer umher, denn alles Große ging durch seine Seele, das er auf der Schule aus den Büchern gelernt: Julius Cäsar, Brutus, Hannibal und der alte Cid. – Da brannte ihm plötzlich sein Gold in der Tasche, auch er hatte sich nicht enthalten können, in dem Goldregen mit seinem Hütlein einige Körner aufzufangen. – „Frei vom Mammon will ich schreiten auf dem Feld der Wissenschaft", sagte er und warf voll Verachtung den Goldstaub in den Sturm, es gab kaum einen Dukaten, aber er fühlte sich noch einmal so leicht.
Unterdes war das Gewitter rasch vorübergezogen, der Wind zerstreute die Wolken wie weiße Nachtfalter in wildem Fluge über den ganzen Himmel, nur tief am Horizont noch schweiften die Blitze, die Nacht ruhte ringsher auf den Höhen aus. Da fühlte Antonio erst die tiefe Einsamkeit, verwirrt eilte er auf den verschlungenen Pfaden durch das Labyrinth der Klippen lange fort. Wie erschrak er aber, als er auf einmal in derselben Gegend herauskam, aus der sie am Morgen entflohen. Der Fels des Königs, mit seinen seltsamen Schluften und Spitzen, stand wieder vor ihm, nur an einem andern Abhange desselben schien er sich zu befinden. Jetzt aber war alles so stumm dort, die Wellen plätscherten einförmig, riesenhaftes Unkraut bedeckte überall wildzerworfenes Gemäuer. – Antonio sah sich zögernd nach allen Seiten um. Schon gestern hatten ihn die Mauertrümmer, die fast wie Leichensteine aus dem

Grün hervorragten, rätselhaft verlockt. Jetzt konnte er nicht länger widerstehen, er zog heimlich seine Schreibtafel hervor, um den kostbaren Schatz von Inschriften und Bilderzeichen, die er dort vermutete, wie im Fluge zu erheben.

Da aber wurde er zu seinem Erstaunen erst gewahr, daß er eigentlich mitten in einem Garten stand. Gänge und Beete, mit Buchsbaum eingefaßt, lagen umher, eine Allee führte nach dem Meere hin, die Kirschbäume standen in voller Blüte. Aber die Beete waren verwildert, Rehe weideten auf den einsamen Gängen, an den Bäumen schlangen sich üppige Ranken wild bis über die Wipfel hinaus, von wunderbaren hohen Blumen durchglüht. Seitwärts standen die Überreste einer verfallenen Mauer, die Sterne schienen durch das leere Fenster, in dem Fensterbogen schlief ein Pfau, den Kopf unter die schimmernden Flügel versteckt.

Antonio wandelte wie im Traum durch die verwilderte Pracht, kein Laut rührte sich in der ganzen Gegend; da war es ihm plötzlich, als sähe er fern am andern Ende der Allee jemand zwischen den Bäumen gehen, er hielt den Atem an und blickte noch einmal lauschend hin, aber es war alles wieder still, es schien nur ein Spiel der wankenden Schatten. Da kam er endlich in eine dunkle Laube, die der Wald sich selber lustig gewoben, das schien ihm so heimlich und sicher, er wollt' nur einen Augenblick rasten und streckte sich ins hohe Gras. Ein würziger Duft wehte nach dem Regen vom Walde herüber, die Blätter flüsterten so schläfrig in der leisen Luft, müde sanken ihm die Augen zu.

Die wunderbare Nacht aber sah immerfort in seinen Schlaf hinein und ließ ihn nicht lange ruhen, und als er erwachte, hörte er mit Schrecken neben sich atmen. Er wollte rasch aufspringen, aber zwei Hände hielten ihn am Boden fest. Beim zitternden Mondesflimmer durchs Laub glaubte er eine schlanke Frauengestalt zu erkennen. – „Ich wußte es wohl, daß du kommen würdest“, redete sie ihn

in spanischer Sprache an. – „So bist du eine Christin?“ fragte er ganz verwirrt. – Sie schwieg. – „Hast du mich denn schon jemals gesehen?“ – „Gestern nachts bei unserem Fest“, erwiderte sie, „du warst allein mit eurem Seekönig.“ – Eine entsetzliche Ahnung flog durch Antonios Seele, er mühte sich in der Finsternis vergeblich, ihre Züge zu erkennen, draußen gingen Wolken wechselnd vorüber, zahllose Johanniswürmchen umkreisten leuchtend den Platz. – Da hörte er fern von den Höhen einen schönen männlichen Gesang. „Wer singt da?“ fragte er erstaunt. – „Still, still“, erwiderte die Unbekannte, „laß den nur in Ruh’. Hier bist du sicher, niemand besucht diesen stillen Garten mehr, sonst war es anders“ –, dann sang sie selber wie in Gedanken:

„Er aber ist gefahren
Weit übers Meer hinaus,
Verwildert ist der Garten,
Verfallen liegt sein Haus.

Doch nachts im Mondenglanze
Sie manchmal noch erwacht,
Löst von dem Perlenkranze
Ihr Haar, das wallt wie Nacht.

So sitzt sie auf den Zinnen,
Und über ihr Angesicht
Die Perlen und Tränen rinnen,
Man unterscheid’t sie nicht.“

Da teilte ein frischer Wind die Zweige, im hellen Mondlicht erkannte Antonio plötzlich die „Frau Venus“ wieder, die sie gestern nachts schlummernd in der Höhle gesehen, ihre eigenen Locken wallten wie die Nacht. – Ein Grauen überfiel ihn, er merkte erst jetzt, daß er unter glühenden Mohnblumen wie begraben lag. Schauernd sprang er empor und schüttelte sich ab, sie wollte ihn halten, aber

er riß sich von ihr los. Da tat sie einen durchdringenden Schrei, daß es ihm durch Mark und Bein ging, dann hörte er sie in herzzerreißender Angst rufen, schelten und rührend flehen.
Aber er war schon weit fort, der Gesang auf den Höhen war verhallt, die Wälder rauschten ihm wieder erfrischend entgegen, hinter ihm versank allmählich das schöne Weib, das Meer und der Garten, nur zuweilen noch hörte er ihre Klagen wie das Schluchzen einer Nachtigall von ferne durch den Wind herüberklingen.

„Du sollst mich doch nicht fangen,
Duftschwüle Zaubernacht!
Es stehn mit goldnem Prangen
Die Stern' auf stiller Wacht
Und machen unterm Grunde,
Wo du verirret bist,
Getreu die alte Runde –
Gelobt sei Jesus Christ!

Wie bald in allen Bäumen
Geht nun die Morgenluft,
Sie schütteln sich in Träumen,
Und durch den roten Duft
Ein' fromme Lerche steiget,
Wenn alles still noch ist,
Den rechten Weg dir zeiget –
Gelobt sei Jesus Christ!"

So sang es im Gebirge, unten aber standen zwei spanische Soldaten fast betroffen unter den Bäumen, denn es war ihnen, als ginge ein Engel singend über die Berge, um den Morgen anzubrechen. Da stieg ein Wanderer rasch zwischen den Klippen herab, sie erkannten zu ihrer großen Freude den Studenten Antonio, er schien bleich und zerstört. – „Gott sei Dank, daß Ihr wieder bei uns seid!" rief ihm der eine Soldat entgegen. „Ihr hättet uns beinah kon-

fus gemacht mit Eurem Gloria“, meinte der andere, „Ihr habt eine gute, geistliche Kehle. Wo kommt Ihr her?“ – „Aus einem tiefen Bergwerke“, sagte Antonio, „wo mich der falsche Flimmer verlockt – wie so unschuldig ist hier draußen die Nacht!“ – „Bergwerk? wo habt Ihr's gefunden?“ fragten die Soldaten mit hastiger Neugier. – „Wie, sprach ich von einem Bergwerk?“ erwiderte Antonio zerstreut, „wo sind wir denn?“ – Die Soldaten zeigten über den Wald, dort läge ihr Landungsplatz. Sie erzählten ihm nun, wie die zersprengte Gesandtschaft unter großen Mühseligkeiten endlich wieder das Lager am Strande erreicht. Da habe der brave Alvarez, da er den Antonio dort nicht gefunden, sie beide zurückgeschickt, um ihn aufzusuchen, und wenn sie jeden Stein umkehren und jede Palme schütteln sollten. Antonio schien wenig darauf zu hören. Die Soldaten aber meinten, es sei diese Nacht nicht geheuer im Gebirge, sie nahmen daher den verträumten Studenten ohne weiteres in ihre Mitte und schritten rasch mit ihm fort.

So waren sie in kurzer Zeit bei ihren Zelten angelangt. Dort stand Alvarez wie ein Wetterhahn auf dem frisch aufgeworfenen Erdwall, vor Ungeduld sich nach allen Winden drehend. Er schimpfte schon von weitem, da er endlich den Verirrten ankommen sah. „Ein Weltentdekker“, sagte er, „muß den Kompaß in den Füßen haben, in der Wildnis bläst der Sturm die Studierlampe aus, da schlägt ein kluger Kopf sich Funken aus den eigenen Augen. Was da Logik und Rhetorik! Sie hätten deinen Kopf aufgefressen mit allen Wissenschaften drin, aber ich hatt's ihnen zugeschworen, sie mußten zum Nachtisch alle unsere bleiernen Pillen schlucken oder meine eignen alten Knochen nachwürgen. Du bist wohl recht verängstigt und müde, armer Junge, Gott, wie du aussiehst!“ – Nun ergriff er den Studenten vor Freude beim Kopf, strich ihm die vollen braunen Locken aus der Stirn und führte ihn eilig ins Lager in sein eigenes Zelt, wo er sich sogleich auf eine Matte hinstrecken mußte. Im Lager aber war schon

ein tiefes Schweigen, die müden Gesellen lagen schlafend wie Tote umher. Nur der Leutnant Sanchez wollte diese Nacht nicht mehr schlafen noch ruhen, er saß auf den zusammengelegten Waffen der Mannschaft; eine Flasche in der Hand, trank er auf eine fröhliche Auferstehung, der Nachtwind spielte mit der roten Hahnfeder auf seinem Hut, der ihm verwegen auf einem Ohr saß; er war wahrhaftig schon wieder berauscht. Antonio mußte nun sein Abenteuer erzählen. Er berichtete verworren und zerstreut, in seinem Haar hing noch eine Traumblume aus dem Garten. Alvarez blieb dabei, das Frauenzimmer sei die Frau Venus gewesen, und jene Höhle, die sie in der Walpurgisnacht entdeckt, der Eingang zum Venusberge. Sanchez aber rückte immer näher, während er hastig ein Glas nach dem andern hinunterstürzte; er fragte wunderlich nach der Lage der Höhle, nach dem Wege dahin, sie mußten ihm alles ausführlich beschreiben. – Auf einmal war er heimlich verschwunden.

Der Abenteurer schlich sich sacht und vorsichtig durch die schläfrigen Posten; über dem Gespräch hatte ihn plötzlich das Gelüsten angewandelt, den dunkeln Vorhang der phantastischen Nacht zu lüften – er wollte die Frau Venus besuchen. Er hatte sich Felsen, Schlünde und Stege aus Alvarez' Rede wohl gemerkt, es traf alles wunderbar zu. So kam er in kurzer Zeit an das stille Tal. Ein schmaler Felsenpfad führte fast unkenntlich zwischen dem Gestrüpp hinab, die Sterne schienen hell über den Klippen, er stieg im trunkenen Übermut in den Abgrund. Da brach plötzlich ein Reh neben ihm durch das Dickicht, er zog schnell seinen Degen. „Hoho, Ziegenbock!" rief er, „hast du die Hexe abgeworfen, die zu meiner Hochzeit ritt! Das ist eine bleiche, schläfrige Zeit zwischen Morgen und Nacht, da schauern die Toten und schlüpfen in ihre Gräber, daß man die Leichentücher durchs Laub streichen hört. Wo sich eine verspätet beim Tanz, ich greif sie, sie soll meine Brautjungfer sein. – Zum Teufel, red' vernehmlicher, Waldeinsamkeit! ich kenn' ja dein Lied aus

alter Zeit, wenn wir auf wilder Freite in Flandern nachts an den Wällen lagen vor mancher schönen Stadt, die von den schlanken Türmen mit ihrem Glockenspiele durch die Luft musizierte. Die Sterne löschen schon aus, wer weiß, wer sie wiedersieht! – Nur leise, sacht zwischen den Werken, in den Laufgräben fort! die Wolken wandern, die Wächter schlafen auf den Wällen, in ihre grauen Mäntel gehüllt, sie tun, als wären sie von Stein. – Verfluchtes Grauen, ich seh dich nicht, was hauchst du mich so kalt an, ich ringe mit dir auf der Felsenwand, du bringst mich nicht hinunter!"

Jetzt stand er auf einmal vor der Kluft, die Alvarez und Antonio in jener Nacht gesehen. Es war die erste geheimnisvolle Morgenzeit, in dem ungewissen Zwielicht erblickte er die junge, schlanke Frauengestalt, ganz wie sie ihm beschrieben worden, auf dem Moosbett in ihrem Schmucke schlummernd, den schönen Leib von ihren Locken verdeckt. Alte halbverwitterte Fahnen, wie es schien, hingen an der Wand umher, der Wind spielte mit den Lappen, hinten in der Dämmerung, den Kopf vornüber gebeugt, saß es wie eine eingeschlafene Gestalt.

„Es ist die höchste Zeit", flüsterte Sanchez ganz verblendet, „sonst versinkt alles wieder, schon hör' ich Stimmen gehn. Wie oft schon sah ich im Wein ihr Bild, das war so schön und wild in des Bechers Grund. Einen Kuß auf ihren Mund, so sind wir getraut, eh' der Morgen graut." – So taumelte der Trunkene nach der Schlummernden hin, er fuhr schauernd zusammen, als er sie anfaßte, ihre Hand war eiskalt. Im Gehen aber hatte er sich mit den Sporen in die Trümmer am Boden verwickelt, eine Rüstung an der Wand stürzte rasselnd zusammen, die alten Fahnen flatterten im Wind, bei dem Dämmerschein war's ihm, als rührte sich alles und dunkle Arme wänden sich aus der Felswand. Da sah er plötzlich im Hintergrunde den schlafenden Wächter sich aufrichten, daß ihn innerlich grauste. An dem irren, funkelnden Blick glaubte er den alten wahnsinnigen Spanier wiederzuerkennen, der warf, ohne

ein Wort zu sagen, seinen weiten Mantel über die Schultern zurück, ergriff das neben ihm stehende Schwert und drang mit solcher entsetzlichen Gewalt auf ihn ein, daß Sanchez kaum Zeit hatte, seine wütenden Streiche aufzufangen. Bei dem Klange ihrer Schwerter aber fuhren große scheußliche Fledermäuse aus den Felsenritzen und durchkreisten mit leisem Fluge die Luft, graue Nebelstreifen dehnten und reckten sich wie Drachenleiber verschlafen an den Wipfeln, dazwischen wurden Stimmen im Walde wach, bald hier, bald dort, eine weckte die andre, aus allen Löchern, Hecken und Klüften stieg und kroch es auf einmal, wilde, dunkle Gestalten im Waffenschmuck, und alles stürzte auf Sanchez zusammen. „Nun, nun, steht's so!" rief der verzweifelte Leutnant, „laß mich los, alter Narr mit deinem verwitterten Bart! Das ist keine Kunst, so viele über einen. Schickt mir euern Meister selber her, es gelüstet mich recht, mit ihm zu fechten! Aber der Teufel hat keine Ehre im Leibe. Ihr höllisches Ungeziefer, nur immer heraus vor meine christliche Klinge! nur immerzu, ich hau' mich durch!" – So, den Degen in der Faust, wich er, wie ein gehetztes Wild, kämpfend von Stein zu Stein, das einsame Felsental hallte von den Tritten und Waffen, im Osten hatte der Morgen schon wie ein lustiger Kriegsknecht die Blutfahne ausgehangen.

Im Lager flatterten unterdes nur noch wenige Wachtfeuer halberlöschend, eine Gestalt nach der andern streckte sich in der Morgenkühle, einige saßen schon wach auf ihrem Mammon und besprachen das künftige Regiment der Insel. Plötzlich riefen draußen die Schildwachen an, sie hätten Lärm im Gebirge gehört. Jetzt vermißte man erst den Leutnant. Alles sprang bestürzt zu den Waffen, keiner wußte, was das bedeuten könnte. Der Lärm aber, als sie so voller Erwartung standen, ging über die Berge wie ein Sturm wachsend immer näher, man konnte schon deutlich dazwischen das Klirren der Waffen unterscheiden. Da, im falben Zwielicht, sahen sie auf einmal den Sanchez droben aus dem Walde dahersteigen, bleich und verstört, mit den

Geistern fechtend. Hinter ihm drein aber toste eine wilde Meute, es war, als ob aller Spuk der Nacht seiner blutigen Fährte folgte. Sein Frevel, wie es schien, hatte das dunkle Wetter, das schon seit gestern grollend über den Fremden hing, plötzlich gewendet, von allen Höhen stürzten bewaffnete Scharen wie reißende Ströme herab, der Klang der Schilde, das Schreien und der Widerhall zwischen den Felsen verwirrte die Stille und bald sahen sich die Spanier von allen Seiten umzingelt. – „Macht dem Leutnant Luft!" rief Alvarez und warf sich mit einigen Soldaten mitten in den dicksten Haufen. Schon hatten sie den Sanchez gefaßt und führten den Wankenden auf einen freien Platz am Meer, aber zu spät, von vielen Pfeilen durchbohrt, brach er neben seinen Kameraden auf dem Rasen zusammen – sein Wort war gelöst, er hatte sich wacker durchgeschlagen. Bei diesem Anblick ergriff alle eine unsägliche Wut, keiner dachte mehr an sich im Schmerz, sie mähten sich wie die Todesengel in die dunklen Scharen hinein, Alvarez und Antonio immer tapfer voran. Da erblickten sie auf einmal ihren wahnsinnigen Landsmann, mitten durch das Getümmel mit dem Schwert auf sie eindringend. Vergebens riefen sie ihm warnend zu – er stürzte sich selbst in ihre Speere, ein freudiges Leuchten ging über sein verstörtes Gesicht, daß sie ihn fast nicht wiedererkannten, dann sahen sie ihn taumeln und mit durchbohrtem Herzen tot zu Boden sinken. – Ein entsetzliches Rachegeschrei erhob sich über dem Toten, die Wilden erneuerten mit verdoppeltem Grimm ihren Angriff, es war, als ständen die Erschlagenen hinter ihnen wieder auf, immer neue scheußliche Gestalten wuchsen aus dem Blut, schon rannten sie jauchzend nach dem Strand, um die Spanier von ihrem Schiffe abzuschneiden. Jetzt war die Not am höchsten, ein jeder befahl sich Gott, die Spanier fochten nicht mehr für ihr Leben, nur um einen ehrlichen Soldatentod. – Da ging es auf einmal wie ein Schauder durch die unabsehliche feindliche Schar, alle Augen waren starr nach dem Gebirge zurückgewandt. Auch Antonio und

Alvarez standen ganz verwirrt mitten in der blutigen Arbeit. Denn zwischen den Palmenwipfeln in ihrem leuchtenden Totenschmucke kam die Frau Venus, die wilden Horden teilend, von den Felsen herab. Da stürzten plötzlich die Eingebornen wie in Anbetung auf ihr Angesicht zur Erde, die Spanier atmeten tief auf, es war auf einmal so still, daß man die Wälder von den Höhen rauschen hörte. Indem sie aber noch so staunend stehn, tritt die Wunderbare mitten unter sie, ergreift Sanchez's Mantel, den sie seltsam um ihren Leib schlägt, und befiehlt ihnen, sich rasch in das Boot zu werfen, ehe der Zauber gelöst. Darauf umschlingt sie Antonio, halb drängt, halb trägt sie ihn ins Boot hinein, die andern, ganz verdutzt, bringen eiligst Sanchez's Leichnam nach, alles stürzt in die Barke. So gleiten sie schweigend dahin, schon erheben sich einzelne Gestalten wieder am Ufer, ein leises Murmeln geht wachsend durch die ganze furchtbare Menge, da haben sie glücklich ihr Schiff erreicht. Dort aber faßt die Unbekannte sogleich das Steuer, die stille See spiegelt ihr wunderschönes Bild, ein frischer Wind vom Lande schwellt die Segel, und als die Sonne aufgeht, lenkt sie getrost zwischen den Klippen in den Glanz hinaus.

Die Spanier wußten nicht, wie ihnen geschehen. Als sie sich vom ersten Schreck erholt, gedachten sie erst ihrer Goldklumpen wieder, die sie auf der Insel zurückgelassen. Da fuhren sie denn wieder so arm und lumpig von dannen, wie sie gekommen. – „Der Teufel hat's gegeben, der Teufel hat's genommen", sagte der spruchreiche Alvarez verdrießlich. – Darüber aber hatten sie den armen Sanchez fast vergessen, der auf dem Verdeck unter einer Fahne ruhte. Alvarez beschloß nun, vor allem andern ihm die letzte Ehre anzutun, wie es einem tapferen Seemann gebührte. Er berief sogleich die ganze Schiffsmannschaft, die einen stillen Kreis um den Toten bildete, dann trat er in die Mitte, um die Leichenrede zu halten. „Seht da, den ge-

wesenen Leutnant", sagte er, „nehmt euch ein Exempel dran, die ihr immer meint, Unkraut verdürb' nicht. Ja, da seht ihn liegen, er war tapfer, oftmals betrunken, aber tapfer" – weiter bracht' er's nicht, denn die Stimme brach ihm plötzlich, und Tränen stürzten ihm aus den Augen, als er den treuen Kumpan so bleich und still im lustigen Morgenrot daliegen sah. Einige Matrosen hatten ihn unterdes in ein Segeltuch gewickelt, andere schwenkten die Flaggen über ihm auf eine gute Fahrt auf dem großen Meere der Ewigkeit – dann ließen sie ihn an Seilen über Bord ins feuchte Grab hinunter. „So ist denn", sagte Alvarez, „sein Leiblied wahr geworden: ‚Ein Meerweib' singt, die Nacht ist lau, da denkt an mich, 's ist meine Frau.' Man soll den Teufel nicht an die Wand malen." – Kaum aber hatte der Tote unten die kalte See berührt, als er auf einmal in seinem Segeltuch mit großer Vehemenz zu arbeiten anfing. „Ihr Narren, ihr", schimpfte er, „was, Wein soll das sein? elendes Wasser ist's!" – Die Matrosen hätten vor Schreck beinah Strick und Mann fallen lassen, aber Alvarez und Antonio sprangen rasch hinzu und zogen voller Freude den Ungestümen wieder über Bord hinauf. Hier drängten sich nun die Überraschten von allen Seiten um ihn herum, und während die einen seine Wunden untersuchten und verbanden, andere jauchzend ihre Hüte in die Luft warfen, glotzte der unsterbliche Leutnant alle mit seinen hervorstehenden Augen stumm und verwogen an, bis sein Blick endlich die wunderbare Führerin des Schiffes traf. Da schrie er plötzlich auf: „Die ist's! ich selber sah sie in den Klüften auf dem Moosbett schlafen!"
Aller Augen wandten sich nun von neuem auf die schöne Fremde, die, auf das Steuer gelehnt, gedankenvoll nach der fernen Küste hinübersah. Keiner traute ihr, Antonio aber erkannte bei dem hellen Tageslicht das Mädchen aus dem wüsten Garten wieder. Da faßte Alvarez sich ein Herz, trat vor und fragte sie, wer sie eigentlich wäre. – „Alma", war ihre Antwort. – Warum sie zu ihnen gekommen? – „Weil sie euch erschlagen wollten", erwiderte sie

in ihrem gebrochenen Spanisch. – Ob sie mit ihnen fahren und ihm als Page dienen wolle? – Nein, sie wolle dem Antonio dienen. – Woher sie denn aber Spanisch gelernt? – Vom Alonzo, den sie erstochen hätten. – „Den tollen Alten“, fiel hier Sanchez hastig ein, „wer war er, und wie kam er zu dir?“ – „Ich weiß nicht“, entgegnete Alma. – „Kurz und gut“, hob Alvarez wieder an, „war die Frau Venus auf Walpurgisnacht auf eurer Insel? Oder bist du gar selber die Frau Venus? Habt ihr beide – wollt’ sagen: du oder die Frau Venus – dazumal in der Felsenkammer geschlafen?“ – Sie schüttelte verneinend den Kopf. – „Nun, so mag der Teufel daraus klug werden! ich will mich heute gar nicht mehr wundern, Frau Venus, Urgande, Megära, das kommt und geht so!“, rief der Hauptmann ungeduldig aus und benannte das Eiland, dessen blaue Gipfel soeben im Morgenduft versanken, ohne weiteres die Venusinsel, von der Frau Venus, die nicht da war.

Die darauffolgende Nacht war schön und sternklar, die Fortuna mit ihren weißen Segeln glitt wie ein Schwan durch die mondlichte Stille. Da trat Antonio leise auf das Verdeck hinaus, er hatte keine Rast und Ruh, es war ihm, als müßte er die schöne Fremde bewachen, die sorglos unten ruhte. Wie erstaunte er aber, als er das Mädchen droben schon wach und ganz allein erblickte; es war alles so einsam in der Runde, nur manchmal schnalzte ein Fisch im Meer, sie aber saß auf dem Boden mitten zwischen wunderlichem Kram, ein Spiegel, Kämme, ein Tambourin und Kleidungsstücke lagen verworren um sie her. Sie kam ihm wie eine Meerfee vor, die, bei Nacht aus der Flut gestiegen, sich heimlich putzt, wenn alle schlafen. Er blieb scheu zwischen dem Tauwerk stehen, wo sie ihn nicht bemerken konnte. Da sah er, wie sie nun einzelne Kleidungsstücke flimmernd gegen den Mond hielt, er erkannte seinen eignen Sonntagsstaat, den er ihr gestern gezeigt:

die gestickte Feldbinde, das rotsamtne, weißgestickte Wämschen. Sie zog es eilig an; Antonio war schlank und fein gebaut, es paßte ihr alles wie angegossen. Darauf legte sie den blendendweißen Spitzenkragen um Hals und Brust und drückte das Barett mit den nickenden Federn auf das Lockenköpfchen. Als sie fertig war, sprang sie auf, sie schien sich über sich selbst zu verwundern, so schön sah sie aus. Da stieß sie unversehens mit den Sporen an das Tambourin am Boden. Sie ergriff es rasch, und den tönenden Reif hoch über sich schwingend, fing sie mit leuchtenden Augen zu tanzen an, fremd und doch zierlich, und sang dazu:

„Bin ein Feuer hell, das lodert
Von dem grünen Felsenkranz,
Seewind ist mein Buhl' und fordert
Mich zum lust'gen Wirbeltanz,
Kommt und wechselt unbeständig.
Steigend wild,
Neigend mild,
Meine schlanken Lohen wend' ich,
Komm nicht nah mir, ich verbrenn' dich!

Wo die wilden Bäche rauschen
Und die hohen Palmen stehn,
Wenn die Jäger heimlich lauschen,
Viele Rehe einsam gehn.
Bin ein Reh, flieg' durch die Trümmer
Über die Höh,
Wo im Schnee
Still die letzten Gipfel schimmern,
Folg' mir nicht, erjagst mich nimmer!

Bin ein Vöglein in den Lüften,
Schwing' mich übers blaue Meer,
Durch die Wolken von den Klüften
Fliegt kein Pfeil mehr bis hierher,

Und die Au'n und Felsenbogen,
Waldeseinsamkeit
Weit, wie weit,
Sind versunken in die Wogen –
Ach, ich habe mich verflogen!"

Bei diesen Worten warf sie sich auf den Boden nieder, daß das Tambourin erklang, und weinte. – Da trat Antonio rasch hinzu, sie fuhr empor und wollte entfliehen. Als sie aber seine Stimme über sich hörte, lauschte sie hoch auf, strich mit beiden Händen die aufgelösten Locken von den verweinten Augen und sah ihn lächelnd an.

Antonio, wie geblendet, setzte sich zu ihr an den Bord und pries ihren wunderbaren Tanz. Sie antwortete kein Wort darauf, sie war erschrocken und in Verwirrung. Endlich sagte sie schüchtern und leise: sie könne nicht schlafen vor Freude, es sei ihr so licht im Herzen. – Gerade so geht mir's auch, dachte er und schaute sie noch immer ganz versunken an. Da fiel ihm eine goldene Kette auf, die aus ihrem Wämschen blinkte. Sie bemerkte es und verbarg sie eilig. Antonio stutzte. „Von wem hast du das kostbare Angedenken?" fragte er. – „Von Alonzo", erwiderte sie zögernd. – „Wunderbar", fuhr er fort, „gesteh es nur, du weißt es ja doch, wer der Alte war und wie er übers Meer gekommen. Und du selbst – wir sahn dich schlummern in der Kluft beim Fackeltanz, und dann an jenem blutigroten Morgen warf sich das Volk erschrocken vor dir hin – wer bist du?" – Sie schwieg mit tiefgesenkten Augen, und wie er so fortredend in sie drang, brach endlich ein Strom von Tränen unter den langen schwarzen Wimpern hervor. „Ach, ich kann ja nicht dafür!" rief sie aus und bat ihn ängstlich und flehentlich, er sollt' es nicht verlangen, sie könnt' es ihm nicht sagen, sonst würde er böse sein und sie verjagen. – Antonio sah sie verwundert an, sie war so schön, er reichte ihr die Hand. Als sie ihn so freundlich sah, rückte sie näher und plauderte so vertraulich, als wären sie jahrelang schon beisammen. Sie erzählte

von der Nacht auf dem Gebirge, wo sie ihn beim flüchtigen Fackelschein zum erstenmal gesehn, wie sie dann traurig gewesen, als er damals im Garten sie so schnell verließ; sie meinte, die Wilden würden ihn erschlagen.
Antonio'n aber war's bei dem Ton ihrer Stimme, als hörte er zur Frühlingszeit die erste Nachtigall in seines Vaters Garten. Die Sterne schienen so glänzend, die Wellen zitterten unter ihnen im Mondenschein, nur von ferne kühlte sich die Luft mit Blitzen, bis endlich Alma vor Schlaf nicht mehr weiter konnte und müde ihr Köpfchen senkte.
Auch Antonio war zuletzt eingeschlummert. Da träumte ihm von dem schönen verwilderten Garten, es war, als wollt' ihm der Vogel in dem ausgebrochenen Fensterbogen im Schlaf von Diego erzählen, der unter den glühenden Blumen sich verirrt. Und als er so, noch halb im Traume, die Augen aufschlug, flog schon ein kühler Morgenwind kräuselnd über die See, er blickte erschrocken umher, da hörte er wieder die Frau Venus neben sich atmen wie damals, und von fern stiegen die Zacken und Felsen der Insel allmählich im Morgengrauen wieder empor, dazwischen glaubte er wirklich den Vogel im Gebirge singen zu hören. Jetzt ruft es auch plötzlich: „Land!" aus dem Mastkorb; verschlafene Matrosen erheben sich, im Innern des Schiffes beginnt ein seltsames Murmeln und Regen. Nun fährt Alma verwirrt aus dem Schlafe empor. Da sie die Wälder, Felsen und Palmen sieht, springt sie voller Entsetzen auf und wirft einen dunkeln, tödlichen Blick auf Antonio. „Du hast mich verraten, ihr wollt mich bei den Meinigen heimlich wieder aussetzen!" ruft sie aus und schwingt sich behende auf den Bord des Schiffes, um sich ins Meer zu stürzen. Aber Antonio faßte sie schnell um den Leib, sie stutzte und sah ihn erstaunt mit ungewissen Blicken an. Unterdes war auch Alvarez auf dem Verdeck erschienen. „Still, still", rief er den Leuten zu, „nur sacht, eh' sie uns drüben merken!" Er ließ die Anker werfen, das Boot wurde leise und geräuschlos heruntergelassen, die Berge und Klüfte breiteten sich immer mächtiger in der

Dämmerung aus. Da zweifelte Antonio selbst nicht länger, daß es auf Alma abgesehn. Ganz außer sich schwang er die arme Verratne auf seinen linken Arm, zog mit der Rechten seinen Degen und rief vortretend mit lauter Stimme: es sei schändlich, treulos und undankbar, das Mädchen wider ihren Willen wieder auf die Insel zu setzen, von der sie alle eben erst mit Gefahr ihres Lebens gerettet. Aber er wolle sie bis zu seinem letzten Atemzuge verteidigen und mit ihr stehn oder fallen, wie ein Baum mit seiner Blüte!

Zu seiner Verwunderung erfolgte auf diese tapfere Anrede ein schallendes Gelächter. „Was Teufel machst du denn für ein Geschrei, verliebter Baccalaureus!" sagte Alvarez, „wir wollen hier geschwind, eh' etwa noch die Wilden erwachen, frisches Wasser holen von den unverhofften Bergen, du siehst ja doch, 's ist ein ganz anderes Land!" Nun sah es Antonio freilich auch, freudig und beschämt, denn die Morgenlichter spielten schon über den unbekannten Gipfeln. Alma aber hatte ihn fest umschlungen und bedeckte ihn mit glühenden Küssen. – Die Sonne vergoldete soeben Himmel, Meer und Berge, und in dem Glanze trug Antonio sein Liebchen hurtig in das Boot, das nun durch die Morgenstille nach dem fremden Lande hinüberglitt.

Alma war die erste, die ans Land sprang, wie ein Kind lief sie erstaunt und neugierig umher. Es blitzte noch alles vom Tau, Menschen waren nirgends zu sehen, nur einzelne Vögel sangen hie und da in der Frische des Morgens. Die praktischen Seeleute hatten indes gar bald eine Quelle, Kokos- und Brotbäume in Menge entdeckt; es ärgerte sie nur, daß die liebe Gottesgabe nicht auch schon gebakken war.

Alvarez aber, da heute eben ein Sonntag traf, beschloß, auf dem gesegneten Eilande einige Tage zu rasten, um das Schiff und die Verwundeten und Kranken wieder völlig instand zu setzen. Währenddes waren mehrere auf den

nächsten Gipfel gestiegen und erblickten überrascht jenseits des Gebirges eine weite, lachende Landschaft. Auf ihr Geschrei kam auch der Hauptmann mit Antonio und Alma herbei. „Das ist ja wie in Spanien", sagte Alvarez erfreut, „hier möcht' ich ausruhn, wenn's einmal Abend wird und die alten Segel dem Sturme nicht mehr halten." – Sie konnten der Versuchung nicht widerstehen, die Gegend näher zu betrachten, sie wanderten weiter, den Berg hinunter und kamen bald in ein schönes, grünes Tal. Auf dem letzten Abhange aber hielten sie plötzlich erschrokken still: ein einfaches Kreuz stand dort unter zwei schattigen Linden. Da knieten sie alle schweigend nieder; Alma sah sie verwundert an, dann sank auch sie auf ihre Knie in der tiefen Sonntagsstille, es war, als zöge ein Engel über sie dahin.

Als sie sich vom Gebet wieder erhoben, bemerkten sie erst einen zierlichen Garten unter dem Kreuz, den die Bäume von oben verdeckt hatten. Voll Erstaunen sahen sie sehr sorgfältig gehaltene Blumenbeete, Gänge und Spaliere, die Bienen summten in den Wipfeln, die in voller Blüte standen, aber der Gärtner war nirgends zu finden. – Da schrie Alma auf einmal erschrocken auf, als hätte sie auf eine Schlange getreten, sie hatte menschliche Fußstapfen auf dem tauigen Rasen entdeckt. – „Den wollen wir wohl erwischen", rief Alvarez, und die Wanderer folgten sogleich begierig der frischen Spur. Sie ging jenseits auf die Berge, sie glaubten den Abdruck von Schuhen zu erkennen. Unverdrossen stiegen sie nun zwischen den Felsen das Gebirge hinan, aber bald war die Fährte unter Steinen und Unkraut verschwunden, bald erschien sie wieder deutlich im Gras; so führte sie immer höher und höher hinauf und verlor sich zuletzt auf den obersten Zacken, wie in den Himmel. – „Es ist heut Sonntag, der Gärtner ist wohl der liebe Gott selber", sagte Alvarez, betroffen in der Wildnis umherschauend.

In dieser Zeit aber war die Sonne schon hoch gestiegen und brannte sengend auf die Klippen; sie mußten die wei-

tere Nachforschung für jetzt aufgeben und kehrten endlich mit vieler Mühe wieder zu den Ihrigen am Strande zurück. Als sie dort ihr Abenteuer erzählten, wollte alles sogleich in das neuentdeckte Tal stürzen. Aber Alvarez schlug klirrend an seinen Degengriff und verbot feierlich allen und jedem, das stille Revier nicht anders als unter seinem eignen Kommando zu betreten. Denn, sagte er, das sei keine Soldatenspelunke, um dort Karten zu spielen, da stecke was ganz Absonderliches dahinter. – Vergebens zerbrachen sie sich nun die Köpfe, was es mit dem Garten für ein Bewenden habe, denn ein Haus war nirgends zu sehen, und soviel hatten sie schon von den Bergen bemerkt, daß das Land eine, wie es schien, unbewohnte Insel von sehr geringem Umfange war. Man beschloß endlich, sich hier an der Küste ein wenig einzurichten und am folgenden Tage gleich in der frühesten Morgenkühle die Untersuchung gemeinschaftlich fortzusetzen.
Unterdes hatten die Zimmerleute schon ihre Werkstatt am Meere aufgeschlagen, rings hämmerte und klapperte es lustig, einige schweiften mit ihren Gewehren umher, andere flickten die Segel im Schatten der überhängenden Felsen, während fremde Vögel über ihnen bei dem ungewohnten Lärm ihre bunten Hälse neugierig aus dem Dickicht streckten.

Mit dem herannahenden Abend versammelte sich nach und nach alles wieder unter den Felsen. Die Jäger kehrten von den Bergen zurück und warfen ihre Beute auf den Rasen; da lag viel fremdes Getier umher, die Schützen an ihren Gewehren müde daneben. Indem kam ein Soldat, der sich auf der Jagd verspätet, ganz erschrocken aus dem Walde und sagte aus, er sei hinter einem schönen, scheuen Vogel weit von hier zwischen die höchsten Felsen geraten, und als er eben auf den Vogel angelegt, habe er plötzlich in der Wildnis ein riesengroßes Heiligenbild auf einer Klippe erblickt, daß ihm die Büchse aus der Hand gesunken. Die

ersten Abendsterne am Firmament hätten das Haupt des Bildes wie ein Heiligenschein umgeben, darauf habe es auf einmal sich bewegt und sei langsam wie ein Nebelstreifen mitten durch den Fels gegangen, er habe es aber nicht wieder gesehen und vor Grauen kaum den Rückweg gefunden. – „Das ist der Gärtner, den wir heut früh schon suchten", rief Alvarez, hastig aufspringend. Dabei traute er nun doch dem unschuldigen Aussehn der Insel nicht und beschloß, noch in dieser Stunde selber auf Kundschaft auszugehen, damit sie nicht etwa mitten in der Nacht unversehens überfallen würden. Das war dem abenteuerlichen Sanchez eben recht, auch Antonio und Alma erboten sich tapfer, den Hauptmann zu begleiten.
Alvarez stellte nun eilig einzelne Posten auf den nächsten Höhen aus; wer von ihnen den ersten Schuß im Gebirge hörte, sollte antworten und auf dieses Signal die ganze Mannschaft nachkommen. Darauf bewaffnete er sorgfältig sich und seine Begleiter, auch Alma mußte einen Hirschfänger umschnallen; jeder steckte aus Vorsicht noch ein Windlicht zu sich, der Soldat aber, der die seltsame Nachricht gebracht, mußte voran auf demselben Wege, den er gekommen. So zog das kleine Häuflein munter in das wachsende Dunkel hinein.
Schon waren die Stimmen unter ihnen nach und nach verhallt, nur manchmal leuchtete das Wachtfeuer noch durch die Wipfel, die Gegend wurde immer kühler und öder. Alma war recht zu Hause hier, sie sprang wie ein Reh von Klippe zu Klippe und half lachend dem steifen Alvarez, wenn ihm vor einem Sprunge graute. Der Soldat vorn aber schwor, daß sie nun schon bald in der Gegend sein müßten, wo er das Bild gesehen. Darüber wurde Sanchez ganz ungeduldig. „Heraus, Nachteule, aus deinem Felsennest!" rief er aus und feuerte schnell sein Gewehr in die Luft ab. Die nahe hohe Felsenwand brach den Schall und warf ihn nach der See zurück, es blieb alles totenstill im Gebirge. – Da glaubten sie plötzlich eine Glocke in der Ferne zu hören, die Luft kam von den Bergen, sie unter-

schieden immer deutlicher den Klang. Ganz verwirrt blieben nun alle lauschend stehen, über ihnen aber brach der Mond durch die Wolken und beleuchtete die unbekannten Täler und Klüfte, als sie auf einmal eine schöne tiefe Stimme in ihrer Landessprache singen hörten:

„Komm, Trost der Welt, du stille Nacht!
Wie steigst du von den Bergen sacht,
Die Lüfte alle schlafen,
Ein Schiffer nur noch, wandermüd,
Singt übers Meer sein Abendlied
Zu Gottes Lob im Hafen.

Die Jahre wie die Wolken gehn
Und lassen mich hier einsam stehn,
Die Welt hat mich vergessen.
Da tratst du wunderbar zu mir,
Wenn ich beim Waldesrauschen hier
In stiller Nacht gesessen.

O Trost der Welt, du stille Nacht!
Der Tag hat mich so müd' gemacht,
Das weite Meer schon dunkelt,
Laß ausruhn mich von Lust und Not,
Bis daß das ew'ge Morgenrot
Den stillen Wald durchfunkelt."

Die Wandrer horchten noch immer voll Erstaunen, als der Gesang schon lange wieder in dem Gewölk verhallt war, das soeben vor ihnen mit leisem Fluge die Wipfel streifte. Alvarez erholte sich zuerst. „Still, still", sagte er, „nur sachte mir nach, vielleicht überraschen wir ihn." – Sie schlichen nun durch das Dickicht leise und vorsichtig immer tiefer in den feuchten Nebel hinein, niemand wagte zu atmen – als plötzlich der Vorderste mit großem Geschrei auf einen Fremden stieß, jetzt schrie wieder einer und noch einer auf, manchmal klang es wie Waffengeras-

sel von ferne. Überwacht und aufgeregt wie sie waren, zog jeder sogleich seinen Degen. Indem sahen sie auch schon mehrere Wilde halbkenntlich zwischen den Klippen herandringen, die unerschrockenen Abenteurer stürzten blind auf sie ein, da klirrte Schwert an Schwert im Dunkeln, immer neue Gestalten füllten den Platz, als wüchse das Gezücht aus dem Boden nach. – In diesem Getümmel bemerkte niemand, wie ein fernes Licht, immer näher und näher, das Laub streifte, auf einmal brach der Widerschein durch die Zweige, den Kampfplatz scharf beleuchtend, und die Fechtenden standen plötzlich ganz verblüfft vor altbekannten Gesichtern – denn die vermeintlichen Wilden waren niemand anders als ihre Kameraden von unten, die verabredetermaßen auf Sanchez's Schuß zu Hülfe gekommen.
„Da ist er!" schrie hier plötzlich der Soldat, der vorhin den Alvarez heraufgeführt. Alle wandten sich erschrocken um: ein schöner riesenhafter Greis mit langem weißen Bart, in rauhe Felle gekleidet, eine brennende Fackel in der Hand, stand vor ihnen und warf dem Sanchez die Fakkel an den Kopf, daß ihn die Funken knisternd umsprühten. „Ruhe da!" rief er; „was treibt euch, hier die Nacht mit wüstem Lärm zu brechen, das wilde Meer murrt nur von fern am Fuß der Felsen, und alle blinden Elemente hielten Frieden hier seit dreißig Jahren in schöner Eintracht der Natur, und die ersten Christen, die ich wiedersehe, bringen Krieg, Empörung, Mord."
Hier erblickte er Alma, deren Gesicht von der Fackel hell beleuchtet war, da wurde er auf einmal still. – Die erstaunten Gesellen standen scheu im Kreise, sie hielten ihn insgeheim für einen wundertätigen Magier. Diese Pause benutzte Alvarez und trat, seinen Degen einsteckend, einige Schritte vor. „Ihr sollt nicht glauben", sagte er, „daß wir loses Gesindel seien, das da ermangelt, einem frommen Waldbruder die gebührende Reverenz zu erweisen; mit dem Lärm vorhin, das war so eine kleine Konfusion." – Der Einsiedler aber schien nicht darauf zu hören, er sah noch immer Alma an, dann, wie in Gedanken in dem

Kreise umherschauend, fragte er, woher sie kämen? – Das wußte nun Alvarez selber nicht recht und berichtete kurz und verworren von der Frau Venus, von Händeln mit den Wilden, von einem prächtigen Reich, das sie entdeckt, aber wieder verloren. – Der Alte betrachtete unterdes noch einmal alle in der Runde. Nach kurzem Schweigen sagte er darauf: es sei schon dunkle Nacht, und seine Klause liege weit von hier, auch habe er oben nicht Raum für so viele unerwartete Gäste, am folgenden Tage aber wollte er sie mit allem, dessen sie zur Reise bedürften, aus dem Überfluß versehen, womit ihn Gott gesegnet. Der Hauptmann solle jetzt die Seinen zum Ankerplatz zurückführen und morgen, wenn sie die Frühglocke hörten, mit wenigen Begleitern wiederkommen.
Die Wanderer sahen einander zögernd an, sie hätten lieber noch heut den Waldbruder beim Wort genommen. Aber in seinem strengen Wesen war etwas Unüberwindliches, das zugleich Gehorsam und Vertrauen erweckte. Er selbst ergriff rasch die Fackel, an der die andern ihre Windlichter anzünden mußten, und zeigte ihnen, voranschreitend, einen von Zweigen verdeckten Felsenweg, der unmittelbar zum Strande führte. Als sie nach kurzem Gange zwischen den Bäumen heraustraten, sahen sie schon das Meer wieder heraufleuchten, tief unter ihnen riefen die zurückgebliebenen Wachen einander von ferne an. – „Mein Gott", sagte der Einsiedler fast betroffen, „das habe ich lange nicht gehört, es ist doch ein herrlich Ding um die Jugend." – Dann grüßt' er alle noch einmal und wandte sich schnell in die Finsternis zurück. Unten aber erschraken die Wachen, da sie ein Licht nach dem andern aus den Klüften steigen und durch die Nacht schweifen sahen, als kämen die verstörten Gebirgsgeister den stillen Wald herab.

Der folgende Tag graute noch kaum, da fuhr Alma schon von ihrem bunten Teppich auf, sie hatte vor Freude auf die bevorstehende Fahrt die ganze Nacht nur leise geschlum-

mert und immerfort von dem Gebirge und dem Einsiedler geträumt. Erstaunt sah sie sich nach allen Seiten um, Antonio lag zu ihren Füßen im Gras. Es war noch alles still, die Wachtfeuer flackerten erlöschend im Zwielicht. Da überfiel Alma ein seltsames Grauen in der einsamen Fremde, sie konnt' es nicht lassen, sie stieß Antonio'n leis und zögernd an. Der verträumte Student richtete sich schnell auf und sah ihr in die klaren Augen. Sie aber wies aufhorchend nach dem Gebirge. Da hörte er hoch über ihnen auch schon die Morgenglocke des Einsiedlers durch die Luft herüberklingen und bei dem Klange fuhren die Langschläfer an den Feuern, einer nach dem andern, empor. Jetzt trat auch Alvarez schon völlig bewaffnet aus dem Zelte und teilte mit lauter Stimme seine Befehle für den kommenden Tag aus. Sanchez sollte heute das Kommando am Strande führen, er mochte ihn nicht wieder auf die Berge mitnehmen, da er ihm überall unverhofften Lärm und Verwirrung anrichtete. Bald wimmelte es nun wieder bunt über den ganzen Platz, und ehe noch die Sonne sich über dem Meere erhob, brach der Hauptmann schon, nur von Alma und Antonio begleitet, zu dem Waldbruder auf.

Alma hatte sich alle Stege von gestern wohl gemerkt und kletterte munter voraus. Antonio trug mühsam ein großes, dickes Buch unter dem Arme, in welchem er mit jugendlicher Wißbegierde und Selbstzufriedenheit merkwürdige Pflanzen aufzutrocknen und zu beschreiben pflegte. Alma meinte, er mache Heu für den Schiffsesel, und brachte ihm Disteln und anderes nichtswürdiges Unkraut in Menge. Das verdroß ihn sehr, er suchte ihr in aller Geschwindigkeit einen kurzen Begriff von dem Nutzen der Wissenschaft beizubringen. Aber sie lachte ihn aus und steckte sich die schönsten frischen Blumen auf den Hut, daß sie selbst wie die Gebirgsflora anzusehen war. – Auf einmal starrten alle überrascht in die Höhe. Denn fern auf einem Felsen, der die andern Gipfel überschaute, trat plötzlich der Einsiedler mitten ins Morgenrot, als wär' er

ganz von Feuer; er schien die Wandrer kaum zu bemerken, so versunken war er in den Anblick des Schiffs, das unten ungeduldig wie ein mutiges Roß auf den Wellen tanzte. Jetzt fiel es dem Alvarez erst aufs Herz, daß er ein verkleidetes Mädchen zu dem frommen Manne mit heraufbringen wolle. Er bestand daher ungeachtet Antonios Fürbitten darauf, daß Alma zurückkehren und ihre Wiederkehr unten erwarten sollte. Sie war betroffen und traurig darüber; als sie aber endlich die Skrupel des Hauptmanns begriff, schien sie schnell einen heimlichen Anschlag zu fassen, sah sich noch einmal genau die Gegend an und sprang dann, ohne ein Wort zu sagen, wieder nach dem Lagerplatz hinab.
Unterdes hatte der Einsiedler oben die Ankommenden gewahrt und wies ihnen durch Zeichen den nächsten Pfad zu dem Gipfel, wo er sie mit großer Freude willkommen hieß. „Laßt uns die Morgenkühle noch benutzen", sagte er dann nach kurzer Rast und führte seine Gäste sogleich wieder weiter zwischen die Berggipfel hinein. Sie gingen lange an Klüften und rauschenden Bächen vorüber, sie erstaunten, wie rüstig ihr Führer voranschritt. So waren sie auf einem hochgelegenen freien Platze angekommen, der nach der Gegend, wo das Schiff vor Anker lag, von höheren Felsen und Wipfeln ganz verschattet war; von der andern Seite aber sah man weit in die fruchtbaren Täler hinaus, während zu ihren Füßen der Garten heraufduftete, den sie schon gestern zufällig entdeckt hatten. – „Das ist mein Haus", sagte der Einsiedler und zeigte auf eine Felsenhalle im Hintergrund. Die Morgensonne schien heiter durch die offene Tür und beleuchtete einfaches Hausgerät und ein Kreuz an der gegenüberstehenden Wand, unter dem ein schönes Schwert hing. Die Ermüdeten mußten sich nun auf die Rasenbank vor der Klause lagern, der Einsiedler aber brachte zu ihrer Verwunderung Weinflaschen und köstliches Obst, schenkte die Gläser voll und trank auf den Ruhm Altspaniens. Unterdessen hatte der Morgen ringsum alles vergoldet und funkelte lustig in den Glä-

sern und Waffen, ein Reh weidete neben ihnen, und schöne bunte Vögel flatterten von den Zweigen und naschten vertraulich mit von dem Frühstück der Fremden.

Hier saßen sie lange zusammen in der erfrischenden Kühle. Der Einsiedler erkundigte sich nach ihrem gemeinschaftlichen Vaterlande, aber er sprach von so alten Zeiten und Begebenheiten, daß ihm fast nur Antonio aus seinen Schulbüchern noch Bescheid zu geben wußte. Da sie ihn aber so heiter sahen, drangen sie endlich in ihn, ihnen seinen eigenen Lebenslauf, und wie er auf diese Insel gekommen, ausführlich zu erzählen. Da besann er sich einen Augenblick. „Es ist mir alles nur noch wie ein Traum", sagte er darauf, „die fröhlichen Gesellen meiner Jugend, die sich daran ergötzen könnten, sind lange tot, andere Geschlechter gehen unbekümmert über ihre Gräber, und ich stehe zwischen den Leichensteinen allein wie in tiefem Abendrote. Doch sei es drum, ich schwieg so lange Zeit, daß mir das Herz recht aufgeht bei den heimatlichen Lauten; ich will euch von allem treulich Kunde geben, vielleicht erinnert sich doch noch jemand meiner, wenn ihr's zu Hause wiedererzählt." So rückten sie denn im Grünen näher zusammen, und der Alte hub folgendermaßen an:

Geschichte des Einsiedlers

„Die letzte Macht der Mohren war zertrümmert, die Zeit war alt und die Waffen verklungen, unsere Burgen standen, einsam über wallenden Kornfeldern, das Gras wuchs auf den Zinnen; da blickte mancher vom Walle übers Meer und sehnte sich nach einer neuen Welt. Ich war damals noch jung, vor meiner Seele dämmerte bei Tag und Nacht ein wunderbares Reich mit blühenden Inseln und goldenen Türmen aus den Fluten herauf – so rüstete ich freudig ein Schiff aus, um es zu erobern.

Was soll ich euch von den ersten Wochen der Fahrt erzählen, von den vorüberfliegenden Küsten, von der Meereseinsamkeit und den weitgestirnten prächtigen Nächten,

ihr kennt's ja so gut wie ich. Es sind jetzt gerade dreißig Jahre, es war des Königs Namenstag, wir fuhren auf offener unbekannter See. Ich hatte zur Gedächtnisfeier des Tages ein Fest auf dem Verdeck bereitet, die Tische waren gedeckt, wir saßen unter bunten Fahnen in der milden Luft, einige sangen spanische Lieder zur Zither, glänzende Fische spielten neben dem Schiff, ein frischer Wind schwellte die Segel. Da, indem wir so der fernen Heimat gedachten, sahen wir auf einmal verflogene Paradiesvögel über uns durch die klaren Lüfte schweifen, alle hielten's für die Verheißung eines nahen Landes. ‚Und was für ein Land muß das sein', rief ich aufspringend, ‚wo der Wind solche Blüten herüberweht!' Wir hofften alle das wunderbare Eldorado zu entdecken. Aber mein Leutnant, ein junger, stiller und finsterer Mann, entgegnete in seiner melancholischen Weise: das Eldorado liege auf dem großen Meere der Ewigkeit, es sei töricht, es unter den Wolken zu suchen. – Das verdroß mich. Ich schenkte rasch mein Glas voll. ‚Wer's hier nicht sucht, der findet's nimmer', rief ich, ‚durch! und wenn's am Monde hinge.' Aber wie ich anstieß, sprang mein Glas mitten entzwei, mir graute – da rief's auf einmal vom Mastkorbe: ‚Land!'
Alles fuhr nun freudig erschrocken auf, wir waren fern von allen bekannten Küsten, es mußte ein ganz fremdes Land sein. Wir sahen erst nur einen Nebelstreif, dann allmählich wuchs und dehnte sich's wie ein Wolkengebirge. Unterdes aber kam der Abend, die Luft dunkelte schläfrig und verdeckte alles wieder. – Wir gingen nun so nah am Strande als möglich vor Anker, um mit Tagesanbruch zu landen. O der schönen, erwartungsvollen Nacht! Es war so still, daß wir die Wälder von der Küste rauschen hörten, ein köstlicher Duft von Kräutern wehte herüber, im Walde sang ein Vogel mit fremdem Schalle, manchmal trat der Mond plötzlich hervor und beleuchtete flüchtig wunderbare Gipfel und Klüfte.
Als endlich der Morgen anbrach, standen wir schon alle wanderfertig auf dem Verdecke vor dem blitzenden Eilan-

de. Ich werde den Anblick niemals vergessen – mir war's, als schlüge die strenge Schöne, die ich oft im Traume gesehen, ihre Schleier zurück und ich säh ihr auf einmal in die wilden, dunkeln Augen. – Wir landeten nun und richteten uns fröhlich am Fuß des Gebirges ein, ich aber machte sogleich mit mehreren Begleitern einen Streifzug ins Land. Wir fanden alles wild und schön, fremde Tiere flogen scheu vor uns in das Dickicht, weiterhin stießen wir auf ein Dorf in einem fruchtbaren Felsentale, die Schmetterlinge flatterten friedlich in den blühenden Bäumen, aber die Hütten waren leer und alles so still in der Einsamkeit zwischen den Klüften und Wasserfällen, als wäre der Morgen der Engel des Herrn, der die Menschen aus dem Paradiese gejagt und nun zürnend mit dem Flammenschwerte auf den Bergen stände.

Als ich zurückkehrte, ließ ich der Vorsicht wegen einige Feldschlangen vom Schiffe bringen und unsern Lagerplatz verschanzen, da ich beschlossen hatte, das Land genau zu durchforschen. So war die Nacht herangekommen. Ich hatte wenig Ruh' vor schweren, seltsamen Träumen, und als ich das eine Mal aufwachte, war unser Wachtfeuer fast schon ausgebrannt, es konnte nicht mehr weit vom Tage sein. Ich begab mich daher zu den äußersten Posten, die ich am Abend ausgestellt, die waren sehr erfreut, mich zu sehen, denn sie hatten die ganze Nacht über eine wunderliche Unruhe im Gebirge bemerkt, ohne erraten zu können, was es gebe. Ich legte mich mit dem Ohr an den Boden, da war's zu meinem Erstaunen, als vernähm' ich den schweren Marsch bewaffneter Scharen in der Ferne. Manchmal erschallte es weit in den Bäumen wie Nachtgeflügel, das aufgeschreckt durch die Zweige bricht, dann war alles wieder still. Indem ich aber noch so lauschte, hör' ich auf einmal ein Flüstern dicht neben mir im Dunkeln. Ich trat einige Schritte zurück, meine Jagdtasche war mit Feuerwerk wohl versehen, ich warf schnell eine Leuchtkugel nach dem Gebirge hinaus. Da bot sich uns plötzlich der wunderbarste Anblick dar: bei dem hellen

Widerschein sahen wir einen furchtbaren Kreis bewaffneter dunkler Gestalten, lauernd an die Palmen gelehnt, hinter Steinen im Dickicht, Kopf an Kopf bis tief in den finstern Wald hinein. Alle Augen folgten dem feurigen Streif der Leuchtkugel, und als sie prasselnd in der Luft zerplatzte, richteten sich mehrere auf und betrachteten erstaunt die funkelnden Sterne, die im Niedersinken die Wipfel vergoldeten. Unterdes waren auf das Feuerzeichen die Unsrigen, die auf meinen Befehl bekleidet und mit den Waffen geruht hatten, erschreckt und noch halbverschlafen herbeigeeilt. Als nun die Wilden das Wirren und ängstliche Hin- und Herlaufen bemerkten, sprangen sie plötzlich aus ihrem Hinterhalt, ein Hagel von Speeren und Steinen flog hinter ihnen drein, ich hatte kaum Zeit, die Meinigen zu ordnen. Ich ließ fürs erste nur blind feuern, die Eingeborenen stutzten, da sie sich aber alle unversehrt fühlten, lachten sie wild und griffen nun um so wütender an. Eine zweite scharfe Ladung empfing die Verwegenen, wir sahen einige von ihnen getroffen sinken, die Hintersten aber gewahrten es nicht und drängten immer unaufhaltsamer über die Gefallenen vor. Mehrere von den Unsrigen wollten unterdes mitten in dem Getümmel ein Weib mit fliegendem Haar gesehen haben, die wie ein Würgengel unter ihren eigenen Leuten die Zurückweichenden mit ihrem Speer durchbohrte, es entstand ein dumpfes, scheues Gemurmel von einer schönen, wilden Zauberin, die Meinigen fingen an zu wanken. Jetzt zauderte ich nicht länger: ich befahl, unsere Feldschlange loszubrennen, der Schuß weckte einen anhaltenden, furchtbaren Widerhall zwischen den Bergen und riß eine breite Lücke in den dichtesten Haufen der Wilden. Das entschied den Kampf; wie vor einer unbegreiflichen übermenschlichen Gewalt standen sie eine Zeitlang regungslos, dann wandte sich auf einmal die ganze Schar mit durchdringendem Geheul; durch den Pulverdampf sahen wir sie ihre Toten und Verwundeten auf den Rücken eilig fortschleppen, und in wenigen Minuten war alles zwi-

schen dem Unkraut und den Felsenritzen wie ein Nachtspuk in der Morgendämmerung verschlüpft, die nun allmählich wachsend das Gebirge erhellte.
Wir standen noch ganz verwirrt, wie nach einem unerhörten Traume. Ich ließ darauf die Verwundeten zurückbringen und sammelte die Frischesten und Kühnsten, um den Saum des Waldes von dem Gesindel völlig zu säubern. So schritten wir eben vorsichtig in die Berge hinein, als plötzlich auf einem Felsen über uns zwischen den Wipfeln eine hohe, schlanke Mädchengestalt von so ausnehmender Schönheit erschien, daß alle, die auf sie zielten, ihre Arme sinken ließen. Sie war in ein buntgeflecktes Pantherfell gekleidet, das von einem funkelnden Gürtel über den Hüften zusammengehalten wurde, mit Bogen und Köcher, wie die heidnische Göttin Diana. Sie redete uns furchtlos und, wie es schien zürnend an, aber keiner verstand die Sprache, und der Klang ihrer Stimme verhallte in den Lüften, bis sie endlich selbst zwischen den Bäumen wieder verschwand.
Mein Leutnant insbesondere war von der wunderbaren Erscheinung ganz verwirrt. Er pflegte sonst nicht viel Worte zu machen, jetzt aber funkelten seine Augen, ich hatte ihn noch nie so heftig gesehn. Er nannte das Mädchen eine teuflische Hexe, man müsse sie tot oder lebendig fangen und verbrennen, er selbst erbot sich, sogleich Jagd auf sie zu machen. Ich verwies ihm seine unsinnige Rede. Wir brauchten, sagte ich, vor allem einige Tage Ruhe und frische Lebensmittel, dazu müßten wir jetzt Frieden halten mit den Eingeborenen. Der Leutnant aber war bei seinem stillen Wesen leicht zum Zorne zu reizen, er hieß mich selber des Teufels Zuhalter und verschwor sich, wenn ihm keiner beistehn wollte, das christliche Werk allein zu vollbringen. Und mit diesen Worten stieg er eilig das Gebirge hinan, ehe wir ihn zurückhalten konnten. Vergebens riefen wir ihm warnend, bittend und drohend nach, ich selbst durchschweifte mit vielen andern fruchtlos die nächsten Berge, es sah ihn niemand wieder.

Dieses ganz unerwartete Ereignis machte mir große Sorge, denn entweder wandte der Unglückliche durch sein Unternehmen das kaum vorübergezogene Ungewitter von neuem auf uns zurück, oder ich verlor, was wahrscheinlich war, einen redlichen und tapferen Offizier. Das letzte schien leider zutreffen zu wollen, denn alle unsere Nachforschungen blieben ohne Erfolg, mehrere Tage waren seitdem vergangen, meine Leute gaben ihn schon auf. Da beschloß ich endlich, mir um jeden Preis Gewißheit über sein Schicksal zu verschaffen. Ich ließ unser Lager abbrechen, lichtete die Anker und segelte, mich immer möglichst dicht zum Lande haltend, weiter an der Küste herab.

Wir fuhren nun abwechselnd an wilden und lachenden Gestaden vorüber, aber wo wir auch ans Land stiegen, sahen wir's verlassen, die Eingebornen flohen scheu vor uns in die Wälder, von dem Leutnant war keine Spur zu entdecken. – So hatten wir uns einmal beim ersten Morgengrauen in einem von Bergen umgebenen Tale gelagert, das mir besonders anmutig und reich bevölkert schien, wie ich aus den vielen Stimmen abnahm, die wir nachts von der Küste gehört hatten. Ich ließ unsern Lagerplatz sogleich mit Zweigen eines Baumes bestecken, von dem ich wußte, daß er in diesen Weltgegenden als Zeichen des Friedens und der Freundschaft angesehen wird, flatternde Bänder und bunte Teppiche wurden ringsum an Stangen ausgehängt, unsere Spielleute mußten dazu musizieren, das klang gar lustig in der Einsamkeit, die nun schon von der schönsten Morgenröte nach und nach erhellt wurde. Ich hatte mich in meiner Erwartung auch nicht getäuscht, denn es währte nicht lange, so erschienen einzelne Wilde neugierig hie und da wie Raben an den Klippen, jetzt erkannten wir auch im steigenden Morgen die Gegend ringsumher, fruchtbare Gründe, Wasserfälle und wunderbar gezackte Felsen, die wie Burgen über den Wäldern hingen.

Bald darauf aber sahen wir es fern am Saum des Waldes in

der Morgensonne schimmern. Ein unübersehbarer Zug von Wilden bewegte sich jetzt unter den Bäumen die nachtkühlen Schlüfte herab, voran schwärmten hohe, schlanke Burschen über den beglänzten Wiesengrund, die gewandt ihre blinkenden Speere in die Luft warfen und wieder auffingen. So im künstlichen Kampfspiel bald sich verschlingend, bald wieder auseinanderfliegend, nahten sie sich langsam unserm Lager, dazwischen sang der Zug dahinter ein rauhes, aber gewaltiges Lied, und sooft sie schwiegen, gaben andere von den Bergen Antwort.

Ich wußte nicht, was ich von dem seltsamen Beginnen halten sollte. Mir war aber alles daran gelegen, mit ihnen in ein friedliches Verständnis zu kommen. Ich hieß daher meine Leute die Feldschlange laden und sich kampffertig halten, während ich selber allein den Ankommenden entgegenging, das grüne Reis hoch über meinem Hute schwenkend. Da gewahrte ich an der Spitze des Zuges mehrere schöne, junge Männer in kriegerischem Schmuck, die über ihren Köpfen breite Schilde wie ein glänzendes Dach emporhielten. Auf diesen aber erblickte ich zu meinem Erstaunen das Wundermädchen wieder, die wir damals auf dem Felsen gesehn. Mit dem schlanken Pantherleib, zu beiden Seiten von den langen dunklen Locken umwallt, ruhte sie in ihrer strengen Schönheit wie eine furchtbare Sphinx auf den Schilden.

Kaum aber hatte sie mich erblickt, als sie sich rasch von ihrem Sitze schwang und auf mich zueilte, die turnierenden Burschen stoben zu beiden Seiten auseinander und senkten ehrerbietig die Lanzen vor ihr – es war die Königin des Landes.

Sie trat, während die andern in einem weiten Halbkreise zurückblieben, mitten unter uns mit einem Anstande, der uns alle erstaunen machte, und betrachtete mich, als den vermeintlichen König der Fremden, lange Zeit mit ernsten Blicken. Ich ließ ihr einen bunten Teppich zum Sitze über den Rasen breiten und überreichte ihr dann ein Geschenk

von Glaskorallen, Tüchern und Bändern. Sie nahm alles wie einen schuldigen Tribut an, ohne sich jedoch, nach einem flüchtigen Blick darauf, weiter darum zu bekümmern, ihre Seele schien von ganz andern Gedanken erfüllt. Unterdes war auch ihr Gefolge nach und nach vertraulicher geworden. Einzelne näherten sich den Unsrigen, einer von ihnen benutzte die Verwirrung, rollte schnell einen Teppich auf und entfloh damit nach dem Walde. Die Königin bemerkte es, rasch aufspringend zog sie einen Pfeil aus ihrem Köcher und durchbohrte den Fliehenden, daß er tot ins Gras stürzte; da hing die ganze Schar wie eine dunkle Wolke wieder unbeweglich am Saume des Waldes.

Mir graute, sie aber wandte sich von neuem zu uns, ihre Blicke spielten umher, sie schien etwas mit den Augen zu suchen. Endlich erblickte sie's: es war unsere Feldschlange. Sie betrachtete sie mit großer Aufmerksamkeit, auf ihr Begehren mußte ich sie wenden und losbrennen lassen. Bei dem Knall stürzten die Eingebornen zu Boden, das Mädchen schauerte kaum und stand wie eine Zauberin in dem ringelnden Dampf. Dann aber flog sie pfeilschnell nach der Gegend, wohin der Schuß gefallen. Ich folgte ihr, denn es schien mir ratsam, ihr die unwiderstehliche Gewalt unseres Geschützes begreiflich zu machen. Es war ein abgelegener Ort tief im Walde, wo die Kugel einen Baum zerschmettert hatte; Stamm, Krone und Äste lagen zerrissen umher, wie vom Blitz gespaltet. – Als sich die Königin von der furchtbaren Wirkung des Schusses überzeugt hatte, wurde sie ganz nachdenklich und traurig; wie vernichtet setzte sie sich auf den Rasen hin. So saß sie lange stumm, ich hatte sie noch nicht so nahe gesehn, nun fesselte mich ihre Schönheit, und ganz verwirrt und geblendet drückte ich flüchtig ihre Hand. Da wandte sie fast betroffen ihr Gesicht nach mir herum und sprang dann plötzlich wild auf, daß ich zusammenschrak. Sie eilte nach unserm Lagerplatz zurück, dort hatte sie, eh' ich's noch hindern konnte, unsere Schiffsfahne ergriffen und

schwenkte sie hoch in der Luft, uns alle auf ihre Berge einladend. Ich hatte kaum noch Zeit genug, die nötigen Wachen am Strande anzuordnen, denn sie flog schon mit dem weißen, flatternden Banner voran. Von Zeit zu Zeit, während wir vorsichtig folgten, erschien sie über den Wipfeln auf überhängenden Felsen, daß uns grauste, und sooft sie oben sichtbar wurde, jauchzten die Eingebornen ihr zu, und ihre Hörner schmetterten dazwischen, daß es weit im Gebirge widerhallte.

Ich übergehe hier unsern Empfang und ersten Aufenthalt auf diesen Felsen, die scheue Gastfreundschaft der Wilden, unser Lagern über den Klüften, die herrlichen Morgen und die wunderbaren Nächte – es ist mir von allem nur noch das Bild der Königin in der Seele zurückgeblieben. Denn sie selber war wie das Gebirge, in launenhaftem Wechsel bald scharf gezackt, bald sammetgrün, jetzt hell und glühend bis in den fernsten, tiefsten Grund, dann alles wieder grauenhaft verdunkelt. Wie oft stand ich damals auf den Bergen und schaute in das blaue Meer! Den Leutnant hatte ich lange aufgegeben, der Wind wehte günstig, alles war zur Abfahrt bereit – und doch mußte ich mich immer wieder zurückwenden in jene wildschöne Einsamkeit.

In dieser Zeit schweifte ich oft mit der Königin auf der Jagd umher. Auf einem solchen Streifzuge war ich eines Tages weit von ihr abgekommen. Vergebens rief ich ihren Namen, die Täler unten ruhten schwül, nur der Widerhall gab Antwort zwischen den Felsen. Auf einmal erblickte ich sie fern im Walde, es war, als ginge jemand unter den Bäumen eilig von ihr fort. Als ich aber hinaufkam, war alles wieder still; dann aber hörte ich sie singen über mir, eine so wunderbare Melodie, daß es mir die Seele wandte. So verlockte sie mich immer weiter in die Wildnis, ihr Lied war auch verklungen, kein Vogel sang mehr in dieser unwirtbaren Höhe – da, wie ich mich einmal plötzlich wende, steht sie auf einer Klippe in der Waldesstille, den Bogen lauernd auf mich angelegt. – Ich starrte sie er-

schrocken an, sie aber lachte und ließ den Bogen sinken, zwischen den Wasserfällen im Widerschein der Abendlichter zu mir herabsteigend. – Es war eine öde Gebirgsebene hoch über allen Wäldern, der Abend dunkelte schon. Sie setzte sich zu mir ins Gras, mir graute, denn um ihren Hals bemerkte ich eine Perlenschnur von Zähnen erschlagener Feinde. Und dennoch wandte ich keinen Blick von ihr, gleichwie man gern in ein Gewitter schaut. So lag ich, den Kopf in meine Hand gestützt, ganz in den Anblick ihrer wunderbaren Erscheinung versunken. Da sie's aber gewahrte, wandte sie sich plötzlich von mir, schwenkte aufspringend ihren Jagdspeer über sich und sang ein seltsames Lied, es waren in unserer Sprache etwa folgende Worte:

‚Bin ein Feuer hell, das lodert
Von dem grünen Felsenkranz,
Seewind ist mein Buhl' und fordert
Mich zum lust'gen Wirbeltanz,
Kommt und wechselt unbeständig.
Steigend wild,
Neigend mild
Meine schlanken Lohen wend' ich,
Komm' nicht nah mir, ich verbrenn' dich!'"

Bei diesen Worten versank Antonio in Nachsinnen, es war offenbar dasselbe Lied, das damals Alma tanzend auf dem Schiffe gesungen. Er mochte aber jetzt den Einsiedler nicht unterbrechen, der in seiner Erzählung folgendermaßen fortfuhr:
„Dieser Abend gab den Ausschlag. Damals tat ich einen heimlichen Schwur, mich selber für die Königin zu opfern. Ich gelobte, Europa zu entsagen für immer, um sie und ihr Volk zum Christentum zu bekehren und dann mit ihr das Eiland zu regieren zu Gottes Ehre. – Ich Tor, ich bildete mir ein, den Himmel zu erobern, und meinte doch nur das schöne Weib! Mein Plan war bald gemacht. Erst

mußte ich sichern Boden haben unter mir. Unter meinen Leuten befanden sich geschickte Werkmeister aller Art; Holz, Steine und was zum Bauen nötig, lag verworren umher, ich ließ rasch zugreifen und auf dem Vorgebirge, welches das ganze Land beherrschte, eine feste Burg errichten zu Schutz und Trutz und pflanzte einen Garten daneben nach unserer Weise.

Nur wenigen von den Meinen hatte ich das eigentliche Vorhaben angedeutet, die andern blendete das Gold, das überall verlockend durch den grünen Teppich der Insel schimmerte. Die Königin wußte nicht, wie ihr geschah, erst wollte sie's hindern, dann stutzte sie und staunte, und während sie noch so zögernd sann und schwankte, wuchsen die Hallen und Bogen und Lauben ihr schon über dem Haupt zusammen, und alles schoß üppig auf und rauschte und blühte, als sollt' es ein ewiger Frühling sein.

Dazumal an einem Sonntage besichtigte ich das neue Werk, meine Leute waren lustig im Grünen zerstreut, ich hatte Wein unter sie verteilen lassen, denn morgen sollten die Kanonen vom Schiff auf die Mauern gebracht, und die Burg feierlich eingeweiht werden. Ich ging durch den einsamen Hof und freute mich, wie die jungen Weinranken überall an den Pfeilern und Wänden hinaufkletterten. Es war ein schwüler Nachmittag, die Bäume flüsterten so seltsam über die Mauer, die Arbeit ruhte weit und breit, nur manchmal schlüpfte eine bunte Schlange durch das Gras, während einzelne Wolken träg und müßig über die Gegend hinzogen. Draußen aber schillerte der junge Garten im Sonnenglanze, wie mit offenen Augen schlafend, als wollt' er mir im Traum etwas sagen. Ich trat hinaus und streckte mich endlich ermattet vor dem Tor unter die blühenden Bäume, wo mich die Bienen gar bald in den Schlummer summten. – So mochte ich lange geschlafen haben, als ich plötzlich Stimmen zu hören glaubte.

Ich bog die Zweige auseinander und erblickte wirklich mehrere Eingeborne im Burghof, sie strichen, heimlich

und scheu umherschauend, an den Mauern hin, ich erkannte die Häuptlinge der Insel an ihrem Schmuck. Im ersten Augenblick glaubte ich, es gelte mir, aber sie konnten mich nicht bemerken. Zu meinem Entsetzen aber gewahre ich nun auch unsern Leutnant mitten unter ihnen mit verworrenem Bart, bleich und verwildert wie ein Gespenst, er redet geläufig ihre Mundart, sie sprechen leise und lebhaft untereinander. Darauf alles auf einmal wieder totenstill – da erblicke ich die Königin am jenseitigen Tor, in ihrem Pantherkleid mit dem Bogen, ganz wie ich sie zum erstenmal gesehen. Sie machte mit ihrem Pfeile wunderliche Zeichen in die Luft und plötzlich, schnell und lautlos, ist alles wieder zerstoben. – Ich rieb mir die Augen, die ganze Erscheinung war mir wie ein Spuk.
Als ich mich ein wenig besonnen, sprang ich hastig auf, da ich aber an den Bergrand trat, stand schon der Abend dunkelrot über der Insel, aus dem Waldgrunde unter mir hörte ich die Meinigen singen. Ich eilte sogleich nach der Gegend des Gebirges hin, wo die Königin mit den Häuptlingen verschwunden war. Da sah ich jemand fern unter den Bäumen sich ungewiß bewegen, bald rasch vortretend, bald wieder zögernd und unschlüssig zurückkehrend. Auf einmal kam er wie rasend auf mich hergestürzt – es war der Leutnant. ‚Fort, fort‘, schrie er, ‚die Nacht bricht schon herein, laßt alles stehen, werft euch auf euer Schiff und flieht, nur fort!‘ – Mir flog eine schreckliche Ahnung durch die Seele. ‚Überläufer!‘ rief ich, meinen Degen ziehend, ‚du hast uns verraten, das Kainszeichen brennt dir blutrot an der Stirn!‘ – ‚Wo, wo brennt's?‘ entgegnete er erschrocken, sich wild nach allen Seiten umsehend. – ‚Aus deinen Augen lodert es versengend‘, sagte ich. – ‚Das ist nicht wahr‘, erwiderte er, ‚im Walde brennt's unter meinen Füßen, in meinem Haar, in meinen Eingeweiden brennt's!‘, und mit diesen Worten ergriff er sein Schwert und drang verzweifelt auf mich ein. ‚Hier, Aug' in Aug', sieh nicht so scheu hinweg!‘ rief ich ihm zu. Ich weiß nicht, täuschte mich die Dämmerung, aber mir war's, als

böte er recht mit Herzenslust die entblößte Brust oft wehrlos meiner Degenspitze – mir graute, ihn zu morden. Da, während wir so fechten, tritt auf einmal die Königin aus dem Walde und mitten zwischen uns. Der Leutnant, da er sie erblickt, taumelt wie geblendet einige Schritte zurück. Dann seinen Degen plötzlich zu ihren Füßen niederwerfend, ruft er aus: ‚Da nimm's, ich *kann* nicht!' Und in demselben Augenblick bricht er zusammen, auf den Boden schlagend. – Die Königin aber neigte sich über ihn und nannte ihn beim Namen so lieblich mit dem wunderbaren Klange ihrer Stimme, daß er verwirrt den Kopf erhob und lauschte. Da setzte sie mutwillig ihren Fuß auf seinen Nacken; ‚geh nur, geh', sagte sie, und ein spöttisches Lächeln flog um ihren Mund. Und zu meinem Erstaunen raffte nun der Leutnant, seinen Degen fassend, sich rasch wieder empor, seine Augen funkelten irr über die hohe Gestalt, die er, ich sah's wohl, tödlich haßte und rasend liebte, er konnte meinen Blick nicht ertragen, seine Kleider waren mit Blut bespritzt von einer leichten Wunde am Arm, aber er bemerkte es nicht. So stürzte er von neuem fort in den Wald, und ein blutiger Streif bezeichnete seine Spur im Grase.
Nun wandte sich die Königin wieder zu mir, ich fragte sie, wo der Leutnant so lange gewesen? Sie schien zerstreut und gab verworren Antwort. Drauf fragte ich, wohin sie ging? – ‚Auf den Anstand', entgegnete sie lachend, ‚der Wind weht vom Gebirge, da wechselt das Wild, es gibt heut ein lustiges Jagen!' Jetzt traten wir droben aus dem Gestrüppe, da sah ich tief unter uns meine gesamte Mannschaft, in buntem Gemisch mit vielen Eingebornen um Becher und Würfelspiel gelagert. Von der einen Seite ragte meine halbfertige Burg über die Wipfel, die Luft dunkelte schon, Vögel schwärmten kreischend um die Mauern. – Ich hatte keine Ruh', es trieb mich zu den Meinen, die Königin führte mich auf dem nächsten Wege hinab. Sie lauschte oft in die Ferne, da hörte ich Stimmen, bald da, bald dort ein Laut, dann sah ich Rauchsäulen im Walde

aufsteigen, ich hielt es für Höhenrauch nach dem schwülen Tage. Unterdes aber kam die Nacht und der Mond, die Bäche rauschten im Dunkeln neben uns, die Königin wurde immer schöner und wilder, sie riß am Wege leuchtende Blumen ab und kränzte sich und mich damit; so stieg sie mit mir von Klippe zu Klippe, selber wie die Nacht. Nun standen wir am letzten Abhange, schon konnte ich die Stimmen der Meinigen im Waldgrunde unterscheiden, da trat sie plötzlich vor mir auf den Fels hinaus und schleuderte ihren Jagdspeer übers Tal. Kaum aber sahen die unten zerstreuten Wilden ihn funkelnd blitzen über sich, so sprangen alle jauchzend auf und warfen sich wie Tigerkatzen über meine Leute, die sich der Tücke nicht versahen. Jetzt wurde mir auf einmal alles schrecklich klar. Ich zog und hieb voll Zorn erst nach der Königin, sie aber flog schon ferne durch den Wald, so stürzt' ich nun den Meinigen zu Hilfe. Diese waren hart bedrängt, nur wenige hatten so schnell zu ihren Waffen gelangen können; ich sammelte, so gut es ging, die Verwirrten, meine unerwartete Gegenwart belebte alle, und in kurzer Zeit war das verräterische Gesindel wieder verjagt.

Aber rings am Saume des Waldes schwoll und wuchs nun die Schar unermeßlich, zahllose dunkle Gestalten mit Feuerbränden wirrten sich kreuzend durch die Nacht und steckten in grauenvoller Geschäftigkeit ringsum die Wälder an. Die Sonne hatte wochenlang gesengt über dem Lande, da griff das Feuer, an den Felswänden auf- und niedersteigend, lustig in die alten Wipfel, der Sturm faßte und rollte die Flammen auf, wie blutige Fahnen; in der entsetzlichen Beleuchtung sah ich die Königin auf ihren Knien, als wollte sie die Lohen auf uns wenden mit ihrem schrecklichen Gebet. Kaum noch vermochten wir zu atmen in dem Rauch, der von Pfeilen schwirrte, von allen Seiten rückt es rasch heran, das Schreien, das sprühende Knistern und Prasseln, nur manchmal von dem Donner stürzender Bäume unterbrochen; schon lief das Feuer in dem verdorrten Heidekraut über den Waldgrund, uns im-

mer enger umzingelnd mit seinem furchtbaren Ringe. Da, in der höchsten Not teilte der Wind auf einen Augenblick den Qualm, und wir gewahrten plötzlich eine dunkle Furt in den Flammenwogen. Ein reißender Waldstrom rang dort mit dem wilden Feuermanne, der zornig Wurzeln, Stämme und Kronen darüber geworfen hatte. Das rettete uns, wir eilten über die lodernden Brücken und erreichten in der allgemeinen Verwirrung glücklich das Meer, eh' uns der große Haufen bemerkte.

Als wir aber an den Strand kamen, sahen wir zu unserm Schrecken unser Boot schon von Eingebornen besetzt. Die Königin war's mit vielen bewaffneten Häuptlingen, sie schienen von unserm Schiffe herzukommen und sprangen soeben leis und heimlich ans Land. Da sie uns erblickten, nicht weniger überrascht als wir, umringten sie eiligst ihre Königin und suchten uns in die Flammen zurückzutreiben. Auf diesem einsamen Platze aber waren wir die Mehrzahl, es entstand ein verzweifelter Kampf, denn unser aller Leben hing an einer Viertelstunde. Vergebens streckte die Königin mit ihrem tödlichen Geschoß meine kühnsten Gesellen zu Boden, die Häuptlinge fochten sterbend noch auf den Knien, und als der letzte sank, schwang ich die Schreckliche gewaltsam auf meinen Arm und stürzte mich mit ihr und den wenigen, die mir geblieben, in das Boot. – Es war die höchste Zeit, denn schon drangen die Eingebornen aus allen Felsenspalten und brennenden Waldtrümmern wie ein Schwarm Salamander auf uns ein, und kaum hatten wir den Bord des Schiffes erklommen, so wimmelte die See von unzähligen bewaffneten Nachen. Ich ließ schnell die Anker lichten, ein frischer Wind schwellte die Segel, die Wilden folgten und bedeckten das Schiff mit einem Pfeilregen.

Nun aber brach auf dem Schiffe selbst der rohe Grimm der verwilderten Soldaten aus. Sie hatten, eh' ich sie zügeln konnte, die Königin gebunden und verhöhnten sie mit gemeinen Spottreden; sie aber saß stolz und schweigend unter ihnen, als wäre sie noch die Herrin hier und wir ihre

Gefangenen. Auf einmal erkannte sie einen Häuptling, der sich auf einem Kahne tollkühn genähert. Sich gewaltsam auf dem Verdeck hoch aufrichtend, fragte sie: ob alle Weißen von der Insel vertilgt seien? und da er's bejahte, winkte sie ihnen zu, unser Schiff zu verlassen. Die Wilden zögerten erschrocken und verwirrt, ein dunkles Gemurmel ging durch den ganzen Schwarm. Da befahl sie ihnen noch einmal mit lauter Stimme, eiligst an den Strand zurückzukehren, und zu unserm Erstaunen wandten sich alle, Boot auf Boot, aber ein wehklagender Abschiedsgesang erfüllte die Luft wie ein Grabeslied.

Mir war das Betragen der Königin unbegreiflich. Noch einmal leuchtete mir die Hoffnung auf, sie wolle alles verlassen und mit uns ziehn, als plötzlich der Schreckensruf: ‚Feuer!' aus dem untern Schiffsraum erscholl. Todbleiche Gesichter, auf das Verdeck stürzend, bestätigten das furchtbare Unheil. Das Feuer hatte die Planken der Pulverkammer gefaßt, an Löschen war nicht mehr zu denken, wir waren alle unrettbar verloren. Mich überflog eine gräßliche Ahnung. Ich sah die Königin durchdringend an; sie flüsterte mir heimlich zu: sie selber habe das Schiff angesteckt, als sie vorhin an Bord gewesen. – Jetzt züngelten die Flammen schon aus allen Luken aufs Verdeck hinauf, da, mitten in der entsetzlichen Verwirrung, zerriß sie plötzlich ihre Banden und freudig und unverwandt nach den brennenden Wäldern schauend, streckte sie beide Arme frei in die sternklare Nacht wie ein Engel des Todes. In demselben Augenblick aber fühlte ich einen dumpfen Schlag, die Bretter wichen unter mir, meine Sinne vergingen, ich sah nur noch einen unermeßlichen Feuerblick, wie tief in die Ewigkeit hinein.

Als ich wieder zu mir selbst kam, war alles still überm Meer, nur dunkle Trümmer des Schiffs und zerrissene Leichname meiner Landsleute trieben einzeln umher. Ich hatte im Todeskampf einen Mastbaum fest umklammert. Jetzt bemerkte ich einen Nachen der Eingeborenen, der verlassen sich neben mir auf den Wellen schaukelte. Ver-

wundet und zerschlagen wie ich war, bot ich meine letzten Kräfte auf und warf mich todmüde hinein. Der Wind trieb mich dicht an dem umbuschten Gestade hin, der Mond schien blaß durch die Rauchwolken, auf der Insel aber hatte unterdes das Feuer auch meine Burg ergriffen, die Flammen schlugen aus allen Fenstern, langsam neigte sich der Turm, und Bogen auf Bogen stürzte alles donnernd in die Glut zusammen. Da sah ich im hellen Widerschein der Flammen fern die Leiche der Königin schwimmen in bleicher Todesschönheit, als schliefe sie auf dem Meere. Auf einem vorspringenden Felsen aber stand der Leutnant, auf sein blutiges Schwert gestützt, ganz allein, vom Feuer verbrannt; er bemerkte mich nicht, mein Schifflein flog um die Klippe – ich sah ihn niemals wieder."

Hier schwieg der Einsiedler, seine Seele schien tief bewegt. Da ihn aber seine Gäste noch immer fragend ansahen, hub er nach einem Weilchen von neuem an: „Was wäre nach jener Nacht noch weiter zu berichten? Ich rang mit Hunger, Sturm und Wogen, ich wünschte mir tausendmal den Tod und haschte doch begierig die zerstreuten Lebensmittel, Werkzeuge und Gerätschaften auf, die der Wind von dem zertrümmerten Schiff an meinen Nachen spülte. So warf die See mich endlich am dritten Tage an dies Eiland. – Hier zwischen diesen Wäldern stieg ich in die Felseneinsamkeit hinauf: meine Jugend, mein Ruhm und meine Liebe waren hinter mir im Meere versunken, und kampfesmüd hing ich mein Schwert an diesen Baum; da seht, da hängt's noch heut, von Blüten ganz verhüllt."

„So seid Ihr Don Diego von Leon!" fuhr hier Antonio plötzlich auf, das Wappen seines Oheims auf dem Degengriff erkennend.

„Der war ich ehemals in der Welt", erwiderte der Einsiedler, „wie kennt Ihr mich?"

Aber der überraschte Antonio lag schon zu seinen Füßen und umklammerte seine Knie, daß ihn des Alten langer, weißer Bart wie Höhenrauch umwallte.

Noch bevor dies an der Klause vorging, war Alvarez unruhig aufgestanden und weiterhin unter die Bäume getreten, denn er glaubte, einen seltsamen Gesang im Walde zu hören. Nun vernahm es auch der Einsiedler. Auf einmal richtete dieser sich gewaltsam aus Antonios Armen auf. „Im Namen Gottes!" rief er nach dem Walde hin, „wende dich ab und gehe ein zur ewigen Ruh!" Antonio und Alvarez schauten erschrocken nach dem Fleck, wohin er starrte, und sahen mit Grauen die Frau Venus von der andern Insel zwischen den wechselnden Schatten über den Bergrücken schweifen. Der Hauptmann zog seinen Degen, man hörte die Flüchtige immer deutlicher und näher durch das Dickicht brechen. Jetzt trat sie unter den Bäumen hervor – es war Alma in der Tracht und dem Schmuck ihrer Heimat, so stand sie scheu und atemlos, sie hatte es unten nicht länger ausgehalten und schon lange Antonio zwischen den Felsen wieder aufgesucht.
Der Einsiedler verwendete keinen Blick von ihr. „Wer bist du?" sagte er endlich. „Du schaust wie sie und bist es doch nicht!" – Alma aber war ganz verwirrt und sah ängstlich einen nach dem andern an. „Ich kann ja nichts dafür", erwiderte sie dann zögernd; „sie sagten's immer, daß ich aussäh' wie meine Muhme, die tote Königin." – „Mein Gott", fiel hier Alvarez ein, „ihr macht mich ganz konfus; so war das also die Insel der wilden Königin, von der wir hergekommen?" – Alma nickte mit dem Köpfchen. „Auch die Meinigen", sagte sie, „hielten mich damals, als wir fortfuhren, für die verstorbene Königin, sonst hätten sie euch sicherlich erschlagen." – Da das Mädchen sah, daß ihr niemand zürnte, wurde sie wieder heiterer und gesprächiger. Sie erzählte nun, daß sie gar oft in ihrer Heimat von alten Leuten gehört, wie die tapfere Königin mit einem spanischen Schiff, das sie selber angezündet, in die Luft geflogen, in jener Schreckensnacht hätten sie dann ihren Leichnam aus dem Meere gefischt und mit den eroberten Fahnen und Waffen der Fremden in die Königsgruft gelegt, wo die besondere eisige Luft die Toten un-

versehrt erhalte. Nur Alonzo allein sei von den Spaniern zurückgeblieben. – „Wie!" rief Alvarez, „so war der wahnsinnige Alte in seinem tollen Ornat derselbe gewesene Schiffsleutnant!" – Alma aber fuhr fort: „Der arme Alonzo bewachte seitdem die tote Königin bei Tag und Nacht und meint', sie schliefe nur, bis er bei unsrer Abfahrt selbst den Tod gefunden." – Der Einsiedler war während dieser Erzählung in tiefes Nachdenken versunken. „Entsetzlich!" sagte er dann halb für sich, „nun ist er abgelöst von seiner schauerlichen Wacht – Gott sei ihm gnädig!"

Unterdes war Alma in die Felsenhalle gegangen und untersuchte dort alles mit furchtsamer Neugier. Alvarez aber rief sie wieder heraus, sie mußte sich zu ihnen vor die Klause setzen, und nun ging es an ein Fragen und Erzählen aus der alten Zeit, daß keiner merkte, wie die Nacht allmählich schon Berg und Tal verschattete.

Tiefer unten aber rumorte es noch immer im Walde. Sanchez machte eifrig die Runde, denn gab es hier auch nichts zu bewachen, den müßigen Gesellen war es in ihrer Langenweile eben nur um den Lärm zu tun. In einzelnen Trupps auf den waldigen Abhängen um die Wachtfeuer gelagert, sangen sie aus der Ferne schöne Lieder, und so oft sie pausierten, hörte man Meer und Wald heraufrauschen. Das hatte die arme Alma lange nicht gehört; sie plauderte froh in ihrer fremden Sprache und sang und tanzte den Kriegstanz ihres Volkes. Diegos Augen aber ruhten bald auf ihr, bald auf dem blühenden Antonio, ihm war, als spiegelte sich wunderbar sein Leben wie ein Traum noch einmal wider.

Die Spanier lagen noch mehrere Tage auf dieser Insel, um günstigen Wind abzuwarten. Don Diego hatte, als er sein Haus im Felsen baute, Gold in Menge gefunden, das lag seitdem vergessen im Schutt. Jetzt fiel's ihm wieder ein, er verteilte den Schatz nach Amt und Würden an seine armen

Gäste. Da war ein Jubilieren, Prahlen und Projektemachen unter dem glückseligen Schwarm, jeder wollte was Rechtes ausbrüten über seinem unverhofften Mammon und ließ allmählich die lustigen Reiseschwingen sinken in der schweren Vergoldung. Den Studenten Antonio aber verlangte wieder recht nach den duftigen Gärten der Heimat, um dort in den blühenden Wipfeln mit seinem schönen, fremden Wandervöglein sich sein Nest zu bauen. So beschlossen sie alle einmütig, die neue Welt vorderhand noch unentdeckt zu lassen und vergnügt in die gute alte wieder heimzukehren. – Diego schüttelte halb unwillig den Kopf. „So", sagte er, „hätte ich nicht getan, als ich noch jung war."

In dieser Zeit erwachte einmal Alma mitten in der schönsten Sommernacht, es war, als hätte sie jemand im Schlafe auf die Stirn geküßt. Sie fuhr erschrocken halb empor und sah soeben Don Diego von dem Platze fortgehn, der zu ihrem Erstaunen ganz still und verlassen war. Als sie sich aber völlig ermunterte, vernahm sie tiefer unten ein verworrenes Getümmel, es war, als sei plötzlich über Nacht der Frühling gekommen: ein Jubel und Rufen und Durcheinanderrennen den ganzen Strand entlang.

Jetzt kamen auch mehrere Soldaten mit gefüllten Schläuchen von den Quellen im Walde herab. „Viktoria!" riefen sie ihr zu, „der Wind hat sich gedreht, nun geht's nach Spanien." – Da sprang Alma pfeilschnell auf, suchte emsig alles zusammen und schnürte ihr Bündel und jauchzte in sich, sie meinte, sie hätte den gestirnten Himmel noch niemals so weit und schön gesehen!

Indem sie aber noch so fröhlich hantierte, sah sie Antonio mit Don Diego eilig und in lebhaftem Gespräch vom Strande kommen. Auf der Klippe über ihr stand Diego plötzlich still. „Nun geh hinab", sagte er zu Antonio, „du beredest mich nicht, ich bleibe hier.

Mein Leben ist wie ein Gewitter schön und schrecklich vorübergezogen, und die Blitze spielen nur noch fern am Horizont wie in eine andere Welt hinüber. Du aber sollst

dir erst die Sporen verdienen, kehre zurück in die Welt und haue dich tüchtig durch, daß du dir einst auch solchen Fels eroberst, der die Wetter bricht – weiter bringt es doch keiner. Fahre wohl!" – Hier umarmte er gerührt den Jüngling und verschwand in der Wildnis. Antonio sah ihm lange in die nachtkühle Einsamkeit nach. – Da erblickte er auf einmal Alma dicht vor sich, schwang sie auf seinen Arm hoch in das aufdämmernde Morgenrot und stürzte mit ihr hinab.

Und als die Sonne aufging, flog das Schiff schon übers blaue Meer, der frische Morgenwind schwellte die Segel. Alma saß vergnügt mit ihrem Reisebündel und schaute in die glänzende Ferne, die Schiffer sangen wieder das Lied von der Fortuna. Auf dem allmählich versinkenden Felsen der Insel aber stand Diego und segnete noch einmal die fröhlichen Gesellen, denen auch wir eine glückliche Fahrt nachrufen.

Das Schloß Dürande

In der schönen Provence liegt ein Tal zwischen waldigen Bergen, die Trümmer des alten Schlosses Dürande sehen über die Wipfel in die Einsamkeit herein; von der andern Seite erblickt man weit unten die Türme der Stadt Marseille; wenn die Luft von Mittag kommt, klingen bei klarem Wetter die Glocken herüber, sonst hört man nichts von der Welt. In diesem Tale stand ehemals ein kleines Jägerhaus, man sah's vor Blüten kaum, so überwaldet war's und weinumrankt bis an das Hirschgeweih über dem Eingang: in stillen Nächten, wenn der Mond hell schien, kam das Wild oft weidend bis auf die Waldeswiese vor der Tür. Dort wohnte dazumal der Jäger Renald, im Dienst des alten Grafen Dürande, mit seiner jungen Schwester Gabriele ganz allein, denn Vater und Mutter waren lange gestorben.

In jener Zeit nun geschah es, daß Renald einmal an einem schwülen Sommerabend, rasch von den Bergen kommend, sich nicht weit von dem Jägerhaus mit seiner Flinte an den Saum des Waldes stellte. Der Mond beglänzte die Wälder, es war so unermeßlich still, nur die Nachtigallen schlugen tiefer im Tal, manchmal hörte man einen Hund bellen aus den Dörfern oder den Schrei des Wildes im Walde. Aber er achtete nicht darauf, er hatte heut' ein ganz anderes Wild auf dem Korn. Ein junger fremder Mann, so hieß es, schleiche abends heimlich zu seiner Schwester, wenn er selber weit im Forst; ein alter Jäger hatte es ihm gestern vertraut, der wußte es vom Waldhüter, dem hatt' es ein Köhler gesagt. Es war ihm ganz unglaublich, wie sollte sie zu der Bekanntschaft gelangt sein? Sie kam nur Sonntags in die Kirche, wo er sie niemals aus den Augen verlor. Und doch wurmte ihn das Gerede, er

konnte sich's nicht aus dem Sinn schlagen, er wollte endlich Gewißheit haben. Denn der Vater hatte sterbend ihm das Mädchen auf die Seele gebunden, er hätte sein Herzblut gegeben für sie.
So drückte er sich lauernd an die Bäume in wechselnden Schatten, den die vorüberfliegenden Wolken über den stillen Grund warfen. Auf einmal aber hielt er den Atem an, es regte sich am Hause, und zwischen den Weinranken schlüpfte eine schlanke Gestalt hervor; er erkannte sogleich seine Schwester an dem leichten Gang; o mein Gott, dachte er, wenn alles nicht wahr wäre! Aber in demselben Augenblick streckte sich ein langer dunkler Schatten neben ihr über den mondbeschienenen Rasen, ein hoher Mann trat rasch aus dem Hause, dicht in einen schlechten grünen Mantel gewickelt, wie ein Jäger. Er konnte ihn nicht erkennen, auch sein Gang war ihm durchaus fremd; es flimmerte ihm vor den Augen, als könnte er sich in einem schweren Traume noch nicht recht besinnen.
Das Mädchen aber, ohne sich umzusehen, sang mit fröhlicher Stimme, daß es dem Renald wie ein Messer durchs Herz ging:

„Ein' Gems auf dem Stein,
Ein Vogel im Flug,
Ein Mädel, das klug,
Kein Bursch holt die ein!"

„Bist du toll!" rief der Fremde, rasch hinzuspringend.
„Es ist dir schon recht", entgegnete sie lachend, „so werd' ich dir's immer machen; wenn du nicht artig bist, sing' ich aus Herzensgrund." Sie wollte von neuem singen, er hielt ihr aber voll Angst mit der Hand den Mund zu. Da sie so nahe vor ihm stand, betrachtete sie ihn ernsthaft im Mondschein. „Du hast eigentlich recht falsche Augen", sagte sie; „nein, bitte mich nicht wieder so schön, sonst sehn wir uns niemals wieder, und das tut uns beiden leid. –

Herr Jesus!" schrie sie auf einmal, denn sie sah plötzlich den Bruder hinterm Baum nach dem Fremden zielen. – Da, ohne sich zu besinnen, warf sie sich hastig dazwischen, so daß sie, den Fremden umklammernd, ihn ganz mit ihrem Leibe bedeckte. Renald zuckte, da er's sah, aber es war zu spät, der Schuß fiel, daß es tief durch die Nacht widerhallte. Der Unbekannte richtete sich in dieser Verwirrung hoch empor, als wär er plötzlich größer geworden, und riß zornig ein Taschenpistol aus dem Mantel; da kam ihm auf einmal das Mädchen so bleich vor, er wußte nicht, war es vom Mondlicht oder vom Schreck. „Um Gottes willen", sagte er, „bist du getroffen?" „Nein, nein", erwiderte Gabriele, ihm unversehens und herzhaft das Pistol aus der Hand windend, und drängte ihn heftig fort. „Dorthin", flüsterte sie, „rechts über den Steg am Fels, nur fort, schnell fort!"

Der Fremde war schon zwischen den Bäumen verschwunden, als Renald zu ihr trat. „Was machst du da für dummes Zeug!" rief sie ihm entgegen und verbarg rasch Arm und Pistol unter der Schürze. Aber die Stimme versagte ihr, als er nun dicht vor ihr stand und sie sein bleiches Gesicht bemerkte. Er zitterte am ganzen Leibe, und auf seiner Stirn zuckte es zuweilen, wie wenn es von fern blitzte. Da gewahrte er plötzlich einen blutigen Streif an ihrem Kleide. „Du bist verwundet", sagte er erschrocken, und doch war's, als würde ihm wohler beim Anblick des Bluts; er wurde sichtbar milder und führte sie schweigend in das Haus. Dort pinkte er schnell Licht an, es fand sich, daß die Kugel ihr nur leicht den rechten Arm gestreift; er trocknete und verband die Wunde, sie sprachen beide kein Wort miteinander. Gabriele hielt den Arm fest hin und sah trotzig vor sich nieder, denn sie konnte gar nicht begreifen, warum er böse sei; sie fühlte sich so rein von aller Schuld, nur die Stille jetzt unter ihnen wollte ihr das Herz abdrükken, und sie atmete tief auf, als er endlich fragte: wer es gewesen? – Sie beteuerte nun, daß sie das nicht wisse, und erzählte, wie er an einem schönen Sonntagsabend, als sie

eben allein vor der Tür gesessen, zum ersten Male von den Bergen gekommen und sich zu ihr gesetzt und dann am folgenden Abend wieder und immer wieder gekommen, und wenn sie ihn fragte, wer er sei, nur lachend gesagt: ihr Liebster.

Unterdes hatte Renald unruhig ein Tuch aufgehoben und das Pistol entdeckt, das sie darunter verborgen hatte. Er erschrak auf das heftigste und betrachtete es dann aufmerksam von allen Seiten. – „Was hast du damit?“ sagte sie erstaunt; „wem gehört es?“ Da hielt er's ihr plötzlich funkelnd am Licht vor die Augen: „Und du kennst ihn wahrhaftig nicht?“

Sie schüttelte mit dem Kopf.

„Ich beschwöre dich bei allen Heiligen“, hob er wieder an, „sag mir die Wahrheit.“

Da wandte sie sich auf die andere Seite. „Du bist heute rasend“, erwiderte sie, „ich will dir gar keine Antwort mehr geben.“

Das schien ihm das Herz leichter zu machen, daß sie ihren Liebsten nicht kannte, er glaubte es ihr, denn sie hatte ihn noch niemals belogen. Er ging nun einige Male finster in der Stube auf und nieder. „Gut, gut“, sagte er dann, „meine arme Gabriele, so mußt du gleich morgen zu unserer Muhme ins Kloster; mach' dich zurecht, morgen, ehe der Tag graut, führ' ich dich hin.“ Gabriele erschrak innerlichst, aber sie schwieg und dachte: kommt Tag, kommt Rat. Renald aber steckte das Pistol zu sich und sah noch einmal nach ihrer Wunde, dann küßte er sie noch herzlich zur guten Nacht.

Als sie endlich allein in ihrer Schlafkammer war, setzte sie sich angekleidet aufs Bett und versank in ein tiefes Nachsinnen. Der Mond schien durchs offene Fenster auf die Heiligenbilder an der Wand, im stillen Gärtchen draußen zitterten die Blätter in den Bäumen. Sie wand ihre Haarflechten auf, daß ihr die Locken über Gesicht und Achseln herabrollten, und dachte vergeblich nach, wen ihr Bruder eigentlich im Sinn habe und warum er vor dem Pistol so

sehr erschrocken – es war ihr alles wie im Traume. Da kam es ihr ein paarmal vor, als ginge draußen jemand sachte ums Haus. Sie lauschte am Fenster, der Hund im Hofe schlug an, dann war alles wieder still. Jetzt bemerkte sie erst, daß auch ihr Bruder noch wach war; anfangs glaubte sie, er rede im Schlaf, dann aber hörte sie deutlich, wie er auf seinem Bett vor Weinen schluchzte. Das wandte ihr das Herz, sie hatte ihn noch niemals weinen gesehen, es war ihr nun selber, als hätte sie was verbrochen. In dieser Angst beschloß sie, ihm seinen Willen zu tun; sie wollte wirklich nach dem Kloster gehen, die Priorin war ihre Muhme, der wollte sie alles sagen und sie um ihren Rat bitten. Nur das war ihr unerträglich, daß ihr Liebster nicht wissen sollte, wohin sie gekommen. Sie wußte wohl, wie herzhaft er war und besorgt um sie; der Hund hatte vorhin gebellt, im Garten hatte es heimlich geraschelt wie Tritte, wer weiß, ob er nicht nachsehen wollte, wie es ihr ginge nach dem Schrecken. – Gott, dachte sie, wenn er noch draußen stünd'! – Der Gedanke verhielt ihr fast den Atem. Sie schnürte sogleich eilig ihr Bündel, dann schrieb sie für ihren Bruder mit Kreide auf den Tisch, daß sie noch heute allein ins Kloster fortgegangen. Die Türen waren nur angelehnt, da schlich sie vorsichtig und leise aus der Kammer über den Hausflur in den Hof, der Hund sprang freundlich an ihr herauf, sie hatte Not, ihn am Pförtchen zurückzuweisen; so trat sie endlich mit klopfendem Herzen ins Freie.

Draußen schaute sie sich tief aufatmend nach allen Seiten um, ja, sie wagte es sogar, noch einmal bis an den Gartenzaun zurückzugehen, aber ihr Liebster war nirgend zu sehen, nur die Schatten der Bäume schwankten ungewiß über den Rasen. Zögernd betrat sie nun den Wald und blieb immer wieder stehen und lauschte; es war alles so still, daß ihr graute in der großen Einsamkeit. So mußte sie nun endlich doch weitergehen und zürnte heimlich im Herzen auf ihren Schatz, daß er sie in ihrer Not so zaghaft verlassen. Seitwärts im Tal aber lagen die Dörfer in tiefer

Ruh. Sie kam am Schloß des Grafen Dürande vorbei, die Fenster leuchteten im Mondschein herüber, im herrschaftlichen Garten schlugen die Nachtigallen und rauschten die Wasserkünste; das kam ihr so traurig vor, sie sang für sich das alte Lied:

„Gut' Nacht, mein Vater und Mutter,
Wie auch mein stolzer Bruder,
Ihr seht mich nimmermehr!
Die Sonne ist untergegangen
Im tiefen, tiefen Meer."

Der Tag dämmerte noch kaum, als sie endlich am Abhange der Waldberge bei dem Kloster anlangte, das mit verschlossenen Fenstern, noch wie träumend, zwischen kühlen, duftigen Gärten lag. In der Kirche aber sangen die Nonnen soeben ihre Metten durch die weite Morgenstille, nur einzelne, früh erwachte Lerchen draußen stimmten schon mit ein in Gottes Lob. Gabriele wollte abwarten, bis die Schwestern aus der Kirche zurückkämen, und setzte sich unterdes auf die breite Kirchhofsmauer. Da fuhr ein zahmer Storch, der dort übernachtet, mit seinem langen Schnabel unter den Flügeln hervor und sah sie mit den klugen Augen verwundert an; dann schüttelte er in der Kühle sich die Federn auf und wandelte mit stolzen Schritten wie eine Schildwacht den Mauerkranz entlang. Sie aber war so müde und überwacht, die Bäume über ihr säuselten noch so schläfrig, sie legte den Kopf auf ihr Bündel und schlummerte fröhlich unter den Blüten ein, womit die alte Linde sie bestreute.

Als sie aufwachte, sah sie eine hohe Frau in faltigen Gewändern über sich gebeugt, der Morgenstern schimmerte durch ihren langen Schleier, es war ihr, als hätt' im Schlaf die Mutter Gottes ihren Sternenmantel um sie geschlagen. Da schüttelte sie erschrocken die Blütenflocken aus dem Haar und erkannte ihre geistliche Muhme, die zu ihrer

Verwunderung, als sie aus der Kirche kam, die Schlafende auf der Mauer gefunden. Die Alte sah ihr freundlich in die schönen frischen Augen. „Ich hab' dich gleich daran erkannt", sagte sie, „als wenn mich deine selige Mutter ansähe!" – Nun mußte sie ihr Bündel nehmen, und die Priorin schritt eilig ins Kloster voraus; sie gingen durch kühle dämmernde Kreuzgänge, wo soeben noch die weißen Gestalten einzelner Nonnen wie Geister vor der Morgenluft lautlos verschlüpften. Als sie in die Stube traten, wollte Gabriele sogleich ihre Geschichte erzählen, aber sie kam nicht dazu. Die Priorin, so lange wie auf eine selige Insel verschlagen, hatte so viel zu erzählen und zu fragen von dem jenseitigen Ufer ihrer Jugend und konnte sich nicht genug verwundern, denn alle ihre Freunde waren seitdem alt geworden oder tot, und eine andere Zeit hatte alles verwandelt, die sie nicht mehr verstand. Geschäftig in redseliger Freude strich sie ihrem lieben Gast die Locken aus der glänzenden Stirn wie einem kranken Kinde, holte aus einem altmodischen, künstlich geschnitzten Wandschrank Rosinen und allerlei Naschwerk, und fragte und plauderte immer wieder. Frische Blumensträuße standen in bunten Krügen am Fenster, ein Kanarienvogel schmetterte gellend dazwischen, denn die Morgensonne funkelte draußen schon durch die Wipfel und vergoldete wunderbar die Zelle, das Betpult und die schwergewirkten Lehnstühle; Gabriele lächelte fast betroffen, wie in eine neue ganz fremde Welt hinein.

Noch an demselben Tage kam auch Renald zum Besuch; sie freute sich außerordentlich, es war ihr, als hätte sie ihn ein Jahr lang nicht gesehn. Er lobte ihren raschen Entschluß von heute nacht und sprach dann viel und heimlich mit der Priorin; sie horchte ein paarmal hin, sie hätte so gern gewußt, wer ihr Geliebter sei, aber sie konnte nichts erfahren. Dann mußte sie auch wieder heimlich lachen, daß die Priorin so geheimnisvoll tat, denn sie merkt' es wohl, sie wußt' es selber nicht. – Es war indes beschlossen worden, daß sie fürs erste noch im Kloster bleiben sollte.

Renald war zerstreut und eilig, er nahm bald wieder Abschied und versprach, sie abzuholen, sobald die rechte Zeit gekommen.

Aber Woche auf Woche verging, und die rechte Zeit war noch immer nicht da. Auch Renald kam immer seltener und blieb endlich ganz aus, um dem ewigen Fragen seiner Schwester nach ihrem Schatze auszuweichen, denn er konnte oder mochte ihr nichts von ihm sagen. Die Priorin wollte die arme Gabriele trösten, aber sie hatt' es nicht nötig, so wunderbar war das Mädchen seit jener Nacht verwandelt. Sie fühlte sich, seit sie von ihrem Liebsten getrennt, als seine Braut vor Gott, der wolle sie bewahren. Ihr ganzes Dichten und Trachten ging nun darauf, ihn selber auszukundschaften, da ihr niemand beistand in ihrer Einsamkeit. Sie nahm sich daher eifrig der Klosterwirtschaft an, um mit den Leuten in der Gegend bekannt zu werden; sie ordnete alles in Küche, Keller und Garten, alles gelang ihr, und wie sie so sich selber half, kam eine stille Zuversicht über sie wie Morgenrot, es war ihr immer, als müßt' ihr Liebster plötzlich einmal aus dem Walde zu ihr kommen.

Damals saß sie eines Abends noch spät mit der jungen Schwester Renate am offenen Fenster der Zelle, aus dem man in den stillen Klostergarten und über die Gartenmauer weit ins Land sehen konnte. Die Heimchen zirpten unten auf den frischgemähten Wiesen, überm Walde blitzte es manchmal aus weiter Ferne. „Da läßt mein Liebster mich grüßen“, dachte Gabriele bei sich. – Aber Renate blickte verwundert hinaus; sie war lange nicht wach gewesen um diese Zeit. „Sieh nur“, sagte sie, „wie draußen alles anders aussieht im Mondschein, der dunkle Berg drüben wirft seinen Schatten bis an unser Fenster, unten erlischt ein Lichtlein nach dem andern im Dorfe. Was schreit da für ein Vogel?“ – „Das ist das Wild im Walde“, meinte Gabriele. –

„Wie du auch so allein im Dunkeln durch den Wald gehen kannst“, sagte Renate wieder; „ich stürbe vor Furcht.

Wenn ich so manchmal durch die Scheiben hinaussehe in die tiefe Nacht, dann ist mir immer so wohl und sicher in meiner Zelle wie unterm Mantel der Mutter Gottes."
„Nein", entgegnete Gabriele, „ich möcht' mich gern einmal bei Nacht verirren recht im tiefsten Wald, die Nacht ist wie im Traum so weit und still, als könnt' man über die Berge reden mit allen, die man lieb hat in der Ferne. Hör' nur, wie der Fluß unten rauscht und die Wälder, als wollten sie auch mit uns sprechen und können nur nicht recht! – Da fällt mir immer ein Märchen ein dabei, ich weiß nicht, hab ich's gehört, oder hat mir's geträumt."
„Erzähl's mir doch, ich bete unterdes meinen Rosenkranz fertig", sagte die Nonne, und Gabriele setzte sich fröhlich auf die Fußbank vor ihr, wickelte vor der kühlen Nachtluft die Arme in ihre Schürze und begann sogleich folgendermaßen:
„Es war einmal eine Prinzessin in einem verzauberten Schlosse gefangen, das schmerzte sie sehr, denn sie hatte einen Bräutigam, der wußte gar nicht, wohin sie gekommen war, und sie konnte ihm auch kein Zeichen geben, denn die Burg hatte nur ein einziges, verschlossenes Tor nach einem tiefen, tiefen Abhang hin, und das Tor bewachte ein entsetzlicher Riese, der schlief und trank und sprach nicht, sondern ging nur immer Tag und Nacht vor dem Tore auf und nieder wie der Perpendikel einer Turmuhr. Sonst lebte sie ganz herrlich in dem Schloß; da war Saal an Saal, einer immer prächtiger als der andere, aber niemand drin zu sehen und zu hören, kein Lüftchen ging und kein Vogel sang in den verzauberten Bäumen im Hofe, die Figuren auf den Tapeten waren schon ganz krank und bleich geworden in der Einsamkeit, nur manchmal warf sich das trockne Holz an den Schränken vor Langeweile, daß es weit durch die öde Stille schallte, und auf der hohen Schloßmauer draußen stand ein Storch, wie eine Vedette, den ganzen Tag auf einem Bein."
„Ach, ich glaube gar, du stichelst auf unser Kloster", sagte Renate. Gabriele lachte und erzählte munter fort:

„Einmal aber war die Prinzessin mitten in der Nacht aufgewacht, da hörte sie ein seltsames Sausen durch das ganze Haus. Sie sprang erschrocken ans Fenster und bemerkte zu ihrem großen Erstaunen, daß es der Riese war, der eingeschlafen vor dem Tore lag und mit solcher grausamen Gewalt schnarchte, daß alle Türen, sooft er den Atem einzog und wieder ausstieß, vor dem Zugwind klappend auf und zu flogen. Nun sah sie auch, sooft die Tür nach dem Saale aufging, mit Verwunderung, wie die Figuren auf den Tapeten, denen die Glieder schon ganz eingerostet waren von dem langen Stillstehen, sich langsam dehnten und reckten; der Mond schien hell über den Hof, da hörte sie zum erstenmal die verzauberten Brunnen rauschen, der steinerne Neptun unten saß auf dem Rand der Wasserkunst und strählte sich sein Binsenhaar; alles wollte die Gelegenheit benutzen, weil der Riese schlief; und der steife Storch machte so wunderliche Kapriolen auf der Mauer, daß sie lachen mußte, und hoch auf dem Dache drehte sich der Wetterhahn und schlug mit den Flügeln und rief immerfort: kick, kick dich um, ich seh ihn gehn, ich sag' nicht wen! Am Fenster aber sang lieblich der Wind: komm mit geschwind! und die Bächlein schwatzten draußen untereinander im Mondglanz, wie wenn der Frühling anbrechen sollte, und sprangen glitzernd und wispernd über die Baumwurzeln: bist du bereit? wir haben nicht Zeit, weit, weit, in die Waldeinsamkeit! – Nun, nun, nur Geduld, ich komm' ja schon, sagte die Prinzessin ganz erschrocken und vergnügt, nahm schnell ihr Bündel unter den Arm und trat vorsichtig aus dem Schlafzimmer; zwei Mäuschen kamen ihr atemlos nach und brachten ihr noch den Fingerhut, den sie in der Eile vergessen. Das Herz klopfte ihr, denn die Brunnen im Hofe rauschten schon wieder schwächer, der Flußgott streckte sich taumelnd wieder zum Schlafe zurecht, auch der Wetterhahn drehte sich nicht mehr; so schlich sie leise die stille Treppe hinab." –

„Ach Gott! wenn der Riese jetzt aufwacht!" sagte Renate ängstlich.

„Die Prinzessin hatte auch Angst genug", fuhr Gabriele fort, „sie hob sich das Röckchen, daß sie nicht an seinen langen Sporen hängenblieb, stieg geschickt über den einen, dann über den anderen Stiefel, und noch einen herzhaften Sprung – jetzt stand sie draußen am Abhang. Da aber war's einmal schön! da flogen die Wolken und rauschte der Strom und die prächtigen Wälder im Mondschein, und auf dem Strom fuhr ein Schifflein, saß ein Ritter darin." –

„Das ist ja gerade wie jetzt hier draußen", unterbrach sie Renate, „da fährt auch noch einer im Kahn dicht unter unserm Garten; jetzt stößt er ans Land."

„Freilich" – sagte Gabriele mutwillig und setzte sich ins Fenster und wehte mit ihrem weißen Schnupftuch hinaus – „und grüß' dich Gott, rief da die Prinzessin, grüß' dich Gott in die weite, weite Fern', es ist ja keine Nacht so still und tief als meine Lieb'!"

Renate faßte sie lachend um den Leib, um sie zurückzuziehen. – „Herr Jesus!" schrie sie da plötzlich auf, „ein fremder Mann, dort an der Mauer hin!" – Gabriele ließ erschrocken ihr Tuch sinken, es flatterte in den Garten hinab. Ehe sie sich aber noch besinnen konnte, hatte Renate schon das Fenster geschlossen; sie war voll Furcht, sie mochte nichts mehr von dem Märchen hören und trieb Gabrielen hastig aus der Tür, über den stillen Gang in ihre Schlafkammer.

Gabriele aber, als sie allein war, riß noch rasch in ihrer Zelle das Fenster auf. Zu ihrem Schreck bemerkte sie nun, daß das Tuch unten von dem Strauche verschwunden war, auf den es vorhin geflogen. Ihr Herz klopfte heftig, sie legte sich hinaus, so weit sie nur konnte, da glaubte sie draußen den Fluß wieder aufrauschen zu hören, darauf schallte Ruderschlag unten im Grunde, immer ferner und schwächer, dann alles, alles wieder still – so blieb sie verwirrt und überrascht am Fenster, bis das erste Morgenlicht die Bergesgipfel rötete.

Bald darauf traf der Namenstag der Priorin, ein Fest, wor-

auf sich alle Hausbewohner das ganze Jahr hindurch freuten; denn auf diesen Tag war zugleich die jährliche Weinlese auf einem nahegelegenen Gute des Klosters festgesetzt, an welcher die Nonnen mit teilnahmen. Da verbreitete sich, als der Morgenstern noch durch die Lindenwipfel in die kleinen Fenster hineinfunkelte, schon eine ungewohnte, lebhafte Bewegung durch das ganze Haus, im Hofe wurden die Wagen von dem alten Staube gereinigt, in ihren besten, blütenweißen Gewändern sah man die Schwestern in allen Gängen geschäftig hin und her eilen; einige versahen noch ihre Kanarienvögel sorgsam mit Futter, andere packten Taschen und Schachteln, als gälte es eine wochenlange Reise. – Endlich wurde von dem zahlreichen Hausgesinde ausführlich Abschied genommen, die Kutscher knallten, und die Karawane setzte sich langsam in Bewegung. Gabriele fuhr nebst einigen auserwählten Nonnen an der Seite der Priorin in einem mit vier alten dicken Rappen bespannten Staatswagen, der mit seinem altmodischen, vergoldeten Schnitzwerk einem chinesischen Lusthaus gleichsah. Es war ein klarer, heiterer Herbstmorgen, das Glockengeläute vom Kloster zog weit durchs stille Land, der Altweibersommer flog schon über die Felder, überall grüßten die Bauern ehrerbietig den ihnen wohlbekannten geistlichen Zug.

Wer aber beschreibt nun die große Freude auf dem Gratialgute, die fremden Berge, Täler und Schlösser umher, das stille Grün und den heitern Himmel darüber, wie sie da in dem mit Astern ausgeschmückten Gartensaal um eine reichliche Kollation vergnügt auf den altfränkischen Kanapees sitzen und die Morgensonne die alten Bilder römischer Kirchen und Paläste an den Wänden bescheint und vor den Fenstern die Sperlinge sich lustig tummeln und lärmen im Laub, während draußen weißgekleidete Dorfmädchen unter den schimmernden Bäumen vor der Tür ein Ständchen singen.

Die Priorin aber ließ die Kinder hereinkommen, die scheu und neugierig in dem Saal umherschauten, in den sie das

ganze Jahr über nur manchmal heimlich durch die Ritzen der verschlossenen Fensterladen geguckt hatten. Sie streichelte und ermahnte sie freundlich, freute sich, daß sie in dem Jahre so gewachsen, und gab dann jedem aus ihrem Gebetbuch ein buntes Heiligenbild und ein großes Stück Kuchen dazu.

Jetzt aber ging die rechte Lust der Kleinen erst an, da nun wirklich zur Weinlese geschritten wurde, bei der sie mithelfen und naschen durften. Da belebte sich allmählich der Garten, fröhliche Stimmen da und dort, geputzte Kinder, die große Trauben trugen, flatternde Schleier und weiße schlanke Gestalten zwischen den Rebengeländern schimmernd und wieder verschwindend, als wanderten Engel über den Berg. Die Priorin saß unterdes vor der Haustür und betete ihr Brevier und schaute oft über das Buch weg nach den vergnügten Schwestern; die Herbstsonne schien warm und kräftig über die stille Gegend, und die Nonnen sangen bei der Arbeit:

„Es ist nun der Herbst gekommen,
Hat das schöne Sommerkleid
Von den Feldern weggenommen
Und die Blätter ausgestreut,
Vor dem bösen Winterwinde
Deckt er warm und sachte zu
Mit dem bunten Laub die Gründe,
Die schon müde gehn zur Ruh'."

Einzelne verspätete Wandervögel zogen noch über den Berg und schwatzten vom Glanz der Ferne, was die glücklichen Schwestern nicht verstanden. Gabriele aber wußte wohl, was sie sangen, und ehe die Priorin sich's versah, war sie auf die höchste Linde geklettert; da erschrak sie, wie so groß und weit die Welt war. – Die Priorin schalt sie aus und nannte sie ihr wildes Waldvöglein. Ja, dachte Gabriele, wenn ich ein Vöglein wäre! Dann fragte die Priorin, ob sie von da oben das Schloß Dürande überm Walde

sehen könne? „Alle die Wälder und Wiesen", sagte sie, „gehören dem Grafen Dürande; er grenzt hier an, das ist ein reicher Herr!" Gabriele aber dachte an *ihren* Herrn, und die Nonnen sangen wieder:

„Durch die Felder sieht man fahren
Eine wunderschöne Frau,
Und von ihren langen Haaren
Goldne Fäden auf der Au
Spinnet sie und singt im Gehen:
Eia, meine Blümelein,
Nicht nach andern immer sehen,
Eia, schlafet, schlafet ein!"

„Ich höre Waldhörner!" rief hier plötzlich Gabriele; es verhielt ihr fast den Atem vor Erinnerung an die alte schöne Zeit. – „Komm schnell herunter, mein Kind", rief ihr die Priorin zu. Aber Gabriele hörte nicht darauf, zögernd und im Hinabsteigen noch immer zwischen den Zweigen hinausschauend, sagte sie wieder: „Es bewegt sich drüben am Saum des Waldes; jetzt seh' ich Reiter; wie das glitzert im Sonnenschein! sie kommen gerade auf uns her."

Und kaum hatte sie sich vom Baum geschwungen, als einer von den Reitern, über den grünen Plan dahergeflogen, unter den Linden anlangte und mit höflichem Gruß vor der Priorin stillhielt. Gabriele war schnell in das Haus gelaufen, dort wollte sie durchs Fenster nach dem Fremden sehen. Aber die Priorin rief ihr nach: der Herr sei durstig, sie solle ihm Wein herausbringen. Sie schämte sich, daß er sie auf dem Baume gesehen, so kam sie furchtsam mit dem vollen Becher vor die Tür mit gesenkten Blicken, durch die langen Augenwimpern nur sah sie das kostbare Zaumzeug und die Stickerei auf seinem Jagdrock im Sonnenschein flimmern. Als sie aber an das Pferd trat, sagte er leise zu ihr: Er sehe *doch* ihre dunkeln Augen im Weine sich spiegeln wie in einem goldnen Brunnen. Bei dem Klang

der Stimme blickte sie erschrocken auf – der Reiter war ihr Liebster – sie stand wie verblendet. Er trank jetzt auf der Priorin Gesundheit, sah aber dabei über den Becher weg Gabrielen an und zeigte ihr verstohlen ihr Tuch, das sie in jener Nacht aus dem Fenster verloren. Dann drückte er die Sporen ein, und flüchtig dankend, flog er wieder fort zu dem bunten Schwarm am Walde, das weiße Tuch flatterte weit im Winde hinter ihm her.

„Sieh nur", sagte die Priorin lachend, „wie ein Falk', der eine Taube durch die Luft führt!"

„Wer war der Herr?" frug endlich Gabriele tief aufatmend. – „Der junge Graf Dürande", hieß es. – Da tönte die Jagd schon wieder fern und immer ferner den funkelnden Wald entlang, die Nonnen aber hatten in ihrer Fröhlichkeit von allem nichts bemerkt und sangen von neuem:

„Und die Vöglein hoch in Lüften
Über blaue Berg' und Seen
Ziehn zur Ferne nach den Klüften,
Wo die hohen Zedern stehn,
Wo mit ihren goldnen Schwingen
Auf des Benedeiten Gruft
Engel Hosianna singen,
Nächtens durch die stille Luft."

Etwa vierzehn Tage darauf schritt Renald eines Morgens still und rasch durch den Wald nach dem Schloß Dürande, dessen Türme finster über die Tannen hersahen. Er war ernst und bleich, aber mit Hirschfänger und leuchtendem Bandelier wie zu einem Feste geschmückt. In der Unruhe seiner Seele war er der Zeit ein gut Stück vorausgeschritten, denn als er ankam, war die Haustür noch verschlossen und alles still, nur die Dohlen erwachten schreiend auf den alten Dächern. Er setzte sich unterdes auf das Geländer der Brücke, die zum Schlosse führte. Der Wallgraben unten lag lange trocken, ein marmorner Apollo mit seltsa-

mer Lockenperücke spielte dort zwischen gezirkelten Blumenbeeten die Geige, auf der ein Vogel sein Morgenlied pfiff; über den Helmen der steinernen Ritterbilder am Tore brüsteten sich breite Aloen; der Wald, der alte Schloßgesell, war wunderlich verschnitten und zerquält, aber der Herbst ließ sich sein Recht nicht nehmen und hatte alles phantastisch gelb und rot gefärbt, und die Waldvögel, die vor dem Winter in die Gärten flüchteten, zwitscherten lustig von Wipfel zu Wipfel. – Renald fror, er hatte Zeit genug und überdachte noch einmal alles: wie der junge Graf Dürande wieder nach Paris gereist, um dort lustig durchzuwintern, wie er selbst darauf mit fröhlichem Herzen zum Kloster geeilt, um seine Schwester abzuholen. Aber da war Gabriele heimlich verschwunden, man hatte einmal des Nachts einen fremden Mann am Kloster gesehn; niemand wußte, wohin sie gekommen. – Jetzt knarrte das Schloßtor, Renald sprang schnell auf, er verlangte seinen Herrn, den alten Grafen Dürande, zu sprechen. Man sagte ihm, der Graf sei eben erst aufgewacht; er mußte noch lange in der Gesindestube warten zwischen Überresten vom gestrigen Souper, zwischen Schuhbürsten, Büchsen und Katzen, die sich verschlafen an seinen blanken Stiefeln dehnten, niemand fragte nach ihm. Endlich wurde er in des Grafen Garderobe geführt, der alte Herr ließ sich soeben frisieren und gähnte unaufhörlich. Renald bat nun ehrerbietig um kurzen Urlaub zu einer Reise nach Paris. Auf die Frage des Grafen, was er dort wolle, entgegnete er verwirrt: seine Schwester sei dort bei einem weitläufigen Verwandten – er schämte sich herauszusagen, was er dachte. Da lachte der Graf. „Nun, nun“, sagte er, „mein Sohn hat wahrhaftig keinen üblen Geschmack. Geh' Er nur hin, ich will Ihm an seiner Fortune nicht hinderlich sein; die Dürandes sind in solchen Affären immer splendid; so ein junger wilder Schwan muß gerupft werden, aber mach' Er's mir nicht zu arg.“ – Dann nickte er mit dem Kopfe, ließ sich den Pudermantel umwerfen und schritt langsam zwischen zwei Reihen von Be-

dienten, die ihn im Vorüberwandeln mit großen Quasten einpuderten, durch die entgegengesetzte Flügeltür zum Frühstück. Die Bedienten kicherten heimlich – Renald schüttelte sich wie ein gefesselter Löwe.
Noch an demselben Tage trat er seine Reise an.
Es war ein schöner blanker Herbstabend, als er in der Ferne Paris erblickte; die Ernte war längst vorüber, die Felder standen alle leer, nur von der Stadt her kam ein verworrenes Rauschen über die stille Gegend, daß ihn heimlich schauerte. Er ging nun an prächtigen Landhäusern vorüber durch die langen Vorstädte immer tiefer in das wachsende Getöse hinein, die Welt rückte immer enger und dunkler zusammen, der Lärm, das Rasseln der Wagen betäubte, das wechselnde Streiflicht aus den geputzten Läden blendete ihn; so war er ganz verwirrt, als er endlich im Wind den roten Löwen, das Zeichen seines Vetters, schwanken sah, der in der Vorstadt einen Weinschank hielt. Dieser saß eben vor der Tür seines kleinen Hauses und verwunderte sich nicht wenig, da er den verstaubten Wandersmann erkannte. Doch Renald stand wie auf Kohlen. „War Gabriele bei dir?“ fragte er gleich nach der ersten Begrüßung gespannt. – Der Vetter schüttelte erstaunt den Kopf, er wußte von nichts. – „Also doch!“ sagte Renald, mit dem Fuß auf die Erde stampfend; aber er konnte es nicht über die Lippen bringen, was er vermute und vorhabe.
Sie gingen nun in das Haus und kamen in ein langes, wüstes Gemach, das von einem Kaminfeuer im Hintergrunde ungewiß erleuchtet wurde. In den roten Widerscheinen lag dort ein wilder Haufe umher: abgedankte Soldaten, müßige Handwerksburschen und dergleichen Hornkäfer, wie sie in der Abendzeit um die großen Städte schwärmen. Alle Blicke aber hingen an einem hohen, hageren Manne mit bleichem, scharfgeschnittenem Gesicht, der, den Hut auf dem Kopf und seinen langen Mantel stolz und vornehm über die linke Achsel zurückgeschlagen, mitten unter ihnen stand. – „Ihr seid der Nährstand“, rief er soeben

aus; „wer aber die andern nährt, der ist ihr Herr; hoch auf, ihr Herren!" – Er hob ein Glas, alles jauchzte wild auf und griff nach den Flaschen, er aber tauchte kaum die feinen Lippen in den dunkelroten Wein, als schlürft' er Blut, seine spielenden Blicke gingen über dem Glase kalt und lauernd in der Runde.
Da funkelte das Kaminfeuer über Renalds blankes Bandelier, das stach plötzlich in ihre Augen. Ein starker Kerl mit rotem Gesicht und Haar, wie ein brennender Dornbusch, trat mit übermütiger Bettelhaftigkeit dicht vor Renald und fragte, ob er dem Großtürken diene? Ein andrer meinte, er habe ja da, wie ein Hund, ein adeliges Halsband umhängen. – Renald griff rasch nach seinem Hirschfänger, aber der lange Redner trat dazwischen, sie wichen ihm scheu und ehrerbietig aus. Dieser führte den Jäger an einen abgelegenen Tisch und fragte, wohin er wolle. Da Renald den Grafen Dürande nannte, sagte er: „Das ist ein altes Haus, aber der Totenwurm pickt schon drin, ganz von Liebschaften zerfressen." – Renald erschrak, er glaubte, jeder müßte ihm seine Schande an der Stirn ansehen. „Warum kommt Ihr gerade auf die Liebschaften?" fragte er zögernd. – „Warum?" erwiderte jener, „sind sie nicht die Herren im Forst, ist das Wild nicht ihre, hohes und niederes? Sind wir nicht verfluchte Hunde und lecken die Schuh', wenn sie uns stoßen?" – Das verdroß Renald; er entgegnete kurz und stolz: Der junge Graf Dürande sei ein großmütiger Herr, er wolle nur sein Recht von ihm und weiter nichts. – Bei diesen Worten hatte der Fremde ihn aufmerksam betrachtet und sagte ernst: „Ihr seht aus wie ein Scharfrichter, der, das Schwert unterm Mantel, zu Gerichte geht; es kommt die Zeit, gedenkt an mich, Ihr werdet der Rüstigsten einer sein bei der blutigen Arbeit." – Dann zog er ein Blättchen hervor, schrieb etwas mit Bleistift darauf, versiegelte es am Licht und reichte es Renald hin. „Die Grafen hier kennen mich wohl", sagte er; er solle das nur abgeben an Dürande, wenn er einen Strauß mit ihm habe, es könnte ihm vielleicht von Nutzen sein. –

„Wer ist der Herr?“ fragte Renald seinen Vetter, da der Fremde sich rasch wieder wandte. – „Ein Feind der Tyrannen“, entgegnete der Vetter leise und geheimnisvoll. Dem Renald aber gefiel hier die ganze Wirtschaft nicht, er war müde von der Reise und streckte sich bald in einer Nebenkammer auf das Lager, das ihm der Vetter angewiesen. Da konnte er vernehmen, wie immer mehr und mehr Gäste nebenan allmählich die Stube füllten; er hörte die Stimme des Fremden wieder dazwischen, eine wilde Predigt, von der er nur einzelne Worte verstand, manchmal blitzte das Kaminfeuer blutrot durch die Ritzen der schlechtverwahrten Tür; so schlief er spät unter furchtbaren Träumen ein.

Der Ball war noch nicht beendigt, aber der junge Graf Dürande hatte dort so viel Wunderbares gehört von den feurigen Zeichen einer Revolution, vom heimlichen Aufblitzen kampffertiger Geschwader, Jakobiner, Volksfreunde und Royalisten, daß ihm das Herz schwoll wie im nahenden Gewitterwinde. Er konnte es nicht länger aushalten in der drückenden Schwüle. In seinen Mantel gehüllt, ohne den Wagen abzuwarten, stürzte er sich in die scharfe Winternacht hinaus. Da freute er sich, wie draußen fern und nah die Turmuhren verworren zusammenklangen im Wind und die Wolken über die Stadt flogen und der Sturm sein Reiselied pfiff, lustig die Schneeflocken durcheinander wirbelnd. „Grüß' mir mein Schloß Dürande!“ rief er dem Sturme zu; es war ihm so frisch zumut, als müßt' er, wie ein lediges Roß, mit jedem Tritte Funken aus den Steinen schlagen.

In seinem Hotel aber fand er alles wie ausgestorben, der Kammerdiener war vor Langeweile fest eingeschlafen, die jüngere Dienerschaft ihren Liebschaften nachgegangen, niemand hatte ihn so früh erwartet. Schauernd vor Frost stieg er die breite, dämmernde Treppe hinauf, zwei tief herabgebrannte Kerzen beleuchteten zweifelhaft das ver-

goldete Schnitzwerk des alten Saales, es war so still, daß er den Zeiger der Schloßuhr langsam fortrücken und die Wetterfahnen im Winde sich drehen hörte. Wüst und überwacht warf er sich auf eine Ottomane hin. „Ich bin so müde", sagte er, „so müde von Lust und immer Lust, langweilige Lust! ich wollt', es wäre Krieg!" – Da war's ihm, als hört' er draußen auf der Treppe gehn mit leisen, langen Schritten, immer näher und näher. „Wer ist da?" rief er. – Keine Antwort. – „Nur zu, mir eben recht", meinte er, Hut und Halbschuhe wegwerfend, „rumor' nur zu, spukhafte Zeit, mit deinem fernen Wetterleuchten über Stadt und Land, als wenn die Gedanken aufstünden überall und schlaftrunken nach den Schwertern tappten. Was gehst du in Waffen rasselnd um und pochst an die Türen unserer Schlösser bei stiller Nacht? Mich gelüstet mit dir zu fechten; herauf, du unsichtbares Kriegsgespenst!"
Da pocht' es wirklich an der Tür. Er lachte, daß der Geist die Herausforderung so schnell angenommen. In keckem Übermut rief er: „Herein!" Eine hohe Gestalt im Mantel trat in die Tür; er erschrak doch, als diese den Mantel abwarf und er Renald erkannte, denn er gedachte der Nacht im Walde, wo der Jäger auf ihn gezielt. – Renald aber, da er den Grafen erblickte, ehrerbietig zurücktretend, sagte: er habe den Kammerdiener hier zu finden geglaubt, um sich anmelden zu lassen. Er sei schon öfters zu allen Tageszeiten hier gewesen, jedesmal aber, unter dem Vorwand, daß die Herrschaft nicht zu Hause oder beschäftigt sei, von den Pariser Bedienten zurückgewiesen worden, die ihn noch nicht kannten; so habe er denn heute auf der Straße gewartet, bis der Graf zurückkäme.
„Und was willst du denn von mir?" fragte der Graf, ihn mit unverwandten Blicken prüfend.
„Gnädiger Herr", erwiderte der Jäger nach einer Pause, „Sie wissen wohl, ich hatte eine Schwester, sie war meine einzige Freude und mein Stolz – sie ist eine Landläuferin geworden, sie ist fort."
Der Graf machte eine heftige Bewegung, faßte sich aber

gleich wieder und sagte halb abgewendet: „Nun, und was geht mich das an?“
Renalds Stirn zuckte wie fernes Wetterleuchten, er schien mit sich selber zu ringen. „Gnädiger Herr“, rief er darauf im tiefsten Schmerz, „gnädiger Herr, gebt mir meine arme Gabriele zurück!“
„Ich?“ fuhr der Graf auf, „zum Teufel, wo ist sie?“
„Hier“ – entgegnete Renald ernst.
Der Graf lachte laut auf, und den Leuchter ergreifend, stieß er rasch eine Flügeltür auf, daß man eine weite Reihe glänzender Zimmer übersah. „Nun“, sagte er mit erzwungener Lustigkeit, „so hilf mir suchen. Horch, da raschelt was hinter der Tapete, jetzt hier, dort, nun sage mir, wo steckt sie?“
Renald blickte finster vor sich nieder, sein Gesicht verdunkelte sich immer mehr. Da gewahrte er Gabrielens Schnupftuch auf einem Tischchen; der Graf, der seinen Augen gefolgt war, stand einen Augenblick betroffen. – Renald hielt sich noch, es fiel ihm der Zettel des Fremden wieder ein, er wünschte immer noch, alles in Güte abzumachen, und reichte schweigend dem Grafen das Briefchen hin. Der Graf, ans Licht tretend, erbrach es schnell, da flog eine dunkle Röte über sein ganzes Gesicht. – „Und weiter nichts?“ murmelte er leise zwischen den Zähnen, sich in die Lippen beißend. „Wollen sie mir drohen, mich schrecken?“ – Und rasch zu Renald gewandt, rief er: „Und wenn ich deine ganze Sippschaft hätt’, ich gäb’ sie nicht heraus! Sag deinem Bettleradvokaten, ich lachte sein und wäre zehntausendmal noch stolzer als er, und wenn ihr beide euch im Hause zeigt, laß ich mit Hunden euch vom Hofe hetzen, das sag’ ihm; fort, fort, fort!“ – Hiermit schleuderte er den Zettel dem Jäger ins Gesicht und schob ihn selber zum Saal hinaus, die eichene Tür hinter ihm zuwerfend, daß es durchs ganze Haus öde erschallte.
Renald stand, wild um sich blickend, auf der stillen Treppe. Da bemerkte er erst, daß er den Zettel noch krampfhaft in den Händen hielt; er entfaltete ihn hastig und las an

dem flackernden Licht einer halbverlöschten Laterne die Worte: „Hütet euch. Ein Freund des Volks." – Unterdes hörte er oben den Grafen heftig klingeln; mehrere Stimmen wurden im Hause wach, er stieg langsam hinunter wie ins Grab. Im Hofe blickte er noch einmal zurück, die Fenster des Grafen waren noch erleuchtet, man sah ihn im Saale heftig auf und nieder gehen. Da hörte Renald auf einmal draußen durch den Wald singen:

„Am Himmelsgrund schießen
So lustig die Stern',
Dein Schatz läßt dich grüßen
Aus weiter, weiter Fern'!

Hat eine Zither gehangen
An der Tür unbeacht',
Der Wind ist gegangen
Durch die Saiten bei Nacht.

Schwang sich auf dann vom Gitter
Über die Berge, über'n Wald –
Mein Herz ist die Zither,
Gibt einen fröhlichen Schall!"

Die Weise ging ihm durch Mark und Bein; er kannte sie wohl. – Der Mond streifte soeben durch die vorüberfliegenden Wolken den Seitenflügel des Schlosses, da glaubte er in dem einen Fenster flüchtig Gabrielen zu erkennen; als er sich aber wandte, wurde es schnell geschlossen. Ganz erschrocken und verwirrt warf er sich auf die nächste Tür, sie war fest zu. Da trat er unter das Fenster und rief leise aus tiefster Seele hinauf, ob sie drin wider ihren Willen festgehalten werde? so solle sie ihm ein Zeichen geben, es sei keine Mauer so stark als die Gerechtigkeit Gottes. – Es rührte sich nichts als die Wetterfahne auf dem Dach. – „Gabriele", rief er nun lauter, „meine arme Gabriele, der Wind in der Nacht weint um dich an den Fen-

stern, ich liebte dich so sehr, ich lieb dich noch immer, um Gottes willen, komm, komm herab zu mir, wir wollen miteinander fortziehen, weit, weit fort, wo uns niemand kennt, ich will für dich betteln von Haus zu Haus, es ist ja kein Lager so hart, kein Frost so scharf, keine Not so bitter als die Schande."

Er schwieg erschöpft, es war alles wieder still, nur die Tanzmusik von dem Ball schallte noch von fern über den Hof herüber, der Wind trieb große Schneeflocken schräg über die harte Erde, er war ganz verschneit. – „Nun, so gnade uns beiden Gott!" sagte er, sich abwendend, schüttelte den Schnee vom Mantel und schritt rasch fort.

Als er zu der Schenke seines Vetters zurückkam, fand er zu seinem Erstaunen das ganzeHaus verschlossen. Auf sein heftiges Pochen trat der Nachbar, sich vorsichtig nach allen Seiten umsehend, aus seiner Tür, er schien auf des Jägers Rückkehr gewartet zu haben und erzählte ihm geheimnisvoll: Das Nest nebenan sei ausgenommen, Polizeisoldaten hätten heute abend den Vetter plötzlich abgeführt, niemand wisse wohin. – Den Renald überraschte und verwunderte nichts mehr, und zerstreut mit flüchtigem Danke nahm er alles an, als der Nachbar nun auch das gerettete Reisebündel des Jägers unter dem Mantel hervorbrachte und ihm selbst eine Zuflucht in seinem Hause anbot.

Gleich am andern Morgen aber begann Renald seine Runde in der weitläufigen Stadt, er mochte nichts mehr von der Großmut des stolzen Grafen, er wollte jetzt nur sein *Recht!* So suchte er unverdrossen eine Menge Advokaten hinter ihren großen Tintenfässern auf, aber die sahen's gleich alle den goldbortenen Rauten seines Rockes an, daß sie nicht aus seiner eigenen Tasche gewachsen waren; der eine verlangte unmögliche Zeugen, der andere Dokumente, die er nicht hatte, und alle forderten Vorschuß. Ein junger, reicher Advokat wollte sich totlachen über die ganze Geschichte; er fragte, ob die Schwester jung, schön, und erbot sich, den ganzen Handel umsonst zu führen

und die arme Waise dann zu sich ins Haus zu nehmen, während ein andrer gar das Mädchen selber heiraten wollte, wenn sie fernerhin beim Grafen bliebe. – In tiefster Seele empört, wandte sich Renald nun an die Polizeibehörde; aber da wurde er aus einem Revier ins andere geschickt, von Pontius zu Pilatus, und jeder wusch seine Hände in Unschuld, niemand hatte Zeit, in dem Getreibe ein vernünftiges Wort zu hören, und als er endlich vor das rechte Bureau kam, zeigten sie ihm ein langes Verzeichnis der Dienstleute und Hausgenossen des Grafen Dürande: seine Schwester war durchaus nicht darunter. Er habe Geister gesehen, hieß es, er solle keine unnützen Flausen machen; man hielt ihn für einen Narren, und er mußte froh sein, nur ungestraft wieder unter Gottes freien Himmel zu kommen. Da saß er nun todmüde in seiner einsamen Dachkammer, den Kopf in die Hand gestützt; seine Barschaft war mit dem frühzeitigen Schnee auf den Straßen geschmolzen, jetzt wußt' er keine Hilfe mehr, es ekelte ihm recht vor dem Schmutz der Welt. In diesem Hinbrüten, wie wenn man beim Sonnenglanz die Augen schließt, spielten feurige Figuren wechselnd auf dem dunkeln Grund seiner Seele: schlängelnde Zornesblicke und halbgeborne Gedanken blutiger Rache. In dieser Not betete er still für sich; als er aber an die Worte kam: „Vergib uns unsere Schuld, als auch wir vergeben unseren Schuldnern", fuhr er zusammen; er konnte es dem Grafen nicht vergeben. Angstvoll und immer brünstiger betete er fort. – Da sprang er plötzlich auf, ein neuer Gedanke erleuchtete auf einmal sein ganzes Herz. Noch war nicht alles versucht, nicht alles verloren, er beschloß, den König selber anzutreten – so hatte er sich nicht vergeblich zu Gott gewendet, dessen Hand auf Erden ja der König ist.

Ludwig XVI. und sein Hof waren damals in Versailles; Renald eilte sogleich hin und freute sich, als er bei seiner Ankunft hörte, daß der König, der unwohl gewesen, heute zum ersten Male wieder den Garten besuchen wolle. Er hatte zu Hause mit großem Fleiß eine Supplik aufgesetzt,

Punkt für Punkt, das himmelschreiende Unrecht und seine Forderung, alles wie er es dereinst vor Gottes Thron zu verantworten gedachte. Das wollte er im Garten selbst übergeben, vielleicht fügte es sich, daß er dabei mit dem König sprechen durfte; so, hoffte er, könne noch alles wieder gut werden.

Vielerlei Volk, Neugierige, Müßiggänger und Fremde hatten sich unterdes schon unweit der Tür, aus welcher der König treten sollte, zusammengestellt. Renald drängte sich mit klopfendem Herzen in die vorderste Reihe. Es war einer jener halbverschleierten Wintertage, die lügenhaft den Sommer nachspiegeln, die Sonne schien lau, aber falsch über die stillen Paläste, weiterhin zogen Schwäne auf den Weihern, kein Vogel sang mehr, nur die weißen Marmorbilder standen noch verlassen in der prächtigen Einsamkeit. Endlich gaben die Schweizer das Zeichen, die Saaltür öffnete sich, die Sonne tat einen kurzen Blitz über funkelnden Schmuck, Ordensbänder und blendende Achseln, die schnell, vor dem Winterhauch, unter schimmernden Tüchern wieder verschwanden. Da schallt' es auf einmal: Vive le roi! durch die Lüfte, und im Garten, soweit das Auge reichte, begannen plötzlich alle Wasserkünste zu spielen, und mitten in dem Jubel, Rauschen und Funkeln schritt der König in einfachem Kleide langsam die breiten Marmorstufen hinab. Er sah traurig und bleich – eine leise Luft rührte die Wipfel der hohen Bäume und streute die letzten Blätter wie einen Goldregen über die fürstlichen Gestalten. Jetzt gewahrte Renald mit einiger Verwirrung auch den Grafen Dürande unter dem Gefolge, er sprach soeben halbflüsternd zu einer jungen schönen Dame. Schon rauschten die taftnen Gewänder immer näher und näher. Renald konnte deutlich vernehmen, wie die Dame, ihre Augen gegen Dürande aufschlagend, ihn neckend fragte, was er drin sehe, daß sie ihn so erschreckten. –

„Wunderbare Sommernächte meiner Heimat", erwiderte der Graf zerstreut. Da wandte sich das Fräulein lachend,

Renald erschrak, ihr dunkles Auge war wie Gabrielens in fröhlichen Tagen – es wollte ihm das Herz zerreißen. Darüber hatte er alles andere vergessen, der König war fast vorüber; jetzt drängte er sich nach, ein Schweizer aber stieß ihn mit der Partisane zurück, er drang noch einmal verzweifelt vor. Da bemerkt ihn Dürande, er stutzt einen Augenblick, dann, schnell gesammelt, faßt er den Zudringlichen rasch an der Brust und übergibt ihn der herbeieilenden Wache. Der König über dem Getümmel wendet sich fragend. – „Ein Wahnsinniger", entgegnet Dürande. –

Unterdes hatten die Soldaten den Unglücklichen umringt, die neugierige Menge, die ihn für verrückt hielt, wich scheu zurück, so wurde er ungehindert abgeführt. Da hörte er hinter sich die Fontänen noch rauschen, dazwischen das Lachen und Plaudern der Hofleute in der lauen Luft; als er aber einmal zurückblickte, hatte sich alles schon wieder nach dem Garten hingekehrt, nur ein bleiches Gesicht aus der Menge war noch zurückgewandt und funkelte ihm mit scharfen Blicken nach. Er glaubte schaudernd den prophetischen Fremden aus des Vetters Schenke wiederzuerkennen.

Der Mond bescheint das alte Schloß Dürande und die tiefe Waldesstille am Jägerhaus, nur die Bäche rauschen so geheimnisvoll in den Gründen. Schon blüht's in manchem tiefen Tal, und nächtliche Züge heimkehrender Störche hoch in der Luft verkünden in einzelnen halbverlornen Lauten, daß der Frühling gekommen. Da fahren plötzlich Rehe, die auf der Wiese vor dem Jägerhaus gerastet, erschrocken ins Dickicht, der Hund an der Tür schlägt an, ein Mann steigt eilig von den Bergen, bleich, wüst, die Kleider abgerissen, mit wildverwachsenem Bart – es ist der Jäger Renald.

Mehrere Monate hindurch war er in Paris im Irrenhause eingesperrt gewesen; je heftiger er beteuerte, verständig

zu sein, für desto toller hielt ihn der Wärter; in der Stadt aber hatte man jetzt Wichtigeres zu tun, niemand bekümmerte sich um ihn. Da ersah er endlich selbst seinen Vorteil, die Hinterlist seiner verrückten Mitgesellen half ihm treulich aus Lust an der Heimlichkeit. So war es ihm gelungen, in einer dunkeln Nacht mit Lebensgefahr sich an einem Seil herabzulassen und in der allgemeinen Verwirrung der Zeit unentdeckt aus der Stadt durch die Wälder, von Dorf zu Dorfe bettelnd, heimwärts zu gelangen. Jetzt bemerkte er erst, daß es von fern überm Walde blitzte, vom stillen Schloßgarten her schlug schon eine Nachtigall, es war ihm, als ob ihn Gabriele riefe. Als er aber mit klopfendem Herzen auf dem altbekannten Fußsteig immer weiter ging, öffnete sich bei dem Hundegebell ein Fensterchen im Jägerhaus. Es gab ihm einen Stich ins Herz; es war Gabrielens Schlafkammer, wie oft hatte er dort ihr Gesicht im Mondschein gesehen. Heut aber guckte ein Mann hervor und fragte barsch, was es draußen gäbe. Es war der Waldwärter, der heimtückische Rotkopf war ihm immer zuwider gewesen. „Was macht Ihr hier in Renalds Haus?" sagte er. „Ich bin müde, ich will hinein." Der Waldwärter sah ihn von Kopf bis zu den Füßen an, er erkannte ihn nicht mehr. „Mit dem Renald ist's lange vorbei", entgegnete er dann, „er ist nach Paris gelaufen und hat sich dort mit verdächtigem Gesindel und Rebellen eingelassen, wir wissen's recht gut, jetzt habe ich seine Stelle vom Grafen." – Drauf wies er Renald am Waldesrand den Weg zum Wirtshause und schlug das Fenster wieder zu. – Oho, steht's so! dachte Renald. Da fielen seine Augen auf sein Gärtchen, die Kirschbäume, die er gepflanzt, standen schon in voller Blüte, es schmerzte ihn, daß sie in ihrer Unschuld nicht wußten, für wen sie blühten. Währenddes hatte sein alter Hofhund sich gewaltsam vom Stricke losgerissen, sprang liebkosend an ihm herauf und umkreiste ihn in weiten Freudensprüngen; er herzte sich mit ihm wie mit einem alten, treuen Freunde. Dann aber wandte er sich rasch zum Hause; die Tür war verschlossen, er stieß

sie mit einem derben Fußtritt auf. Drin hatte der Waldwärter unterdes Feuer gepinkt. „Herr Jesus!“ rief er erschrocken, da er, entgegentretend, plötzlich beim Widerschein der Lampe den verwilderten Renald erkannte. Renald aber achtete nicht darauf, sondern griff nach der Büchse, die überm Bett an der Wand hing. „Lump“, sagte er, „das schöne Gewehr so verstauben zu lassen!“ Der Waldwärter, die Lampe hinsetzend und auf dem Sprunge, durchs Fenster zu entfliehen, sah den furchtbaren Gast seitwärts mit ungewissen Blicken an. Renald bemerkte, daß er zitterte. „Fürcht' dich nicht“, sagte er, „dir tu ich nichts, was kannst du dafür! ich hol' mir nur die Büchse, sie ist vom Vater, sie gehört mir und nicht dem Grafen, und so wahr der alte Gott noch lebt, so hol' ich mir auch mein *Recht,* und wenn sie's im Turmknopf von Dürande versiegelt hätten, das sag' dem Grafen und wer's sonst wissen will.“ – Mit diesen Worten pfiff er dem Hunde und schritt wieder in den Wald hinaus, wo ihn der Waldwärter bei dem wirren Wetterleuchten bald aus den Augen verloren hatte.

Währenddes schnurrten im Schloß Dürande die Gewichte der Turmuhr ruhig fort, aber die Uhr schlug nicht, und der verrostete Weiser rückte nicht mehr von der Stelle, als wäre die Zeit eingeschlafen auf dem alten Hofe beim einförmigen Rauschen der Brunnen. Draußen, nur manchmal vom fernen Wetterleuchten zweifelhaft erhellt, lag der Garten mit seinen wunderlichen Baumfiguren, Statuen und vertrockneten Bassins wie versteinert im jungen Grün, das in der warmen Nacht schon von allen Seiten lustig über die Gartenmauer kletterte und sich um die Säulen der halbverfallenen Lusthäuser schlang, als wollt' nun der Frühling alles erobern. Das Hausgesinde aber stand heimlich untereinander flüsternd auf der Terrasse, denn man sah es hie und da brennen in der Ferne; der Aufruhr schritt wachsend schon immer näher über die stillen Wälder von Schloß zu Schloß. Da hielt der kranke alte Graf um die gewohnte Stunde einsam Tafel im Ahnensaal, die hohen

Fenster waren fest verschlossen, Spiegel, Schränke und Marmortische standen unverrückt umher wie in der alten Zeit, niemand durfte, bei seiner Ungnade, der neuen Ereignisse erwähnen, die er verächtlich ignorierte. So saß er, im Staatskleide, frisiert, wie eine geputzte Leiche, am reichbesetzten Tisch vor den silbernen Armleuchtern und blätterte in alten Historienbüchern, seiner kriegerischen Jugend gedenkend. Die Bedienten eilten stumm über den glatten Boden hin und her, nur durch die Ritzen der Fensterladen sah man zuweilen das Wetterleuchten, und alle Viertelstunden hakte im Nebengemach die Flötenuhr knarrend ein und spielte einen Satz aus einer alten Opernarie.

Da ließen sich auf einmal unten Stimmen vernehmen, drauf hörte man jemand eilig die Treppe heraufkommen, immer lauter und näher. „Ich muß herein!" rief es endlich an der Saaltür, sich durch die abwehrenden Diener drängend, und bleich, verstört und atemlos stürzte der Waldwärter in den Saal, in wilder Hast dem Grafen erzählend, was ihm soeben im Jägerhaus mit Renald begegnet. –

Der Graf starrte ihn schweigend an. Dann, plötzlich einen Armleuchter ergreifend, richtete er sich zum Erstaunen der Diener ohne fremde Hilfe hoch auf. „Hüte sich, wer einen Dürande fangen will!" rief er, und gespenstisch wie ein Nachtwandler mit dem Leuchter quer durch den Saal schreitend, ging er auf eine kleine eichene Tür los, die zu dem Gewölbe des Eckturms führte. Die Diener, als sie sich vom ersten Entsetzen über sein grauenhaftes Aussehen erholt, standen verwirrt und unentschlossen um die Tafel. „Um Gottes willen", rief da auf einmal ein Jäger herbeieilend, „laßt ihn nicht durch, dort in dem Eckturm hab' ich auf sein Geheiß heimlich alles Pulver zusammentragen müssen; wir sind verloren, er sprengt uns alle mit sich in die Luft!" – Der Kammerdiener, bei dieser schrecklichen Nachricht, faßte sich zuerst ein Herz und sprang rasch vor, um seinen Herrn zurückzuhalten, die andern folgten seinem Beispiel. Der Graf aber, da er sich

so unerwartet verraten und überwältigt sah, schleuderte dem nächsten den Armleuchter an den Kopf, darauf, krank wie er war, brach er selbst auf dem Boden zusammen.

Ein verworrenes Durcheinanderlaufen ging nun durch das ganze Schloß; man hatte den Grafen auf sein seidenes Himmelbett gebracht. Dort versuchte er vergeblich, sich noch einmal emporzuziehen, zurücksinkend rief er: „Wer sagte da, daß der Renald nicht wahnsinnig ist?" – Da alles still blieb, fuhr er leiser fort: „Ihr kennt den Renald nicht, er kann entsetzlich sein, wie fressend Feuer – läßt man denn reißende Tiere frei aufs Feld? – Ein schöner Löwe, wie er die Mähnen schüttelt – wenn sie nur nicht so blutig wären!" – Hier, sich plötzlich besinnend, riß er die müden Augen weit auf, und starrte die umherstehenden Diener verwundert an.

Der bestürzte Kammerdiener, der seine Blicke allmählich verlöschen sah, redete von geistlichem Beistand, aber der Graf, schon im Schatten des nahenden Todes, verfiel gleich darauf von neuem in fieberhafte Phantasien. Er sprach von einem großen, prächtigen Garten und einer langen, langen Allee, in der ihm seine verstorbene Gemahlin entgegenkäme, immer näher und heller und schöner. – „Nein, nein", sagte er, „sie hat einen Sternenmantel um und eine funkelnde Krone auf dem Haupt. Wie rings die Zweige schimmern von dem Glanz! – Gegrüßt seist du, Maria, bitt' für mich, du Königin der Ehren!" – Mit diesen Worten starb der Graf.

Als der Tag anbrach, war der ganze Himmel gegen Morgen dunkelrot gefärbt; gegenüber aber stand das Gewitter bleifarben hinter den grauen Türmen des Schlosses Dürande, die Sterbeglocke ging in einzelnen abgebrochenen Klängen über die stille Gegend, die fremd und wie verwandelt in der seltsamen Beleuchtung heraufblickte. – Da sahen einige Holzhauer im Walde den wilden Jäger Renald mit seiner Büchse und dem Hunde eilig in die Morgenglut hinabsteigen; niemand wußte, wohin er sich gewendet.

Mehrere Tage waren seitdem vergangen, das Schloß stand wie verzaubert in öder Stille, die Kinder gingen abends scheu vorüber, als ob es drin spuke. Da sah man eines Tages plötzlich droben mehrere Fenster geöffnet, buntes Reisegepäck lag auf dem Hof umher, muntere Stimmen schallten wieder auf den Treppen und Gängen, die Türen flogen hallend auf und zu, und vom Turm fing die Uhr trostreich wieder zu schlagen an. Der junge Graf Dürande war, auf die Nachricht vom Tode seines Vaters, rasch und unerwartet von Paris zurückgekehrt. Unterwegs war er mehrmals verworrenen Zügen von Edelleuten begegnet, die schon damals flüchtend die Landstraßen bedeckten. Er aber hatte keinen Glauben an die Fremde und wollte ehrlich Freud' und Leid mit seinem Vaterlande teilen. Wie hatte auch der erste Schreck aus der Ferne alles übertrieben! Er fand seine nächsten Dienstleute ergeben und voll Eifer und überließ sich gern der Hoffnung, noch alles zum Guten wenden zu können.
In solchen Gedanken stand er an einem der offenen Fenster, die Wälder rauschten so frisch herauf, das hatte er so lange nicht gehört, und im Tale schlugen die Vögel und jauchzten die Hirten von den Bergen, dazwischen hörte er unten im Schloßgarten singen:

„Wär's dunkel, ich läg' im Walde,
Im Walde rauscht's so sacht,
Mit ihrem Sternenmantel
Bedeckt mich da die Nacht,
Da kommen die Bächlein gegangen:
Ob ich schon schlafen tu?
Ich schlaf' nicht, ich hör' noch lange
Den Nachtigallen zu,
Wenn die Wipfel über mir schwanken,
Es klinget die ganze Nacht,
Das sind im Herzen die Gedanken,
Die singen, wenn niemand wacht."

Jawohl, gar manche stille Nacht, dachte der Graf, sich mit der Hand über die Stirn fahrend. – „Wer sang da?" wandte er sich dann zu den auspackenden Dienern; die Stimme schien ihm so bekannt. Ein Jäger meinte, es sei wohl der neue Gärtnerbursch aus Paris, der habe keine Ruhe gehabt in der Stadt; als sie fortgezogen, so sei er ihnen zu Pferde nachgekommen. „Der?" – sagte der Graf – er konnte sich kaum auf den Burschen besinnen. Über den Zerstreuungen des Winters in Paris war er nicht oft in den Garten gekommen; er hatte den Knaben nur selten gesehn und wenig beachtet, um so mehr freute ihn seine Anhänglichkeit. Indes war es beinahe Abend geworden, da hieß der Graf noch sein Pferd satteln, die Diener verwunderten sich, als sie ihn bald darauf so spät und ganz allein noch nach dem Walde hinreiten sahen. Der Graf aber schlug den Weg zu dem nahen Nonnenkloster ein, und ritt in Gedanken rasch fort, als gält' es, ein lange versäumtes Geschäft nachzuholen; so hatte er in kurzer Zeit das stille Waldkloster erreicht. Ohne abzusteigen, zog er hastig die Glocke am Tor. Da stürzte ein Hund ihm entgegen, als wollt' er ihn zerreißen, ein langer, bärtiger Mann trat aus der Klosterpforte und stieß den Köter wütend mit den Füßen; der Hund heulte, der Mann fluchte, eine Frau zankte drin im Kloster, sie konnte lange nicht zu Worte kommen. Der Graf, befremdet von dem seltsamen Empfang, verlangte jetzt schleunig die Priorin zu sprechen. – Der Mann sah ihn etwas verlegen an, als schämte er sich. Gleich aber wieder in alter Roheit gesammelt, sagte er, das Kloster sei aufgehoben und gehöre der Nation; er sei der Pächter hier. Weiter erfuhr nun der Graf noch, wie ein Pariser Commissair das alles so rasch und klug geordnet. Die Nonnen sollten nun in weltlichen Kleidern hinaus in die Städte, heiraten und nützlich sein; da zogen alle in einer schönen stillen Nacht aus dem Tal, für das sie so lange gebetet, nach Deutschland hinüber, wo ihnen in einem Schwesterkloster freundliche Aufnahme angeboten worden.

Der überraschte Graf blickte schweigend umher, jetzt bemerkte er erst, wie die zerbrochenen Fenster im Winde klappten; aus einer Zelle unten sah ein Pferd schläfrig ins Grün hinaus, die Ziegen des Pächters weideten unter umgeworfenen Kreuzen auf dem Kirchhof, niemand wagte es, sie zu vertreiben; dazwischen weinte ein Kind im Kloster, als klagte es, daß es geboren in dieser Zeit. Im Dorfe aber war es wie ausgekehrt, die Bauern guckten scheu aus den Fenstern, sie hielten den Grafen für einen Herrn von der Nation. Als ihn aber nach und nach einige wiedererkannten, stürzte auf einmal alles heraus und umringte ihn, hungrig, zerlumpt und bettelnd. Mein Gott, mein Gott, dachte er, wie wird die Welt so öde! – Er warf alles Geld, das er bei sich hatte, unter den Haufen, dann setzte er rasch die Sporen ein und wandte sich wieder nach Hause.

Es war schon völlig Nacht, als er in Dürande ankam. Da bemerkte er mit Erstaunen im Schlosse einen unnatürlichen Aufruhr, Lichter liefen von Fenster zu Fenster, und einzelne Stimmen schweiften durch den dunklen Garten, als suchten sie jemand. Er schwang sich rasch vom Pferde und eilte ins Haus. Aber auf der Treppe stürzte ihm schon der Kammerdiener mit einem versiegelten Blatte atemlos entgegen: es seien Männer unten, die es abgegeben und trotzig Antwort verlangten. Ein Jäger, aus dem Garten hinzutretend, fragte ängstlich den Grafen, ob er draußen dem Gärtnerburschen begegnet? Der Bursch habe ihn überall gesucht, der Graf möge sich aber hüten vor ihm, er sei in der Dämmerung verdächtig im Dorf gesehen worden, ein Bündel unterm Arm, mit allerlei Gesindel sprechend, nun sei er gar spurlos verschwunden.

Der Graf, unterdes oben im erleuchteten Zimmer angelangt, erbrach den Brief und las in schlechter, mit blauer Tinte mühsam gezeichneter Handschrift: „Im Namen Gottes verordne ich hiermit, daß der Graf Hippolyt von Dürande auf einem, mit dem gräflichen Wappen besiegelten Pergament die einzige Tochter des verstorbenen För-

sters am Schloßberg, Gabriele Dubois, als seine rechtmäßige Braut und künftiges Gemahl bekennen und annehmen soll. Dieses Gelöbnis soll heute bis elf Uhr nachts in dem Jägerhause abgeliefert werden. Ein Schuß aus dem Schloßfenster aber bedeutet: Nein. *Renald.*"
„Was ist die Uhr?" fragte der Graf. – „Bald Mitternacht", erwiderten einige, sie hätten ihn so lange im Walde und Garten vergeblich gesucht. – „Wer von euch sah den Renald, wo kam er her?" fragte er von neuem. Alles schwieg. Da warf er den Brief auf den Tisch. „Der Rasende!" sagte er und befahl für jeden Fall die Zugbrücke aufzuziehen, dann öffnete er rasch das Fenster und schoß ein Pistol, als Antwort, in die Luft hinaus. Da gab es einen wilden Widerhall durch die stille Nacht, Geschrei und Rufen und einzelne Flintenschüsse, bis in die fernsten Schlünde hinein, und als der Graf sich wieder wandte, sah er in dem Saal einen Kreis verstörter Gesichter lautlos um sich her.
Er schalt sie Hasenjäger, denen vor Wölfen graute. „Ihr habt lange genug Krieg gespielt im Walde", sagte er, „nun wendet sich die Jagd, wir sind jetzt das Wild, wir müssen durch. Was wird es sein! Ein Tollhaus mehr ist wieder aufgeriegelt, der rasende Veitstanz geht durchs Land, und der Renald geigt ihnen vor. Ich hab' nichts mit dem Volk, ich tat ihnen nichts als Gutes, wollen sie noch Besseres, sie sollen's ehrlich fordern, ich gäb's ihnen gern, abschrecken aber laß ich mir keine Hand breit meines alten Grund und Bodens; Trotz gegen Trotz!"
So trieb er sie in den Hof hinab, er selber half die Pforten, Luken und Fenster verrammen. Waffen wurden rasselnd von allen Seiten herbeigeschleppt, sein fröhlicher Mut belebte alle. Man zündete mitten im Hofe ein großes Feuer an, die Jäger lagerten sich herum und gossen Kugeln in den roten Widerscheinen, die lustig über die stillen Mauern liefen – sie merkten nicht, wie die Raben, von der plötzlichen Helle aufgeschreckt, ächzend über ihnen die alten Türme umkreisten. – Jetzt brachte ein Jäger mit gro-

ßem Geschrei den Hut und die Jacke des Gärtnerburschen, die er zu seiner Verwunderung beim Aufsuchen der Waffen im Winkel eines abgelegenen Gemaches gefunden. Einige meinten, das Bürschchen sei vor Angst aus der Haut gefahren, andere schworen, er sei ein Schleicher und Verräter, während der alte Schloßwart Nicolo, schlau lächelnd, seinem Nachbar heimlich etwas ins Ohr flüsterte. Der Graf bemerkte es. „Was lachst du?“ fuhr er den Alten an; eine entsetzliche Ahnung flog plötzlich durch seine Seele. Alle sahen verlegen zu Boden. Da faßte er den erschrockenen Schloßwart hastig am Arm und führte ihn mit fort in einen entlegenen Teil des Hofes, wohin nur einige schwankende Schimmer des Feuers langten. Dort hörte man beide lange Zeit lebhaft miteinander reden, der Graf ging manchmal heftig an dem dunkeln Schloßflügel auf und ab, und kehrte dann immer wieder fragend und zweifelnd zu dem Alten zurück. Dann sah man sie in den offenen Stall treten, der Graf half selbst eilig den schnellsten Läufer satteln, und gleich darauf sprengte Nicolo quer über den Schloßhof, daß die Funken stoben, durchs Tor in die Nacht hinaus. „Reit zu“, rief ihm der Graf noch nach, „frag’, suche bis ans Ende der Welt!“

Nun trat er rasch und verstört wieder zu den andern, zwei der zuverlässigsten Leute mußten sogleich bewaffnet nach dem Dorf hinab, um den Renald draußen aufzusuchen; wer ihn zuerst sähe, solle ihm sagen: er, der Graf, wolle ihm Satisfaktion geben wie einem Kavalier, und sich mit ihm schlagen. Mann gegen Mann – mehr könne der Stolze nicht verlangen.

Die Diener starrten ihn verwundert an, er aber hatte unterdes einen rüstigen Jäger auf die Zinne gestellt, wo man am weitesten ins Land hinaussehen konnte. „Was siehst du?“ fragte er, unten seine Pistolen ladend. Der Jäger erwiderte: die Nacht sei zu dunkel, er könne nichts unterscheiden, nur einzelne Stimmen höre er manchmal fern im Feld und schweren Tritt, als zögen viele Menschen lautlos durch die Nacht, dann sei alles wieder still. „Hier ist’s lu-

stig oben“, sagte er, „wie eine Wetterfahne im Wind – was ist denn das?“
„Wer kommt?“ fuhr der Graf hastig auf.
„Eine weiße Gestalt, wie ein Frauenzimmer“, entgegnete der Jäger, „fliegt unten dicht an der Schloßmauer hin.“ – Er legte rasch seine Büchse an. Aber der Graf, die Leiter hinauffliegend, war schon selber droben und riß dem Zielenden heftig das Gewehr aus der Hand. Der Jäger sah ihn erstaunt an. „Ich kann auch nichts mehr sehen“, sagte er dann halb unwillig und warf sich nun auf die Mauer nieder, über den Rand hinausschauend: „Wahrhaftig, dort an der Gartenecke ist noch ein Fenster offen, der Wind klappt mit den Laden, dort ist's hereingehuscht.“
Die Zunächststehenden im Hofe wollten eben nach der bezeichneten Stelle hineilen, als plötzlich mehrere Diener wie Herbstblätter im Sturm über den Hof daherflogen; die Rebellen, hieß es, hätten im Seitenflügel eine Pforte gesprengt, andere meinten, der rotköpfige Waldwärter habe sie mit Hilfe eines Nachschlüssels heimlich durch das Kellergeschoß hereingeführt. Schon hörte man Fußtritte hallend auf den Gängen und Treppen und fremde, rauhe Stimmen da und dort, manchmal blitzte eine Brandfackel vorüberschweifend durch das Fenster. – „Hallo, nun gilt's, die Gäste kommen, spielt auf zum Hochzeitstanze!“ rief der Graf, in niegefühlter Mordlust aufschauernd. Noch war nur erst ein geringer Teil des Schlosses verloren; er ordnete rasch seine kleine Schar, fest entschlossen, sich lieber unter den Trümmern seines Schlosses zu begraben, als in diese rohen Hände zu fallen.
Mitten in dieser Verwirrung aber ging auf einmal ein Geflüster durch die Leute: der Graf zeigte sich doppelt im Schloß, der eine hatte ihn zugleich im Hof und am Ende eines dunkeln Ganges gesehen, einem andern war er auf der Treppe begegnet, flüchtig und auf keinen Anruf Antwort gebend, das bedeutete seit uralter Zeit dem Hause großes Unglück. Niemand hatte jedoch in diesem Augenblick das Herz und die Zeit, es dem Grafen zu sagen, denn

soeben begann auch unten der Hof sich schon grauenhaft zu beleben; unbekannte Gesichter erschienen überall an den Kellerfenstern, die Kecksten arbeiteten sich gewaltsam hervor und sanken, ehe sie sich draußen noch aufrichten konnten, von den Kugeln der wachsamen Jäger wieder zu Boden, aber über ihre Leichen weg kroch und rang und hob es sich immer wieder von neuem unaufhaltsam empor, braune verwilderte Gestalten, mit langen Vogelflinten, Stangen und Brecheisen, als wühlte die Hölle unter dem Schlosse sich auf. Es war die Bande des verräterischen Waldwärters, der ihnen heimtückisch die Keller geöffnet. Nur auf Plünderung bedacht, drangen sie sogleich nach dem Marstall und hieben in der Eile die Stränge entzwei, um sich der Pferde zu bemächtigen. Aber die edlen schlanken Tiere, von dem Lärm und der gräßlichen Helle verstört, rissen sich los und stürzten in wilder Freiheit in den Hof; dort mit zornig-funkelnden Augen und fliegender Mähne, sah man sie bäumend aus der Menge steigen und Roß und Mann verzweifelnd durcheinander ringen beim wirren Wetterleuchten der Fackeln, Jubel und Todesschrei und die dumpfen Klänge der Sturmglocken dazwischen. Die versprengten Jäger fochten nur noch einzeln gegen die wachsende Übermacht; schon umringte das Getümmel immer dichter den Grafen, er schien unrettbar verloren, als der blutige Knäuel mit dem Aufruf: dort ist er! sich plötzlich wieder entwirrte und alles dem andern Schloßflügel zuflog.

Der Graf, in einem Augenblick fast alleinstehend, wandte sich tiefaufatmend und sah erstaunt das alte Banner des Hauses Dürande drüben vom Balkon wehen. Es wallte ruhig durch die wilde Nacht, auf einmal aber schlug der Wind wie im Spiel die Fahne zurück – da erblickte er mit Schaudern sich selbst dahinter, in seinen weißen Reitermantel tief gehüllt, Stirn und Gesicht von seinem Federbusch umflattert. Alle Blicke und Rohre zielten auf die stille Gestalt, doch dem Grafen sträubte sich das Haar empor, denn die Blicke des furchtbaren Doppelgängers wa-

ren mitten durch den Kugelregen unverwandt auf ihn gerichtet. Jetzt bewegte er die Fahne, er schien ihm ein Zeichen geben zu wollen, immer deutlicher und dringender ihn zu sich hinaufwinkend.

Eine Weile starrt er hin, dann, von Entsetzen überreizt, vergißt er alles andere, und unerkannt den Haufen teilend, der wütend nach dem Haupttor dringt, eilt er selbst dem gespenstischen Schloßflügel zu. Ein heimlicher Gang, nur wenigen bekannt, führt seitwärts näher zum Balkon, dort stürzt er sich hinein; schon schließt die Pforte sich schallend hinter ihm, er tappt am Pfeiler einsam durch die stille Halle, da hört er atmen neben sich, es faßt ihn plötzlich bei der Hand, schauernd sieht er das Banner und den Federbusch im Dunkeln wieder schimmern. Da, den weißen Mantel zurückschlagend, stößt es unten rasch eine Tür auf nach dem stillen Feld, ein heller Mondblick streift blendend die Gestalt, sie wendet sich. – „Um Gottes willen, *Gabriele!*“ ruft der Graf und läßt verwirrt den Degen fallen.

Das Mädchen stand bleich, ohne Hut vor ihm, die schwarzen Locken aufgeringelt, rings von der Fahne wunderbar umgeben. Sie schien noch atemlos. „Jetzt zaudere nicht“, sagte sie, den ganz Erstaunten eilig nach der Tür drängend, „der alte Nicolo harrt deiner draußen mit dem Pferde. Ich war im Dorf, der Renald wollte mich nicht wiedersehn, so rannte ich ins Schloß zurück, zum Glück stand noch ein Fenster offen, da fand ich dich nicht gleich und warf mich rasch in deinen Mantel. Noch merken sie es nicht, sie halten mich für dich; bald ist’s zu spät, laß mich und rette dich, nur schnell!“ Dann setzte sie leiser hinzu: „Und grüße auch das schöne Fräulein in Paris, und betet für mich, wenn’s euch wohlgeht.“

Der Graf aber, in tiefster Seele bewegt, hatte sie schon fest in beide Arme genommen und bedeckte den bleichen Mund mit glühenden Küssen. Da wand sie sich schnell los. „Mein Gott, liebst du mich denn noch, ich meinte, du freitest um das Fräulein?“ sagte sie voll Erstaunen, die großen Augen fragend zu ihm aufgeschlagen. – Ihm war’s

auf einmal, wie in den Himmel hineinzusehen. „Die Zeit fliegt heut entsetzlich", rief er aus, „dich liebte ich immerdar, da nimm den Ring und meine Hand auf ewig, und so verlaß' mich Gott, wenn ich je von dir lasse!" – Gabriele, von Überraschung und Freude verwirrt, wollte niederknien, aber sie taumelte und mußte sich an der Wand festhalten. Da bemerkte er erst mit Schrecken, daß sie verwundet war. Ganz außer sich riß er sein Tuch vom Halse, suchte eilig mit Fahne, Hemd und Kleidern das Blut zu stillen, das auf einmal unaufhaltsam aus vielen Wunden zu quellen schien. In steigender unsäglicher Todesangst blickte er nach Hilfe ringsumher, schon näherten sich verworrene Stimmen, er wußte nicht, ob es Freund oder Feind. Sie hatte währenddes den Kopf müde an seine Schulter gelehnt. „Mir flimmert's so schön vor den Augen", sagte sie, „wie dazumal, als du durchs tiefe Abendrot noch zu mir kamst; nun ist ja alles, alles wieder gut." Da pfiff plötzlich eine Kugel durch das Fenster herein. „Das war der Renald!" rief der Graf, sich nach der Brust greifend; er fühlte den Tod im Herzen. – Gabriele fuhr hastig auf. „Wie ist dir?" fragte sie erschrocken. Aber der Graf, ohne zu antworten, faßte heftig nach seinem Degen. Das Gesindel war leise durch den Gang herangeschlichen, auf einmal sah er sich in der Halle von bewaffneten Männern umringt. – „Gute Nacht, mein liebes Weib!" rief er da; und mit letzter, übermenschlicher Gewalt das von der Fahne verhüllte Mädchen auf den linken Arm schwingend, bahnt' er sich eine Gasse durch die Plünderer, die ihn nicht kannten und verblüfft von beiden Seiten vor dem Wütenden zurückwichen. So hieb er sich durch die offene Tür glücklich ins Freie hinaus, keiner wagte ihm aufs Feld zu folgen, wo sie in den schwankenden Schatten der Bäume einen heimlichen Hinterhalt besorgten.

Draußen aber rauschten die Wälder so kühl. „Hörst du die Hochzeitsglocken gehn?" sagte der Graf; „ich spür' schon Morgenluft." – Gabriele konnte nicht mehr sprechen, aber sie sah ihn still und selig an. – Immer ferner und

leiser verhallten unterdes schon die Stimmen vom Schlosse her, der Graf wankte verblutend, sein steinernes Wappenschild lag zertrümmert im hohen Gras, dort stürzt' er tot neben Gabrielen zusammen. Sie atmeten nicht mehr, aber der Himmel funkelte von Sternen, und der Mond schien prächtig über das Jägerhaus und die einsamen Gründe; es war, als zögen Engel singend durch die schöne Nacht. –

Dort wurden die Leichen von Nicolo gefunden, der vor Ungeduld schon mehrmals die Runde um das Haus gemacht hatte. Er lud beide mit dem Banner auf das Pferd, die Wege standen verlassen, alles war im Schloß, so brachte er sie unbemerkt in die alte Dorfkirche. Man hatte dort vor kurzem erst die Sturmglocke geläutet, die Kirchtür war noch offen. Er lauschte vorsichtig in die Nacht hinaus, es war alles still, nur die Linden säuselten im Winde, vom Schloßgarten hörte er die Nachtigallen schlagen, als ob sie im Traume schluchzten. Da senkte er betend das stille Brautpaar in die gräfliche Familiengruft und die Fahne darüber, unter der sie noch heut' zusammen ausruhn. Dann aber ließ er mit traurigem Herzen sein Pferd frei in die Nacht hinauslaufen, segnete noch einmal die schöne Heimatsgegend und wandte sich rasch nach dem Schloß zurück, um seinen bedrängten Kameraden beizustehn; es war ihm, als könnte er nun selbst nicht länger mehr leben.

Auf den ersten Schuß des Grafen aus dem Schloßfenster war das raubgierige Gesindel, das durch umlaufende Gerüchte von Renalds Anschlag wußte, aus allen Schlupfwinkeln hervorgebrochen, er selbst hatte in der offenen Tür des Jägerhauses auf die Antwort gelauert und sprang bei dem Blitz im Fenster wie ein Tiger allen voraus, er war der erste im Schloß. Hier, ohne auf das Treiben der andern zu achten, suchte er mitten zwischen den pfeifenden Kugeln in allen Gemächern, Gängen und Winkeln unermüdlich den Grafen auf. Endlich erblickt' er ihn durchs Fen-

ster in der Halle, er hört' ihn drin sprechen, ohne Gabrielen in der Dunkelheit zu bemerken. Der Graf kannte den Schützen wohl, er hatte gut gezielt. Als Renald ihn getroffen taumeln sah, wandte er sich tiefaufatmend – sein Richteramt war vollbracht.

Wie nach einem schweren, löblichen Tagewerke durchschritt er nun die leeren Säle in der wüsten Einsamkeit zwischen zertrümmerten Tischen und Spiegeln, der Zugwind strich durch alle Zimmer und spielte traurig mit den Fetzen der zerrissenen Tapeten.

Als er durchs Fenster blickte, verwunderte er sich über das Gewimmel fremder Menschen im Hofe, die ihm geschäftig dienten wie das Feuer dem Sturm. Ein seltsam Gelüsten funkelte ihn da von den Wänden an aus dem glatten Getäfel, in dem der Fackelschein sich verwirrend spiegelte, als äugelte der Teufel mit ihm. – So war er in den Gartensaal gekommen. Die Tür stand offen, er trat in den Garten hinaus. Da schauerte ihn in der plötzlichen Kühle. Der untergehende Mond weilte noch zweifelnd am dunkeln Rand der Wälder, nur manchmal leuchtete der Strom noch herauf, kein Lüftchen ging, und doch rührten sich die Wipfel, und die Alleen und geisterhaften Statuen warfen lange, ungewisse Schatten dazwischen, und die Wasserkünste spielten und rauschten so wunderbar durch die weite Stille der Nacht. Nun sah er seitwärts auch die Linde und die mondbeglänzte Wiese vor dem Jägerhause; er dachte sich die verlorne Gabriele wieder in der alten unschuldigen Zeit als Kind mit den langen dunkeln Locken, es fiel ihm immer das Lied ein: „Gute Nacht, mein Vater und Mutter, wie auch mein stolzer Bruder" – es wollte ihm das Herz zerreißen, er sang verwirrt vor sich hin, halb wie im Wahnsinn:

„Meine Schwester, die spielt an der Linde. –
Stille Zeit, wie so weit, so weit!
Da spielten so schöne Kinder
Mit ihr in der Einsamkeit.

Von ihren Locken verhangen,
Schlief sie und lachte im Traum,
Und die schönen Kinder sangen
Die ganze Nacht unterm Baum.

Die ganze Nacht hat gelogen,
Sie hat mich so falsch gegrüßt,
Die Engel sind fortgeflogen
Und Haus und Garten stehn wüst.

Es zittert die alte Linde
Und klaget der Wind so schwer,
Das macht, das macht die Sünde –
Ich wollt', ich läg' im Meer. –

Die Sonne ist untergegangen
Und der Mond im tiefen Meer,
Es dunkelt schon über dem Lande;
Gute Nacht! seh' dich nimmermehr."

„Wer ist da?" rief er auf einmal in den Garten hinein. Eine dunkle Gestalt unterschied sich halb kenntlich zwischen den wirren Schatten der Bäume; erst hielt er es für eins der Marmorbilder, aber es bewegte sich, er ging rasch darauf los, ein Mann versuchte sich mühsam zu erheben, sank aber immer wieder ins Gras zurück. „Um Gott, Nicolo, du bist's!" rief Renald erstaunt; „was machst du hier?" – Der Schloßwart wandte sich mit großer Anstrengung auf die andere Seite, ohne zu antworten.
„Bist du verwundet?" sagte Renald, besorgt nähertretend, „wahrhaftig, an dich dacht' ich nicht in dieser Nacht. Du warst mir der liebste immer unter allen, treu, zuverlässig, ohne Falsch; ja, wär' die Welt wie du! Komm nur mit mir, du sollst herrschaftlich leben jetzt im Schloß auf deine alten Tage, ich will dich über alle stellen."
Nicolo aber stieß ihn zurück: „Rühre mich nicht an, deine Hand raucht noch von Blut."

„Nun", entgegnete Renald finster, „ich meine, ihr solltet mir's alle danken, die wilden Tiere sind verstoßen in den wüsten Wald, es bekümmert sich niemand um sie, sie müssen sich ihr Futter selber nehmen – bah, und was ist Brot gegen Recht?"
„Recht?" sagte Nicolo, ihn lange starr ansehend, „um Gottes willen, Renald, ich glaube gar, du wußtest nicht" –
„Was wußt' ich nicht?" fuhr Renald hastig auf.
„Deine Schwester Gabriele –"
„Wo ist sie?"
Nicolo wies schweigend nach dem Kirchhof; Renald schauderte heimlich zusammen. „Deine Schwester Gabriele", fuhr der Schloßwart fort, „hielt schon als Kind immer große Stücke auf mich, du weißt es ja; heut abend nun in der Verwirrung, eh's noch losging, hat sie in ihrer Herzensangst mir alles anvertraut."
Renald zuckte an allen Gliedern, als hinge in der Luft das Richtschwert über ihm. „Nicolo", sagte er drohend, „belüg' mich nicht, denn dir, gerade dir glaube ich."
Der Schloßwart, seine klaffende Brustwunde zeigend, erwiderte: „Ich rede die Wahrheit, so wahr mir Gott helfe, vor dem ich noch in dieser Stunde stehen werde! – Graf Hippolyt hat deine Schwester nicht entführt."
„Hoho", lachte Renald, plötzlich wie aus unsäglicher Todesangst erlöst, „ich sah sie selber in Paris am Fenster in des Grafen Haus."
„Ganz recht", sagte Nicolo, „aus Lieb' ist sie bei Nacht dem Grafen heimlich nachgezogen aus dem Kloster." –
„Nun siehst du, siehst du wohl? ich wußt's ja doch. Nur weiter, weiter", unterbrach ihn Renald; große Schweißtropfen hingen in seinem wildverworrenen Haar.
„Das arme Kind", erzählte Nicolo wieder, „sie konnte nicht vom Grafen lassen; um ihm nur immer nahe zu sein, hat sie verkleidet als Gärtnerbursche sich verdungen im Palast, wo sie keiner kannte."
Renald, aufs äußerste gespannt, hatte sich unterdes neben dem Sterbenden, der immer leiser sprach, auf die Knie

hingeworfen, beide Hände vor sich auf die Erde gestützt. „Und der Graf", sagte er, „der Graf, aber der Graf, was tat der? Er lockte, er kirrte sie, nicht wahr?"
„Wie sollt' er's ahnen?" fuhr der Schloßwart fort; „er lebte wie ein loses Blatt im Sturm, von Fest zu Fest. Wie oft stand sie des Abends spät in dem verschneiten Garten vor des Grafen Fenstern, bis er nach Hause kam, wüst, überwacht – er wußte nichts davon bis heute abend. Da schickt' er mich hinaus, sie aufzusuchen; sie aber hatte sich dem Tode schon geweiht, in seinen Kleidern euch täuschend wollte sie eure Kugeln von seinem Herzen auf ihr eigenes wenden – o jammervoller Anblick – so fand ich beide tot im Felde Arm in Arm – der Graf hat ehrlich sie geliebt bis in den Tod – sie beide sind schuldlos – rein – Gott sei uns allen gnädig!"
Renald war über diese Worte ganz still geworden, er horchte noch immer hin, aber Nicolo schwieg auf ewig, nur die Gründe rauschten dunkel auf, als schauderte der Wald.
Da stürzte auf einmal vom Schloß die Bande siegestrunken über Blumen und Beete daher, sie schrien vivat und riefen den Renald im Namen der Nation zum Herrn von Dürande aus. Renald, plötzlich sich aufrichtend, blickte wie aus einem Traum in die Runde. Er befahl, sie sollten schleunig alle Gesellen aus dem Schlosse treiben und keiner, bei Lebensstrafe, es wieder betreten, bis er sie riefe. Er sah so schrecklich aus, sein Haar war grau geworden über Nacht, niemand wagte es, ihm jetzt zu widersprechen. Darauf sahen sie ihn allein rasch und schweigend in das leere Schloß hineingehen, und während sie noch überlegen, was er vorhat und ob sie ihm gehorchen oder dennoch folgen sollen, ruft einer erschrocken aus: „Herr Gott, der rote Hahn ist auf dem Dach!" Und mit Erstaunen sahen sie plötzlich feurige Spitzen bald da, bald dort aus den zerbrochenen Fenstern schlagen und an dem trocknen Sparrwerk hurtig nach dem Dache klettern. Renald, seines Lebens müde, hatte eine brennende Fackel er-

griffen und das Haus an allen vier Ecken angesteckt. – Jetzt, mitten durch die Lohe, die der Zugwind wirbelnd faßte, sahen sie den Schrecklichen eilig nach dem Eckturme schreiten, es war, als schlüge Feuer auf, wohin er trat. Dort in dem Turme liegt das Pulver, hieß es auf einmal, und voll Entsetzen stiebte alles über den Schloßberg auseinander. Da tat es gleich darauf einen furchtbaren Blitz, und donnernd stürzte das Schloß hinter ihnen zusammen. Dann wurde alles still; wie eine Opferflamme, schlank, mild und prächtig stieg das Feuer zum gestirnten Himmel auf, die Gründe und Wälder ringsumher erleuchtend – den Renald sah man nimmer wieder.

Das sind die Trümmer des alten Schlosses Dürande, die weinumrankt in schönen Frühlingstagen von den waldigen Bergen schauen. – Du aber hüte dich, das wilde Tier zu wecken in der Brust, daß es nicht plötzlich ausbricht und dich selbst zerreißt.

Die Entführung

Der Abend senkte sich schon über der fruchtbaren Landschaft, welche die Loire durchströmt, als ein junger Mann, jagdmüde und mit der Büchse über dem Rücken aus dem Walde tretend, unerwartet zwischen den grünen Bergen in der schönsten Einsamkeit ein altes Schloß erblickte. Er konnte durch die Wipfel nur erst Dach und Türme sehen, von Efeu überwachsen, mit geschlossenen Fenstern, halb wie im Schlafe. Neugierig drang er durch das verworrene Gebüsch die Anhöhe hinan, es schien der ehemalige Schloßgarten zu sein, denn künstliche Hecken durchschnitten oben den Platz, weiterhin schimmerte noch eine weiße Statue durch die Zweige, aber rings aus den Tälern ging der Frühling, mit Waldblumen funkelnd, lustig über die gezirkelten Beete und Gänge, alles prächtig verwildernd.

Jetzt, um eine Ecke biegend, sah er auf einmal das ganze Schloß vor sich, mitten im Grün, als wollt's in alle Fenster steigen; auf der steinernen Rampe vor der Saaltür, vom Abendrot beschienen, saßen eine ältliche Dame und eine schlanke Mädchengestalt am Stickrahmen, ein zahmes Reh graste neben ihnen in der schönen Wildnis, alle drei den Ankommenden erstaunt betrachtend.

Dieser stutzte überrascht, aber schnell entschlossen näherte er sich den Frauen und entschuldigte mit vielem Anstand seinen unwillkürlichen Überfall, er kenne hier die Waldgrenzen noch zu wenig, so sei er in dies fremde Revier geraten und lege nun als Wildschütz sein Geschick in ihre Hände. Die alte Dame, ohne seine Entschuldigung besonders zu beachten und ihn vom Kopf bis zu den Füßen mit den Blicken messend, bat ihn, da er fein gekleidet erschien, ziemlich kalt, neben ihnen Platz zu nehmen, in-

dem sie auf einen Lehnstuhl wies, den auf ihren Wink ein bejahrter Diener in etwas verschossener Livre soeben aus dem Gartensaal brachte.

Die Unterhaltung stockte einen Augenblick, aber der Fremde, der sich in der maskenhaften Freiheit eines Unbekannten zu gefallen schien, wußte bald mit großer Gewandtheit das Gespräch zu ergreifen und zu beleben. Sie sprachen demnächst von der Räuberbande, die sich in diesem Frühjahr hier zwischen den Bergen eingenistet und durch ihre verwegenen Züge die ganze Gegend in Furcht und Schrecken setzte. Der Gast sagte lachend, das komme von der langen Friedenszeit, da spiele der Krieg, der sich sein Recht nicht nehmen lasse, auf seine eigne Hand im Lande. Der Mensch verlange immer etwas Außerordentliches, und wenn es das Entsetzlichste wäre, um nur dem unerträglichsten Übel, der Langenweile, zu entkommen. – Die neueste Zeitung lag soeben auf dem Tischchen vor ihnen, sie enthielt eine ungefähre Personenbeschreibung des vermutlichen Hauptmannes der Bande. Der Fremde las sie mit großer Aufmerksamkeit, und es fiel der Dame auf, da er darauf um die Erlaubnis bat, das Blatt mitzunehmen, und es hastig einsteckte.

Währenddes war Frenel, der alte Diener, mit sichtbaren Zeichen von Bestürzung wieder hinzugetreten. Er schien aus dem Hofe zu kommen, und, der Dame einen heimlichen Wink gebend, sprach er lange leise und lebhaft mit ihr im Hintergrunde des Saales. Er meldete, daß sich im Walde, unweit des Schlosses, unbekannte, bewaffnete Männer zu Pferde gezeigt, sie hielten ein lediges Roß, das schöner und kostbarer gezäumt als die andern. Der Waldhirte, der unbemerkt in ihrer Nähe gewesen, habe deutlich vernommen, wie sie von ihrem Herrn geredet, mehrmals ungeduldig nach dem Schlosse schauend, als ob sie jemanden von hier erwarteten. – Die alte Dame, bei dieser seltsamen Nachricht einen Augenblick nachsinnend, überflog unwillkürlich in Gedanken die Beschreibung des Räuberhauptmannes aus der Zeitung, er war als ein jun-

ger, schöner, weltgewandter Mann geschildert – es fuhr ihr auf einmal wie ein Blitz durch die Seele, wie alles gar wohl auf ihren rätselhaften Gast bezogen werden konnte.
Indem sie so in großer Bewegung mit sich selber schnell beriet, wie sie in dieser sonderbaren Lage sich zu benehmen habe, schien der Fremde von alledem nichts zu bemerken. Er unterhielt sich heiter und angelegentlich mit dem Fräulein, während der Abend über dem wilden Garten schon immer tiefer hereindunkelte. Da fiel plötzlich ein Schuß unten im Walde. Die Dame trat entschlossen einige Schritte auf den Fremden zu. „Das sind meine Leute", sagte dieser, rasch aufspringend. – „Ihre Leute?" – „Gewiß", erwiderte er. – Da er aber auf einmal den Schreck der erbleichten Dame bemerkte, entschuldigte er sich abermals wegen dieser Unruhe, versprach den Frevler ernstlich zu bestrafen und nahm sogleich Abschied, indem er, flüchtig seinen Namen nennend, noch um die Erlaubnis bat, wiederkommen zu dürfen. Aber niemand hörte oder antwortete ihm in der Verwirrung; so flog er den Schloßberg hinab. Der Abend tat noch einen roten, falschen Blick über die Bergkuppen, unten war schon alles finster und still, man hörte nur den Hufschlag von mehreren Rossen den Waldgrund entlang. Das Fräulein, das nun auch den entsetzlichen Verdacht vernommen, rief aufs tiefste erschrocken: „O Gott, o Gott, er kommt gewiß wieder!"
Wirklich konnte die Lage der verwitweten Marquise Astrenant – so hieß die Dame – gerechte Besorgnis erregen. Die Erinnerung an den alten Glanz und den verschwenderischen Aufwand ihres verstorbenen Gemahls war in der Gegend noch frisch genug, um die Anschläge des Raubgesindels auf das abgelegene Schloß zu lenken, und doch war sie in der Tat so verarmt, daß sie nicht daran denken konnte, in diesem Augenblick mit ihrer Tochter Leontine diese gefährliche Einsamkeit zu verlassen. In dieser Not fiel ihr ein, daß der Graf Gaston, wie sie von ihren Leuten gehört, soeben auf kurze Zeit auf einem sei-

ner benachbarten Jagdschlösser angekommen war. Diesen glücklichen Umstand benutzend, stellte sie dem Grafen, obgleich sie ihn noch nicht persönlich kannte, schriftlich in wenigen Worten ihre Abgeschiedenheit und Gefahr vor und beschwor ihn, als Nachbar sie in ihrer hilflosen Lage zu beschützen. Mit diesem Briefe wurde noch denselben Abend ein reitender Bote nach dem Jagdschlosse gesandt.

So war die Nacht allen unter mancherlei Vorsichtsmaßregeln schlaflos vergangen. Schon am folgenden Morgen aber erhielten sie die Antwort: der Graf werde nicht ermangeln, ihren Wünschen nach Kräften zu entsprechen, und womöglich heute noch selbst seine Aufwartung machen. Diese Zusage und das tröstliche Morgenlicht hatten alle Sorge gewendet. Sie schämten sich fast und lachten über die übertriebene Furcht und Besorgnis, womit die Wälder ringsumher im Dunkeln sie geschreckt. Und wie nach Gewittern oft ein heiterer Glanz über die Landschaft fliegt, so brachte auch hier der angekündigte Besuch des Grafen Gaston sehr bald das ganze stille Haus in eine ungewohnte fröhliche Bewegung. Die gläsernen Kronleuchter, die so lustig funkelten, wurden sorgfältig geputzt, die verstaubten Tapeten ausgeklopft und Teppiche gelüftet, der Morgen glänzte durch die verbleichten, rotseidenen Gardinen seltsam auf dem getäfelten Boden der Zimmer, während draußen über dem sonnigen Rasenplatz vor dem Hause die Schwalben jauchzend hin und her schossen. Leontine erschien besonders fleißig, sie war aufgewachsen zwischen diesen Trümmern des früheren Glanzes, nun schien ihr alles so prächtig, weil es ins Morgenrot ihrer Kindheit getaucht. Die Marquise lächelte schmerzlich, aber sie mochte die Freude der Tochter nicht stören.

Die Sonne stieg indes und senkte sich schon wieder nach den Tälern, und der Graf war zu ihrem Befremden noch immer nicht angekommen, noch hatte er den ganzen Tag über etwas von sich hören lassen. Sie mußten seinen Besuch für heute schon aufgeben, und als endlich der Abend

von neuem die Wälder färbte, saßen beide Frauen, durch die Geschäftigkeit des Tages zerstreut und zuversichtlicher geworden, wie sonst wieder auf der steinernen Rampe vor dem Garten an ihrer Arbeit, als wäre eben nichts vorgefallen. Leontine, in vergeblicher Erwartung des Grafen, war geschmückt wie eine arme Braut, die nicht weiß, wie schön sie in ihrer Armut ist. Aber die Abendsonne blitzte über ihre frischen Augen und hüllte sie ganz in ihr schönstes, goldnes Kleid, und ihr Reh sah von fern verwundert nach der prächtigen Herrin, es war, als hätt' es alle seine Spielkameraden mit herbeigerufen, so neugierig wimmelten die Waldvögel im Garten und guckten durch die Zweige und schwatzten vergnügt untereinander. Vor dem Hause aber ging die Abendluft lind durch die Blumen unter ihnen. Leontine sah oft in Gedanken über ihre Arbeit ins Tal hinaus und sang:

„Überm Lande die Sterne
Machen die Runde bei Nacht,
Mein Schatz ist in der Ferne,
Liegt am Feuer auf der Wacht."

Die Marquise sagte: „Das hast du von unserm alten Frenel, als er noch Soldat war; sollte man doch glauben, du hätt'st einen Offizier zum Liebsten." Leontine lachte und sang weiter:

„Übers Feld bellen Hunde,
Wenn der Mondschein erblich,
Rauscht der Wald auf dem Grunde:
Reiter, jetzt hüte dich!"

„Ist's denn schon so spät?" unterbrach sie sich selbst, „sie läuten ja schon die Abendglocken, der Wind kommt über den Wald her, wie schön das klingt aus der Ferne herüber." Sie sang von neuem:

„Um das Lager im Dunkeln
Jetzt schleichen sie sacht,
Die Gewehre schon funkeln –
So falsch ist die Nacht!"

„Was steigt denn da für ein Rauch auf im Walde?" fragte hier die Mutter. – „Es wird wohl der Köhler sein", erwiderte Leontine, aber sie sah doch gespannt hin und sang zögernd:

„Ein Gesell durchs Gesteine
geht sacht in ihrer Mitt',
Es rasseln ihm die Beine –
Hat einen leisen, leisen Tritt. –"

„Nein!" sprang sie auf, „das ist ein Brand, da schlägt ja die helle Flamme auf, horch, sie läuten die Sturmglocken drüben!"
Indem nun beide sich erhoben, hörten sie in derselben Richtung ein paarmal schießen, dann war alles wieder still. „Da haben gewiß die Nachbarn großes Jagen", sagte die Marquise, „sie können nun einmal nicht fröhlich sein ohne Lärm." Da sie aber jetzt das Schloßgesinde am Abhange des Gartenberges versammelt sah, in großer Aufregung untereinander redend und nach jener Gegend hinausschauend, rief sie hinab: was es gebe? – „Blutige Köpfe!" hieß es zurück, der Waldwärter sei eben aus den Bergen gekommen, der Graf Gaston habe vor Tagesanbruch heimlich alle seine Bauern und Jäger bewaffnet und die Räuberbande aufgespürt und treibe sie von einem brennenden Schlupfwinkel zum andern durch den Wald, es gehe scharf her da drüben! – Da wandte sich Leontine, die bisher wie im Traume gestanden, plötzlich herum, sie sagte: es sei schändlich und gottlos, die Schlafenden zu überfallen und Menschen zu hetzen wie die wilden Tiere! – Die Mutter sah sie erstaunt an. Aber sie hatte keine Zeit, dem sonderbaren Betragen der Tochter nachzudenken, denn der alte Frenel trat soeben voll Eifer aus dem Hause, er

hatte hastig seine Büchse geladen und wollte mit hinunter. Die Marquise beschwor ihn, zum Schutze bei ihnen zu bleiben, wenn etwa einzelne versprengte Räuber hier vorüberschweiften, die andern sollten das Hoftor schließen, sich mit Beilen und Sensen versehen und den offenen Garten umstellen.

Leontine aber war indes schon in das obere Stockwerk gestiegen, die Fledermäuse in den wüsten Sälen schossen verstört aus den offenen Fenstern; sie schaute aus einem Erker angestrengt in die Waldgründe hinaus, als wollte sie durch die Wipfel sehen. Es dunkelte schon über den Tälern, die Schüsse schienen näher zu kommen, manchmal brachte der Wind einen wilden Schrei aus der Ferne herüber, vom Walde sah sie ein Reh von dem Lärm erschrokken unten über die Wiese fliehen. O wäre ich doch ein Mann! dachte sie tausendmal; dazwischen betete sie wieder still im Herzen vor der aufsteigenden Nacht, dann lehnte sie sich weit aus dem Fenster und winkte mit ihrem weißen Schnupftuch über die dunklen Wälder, sie wußte selbst nicht, was sie tat.

Jetzt hörte sie, wie unten im Garten nach und nach mehrere Boten zurückkamen, die die Mutter auf Kundschaft ausgeschickt; sie konnte in der Stille jedes Wort vernehmen. Die Bande, hieß es, sei völlig geschlagen, gefangen oder zerstreut. Ein anderer erzählte von der außerordentlichen Kühnheit des Grafen Gaston, wie er, überall der Erste voran, den Hauptmann selber aufs Korn genommen. Auf der Felsenkante im Walde seien sie endlich aneinander geraten, da habe der Graf ihn, immerfort fechtend, samt dem Pferde über den Abhang hinabgestürzt. Aber Unkraut verdirbt nicht, unten sich überkugelnd, seien Roß und Reiter, wie die Katzen, wieder auf die Beine gekommen; nun jagen sie alle den Räuber hier nach dem Schlosse zu, aber er sei ganz umzingelt, er könne nicht mehr entwischen. – „Gott segne den tapfern Grafen!“ rief die Marquise bei diesem Berichte aus, „er hat ritterlich sein Wort gelöst.“

Leontine aber sah wieder unverwandt nach dem Walde, denn draußen hatte die wilde Jagd sich plötzlich gewendet, ein Schuß fiel ganz nah, darauf mehrere, immer näher und näher, man sah die einzelnen Schüsse blitzen im Dunkeln. Auf einmal glaubte sie einen Reiter in verzweifelter Flucht längs dem Saume des Waldes flimmern zu sehen; die Jäger des Grafen, eine andere Fährte einschlagend, schienen ihn nicht zu bemerken, er flog gerade nach dem Schlosse her. Da, in wachsender Todesangst sich plötzlich aufraffend, stürzte sie pfeilschnell über die steinernen Treppen durch das stille Haus hinab und unten an dem alten Walle durch eine geheime Pforte, den Riegel sprengend, ins Freie. Als sie aber am Fuß des Schloßberges atemlos anlangt, vor Ermattung fast in die Knie sinkend, kommt auch der Reiter schon durch die dunkelnde Luft daher – es war, wie sie geahnt, der Fremde von gestern, verstört, mit fliegenden Haaren, sein Pferd ganz von Schaum bedeckt.
„Was wollen Sie hier?" rief sie ihm schon von fern entgegen. – Er, bei ihrem Anblick stutzend, hielt schnell an und, sich vom Pferde schwingend, erwiderte er höflich: er wolle, seinem Versprechen gemäß, sie und die Marquise noch einmal begrüßen. – „Um Gottes willen, sind Sie rasend? heut, in dieser Stunde?" – Der Reiter entschuldigte sich, der Kampf sei ernster geworden und habe ihn länger aufgehalten, als er gedacht, es sei der einzige noch übrige Augenblick, er müsse sogleich wieder weiter. – „O Gott, ich weiß", fiel Leontine ein. – „Sie wissen?" –
Leontine schauderte, da er, dicht vor ihr, sie auf einmal so durchdringend ansah. – „Sie bluten", sagte sie dann erschrocken. – „Nur ein Streifschuß", entgegnete er; „doch Sie haben recht", fuhr er lächelnd fort, „es ziemt sich nicht, in diesem Zustande bei Damen Besuche abzustatten." Aber Leontine hörte kaum mehr, was er sprach, sie stand in tiefen Gedanken. „Ich wüßte wohl einen verborgenen Ort für diese Nacht", sagte sie darauf schnell und leise, „wenn nur – nein, nein, es ist unmöglich! Das

Schloß ist voll Leute, vielleicht kommt der Graf selbst noch." – Und den Fremden in steigender, höchster Angst fortdrängend, wies sie ihm einen abgelegenen Fußsteig, der führe zu einer Furt des Flusses, da solle er hinüber, dann den Pfad rechts einschlagen – „nur schnell, schnell", flehte sie, „da kommen schon Leute zwischen den Bäumen, Sie suchen." – „Wen?" fragte der Reiter, sich rasch umsehend. – „O mein Gott", rief Leontine fast weinend, „Sie selbst, den unglücklichen Hauptmann!" – Der Fremde, bei diesen Worten plötzlich wie aus einem Traume erwachend, schlug schnell den Mantel zurück und nahm sie in beide Arme: „Kind, Kind, wie liebst du mich so schön! Das werde ich dir gedenken mein Leben lang, du sollst noch von dem Räuberhauptmann hören. – Jetzt drängt die Zeit. Grüße die Mutter oben, sag ihr, das Land sei frei, sie könne ohne Sorgen schlafen, leb wohl!" – Noch vom Pferde aber bat er sie um ihr weißes Tuch, sie reicht' es ihm zögernd; das wollte er um seine Wunde schlagen, da heilt' es über Nacht. – So ritt er fort.
Jetzt bemerkte sie erst, daß ihr Handschuh blutig geworden von seinem Arm, sie verbarg ihn, heftig an allen Gliedern zitternd. Im Walde indes und droben im Schlosse gingen verworrene Stimmen, sie sah noch immer dem Reiter nach und atmete tief auf, als er endlich in der schirmenden Wildnis verschwunden. Dann setzte sie sich auf den Rasen, den Kopf in beide Hände gestützt, und weinte bitterlich.

Noch in derselben Nacht brach auch Graf Gaston von seinem Jagdschlosse wieder auf, wohin er nur erst vor wenigen Tagen mit dem Ruhme eines ausgezeichneten Offiziers aus fremdem Kriegsdienst zurückgekehrt, um sich in der Einsamkeit zu erholen. Aber der Ruf seiner Tapferkeit war ihm längst nach Paris vorangeeilt, und fast gleichzeitig mit der Bitte der Marquise um seinen Schutz vor den Räubern erhielt er den unerwarteten Befehl des Königs,

sich unverzüglich an den Hof zu begeben, wo man bei den damaligen heimlichen Kriegsrüstungen seine Erfahrung benutzen wollte. So war es gekommen, daß er, um sein Wort gegen die besorgte Dame zu lösen, die Räuberjagd auf das gewaltsamste beschleunigt, dann aber keine Zeit mehr übrig hatte, bei der Marquise noch den versprochenen Besuch abzustatten.

In Paris zog er wie im Triumphe ein. Der frische Lorbeerkranz stand der hohen, schlanken Gestalt gar anmutig zu dem gebräunten Gesicht. Nun folgte ihm auch noch das vergrößernde Gerücht der Kühnheit, womit er soeben die lange vergeblich aufgesuchte Räuberbande wie im Fluge zwischen den Bergen vernichtet. Der König selbst hatte ihn ausgezeichnet empfangen, jedermann wollte ihn kennenlernen, und die Damen sahen scheu und neugierig durch die Fenstergardinen, wenn er im vollen Schmuck soldatischer Schönheit die Straßen hinabritt. – Unter ihnen aber zog nur *eine* seine Aufmerksamkeit auf sich, und diese hatte er bis jetzt noch nirgend erblickt.

Ganz Paris sprach damals von der jungen, reichen Gräfin *Diana,* einer amazonenhaften, spröden Schönheit mit rabenschwarzem Haar und dunklen Augen. Einige nannten sie ein prächtiges Gewitter, das über die Stadt fortzöge, unbekümmert, ob und wo es zünde; andere verglichen sie mit einer zauberischen Sommernacht, die, alles verlokkend und verwirrend, über seltsame Abgründe scheine. So fremd und märchenhaft erschien diese wilde Jungfräulichkeit an dem sittenlosen Hofe.

Über ihr früheres Leben konnte Graf Gaston nur wenig erfahren. Schon als Kind elternlos und auf dem abgelegenen Schlosse ihres Vormunds ganz männlich erzogen, soll sie diesen in allen Reiter- und Jagdkünsten sehr bald übertroffen haben. Da verliebte sich, so hieß es, der unkluge Vormund sterblich in das wunderbare Mädchen, dem schon längst der benachbarte Graf Olivier mit aller schüchternen Schweigsamkeit der ersten Liebe heimlich zugetan war. Um den Vormund zu vermeiden, hatte er,

wie von einem Spazierritt oder vom Jagen zurückkehrend, sich fast jeden Abend, wenn im Schlosse schon alles schlief, unter ihren Fenstern eingefunden, wo sie in der Stille der Nacht, da sie seine zärtlichen Blicke nicht verstand, sorglos und fröhlich mit ihm zu plaudern pflegte. – Jetzt aber, da er eines Abends spät wiederkommt, trifft er zu seinem Erstaunen die Gräfin reisefertig draußen im Garten. Sie verlangt ein Pferd von ihm, sie könne mit dem Vormund nicht länger zusammen wohnen. Überrascht und einen Augenblick ungemessenen Hoffnungen Raum gebend, bietet er ihr sein eigenes Roß an und schwingt sich freudig auf das seines Dieners, der unter den hohen Bäumen am Garten hielt. So reiten sie lange schweigend durch den Wald. Da öffnet ihm die schöne Einsamkeit das Herz, er spricht zum erstenmal glühend von seiner Liebe zu ihr, während sie eben an einem tiefen Felsenriß dahinziehn. Diana, bei seinen Worten erschrocken auffahrend, sieht ihn verwundert von der Seite an, drauf, nach kurzem Besinnen plötzlich ihr Pferd herumwerfend, setzt sie grauenhaft über die entsetzliche Kluft – sein störrisches Pferd bäumt und sträubt sich, er kann nicht nach. Drüben aber hört er sie lachen, und eh' sie im Walde verschwunden, blitzt noch einmal die ganze Gestalt seltsam im Mondlicht auf; es war ihm, als hätt' er eine Hexe erblickt. – So kam sie mitten in der Nacht ohne Begleitung auf dem Landhaus ihrer Tante bei Paris an. Olivier aber hatte wenige Tag darauf seine Güter verlassen und fiel im Auslande im Kriege; man sagt, er habe sich selbst in den Tod gestürzt. Der Tor! dachte Gaston, wer schwindelig ist, jage nicht Gemsen! Es war ihm recht wie Alpenluft bei der Erzählung von der schönen Gräfin, und er freute sich auf das bevorstehende Hoffest, wo er ihr endlich zu begegnen hoffte.

Der Ball bei dem Hofe war schon halb verrauscht, als Gaston, den Besuche, Freunde und alte Erinnerungen auf jedem Schritte aufgehalten hatten, in seinen Domino gewik-

kelt, die Treppen des königlichen Schlosses hinaufeilte. Betäubt, geblendet trat er mitten in der Nacht in das erschreckende Gewirr der Masken, die sich gespenstisch schrillend kreuzten, durchblitzt vom grünen Gefunkel der Kronleuchter und in den Spiegelwänden tausendfach verdoppelt, wie wenn das heidnische Gewimmel von den gemalten Decken der Gemächer plötzlich lebendig geworden und herabgestiegen wäre.

Als er, sich mühsam durchdrängend, endlich den großen Saal erreicht, fiel eben die Musik majestätisch in ein Menuett ein, die tanzfertigen Paare, einander an den Fingerspitzen haltend, verneigten sich feierlich gegen den Eingang, als wollten sie den Eintretenden bewillkommnen, der sich nicht enthalten konnte, die Begrüßung mit einem tiefen Kompliment zu erwidern. Da schwang der Kapellmeister auf dem goldverschnörkelten Chor seine Rolle wieder: ein neuer Akkord, und wie auf einen Zauberschlag mit den taftenen Gewändern auseinanderrauschend, auf den Zehen sich zierlich wendend und wieder verschlingend, wogt' es auf einmal melodisch den ganzen kerzenhellen Saal entlang.

Gaston aber sah wie ein Falk durch die duftende Tanzwolke, denn so oft sie sich teilte, erblickte er im Hintergrunde mitten zwischen den fliegenden Schößen und Reifröcken, gleich einer Landschaft durch Nebelrisse, eine prächtige Zigeunerfürstin, hoch, schlank, mit leuchtendem Schmuck, die Locken aufgeringelt über die glänzenden Schultern.

Und wie er noch so hinstarrend stand, kam sie selber quer durch den Saal und ein Kometenschweif galanter Masken hinter ihr, die ihr eifrig den Hof zu machen schienen. Sie war in seltsamer Geschäftigkeit. Aus ihrem Handkörbchen ein Band aufrollend, schwang sie es plötzlich wie einen Regenbogen über die Verliebten, jeder griff und haschte graziös darnach. Drauf hier und dort durch den Haufen sich schlingend und alle wie mit Zaubersprüchen rasch umgehend, das eine Ende des Bandes fest in der

Hand, schlang sie's behend dem einen um den Hals, dem andern um Arm und Füße, immer schneller, dichter und enger. Die überraschten Liebhaber, Ritter, Chinesen und weise Ägyptier, als sie die unverhoffte Verwickelung gewahr wurden, wollten nun schnell auseinander, aber je zierlicher sie sich wanden und reckten, je unauflöslicher verwirrte sich der Knäul; auf dem glatten Boden ausglitschend, verloren sie Larven, Helme und phrygische Mützen, daß die Haarbeutel zum Vorschein kamen und der Puder umherstob, das Menuett selbst kam aus seiner Balance, man hörte im Saale ein kurzes, anständiges Lachen – die Zigeunerin aber war unterdes in dem Getümmel verschlüpft.

Gaston aber, eh' sich die andern besannen, flog ihr schon nach, aus dem Saal, durch mehrere anstoßende Zimmer. Dort in den Spiegeln ihn hinter sich gewahrend, wandte sie sich einmal nach ihm herum, daß er vor den Augen erschrak, die aus der Larve funkelten. Dann sah er sie durch den Gartensaal schweifen, jetzt trat sie aus der Tür auf die Terrasse und schien plötzlich draußen in der Nacht zu verschwinden, wie ein Elfe, der nur neckend zum flüchtigen Besuch gekommen.

Gaston wollte dennoch seine Jagd nicht aufgeben, wurde aber durch einen ungewöhnlichen Aufruhr der Gesellschaft aufgehalten. Die Masken traten rasch auseinander, ehrfurchtsvoll eine Gasse bildend; der König mit seiner vertrautesten Umgebung nahte, nach allen Seiten sprechend und lachend, unmaskiert in bürgerlicher Kleidung, ein schöner Jüngling voll lebensfrohen Mutwillens, wie damals Ludwig der Fünfzehnte war. „Hütet Euch, Gaston" – sagte er, diesen sogleich an Größe und Haltung erkennend – „dies ist eine gefährliche Räubernacht, es wird mit Augen um Herzen gefochten."

Alle Blicke waren auf den Grafen gerichtet, der nun, die Larve abnehmend, dem König folgen mußte. Sie traten, um sich zu erfrischen, vor den Gartensaal hinaus. Es war eine schwüle Sommernacht, der Himmel halb verdunkelt

von finstern Wolken, aus denen sich die weißen Statuen fast gespenstisch abhoben, tiefer im Garten hörte man eine Nachtigall schlagen, zuweilen blitzte es von fern über den hohen schwarzen Bäumen.
Der König, indem er sich tanzmüde und gähnend unter den Orangenbäumen auf der Terrasse niederließ, wollte zur Unterhaltung von Gaston irgendein Abenteuer seiner Fahrten hören. Diesem, der noch immer zerstreut und unruhig in den Garten schaute, wo die Zigeunerin verschwunden, war bei dem plötzlichen Anblick der stillen Nacht soeben ein seltsamer Vorfall wieder ganz lebendig geworden, und ohne sich lange zu besinnen, erzählte er, wie er auf seiner jetzigen Reise hierher eine alte, verfallene Burg, in der es der Sage nach spuken sollte, aus Neugier besucht und, da es gerade schwüle Mittagszeit, unter den Trümmern im hohen Grase rastend eingeschlummert.
„Gute Nacht, gute Nacht!“ unterbrach ihn der König, „das ist ein schläfriges Abenteuer.“
„Es wird gleich wieder munter, Sire“, entgegnete Gaston, „denn auf einmal, mitten in dieser Einsamkeit, fiel ein Schuß ganz in der Nähe, traumtrunken sehe ich ein Reh getroffen vor mir in den Abgrund stürzen, und wie ich erschrocken aufspringe, steht über mir zwischen den wilden Nelken im zerbrochenen Fensterbogen der Burg eine unbekannte, wunderschöne Frauengestalt auf ihr Gewehr gestützt, die wandte sich nach mir, – den Blick vergesse ich nimmer, gleichwie das Wetterleuchten überm Garten dort!“
Der König lachte: das sei eine Waldfrau gewesen mit dem Zauberblick, von dem die Jäger sprechen, die hab' es ihm angetan.
„Und Sie setzten ihr nicht nach?“ riefen die andern.
„Wohl tat ich das“, erwiderte Gaston, „aber ich konnte so bald über das Gemäuer und Geröll nicht den Eingang finden, und als ich endlich in die Hallen eintrat, war alles still und kühl, nur ein wilder Apfelbaum blühte im leeren Ho-

fe, die Bienen summten drin, kein Vogel sang den weiten Wald entlang – Herr Gott, das ist sie!"
„Wie, unsere Amazone?" rief der König überrascht herumgewendet.
Die Zigeunerin, ihre Larve am Gürtel und vom Streiflicht der Fenster getroffen, trat aus einer der Alleen zu ihnen auf die Terrasse. Gaston war ganz verwirrt, da sie ihm gleich darauf als die Gräfin Diana vorgestellt wurde.
Sie aber, als sie seinen Namen nennen hörte, der so tapfern Klang hatte, sah ihn mit großer, fast scheuer Aufmerksamkeit an. „Wenn ich nicht irre", sagte sie, „so traf ich schon letzthin auf der alten Burg –"
„Ein edles Wild mit Zauberblicken", fiel rasch der König ein. –
„Also auch schon lahm!" erwiderte sie halb für sich und wandte plötzlich dem Grafen verächtlich den Rücken. – Die Umstehenden blickten ihn schadenfroh an, Gaston aber lachte wild und kurz auf und verschwor sich innerlich, die Stolze zu demütigen, und sollt' er auf den Zinnen von Notre-Dame mit ihr den Tanz wagen!
Über des Königs Stirn aber flog eine leichte Röte, denn er hegte seit Gastons Anwesenheit in Paris insgeheim den Wunsch, ihn mit Diana zu verbinden. Etwas verstimmt, um nur die plötzlich eingetretene peinliche Stille zu unterbrechen, fragte er Diana: ob sie denn so allein im Garten nicht fürchte, daß sie entführt werde? – Sie lachte: der König habe alles zahm gemacht, sie hätte nur Grillen gefunden in den Hecken; die zirpten lieblich, dort wie hier. – Gaston meinte: die Gräfin habe ganz recht, solche Grillenhaftigkeit sei nicht gefährlich, und, mache auch manche noch so weite Sprünge, jeder wackere Bursch überhole sie leicht. – Diana schüttelte die Locken aus der Stirn; es verdroß sie doch gerade von ihm, daß er ihr trotzte. Und da einer der Kammerherren, um wieder einzulenken, soeben zirpte: selbst die Heimchen brächten ihr Ständchen, wenn sie träumend durch den nächtlichen Garten ginge, erwiderte sie rasch in heimlicher Aufregung: „Wahrhaf-

tig, mir träumte, der Tag mache der Nacht den Hof, er duftete nach Jasmin und Lavendel, blond, artig, lau, etwas lispelnd, mit kirschblütenen Manschetten und Hirtenflöte, ein guter, langweiliger Tag." – Man lachte, keiner bezog es auf sich; ein Vicomte, als Troubadour die Zither im Arme, sagte zierlich: „Aber die keusche Nacht wandelte unbekümmert fort, ihren Elfenreihen ätherisch dahinschwebend." – „Nein", entgegnete Diana, indem sie ihm in ihrer wunderlichen Laune die Zither nahm und, sich auf das Marmorgeländer der Terrasse setzend, zur Antwort sang:

„Sie steckt' mit der Abendröte
In Flammen rings das Land
Und hat samt Manschetten und Flöte
Den verliebten Tag verbrannt.

Und als nun verglommen die Gründe!
Sie stieg auf die stillen Höh'n,
Wie war da rings um die Schlünde
Die Welt so groß und schön!

Waldkönig zog durch die Wälder
Und stieß ins Horn vor Lust,
Da klang über die stillen Felder,
Wovon der Tag nichts gewußt. –

Und wer mich wollt' erwerben,
Ein Jäger müßt's sein zu Roß,
Und müßt' auf Leben und Sterben
Entführen mich auf sein Schloß!"

Hier gab sie lachend die Zither zurück. Gaston aber bei der plötzlichen Stille erwachte wie aus tiefen Gedanken. „Und wenn es wirklich einer wagte?" sagte er rasch in einem seltsamen Tone, daß es allen auffiel. – „Wohlan, es gilt", fiel da der junge König ein, „ich trete der Herausfor-

derung der Gräfin als Zeuge und Kampfrichter bei, ihr alle habt's gehört, welchen Preis sie dem Entführer aussetzt."

Diana stand einen Augenblick überrascht. „Und verspielt der Vermessene?" fragte sie dann ernst. – – „So wird er tüchtig ausgelacht", erwiderte der König, „wie ein Nachtwandler, der bei Mondschein verwegen unternimmt, wovor ihm bei Tage graut."

Mit diesen Worten erhob er sich, und im Vorbeigehen dem Grafen noch leise zuflüsternd: „Wenn ich nicht der König wär', jetzt möcht' ich Gaston sein!" wandte er sich, wie über einen herrlich gelungenen Anschlag lebhaft die Hände reibend, durch den Gartensaal in die innern Gemächer. – Diana aber schien anders bei sich zu beschließen, sie folgte zürnend.

Jetzt umringten die Hofleute von allen Seiten den Grafen, ihm zu dem glänzenden Abenteuer, wie einem verzauberten Prinzen und Feenbräutigam, hämisch Glück wünschend. Die übrige Gesellschaft unterdes, da der König sich zurückgezogen, strömte schon eilig nach den Türen, die Masken hatten ihre Larven abgenommen und zeigten überwachte, nüchterne Gesichter, durch die Säle zwischen den wenigen noch wankenden Gestalten strich die Langeweile unsichtbar wie ein böser Luftzug.

Gaston blieb nachdenklich am offenen Fenster, bis alles zerstoben. Er sah sich hier unerwartet durch leichtsinnige Reden, die anfänglich nur ein artiges Spiel schienen, plötzlich seltsam und unauflöslich verwickelt. Es war ihm wie eine prächtige Nacht, vor der eine marmorkalte Sphinx lag, er mußte ihr Rätsel lösen oder sie tötete ihn.

Währenddes war Diana schon in ihrem Schlafgemache angelangt. Als sie in dem phantastischen Ballschmuck eintrat, erstaunte die Kammerjungfer von neuem und rief fast erschrocken aus: wie sie so wunderschön! Die Gräfin verwies es ihr unwillig, das sei ein langweiliges Unglück. Und da das Mädchen drauf ihr Befremden äußerte, daß sie durch solche Härte so viele herrliche Kavaliere in Gefahr

und Verzweiflung stürze, erwiderte Diana streng: „Wer nimmt sich meiner an, wenn diese Kavaliere bei Tag und Nacht mit Listen und Künsten bemüht sind, mich um meine Freiheit zu betrügen?" –
Draußen aber rollten indes die Wagen noch immer fort, jetzt flog das rote Licht einer Fackel über die Scheiben, in dem wirren Widerschein der Windlichter unten erblickte sie noch einmal flüchtig den Gaston, wie er eben sein Pferd bestieg, die Funken stoben hinter den Hufen, sie sah ihm gedankenvoll nach, bis er in der dunkeln Straße verschwunden. Dann, vor den Wandspiegel tretend, löste sie die goldne Schlange aus dem Haar, die schwarzen Locken rollten tief über die Schultern hinab, ihr schauerte vor der eigenen Schönheit.

Kurze Zeit nach diesem Feste war der Hof fern von Paris zum Jagen versammelt. Da ging das Rufen der Jäger, Hundegebell und Waldhornsklang wie ein melodischer Sturmwind durch die stillen Täler, breite, ausgehauene Alleen zogen sich geradlinig nach allen Richtungen hin, jede an ihrem Ende ein Schloß oder einen Kirchturm in weiter Ferne zeigend. Jetzt brachte die Luft den verworrenen Schall immer deutlicher herüber, immer näher und häufiger sah man geschmückte Reiter im Grün aufblitzen, plötzlich brach ein Hirsch, das Geweih zurückgelegt, aus dem Dickicht in weiten Sätzen quer über eine der Alleen und ein Reiter leuchtend hinterdrein, mit hohen, steifen Jagdstiefeln, einen kleinen dreieckigen Tressenhut über den gepuderten Locken, in reichgesticktem grünen Rock, dessen goldbordierte Schöße weit im Winde flogen – es war der junge König. – „Das ist heute gut' Jagdwetter, man muß es rasch nutzen!" rief er flüchtig zurückgewandt zu Gaston herüber, der im Gefolge ritt. Gaston erschrak, er wußte wohl, was der König meinte.
Diana aber fehlte im Zuge, sie war zuletzt auf einer der entfernteren Waldhöhen gesehen worden. Des Treibens

müde und ohne jemandem von ihrem Vorhaben zu sagen, hatte sie sich mitten aus dem Getümmel nach einem nahegelegenen, ihr gehörigen Jagdschloß gewendet; denn sie kam sich selber als das Wild vor auf dieser Jagd, auf das sie alle zielten. Es war das Schloß, wo sie als Kind gelebt, sie hatte es lange nicht mehr besucht. Die Nacht war schon angebrochen, als sie anlangte, niemand erwartete sie dort, alle Fenster waren dunkel im ganzen Hause, als ständ' es träumend mit geschlossenen Augen. Und da endlich der erstaunte Schloßwart, mit einem Windlicht herbeigeeilt, die alte schwere Türe öffnete, gab es einen weiten Schall durch den öden Bau, draußen schlug soeben die Uhr vom Turme, als wolle sie mit dem wohlbekannten Klange grüßen.

Diana, fast betroffen oben im Saale umherblickend, öffnete rasch ein Fenster, da rauschten von allen Seiten die Wälder über den stillen Garten herauf, daß ihr das Herz wuchs. Mein Gott, dachte sie, wo bin ich denn so lange gewesen! O wunderschöne Einsamkeit, wie bist du kühl und weit und ernst und versenkst die Welt und baust dir in den Wolken drüber Schlösser kühn wie auf hohen Alpen. Ich wollt', ich wäre im Gebirg', ich stieg' am liebsten auf die höchsten Gipfel, wo ihnen allen schwindelte, nachzukommen – ich tu's auch noch, wer weiß wie bald!

Unterdes war das Nötigste zu ihrer Aufnahme eingerichtet, jetzt wurde nach und nach auch im Schlosse alles wieder still, sie aber konnte lange nicht einschlafen, denn die Nacht war so schwül, und in den Fliederbüschen unter den Fenstern schlugen die Nachtigallen und das Wetter leuchtete immerfort von fern über dem dunkeln Garten.

Als Diana am folgenden Morgen erwachte, hörte sie draußen eine kindliche Stimme lieblich singen. Sie trat rasch ans Fenster. Es war noch alles einsam unten, nur des Schloßwarts kleines Töchterlein ging schon geputzt den stillen Garten entlang, singend, mit langem blondem

Haar, wie ein Engel, den der Morgen auf seinem nächtlichen Spielplatz überrascht. Bei diesem Anblick flog eine plötzliche Erinnerung durch ihre Seele, wie einzelne Klänge eines verlorenen Liedes, es hielt ihr fast den Atem an, sie bedeckte die Augen mit beiden Händen und sann und sann – auf einmal rief sie freudig: „Leontine!"

Da sprang sie schnell auf, es fiel ihr ein, daß die Marquise Astrenant mit ihrer Tochter ja nur wenige Meilen von hier wohnte. Sie setzte sich gleich hin und schrieb an Leontine. Sie erinnerte sie an die schöne Morgenstille ihrer gemeinschaftlichen Jugendzeit, wo sie immer die kleine Elfe genannt wurde wegen ihrer langen blonden Locken, wie sie da in diesem Garten hier als Kinder wild und fröhlich miteinander gespielt und seitdem eines das andere nicht wiedergesehen. Sie werde sie auch nicht mehr schlagen oder im Sturm auf dem Flusse unterm Schlosse mit ihr herumfahren wie damals. Sie solle nur eilig herüberkommen, so wollten sie wieder einmal ein paar Tage lang zusammen sich ins Grüne tauchen und nach der großgewordenen Welt draußen nichts fragen. – Diese Aussicht hatte sie lebhaft bewegt. Sie klingelte und schickte noch in derselben Stunde einen Boten mit dem Brief nach dem Schlosse der Marquise ab.

Darauf ging sie in den Garten hinab. Sie hätte ihn beinahe nicht wiedererkannt, so verwildert war alles, die Hecken unbeschnitten, die Gänge voll Gras, weiterhin nur glühten noch einige Päonien verloren im tiefen Schatten. Da fiel ihr ein Lied dabei ein:

„Kaiserkron' und Päonien rot,
Die müssen verzaubert sein,
Denn Vater und Mutter sind lange tot,
Was blühn sie hier so allein?"

Jetzt sah sie sich nach allen Seiten um, sie kam sich selber wie verzaubert vor zwischen diesen stillen Zirkeln von Buchsbaum und Spalieren. Die Luft war noch immer

schwül, in der Ferne standen Gewitter, dazwischen stach die Sonne heiß, von Zeit zu Zeit glitzerte der Fluß, der unten am Garten vorüberging, heimlich durch die Gebüsche herauf. Es war ihr, als müßte ihr heut was Seltsames begegnen, und die stumme Gegend mit ihren fremden Blikken wollte sie warnen. Sie sang das Lied weiter:

„Der Springbrunnen plaudert noch immerfort
Von der alten schönen Zeit,
Eine Frau sitzt eingeschlafen dort,
Ihre Locken bedecken ihr Kleid.

Sie hat eine Laute in der Hand,
Als ob sie im Schlafe spricht,
Mir ist, als hätt' ich sie sonst gekannt –
Still, geh' vorbei und weck' sie nicht!

Und wenn es dunkelt das Tal entlang,
Streift sie die Saiten sacht,
Da gibt's einen wunderbaren Klang
Durch den Garten die ganze Nacht."

„Ich weckte sie doch", sagte sie, „wenn ich sie so im Garten fände, und spräch' mit ihr."
Unterdes aber waren die Wolken von allen Seiten rasch emporgestiegen, es donnerte immer heftiger, die Bäume im Garten neigten sich schon vor dem voranfliegenden Gewitterwinde. Die schwülen Traumblüten schnell abschüttelnd, blickte sie freudig in das Wetter. Da gewahrte sie erst dicht am Abhang den alten Lindenbaum wieder, auf dem sie als Kind so oft gesessen und vom Wipfel die fernen weißen Schlösser weit in der Runde gezählt. Er war wieder in voller Blüte, auch die Bank stand noch darunter, deren künstlich verflochtene Lehne fast bis an die ersten Äste reichte. Sie stieg rasch hinauf in die grüne Dämmerung, der Wind bog die Zweige auseinander. Da rollte sich plötzlich rings unter ihr das verdunkelte Land auf, der

Strom, wie gejagt von den Blitzen, schoß pfeilschnell daher, manchmal klangen von fern die Glocken aus den Dörfern, alle Vögel schwiegen, nur die weißen Möwen über ihr stürzten sich jauchzend in die unermeßliche Freiheit – sie ließ vor Lust ihr Tuch im Sturme mit hinausflattern.

Auf einmal aber zog sie es erschrocken ein. Sie hatte einen fremden Jäger im Garten erblickt. Er schlich am Rande der Hecken hin; bald sachte vorgebogen, bald wieder verdeckt von den Sträuchern, keck und doch vorsichtig, schien er alles ringsumher genau zu beobachten. Sie hielt den Atem an und sah immerfort unverwandt hin, wie er, durch die Stille kühn gemacht, nun hinter dem Gebüsch immer näher und näher kam; jetzt, schon dicht unter dem Baume, trat er plötzlich hervor – sie konnte sein Gesicht deutlich erkennen. In demselben Augenblick aber hörte er eine Tür gehn im Schlosse und war schnell im Grünen verschwunden.

Diana aber, da alles wieder still geworden, glitt leise vom Baume; darauf, ohne sich umzusehen, stürzte sie durch den einsamen Garten die leeren Gänge entlang nach dem Schlosse, die eichene Tür hinter sich zuwerfend, als käme das Gewitter hinter ihr, das nun in aller furchtbaren Herrlichkeit über den Garten ging.

Sie achtete aber wenig darauf. In großer Aufregung im Saale auf und nieder gehend, schien sie einem Anschlag nachzusinnen. Manchmal trat sie wieder ans Fenster und blickte in den Garten hinab. Da sich aber unten nichts rührte als die Bäume im Sturn, nahm sie ein paar Pistolen von der Wand, die sie sorgfältig lud; dann setzte sie sich an den goldverzierten Marmortisch und schrieb eilig mehrere Briefe. Und als das Wetter draußen kaum noch gebrochen, wurden im Hofe gesattelte Pferde aus dem Stalle geführt, und bald sah man reitende Boten nach allen Richtungen davonfliegen.

Gleich darauf aber rief sie ihr ganzes Hausgesinde zusammen. Sie mußten schnell herbeischaffen, was die Vorräte

vermochten, Wild, Früchte, Wein und Geflügel. Einer der Jäger, dessen Vater einst Küchenmeister gewesen, verstand sich noch am besten unter ihnen auf den guten Geschmack und mußte, zu allgemeinem Gelächter, eine weiße Schürze vorbinden und den Kochlöffel statt des Hirschfängers führen. Bald loderte ein helles Feuer im Kellergeschoß, die halb verrosteten Bratspieße drehten sich knarrend in der alten verödeten Küche, überall war ein lustiges Plaudern und Getümmel. Alle guten Stühle und Kanapees aber ließ die Gräfin oben in den großen Saal zusammentragen, Spieltische wurden zurechtgerückt und in der Mitte des Saales eine lange Tafel gedeckt. Die feierlichen Anstalten hatten fast etwas Grauenhaftes in dieser Einsamkeit, als sollten die Ahnenbilder, die mit ihren Kommandostäben ernst von den Wänden schauten, sich zu Tische setzen, denn niemand wußte sonst, wer die Gäste sein sollten.

So war in seltsamer Unruhe der Abend gekommen und das Gewitter lange vorbei, als Diana allein mit ihrer Kammerjungfer unten in das Gartenzimmer trat, die sich beim Hereintreten rasch und verstohlen nach allen Seiten umsah. Sie hatte, ohne zu wissen zu welchem Zweck, das schöne Kleid anziehen müssen, das die Gräfin heute getragen; das hinderte sie, es war überall zu knapp und zu lang. Sie ging vor den Spiegel, als wollte sie sich's zurechtrükken, ihre Blicke aber schweiften seitwärts durchs Fenster, und als Diana sich einmal wandte, benutzte sie's schnell und schien zornig jemanden in den Garten hinauszuwinken. Die Gräfin, sie an ihre Verabredung erinnernd, hieß sie vom Fenster wegtreten, ordnete rasch noch die Locken des Mädchens und setzte ihr ihren eigenen Jagdhut auf. Dann, die Verkleidete von allen Seiten zufrieden musternd, schärfte sie ihr nochmals ein, sich in diesem Zimmer still zu verhalten und nicht in den Garten zu gehen, bis sie draußen dreimal leise in die Hände klatschen höre, denn es dunkele schon, und die Nacht habe wilde Augen. – „Wo?" rief das ganz zerstreute Mädchen heftig er-

schrocken. – Aber Diana, eilig wie sie war, bemerkte es nicht mehr; heftig einen Jägermantel umwerfend, der über dem Stuhle lag, und einen Männerhut tief in die Augen drückend, flog sie in den dämmernden Garten hinaus.

Kaum aber war sie verschwunden, so sprang die Kammerjungfer geschwind ans Fenster. „Aber, Robert, bist du denn ganz toll!“ rief sie einem fremden Jäger entgegen, der schon längst draußen im Gebüsch steckte und nun rasch hinzutrat. – „I Gott bewahre, hast du mich doch erschreckt!“ entgegnete dieser, sie erstaunt vom Kopf bis zu den Füßen betrachtend, „das ist ja ganz wie deine Gräfin!“ – Das Mädchen aber nannte ihn einen Unverschämten, daß er sie hier auf dem Lande besuche; wenn die Gräfin ihn sähe, sei es um ihren Dienst geschehen, er solle auf der Stelle wieder fort. – „Nicht eher“, erwiderte der eifersüchtige Liebhaber, „bis ich weiß, wer der Mann war, der soeben von dir ging.“ – Da lachte sie ihn tüchtig aus: er sei ein rechter Jäger, der auf dem Anstand das Wild verwechsele, es sei ja die Gräfin selber gewesen. – „So?“ – sagte Robert sehr überrascht und einen Augenblick in Nachsinnen versunken. Dann plötzlich mit leuchtenden Blicken fragte er hastig: warum denn die Gräfin sich verkleidet, wohin sie ginge, ob sie diesen Abend in dem Mantel bleibe? Aber das ungeduldige Mädchen, in wachsender Furcht, drängte ihn statt aller Antwort schon von der Schwelle über die Stufen hinab. Er gab ihr noch schnell einen Kuß, dann sah sie ihn freudig über Beete und Sträucher fortspringen.

Als sie wieder allein war, fiel ihr erst die seltsame Hast und Neugierde des Jägers aufs Herz; es überflog sie eine große Angst, daß sie in der Verwirrung die Verkleidung der Gräfin ausgeplaudert. Auch schreckte sie nun in dieser Stille die aufsteigende Nacht im Garten, es war ihr, als blickten wirklich überall wilde Augen aus dem Dunkel auf sie, manchmal glaubte sie gar Stimmen in der Ferne zu hören. Sie konnte durchaus nicht erraten, was es geben soll-

te, und verwünschte tausendmal ihre Liebschaften und die unbegreiflichen Einfälle der Gräfin und das ganze dumme Landleben mit seiner spukhaften Einsamkeit.

Ein tiefes Schweigen bedeckte nun schon alle Gründe, nur fern im Garten war noch ein heimlich Knistern und Wispern überall zwischen den Büschen, als zög' eine Zwerghochzeit unsichtbar über die stillen Beete hin, von Zeit zu Zeit funkelte es aus den Hecken herüber wie Waffen oder Schmuck. Dann hörte man von der andern Seite eine Zither anschlagen und eine schöne Männerstimme sang:

„Hörst du die Gründe rufen
In Träumen halb verwacht?
O, von des Schlosses Stufen
Steig nieder in die Nacht! –"

Drauf alles wieder still, nur eine Nachtigall schlug in dem blühenden Lindenbaum am Abhange. Auf einmal raschelt was, eine schlanke Gestalt schlüpft droben aus dem Gebüsch. Es war Diana, in ihren Jägermantel dicht verhüllt, die über den Rasen nach dem Schlosse ging. Tiefer im Garten sang es von neuem:

„Die Nachtigallen schlagen,
Der Garten rauschet sacht,
Es will dir Wunder sagen
Die wunderbare Nacht."

Jetzt stand Diana vor der Tür des Gartenzimmers und klatschte dreimal leise in die Hand. In demselben Augenblick aber sieht sie auch schon zwei dunkle Gestalten zwischen den Bäumen vorsichtig hervortreten, – „Bist du es, Robert?" und „wo ist sie?" flüsterte der eine dem andern leise zu.
Sie zogen sich tiefer in den Garten zurück. Da sah sie, wie

die Kammerjungfer auf das verabredete Zeichen oben aus dem Hause getreten, die eine Gestalt schien sich ihr zu nähern. – Diana triumphierte schon im Herzen, als jetzt plötzlich der andere gerade auf ihr Versteck losschritt. Bei dieser unerwarteten Wendung flog sie erschrocken über den Rasenplatz den Gartenberg hinab, seitwärts sah sie den Fremden bei ihrem Anblick rasch durch die Hecken brechen, als wollt' er ihr den Vorsprung abgewinnen, sie verdoppelte ihre Eile, schon glaubte sie unten Bekannte zwischen den Bäumen zu erblicken, jetzt trat sie atemlos am Fuß des Berges aus dem Garten, zu gleicher Zeit aber war auch der Fremde angelangt und vor ihr stand Graf Gaston. Hut und Mantel waren ihr im Gebüsch entfallen, Gaston, rasch die Zither wegwerfend, blickte ihr lächelnd in die Augen. – „Ihr seid der kühnste Freier, den ich jemals sah", sagte sie nach einem Weilchen finster. Gaston küßte feurig ihre Hand, die er nicht wieder losließ. Vor ihnen aber, vom Gesträuche halb verdeckt, stand ein leichter Wagen mit vier Pferden, die Kutscher in den Sätteln, die Pferde schnaubend, scharrend, alles wie ein Pfeil auf gespanntem Bogen, der eben losschnellen will.

Indem aber, wie Gaston, den Kutschern winkend und ihr ehrerbietig den Arm reichend, sie in den Wagen heben will, sieht er, daß sie, einige Schritte zurückgetreten, mit einem Pistol nach ihm zielt. Er stutzt, sie aber lacht und feuert das Pistol in die Luft. Da bei dem Knall, wie ein Schwarm verstörter Dohlen, brechen plötzlich seitwärts aus allen Hecken Gestalten mit Haarbeuteln, Staubmänteln und gezückten Stahldegen. Gaston erkennt sogleich mit Erstaunen die alten Gesichter aus der Residenz, alles jubelfröhlich, siegesgewiß.

„Fahrt zu!" ruft er da, ohne sich zu bedenken, den Kutschern zu, die nun, ihre Peitschen schwingend, gerade in den glänzenden Schwarm hineinjagen, der sogleich von allen Seiten lachend den Wagen umringt, um die vermeintlich Entführte daraus zu erlösen. Gaston und Diana aber standen währenddes dicht am Bergstrom, der unter

dem Garten vorbeischoß, ein Kahn lag dort am Ufer angebunden. Der Graf, eh' Diana sich besinnt, schwingt sie hoch auf dem Arm in den Nachen, zerhaut mit seinem Hirschfänger das Tau und lenkt rasch mitten ins Fahrwasser; so flogen sie, bevor noch die am Wagen es gewahr wurden, in der entgegengesetzten Richtung pfeilschnell den Fluß hinab.

Er selbst war es gewesen, den Diana am Morgen vom Lindenbaum umherspähend erblickt. Da zweifelte sie keinen Augenblick länger, daß er sein verwegenes Vorhaben in der folgenden Nacht auszuführen gedenke. Ihr Anschlag war schnell gefaßt. Voll Übermut lud sie durch vertraute Boten sogleich das ganze Hoflager zu Entführung und Abendbrot herüber, die einzeln und ohne Aufsehen eingetroffenen Hofleute wurden am Wege versteckt, Gaston in der Verwirrung und Dunkelheit sollte, statt ihrer, das verkappte Kammermädchen entführen und so vor den Augen des hervorbrechenden Hinterhalts doppelt beschämt werden. – Nun aber hatte die unzeitige Liebschaft des Mädchens und Dianas eigene Unbesonnenheit im entscheidenden Augenblick plötzlich alles anders gewendet! Schon waren Schloß und Garten hinter den Fortschiffenden dämmernd versunken, immer ferner und schwächer nur hörte man von dorther noch verworrenes Rufen, Schüsse und Hörnersignale der bestürzten Hofleute, die sich wie durch eine unbegreifliche Verzauberung auf einmal in allen Plänen gekreuzt sahen und nun die auf Gaston geladenen Witze verzweifelt gegeneinander selbst abschossen.

Der Fluß indes ging rasch durch wüsten Wald. Diana wußte recht gut, daß hier kein Haus und keine menschliche Hilfe in der Nähe war; so saß sie still am Rande des Kahns und schaute vor sich in die Flut, die von Zeit zu Zeit in Wirbeln dunkel aufrauschte. Gaston aber, wohl fühlend, daß in dieser unerhörten Lage alle gewöhnliche Galanterie und Entschuldigung nur lächerlich und in den Wind gesprochen sei, blieb gleichfalls stumm; und so glit-

ten sie lange Zeit schweigend zwischen stillen Wäldern und Felsenwänden durch die tiefe Einsamkeit der Nacht, während der Graf immerfort Dianas Spiegelbild im mondbeschienenen Wasser vor sich sah, als zöge eine Nixe mit ihnen neben dem Schiffe.

Endlich, um nur die unerträgliche Stille zu brechen, sagte er, als wäre nichts geschehen: alles hier erinnere ihn wunderbar an eine Sage seiner Heimat. Da stehe im Schloßgarten ein marmornes Frauenbild und spiegele sich in einem Weiher. Keiner wage es, in stiller Mittagszeit vorbeizugehen, denn wenn die Luft linde kräuselnd übers Wasser ging und das Spiegelbild bewegte, da sei's, als ob es sachte seine Arme auftät.

Diana, ohne ein Wort zu erwidern, fuhr unwillig mit der Hand über das Wasser, daß alle Linien ihres Bildes drin durcheinanderlaufend im Mondesflimmer sich verwirrten.

„Von diesem Bilde", fuhr Gaston fort, „geht die Rede, daß es in gewissen Sommernächten, wenn alles schläft und der Vollmond, wie heut, über die Wälder scheint, von seinem Steine steigend, durch den stillen Garten wandle. Da soll sie mit den alten Bäumen und den Wasserkünsten in fremder Sprache reden, und wer sie da zufällig erblickt, der muß in Liebesqual verderben, so schön ist die Gestalt."

„Was ist das für ein Turm dort überm Walde?" rief hier Diana, sich plötzlich aufrichtend, daß er zusammenschrak, als hätt' er selbst das Marmorbild erblickt, von dem er sprach – es waren ihre ersten Worte. Er sah sich verwundert nach allen Seiten um, weiterhin schien sich die Schlucht zu öffnen, durch eine Waldlichtung erblickte er wirklich schon flüchtig den Turm seines Jagdschlosses, tiefer unten den Fahrweg, der in weiten Umkreisen um das Gebirge ging; dort hatte er seine Leute vom Schloß zum Empfange hinbestellt. Gleich darauf aber verdeckten Felsen und Bäume alles wieder, und der Fluß wandte sich von neuem. Gaston, der das abgelegene Schloß selten be-

sucht, kannte die Umgebung nur wenig, er stand einen Augenblick verwirrt, und wußte nicht, an welchem Ufer er landen sollte.

Da bemerkte er rechts den Schimmer eines kleinen Feuers ungewiß durch die Büsche. Da sind sie, dachte er, und lenkte darauf hin. Der Kahn stieß hart ans Land; indem er aber, schon am Ufer, das Gestrüpp auseinanderbog, um der Gräfin Platz zu schaffen, stieß diese, eh' er's hindern konnte, im Heraussteigen den Nachen weit hinter sich ab, der nun unwiederbringlich mit dem reißenden Strom forttrieb. Gaston sah sie überrascht an, sie blickte funkelnd nach allen Seiten in der schönen Nacht umher.

So standen sie an einem wildumzirkten Platz, Bäume, Fels und altes Bauwerk wirr durcheinander gewachsen. Es war, wie er beim Mondlicht erkannte, eine verfallene, unbewohnte Wassermühle, hinten, wie ein Schwalbennest, an die hohe, unersteigliche Felsenwand gehängt, von zwei andern Seiten vom schäumenden Fluß umgeben. Von dort zwischen Unkraut und Gebälk kam der Lichtschein her, den er vom Strom gesehen; er trat eilig mit Diana in das wüste Gehöft, voll Zuversicht, die Seinigen zu treffen. Wie groß aber war sein Erstaunen, da er den Platz leer fand, nur einzelne blaue Flämmchen zuckten noch aus der halb verloschenen Brandstätte, als wäre sie eben von Hirten verlassen worden. –

„Ist das Ihr Schloß?" fragte Diana höhnend. Gaston aber, der einen zerbrochenen Fensterladen im Winde klappen hörte, war schon ins Haus gegangen. Dort durch die Öffnung schauend, gewahrte er zu seinem Schrecken erst, daß er auf dem falschen Ufer gelandet, drüben hinter den dunkeln Wipfeln lag sein Jagdschloß im prächtigen Mondschein – nun wußte er's auf einmal, warum Diana den Nachen zurückgestoßen!

In dieser Verlegenheit zog er schnell ein Pistol unter seinem Mantel hervor und feuerte es in die Nacht ab, ein Reh fuhr nebenan aus dem Dickicht, man konnte seinen Hufschlag noch weit durch den stillen Waldgrund hören. Zu-

gleich aber gab zu seiner großen Freude ein Schuß drüben Antwort, bald wieder einer, und drauf ein Schreien und Rufen vom Felde, daß fern in den Dörfern die Hunde anschlugen. Schon glaubte er einige der Stimmen zu erkennen und wollte eben ein zweites Pistol abschießen, als er auf einmal ein seltsames Knistern und Blinken in allen Ritzen des alten Hauses bemerkte. „Um Gottes willen, da schlagen Flammen auf!" schrie er, entsetzt hinausstürzend, der einzige Ausgang zum Walde brannte schon lichterloh – Diana, da sie bei dem Herannahen der Signale und Stimmen keine Rettung mehr sah, hatte das Haus an allen vier Ecken angezündet. Jetzt erblickte er die Schreckliche selbst hoch auf dem hölzernen Balkon der Mühle, gerade über dem Strom. Da sie ihn gewahrte, wandte sie sich schnell herum, es war wieder jenes Wetterleuchten des Blicks, das ihn schon einmal geblendet. – „Komm nun und hol' die Braut!" rief sie ihm wild durch die Nacht zu, „das Brautgemach ist schon geschmückt, die Hochzeitsfackeln brennen."

Unterdes aber züngelten einzelne Flammenspitzen schon hier und da durch die Fugen, der heiße Sommer hatte alles gedörrt, das Feuer, im Heidekraut fortlaufend, kletterte hurtig in dem trockenen Gebälk hinauf und der Wind faßte lustig die prächtigen Lohen, und von drüben kam das Rufen und Schießen rasch immer näher und lauter, und: „hol' deine Braut!" frohlockte Diana wieder dazwischen. – Da, ohne hinter sich zu blicken, stürzte Gaston durch den wirbelnden Rauch die brennende Treppe hinan. „Zurück, rühr' mich nicht an!" rief ihm Diana entgegen, „wer hieß dich mit Feuer spielen, nun ist's zu spät, wir beide müssen drin verderben!"

Aber die Funken von den Kleidern stäubend, stand er schon droben dicht bei ihr; am Ufer brannte ein schlanker Tannenbaum vom Wipfel bis zum Fuß, die schöne Gestalt und die stille Gegend beleuchtend. Gaston blickte ratlos in der Verwüstung umher, es schien keine Hilfe möglich, die Balken stürzten rings schon krachend in die Glut zu-

sammen, hinten die steile Felswand und unter ihnen der Strom, in dem der Brand sich gräßlich spiegelte.
Indem aber hat das Feuer die dürren Wurzeln der Tanne zerfressen und, wie das Gerüst eines abgebrannten Feuerwerks allmählich verdunkelnd und sich neigend, sinkt der Baum prasselnd quer über den wütenden Felsbach. Da faßt Gaston, der alles ringsher scharf beachtet, plötzlich Dianas Hand, schwingt sie selbst, ehe sie sich das versieht, auf seinen Arm, und, seinen Mantel um sie schlagend, mit fast übermenschlicher Gewalt, trägt er die Sträubende mitten durch die Flammen über die grauenvolle Brücke, unter der der Fluß wie eine feurige Schlange dahinschoß. Jetzt hat er, aus dem furchtbaren Bezirk tretend, glücklich das jenseitige Ufer erreicht und schleudert den brennenden Mantel hinter sich in den Fluß. Diana, plötzlich Stirn und Augen enthüllt, wandte sich von ihm ab in die Nacht. „Sieh mich nicht so an", sagte sie, „du verwirrst mir der Seele Grund." – Da hörte er auf einmal auch die Stimmen wieder im Felde, mehrere Gestalten schwankten fern durch den Mondschein; es waren seine Leute, die, der Verabredung gemäß, am Fahrwege auf ihn gewartet und nun ganz erstaunt herbeieilten, da sie den Herrn auf dem Wege vom Fluß erkannten. „Zum Schloß!" rief ihnen Gaston zu, und alle Kräfte noch einmal zusammenraffend, trug er seine Beute rasch den Gartenberg hinan; schon schimmerten rechts und links ihm altbekannte Plätze entgegen, jetzt teilten sich die alten Bäume, und vor ihnen ernst und dunkel lag das stille Haus; da ließ er erschöpft die Gräfin auf den steinernen Stufen vor der Schloßtür nieder. Von drüben aber beleuchtete der Brand taghell Garten und Schloß und Dianas grausame Schönheit; Gaston schüttelte sich heimlich vor Grausen.
Indem waren auch die Diener, entschuldigend, fragend und erzählend, von allen Seiten herbeigekommen. Der Graf, ohne ihrer Neugier Rede zu stehen, befahl ihnen, rasch die Türen zu öffnen und die Kerzen anzuzünden, er schien in seinem ganzen Wesen auffallend verändert, daß

sie sich fast vor ihm fürchteten. Darauf der Gräfin seinen Arm reichend, indem er sie in das unterdes geöffnete Schloß führte, sagte er mit glatter, seltsamer Kälte zu ihr: die Aufgabe sei gelöst und die wunderliche Wette entschieden, sie möge nun ausruhen und Schloß, Garten, Diener und Wildbahn hier ganz als die ihrigen betrachten. Und so, ohne ihre Antwort abzuwarten, ließ er sie im kerzenhellen Saale allein.

Draußen aber, in großer Aufregung, hieß er schnell alle Gemächer reinigen und schmücken und ordnete, zu allgemeiner Verwunderung der Diener, sogleich alles zu einem glänzenden Feste an. Die Jäger flüsterten mit verbissenem Lachen heimlich untereinander, der eine winkte schlau mit den Augen nach der schönen Fremden im Saale. Gaston, der es bemerkte, faßte ihn zornig an der Brust und schwor jedem den Tod, der der Gräfin drin, als ihrer Herrin, nicht ehrfurchtsvoll und pünktlich wie ihm selber diente.

Drauf ließ er sein Pferd satteln und ritt noch dieselbe Stunde fort, niemand wußte wohin.

Auf dem Schlosse der Marquise Astrenant ging seit jener Räuberjagd gar mancherlei Gerede. Den Anführer der Räuber, hieß es, habe von dem Augenblick, da Graf Gaston ihn vom Felsen gestürzt, niemand mehr wiedergesehen, nur eine blutige Fährte hätten sie beim Verfolgen bemerkt, die führte endlich zwischen ungangbaren Klippen in einen Abgrund, wo keiner hinabkommt, da habe er ohne Zweifel in dem Felsstrom unten seinen wohlverdienten Tod gefunden. – Leontine wußt' es wohl besser, aber das Geheimnis wollt' ihr das Herz abdrücken.

In den Wäldern war es unterdes schon lange wieder still geworden, über den wilden Garten vor dem Schlosse schien soeben die untergehende Sonne, die Luft kam vom Tale, man hörte die Abendglocken weither durch die schöne Einsamkeit herüberklingen. Da stand Leontine,

wie damals, zwischen den Hecken und fütterte wieder ihr Reh und streichelte es und sah ihm in die klaren, unschuldigen Augen. „Deine Augen sind ohne Falsch", sagte sie schmeichelnd zu ihm, „du bist mir treu, wir wollen auch immer zusammenbleiben hier zwischen den Bergen, es fragt ja doch niemand draußen nach uns." Und da die Vögel so schön im Walde sangen, fiel ihr dabei ein Lied wieder ein, an das sie lange nicht gedacht, und sie sang halb traurig:

„Konnt' mich auch sonst mit schwingen
Übers grüne Revier,
Hatt' ein Herz zum Singen
Und Flügel wie ihr.

Flog über die Felder,
Da blüht' es wie Schnee,
Und herauf durch die Wälder
Spiegelt' die See.

Ein Schiff sah ich gehen
Fort über das Meer,
Meinen Liebsten drin stehen. –
Dacht' meiner nicht mehr.

Und die Segel verzogen,
Und es dämmert' das Feld,
Und ich hab' mich verflogen
In der weiten, weiten Welt –"

„Leontine!" rief da die Marquise an der Gartentür des Schlosses, „sieh doch einmal, was wirbelt denn dort für Staub auf vom Wege?" Leontine trat an den Abhang des Gartens und, die Hand vor dem Glanz über die Augen haltend, sagte sie: „Ein Reiter kommt, die Sonne glitzert nur zu sehr, ich kann nichts deutlich erkennen." – Gott, dachte sie heimlich, wenn er es wäre! – jetzt biegt er schon

um den Weidenbusch, wie das fliegt! – ach nein, ein fremder Jäger ist's, was der nur noch bringen mag?
Die Mutter aber, voll Neugier und Verwunderung, war dem Reiter schon entgegengegangen und kam gleich darauf mit einem geöffneten Briefe zurück. Es war Dianas Einladung; sie beschwor das Fräulein in wenigen Zeilen herzlich und ungestüm, doch ja sogleich zu ihr hinüberzukommen, da sie nur eben ein paar Tage für sich habe und sich selbst dort nicht losmachen könne. – Die Marquise stand einen Augenblick sinnend. „Daran hatt' ich am wenigsten gedacht", sagte sie dann. „Diana ist übermütig, herrisch und gewaltsam, ihre Art ist mir immer zuwider gewesen, aber sie hat wie ein prächtiges Feuerwerk mit ihren Talenten, die sie selbst nicht kennt, den Hof und ganz Paris geblendet, du mußt ja doch endlich auch in die Welt hinaus, es ist wie ein Fingerzeig Gottes, sein Wille geschehe." – Leontinen aber flimmerten die Zeilen lustig im Abendrot, es blitzte ihr plötzlich alles wieder auf daraus: die schöne Jugendzeit, die wilden Spiele und kindischen Zänkereien mit Diana, alle ihre Gedanken waren auf einmal in die schimmernde Ferne gewendet, die sich so unerwartet aufgetan.
Es wurde nun nach kurzer Beratung beschlossen, daß sie, um keine Zeit zu verlieren und die angenehme Kühle zu benutzen, noch heute abreisen und die schöne Sommernacht hindurchfahren sollte; der alte Frenel sollte sie begleiten. Und nun ging es sogleich herzhaft an die nötigen Vorbereitungen, treppauf, treppab, die Türen flogen, Frenel klopfte seine alte Staatslivree aus, aus dem Schuppen wurde der verstaubte Reisewagen geschoben, der Hund bellte im Hofe und der Truthahn kollerte in dem unverhofften Rumor.
Oben aber in der Stube saß Leontine mit untergeschlagenen Beinen fröhlich plaudernd auf dem glänzenden Getäfel des Fußbodens vor ihrem Koffer, Kleider und Schuhe und Schals in reizender Verwirrung um sie her, und die Mutter half ihr einpacken, das Schönste, das sie

hatt'. Dann brachte sie ihr das Reisekleid und strich ihr die Locken aus der Stirn und putzte sie auf vor dem Spiegel. Und von draußen sah der Abend durchs offene Fenster herein und füllte das ganze Zimmer mit Waldhauch, und unten sangen die Vögel wieder so lustig zum Valet, und Leontine war so schön in ihrem neuen Reisehut; es war lange nicht solche Freude gewesen in dem stillen Hause.

Endlich fuhr unten der Wagen vor, es war alles bereit, vor der Haustür stand das ganze Hofgesinde versammelt, um ihr Fräulein fortfahren zu sehen. Beim Hinabsteigen sagte die Marquise: „Ich weiß nicht, jetzt ängstigt mich ein Traum von heute nacht, ich sah dich prächtig geschmückt die große Allee hinuntergehen, da war's, als würde sie immer länger und länger und hinten eine ganz fremde Gegend, ich rief dir nach, aber du hörtest mich nicht mehr, als wärst du nicht mehr mein." – Leontine lachte: der Schmuck bedeute große Ehre und Freude, wer weiß, was für ein Glück sie in der Fremde erwarte. Damit küßte sie noch einmal herzlich die Mutter und sprang in den Wagen. Aber es war ihr doch wehmütig, als nun die Wagentür wie ein Sargdeckel hinter ihr zuschlug und die Mutter, die immer noch mit dem Tuche nachwinkte, im Dunkel verschwand und Schloß und Garten allmählich hinter den schwarzen Bäumen versanken.

Jetzt rollte sie schon im Freien durch die einsame Gegend hin, der Mondschein wiegte sich auf den leise wogenden Kornfeldern, der Kutscher knallte lustig, daß es weit in den Wald schallte, manchmal schlugen Hunde an, fern in den Dörfern, und Frenels Tressenhut blinkte immerfort vom hohen Kutschbock. Leontine hatte das Wagenfenster geöffnet, sie war noch niemals zu dieser Stunde im Felde gewesen; nun war sie ganz überrascht, so wunderbar ist die ernste Schönheit der Nacht, die nur in Gedanken spricht und das Entfernteste wie im Traum zusammenfügt. Sie hatte auch Leontinen gar bald in sich versenkt. Im Fahren durch die stille Einsamkeit dachte sie sich den

Räuberhauptmann hoch im Gebirge am Feuer zwischen Felswänden, wie sie neben ihm auf dem Rasen schlief und er sie bewachte, tief unten aber durch den Felsenriß die Täler unermeßlich im Mondschein heraufdämmernd, Städte, Felder, gewundene Ströme und ihrer Mutter Schloß weit in der Ferne, und das Feuer, mit dem die Luft spielte, spiegelte sich flackernd an den feuchten Felsenwänden, und die Nachtigallen schlugen tief unten in den stillen Gärten, wo die Menschen wohnten, und die Wälder rauschten darüber hin, bis allmählich Wald und Strom und Flammen sich seltsam durcheinanderwirrten und sie wirklich einschlummerte.

Sie mochte lange geschlafen haben, denn als sie erwachte, hielt der Wagen still mitten in der Nacht, Frenel und der Kutscher waren fort, seitwärts stand eine einzelne Hütte, man sah das Herdfeuer durch die kleinen Fenster schimmern, im Hause hörte sie den Frenel sprechen, er schien nach dem Wege zu fragen. Sie lehnte sich an das Kutschenfenster, ein finstrer Wald lag vor ihnen und drüben auf einer Höhe ein Schloß im Mondschein. Wie sie aber so, nicht ohne heimliches Grauen, mit ihren Augen noch die Öde durchmißt, hört sie auf einmal Pferdetritte fern durch die Stille der Nacht. Es schallt immer näher und näher, jetzt sieht sie einen Reiter, in seinen Mantel gehüllt, im scharfen Trabe auf demselben Wege vom Walde rasch daherkommen. Sie fährt erschrocken zurück und drückt sich in die Ecke des Wagens. Der Reiter aber, da er den verlassenen Wagen bemerkt, hält plötzlich an.

„Wer ist da", rief er, „wo wollen Sie hin?" – „Nach St. Lüc", erwiderte Leontine, ohne sich umzusehen. – „St. Lüc? das ist das Schloß der Gräfin Diana", sagte der Reiter; „wenn Sie die Gräfin sehen wollen, die ist seit einigen Stunden schon auf des Grafen Gaston Schloß dort überm Wald." – „Unmöglich", versetzte das Fräulein, sich lebhaft aufrichtend bei der unerwarteten Nachricht.

„Leontine!" – rief da auf einmal der Fremde, ganz dicht an den Wagenschlag heranreitend, daß sie zusammenfuhr,

ein Mondblick durch die Wipfel funkelte über Reiter und Roß – es war der Räuberhauptmann.
Er zog, da er sie nun erkannte, schnell das weiße Tuch hervor, das sie ihm damals gegeben, und es ihr vorhaltend, fragte er: ob sie das kenne und seiner manchmal noch gedacht? – Leontine, auf das heftigste erschrocken und an allen Gliedern zitternd, hatte doch die Besinnung, nicht um Hilfe zu schreien. „Um Gottes willen", rief sie, „nur jetzt nicht, reiten Sie fort!" – Er aber, sich verbeugend in sichtlicher Spannung, als hing' die Welt an ihrer Antwort, fragte noch einmal dringender: ob sie ihn und jene wildschöne Nacht vergessen oder nicht? – „Rasender, was tun Sie!" erwiderte sie mit einiger Heftigkeit, „meine Leute sind nur wenige Schritte von hier, verlassen Sie mich auf der Stelle!" – Da ließ er langsam Arm und Tuch sinken und vor sich sehend, sagte er finster: „Was tut's, ich bin des Lebens müde." –
Jetzt hörte sie plötzlich die Tür gehen im Hause und Frenels Stimme. „Sie kommen", rief sie in Todesangst und fast in Weinen ausbrechend; „o, ich beschwöre dich, reite eilig fort, sie fangen dich, ich überlebt' es nicht!"
„Das war der alte Klang, du liebst mich noch!" jubelte da plötzlich der Reiter auf, sein Pferd lustig herumwerfend. Nun traten auch Frenel und der Kutscher wieder aus dem Hause. „Dort hinaus, immer den Wald entlang!" rief er ihnen im Vorübersprengen zu und verschwand im Dunkel vor ihnen.
„Wer war denn das?" fragte Frenel, ihm erstaunt nachsehend. Aber Leontine, noch ganz verwirrt, atmete erst tief auf, als die letzten Roßtritte verhallt und sie den Reiter in der Freiheit der Nacht wieder geborgen wußte. Darauf befahl sie, sogleich nach dem Schloß des Grafen Gaston zu fahren, das sie dort über dem Walde sähen, die Gräfin Diana sei dort, sie habe es soeben von jenem Reiter gehört, einem reisenden Herrn, setzte sie zögernd hinzu, der von dorther gekommen. – Frenel, sehr verwundert, wollte noch mancherlei fragen, aber sie trieb ihn in großer

Hast. – „Nun, nun, es wird auch ganz finster, der Mond geht schon unter, wir mußten ohnedies an dem Schlosse vorüber", sagte er, mühsam seinen Sitz besteigend, der Kutscher schwang die Peitsche, und sie flogen dem Walde zu; es war derselbe Weg, den ihnen der Reiter gewiesen. So fuhren sie rasch an den Tannen hin, von der andern Seite schwebten Wiesen, Felder und Hecken leise wechselnd vorüber, das Schloß trat immer deutlicher über den Wipfeln heraus, man hörte fern schon Nachtigallen in den Gärten schlagen. Leontine, in Nachsinnen versunken, sah sich noch manchmal scheu nach allen Seiten um, es war ihr alles wie ein Traum.

„Da blitzt es von weitem", sagte sie nach einem Weilchen zu Frenel, um in der Angst nur etwas zu sprechen. Aber Frenel, der von seiner hohen Warte freier ins Land schauen konnte, schüttelte den Kopf: er sehe schon lange hin, das sei kein Wetterleuchten, sondern Raketen oder Leuchtkugeln, die sie vom Schlosse würfen, jetzt habe er's ganz deutlich gesehen, sie müßten droben heut ein Fest haben.

Während sie aber noch so sprachen, kam plötzlich ein Lakai zu Pferde, in prächtiger Liverei und von Golde flimmernd, ihnen durch die Nacht entgegen. Frenel, ganz überrascht, zog ehrerbietig seinen Tressenhut. Jener aber ritt dicht an den Wagen, das Fräulein begrüßend, indem er sich als einen Diener aus dem Schlosse ankündigte, wohin er die Herrschaft geleiten solle. Und mit diesen Worten, ohne eine Antwort abzuwarten, drückte er die Sporen wieder ein und setzte sich rasch an die Spitze, in der hohen dunkeln Kastanienallee dem Wagen vorreitend. – Frenel hatte sich von seinem Bocke ganz zurückgebogen und sah durch die Scheiben erstaunt und fragend das Fräulein an. Leontine zuckte nur mit den Achseln, sie wußte durchaus nicht mehr, was sie davon denken sollte. Ihre Verwirrung wurde aber noch größer, als sie bald darauf an mehreren kleinen Häusern vorüberkamen, wo ungeachtet der weit vorgerückten Nacht alles noch in seltsamer Erwartung

und Bewegung schien. Überall brannte Licht, daß man weit in die reinlichen Zimmer hineinsehen konnte, Mädchen und Frauen lagen neugierig in den offenen Fenstern. „Da kommt sie, das ist sie!" hörte Leontine im Vorüberfahren ausrufen. „Mein Gott", sagte sie zu Frenel, „das muß hier irgendein Mißverständnis sein."

In diesem Augenblick aber bogen sie rasch um eine Ecke, der Wagen rollte über eine steinerne Brücke und gleich darauf in das hohe, dunkle, lange Schloßtor hinein. Jetzt flog rotes Licht spielend über die alten Mauern und Erker, Leontine, als hätte sie plötzlich ein Gespenst erblickt, starrte mit weit offenen Augen in die Blendung, denn der ganze Hof wimmelte von Windlichtern und reichgeschmückten Dienern, und auf den Stufen des Schlosses mitten im wirren Widerschein der Fackeln stand schon wieder der Räuberhauptmann!

Er schien selbst auch erst angelangt, sein Pferd, noch rauchend, wurde eben abgeführt. Als der Wagen anhielt, stieg er rasch hinab, alles wich ihm ehrerbietig aus. Er hob die ganz Verstummte aus dem Wagen und führte sie, wie einen längst erwarteten Besuch, durch die Reihe von Dienern mit höfischem Anstand die Treppe hinan, ohne ein Wort oder Mienen anzudeuten, was zwischen ihnen vorgefallen. So gingen sie durch mehrere Gemächer, alle waren hell erleuchtet, eine seltsame Ahnung flog durch Leontinens Seele, sie wagt' es kaum zu denken. Jetzt traten sie in den Saal. „Mein Gott", sagte sie, „Sie sind –"

„Graf Gaston", erwiderte ihr Begleiter, „vergeben Sie die Täuschung, sie war so schön!"

Drauf blickte er rasch im Saal umher. „Wo ist die Gräfin Diana?" fragte er die Diener. Man sagte ihm, die Gräfin habe gleich, nachdem er das Schloß verlassen, Pferd und Wagen verlangt, so sei sie mitten in der Nacht fortgefahren, der Kutscher selbst habe noch nicht gewußt, wohin es ginge. – Gastons Stirn verdunkelte sich bei dieser Nachricht, er sah nachsinnend vor sich nieder.

Leontine aber hatte unterdes schnell noch einmal alles

überdacht: den ersten Besuch des Unbekannten, seine flüchtige Erscheinung, dann unten vor dem Schloß die verworrenen Gerüchte von dem Tode des Räubers – wie hatte Schreck und Zufall alles wunderbar verwechselt! Sie stand verwirrt mit niedergeschlagenen Augen, tiefbeschämt, daß er nun alles, alles wußte, wie sehr sie ihn geliebt.

Da wandte sich Gaston, nach kurzem Überlegen, lächelnd wieder zu ihr. „Das Spiel ist aus", sagte er, „ein todwunder Räuber steht vor Ihnen und gibt sich ganz in Ihre Hand. Morgen geleit' ich Sie zurück zur Mutter, da sollen Sie richtend entscheiden über ihn auf Leben oder Tod." –

Drauf, als wollte er schonend die Überraschte heut nicht weiter drängen, klingelte er rasch; weibliche Dienerschaft trat herein zu des Fräuleins Aufwartung. Und ihre Hand küssend, eh' er schied, flüsterte er ihr noch leise zu: „Ich kann nicht schlafen, ich zieh' heut mit den Sternen auf die Wacht und mach' die Runde um das Schloß die ganze schöne Nacht, es ist ein heimlich Klingen draußen in der stillen Luft, als zög' eine Hochzeit ferne an den Bergen hin." –

Leontine stand noch lange am offnen Fenster über dem fremden Garten, Johanniswürmchen schweiften leuchtend durch Blumen und Sträucher, manchmal schlug eine Nachtigall fern im Dunkel. „Es ist nicht möglich", sagte sie tausendmal still in sich, „es ist nicht möglich!" –

Unten im Hofe aber erkundigte sich Gaston jetzt noch genauer, wiewohl vergeblich, nach der Richtung, die Diana genommen. Verblendet, wie er war, von ihrer zauberischen Schönheit, hatte sich, als er in den Flammen dieser Nacht sie plötzlich in allen ihren Schrecken erblickt, schaudernd sein Herz gewendet, und, wie eine schöne Landschaft nach einem Gewitter, war in seiner Seele Leontines unschuldiges Bild unwiderstehlich wieder aufgetaucht, das Diana so lange wetterleuchtend verdeckt. Dieser hatte er nun auf dem Schlosse hier Leontinen als

seine Braut vorstellen wollen; das sollte seine Rache sein und ihre Buße. Nun aber war unerwartet alles anders gekommen.

Wenige Wochen darauf ging an dem Schloß der Marquise ein fröhliches Klingen durch die stille Morgenluft, eine Hochzeit zog an den Waldbergen hin: glänzende Wagen und Reiter, Leontine als Braut auf zierlichem Zelter voran, heiter plaudernd an Gastons Seite. Die Vögel sangen ihr nach aus der alten schönen Einsamkeit, das treue Reh folgte ihr frei, manchmal am Wege im Walde grasend. Sie zogen nach Gastons prächtigem Schloß an der Loire.
Hier lebte er in glücklicher Abgeschiedenheit mit seiner schönen Frau. Nur manchmal überflog ihn eine leise Wehmut, wenn bei klarem Wetter die Luft den Klang der Abendglocken von dem Kloster herüberbrachte, das man aus dem stillen Schloßgarten fern überm Walde sah. Dort hatte Diana in der Nacht nach ihrer Entführung sich hineingeflüchtet und gleich darauf, der Welt entsagend, den Schleier genommen. Als Oberin des Klosters furchtbare Strenge gegen sich und die Schwestern übend, wurde sie in der ganzen Gegend fast wie eine Heilige verehrt. Den Gaston aber wollte sie nie wiedersehen.

Die Glücksritter

1. Suppius und Klarinett

Der Abend funkelte über die Felder, eine Reisekutsche fuhr rasch die glänzende Straße entlang, der Staub wirbelte, der Postillion blies, hinten auf dem Wagentritte aber stand vergnügt ein junger Bursch, der im Wandern aufgestiegen, bald auf den Zehen lang gestreckt, bald sich dukkend, damit die im Wagen ihn nicht bemerkten. Und hinter ihm ging die Sonne unter und vor ihm der Mond auf, und manchmal, wenn der Wald sich teilte, sah er von ferne Fenster glitzern im Abendgold, dann einen Turm zwischen den Wipfeln und weiße Schornsteine und Dächer immer mehr und mehr, es mußte eine Stadt ganz in der Nähe sein. Da zog er geschwind die Ärmel seines Rockes tiefer über die Handgelenke, denn er hatte ihn ausgewachsen, auch war derselbe etwas dünn und spannte über dem Rücken. Im Walde neben ihm aber war ein großes Gefunkel und Zwitschern und Hämmern von den Spechten, bald da bald dort, als wollten sie ihn necken, und die Eichkätzchen guckten um die Stämme nach ihm, und die Schwalben kreuzten jauchzend über den Weg: „Kiwitt, kiwitt, was hat dein Rock für einen schönen Schnitt!"

So ging's wie im Fluge fort, es wurde allmählich dunkel, jetzt klangen schon deutlich die Abendglocken über den Wald herüber. „Sind wir bald dort?" fragte eine wunderliebliche Stimme aus dem Wagen. – „Gleich, gleich", antwortete rasch der Bursch, der sich in der Freude vergessen; da bemerkten sie ihn erst alle. „Wart', ich will dir herunterhelfen!" rief der Postillion und hieb mit der Peitsche zurück nach ihm, eine Hand haspelte eifrig von innen am Wagenfenster. Indem aber fuhren sie eben an einer Gartenmauer hin, über die der Ast eines Apfelbaumes weit hinauslangte, der Bursch hatte ihn schon gefaßt und

schwang sich behend auf die Mauer und von der Mauer auf den Baum. Darüber öffnete sich das Glasfenster der Kutsche, ein junges Mädchengesichtchen guckte neugierig hervor. „Gott, wie ist die schön!" rief der Bursch und schüttelte aus Leibeskräften den Baum vor Lust, daß der Wagen im Vorbeifliegen ganz von Blüten verschneit war. Über dem Schütteln aber flog ihm droben der Hut vom Kopf, er wollte ihn haschen, darüber verlor er sein Bündel, und eh' er sich's versah, fuhren Hut und Bündel und Bursch prasselnd zwischen den Zweigen in den fremden Garten hinab.

Jetzt tat's plötzlich unten einen lauten Schrei, er aber erschrak am allermeisten, denn als er aufblickte, bemerkte er in der Dunkelheit eine Dame und einen Herrn dicht vor sich, die dort zu lustwandeln schienen. Da ruft ihm aber zu seinem großen Erstaunen auch schon der Herr lachend entgegen: „Nun, endlich, endlich, willkommen!" und: „Wir haben schon recht auf Sie gewartet", sagte die Dame. Der Bursch, ohne sich in der Konfusion lange zu besinnen, macht ein Kompliment und erwidert: sein Kurier wäre an allem schuld, der hätte zur Unzeit mit der Peitsche geschnalzt, da habe sein Roß einen erstaunlichen Satz gemacht, daß er mit der Frisur am Aste hängengeblieben; so habe er in der Geschwindigkeit die Gartentür verfehlt – „und den rechten Ton getroffen", meinte die Dame, „Sie spielen zum Entzücken." – „Bloß das Klarinett ein wenig", sagte der Bursch verwundert. – „Aber wo bleibt denn dein Schatz?" fragte der Herr wieder. – „Schatz?" – entgegnete der Bursch – „o, die kommt mit der Extrapost nachgefahren wie eine Ananas im Glaskasten." – Und wahrhaftig, als er unter den dunkeln Bäumen umherschaute, sah er seitwärts am Gartentor den Wagen, den er kaum verlassen, soeben im hellen Mondschein stillhalten. Aber die andern bemerkten es nicht mehr, sie waren schon lachend vorausgeeilt. „Er ist da, Herr Klarinett ist da!" riefen sie und sprangen nach dem Hause im Garten, daß der taftene Reifrock der Dame im Winde rauschte.

Indem aber hüpft auch das hübsche Frauenzimmer am Tore schon aus dem Wagen und gleich hinter ihr ein junger Mensch, schlank, gesellenhaft, ein Bündel auf dem Rücken; die streichen im Dunkel an dem Burschen, der nicht weiß, wie ihm geschieht, schnell vorüber, gerade nach dem Hause hin, und wie sie ankommen, geht eben die Haustür auf, ein Glanz von Lichtern schlägt blendend heraus, drin sumst und wimmelt es ordentlich vor Gesellschaft. „Da, Herr Klarinett und sein Schatz – und superb und tausendwillkommen", hört der Bursch von dem Hause, drauf noch ein großes Scharren und Komplimentieren auf der Schwelle, dann klappt auf einmal die Saaltür hinter dem ganzen Jubel zu, und der Bursch stand wieder ganz allein draußen in der Nacht.
Das ärgerte ihn sehr, denn wußt' er gleich in der Finsternis nicht recht, wo eigentlich Fortunas Haarzopf hier flatterte, so hatte er ihn doch fast schon erwischt und sah nun unschlüssig zwischen einem Holunderstrauch hervor. Da eilt plötzlich ein gallonierter Bedienter dicht an ihm vorüber, und in demselben Augenblick öffnet sich leise seitwärts ein Fensterchen, und: „Pst, pst, bist du's?" reicht ein weißer Arm fix eine Flasche Wein heraus. Der Bursch, nicht zu faul, langt schnell nach der Flasche, der Bediente, der soeben der prächtigen Felsentorte, die er nach dem Hause trug, heimlich zugesprochen, hatte beide Backen voll und konnte weder gleich reden noch zugreifen. Und eh' er sich noch besinnt, hat der Bursch auch schon der Torte das Dach eingeschlagen und schiebt sie zur Flasche in den Schubsack, das ging alles so still und rasch hintereinander, daß man's nicht so geschwind erzählen kann. Nun aber bekam der Bediente endlich Luft und schrie: „Diebe, Spitzbuben!" Das Frauenzimmer am Fensterchen kreischte, ein Hund schlug im Garten an, mehrere Türen im Hause flogen heftig auf. Der Bursch indes war quer durchs Gesträuch schon am andern Ende des Gartens. Kaum aber hatte er beide Beine über den Zaun geschwungen, so schreit's schon wieder draußen: „Wer da!"

neben ihm. Er, ohne Antwort zu geben, mit den dickgeschwollenen Rocktaschen über ein frischgeackertes Feld immerfort, daß der Staub flog, zwei Kerls mit langen Stangen hinter ihm: „Hallo!“ und „fangt den Schnappsackspringer!“ Und Gärten rechts und Gärten links, so stürzten endlich alle miteinander durch ein altes Tor unverhofft mitten in eine Stadt herein.
Hier wäre er ihnen um ein Haar entwischt, denn er hatte einen guten Vorsprung und flog eben in ein abgelegenes Seitengäßchen, aber das war zum Unglück eine Sackgasse, dort trieben sie ihn hinein und warfen ihm ihre Stangen nach den Füßen, worüber in der Gegend ein großes Verwundern und Tür- und Fensterklappen entstand. Da trat aber plötzlich ein langer Mann in einem zottigen Mantel um die Ecke, wie ein Tanzbär in Stiefeln, der faßte, ohne ein Wort zu sagen, den einen Häscher am Genick, den andern an der Halsbinde, warf den dahin, den dorthin, riß dem dritten seine Stange aus der Hand und versetzte damit dem vierten, der etwas dick war und nicht so geschwind entspringen konnte, einen Schlag über den breiten Rükken, und in einem Augenblick war alles auseinandergestoben und der Platz leer. Nun wetzte er die eroberte Stange, die unten mit Eisen beschlagen war, kreuzweis auf dem Pflaster, daß es Funken gab, und rief zu wiederholten Malen: „Hoho, sind noch mehrere da, die Prügel haben wollen?“ Da sich aber niemand weiter meldete, so nahm er die Stange, die er einen Bleistift nannte, unter den einen Arm und den Burschen unter den andern und führte ihn über die Straße fort. Unterwegs, als dieser sich wieder etwas erholt und nach allen Seiten umgesehen hatte, fragte er endlich, was denn das für eine Stadt sei? – „Das wird Halle geheißen“, erwiderte jener.
So kamen sie an ein kleines Haus und über eine enge Treppe, wo der Graumantel mit seinen ungeheuren Reiterstiefeln mehrmals stolperte, in eine große wüste Stube, in der eine Öllampe verwirrte Scheine über die kahlen Wände und in die staubigen Winkel umherwaf. Der alte Student (denn

das war der im Mantel) warf, wie er eintrat, seinen Bleistift mitten in die Stube und zog mühsam den Docht der halbverloschenen Lampe zurecht; da tauchte nach und nach allerlei Gerümpel ringsumher aus der Dämmerung; ein ausgetrocknetes Tintenfaß, leere Bierflaschen, die als Leuchter gedient, Rapiere und ein alter Stiefel daneben, da hatt' er seine Wäsche drin. Er selbst aber nahm sich, so bei Licht besehen, ziemlich graulich aus: große, weit herausstechende Augen, eine lederne Kappe auf dem zerzausten Kopf, ein Strick um den Leib und lauter Bart, wie ein Eremit.

Als er mit der Lampe fertig war, reckte er sich zufrieden, daß ihm alle Glieder knackten. „Ach", sagte er, „solche Motion tut not, wenn man so den ganzen Tag über den Büchern hockt." – Der Bursch sah sich überall um, aber es war kein Buch zu sehen. – Drauf wandte der Student sich zu ihm: „Aber Fuchs, bist du denn des Teufels", sagte er, „gleich zwischen Spießen und Stangen hier mit der Tür ins Haus zu brechen!" – „Zerbrochen?" entgegnete der Bursch, erschrocken nach seinem Schubsacke greifend, „nein, da ist die ganze Bescherung."

Mit diesen Worten brachte er Flasche und Torte aus den Taschen hervor. Als der Student das sah, fragte er nicht weiter nach dem Herkommen, sondern verbiß sich, obgleich es fast über Mitternacht war, sogleich mit so erstaunlichem Appetit in die Felsentorte, daß ihm die Trümmer über den Bart herabkollerten. „Wie heißt du denn?" fragte er dazwischen. – Der Bursch, ohne sich lange zu bedenken, erwiderte: „Klarinett." – „Hm, ein guter Klang", meinte der Student. Dann griff er nach dem Wein, und da kein Glas da war, trank er ihm aus der Flasche zu: „Daß dich der Donner erschlag', Klarinett, wenn du nicht ein ordentlicher Kerl wirst! Überhaupt", fuhr er, sich den Bart wischend, fort, „wenn du studieren willst, da mußt du die Bücher in die Nase – wollt' sagen die Nase in die Bücher stecken und Cajus, Cujacius und allen den schweinsledernen Kerls auf den Leib gehen, und wenn sie noch so dick wären!"

„Aber", fiel ihm hier der Bursch ins Wort, „ich bin ja gar kein Student, sondern eigentlich ein wandernder Musikus."

„Was, ein Musikant?" rief der Student, „was spielst du?" – „Das Klarinett." – „Oho", sagte er, „du pfeifst also deinen eigenen Namen wie der Kuckuck." Hier ging er, wie in reiflicher Überlegung, mit langen Schritten ein paarmal im Zimmer auf und nieder, dann blieb er plötzlich vor dem Burschen stehen und vertraute ihm, wie er eine große, heimliche Lieb' gefaßt hätte seit langer Zeit zu einer vornehmen Dame hier am Ort; er wüßte aber nicht, wie sie hieße, sondern ginge nur zuweilen an ihrem Hause vorüber, wo sie mit ihrem dicken Kopfzeug wie eine prächtige Hortensia am Fenster säße, aber so oft er unter die Fenster käme, hörte er bloß ein angenehmes Flüstern droben und sähe nichts als weiße Arme flimmern und Augen funkeln durch die Blumen. –

Der Bursch versetzte darauf, er sollte sich nur etwas besser herausputzen bei solchen Gelegenheiten. – Der Student sah an sich herunter, schüttelte den Kopf und schien ganz zufrieden mit seinem Aufzuge. Dann sagte er, er hätte schon lange die Intention gehabt, vor ihren Fenstern eine Serenade aufzuführen, aber seine Kommilitonen könnte er dazu nicht brauchen, die würden ihn auszustechen suchen bei ihr; nun aber wolle er ihr morgen abend das Ständchen bringen, da sollte der Bursch mit blasen helfen.

Dieser war damit zufrieden, und nun sollte auch sogleich die Serenade eingeübt werden. Der Student nahm voll Eifer ein Waldhorn von der Wand, staubte es erst sorgfältig ab, setzte ein wackeliges Notenpult unter Zorn und Fluchen, weil es nicht fest stehen wollte, mitten in der Stube zurecht, legte die Notenbücher drauf, und beide stellten sich nun einander gegenüber und fingen mit großer Anstrengung ein sehr künstliches Stück zu blasen an. Darüber aber war bei der nächtlichen Stille nach und nach die ganze Nachbarschaft in Aufruhr geraten. Ein Hund fing

im Hofe zu heulen an, drauf tat sich erst bescheiden ein Fenster gegenüber auf, dann wieder eins, und endlich unaufhaltsam immer mehrere vom Keller bis zum Dach, und dicke und dünne Stimmen durcheinander, alles schimpfte und zankte auf die unverhoffte Nachtmusik. Zuletzt wurde es doch dem Studenten zu toll, er warf voller Wut das Horn weg, ergriff ein altes, verrostetes Pistol vom Tisch und drohte zum offenen Fenster hinaus, den Zipfel von jeder Schlafmütze herabzuschießen, die sich ferner am Fenster blicken ließe. Da duckten auf einmal alle Mausköpfe unter und es wurde wieder stille draußen, nur der Hund bellte noch ein Weilchen den Mond an, der prächtig über die alten Dächer schien.

Der Student aber, sich den Schweiß von der Stirn wischend, streckte sich nun ganz ermüdet der Länge nach auf das zerrissene Sofa hin, Klarinett sollte sich's auch kommode machen, aber es war nur ein einziger Stuhl in der Stube, und als er ihn ergriff, ging die Lehne auseinander. Da wies der Student auf einen leeren Koffer neben dem Kanapee, dann verlangte er gähnend, Klarinett sollte ihm seinen Lebenslauf erzählen, damit er ihm darnach gute Ratschläge für sein weiteres Fortkommen erteilen könnte.

Der Bursch schoß einen seltsamen scharfen Blick herüber, als wollt' er erst prüfen, wieviel er hier vertrauen dürfte, dann rückte er sich auf seinem Koffer zurecht und begann nach kurzem Besinnen:

„Ich weiß nicht, ob mein Vater Müller war, aber er wohnte in einer verfallenen Waldmühle, da rauschten die Wasser lustig genug, aber das Rad war zerbrochen und das Dach voller Lücken, in den klaren Winternächten sahen oft die Wölfe durch die Löcher ins Haus herein."

„Was lachst du denn?" unterbrach ihn hier der Student. – „Wahrhaftig", erwiderte der Bursch, „Ihr gemahnt mich heut ganz an meinen seligen Vater, wie ihn mir die Mutter einmal beschrieben hat." – „Was geht mich dein seliger Vater an", meinte der Student. Aber der Bursch fuhr von

neuem lachend fort: „Es war nämlich gerade den Abend nach einer Schlacht, man hatte den ganzen Tag in der Ferne schießen hören, da ging mein seliger Vater eilig ins Feld hinaus, denn die Mühle lag seitwärts im Grunde tief verschneit; so war der Krieg darüber weggegangen. Draußen aber hatte er mancherlei Plunder im Schnee verstreut, zerhauene Wämser, Fahnen, Pickelhauben und Waffen; mein Vater konnte alles brauchen, er fuhr sogleich in ein Paar ungeheure Reiterstiefel hinein, zog hastig Pappenheimsche Kürasse, schwedische Koller und Kroatenmäntel an, eins über das andere, dabei war er in der Geschwindigkeit mit beiden Armen in ein Paar spanische Pluderhosen geraten, der Wind blies den Kroatenmantel im Freien weit auf, je mehr er zuckte und reckte, je verwickelter wurde die Konfusion von Schlitzen, Falten, flatternden Zipfeln und Quasten, und als nun meine Mutter, die eben guter Hoffnung war, ihn so haspelnd und fluchend mit ausgespreizten Armen wie einen fliegenden Wegweiser daherstreichen sah, mußte sie so darüber lachen, daß sie plötzlich meiner genas. Und in demselben Augenblick, wo ich zur Welt kam, ging draußen klingendes Spiel durch die stille Luft, die Kaiserlichen bliesen noch im Fortziehn Viktoria weit auf den Bergen, daß es lustig über den Schnee herüberklang, mein Vater meinte, das wäre ein gutes Zeichen, ich würde ein glücklicher Soldat werden. Ich selber aber weiß mich von allem den nur noch dunkel so viel zu erinnern, daß ich so recht still und warm in der wohlgeheizten Stube in meinen Kissen lag und verwundert die spielenden Ringe und Figuren betrachtete, welche die Nachtlampe an der Stubendecke abbildete. Das zahme Rotkehlchen war von dem ungewohnten Licht und Nachtrumor aufgewacht, schüttelte die Federn, wie wenn es auch sein Bettlein machen wollte, setzte sich dann neugierig auf die Bettlade vor mir und sang ganz leise, als wollt' es mir zum Geburtstag gratulieren. Meine Mutter aber neigte sich mit ihrem schönen, bleichen Gesicht und den großen Augen freundlich über mich, daß ihre Locken

mich ganz umgaben, zwischen denen ich draußen die Sterne und den stillen Schnee durchs kleine Fenster hereinfunkeln sah. Seitdem, so oft ich eine klare, weitgestirnte Winternacht sehe, bin ich immer wieder wie neugeboren." –

Hier hielt er plötzlich inne, denn er hörte soeben Herrn Suppius (so hieß der Student) auf dem Kanapee schon tüchtig schnarchen. Der Mondschein lag wie Schnee auf den Dächern, da war's ihm in dieser Stille, wie der Lampenschein so flatternd an der Decke spielte, als hörte er draußen die Wasser und den Wind wieder gehen durch die Wipfel im Walde und das Rotkehlchen wieder dazwischen singen.

2. *Die Serenaden*

Am folgenden Tage durchstrich Klarinett neugierig alle Gassen und Plätze, die der dreißigjährige Kriegssturm übel zugerichtet. Aber es gefiel ihm doch sehr, denn die ganze Stadt war jetzt wie ein lustiges Feldlager, die Studenten in schönen, unerhörten Trachten schwärmten plaudernd durch die Straßen, überall Lachen, Waffengeklirr und der fröhliche Klang der Jugend, als hätte sich mitten aus dem neuen Frieden, der nun allmählich draußen die müde Welt überzog, ein Haufen Holkscher Jäger hierhergeworfen, um die Wissenschaften zu erstürmen.

Als er endlich, nach vielem Umherirren und Fragen, ziemlich spät die Sackgasse wiedergefunden, traf er Herrn Suppius schon unten an der Haustür voller Unruhe wegen der verabredeten Serenade. Er hätte ihn beinah nicht wiedererkannt, denn er hatte einen gestickten Modefrack mit steifen Schößen angezogen und eine große Wolkenperükke auf dem Kopf, wie ein Gesandter. Er quälte sich soeben voll Zorn und Eifer, einen alten Degen, der nicht passen wollte, galant anzustecken, darüber waren mehrere Lok-

ken der Perücke aufgegangen, da und dort kam sein eignes struppiges Haar darunter hervor, aber er fragte nichts danach und stülpte einen dreieckigen Tressenhut drauf, daß es staubte, der saß ihm ganz hintenüber recht im Genick. Klarinett mußte nun auch geschwind seine besten Kleider anlegen, und als die balsamische Nacht über die verräucherten Dächer daherkam, wanderten schon beide vergnügt mit ihren Instrumenten durch die finstere Stadt. Ihre Tritte hallten in der abgelegenen Einsamkeit, nur ein Student sang noch am offenen Fenster zur Zither, mehrere Uhren schlugen verworren durch den Wind, der Nachtwächter rief eben die elfte Stunde, einige Stimmen ahmten ihn verhöhnend nach, man hörte Lärm und Gezänke in der Ferne, dann plötzlich alles wieder still. Auf einmal winkte Suppius, sie schlüpften durch eine Lücke der Stadtmauer ins Freie, und standen vor einem schönen, großen Hause. Klarinett betrachtete verwundert Dach, Erker und den mondbeschienenen Garten zur Seite, er glaubte nach und nach dieselbe Villa wiederzuerkennen, wo er gestern abends angekommen; da dacht' er sich's gleich, daß es wieder nicht gut ablaufen würde.

Aber alles erschien heute von einer andern Seite, sie waren in einen kleinen, winkeligen Hof geraten voll Gerümpel und alter Tonnen, die Fenster im Hause waren fest verschlossen, nur die Wetterfahne drehte sich manchmal knarrend auf dem Dach, eine Katze unten funkelte sie mit ihren grünfeurigen Augen an und wand sich mit gebogenem Buckel spinnend um ihre Stiefel. „Hier heraus muß sie schlafen, halt' dich nur dicht hinter mir", sagte Suppius, sein Waldhorn leise zurechtsteckend.

Kaum aber hatten sie sich zwischen den Tonnen zum Blasen zurechtgestellt, so war's ihnen, als hörten sie von der einen Seite draußen ein Pferd schnauben. Sie setzten die Instrumente ab und horchten ein Weilchen, da ließ sich gleich darauf ein heimliches Knistern im Hause vernehmen, in demselben Augenblick tat sich ein Hinterpförtchen leise auf, ein Mann, vorsichtig nach allen Seiten um-

schauend, trat hervor und führte ein Frauenzimmer, die zögernd folgte, schnell bei der Hand an den blühenden Sträuchern fort. Der Mond schien bald hell, bald dunkel zwischen wechselnden Wolken, da sahen sie deutlich, wie der Mann jetzt unter den hohen Bäumen die Dame auf ein Pferd hob, sich selber hinter ihr hinaufschwang, einen weiten weißen Mantel um beide schlug und sacht und lautlos davonritt. Da warf Suppius plötzlich die leeren Tonnen auseinander, und mit einem Satz sich über den Zaun schwingend, rannte er unaufhaltsam mit entsetzlichem Geschrei übers Feld an den letzten Häusern vorüber, daß alle Hunde erwachten und die Leute erschrokken an die Fenster fuhren. Der Herr auf dem Pferde aber, da er ihn unverhofft mit seinen großen Stiefeln hinter sich so hohe, weite Sprünge machen sah, setzte die Sporen ein und es dauerte nicht lange, so waren Roß und Reiter verschwunden.

Der Student nun, als er sie im Dunkel verloren, blieb atemlos mitten im Felde stehen und schimpfte auf die Nacht, die alles bemäntelte, und auf den Mond, der wie eine Spitzbubenlaterne dazu leuchtete, und auf den Wind, der ihm die Wolkenperücke zerzaust, und auf Klarinett, der darüber lachte. – „Aber um Gottes willen, was gibt's denn eigentlich?" fragte dieser endlich ganz erstaunt. – „Was es gibt?" erwiderte Suppius zornig, „Mord, Totschlag, Entführung gibt's, hast du nicht den Reiter gesehen?" – „Ja, und eine Dame." – „Und das war just meine Liebste!" rief Suppius.

Klarinett aber, da er diese unerwartete Nachricht vernommen, lag schon der Länge nach im Grase und legte das Ohr an den Boden. „Die Luft kommt von dorther", sagte er eifrig, „ich höre noch den Klang der Huftritte von fern, jetzt schlagen die Hunde an drüben im Dorf, dort sind sie hin." – „Gut, so steh' nur rasch wieder auf", sagte Suppius und beschloß sogleich, dem Entführer weiter nachzusetzen, Klarinett sollte auch mit, er selber habe alles von Wert bei sich und in der Stadt nichts zurückgelassen, als

ein paar lumpige Schulden, den Weg aber, den der Räuber eingeschlagen, kenne er wie seine Tasche und wisse recht gut, wohin er führe, sie brauchten nur schnell auf der Saale sich in einen Kahn zu werfen, so kämen sie ihnen noch vor Tagesanbruch ein gut Stück voraus.

Das war dem Klarinett eben recht, und so gingen sie rasch miteinander nach dem Ufer zu. Dort fanden sie bald unter dem Weidengebüsch einen angebundenen Nachen, ein Fischer lag drin voller Gedanken auf dem Rücken, der machte große Augen, als er Herrn Suppius, den hier in der Gegend alle kannten, so martialisch auf sich zukommen sah. Suppius sagte ihm, wo sie hinauswollten, der Fischer griff stumm und verschlafen nach den Rudern, und nach einigen Minuten fuhren sie alle schon lustig die Saale hinunter. Der Wind hatte unterdes die Wolken zerstreut, da legte Suppius, der sich in der Nachtkühle wieder ein wenig beruhigt, dem Fischer gelehrt den ganzen Himmelsplan aus mit lateinischen Skorpionen, Krebsen und Schlangen und geriet, da der ungläubige Fischer von dem allen nichts wissen wollte, immer tiefer und eifriger in den Disput. Klarinett aber saß in der Einsamkeit ganz vorn im Kahn; das war eine prächtige Nacht! Sternschnuppen am Himmel, und Berge, Wälder und Dörfer am Ufer flogen wie im Traum vorüber, manchmal rauscht' es leise im Wasser auf, als wollte eine Nixe auftauchen in der großen Stille, von beiden Seiten hörte man Nachtigallen fern in den Gärten. Da sang Klarinett:

„Möcht' wissen, was sie schlagen
So schön bei der Nacht,
's ist in der Welt ja doch niemand,
Der mit ihnen wacht.

Und die Wolken, die reisen,
Und das Land ist so blaß,
Und die Nacht wandert leise,
Man hört's kaum, durchs Gras.

Nacht, Wolken, wohin sie gehen,
Ich weiß es recht gut,
Liegt ein Grund hinter den Höhen,
Wo meine Liebste jetzt ruht.

Zieht der Einsiedler sein Glöcklein,
Sie höret es nicht,
Es fallen ihr die Löcklein
Übers ganze Gesicht.

Und daß sie niemand erschrecket,
Der liebe Gott hat sie schier
Ganz mit Mondschein bedecket,
Da träumt sie von mir."

Jetzt glitt der Nachen durch das säuselnde Schilf ans Ufer, ein erleuchtetes Fenster spiegelte sich im Fluß, Klarinett erkannte nach und nach alte Mauern und Türme und eine Stadt im Mondschein. Suppius aber hatte ihn schon am Arme gefaßt und sprang mitten aus seinem Diskurse ans Land. „Dort am Galgen geht der Feldweg vorbei, den sie kommen müssen", sagte er und bezahlte rasch den Schiffer, der gähnend wieder in die schöne Nacht hinausstieß. Die beiden aber schritten nun sogleich durch das alte Tor, da hatte der Krieg das Stadtwappen ausgebissen, bei der angenehmen Friedenszeit lag der Nachtwächter schnarchend auf der steinernen Bank daneben, der Mond beschien hell die stille Straße mit ihren spitzen, finstern Giebeln, draußen vom Felde hörte man fern eine Wachtel schlagen. Als sie auf den Markt kamen, machte Suppius plötzlich halt. „Die Stadt hat nur zwei Tore", sagte er, „von dem Brunnen hier kann man von einem Tor zum andern sehen, die Nacht ist klar, sie mögen nun erst ankommen oder schon drin sein, hier können sie uns nicht entwischen." Mit diesen Worten postierte er den Klarinett an die eine Seite des Brunnens und setzte sich selbst von der andern auf die steinerne Rampe, die Arme über der Brust

verschlungen und unverwandt in die Straße hineinschauend. Indem bemerkte Klarinett noch Licht in einem schönen, großen Hause, ein tief heruntergebrannter Kronleuchter drehte sich, wie verschlafen hinter den Scheiben, man schien soeben nach einem Tanze die Kerzen auszupusten von einem Fenster zum andern, und bald war das ganze Haus ebenfalls dunkel bis auf ein einziges Zimmer. Da tat sich plötzlich unten eine Tür auf, und laut plaudernd, scherzend und lachend, brach ein dunkles Häuflein in die kühle Stille heraus, es waren Schüler oder Musikanten mit überwachten Gesichtern, ihre Instrumente unter den Mänteln. Als sie noch das Licht oben sahn, traten sie schnell wieder zusammen, stellten sich unter das erleuchtete Fenster und fingen sogleich ein Ständchen zu blasen an, das zog wie ein goldener Traum über die schlafende Stadt. Auf einmal aber öffnete sich oben das Fenster, zwischen den rotseidenen Gardinen erschien eine schöne, schlanke Mädchengestalt und bog sich weit heraus in den Mondschein, als wollte sie zu ihnen sprechen.

„Das ist *sie!*“ rief hier plötzlich Suppius, von dem Rande des steinernen Brunnens aufspringend. In demselben Augenblick aber faßte von hinten ein dunkler Arm das Mädchen schnell um den Leib, zog sie in das Zimmer zurück, und warf hastig das Fenster zu, dann sah man noch drin an den Wänden lange Schatten wie Windmühlflügel verworren durcheinander arbeiten, und gleich darauf war auch das Licht oben ausgelöscht und alles wieder still.

Die unverhoffte Erscheinung des Suppius brachte die erschrockenen Musikanten unten ganz aus dem Konzept, einer sah den andern verwundert an, nur hier und da fuhr ein verlegener Ton aus, wie bei einer Orgel, der der Wind ausgegangen. Zu beiden Seiten ehrerbietig ausweichend, antworteten alle eifrig durcheinander: „Wir sind's, wir sind's, wir wollten ihnen, da sie oben noch Licht hatten, einen Willkommen blasen.“ – „Wem denn?“ – „Nun Ihr

wißt's ja, die vorhin ankamen, als wir drin zum Tanze aufspielten, der fremde Herr mit der Dame." – „Zu Pferd im langen Mantel?" – „Ja, die Euch so höflich grüßten, Ihr saht eben auch zum Fenster heraus." – „Ich?" – „Freilich, und ‚Ha, das faule Hofgesind'!' rief der fremde Kavalier im Hofe, ‚wo bleibt meine Leibkarosse?' Und als Ihr eben droben den Kehraus tanztet" – „Da möcht' man ja gleich des Teufels werden!" – „kam auch die Karosse wirklich nach, Ihr rieft noch dem Kutscher aus dem Fenster zu, er sollt' nach dem Hof fahren." – „Wer ist hier betrunken, ich oder ihr?" – „Ich und ihr und wir alle für unseren Herrn Burgemeister, vivat hoch!" schrien da auf einmal die berauschten Musikanten und wollten nun den Suppius, den sie in seinen höfischen Staatskleidern im Dunkeln für den Burgemeister hielten durchaus mit Musik nach Hause bringen. Vergebens sträubte sich der entrüstete Student, sie ließen sich's nicht nehmen, und eh' er sich's versah, setzten sie sich paarweise in Ordnung und schritten, einen feierlichen Marsch spielend, quer über den Markt voran, als wollten sie die Sterne am Himmel ausblasen. In ihrem Eifer merkten sie's gar nicht, daß Suppius an einer Straßenecke hinter ihnen entwischt war; immerfort blasend, bogen sie in die finstre Gasse hinein, da wurden von allen Seiten über dem Lärm die Hunde wach, dann hörte man sie noch mit dem Nachtwächter um den verlornen Burgemeister zanken, immer weiter und weiter, bis endlich alles zwischen den dunkeln Häusern nach und nach vertoste.

Unterdes aber hatten Supius und Klarinett, der eine schimpfend, der andre lachend, schon den offenen Hof des Wirtshauses erreicht, als ihnen eine ausgespannte Reisekutsche mit Glasfenstern und vergoldeten Schnörkeln im Mondschein prächtig entgegenglitzerte. Suppius, bei dem erfreulichen Anblick, ohne ein Wort zu sprechen, öffnete sogleich die Tür der verlassenen Kutsche, schob den verwunderten Klarinett in den Wagen und schwang sich selber hurtig nach. „So", sagte er, nachdem er das

Glasfenster hinter ihnen behutsam wieder geschlossen hatte, „jetzt sitzen wir mitten in der Entführung drin, wie der fromme Äneas im hölzernen Pferde, um die geraubte Helena zu retten; der Kavalier kann nicht fahren ohne Wagen, der Wagen nicht ohne mich, und ich nicht, ohne den Kavalier und den Wagen und ganz Troja umzuwerfen." – „Amen, Gott weiß, wer dabei zu oberst oder zu unterst zu liegen kommt", erwiderte Klarinett, dem die Bündigkeit des trojanischen Anschlages noch nicht recht einleuchten wollte. Eigentlich aber freute er sich selber auf die Konfusion, die nun jeden Augenblick ausbrechen konnte.

Suppius hatte sich indes in der Finsternis des Wagens unverhofft in die seidnen Fransen und Quasten, die überall herumbommelten, verhaspelt und kam nicht aus dem Ärger. Dabei unterließ er aber doch nicht, von Zeit zu Zeit die Gardinen am Wagenfenster zurückzuschlagen und aus seinem Kastell Beobachtungen anzustellen. Das ganze Haus lag in tiefem Schlaf, nur von der einen Seite stand die Stalltür halb offen, sie hörten drin zuweilen Pferde stampfen und schnauben und einzelne Fußtritte, der Kutscher schien schon wach zu sein. Auf einmal stieß er Klarinett an. „Sieh doch", sagte er, „was ist das für ein großer Pilz da auf der Hofmauer?"

„Das wackelt ja", entgegnete Klarinett, scharf hinblikkend, „ein breiter Klapphut ist's, den Wind und Wetter so zerknattert haben, seht Ihr nicht die Augen darunter hervorfunkeln?"

„Wahrhaftig", bemerkte Suppius wieder, „nun hampelt's und hebt sich's, Haare, Bart und Mantel verworren durcheinandergefilzt, jetzt kommt ein Bein über die Mauer."

„Und ein Ellbogen aus dem Ärmel", meinte Klarinett.

Indem aber schwang sich die ganze Figur plötzlich von der Mauer in den Hof hinab, eine zweite folgte, lange, bärtige, soldatische Gesellen.

Beide, erst nach allen Seiten umherspähend, schlichen an

die Haustür und versuchten vorsichtig zu öffnen, fanden aber alles fest verschlossen. Suppius und Klarinett verwandten kein Auge von ihnen. Jetzt bemerkten sie, wie die Fremden, an der Stalltür vorbei, quer über den Hof gingen und in der Gaunersprache miteinander redeten. „Schau", sagte der eine, „haben die schöne Klebis (Pferde), werden Santzen (Edelleute) sein, oder vornehme Kummerer (Kaufleute), die nach Leipzig schwänzen (reisen)." – „Eine gute Schwärze (Nacht)", versetzte der andere, „es schlunt (schläft) noch alles im Schöcherbeth (Wirtshaus), kein Quin (Hund) bellt, und kein Strohbohrer (Gans) raschelt. Alch' (troll dich), wollen die Karosse zerlegen, hat vielleicht Messen (Gelder) in den Eingeweiden."
„Das sind verlaufne Lenninger (Soldaten)", flüsterte Klarinett, „die kommen bracken (stehlen), ich wollt', ich könnt' den Mausköpfen grandige Kuffen stecken (schwere Schläge geben)!" – „Was Teufel, verstehst du denn auch das Rotwelsch?" fragte Suppius erstaunt.
Aber da war keine Zeit mehr zu Erklärungen, denn die Lenninger kamen jetzt gerade auf den Wagen los; der eine schnupperte ringsherum, ob er nicht einen Koffer oder Mantelsack fände, der andre aber griff geschwind, damit es sein Gesell nicht merken sollte, nach der Wagentür. Suppius und Klarinett hielten sie von innen fest, er konnte sie mühsam nur ein wenig öffnen, wunderte sich, daß es so schwer ging, und tappte sogleich mit der Hand hinein. „Aha, ein Paar Stiefeln!" sagte er vergnügt in sich, des überraschten Suppius Füße fassend. Indem aber schnappt Klarinett die Tür, wie eine Auster, rasch wieder zu, der Dieb hatte kaum so viel Zeit, die gequetschte Hand zurückzuziehn, er meinte in der Finsternis nicht anders, sein Kamerad hätt' ihn geklemmt, weil er ihm den ersten Griff nicht gönnte. „Was ist das!" rief er zornig und böse diesem zu, „bist ein Hautz (Bauer) und kein ehrlicher Gleicher (Mitgesell), möchtest alles allein schöchern (trinken) und mir den leeren Glestrich (Glas) lassen!" – Der andre,

der gar nicht wußte, was es gab, erwiderte ebenso: „Was barlest (sprichst) du so viel, wenn wir eben was auf dem Madium (Ort) haben, komm nur her, sollst mir den Hautz wie gefunkelten Johann (Branntwein) hinunterschlingen!" Da trat plötzlich der Mond aus den Wolken und der Kutscher in die Stalltür, und die erschrockenen Schnapphähne flogen, wie Eidechsen unter dem Schatten des Hauses zwischen Steinen und Ritzen durch den Hof und über die Mauer wieder in die alte Freiheit hinaus.

„Nun, die bleiben auch noch draußen am Galgen hängen", meinte Suppius aufatmend. Der schlaftrunkene Kutscher aber, der von allem nichts bemerkt hatte, siebte im Mondschein den Hafer für seine Pferde, gähnte laut und sang:

„Wann der Hahn kräht auf dem Dache,
Putzt der Mond die Lampe aus,
Und die Stern' ziehn von der Wache,
Gott behüte Land und Haus."

Darauf ging der Knecht an den Brunnen im Hofe, pumpte Wasser in den Eimer und kämmte und wusch sich umständlich mit vielem Gegurgel und Geräusch, zu großem Ärger des Suppius, der gerne gesprochen hätte. Endlich kehrte er in den Stall zurück, auch die Schnapphähne ließen sich nicht wieder blicken, und da nun alles still blieb, sagte Suppius ernst zu Klarinett gewendet: „Hör', junger Gesell, es ist ein löblicher Brauch, Verirrte auf den rechten Weg zu weisen. Du redetest vorhin ziemlich geläufig eine gewisse Sprache – Ex ungue leonem – also glaube ich –"

„Was denn?" unterbrach ihn Klarinett etwas betroffen; „unter den Römern gab's Schnapphähne genug und Ihr redet doch auch Lateinisch." Aber Suppius, den der Tiefsinn der Nacht angeweht, ließ sich nicht aus seiner feierlichen Verfassung bringen. Er hatte sich in das Wagenfenster gelehnt, den Kopf in die rechte Hand gestützt, die Sterne funkelten durch den Lindenbaum vor dem Hause,

von den Bergen rauschte der Wald über die Dächer herein. „Da nimm dir ein Exempel dran", fuhr er fort, „Wälder und Berge stehn nachts in Gedanken, da soll der Mensch sich auch bedenken. Alle weltliche Lust, Hoheit und Pracht, die Nacht hat alles umgeworfen, die wunderbare Königin der Einsamkeit, denn ihr Reich ist nicht von dieser Welt. Sie steigt auf alle Berge und stellt sich auf die Zinnen der Schlösser und schlägt mahnend die Glocken an, aber es hört es niemand als die armen Kranken, und niemand hört die Gewichte der Turmuhr schnurren und den Pendel der Zeit gehn in der stillen Stadt. Der Schlaf probiert heimlich den Tod und der Traum die Ewigkeit. Da hab' ich immer meine schönsten –"
Hier überwältigte ihn unversehens der Schlaf, er nickte ein paarmal mit seinem dreieckigen Tressenhut; dann plötzlich ein Weilchen wieder hinausstarrend, in abgebrochenen Sätzen wie eine abgelaufene Spieluhr: „Meine schönsten Gedanken", hub er noch einmal an – „in der Nacht, wo Laub und Fledermaus und Igel und Iltis verworren miteinander flüstern – und der Mensch im Traume – ihre Sprache versteht." –
Jetzt aber hatte die Nacht ihn selber umgeworfen. Klarinett horchte noch immer hin, denn es war ihm wirklich bei den Worten, als hört' er des Einsiedlers Glöcklein fern überm Wald. Er zog, da Suppius nun fest schlief, das Wagenfenster vorsichtig auf; dann lehnt' er in Gedanken die Stirn an die Scheibe, da hörte er vom Stall her wieder das einförmige Schnurzen der Pferde beim Futter, und über ihm rauschte der Baum und seitwärts die Saale hinter dem Hause fort und immerfort, bis auch er endlich vor großer Ermüdung einschlummerte. –
Ruck! – stießen da auf einmal beide so hart mit den Köpfen aneinander, daß es dröhnte. Suppius blickte wild nach allen Seiten um sich und wußte durchaus nicht, wo er war. Als er sich aber endlich auf seine Liebste und die ganze Entführungsgeschichte wieder besonnen hatte, sagte er verwirrt: „Was ist das, Klarinett? wir fahren ja, ich glaube

gar, nun werden wir selbst entführt." – „Ja, und gerade in einen Wald hinein", erwiderte Klarinett nicht weniger verwundert, „seht nur, vier prächtige Rosse vor dem Wagen und der fromme Kutscher drauf." – „Mit einem goldbordierten Hut", sagte Suppius wieder, „und hinter uns aus der Stadt krähen uns die Hähne nach, als wollten sie uns foppen, mir scheint, ich wittre schon Morgenluft." – „Freilich, aber die Fledermäuse schwirren noch durch die Dämmerung", versetzte Klarinett, plötzlich aufmerksamer zur Seite blickend, „da schaut nur zwischen die Bäume, da noch einer, dort wieder einer: bei Gott, das sind die Bärenhäuter von heute nacht, die halten Euch gewiß für den reisenden Kavalier."

Indem aber fiel auch schon ein Schuß aus dem Walde und gleich darauf noch ein zweiter. Der Kutscher duckte sich, die Kugel pfiff über ihn weg, er peitschte heftig in die Pferde, Suppius schrie voll Wut aus dem Wagen: „Fehlgeschossen, ihr Narren! ich bin's ja nicht!" Der Kutscher, da er zu seinem großen Erstaunen auf einmal fremde Leute im Wagen bemerkte, die er gleichfalls für Strauchdiebe hielt, warf sich nun ohne weiteres aus dem Sattel, überkugelte sich ein paarmal im Graben und war dann schnell im Dickicht verschwunden. Über dem Lärm aber wurden die ledigen Pferde ganz wild, die Räuber fluchten, die Kugeln pfiffen, Suppius drohte, so sausten sie unaufhaltsam dahin, man hört' es noch lange durch die heitere Morgenstille rumpeln und schimpfen.

3. Waldesrauschen

In einer warmen Sommernacht schlief ein Mädchen im Wald, sie hatte den Kopf über den rechten Arm auf ihr Tamburin gelegt und das Gesicht gegen den Tau mit der Schürze bedeckt, ein Pferd weidete daneben, weiterhin lag ein junger Bursch, der wendete sich manchmal und re-

dete unverständlich im Schlaf. Zwischen den Bäumen aber flog das erste halbe Morgenlicht schon schräg über den luftigen Rasen, ein paar Rehe, die in der Nacht mit den Pferden geweidet, schlüpften raschelnd durch die Dämmerung tiefer in den Wald zurück, sonst war noch alles still.

Auf einmal ertönte ein gellender Wachtelschlag, das Mädchen erhob sich rasch, daß die Glöcklein am Tamburin klangen. Es war der Vater, der mit seinem Pfeifchen die Schlafenden weckte. Er stand schon in voller Reisetracht: knappe blaue Beinkleider mit rotem Paß und eine grüne ungersche Jacke mit gelben Schnüren und blinkenden Knöpfchen nachlässig über die Schulter geworfen, ein ehemaliger Soldat, der nun als Puppenspieler und starker Mann mit den Kindern durchs Land zog.

„Horch", sagte er, „da krähen Hähne in weiter Ferne nach jener Seite hin, die Luft kommt von drüben, da muß ein Dorf sein, der Wald liegt hoch, besteig einmal den Tannenbaum, Seppi, und sieh dich um." Der Bub reckte und dehnte sich mit beiden Armen in die ungewisse Luft und schüttelte die Locken aus der Stirn, dann kletterte er schnell in den höchsten Wipfel hinauf. Nach einem Weilchen rief er herab: „Da unten ist noch alles nachtkühl und still, es liegt alles durcheinander im tiefen Grund, da haben sie wieder ein Dorf verbrannt." – „Ja, ja", versetzte der Vater, „der große Schnitter Krieg mäht uns tapfer voran, man hört seine Sense bei Tag und Nacht klingen durchs Land, wir geringen Leut' haben die Nachlese auf den Stoppeln. Siehst du sonst nichts?" – „In der Ferne ein schönes Schloß überm Wald, die Fenster glitzern herüber." – „Raucht der Schornstein?" – „Ja, kerzengrad' aus den Wipfeln." – „Gut", versetzte der Vater, „so komm nur wieder herunter, da wollen wir hin." – Aber im Herabsteigen zögernd, rief der Bursch noch einmal: „Ach, aber da drüben, da liegt das ganze Tal schon im Sonnenschein, jetzt blitzen drunten Hellebarden aus den Kornfeldern, Landsknechte ziehen nach dem Walde zu, wie

schön sie singen!" – „Da ist der Siglhupfer dabei!" sagte das Mädchen freudig. – Der Vater blickte rasch nach ihr herüber, man wußte niemals recht, ob er lächelte oder heimlich schnappen und beißen wollte, so scharf blitzten manchmal seine Zähne unter dem langen, gewichsten Schnurrbart hervor. „Rauch und Wind!" sagte er, „wer weiß, wo der Siglhupfer schon zerhauen im Graben liegt." – Das Mädchen aber lachte: „Ihr sprecht immer so barsch, er denkt doch an mich, er ist ein Soldat von Fortüne und kommt wohl wieder, eh' wir's denken, als Offizier zu Pferde mit hohen Federn auf dem Hut."
Währenddes hatte sie ein Stück von einem zerschlagenen Spiegel vor sich an den Baum gelehnt, setzte sich davor ins Gras und flocht ihr langes schwarzes Haar auf zigeunerisch in zierliche Zöpfchen, dabei biß sie von Zeit zu Zeit in eine Wecke und streute einzelne Krümelchen über den Rasen für die Vögel, die ihr neugierig aus dem Laube zusahen. Der Vater und Seppi aber zäumten und packten schon das Saumroß, unverdrossen bald einen Königs-, bald einen Judenbart zurückschiebend, die, in schmählicher Gleichheit durcheinander geworfen, aus dem löcherigen Puppensack herausdrängten. Dann hauchte der Vater ein paarmal auf ein großes schwarzes Pflaster, das er über das linke Auge und Backe legte, damit er martialischer aussäh' und die Leute sich vor ihm fürchteten. Und als endlich alles reisefertig war, schwang er die Tochter in den Sattel, Seppi mußte vorausgehen, er aber führte das Pferd über die Wurzeln und Steine vorsichtig hinter sich am Zügel, und droben auf ihrem luftigen Sitze, das Tamburin neben sich gehängt, baumelte das Mädchen vergnügt mit den Füßen und freute sich über ihre neuen roten Halbstiefeln; manchmal streifte ihr ein Zweig Stirn und Wange, daß sie wie eine Blume ganz voll Tauperlen hing. Da stimmte Seppi vorne lustig an:

„Der Wald, der Wald, daß Gott ihn grün erhalt',
Gibt gut Quartier und nimmt doch nichts dafür!"

Und das Mädchen antwortete sogleich:

„Zum grünen Wald wir Herberg' halten,
Denn Hoffahrt ist nicht unser Ziel,
Im Wirtshaus, wo wir nicht bezahlten,
Es war der Ehre gar zu viel,
Der Wirt, er wollt' uns gar nicht lassen,
Sie ließen Kanne und Kartenspiel,
Die ganze Stadt war in den Gassen.
Und von den Bänken mit Gebraus
Stürzt' die Schule heraus,
Wuchs der Haufe von Haus zu Haus,
Schwenkt' die Mützen und jubelt' und wogt',
Der Hatschier, die Stadtwacht, der Bettelvogt,
Wie wenn ein Prinz zieht auf die Freit',
Gab alles, alles uns fürstlich Geleit.
Wir aber schlugen den Markt hinab
Uns durch die Leut' mit dem Wanderstab
Und hoch mit dem Tamburin, daß es schallt' –"

Und der Puppenspieler und Seppi fielen jubelnd ein:

„Zum Wald, zum Wald, zum schönen grünen Wald!"

Das Mädchen sang wieder:

„Und da nun alle schlafen gingen,
Der Wald steckt seine Irrlicht' an,
Die Frösche tapfer Ständchen bringen,
Die Fledermaus schwirrt leis voran,
Und in dem Fluß auf feuchtem Steine
Gähnt laut der alte Wassermann,
Strählt sich den Bart im Mondenscheine
Und frägt ein Irrlicht, wer wir sind?
Das aber duckt sich geschwind,
Denn über ihn weg im Wind
Durch die Wipfel der wilde Jäger geht,

Und auf dem alten Turm sich dreht
Und kräht der Wetterhahn uns nach:
Ob wir nicht einkehr'n unter sein Dach?
O Gockel, verfallen ist ja dein Haus,
Es sieht die Eule zum Fenster heraus,
Und aus allen Toren rauschet der Wald,
Der Wald, der Wald, der schöne grüne Wald!

Und wenn wir müd' einst, sehn wir blinken
Eine goldne Stadt still überm Land,
Am Tor Sankt Peter schön tut winken:
‚Nur hier herein, Herr Musikant!'
Die Engel von den Zinnen fragen,
Und wie sie uns erst recht erkannt,
Sie gleich die silbernen Pauken schlagen,
Sankt Peter selbst die Becken schwenkt,
Und voll Geigen hängt
Der Himmel, Cäcilia an zu streichen fängt,
Dazwischen hoch vivat! daß es prasselt und pufft,
Werfen die andern vom Wall in die Luft
Sternschnuppen, Kometen,
Gar prächtige Raketen,
Versengen Sankt Peter den Bart, daß er lacht,
Und wir ziehen heim, schöner Wald, gute Nacht!"

Und zum Chor machte der Puppenspieler mit dem Munde prasselnd das Feuerwerk nach, und Seppi schmetterte mit einem Pfeifchen wie eine Nachtigall, und die Tochter schwang ihr Tamburin schwirrend dazwischen; so zogen sie wie eine Bauernhochzeit durch den Wald in den aufblitzenden Morgen hinunter, als zögen sie schon ins Himmelreich hinein.
Als sie aber am Rande des Waldes zu sein vermeinten, fing jenseits der Wiese schon wieder ein andrer an, die Heiden waren ohne Weg, die Bäche ohne Steg, manchmal war's ihnen, wie wenn sie Hunde bellen hörten aus der Ferne und Stimmen gehn im Grund, das Schloß aber, wohin sie

zielten, stand bald drüben, bald dort, immer neue Schluchten dazwischen, als wollt' es sie foppen. Und so war es fast schon wieder Abend geworden, als sie endlich, aus einem verworrenen Gebüsch tretend, auf einmal die Burg ganz nahe vor sich sahen.

Sie schauten sich erst nach allen Seiten um, eine Allee von wilden Kastanien führte nach dem Tor, man konnte bis in den gepflasterten Hof und im Hofe einen Brunnen und Galerien rings an dem alten Hause sehen, es rührte sich aber nichts darin. „Ich weiß nicht, Denkeli", sagte der Puppenspieler nach einem Weilchen zur Tochter, „das kommt mir doch kurios vor mit dem Schloß, das hängt ja alles so liederlich, die Sparren vom Dach und die Laden aus den Fenstern, als wär' auch schon der Kriegsbesen darübergefahren." – Indem schlug die Uhr vom Turme langsam durch die große Einsamkeit. – „Da muß aber doch jemand wohnen, der die Uhr aufzieht", sagte Denkeli. – „Das tun die Toten bei Nacht in solchen Schlössern", erwiderte der Vater verdrießlich.

Darüber waren sie an ein altes Gittertor gekommen und blickten durch die ehemals vergoldeten Stäbe in den Schloßgarten hinein. Da lag alles einsam und schattigkühl, Regen, Wind und Sonnenschein waren, wie es schien, schon lange die Gärtner gewesen, die hatten einen steinernen Neptun aufs Trockne gesetzt und ihm eine hohe grüne Mütze von Ginster bis über die Augen gezogen, wilder Wein, Efeu und Brombeer kletterten von allen Seiten an ihn heran, eine Menge Sperlinge tummelte sich lärmend in seinem Bart, er konnt' sich mit seinem Dreizack vor dem Gesindel gar nicht mehr erwehren. Und wie er so sein Regiment verloren, reckten und dehnten sich auch die künstlich verschnittenen Laubwände und Baumfiguren aus ihrer langen Verzauberung phantastisch mit seltsamen Fühlhörnern, Kamelhälsen und Drachenflügeln in die neue Freiheit hinaus, und mitten unter ihnen auf dem Dach eines halbverfallenen Lusthauses saß melancholisch ein Pfau noch aus der vorigen Pracht und rief der unterge-

henden Sonne nach, als hätte sie ihn hier in der Wildnis vergessen. Auf einmal aber tat es einen leuchtenden Blitz durchs Grün, eine wunderschöne Dame erschien tiefer im Garten, durch die stillen Gänge nach dem Schlosse zu wandelnd, ganz allein in prächtigem Gewande, ihr langes Haar wallte ihr wie ein goldener Mantel über die Schultern, die Abendsonne blitzte noch einmal leuchtend über das kostbare Geschmeide auf Stirn und Gürtel. Denkeli blickte sie scheu, doch unverwandt an, sie dachte an die vorigen Reden des Vaters, es war ihr, als ginge die Zauberin dieser Wildnis vorüber. Die Dame aber bemerkte die Wanderer nicht, sie sah ein paarmal zurück nach ihrer taftenen Schleppe, die schlängelnd hinter ihr herrauschte, und verlor sich dann wieder zwischen den Bäumen.

Jetzt hörten sie zu ihrem Erstaunen plötzlich auch Stimmen am Schloß, sie gingen eilig hin und bemerkten nach langem Umherirren endlich einen Balkon zwischen den Wipfeln, der nach dem Walde herausging. Dort sahen sie einige Herren an dem steinernen Geländer stehen, die Dame aus dem Garten schien auch bei ihnen zu sein; aber sie konnten nichts deutlich erkennen, denn die Linde, die in voller Blüte stand, reichte bis an den Balkon, und die Abendsonne funkelte blendend dazwischen. Der Puppenspieler war auf alle Glücksfälle vorbereitet, er zog schnell seine Orgelpfeife, die er vor den Mund band, und eine Geige hervor, Seppi einen Triangel und Denkeli ihr Tamburin, und so stellten sie sich unter die Bäume und brachten gleich den Herrschaften ein Ständchen. Denkeli sah dabei öfters scharf hinauf; auf einmal ließ sie, mitten in dem Geschwirre abbrechend, Arm und Tamburin sinken, sie hatte in größter Verwirrung in dem einen Kavalier droben den Siglhupfer erkannt, sie sah, wie er galant und scharmant sich neigte und beugte und mit der Dame parlierte, sie konnt' es gar nicht begreifen. Der Vater stieß sie ein paarmal mit dem Ellbogen an, sie sollte zu singen anfangen, aber sie warf das Köpfchen trotzig empor und

wollte durchaus nicht, und dem Vater mochte sie die Ursach' nicht sagen, denn er lachte sie immer aus mit ihrer Liebschaft. Während des Hin- und Herwinkens aber kam auch schon eine Kammerjungfer schnell aus dem Schloß herunter und brachte ihnen einen Krug Wein und jedem einen Rosenobel sauber in Papier gewickelt mit der Botschaft, ihre Herrschaft sei heute gar nicht wohl und zu müde, um die Musik anzuhören, auch sei im ganzen Hause kein Unterkommen für sie zur Nacht.

„Seht Ihr, sie mögen meinen Gesang ja nicht", sagte Denkeli zum Vater; sie dachte bei sich, Siglhupfer habe sie erkannt und wollte sie nur los sein, weil er sich ihrer schäme vor der vornehmen Dame.

Der Puppenspieler zuckte, ohne zu antworten, ein paarmal zornig mit den buschigen Augenbrauen, trank aber doch auf die Gesundheit der Dame und reichte darauf den Krug der Tochter, die ihn mit der Hand von sich stieß. So stritten sie heimlich untereinander, der Vater zankte noch immer über Denkelis Eigensinn, dann packte er heftig seine Instrumente zusammen, um weiterzuziehn, sie wußten nicht wohin in der fremden Gegend. Über ihnen summten die Bienen im Wipfel, und hinter den Blüten droben plauderten und lachten die Herrschaften in der schönen Abendkühle und machten sich lustig über die Bettelmusikanten, Denkeli erkannte Siglhupfers Stimme darunter recht gut, das schnitt ihr durch die Seele! Manchmal sah sie auch seinen Federhut und die Lokken und den Schmuck der Dame durch die Zweige schimmern, es war ihr alles wie ein Traum. Im Weggehn fragte sie die Jungfer noch: „Wer ist denn der junge Herr da droben?"

„Ei, Ihr kommt wohl von weit her?" erwiderte diese, „das ist ja der Herr Rittmeister von Klarinett, der Bräutigam des gnädigen Fräuleins."

4. *Das verzauberte Schloß*

Der Schall einer Trompete gab das Zeichen zur Tafel. Eine Flügeltür tat sich plötzlich auf, und Suppius, in goldbrokatenem Staatskleid leuchtend, einen Federhut in der einen Hand, führte an der andern eine prächtige Dame, von kostbaren Armbändern, Halsketten und Ohrgehängen umblitzt und umbommelt, daß man nicht hinsehen konnte, wenn die Sonne darauf schien. So stiegen beide feierlich eine steinerne Treppe in den großen alten Gartensaal hinab, ein Hündchen mit silbernen Schellen um den Hals trat oft der Dame auf die schwere Schleppe, die von Stufe zu Stufe hinter ihnen herrauschte. Klarinett folgte in reicher Offizierskleidung: in dunkelgrünem Samt mit geschlitzten Ärmeln, einem Kragen von Brüßler Kanten darüber und den Hut mit goldner Spange und nickenden Federn schief auf den Kopf gedrückt, es paßte ihm alles prächtig. Er spielte vornehm mit einer Reitgerte und nickte kaum, als ihm der Diener der Dame meldete, daß sein Reisegepäck gehörig untergebracht sei.
Im Saale aber war der Tisch schon gedeckt, sie nahmen mit großem Geräusch und unter vielen Komplimenten Platz auf den schweren, rotsamtenen Sesseln mit hohen künstlich geschnitzten Lehnen. Klarinett überblickte unterdes erstaunt die Tafel, da gab's so wunderliche Pracht, abenteuerlich gehenkelte Krüge, hohe, altmodisch geschliffene Stengelgläser von den verschiedensten Farben und Gestalten, seltsam getürmte Speisen und Schaugerichte und heidnische Götter von Silber dazwischen, die Pomeranzen in den Händen hielten. Seitwärts aber stand die Tür auf, daß man weit in den Garten sehn konnte, die Sonne funkelte in den Gläsern, der Diener eilte mit Schüsseln und vergoldeten Aufsätzen flimmernd hin und her, und draußen sangen die Vögel dazu und vor der Tür saß ein Pfau auf der marmornen Rampe und schlug sein prächtiges Rad.
So saßen sie lange in freudenreichem Schalle, da hub Fräulein Euphrosine (so war die Dame genannt) mit freundli-

cher Gebärde an: sie könne sich noch immer nicht drein finden, denn es käme selten ein Fremder in die Einsamkeit, und keiner so seltsam als ihre beiden Gäste, die, wie sie versicherte, heut beim ersten Morgengrauen vom Walde quer übers Feld plötzlich mit vier schäumenden Rossen ohne Kutscher mitten in den Schloßhof und gewiß auch am andern Ende wieder hinausgeflogen wären, hätten sie nicht am Torpfeiler Achse und Deichsel gebrochen. – Klarinett, mit zierlichen Reden den verursachten Schreck entschuldigend, erzählte nun, sie seien fremde Kavaliere, die, vom Westfälischen Frieden nach ihren Herrschaften reisend, in jenem Wald von Räubern überfallen worden, Haushofmeister, Kutscher, Leibhusar, alles sei erschossen; und da das Fräulein auf die Frage: ob sie in Tztschneß hinter Tzquali in Mingrelim bekannt? mit dem Kopf schüttelte, bedauerte er das sehr, denn gerade von dort seien sie her.
Suppius stürzte ein Glas Ungarwein so eilig aus, daß er sich den gestickten Zipfel seiner Halsbinde begoß; es war, als hätte Klarinett mit seinen Lügen ihn plötzlich in einen Strom gestoßen, nun mußte er mit durch oder schmählich vor den Augen der Dame untergehn. Dabei sah er oft das Fräulein bedenklich von der Seite an, sie kam ihm schon wieder auf ein Haar wie seine entführte Geliebte vor, aber er traute sich doch nicht recht, er hatte seine Liebste so selten und immer nur flüchtig am Fenster hinter den Blumen gesehen; so wurde er ganz konfus und wagte es nicht, von der Entführung zu reden. Und als er darauf dennoch mit großer Feinheit die Sommerkühle der vergangenen Nacht pries, gelegentlich einen Seitenblick über jenes mondbeschienene Städtchen warf, und endlich leise über den Marktplatz am steinernen Brunnen vorbei zu dem Wirtshaus kam, auf das Fenster zielend, wo ihnen damals der lieblichste Stern erschienen: sah die Dame ihn befremdet an und wußte durchaus nicht, was er wollte. Aber Suppius war einmal im Zuge ausbündiger Galanterie. „Was frag' ich noch nach Sternen!" rief er aus, „flogen wir doch auf

vergoldeten Rädern Fortunas aus Nacht zu Aurora, daß ich vor Blendung noch nicht aufzublicken vermag.“ – Da schlug das Fräulein mit einem angenehmen Lächeln die schönen Augen nieder, Suppius, entzückt, griff hastig nach ihren Fingerspitzen, um sie zu küssen, warf aber dabei mit dem breiten Aufschlage seines Ärmels dem silbernen Cupido die Pomeranze aus der Hand, und wie er sie haschen wollte, verwickelte er sich mit Sporen und Degenspitze unversehens ins Tischtuch, alle Gläser stießen auf einmal klirrend an, als wollten sie seine Gesundheit ausbringen, der Cupido stürzte und riß einen Weinkrug mit, das Hündchen bellte, der Pfau draußen schrie. Euphrosine aber mit flüchtigem Erröten stand rasch auf, die Tafel aufhebend, indem sie dem Klarinett ihren Arm reichte.

Sie traten vor die Saaltür auf die Terrasse, von der eine breite Marmortreppe nach dem Garten führte. Eine Eidechse, als sie herauskamen, fuhr erschrocken zwischen die Ritzen der Stufen, aus denen überall das Gras hervordrang, seitwärts stand ein alter Feldstuhl, eine Zither lehnte daran. Als Suppius, der noch immer den Aufruhr an der Tafel mit seinen weiten Alamodeärmeln ausführlich zu entschuldigen beflissen war, das Instrument erblickte, stockt' er auf einmal und entfernte sich schnell wie einer, der plötzlich einen guten Einfall hat. Das Fräulein aber ließ sich in der Tür auf dem Feldstuhl nieder, Klarinett, die Zither auf den Knien prüfend und stimmend, setzte sich auf die Stufen zu ihren Füßen, daß der Pfau von dem steinernen Geländer ihm mit seinem schlanken Hals über die Schulter sah. Draußen aber war es unterdes kühl geworden, der ganze Garten stand tief in Abendrot, während die Täler schon dunkelten, auch der Pfau steckte jetzt den Kopf unter die Flügel zum Schlaf, die Luft kam über den Garten und brachte den Schall einer Abendglocke aus weiter Ferne. Da fiel dem Klarinett in dieser Abgeschiedenheit eine Sage ein, die er unten in den Dörfern gehört, und da das Fräulein sie wissen wollte, erzählte er von

einem verzauberten Schloß des Grafen Gerold; da wüchse auch das Gras aus den Steinen, da sänge kein Vogel ringsum, und kein Fenster würde jemals geöffnet; man höre nichts als den Wetterhahn sich drehn und den Zugwind flüstern und zuweilen bei großer Trockne das Getäfel krachen im Schloß; so stünd' es öde seit hundert Jahren, als redet' es mit geschlossenen Augen im Traum. – Jetzt hatte er die Zither in Ordnung gebracht – „Es gibt auch eine Weise darauf", sagte er und sang:

„Doch manchmal in Sommertagen
Durch die schwüle Einsamkeit
Hört man mittags die Turmuhr schlagen
Wie aus einer fremden Zeit.

Und ein Schiffer zu dieser Stunde
Sah einst eine schöne Frau
Vom Erker schaun zum Grunde –
Er ruderte schneller vor Graun.

Sie schüttelt' die dunklen Locken
Aus ihrem Angesicht:
‚Was ruderst du so erschrocken,
Behüt' dich Gott, dich mein' ich nicht.'

Sie zog ein Ringlein vom Finger,
Warf's tief in die Saale hinein:
‚Und der mir es wiederbringet,
Der soll mein Liebster sein!'"

Hier gewahrte Klarinett auf einmal, daß das Fräulein, wie im tiefsten Nachsinnen versunken, aufmerksam den kostbaren Demantring betrachtete, den er mit dem andern Staat in der fremden Karosse gefunden und leichtsinnig angesteckt. Er stutzte einen Augenblick, das Fräulein aber, als hätte sie nichts bemerkt, fragte mit seltsamem Lächeln nach dem Ausgang der Sage. Klarinett, etwas ver-

wirrt, erzählte weiter: „Und wenn nun der Rechte mit dem Ringe kommt, hört die Verzauberung auf, aus den Winkeln der stillen Gemächer erheben sich überall schlaftrunken Männer und Frauen in seltsamen Trachten, das öde Schloß wird nach und nach lebendig. Diener rennen, die Vögel singen wieder draußen in den Bäumen, und dem Liebsten gehört das Land, so weit man vom Turme sehn kann."

Bei diesen Worten fiel auf einmal draußen ein Waldhorn ein; der galante Suppius war es, er zog in seinem Goldbrokat wie ein ungeheurer Johanniswurm durch den finstern Garten, als wollt' er mit seinen Klängen die Nacht anbrechen, die nun von allen Seiten prächtig über die Wälder heraufstieg. Schloß, Büsche und Garten wurden immer wunderbarer im Mondschein, und wenn die Luft die Zweige teilte, blinkte aus der Tiefe unterm Schloß die Saale herauf und das Geschmeide und die Augen des Fräuleins blitzten verwirrend dazwischen. – Da hob plötzlich die Uhr vom Turme zu schlagen an. Klarinett fuhr unwillkürlich zusammen, in demselben Augenblick glaubte er einen flüchtigen Händedruck zu fühlen, und als er verwundert aufsah, traf ihn ein funkelnder Blick der Dame. Indem aber trat der Diener mit einer Kerze hinter ihnen in den Saal, um die Fremden ins Schlafgemach zu geleiten, die Dame erhob sich zierlich und gemessen wie sonst und war nach einer freundlichen Verbeugung schnell durch eine innere Tür des Saals verschwunden. Doch als Klarinett sich betroffen wandte, ging eben der Mond aus einer Wolke und beschien das steinerne Bildwerk über der Tür: es war wirklich das ihm wohlbekannte Wappen des Grafen Gerold. – „Was ist denn das?" dachte er erschrocken, „am Ende hab' ich da selber den Ring." –

Am folgenden Tage hielt er's fast für einen Traum, so ganz anders sah die Welt aus, der Morgen hatte alles wieder mit Glanz und Vogelschall verdeckt, nur das unheimliche Wappen über der Tür blieb aus jener Nacht und der Zauberblick der Dame. Er hatte sich in dem Wetterleuchten ihrer Augen nicht geirrt, sie spielten munter fort, ihre Liebe

zu Klarinett brach rasch aus wie der Frühling nach einem warmen Gewitterregen. Und so ließ er denn auch alles gut sein und wollte mit Grübeln das Glück nicht versuchen, das ihm so unversehens über den Kopf gewachsen.
Dem Suppius aber ging es über den seinigen weg, ohne daß er's merkte. Jeden Morgen putzte er sich, mit Rat und Beistand des mutwilligen Klarinett, auf das sorgfältigste heraus und probierte vor dem Wandspiegel insgeheim artige Stellungen. Aber bis Mittag war doch alles wieder schief und verschoben, das vornehme Kleid der guten Lebensart saß ihm, als wär' er in der Eile mit einem Arm in den falschen Ärmel gefahren. Manchmal fielen ihm auch plötzlich die Wissenschaften wieder ein, da erschrak er sehr und verwünschte alle Abenteuer, die er doch immer selber wieder anzettelte. Dann ergriff er hastig das dicke Buch, das in der Tasche seines Serenadenrockes mitgekommen, damit setzte er sich in die abgelegensten Winkel des Gartens ins Gras und schlug das Kapitel auf, wo er in Halle stehn geblieben. Aber der alte Ungarwein aus dem Schloßkeller war stärker als er, der ließ die Buchstaben auf magyarisch vor ihm tanzen und drückte ihm jedesmal die Augen zu und die Nase ins Buch. Und wenn er aufwachte, steckte zu seinem Erstaunen das Zeichen im Buch immer beim unrechten Paragraphen, auch glaubte er auf dem Rasen Spuren von Damenschuhen zu bemerken, als hätten ihn Elfen im Schlafe besucht, ja das eine Mal lag, statt des Zeichens, ein ganzer Strauß brennender Liebe zwischen den Blättern. Da steckt' er ihn triumphierend vorn an die Brust und sprach den ganzen Tag durch die Blume zu Euphrosine von heimlicher Liebe und Hochzeit. Er zweifelte und verwunderte sich nicht, daß sie in ihn verliebt, und ließ oft gegen Klarinett fallen, wie er darauf bedacht sein werde, ihn hier als seinen Kapellmeister oder Fasanengärtner anzustellen.
Klarinett aber wußt' es wohl besser, es kam alles bald zum Ausgang. Denn als er eines Morgens bei einem Spaziergange mit Euphrosine und ihrem Diener auf eine Anhöhe

gestiegen, von der man weit ins Land hinaussehen konnte, wies ihm der Diener rings in die Runde die Schlösser, Wälder, Teiche, weidende Herden und Untertanen, die alle seinem Fräulein gehörten. Der Morgen funkelte darüber, die Teiche blickten wie Augen aus dem Grün, alle Wälder grüßten ehrerbietig rauschend herauf, Klarinett war wie geblendet. Da sagte Euphrosine rasch: „Und alles ist dein – wenn du diese Hand nicht verschmähst", setzte sie mit gesenkten Augen kaum hörbar hinzu. Klarinett aber, ganz verblüfft, stürzte auf ein Knie nieder und schwor, so wahr er Kavalier und Rittmeister sei, wolle er sie nimmer verlassen, und ein Kuß auf ihre Hand besiegelte den schönen Bund, und in dem Auge des grauen Dieners zitterte eine Freudenträne.

Nun aber lebten sie alle vergnügt von einem Tag zum andern, da war nichts als Schmausen und Musizieren und Umherliegen über Rasenbänken und Kanapees. Täglich zur selben Zeit lustwandelten sie rauschend in vollem Staate vor dem Schloß, gleichsam leuchtende Zirkel und Namenszüge durch den Garten beschreibend, der mit seinen Schnörkeln von bunten Scherben wie ein Hochzeitskuchen im Sonnenschein lag, im Hofe hatte der blühende Holunderbusch ihre Staatskarosse schon beinah ganz überwachsen, auf der Marmortreppe schlug der Pfau täglich dasselbe Rad, die Vögel sangen immer dieselben Lieder in denselben Bäumen. Und an einem prächtigen Morgen, den er halb verschlafen, dehnte sich Klarinett, daß ihm die Glieder vor Nichtstun knackten; „nein", sagte er, „nichts langweiliger als Glück!"

5. Fortunas Schildknappen

Zur selben Zeit lag das Dorf, das einst zu dem Schlosse gehört, fern unterm Berg in Trümmern. Es war seit dem letzten Durchzug der Schweden zerstört und verlassen,

nun rückte der Wald, den die Bauern so lange tapfer zurückgedrängt, über die verrasten Beete unter Vogelschall mit Stacheln, Disteln und Dornen wieder ein und hatte sich das verbrannte Gebälk schon mit Efeu und wilden Blumen prächtig ausgeschmückt und auf dem höchsten Aschenhaufen einen blühenden Strauch als Siegesfahne ausgesteckt, nur einzelne Schornsteine streckten noch, wie Geister, verwundert die langen weißen Hälse aus der verwilderten Einsamkeit. Heute aber fing auf einmal der eine Schornstein wieder zu rauchen an, ein helles Feuer knisterte unter demselben, und so oft der Wind den Rauch teilte, sah man in der Glut des Widerscheins wilde dunkle Gestalten, wie Arbeiter in einem Eisenhammer, mit aufgestreiften Ärmeln vor dem Feuer hantieren, kochen und Bratspieße drehen; einer saß im Grase und flickte sein Wams, ein andrer lag daneben und sah ihm verächtlich zu, den Arm stolz in die Seite gestemmt, daß ihm im Mondschein der Ellbogen aus dem Loche im Ärmel glänzte, während weiterhin zwei Holkische Jäger soeben durch das Dickicht brachen und ein frischgeschossenes Reh herbeischleppten. Es waren versprengte Landsknechte, die das Ende des Dreißigjährigen Krieges plötzlich vom Pferd auf den Friedens- und Bettelfuß gesetzt. In solchem Schimpf hatten sie beschlossen, den Krieg auf ihre eigene Faust fortzusetzen und sich mitten durch ihren gemeinschaftlichen Feind, den Frieden, nach Ungarn durchzuschlagen, wo sie gegen den Türken neue Ehre und Beute zu gewinnen hofften.

„Hartes Bett, gemeines Bett!“ sagte der Stolze mit dem Loch im Ärmel, „heute ist's gerade ein Jahr, es war auch so eine blanke Nacht, da hing's nur von mir ab, ich konnte auf kostbaren Teppichen liegen mit eingewirkten Wappen, in jedem Zipfel mein Namenszug in Gold.“ –

Da kniff ein grauer Kerl seitwärts den neben ihm liegenden Dudelsack, der plötzlich schnarrend einfiel. – „Ruhe da!“ rief ein breiter Landsknecht hinüber, und mehrere Schalke rückten zum Feuer, um den Schreckenberger (so

hieß der Stolze) besser zu hören. Dieser warf dem Dudelsack einen martialischen Blick zu und fuhr fort:
„Denkt ihr noch dran, nach der Schlacht bei Hanau, wie wir da querfeld mit der Regimentskasse retirierten, nichts als Rauchwirbel in der Ferne und Rabenzüge über uns, in den Dörfern guckten die Wölfe aus den Fenstern, und die Bauern grasten im Wald." – „Freilich", versetzte der schlaue Landsknecht, „und eine Dame auf kostbarem Zelter, einen Pagen hinter sich, immer neben uns her, und als wir am Abend an einem verbrannten Dorfe Halt machten, kehrte sie auch über Nacht ein in dem wüsten Gartenschloß daneben." – „Ja, und die Augen", sagte Schreckenberger, „spielten ihr wie zwei Spiegel im Sonnenschein, dich und die andern hat's geblendet, ihr war't alle vernarrt in sie. Nun denk' ich an nichts und gehe abends am Schloß vorüber, da schreibt sie euch aus dem Fenster ordentlich: Vivat Schreckenberger! mit den feurigen Blicken in die Luft, und wie ich mich wende, ruft sie: Ach! und fällt in Ohnmacht vor großer Lieb' zu mir. So was war mir schon oft passiert, ich fragte wenig darnach, da ich aber tiefer im Garten bin, kommt plötzlich der Page im Dunkel daher mit einem Briefe an mich auf rosenfarbenem Papier."
Hier zog Schreckenberger ein Brieflein aus dem Wams und reichte es mit vornehm zugekniffenen Augen über die Achsel dem andern hin. Der Landsknecht nahm es hastig und las: „Im Garten bei Nacht – Das Lusthaus ohne Wacht – Sturmleitern daran, – Cupido führt an – Um Mitternacht Runde – Parol': ‚Adelgunde'."
„Das klappt ja wie ein Trommelwirbel", sagte der Landsknecht, indem er, den Brief zurückgebend, neugierig noch näher rückte, „ja, Cupido hat schon manchen angeführt, nur weiter, weiter!"
„Kurz: Um Mitternacht bin ich auf meinem Posten", hub Schreckenberger wieder an, „im Garten nichts als Mondschein, große Stille, das Lusthaus, wie's im Briefe steht, droben ein offnes Fenster auf dem Dache, drunten eine

Leiter, ich weiß nicht mehr, ob von Sandelholz oder Seide oder Frauenhaaren. Ich fackle nicht lange, die Büchse auf dem Rücken, in jeder Hand ein Pistol, den blanken Säbel zwischen den Zähnen, so klettr' ich hinauf –"

„Also du warst es doch!" fiel hier der Landsknecht verwundert ein.

„Nun wer denn sonst?" erwiderte Schreckenberger, „und Jasmin, wie ich hinaufsteige, Rose von Jericho, Holunder, Jelängerjelieber, alles umhalst und umschlingt mich vor Freuden, das riß sich ordentlich um mich, daß ich die Sporen nicht nachbringen konnte, und vom Fenster droben hoben mich plötzlich zwei alabasterne Schwanenarme aus dem Brunnen der Nacht, und über mir ein prächtiges Gewitter von Locken, da blitzten Augen und Juwelen draus, und in dem Brunnen gehn immerfort goldne Eimer auf und nieder mit Muskateller und Konfekt, und die Gräfin Adelgunde sitzt neben mir auf einem mit Diamanten gesprenkelten Kanapee, und: ‚Langen Sie zu', sagte sie, und ‚O, ich bitte sehr', sag' ich – da hör' ich auf einmal unter uns in dem Lustpalaste inwendig ein Gesumse wie in einem Bienenstock. ‚Was war das?' ruf' ich –"

Jetzt brach plötzlich ein Lachen aus. „*Wir* waren's", sagte einer der Zuhörer, „denn wir steckten ja alle drin, der Page hatte uns alle nacheinander auch in das Lusthaus geladen und drauf die Tür hinter uns verriegelt."

Aber Schreckenberger, einmal im Strom der Erzählung, ließ sich nicht irre machen; „ich springe auf", fuhr er fort, „ha, Verrat! schrei' ich –"

Nun sprachen alle rasch durcheinander: „Ja, du machtest einen Teufelslärm auf dem Dache, denn sie hatten hinter dir die Leiter weggenommen und das Fenster oben war verschlossen."

„Und die Gräfin in dem einen Arm, den Säbel im andern, und unter mir kocht's und zischt's und rumpelt's –"

„Freilich, im dunkeln Lusthaus stießen wir einer auf den andern, und einer fragte den andern trotzig, was er hier suche, und jeder hatte seine Parole Adelgunde, bis wir zu-

letzt alle aneinander gerieten und aus der Parole ein großes Feldgeschrei und Geraufe wurde."

„Und ich steche links, steche rechts, die Gräfin ohnmächtig, ruft: ‚Genug des Gemetzels!' Aber ich laß mich nicht halten und feure prasselnd alle meine Pistolen ab nach allen Seiten wie ein Feuerwerk –"

„Das hörten wir wohl", fiel nun der Landsknecht wieder ein, „und hielten's für einen ordentlichen Überfall, da arbeiteten wir und stemmten uns an die verriegelte Tür und die Wände, bis das ganze morsche Lusthaus über uns in Stücken auseinanderging. So kamst du auch kopfüber mit herunter – du machtest einmal Sprünge quer übers Feld fort, ohne dich umzusehn! wir erkannten dich nicht in der Verwirrung und wußten dann gar nicht, wo du auf einmal hingekommen; später hieß es, du wärst zu den Kaiserlichen desertiert in dieser Nacht."

„Nacht?" fuhr der unverwüstliche Schreckenberger noch immer fort, „ja recht mitten durch die Nacht auf einem schneeweißen Zelter, sich die Tränen wischend mit dem goldbordierten Schleier und mir zuwinkend, flog die dankbar gerettete Gräfin –"

„Mit eurer verlaßnen Regimentskasse in die weite Welt", versetzte einer der Holkischen Jäger, „denn es war unsere Marketenderin, die schöne Sinka, die hat's euch allen angetan, das merkte sie wohl und vexierte euch von der Feldwacht fort."

Schreckenberger schwieg und warf wieder einen martialischen Blick rings in die Runde. Aber der Jäger fuhr fort: „Und gleich am andern Morgen, da wir bei unserem Regiment sie alle kannten, wurden wir kommandiert, ihr nachzusetzen. Das war eine lustige Jagd, wir strichen wie die Füchse auf allen Diebswegen und schüttelten jeden Baum, ob das saubre Früchtchen nicht herabfiel. So kamen wir am folgenden Abend – es war gerade ein Sonntag – in ein kleines Städtchen; da war großes Gewirr auf dem Platz, ein Stoßen und Drängen und Lärm von Trommeln und Pfeifen, in allen Fenstern lagen Damen wie ein Blu-

mengeländer bis an die Dächer herauf, wo die Schornsteinfeger aus den Rauchfängen guckten und vor Lust ihre Besen schwangen. An des Burgemeisters Hause aber war vom Balkon ein Seil gespannt über die Stadt und die Gärten weg bis zum Waldberg jenseits überm Fluß. Ein schlanker Bursch stand auf dem Geländer des Balkons in flimmernder spanischer Tracht mit wallenden Locken. Der alte Burgemeister schien wie vernarrt in das blanke Püppchen, plauderte und nickte ihm freundlich zu, daß die Sonne in den Edelsteinen seines kostbaren Hutes spielte, der Bursch reckte ihm lachend den Fuß hin, er mußte ihm mit einem großen Stücke Kreide die Sohlen einreiben. Auf einmal wendet er sich herum – ‚Das ist Sinka!' ruf' ich erstaunt meinen Kameraden zu. – Aber sie hatte uns auch schon bemerkt, und eh' wir uns durchdrängen können, nimmt sie rasch dem Burgemeister den kostbaren Hut von der Glatze, drückt sich ihn auf die Locken, und zierlich mit zwei bunten Fähnchen schwenkend und grüßend, schreitet sie unter großem Jubelgeschrei über Köpfe, Dächer und Gärten fort. Der Abend dunkelte schon, das Seil wurde unkenntlich aus der Ferne, es war, als ginge sie durch die leere Luft, die untergehende Sonne blitzte noch einmal in den Steinen am Hut, so verschwand sie wie eine Sternschnuppe jenseits überm Walde; niemand hat sie wiedergesehn."

„Meinetwegen, Stern oder Schnuppe!" fiel hier Schrekkenberger ein, tat einen Zug aus seiner Feldflasche und sang:

„Aufs Wohlsein meiner Dame,
Ein Windfahn' ist ihr Panier,
Fortuna ist ihr Name,
Das Lager ihr Quartier.

Und wendet sie sich weiter,
Ich kümmre mich nicht drum,
Da draußen ohne Reiter
Da geht die Welt so dumm.

Statt Pulverblitz und Knattern:
Aus jedem wüsten Haus
Gevattern sehn und schnattern
Alle Lust zum Land hinaus.

Fortuna weint vor Ärger,
Es rinnet Perl' auf Perl'.
,Wo ist der Schreckenberger?
Das war ein andrer Kerl!'

Sie tut den Arm mir reichen,
Fama bläst das Geleit,
So zu dem Tempel steigen
Wir der Unsterblichkeit."

Nun schwenkten die andern die Hüte, und „Vivat das hohe Brautpaar", schrien sie jubelnd, „hoch lebe unser Tempelherr der Unsterblichkeit!" Und der Dudelsack schnarrte wieder einen Tusch dazu.
Da schlugen plötzlich die großen Hunde an, die jede Nacht um ihr Lager die Runde machten, die Gesellen horchten auf, es war auf einmal alles totenstill. Man hörte in der Ferne Äste knacken, wie wenn jemand durchs Dikkicht bräche, es kam immer näher, jetzt vernahmen sie deutlich Fußtritte und Stimmen, die Wipfel der Sträucher bewegten sich schon, Schreckenberger nahm schnell seine Muskete und zielte nach der Gegend hin.
Plötzlich aber ließ er Arm und Flinte wieder sinken: „Ih, Pamphil, wo kommst denn du hergezigeunert!" rief er ganz verwundert aus. Der Puppenspieler trat aus dem Gebüsch, Seppi und Denkeli hinter ihm, die großen Hunde, denen sie Brocken zuwarf, gaben ihnen frei Geleit. Der Puppenspieler visierte erst die ganze Gesellschaft rings im Kreise scharf mit dem einen Auge, dann, da er lauter bekannte Gesichter bemerkte, nahm er das schwarze Pflaster vom andern, „Hast du wieder Mondfinsternis gemacht, um besser zu mausen?" fragte lachend der Lands-

knecht. – „Wir sind alle im abnehmenden Monde bei dem wachsenden Frieden", erwiderte Pamphil, „wir haben den faulen Bauern die Felder mit Blut gedüngt, nun schießt alles in Kraut und Rüben, die Welt wird noch ersticken vor Langerweile. Aber was treibt ihr hier, ihr alten Kriegsgurgeln, man hört euch ja eine halbe Meile weit durch die stille Nacht, ich konnt' nicht fehlen."

Nun raschelte es in allen Winkeln, immer mehr wilde Gestalten richteten sich aus dem Dunkel empor, da war des Begrüßens, Händeschüttelns und Fragens kein Ende. Wie sie aber hörten, daß Pamphil soeben von dem Schlosse kam, das sie unterwegs von fern überm Walde gesehen, trat alles um ihn herum, und da er von zwei Kavalieren droben erzählte und von einem schönen Reisewagen im Hofe, mußte er ihnen alles ausführlich beschreiben; sie zweifelten nicht, daß es die beiden Edelleute mit der Karosse seien, die sie vor einiger Zeit bei Nacht in dem Städtchen gesehen, und die ihnen dann im Walde mitten durchs Kreuzfeuer ihrer Pistolen so schnöde entwischt.

Unterdes saß Denkeli seitwärts auf einem Baumsturz, den Kopf in die Hand gestützt und ohne sich um die andern zu bekümmern; man wußte nicht, ob sie müde oder traurig. Das stach den Gesellen in die Augen, einige wollten sich galant zeigen und scharrten und kollerten wie aufgeblasene Truthähne um sie herum. Der Holkische Jäger, kecker als die andern, schlich sich leis von hinten heran, um das Mädchen zu küssen, da wandt' sie sich und gab ihm unversehens eine Ohrfeige, daß es laut klatschte. Der Überraschte griff wütend nach seinem Hirschfänger, aber der Puppenspieler, der alles bemerkt, hatte ihn schon von unten an dem einen Bein gefaßt und hob ihn so, zu allgemeinem Gelächter, mit ausgestrecktem Arm hoch über sich in die Luft. „Bleibt meiner Denkeli vom Leib", rief er mit martialischen Mienen, „oder ich mach' meine schönsten Kunststücke an eueren eignen Knochen durch." – „Laßt sie nur", sagte Denkeli, „ich werde schon allein mit ihnen fertig, heute kommen sie mir gerade recht." – Der Jäger,

da er wieder auf dem Boden war, sah den Puppenspieler halb verwundert, halb trotzig vom Kopf bis zu den Füßen an, wie ein Mops, der unverhofft auf einen Bullenbeißer gestoßen.

Denkeli aber blickte scharf zur Seite zwischen die dunklen Bäume, dort waren die andern unterdes wieder zusammengetreten und redeten heimlich untereinander in der Spitzbubensprache. Eine entsetzliche Ahnung stieg plötzlich in ihrer Seele auf, denn sie hörte von Zeit zu Zeit des reichen Fräuleins auf dem Schloß und der beiden Kavaliere erwähnen. Ihr Herz klopfte; scheinbar gleichgültig am Feuer kauernd und die Flamme schürend, horchte sie mit wachsender Angst hinüber, da erfuhr und erriet sie nach und nach alles: wie sie noch heute den Berg hinaufschleichen, das schlechtverwahrte Schloß im ersten Schlafe überfallen und die Beraubten auf ewig still machen wollten. Auch der Vater trat nun hinzu und schien mancherlei guten Rat zu erteilen.

Denkeli dachte mit Schrecken an Siglhupfer, den sie oben gesehn. Sonst achtete sie wenig auf die Anschläge der Männer, sie war von Jugend dran gewöhnt; jetzt kam ihr auf einmal alles ganz anders und unleidlich vor. Aber zu verhindern war's nicht mehr, das wußt' sie wohl, eher hätte sie den Sturmwind im Fluge wenden können. So suchte sie nach kurzem Bedenken unbemerkt die Pistolen des Vaters hervor, lud sie und legte drauf hastig ihren schönsten Putz an, ihre Augen funkelten, und wie sie auf einmal, von den schwarzen Locken umringelt, sich in ihrem Schmuck am Feuer aufrichtete, erschrak alles, so prächtig war sie. Der Vater lobte sie, daß sie etwas auf sich hielt vor den Leuten. Sie erwiderte rasch: sie wisse schon alles, sie habe sich die Gegend wohl gemerkt und wolle nach dem Schloß vorausgehn, um auszukundschaften, ob der Wald sicher, eh' die andern nachkämen. Es fiel dem Vater nicht auf, er kannte sie, wie beherzt sie war. Da stand sie noch einen Augenblick zögernd. „Lebt wohl“, sagte sie dann aus tiefstem Herzensgrund. Der Vater stutzte bei dem un-

gewöhnlich bewegten Klang der Stimme und sah ihr in Gedanken nach, aber, ihr Tamburin schwingend, war sie schon im Walde verschwunden.

6. Viel Lärmen um Nichts

Währenddes ruhte schon alles im Schloß, nur Klarinett konnte von den vielen schlagenden Nachtigallen im Garten nicht einschlafen. Der Mond schien hell durchs ganze Zimmer, manchmal bewegte die Zugluft die alten Tapeten, und wo sie zerrissen, waren auf den kahlen Wänden, dem Stammbuch müßiger Soldaten, überall Gesichter und Figuren ungeschickt mit Kohle gemalt. Seitwärts in einen weiten damastenen Schlafrock gehüllt, saß er auf dem schweren Himmelbett, an dem Himmel und Betten fehlten, und dachte über seine immer näher heran rückende Vermählung nach. – Jetzt öffnete er ungeduldig ein Fenster, der frische Waldhauch wehte ihn plötzlich über die Dächer an, da war's, als wollten die rauschenden Wipfel ihn an ein Lied erinnern, das er früher gar oft in solcher nächtlichen Einsamkeit gesungen. Er besann sich lange, dann stimmte er, halb singend, halb sprechend, leise vor sich an:

„Es ist ein Klang gekommen
Herüber durch die Lúft."

Die Weise wollte ihm durchaus nicht einfallen –

„Der Wind hat's gebracht und genommen –"

Er ärgerte sich, daß er hier alles verlernt, was ihm sonst lieb gewesen, es wurde ihm so heiß und angst, er schob's auf den ungewohnten Ungarwein und eilte endlich aus dem schwülen Gemach die stille Treppe hinab, durch ein

verborgenes Pförtchen ins Freie. Er ging so eilig durch den Garten, daß er sich alle Augenblicke in die weiten Falten des Schlafrockes verwickelte, die Mücken stachen ihn, die Gedanken jagten sich ihm durch die Seele wie die Wolken am Himmel, er wußt' sich gar nicht zu retten. „Sei kein Narr, sei kein Narr", sagte er hastig zu sich selbst, „ein Schloß, drei Weiler, vier Teiche und fette Karpfen und Untertanen und Himmelbett – Und was macht die Frau Liebste? – Danke für höfliche Nachfrage, sie wiegt – Ach und die lieben Kleinen? – sie schreien und die Wiegen rumpeln – und derweil rauscht der Wald draußen und schilt mich, und die Rehe gucken durch den Gartenzaun und lachen mich aus – ja Wald und Rehe, als wenn das alles nur so zum Einheizen und Essen wär'!"

So war er in seinem Eifer mit dem langen Schlafrock mitten ins Dickicht zwischen Dornen und Nesseln geraten, und als er sich umsah, erblickte er wahrhaftig die wunderbare Fei in einem Fensterbogen über sich. Er starrte betroffen hin, denn dieser Teil des Schlosses war völlig wüst und unbewohnt, auch kam die Gestalt ihm jetzt schlanker und ganz anders vor als Euphrosine, sie bog sich weit herüber, als säh' sie sich nach jemand aus, ihn schauerte. – Da schien sie ihn zu bemerken und verschwand schnell wieder am Fenster.

Jetzt aber hörte er zu seinem Erstaunen eine wunderschöne Stimme singen, bald näher bald ferner, wie in goldenen Kreisen um das ganze stille Haus. Er stutzte und hielt den Atem an, das Herz wurde ihm so leicht und fröhlich bei dem Klange, die Luft kam vom Schloß, er meinte die Weise zu kennen aus alter Zeit. Da schlug er sich plötzlich vor die Stirn, jetzt wußte er auf einmal das Lied, auf das er sich niemals besinnen konnte, und sang jauchzend aus frischer Brust:

„Es ist ein Klang gekommen
Herüber durch die Luft,
Der Wind hat's gebracht und genommen,
Ich weiß nicht, wer mich ruft.

Es schallt der Grund von Hufen,
In der Ferne fiel ein Schuß –
Das sind die Jäger, die rufen,
Daß ich hinunter muß!"

Und auf einmal ganz nahe unter dem Garten antwortete die Stimme:

„Das sind nicht die Jäger – im Grunde
Gehn Stimmen hin und her,
Hüt' dich zu dieser Stunde!
Mein Herz ist mir zu schwer,
Wer dich lieb hat, macht die Runde,
Steig' nieder und frag' nicht wer?
Ich führ' dich aus diesem Grunde –
Dann siehst du mich nimmermehr."

Aber Klarinett hatte schon den Schlafrock abgeworfen, er fühlt' sich auf einmal so leicht in dem alten Wanderkleide und schaute in das stille Meer der Nacht, als hört' er die Glocken gehen von den versunkenen Städten darunter, und aus dem Waldgrund tönte der Gesang immerfort dazwischen:

„Ich weiß einen großen Garten,
Wo die wilden Blumen stehn,
Die Engel frühmorgens sein' warten,
Wenn alles noch still auf den Höhn,
Manch zackiges Schloß steht darinne,
Die Rehe grasen ums Haus,
Da sieht man weit von der Zinne
Weit über die Länder hinaus –"

Klarinett erkannte die Stimme recht gut, und ganz verwirrt, zwischen den wankenden Schatten der Bäume stieg er durch den Garten in die mondbeglänzte Einsamkeit hinab, immer tiefer, tiefer, das Schloß war hinter ihm schon versunken.

Nun wurde oben alles wieder totenstill, nur der Wetterhahn auf dem Turm drehte sich unruhig im Winde hin und her, als traute er der falschen Nacht nicht und wollte die Schlafenden warnen. Da raschelt plötzlich etwas in der Ferne, lockeres Steingeröll, wie hinter Fußtritten, rollt schallend in den Abgrund, drauf wieder die alte unermeßliche Stille. Allmählich aber schien das heimliche Geknister ringsum sich zu nähern, manchmal fuhr ein verstörter Waldvogel aus dem Gebüsch, sich erschrocken in wildem Zickzack in die Nachtluft stürzend, da und dort blinkte es wie Stahl auf und funkelten wilde Augen durchs Gesträuch. Jetzt trat eine fremde Gestalt vorsichtig aus den Hecken hervor, ein zweiter und mehrere folgten von allen Seiten, die ganze Bande mit Blendlaternen, Brecheisen, Stricken und Leitern schritt sacht und lautlos dem Schlosse zu. – „Nur immer mir nach hier die Marmorstufen hinauf", flüsterte der Puppenspieler zurück. Sie arbeiteten nun, daß ihnen die Schweißtropfen aus dem struppigen Haar rannen, an der verschlossenen Tür, um sie unbemerkt zu öffnen. Andere hoben ungeduldig indes die Scheiben aus den Fenstern und legten die Leitern an, eifrig hinansteigend. Indem aber tut auch die Tür sich schon mit Krachen auf, und das ganze Gesindel durch Fenster und Tür stürzt auf einmal mitten in den Gartensaal. – „Das Fräulein!" schreit plötzlich der Puppenspieler. Euphrosine, von ihrem Diener begleitet, erschrocken, mit fliegendem Haar im Widerschein eines Windlichts tritt ihnen rasch entgegen. – „Was Teufel, die tolle Sinka", ruft da der Holkische Jäger, und alle stehn wie verzaubert.

Pamphil war der erste, der sich von seinem Erstaunen wieder erholte. „Was ist das, wie kommt ihr hierher?" fragte er den Diener, „ich traf dich doch erst vor kurzem in Halle, es war gerade Geburtstag, glaub' ich, und Maskerade in des Grafen Gerold Haus an der Stadtmauer; da sagtest du, du hättest einen Schatz drin." – „Und den hab' ich auch in der folgenden Nacht gehoben aus der Jungfernkammer auf mein Roß", entgegnete der Diener, „denn Sinka war

Kammerjungfer im Haus und ich entführte sie die Nacht nach dem Feste." – Wie die andern so viel von Schätzen hörten, schrien alle durcheinander: da stecke was dahinter, sie wüßten's wohl, Sinka hätte hier auf dem Schloß wie eine Prinzessin gelebt und aus dem gräflichen Hause mehr als ihren Abschied genommen, auch sei sie ihnen noch ihre Regimentskasse schuldig, sie sollte ihnen zur Goldtruhe vorleuchten, oder sie würden ihr das Schloß überm Kopfe anzünden.

Sinka blickte ratlos umher, wie nach einem guten Einfall, denn sie gedachte des in Halle gestohlenen Schmuckkästchens droben unter ihrem Bett und verwünschte im Herzen die beiden Kavaliere und ihr Heiratsprojekt, das sie so lange hier im Schlosse aufgehalten. Doch die Gesellen ließen keine Bedenkzeit, überwacht und in der übelsten Laune stürmten die einen schon die innere Saaltür, die andern wollten das Schlafzimmer der beiden Edelleute aufsuchen, wieder andere verrannten diesen wie jenen den Weg, um die ersten zu sein beim Fange, und jeder zankte auf den Puppenspieler, daß er sie mit seinem falschen Schloßfräulein vexiert. So gerieten endlich alle, lärmend, stoßend und über die Marmorstufen sich wieder hinabdrängend, auf dem Gartenplatz vor dem Schlosse wütend aneinander. Vergebens warf sich Sinka dazwischen und schimpfte sie wilde Gänse, die ihr ins Netz fielen und alle Maschen zerrissen, da sie eben einen jungen Goldfasan fangen wollte, morgen sei die Hochzeit mit dem Rittmeister, sie wollte ehrlich mit ihnen teilen. Keiner hörte mehr, alles stach, hieb und raufte in der stockfinstern Nacht, daß die Fetzen flogen und die Funken von den Klingen sprühten.

Da schrie plötzlich Sinka durchdringend auf, mit Entsetzen bemerkten sie auf einmal mitten unter sich ein fremdes Gesicht, jetzt wieder eins, bald da bald dort beim Streiflicht des Mondes immer mehr unbekannte Gestalten, die schweigend mit kämpften, die eine von furchtbarem Aussehn ingrimmig durch den dicksten Haufen mähend, als

föchte der Teufel mit ihnen. Da faßt alle ein unwiderstehliches Grauen, und Sinka voran, stiebt plötzlich der ganze verbissene Knäuel wie ein Nachtspuk in die Waldschluchten auseinander.
Nur der grimmige Fechter, mit zerhauenem Hute blutend auf ein Knie gesunken, verteidigte sich noch immer gegen die geisterhafte Runde der Unbekannten, die nun allein auf dem Platz zurückgeblieben. Der eine leuchtete ihm mit seiner Fackel unter die herabhängende Hutkrempe – „Ei, Herr Suppius, was machen Sie denn hier!“ rief er erschrocken zurückprallend.
Suppius – der bei dem ersten Lärm sich sogleich aus seinem Schlafgemach in das Getümmel gestürzt hatte – blickte im Kreise herum und erkannte nun mit großem Erstaunen einige reichgekleidete Jäger des Grafen Gerold aus Halle, die er damals öfters gesehn, wenn er unter den Fenstern seiner eingebildeten Geliebten vorbeistrich. Sie halfen ihm sogleich wieder auf die Beine, und da sie seine umherschweifenden fragenden Blicke bemerkten, erzählten sie ihm in aller Geschwindigkeit, wie ihrem Herrn vor kurzem, da er mit seiner Tochter im nächsten Städtchen übernachtet, eine Karosse nebst Effekten, die er auf der Reise vorausgeschickt, verwegen weggeschnappt worden, da seien sie endlich der Diebesbande auf die Spur gekommen und ihr immer dicht auf den Fersen bis hier zu des Grafen wüstem Jagdschloß gefolgt.
„Des Grafen Schloß“, fragte Suppius ganz verwirrt. Aber er hatte nicht Zeit, sich lange zu verwundern. „Wo ist der Samson, der die Philister geschlagen?“ rief ein stattlicher Herr im Garten. Es war der Graf Gerold selbst, der, sich rasch vom Pferde schwingend, herzutrat und den abenteuerlichen Studenten mit heimlichem Lächeln betrachtete. Hinter ihm hielt seine Tochter, im ersten Morgenlicht mit den wallenden Federn vom Zelter nickend. – „Das ist sie wirklich und leibhaftig!“ dachte Suppius überrascht.
Nun war unter den Schalken ringsum viel Rühmens von dem wütenden Studenten, der wie ein Sturmwind das Ge-

sindel auseinandergeblasen. Indem hatten die Jäger im Schloßhofe auch die verschwundene Karosse entdeckt, andre brachten soeben den verlorenen Reisekoffer mit den Staatskleidern und das gestohlene Schmuckkästchen herbei. Der lustige Graf, ohne lange zu kramen, zog sogleich eine schwere goldne Kette hervor, aus lauter St. Jürgen und Lindwürmern künstlich zusammengefügt, und reichte sie seiner Tochter, die mußte sie feierlich dem tapfern Retter des Schlosses um den Hals hängen. Dann gab er seinen Leuten einen Wink. Da setzten sie rasch die Trompeten an und bliesen dem Suppius zu Ehren einen schmetternden Tusch, während die andern, eh' er sich's versah, ihn auf ihre Schultern schwangen und so im Triumph ins Schloß zum Frühstück trugen.

Unterdes war der Tag schon angebrochen, Suppius konnte von seinem luftigen Sitz weit über die Hecken weg ins Tal schauen. Da sah er, zu neuem Erstaunen, unten seinen Gefährten Klarinett zu Roß, seine Denkeli vor sich im Sattel, wie einen Morgenblitz am Saum des Waldes dahinfliegen. Siglhupfer (denn niemand anders war Klarinett) hatte sich nicht getäuscht: Denkeli, entschlossen, mit Gefahr ihres eigenen Lebens ihn zu warnen und zu retten, war die singende Fei im Fenster gewesen – nun verstand er erst die Sage; so weit man vom Turm des Schlosses sehen konnte, es war ja alles, alles wieder sein!

Oben aber schmetterten jetzt von frischem die Trompeten, Vivat und Jubelgeschrei, und hinter sich sah Suppius die Hüte schwenken und Weinflaschen blinken und die schönen Augen der jungen Gräfin dazwischen funkeln. – So hatte er, wie man die Hand umdreht, sein Glück gemacht. – Siglhupfer aber blieb fortan in den Wäldern selig verschollen.

Libertas und ihre Freier

Ein Märchen

Es war einmal ein Schloß in Deutschland mit dicken Pfeilern, Bogentor und Türmchen, von denen Wind und Regen schon manchen Schnörkel abgebissen hatten. Das Schloß lag mitten im Walde und war sehr verrufen in der ganzen Gegend, denn man wußte nicht, wer eigentlich darin wohnte. Jemand konnte es nicht sein, sonst hätte man ihn doch manchmal am Fenster erblicken müssen; und niemand auch nicht, denn in dem Schlosse hörte man bei Tag und Nacht beständig ein entsetzliches Rumoren, Seufzen, Stöhnen und Zischen, als würde drin die Welt von neuem erschaffen; ja des Nachts fuhr bald da, bald dort ein Feuerschein aus einem der langen Schornsteine oder Fenster heraus, als ob gequälte Geister plötzlich ihre lechzenden Zungen ausstreckten. Über dem Schloßportal aber befand sich eine überaus künstliche Uhr, die mit großem Geknarre Stunden, Minuten und Sekunden genau angab, aber aus Versehen rückwärts fortrückte und daher jetzt beinahe schon um fünfzig Jahre zu spät ging; und jede Stunde spielte sie einen sinnigen Verein gebildeter Arien zur Veredelung des Menschengeschlechts, zum Beispiel:

„In diesen heil'gen Hallen
Kennt man die Rache nicht –
Und Ruhe ist vor allen
Die erste Bürgerpflicht usw."

Die benachbarten Hirten, Jäger und andere gemeinen Leute aber waren das schon gewöhnt und fragten nicht viel darnach, denn sie wußten ohnedem von der Sonne schon besser, was es an der Zeit war, und sangen unbe-

kümmert ihre eigenen Lieder. Wer aber recht genau aufpaßte, der konnte wirklich zuweilen zur Nachtzeit oder in der schwülen Mittagstille den Schloßherrn aus dem großen Uhrportal hervortreten und auf den einsamen Kiesgängen des Ziergartens lustwandeln sehen; einen hagern, etwas schiefbeinigen Herrn mit gebogener Nase und langem Schlafrock, der war von oben bis unten mit allerlei Hieroglyphen und Zaubersprüchen verblümt und punktiert und hatte unten einige Zimbeln am Saume, die aber immer gedämpft waren, um ihn nicht im Nachdenken zu stören. Das war aber niemand anders als der Baron Pinkus, der große Nekromant, und die Sache verhielt sich folgendermaßen:

Vor geraumer Zeit und bevor er noch Baron war, hatte der Staatsbürger Pinkus auf dem Trödelmarkte in Berlin den ganzen Nachlaß des seligen Nicolai (der damals gerade altmodisch geworden, weil soeben die Romantik aufgekommen war) für ein Lumpengeld erstanden und machte in Ideen. Er war ein anschlägiger Kopf und setzte die Ware ab, wo sie noch rar war. So war er denn eines Tages an das abgelegene Schloß eines gewissen Reichsgrafen gekommen. Der Graf saß gerade in freudenreichem Schalle an der Mittagstafel mit seinem Stallmeister, Hofmarschall und dem anderen Hofgesind. Da riß es plötzlich so stark an der Hausglocke, daß die Kanarienvögel, Papageien und Pfauen vor Schreck zusammen schrien und die Puthähne im Hofe zornig zu kollern anfingen. Der Graf rief: „Wer ist da draußen vor dem Tore?" Der Page rief: „Was wollen Sie, mein Herr?" – „Menschenwohl, Jesuiten wittern und Toleranzen." – Der Page kam: „Dem Menschen ist nicht wohl, er will einen Bitteren oder Pomeranzen." – „Das verdenk' ich ihm nicht", entgegnete der Graf, „aber geh und frag' noch einmal genauer, wer er sei." – Der Page ging: „Ihr Charakter, mein Herr?" – „Kosmopolit!" – Der Page kam: „Großhofpolyp." – Das Brockhausische Konversationslexikon war damals noch nicht erfunden, um darin nachschlagen zu können, es entstand daher ein

allgemeines Schütteln des Kopfes, und der Graf war sehr neugierig, die neue Hofcharge kennen zu lernen. So wurde nun Pinkus eingelassen und trat mit stolzer Männerwürde in den Saal, und nachdem die notwendigen Bewillkommnungskomplimente zu beiderseitiger Zufriedenheit glücklich ausgewechselt waren, begann er sogleich eine wohlstilisierte Rede von der langen Nacht, womit die schlauen Jesuiten das Land überzogen, kam dann auf den großen Nicolai, wie derselbe, da in dem Stichdunkel alle mit den Köpfen aneinanderrannten, in edler Verzweiflung seinen unsterblichen Zopf ergriff, ihn an seiner Studierlampe anzündete und mit dieser Fackel das Volk der Tugendusen, die bloß von Moral lebten, siegreich bis mitten in die Ultramontanei führte. – Hier nahm der Hofmarschall verzweiflungsvoll eine Prise, und verschiedene Kavaliere gähnten heimlich durch die Nase. Aber Pinkus achtete nicht darauf, sondern fing nun an, den besagten Nicolaischen Zopf ausführlich in seine einzelnen philosophischen Bestandteile zu entwickeln. „Das ist ja nicht auszuhalten!" rief der Oberstallmeister mit schwacher, kläglicher Stimme, die anderen stießen schon schlummernd mit ihren Frisuren gegeneinander, daß der Puder stob, die Pfauen draußen hatten längst resigniert die Köpfe unter die Flügel gesteckt, im Vorzimmer schnarchte die umgefallene Dienerschaft fürchterlich auf Stühlen und Bänken. Es half alles nichts, der unaufhaltsame Pinkus zog immer neue, lange, vergilbte Papierstreifen aus dem erstandnen Nachlaß, rollte sie auf und murmelte fort und immerzu von Aufklärung, Intelligenz und Menschenbeglückung. – „Sapperment!" schrie endlich der Graf voll Wut und wollte aufspringen, aber er konnte nicht mehr, sondern versank mit dem ganzen Hofstaat in einen unendlichen Zauberschlaf, aus dem sie alle bis heute noch nicht wieder erwacht sind.

„Man muß nur haben Verstand!" rief da der böse Nekromant und rieb sich vergnügt die Hände, legte sie aber nicht müßig in den Schoß, denn durch die offnen Türen, da nie-

mand mehr da war sie zuzumachen, kam der Wind dahergepfiffen und griff unverschämt nach seinen Papieren; aus der großen Kristallflasche, die der Hofmarschall beim Einschlafen umgeworfen, war ihm das Wasser in die Schnallenschuhe gestürzt, und die Kerze, woran sie ihre Pfeifen anzuzünden pflegten, flackerte unordentlich und wollte durchaus die seidene Gardine anstecken. Pinkus aber hatte sie alle schon lange auf dem Korn und eine gründliche Verachtung vor der Luft, dem landstreicherischen Windbeutel, sowie vor dem Wasser, das keine Balken hat und immer nur von Stein zu Stein springen, glitzern, schlängeln und die unnützen Vergißmeinnichts küssen möchte, und vor dem Fenster, das nichts tut, als vertun und verzehren. Er trat daher entrüstet in den Garten hinaus, zivilisierte ohne Verzug jene ungeschlachten Elemente durch seine weitschweifigen Zaubersprüche, die keine Kreatur lange aushält, und stellte sie dann in dem verstorbenen Schlosse an. In demselben Schlosse aber legte er sofort eine Gedankendampffabrik an, die ihre Artikel zu Benjowskys Zeiten bis nach Kamtschatka absetzte und eben den außerordentlichen Lärm machte, den sich die dummen Leute in der Umgegend nicht zu deuten wußten.
So war also der Staatsbürger Pinkus ein überaus reicher Mann und Baron geworden und befand, daß alles gut war.

Seitdem waren viele Jahre vergangen, da gewahrte man in einer schönen Nacht dort in der Gegend ein seltsames Zittern und Blinkern in der Luft, als würde am Himmel ganz was Absonderliches vorbereitet. Die Vögel erwachten darüber und reckten und dehnten noch verschlafen ihre Flügel, da sahen sie droben auch den Adler schon wach und fragten erstaunt:

„Was gibt's, daß vom Horste
An der zackigen Kluft

Der Adler schon steigt
Und hängt überm Forste
In der stillen Luft,
Wenn alles noch schweigt?“

Der Adler aber vernahm es und rief hinab:

„Ich hörte in Träumen
Ein Rauschen gehn,
Sah die Gipfel sich säumen
Von allen Höhn –
Ist's ein Brand, ist's die Sonne,
Ich weiß es nicht,
Aber ein Schauer voll Wonne
Durch die Wälder bricht.“

Jetzt schüttelten die Vögel geschwind den Tau von den bunten Wämschen und hüpften und kletterten nun selber in ihrem grünen Hause bis in die allerhöchsten Wipfel hinaus, da konnten sie weit ins Land hinaussehen, und sangen:

„Sind das Blitze, sind das Sterne?
Nein, der Aar hat recht gesehn,
Denn schon leuchtet's aus der Ferne,
Daß die Augen übergehn.

Und in diesen Morgenblitzen
Eine hohe Frau zu Roß,
Als wär' mit den Felsenspitzen
Das Gebirge dort ihr Schloß.

Geht ein Klingen in den Lüften,
Aus der Tiefe rauscht der Fluß,
Quellen kommen aus den Schlüften,
Bringen ihr der Höhen Gruß.

Und die grauen Schatten sinken,
Wie sie durch die Dämmrung bricht,
Und die Kreaturen trinken
Dürstend alle wieder Licht.

Ja, sie ist's, die wir da schauen,
Unsre Königin im Tal!
O Libertas! schöne Fraue,
Grüß' dich Gott vieltausendmal!"

„Habt Dank, meine lustigen Kameraden!" rief da eine wunderliebliche Stimme, die wie ein Glöcklein durch die Einsamkeit klang, und die Lerche stieg sogleich kerzengerade in die Höh' und jubilierte: „Die Libertas ist da, die Libertas ist da!" – es wollt's niemand glauben. Sie war's aber wirklich, die soeben zwischen dem Gesträuche auf den Schloßberg heraustrat. Sie ließ ihr Rößlein frei neben sich weiden und schüttelte die langen, wallenden Locken aus der Stirn; die Bäume und Sträucher hatten sie ganz mit funkelndem Tau bedeckt, daß sie fast wie eine Kriegsgöttin in goldner Rüstung anzusehen war. Hinter ihr aber, wo sie geritten, zog sich's wie eine leuchtende Furt durchs Land, denn sie war über Nacht gekommen, der Mond hatte prächtig geschienen und die Wälder seltsam dazu gerauscht, in den Tälern aber schlief noch alles, nur die Hunde bellten erschrocken in den fernen Dörfern, und die Glocken auf den Türmen schlugen von selbst an, wo sie vorüberzog.

„Ich wollte doch auch wieder einmal meine Heimat besuchen", sagte sie jetzt, „die schönen Wälder, wo ich aufgewachsen. Da ist viel abgeholzt seitdem, das wächst sobald nicht wieder nach auf den kahlen Bergen." Nun erblickte sie erst das geheimnisvolle Schloß und den Ziergarten. „Aber wo bin ich denn hier hingeraten?" fragte sie erstaunt. Es schwieg alles; was wußten die Vögel von dem Baron Pinkus! Es war ihr alles so fremd, sie konnte sich gar nicht zurechtfinden. „Das ist die Burg nicht mehr, wo

sonst meine liebsten Gesellen gewohnt. Mein Gott! wo sind die alten Linden hin, unter denen wir damals so oft zusammengesessen?" – Darüber wurde sie auf einmal ganz ernsthaft, trat an den Abhang und sprach laut in die Tiefe hinaus:

„Die gebunden da lauern,
Sprengt Riegel und Gruft,
Du ahnend Schauern
Der Felsenkluft,
Unsichtbar Ringen
In der stillen Luft,
Du träumend Singen
Im Morgenduft!
Brecht auf! schon ruft
Der webende blaue
Frühling durchs Tal!"

Und die Vögel jubelten wieder:

„O Libertas, schöne Fraue,
Grüß' dich Gott vieltausendmal!"

Da ging erst ein seltsames Knistern und Flüstern durch die Buchsbäume und Spaliere, fast grauenhaft, wie wenn sie heimlich miteinander reden wollten in der großen Einsamkeit, drauf kam von den Waldbergen auf einmal ein Rauschen immerfort wachsend über den ganzen Garten, es war, als stiege über die Hecken und Gitter von allen Seiten verwildernd der Wald herein, die Fontäne fing wie eine Fee mit kristallenen Gewändern zu tanzen an, und Krokus, Tulipanen, Königskerzen und Kaiserkronen kicherten lustig untereinander; im Schloß aber entstand zu gleicher Zeit ein entsetzliches Krachen und Tosen, daß alle Türen und Fenster aufsprangen. Da kam plötzlich Pinkus, ganz verstört und zerzaust, aus dem Haupttore mit solcher Vehemenz dahergeflogen, daß die Schöße seines

punktierten Schlafrocks weit hinter ihm drein rauschten. Er wollte vernünftig reden, aber der Frühlingssturm hatte ihn mit erfaßt, er mußte zu seinem großen Ärger in lauter Versen sprechen und schrie ingrimmig:

„Bin ich selber von Sinnen?
Im Schlosse drinnen
Ein Brausen, Rumoren,
Alles verloren!
Die Wasser, die Winde,
Das Feuer, das blinde,
Die ich besprochen,
Wild ausgebrochen,
Die rasen und blasen
Aus feurigen Nasen,
Mit glühenden Blicken,
Brechen alles in Stücken!"

Hier stutzte er auf einmal, er hatte die Libertas erblickt, da schoß ihm plötzlich das Blatt. Er kannte sie zwar nicht von Person, aber der schlaue Magier wußte nun sogleich, wer die ganze Verwirrung angerichtet. Ohne Verzug schritt er daher auf sie los und forderte ihren Paß. Sie betrachtete ihn von oben bis unten, er sah vom Schreck so windschief und verschoben aus; sie mußte ihm hellaut ins Gesicht lachen. Da wurde er erst recht wild und rief die bewaffnete Macht heraus, die sich nun von allen Seiten mit großer Anstrengung mobil machte, denn der Friedensfuß, auf dem sie solange gestanden, war ihr soeben etwas eingeschlafen. Libertas stand unterdessen wie in Gedanken und wußte gar nicht, was die närrischen Leute eigentlich wollten. Doch sie sollte es nur zu bald erfahren. Pinkus befahl, die gefährliche Landstreicherin im Namen der Gesittung zu verhaften. Sie ward eiligst wie ein Wickelkind mit Stricken umwunden und ihr, in gerechter Vorsicht, darüber noch die Zwangsjacke angelegt. Da hätte man sehen sollen, wie bei dieser Arbeit manchem würdigen Krie-

ger eine Träne in den gewichsten Schnurrbart herabperlte; aber der Patriotismus war groß, und Stockprügel tun weh. So wurde Libertas unter vielem Lärm in das mit dem Schlosse verbundene Arbeitshaus abgeführt.
Pinkus aber, nachdem er sich von der Alteration einigermaßen wieder erholt hatte, schrieb sogleich ein großes Renaissancefest aus, das in einem feierlichen Aufzuge aus dem chinesischen Lusthause nach dem Schloß bestand und wohl einer würdigeren Feder wert wäre. Da sah man nämlich zuerst zwölf weißgekleidete Mädchen, eine hinter der anderen vorschreitend, in den chinesischen Saal hereinschweben, sie trugen auf ihren Achseln eine wunderliche Festgabe, die wie eine lange Wurst oder wie ein gräulicher Wurm aussah. Damit traten sie in einer Reihe vor Pinkus, stellten sich auf das eine Bein und streckten das andere anmutsvoll in die Luft, während eine jede die rechte Hand auf ihr Herz legte, mit der linken aber das langschweifige Weihopfer hoch in die Höhe hob und alle lieblich dazu sangen:

„Wir bringen dir der Treue Zopf
Von eigner Locken Seide,
Lang' trag ihn dein erhabner Kopf
Zu deines Landes Freude,
Kopf, Zopf und Lockenseide!"

Es war wirklich ein ungeheurer Zopf, den sie eiligst aus ihren eigenen Locken zusammengewunden hatten. Der gerührte Pinkus riß sich sofort den Haarbeutel vom Haupt, verehrte ihn unter angemessenen Worten den Jungfrauen, um ihn als teures Andenken in dem Prüfungssaale ihrer Pensionsanstalt aufzuhängen, und ließ sich dann den patriotischen Zopf im Genick befestigen, was sich sehr feierlich ausnahm, denn er schleppte ihn hinten etwas nach, so daß ihm jeder drei Schritt vom Leibe bleiben mußte, um nicht unversehens darauf zu treten. Jetzt aber begann der Zug durch den Garten. Voran schritten,

wie eine Schar schneeweißer Gänse, die glücklichen Jungfrauen mit dem Haarbeutel auf sammetnem Kissen, ihnen folgte der Haushofmeister, an dessen Allongeperücke in der feuchten Abendluft die Locken aufgegangen waren und wie ein Fürstenmantel fast bis an die Fersen herabfielen, endlich kam Pinkus selbst, dem der Kammerdiener den Zipfel des Opferzopfes ehrerbietig nachtrug. Auch der Ziergarten, der seit Libertas gebunden war, hatte unterdes seine vorige würdige Haltung wiedergewonnen, und wo Pinkus vorüberschritt, präsentierte der marmorne Herkules mit seiner Keule, der geigende Apollo salutierte mit dem Fiedelbogen, und die Tritonen in den steinernen Becken bliesen auf ihren Muscheln aus Leibeskräften: Heil dir im Siegerkranz!

Die Geschichte machte damals großes Aufsehn in Deutschland. Die Schwalbe schoß ängstlich hin und her und schwatzte, und schrie von allen Dächern und Zäunen: „Weh, weh, Frau Libertas ist gefangen!“ Die Lerche stieg sogleich wieder kerzengerade in die Höh' und meldete es dem Adler, die Nachtigall schluchzte und konnt' sich gar nicht erholen, selbst der Uhu seufzte einigemale tief auf; die Rohrdommel aber trommelte sofort Alarm, und der Storch marschierte im Paradeschritt durch alle Wiesen und Felder und klapperte unablässig zum Appell. Bald wurde es auch weiter im Walde lebendig; der Hase duckte sich im Kohl und mochte von der ganzen Sache nichts wissen, der Fuchs wollte erst abwarten, welche Wendung sie nehmen würde; der biedere Bär dagegen ging schnaubend um und wurde immer brummiger, und die Hirsche rannten verzweiflungsvoll mit ihren Geweihen gegen die dicksten Eichen, oder fochten krachend miteinander, um sich in den Waffen zu üben.
Da kam zur selben Stunde der Doktor Magog dahergewandert, der seinen Verleger nicht finden konnte und daher soeben in großer Verlegenheit war. Der hörte mit Ver-

wunderung das ungewöhnliche Geschrei der Vögel; durch einen entflogenen Star, der reden gelernt, erfuhr er alles, was geschehen, und wollte aus der Haut fahren über diese Nachricht. „Ha!“ rief er, und dabei fuhr ihm wirklich der Ellbogen aus dem Ärmel. Aber sein Entschluß war sogleich gefaßt: er wandte sich eiligst seitwärts nach dem Walde hin. Da erblickte ihn ein Köhler von fern und rief ihm zu, wohin er ginge. – „Zum Urwald“, erwiderte Magog. – „Seid Ihr toll?“ schrie der Köhler wieder herüber:

„Kehrt um auf der Stelle,
Dort steht ein Haus,
Da brennt die Hölle
Zum Schornstein heraus,
Und auf der Schwelle
Tanzt der Teufel Kehraus.“

„Laßt ihn tanzen!“ entgegnete Magog und schritt stolz weiter. Der fromme Köhler sah ihm nach, bis er im Walde verschwunden war. „So gnad’ ihm Gott“, sagte er dann und schlug ein Kreuz. Magog aber räsonierte noch lange innerlich: „Abergläubisches Volk, das im Mittelalter und in der Religion steckengeblieben! Darum wächst auch der Wald hier so dumm ins Blaue hinein, daß man keinen vernünftigen Fortschritt machen kann.“
So war er eine Weile durch das Dickicht vorgedrungen, als er unverhofft eine dünne Gestalt sehr eilfertig auf sich zukommen sah. Es war eine lange, hagere alte Dame in ganz verschossenem altmodischem Hofstaat, das graue Haar in lauter Papilloten gedreht, wie ein gespickter Totenkopf, die hatte unter jedem Arm eine große Pappschachtel, hielt mit der einen Hand ein zerrissenes Parasol über sich und stützte sich mit der anderen auf einen Haubenstock. – „Ist das der rechte Weg zum Urwald?“ fragte Magog. – „Gewiß, leider, mein Herr“, erwiderte die Dame, sich feierlich verneigend. „Ja“, setzte sie dann mit außerordentli-

cher Geschwindigkeit in *einem* Striche fortredend hinzu – „ja, diese bäuerische ungesittete Nachbarschaft macht sich von Tag zu Tag breiter, besonders seit einigen Tagen, man sagt, die famose Libertas sei wieder einmal in der Luft, es ist nicht mehr auszuhalten in dieser gemeinen Atmosphäre, keine Gottesfurcht mehr vor alten Familien, aber ich hab' es meinem hochseligen Herrn Neveu immer vorausgesagt, das war auch so ein herablassender Volksfreund, wie sie es nennen, ja das eine Mal embrassierte er sich gar mit dem Pöbel, da haben sie ihn jämmerlich erdrückt, und nun gar wir Jungfrauen sind beständigen Attacken ausgesetzt, und so sehe ich mich soeben bemüßiget zu emigrieren; o Sie glauben gar nicht, mein Herr, was so eine arme Waise von Distinktion sich zerärgern muß in der gegenwärtigen Abwesenheit aller Tugenden von Stande!" Hier kam sie vor großem Eifer ins Singen und machte plötzlich einen langen, feinen Triller wie eine verdorbene Spieluhr, bis sie sich endlich ganz verhustete. Magog, der ihr voll Erstaunen zugehört, brach in ein schallendes Gelächter aus. Darüber geriet die Dame in solchen Zorn, daß sie verächtlich und ohne Abschied zu nehmen eiligst weiter emigrierte. – „Ohne Zweifel die Urtante, da kann ich nicht mehr weit haben", dachte Magog und schritt getrost wieder vorwärts.

Bald aber verlor sich der Fußsteig vor seinen Füßen, der Forst wurde immer wilder und dichter, von fern nur sah er eine seltsame Rauchsäule über die Wipfel aufsteigen; da gedachte er der Warnung des Köhlers und des wüsten Hauses, aus dem die Hölle brennen sollte. Aber ein rauchender Schornstein war ihm von jeher ein anziehender Anblick, und so klomm er mühsam eine Anhöhe hinan, um das ersehnte Haus zu entdecken. Doch zu seinem Schrecken bemerkte er, daß es ringsum bereits zu dunkeln anfing. Jetzt begann es auch unten am Boden schon sich geheimnisvoll zu rühren, Eidechsen raschelten durch das trockene Laub, die Fledermäuse durchkreuzten mit leisem Flug die Dämmerung, aus den feuchten Wiesen kro-

chen und wanden sich überall trägringelnd lange Nebelstreifen und hingen sich an die Tannenäste wie Trauerflöre, und als Magog endlich droben ins Freie trat, stieg die kühle stille Nacht über die Wälder herauf und bedeckte alles mit Mondschein. Auch die Rauchsäule konnte er nicht mehr bemerken, es war, als hätte die fromme Nacht die Hölle ausgelöscht. Da beschloß er, hier oben den Morgen abzuwarten, streckte sich auf das weiche Moos hin, schob sein mit Manuskripten vollgepfropftes Reisebündel unter den Kopf, betrachtete dann noch eine Zeitlang die zerrissenen Wolken, die über ihm dahinjagten und manchmal wie Drachen nach dem Monde zu schnappen schienen, und war endlich vor großer Müdigkeit fest eingeschlafen.
So mochte er eine geraume Zeit geruht haben, da meinte er mitten durch den Schlummer ein Geflüster zu vernehmen und dazwischen ein seltsames Geräusch, wie wenn ein Messer auf den Steinen gewetzt würde. Die Stimmen kamen immer näher und näher. „Er schläft", sagte die eine, „jetzt ist's die rechte Zeit." – „Ein schlechter Braten", entgegnete eine andere tiefe Stimme, „er ist sehr mager, hab' seinen Futtersack untersucht, den er unterm Kopfe hat, er lebt bloß von Papier." – Nun schien es dem Magog, als hörte er auch die emigrierte Tante leise und eifrig dazwischenreden in verschiedenen unbekannten Sprachen, die anderen antworteten ebenso, die Wipfel rauschten verworren drein, auf einmal schlug sie wieder ihren schrillen Triller. Da sprang Magog ganz entsetzt auf – es war ein heiserer Hahn, der fern im Tale krähte. Verstört blickte er um sich, der Morgen blitzte zu seinem Erstaunen schon über die Wälder, er wußte nicht, ob ihm das alles nur geträumt oder sich wirklich ereignet hatte.
Jetzt sah er auch die Rauchsäule von gestern wieder emporwirbeln, er hielt es für einen unverhofften feuerspeienden Berg. Als er indes näher kam, erkannte er, daß es nur eine ungeheure Lehmhütte war, in welcher wahrscheinlich das Frühstück gekocht wurde. In diesen tröstlichen Gedanken ging er also unaufhaltsam darauf los. Auf ein-

mal aber blieb er ganz erschrocken stehen. Denn auf dem Rasenplatze vor der Hütte war ein Riesenweib wahrhaftig soeben damit beschäftigt, ein großes Schlachtmesser zu wetzen. Sie schien ihn nicht zu bemerken oder weiter nicht zu beachten, weil er so klein war, und in demselben Augenblick brachen auch mehrere Riesenkinder mit großem Geschrei aus der Hütte und zankten und würgten und rauften untereinander, daß die Haare davonflogen. Über diesem Lärme aber erhob sich plötzlich eine wunderbare, baumlange Gestalt und gähnte, daß ihr die Morgensonne bis tief in den Schlund hineinschien. Der Mann war greulich anzusehen, ungewaschen und ungekämmt, wie ein zerzaustes Strohnest, und hatte eine ungeheure Wildschur an, die war aus lauter Lappen und Fetzen von Fuchsbalg, wilden Schweinshäuten und Bärenfellen zusammengeflickt. – „Herr *Rüpel?!*" rief da Magog in freudigem Erstaunen. – „Wer ruft mich?" erwiderte der Riese noch halb im Schlafe und sah den Fremden verwundert an. – „*Sie* eben hab' ich aufgesucht", entgegnete Magog, „eine höchst wichtige Angelegenheit." – Aber Rüpel hatte gerade mit der Kindererziehung zu tun. „Hetzoh!" schrie er den Jungens zu, die noch immer fortrauften, „du da wirst dich doch nicht unterkriegen lassen, frisch drauf!" Dann streckte er unversehens sein langes Bein vor, da stürzten und kollerten die Verbissenen plötzlich verworren übereinander, während die Riesenmutter voller Zorn ihren Kehrbesen mitten in den Knäuel warf. Darüber kamen alle in ein so herzhaftes Lachen, daß der Wald zitterte.

Da nun Magog die Familie in so guter Laune sah, faßte er sich ein Herz und rückte sogleich mit seinem eigentlichen Plane heraus. „Herr Rüpel", sagte er, „ich bin ein Biedermann und kenne kein Hofieren und keinen Hof, als den Hühnerhof meiner Mutter, aber das muß ich Ihnen rund heraussagen: Ihre Macht und Gesinnungstüchtigkeit ist durch ganz Europa ebenso berühmt als geschätzt und ebenso geschätzt als gefürchtet. Darum wende ich mich

vertrauensvoll an Ihr großes Herz und rufe: Wehe und abermals wehe! die Libertas ist geknechtet! – wollen wir das dulden?" – „Libertas? wer ist die Person?" fragte Rüpel. – „Libertas?" erwiderte Magog, „Libertas ist die Schutzpatronin aller Urwälder, die Patronin dieses langweiligen – wollt' sagen: altheiligen Waldes." – „I bewahre", fiel ihm hier die Riesin ins Wort, „unsere Grundherrschaft ist das gnädige Fräulein Sibylla da draußen." – „Was? die mit den Papilloten und großen Haubenschachteln?" rief Magog, den dieser unerwartete Einwurf ganz aus dem Konzept gebracht hatte. Aber er faßte sich bald wieder. „Grundherrschaft!" fuhr er fort, „schützt die Grille Krokodile, der Frosch das Rhinozeros, der Weißfisch den Haifisch? – Wer die Macht hat, ist der Herr, und Ihr habt die Macht, wenn die Libertas regiert, und habt die Macht nicht, wenn die Libertas gefangen ist, und die Libertas ist gefangen – ich frage also nochmals, wollen wir das dulden?"

Hier aber wurde er, da er eben im besten Zuge war, durch einen seltsamen Auftritt unterbrochen. Ein Reiher kam nämlich pfeilschnell dahergeschossen, setzte sich gerade auf seinen zerknitterten Kalabreser, drehte ein paarmal mit dem dünnen Halse, verneigte sich dann feierlich vor der Gesellschaft und sagte: „Sie lassen alle ihren Respekt vermelden und es tut ihnen sehr leid, aber sie können heut und morgen nichts bringen, wir haben alle außerordentlich Wichtiges zu tun; schönen guten Morgen!" Und damit sich abermals höflich verneigend, schwang er sich wieder in die Lüfte. – „Guten Morgen, Herr Fischer", erwiderte Rüpel, ihm ganz verblüfft und mit einer verzweifelten Resignation nachschauend. Jetzt sah man auf einmal auch einen ungeheuren Schwarm wilder Gänse über den Wald fortziehen, einen alten gewiegten Gänserich voran, alle die Hälse wie Lanzen weit vorgestreckt und in einem spitzen Keile dahinstürmend, als wollten sie den Himmel durchbrechen, und dabei machten sie ein so entsetzliches kriegerisches Geschreie, daß man sein eigenes

Wort nicht hören konnte. Währenddes aber hatte der eine Riesenknabe sich mit dem Ohre auf den Boden gelegt und sagte: „Draußen im Grunde hör' ich ein groß' Getrampel, man kann die Tritte deutlich unterscheiden: Hirsche, Auerochsen, Bären, Damhirsche, Rehe, zieht alles wild durcheinander den großen See entlang." – „Die Tollköpfe!" rief die Riesenmutter aus, „da haben sie gewiß wieder Verdruß gehabt mit dem gnädigen Fräulein und haben unseren guten Wald in Verruf getan und wandern aus; denn das Fräulein ist ihnen immer spinnefeind gewesen und ließ sie mit Hunden hetzen und schinden und braten obendrein."

„Nein, nein, die alte Spinne ist ja selber ausgewandert, ich bin ihr gestern begegnet", sagte Magog voll Verwunderung, „aber warum nehmen Sie sich denn die Sache so sehr zu Herzen, teuerste Frau von Rüpel?"

„Wie sollt' ich nicht!" erwiderte die Riesin, „ach wir armen Waldleute müssen uns gar kümmerlich durchhelfen mit der großen Familie. Sehen Sie, lieber Herr, ich und mein Mann arbeiten hier für die vornehmen Tiere: Hirsche, Rehe und anderes Hochwild um Tageslohn, den wir von ihnen in Naturalien beziehen. Des Abends spricht mancher Edelhirsch bei uns ein, wenn er nachts auf die Freite gehen will, da muß ihm mein Mann die Pelzstiefelchen putzen, dafür erhalten wir denn die Felle der verunglückten Kameraden und die abgeworfenen Geweihe in die Wirtschaft. Alle Morgen aber kommen die Bären und lassen sich ihre Pelze ausklopfen und bringen uns große Honigfladen, oder ein paar wilde Schweine lassen sich ihre Hauer schleifen und werfen uns zum Dank einen fetten Frischling auf die Schwelle, denn die Zeiten sind schlecht, da kommt es ihnen auf ein Kind mehr oder weniger nicht an. Ich aber flechte Nester für die Adler, Habichte und Auerhühner, und die lassen uns dann im Vorüberfliegen einen Hasen oder ein Zicklein herunterfallen oder legen uns nachts einige Schock Eier vor die Tür, wenn sie eben nicht Lust haben, alle auszubrüten. Und

nun – ach das große Unglück! jetzt haben wir unsere Kundschaft verloren und stehen ganz verlassen in der Welt, o! o!" – Und hier fing sie jämmerlich zu heulen an, und der Riese, der sich lange gehalten, stimmte plötzlich furchtbar mit ein.

Da trat Magog mannhaft mitten unter sie. „Das soll bald anders werden!" rief er; „kennt ihr das Schloß des Baron Pinkus?" Der Riese entgegnete, er habe es wohl von fern gesehen, wenn er manchmal zur Unterhaltung bis an den Rand des Waldes gegangen, um die Köhler und andere kleine Leute zu schrecken. – „Nun gut", fuhr Magog fort, „dort eben sitzt die Libertas gefangen. Seht, mich hat auch die Welt nur auf elende Lorbeeren gebettet, daß ich mir an dem stacheligen Zeug schon den ganzen Ärmel am Ellbogen durchgelegen; darum habe ich ein Herz für das arme Riesenvolk. Die Libertas ist eine reiche Partie, wir müssen sie befreien! Dabei kann es vielleicht einige Püffe setzen, was frag' ich darnach! Ihr habt ja ein dickes Fell, alles für meine leidenden Brüder! Mit *einem* Wort: Ihr befreit sie und ich heirate sie dann und Ihr seid auf dem Schlosse Portier und Schloßwart und Haushofmeister, eh' man die Hand umdreht. Topp, schlagt ein – aber nicht zu stark, wenn ich bitten darf."

Darüber war Rüpel ganz wild geworden und schritt, ohne ein Wort zu sagen, so eilig in die Hütte, daß Magog nur mühsam und mit vorgehaltenen Händen tappend folgen konnte. Denn sie stiegen über viele ungeschickte Felsenstufen in eine große Höhle hinab, über welcher der Berg, den Magog für die Hütte gehalten, nur das Dach und den Schornstein bildete. Im Hintergrunde der Höhle hing ein Kessel über dem Feuer, ein zahmer Uhu mit großen funkelnden Augen saß in einem Felsenspalt daneben und fachte mit seinen Flügeln die Flamme an und schnappte manchmal nach den Fledermäusen, die geblendet nach dem Feuer flogen. Die Flamme warf ein ungewisses Licht über die rauhen und wunderlichen Steingestalten umher, die bei den flackernden Widerscheinen sich heimlich zu

bewegen schienen, und mächtige Baumwurzeln drängten sich überall wie Schlangen aus den Wänden, in der Tiefe aber hörte man ein Picken und Hämmern und unterirdische Wasser verborgen gehen, und dazwischen rauschte der Wald immerfort durch die offene Tür herein. Rüpel aber rumorte eifrig in der Höhle herum, er schien allerlei zusammenzusuchen. Auf einmal wandte er sich zu Magog: „Und damit Punktum, ich geh' mit auf die Befreiung!"

Da nun die Riesin merkte, wo das alles eigentlich hinauswollte, wurde sie plötzlich ganz empfindlich und nannte ihren Mann einen alten Bummler und den Magog einen verlaufenen Schnappsackspringer, der nur gekommen, das häusliche Familienglück zu stören. Vergebens hielt ihr Magog den Patriotismus und den gebieterischen Gang der neuen Weltgeschichte entgegen. Sie behauptete, sie hätten schon hier im Hause Geschichten genug und nicht nötig, noch neue zu machen, und die ganze Geschichte ging' die Welt gar nichts an! So entspann sich unversehens ein bedenklicher Streit. Rüpel fluchte, die Riesin zankte, die Kinder schrien, und draußen war von dem Lärm das Echo aus dem Morgenschlummer erwacht und schimpfte immerfort mit drein, man wußte nicht, ob auf Rüpel, auf Magog oder auf die Riesin.

Da hob sich auf einmal im Boden ein Stein dicht neben Magog, der erschrocken die Beine einzog, denn er meinte, es wollte ihn ein Riesenmaulwurf in die Zehen beißen. Es war aber nur eine heimliche Falltür, und aus dieser fuhr mit halbem Leibe ein winziges Kerlchen mit altem Gesicht und spitzer Mütze zornig empor: „Was macht ihr heute hier oben wieder für ein greuliches Spektakel", sagte er mit seiner dünnen Stimme, „wenn ihr nicht manierlich seid, kündigen wir euch die Miete auf!" Dabei tat es einen glühenden Blick aus der Tiefe herauf, und Magog konnte durch die Öffnung weit hinabschauen. Da sah er unzählige kleine Wichte, jedes eine Grubenlampe auf dem Kopf, in goldenen Eimern wundersam singend auf und nieder

schweben, und ganz unten blitzte und funkelte es bei den vielen irrenden Lichtern von Diamanten, Kristallen und Saphiren wie ein prächtiger Garten. – „Um Gottes willen“, rief die Riesin ihm leise und ängstlich zu, „schaut nicht so hin, man wird wahnsinnig, wenn man lange da hinuntersieht; das sind unsere Hausherren, die Zwerge und Grubenleute, die unter uns wohnen und uns diese Dachkammer für ein Billiges überlassen haben.“ Aber Rüpel, dem noch der vorige Zank in den Gliedern steckte, hatte schon mit dem Fuße nach dem Zwerglein gestoßen und hätte es sicherlich zertreten, wenn es nicht fix wieder untergeduckt und den Stein hinter sich zugeklappt hätte. Sodann ergriff Rüpel rasch seinen knotigen Wanderstab, warf einen Sack über die Schultern und stand in seinen Pelzhäuten wie eine Kürschnerbude reisefertig in der Tür. Da hätte man nun die feierliche Abschiedszene sehen sollen, die wohl geeignet war, ein fühlendes Herz mit den sanftesten Regungen zu erfüllen! Die Riesin hing mit aufgelöstem Haar am Halse des geliebten Mannes und schluchzte außerordentlich: auch von seinem gerechten Schmerze zeugte eine ungeheure Träne im Auge, die lieben Kleinen umklammerten kindlich lallend die Knie ihres verehrten Erzeugers, da hörte man nichts, als die süßen Namen: Papa und teurer Gatte und treue Lebensgefährtin! Aber Rüpel zerdrückte die Träne und riß sich los wie ein Mann. „Weib, du sollst von mir hören!“ rief er und schritt majestätisch in den Wald hinein, und Magog versäumte nicht, ihm auf das allereilfertigste nachzufolgen, denn hinter ihnen hörte er noch immer die Stimme der verwaisten Familienmutter und konnte nicht recht unterscheiden, ob sie noch immer weinte oder etwa von neuem schimpfte.

Endlich war alles verhallt, man vernahm nur noch den Tritt der einsamen Wanderer. Magog bemerkte mit vieler Genugtuung den langen Fortschritt seines Reisekumpans,

und da er seinen Rücken recht betrachtete, freute er sich dieser breitesten Grundlage und lud ihm auch noch sein eigenes Ränzel mit auf, das freilich nicht sonderlich schwer war. Durch die Wildnis aber wehte ihnen ein kräftiger Waldhauch entgegen, da wurden beide ganz lustig. Rüpel erzählte, wie er eigentlich von dem berühmten deutschen Bärenhäuter abstamme. Magog aber stimmte sein Lieblingslied an:

„Von des Volkes unverjährbaren Rechten
Und der Tyrannen Attentaten,
Die die Völker verdummen und knechten,
Fürsten und Pfaffen und Bureaukraten."

„Und Bier und Braten!" fiel hier Rüpel jubelnd mit ein. – „Haben Sie etwas mit?" wandte sich Magog rasch herum. Rüpel schüttelte mit dem Kopfe. – „Ha, also nur immer vorwärts, vorwärts!" ermutigte Magog.

Über dem Singen und den vergnügten Gesprächen aber hatte Rüpel unvermerkt den rechten Weg verloren. Vergebens bestieg er nun jeden Berg, dem sie begegneten, um sich wieder zurechtzufinden; man sah nichts als Himmel und Wald, der wie ein grünes Meer im frischen Winde Wellen schlug, so weit die Blicke reichten. Und fragen konnten sie auch niemand. Denn der Lärm, den sie unterwegs machten, war groß, und wo sie etwa ein einsamer Hirt oder Jäger hörte und des erschrecklichen Riesen ansichtig wurde, entfloh er sogleich oder verbarg sich im dicksten Gebüsch, bis sie vorüber waren. So irrten sie den ganzen Tag umher.

Des Abends, da sie schon sehr hungrig waren, kamen sie endlich an eine anmutige Anhöhe, an der unten ein Fluß vorüberging. Jenseits des Flusses aber lag ein weiter wüster Platz, rings vom finstern Walde eingeschlossen, und auf dem Platze lagen einzelne Felsblöcke zerstreut, wie Trümmer einer verfallenen Stadt, was sehr einsam anzusehen war. Auf dieser Höhe machte Rüpel plötzlich Halt

und ließ den Magog seitwärts zwischen das Gebüsch treten und sich dort ganz still verhalten. Er selbst aber setzte sich mitten auf die Höhe, zog sein haariges Wams, gleich einer Nebelkappe, aus der nur seine großen Augen hervorfunkelten, bis über den Kopf herauf, kniff aus den Fellen ein paar seltsame Ohren darüber und breitete mit beiden Armen den Pelzmantel aus wie zwei Flügel, so daß er wie eine ungeheuere Nachteule aussah. Es dauerte auch nicht lange, so kamen von allen Seiten die schreckhaftesten Vögel, wilde Auerhühner, Birkhähne und Fasanen mit großem Geschrei herbei und stießen und hackten auf das Ungetüm; und als der Schwarm am dicksten, schlug er rasch beide Pelzflügel über ihnen zusammen und schob alles in seine weitläufigen Manteltaschen. – „Das hab' ich von meinem Urgroßvater Kauzenweitel gelernt", rief er sehr zufrieden aufstehend zu Magog hinüber. Dann ging er zu dem Fluß hinab und streckte sich unter dem hohen Schilfe platt auf den Leib am Ufer hin. Magog meinte, er sei durstig und wolle den Fluß austrinken; aber Rüpel ließ bloß seinen verworrenen Bart ins Wasser gleiten, den hielten die klügsten Hechte und die breitmauligsten Karpfen für spielendes Gewürm, und so oft sie danach schnappten, schnappte Rüpel auch nach ihnen und hatte gar bald mehrere Mund voll auserlesene Fische aufs Trockne gebracht. Darauf kehrte er wieder zu Magog zurück, holte aus seinem Reisesack einen Feldkessel, Bratspieß, Messer und Gabeln hervor und schlug sich mit der Faust auf beide Augen, daß es Funken gab. Daran zündete er ein großes Feuer an und fing sogleich mit vielem Eifer zu kochen und zu braten an; und eh' es noch dunkel wurde, saßen beide Wanderer um die lustige Flamme gelagert und schmausten in freudereichem Schalle.

Unterdes war die Nacht herangekommen, in dem Feuer neben ihnen flackerte nur noch manchmal ein blaues Flämmchen auf; sie richteten sich daher in dem trocknen Laube, so gut es gehen wollte, zur Ruhe ein und waren auch beide sehr bald eingeschlafen. Es mochte aber noch

lange nicht Mitternacht sein, als Magog, wie in seiner ersten Reisenacht, wieder ein seltsames Rauschen und Murmeln vernahm, das bald schwächer, bald wieder lauter wurde, fast wie das verworrene Brausen einer fernen Stadt. Er richtete sich mit halbem Leibe auf, aber diesmal war es kein bloßer Traum. Denn obgleich der Mond zwischen vorüberjagendem Gewölk den wüsten Platz jenseits des Flusses nur flüchtig beleuchtete, so konnte er doch zu seinem Erstaunen deutlich bemerken, daß der Platz jetzt ganz belebt war. In einem weiten Halbkreise am Waldrande drüben lagen nämlich, dicht Kopf an Kopf gereiht, zahllose Auerochsen, zunächst hinter ihnen standen Rehe und Damhirsche, über diese hinweg starrte dann ein ganzer Wald von Hirschgeweihen, und weiterhin noch bis tief in die Schatten des Waldes schien es verworren zu wimmeln und zu drängen, denn sooft ein Mondstrahl das Dunkel streifte, sah man da und dort den Kopf eines Einhorns oder bärtigen Elens sich abenteuerlich hervorstrekken, und zwischen ihren Beinen Marder, Iltis und andere geringe Tiere geschäftig hin und her schlüpfen. Selbst die Bäume, die den Platz von der einen Seite umschlossen, waren von allerlei großen und kleinen Vögeln bedeckt, daß sie aussahen wie Weinstöcke im Herbst, und man nicht wußte, was Blatt oder Vogel war, rings um den Platz aber machten Störche ernsthaft die Runde und hoben die langen Schnäbel gegen den Wind, ob etwa von fern ein Feind nahe.
„Aha, das sind gewiß die Tiere, die der Riesenknabe schon heute früh in der Ferne hat marschieren gehört“, dachte Magog und wollte, als er sich vom ersten Erstaunen ein wenig erholt, geschwind den Rüpel wecken und rüttelte und schüttelte ihn mit großer Anstrengung aus Leibeskräften. Der tat aber nach der guten Mahlzeit einen schweren Schlaf, er hob bloß den Kopf in die Höh’ und glotzte ihn an, ohne etwas zu sehen, dann wälzte er sich auf die andere Seite und schnarchte so schrecklich weiter, daß von dem Atem die nächsten Bäume sich auf und nieder bogen.

Nun schaute Magog still und unverwandt nach dem Platze hinüber, denn er war sehr neugierig, was die Tiere in dieser Einsamkeit eigentlich vorhätten. Da sah er, wie ein Auerochs plötzlich aus der vordersten Reihe brach, mit einem gewaltigen Satze auf einen der umherliegenden Steinblöcke sprang und, nachdem er mit seinem zottigen Haupte sich dreimal vor der Versammlung verneigt, sofort eine donnernde Rede begann. Dabei brüllte er mitten im Sprechen oft plötzlich furchtbar auf, scharrte mit dem einen Vorderfuß, ringelte wütend den Schweif in die Luft und schüttelte die Mähne, daß man beim Mondschein seine rotglühenden Augen rollen sah. Magog konnte nichts davon verstehen, aber die Rede mußte sehr hinreißend sein, denn als er endlich von dem Steine wieder zu seinen Kameraden zurücksprang, ging ein freudiges Brüllen, Schnurren und Scharren durch die ganze Versammlung, und alle Hirsche schlugen mutig mit ihren Geweihen zusammen. Darauf hatte ein Bär das Wort erhalten. Auch dieser kletterte bedächtig auf einen der Steine herauf, stellte sich auf die Hinterbeine und streckte während seiner Ansprache bald das eine, bald das andere Vorderbein weit vor sich aus, dann legte er die eine Tatze an sein Herz – er konnte vor Rührung nicht weiter und mußte abtreten. Jetzt ließ sich unerwartet aus irgend einem dunkeln Winkel ein Uhu auf dem Steine nieder. Das wollten die andern Vögel durchaus nicht leiden, ja ein kecker Nußhäher schoß plötzlich hervor und hackte nach ihm, aber die wachthabenden Störche stellten klappernd sogleich die Ruhe wieder her. Nun schüttelte der Uhu seine Federn auf, daß er aussah wie eine Allongeperücke, klappte zum Gruß dreimal mit dem Schnabel, setzte eine Brille auf und fing aus einem Blatte, das er mit der einen Klaue vor sich hielt, zu lesen an. Er schien alles sehr weitläufig und gründlich auseinanderzusetzen, denn die ganze Gesellschaft hörte dem gelehrten Redner so aufmerksam zu, daß man dazwischen das Wiederkäuen der Ochsen vernehmen konnte; nur die ungeduldigen Vögel in den Bäumen, die

nun einmal ärgerlich geworden, störten leider zuweilen die feierliche Stille durch plötzliches ungebührliches Schreien und Raufen. Unterdes aber ging die Vorlesung ohne Komma und ohne Punktum in einem Tone immer fort und fort, wie murmelnde Bäche und spinnende Kater, und Magog wußte nicht, wie lange die Rede gedauert, denn ehe sie noch ihr Ende erreicht hatte, war er über dem einförmigen Gemurmel, so sehr er sich auch dagegen sträubte, unaufhaltsam eingeschlummert.

Er hätte auch wahrscheinlich bis in den Tag hinein geschlafen, wenn ihn nicht mitten in der Nacht Rüpel auf einmal durch unablässiges Rufen geweckt hätte. Sein erster Blick fiel auf den geheimnisvollen Platz drüben, der war aber, als wäre eben nichts geschehen, wieder so still und einsam wie gestern. Rüpel aber verzehrte bereits mit großem Appetit die Überbleibsel vom gestrigen Mahle und hatte auch ein gut Stück davon für Magog zurückgelegt. Da dieser ihm nun erzählt, was er in der Nacht jenseits des Flusses gesehen, gab Rüpel wenig darauf und meinte, das sei ohne Zweifel eine geheime Verschwörung, da kümmere er sich nicht darum, wenn er nur sein Auskommen habe. Mit dem Auskommen aber stehe es heute gerade sehr schlimm. Er habe nämlich jetzt erst an den Gestirnen die rechte Richtung erkannt, sie seien ganz auf den Holzweg geraten und hätten noch weit zu gehen. In dieser Richtung gebe es jedoch keinen Fluß, um darin zu fischen, und mit dem vom seligen Kauzenweitel ererbten Kunststück sei es auch nichts, weil die verschwornen Vögel heut alle nicht zu Hause seien. Sie mußten daher eilen, um womöglich noch in der Nacht ihr Ziel zu erreichen.

So geschah es also, daß sie noch zur selben Stunde, nachdem sie sich gehörig gestärkt hatten, ihren Befreiungszug unverdrossen wieder fortsetzten. War aber schon der Anfang dieser Nacht schön gewesen, so war sie jetzt noch viel tausendmal schöner. Die Sterne blinkten durch das dunkle Laub, als ob die Bäume silberne Blüten trügen, und der Mond ging wie ein Einsiedler über die stillen Wälder und

spielte melancholisch mit der schlummernden Erde, indem er bald einen Felsen beleuchtete, bald einen einsamen Grund in tiefen Schatten versenkte und Berg und Wald und Tal verworren durcheinanderstellte, daß alles fremd und wunderbar aussah. Auf einmal blieb Rüpel stehen, denn ein seltsam schweifendes Licht streifte die Spitzen des Gebüsches vor ihnen. Sie bogen die Zweige vorsichtig auseinander und erblickten nun mehrere schöne schlanke Mädchengestalten in leuchtenden Gewändern, die sich bei den Händen angefaßt hatten und dort einen Ringeltanz hielten. Ihre langen blonden Haare flogen in der leisen Luft, daß es wie ein Schleier von Mondschein um sie her wehte, und doch sahen sie aus wie Kinder und berührten mit den zierlichen Füßchen kaum den Boden, und wo sie ihn berührten, schimmerte das Gras von goldnem Glanze. Dabei sangen sie überaus lieblich:

„Luft'ge Kreise, lichte Gleise
Von Gesang und Mondenschein
Ziehn wir leise dir zur Reise,
Kehre bei uns Elfen ein!"

Das ließen sich die Reisenden nicht zweimal sagen und eilten sehr erfreut über die große Höflichkeit aus ihrem Versteck hervor. Kaum waren sie indes auf den freien Platz herausgekommen, so war plötzlich die ganze Erscheinung lautlos verschwunden, und sie schwankten auf einem mit trügerischem Rasen bedeckten Moorgrund, in welchem Rüpel sogleich bis über die Knie versank. Dabei glaubten sie hier und da heimlich lachen zu hören, konnten jedoch durchaus niemand mehr entdecken. Rüpel aber, um sich zu helfen, griff wütend um sich, erwischte den Magog, der soeben schon wieder aufs Trockne sprang, beim Rockzipfel und riß ihm einen Schoß seines alten Frackes glatt weg, worüber der Doktor höchst entrüstet wurde und beide in einen sehr unangenehmen und lauten Wortwechsel gerieten.

Nachdem sie sich endlich herausgearbeitet und an dem Moose möglichst wieder gesäubert hatten, sagte Rüpel: „Ja, in dieser Gegend ist's nicht recht geheuer, hier nahebei muß auch der stille See liegen mit dem versunkenen Schlosse; man kann, wenn's windstill ist, tief im Grunde noch die Türme sehen, und manchmal in schönen Sommernächten taucht es herauf, bis die ersten Hähne krähen." Und in der Tat, der unheimliche Spuk wollte gar nicht aufhören, je weiter sie in der verrufenen Gegend fortschritten. Irrlichter hüpften überall über den Weg vor ihnen und spielten und wandten sich untereinander wie junge Kätzchen; dann fuhren sie neckend nach Rüpels Bart, setzten sich auf Magogs Hut oder haschten von hinten nach ihm, als wollten sie ihm den noch übriggebliebenen Frackschoß abreißen. Rüpel sagte: „Die närrischen Dinger werden mir noch meine Wildschur anzünden", und suchte immerfort eines zu greifen, und da es jedesmal mißlang, brach er endlich in ein so herzhaftes Lachen aus, daß es weit durch den Wald schallte und die Irrlichter erschrocken nach allen Seiten auseinanderfuhren.

„Hab ich's nicht gesagt?!" rief dann Rüpel, indem er plötzlich ganz erschrocken stillstand und mit dem Finger in die Nacht hinauswies. Magog wandte sich rasch herum und erblickte in der Waldeinsamkeit einen großen klaren See, und mitten in dem See ein schneeweißes Schloß mit goldnen Zinnen, das sich wie ein schlummernder Schwan im Wasser spiegelte, und rings um das Schloß herum schien ein Garten mit Myrten, Palmen und andern wunderbaren Bäumen gleichfalls zu schlummern, so still war es dort. Jetzt aber erhoben sich auf einmal einige Elfen, die unter den Palmen geschlafen hatten, dann immer mehrere, und gleich darauf sah man sie alle wie Johanniswürmchen geschäftig hin und her irren, als würde dort ein großes Fest vorbereitet. Dabei streiften sie im Vorüberschweben mit ihren Fingerspitzen Bäume, Blumen und Sträucher, die von der flüchtigen Berührung allmählich in hun-

dertfarbigem Glanze, wie lauter Bergkristalle, Rubinen, Smaragden und Saphire zu leuchten anfingen, und wenn die Luft durch den Garten ging, gab es einen wunderbaren Klang, als ob der Mondschein selber sänge. – „Das ist ihr Traumschloß", flüsterte Rüpel dem Magog zu und wandte kein Auge von der prächtigen Illumination. Magog aber warf stolz den Kopf zurück. „Einfältiges Waldesrauschen, alberne Kobolde, Mondenschein und klingende Blumen", sagte er mit außerordentlicher Verachtung, „nichts als Romantik und eitel Märchen, wie sie müßige Ammen sonst den Kindern erzählten. Aber der Menschengeist ist seitdem mündig geworden. Vorwärts! die Weltgeschichte wartet draußen auf uns." Mit diesen Worten drängte er den kindischen Riesen fort zu verdoppelter Eile und ruhte nicht, bis der Blumengesang und der schimmernde Garten hinter ihnen verklungen und versunken.

Das war aber nun einmal eine wahre Hexennacht, denn sie mochten kaum noch eine Stunde lang gegangen sein, so hörten sie schon wieder ein seltsames Geräusch vor sich, ein Schwanken und Knistern in den Zweigen und Hufklang dazwischen, immer näher und näher, wie wenn jemand rasch und heimlich durch das Dickicht bräche. Und es war auch wirklich ein flüchtiger Zug, der gerade auf sie zukam. Voran eilten viele Irrlichter in lustigen Sprüngen, um unter den Eichenschatten den Weg zu zeigen, dann folgte ein Hirsch, und auf dem Hirsche saß eine sehr schöne Dame, von ihren Locken, wie von einem goldnen Mantel, durch den die Sterne schienen, rings umwallt und einen Kranz ums Haupt, der in grüngoldnem Feuer funkelte. Als sie die beiden Wandrer gewahrte, stutzte sie, und auf einen Wink von ihr hielten Hirsch und Irrlichter plötzlich an. Rüpel verneigte sich, so tief er's vermochte, und wagte kaum verstohlen aufzublinzeln, während die Irrwische, die keinen Augenblick ruhig bleiben konnten, sich schon wieder mit Magogs verwitwetem Rockschoß zu schaffen machten. „Was sucht ihr hier?" fragte die Rei-

terin, die Fremden mit einem strengen und durchdringenden Blick betrachtend. – „Die Libertas", entgegnete Magog stolz. Da lachte die Dame und winkte wieder, und wieder eilten die Irrlichter voran und flog der Hirsch mit seiner schönen Herrin über den Rasen fort – sie schienen nach dem Traumschlosse hinzuziehn.

Jetzt erst richtete sich Rüpel mühsam aus seiner Devotion wieder auf; „gewiß Ihre Majestät die Elfenkönigin", rief er, dem Zuge noch lange nachsehend. „Das wäre mir eine schöne Königin", erwiderte Magog, „ihr Diadem war nicht einmal echt, nichts als leuchtende Johanniswürmchen."

Der Morgen fing endlich an zu dämmern, in der Ferne krähte schon ein Hahn; da bog Rüpel bald da, bald dort die Wipfel auseinander und spähte unruhig nach allen Seiten umher. „Jetzt hab' ich's!" rief er auf einmal, „dort ist das Schloß des Baron Pinkus." – „Das trifft sich ja vortrefflich", entgegnete Magog, „es scheint noch alles zu schlafen droben, wir müssen das Schloß überrumpeln. Der Star hat mir alles ausführlich beschrieben; dort in dem Eckturm sitzt die Libertas gefangen. Sie, lieber Herr Rüpel, haben gerade die gehörige Leibeslänge, Sie langen also ohne weiteres in das Turmfenster hinein und heben die Gefangene in meine Arme. Ja, jetzt gilt's: Entführung, Hochzeit, Tod oder Haushofmeister!" Nun aber hatte er seine Not mit dem Riesen, der nicht so leise auftreten konnte, wie es die Wichtigkeit des entscheidenden Augenblicks erheischte und überdies bald Eicheln knackte, bald wieder einen Ast abbrach, um sich die Zähne zu stochern. Jetzt glaubten sie in dem Schloßhofe einen Hund anschlagen zu hören. „Um des Himmels willen", flüsterte Magog seinem Gefährten zu, „nur still jetzt, sachte, sachte!" – So zogen sie sich vorsichtig am Rande des Waldes hin, als ob sie ein Eulennest beschleichen wollten.

Da sahen sie zu ihrer nicht geringen Verwunderung auf

einmal einen glänzenden Punkt sich wie eine Sternschnuppe übers Feld bewegen. Es kam immer näher, und bald konnten sie deutlich unterscheiden, daß es eine Frauengestalt und die Sternschnuppe eine glimmende Zigarre war, die sie im Munde hielt. Sie kam, wie es schien, in großer Angst vom Schlosse gerade auf sie dahergeflogen; eine prächtige Amazone mit Schärpe, Reitgerte und klingenden Sporen, ein zierliches Reisebündel unter dem Arm. Jetzt stand sie atemlos dicht vor Magog, den sie beinahe umgerannt hätte. – „Mein Ideal!" rief sie da plötzlich aus, und „Libertas!" schallte es aus Magogs entzücktem Munde herüber. Sie hatten einander im Augenblick erkannt, ein geheimnisvoller Zug gleichgestimmter Seelen riß Herz an Herz, und in einer langen stummen Umarmung ging ihnen die Welt unter und die Ewigkeit auf. – Unterdes war auch Rüpel neugierig zwischen den Bäumen hervorgetreten, da erschrak die Dame sehr und sah ihn scheu von der Seite an. Rüpel aber, dem ihr neckisches Wesen gefiel, wurde auf einmal sehr galant, wollte ihr seine Bärenhaut unterbreiten und sie in seinem Futtersack durch den Wald tragen, ja er versuchte sogar in seiner Lustigkeit auf dem Rasen eine Menuett auszuführen, die er einst die alte Urtante hatte tanzen gesehen. Nun wurde auch die Dame wieder ganz vertraulich und erzählte, wie sie es auf dem barbarischen Schlosse nicht länger habe aushalten können; dann geriet sie immer mehr in sichtbare Begeisterung und sprach von Tyrannenblut, von Glaubens-, Rede-, Preß- und allen erdenklichen Freiheiten. Da hielt sich Magog nicht länger, reckte zum Treuschwur den Arm hoch zu den Göttern empor, reichte ihr darauf die Rechte und verlobte sich sogleich mit ihr, und Rüpel schrie in einem fort Vivat! dazu.

Über diesem Freudengeschrei aber entstand nach und nach ein bedenkliches Rumoren im Schlosse. Die Verliebten draußen merkten es gar nicht, wie erst einzelne Wachen verdächtig über das stille Feld fast bis zum Walde streiften und dann eiligst wieder zum Schloß zurückkehr-

ten. Auf einmal aber tat sich das Schloßtor auf, und die ganze bewaffnete Macht schritt mit dem Feldgeschrei: „Die Libertas ist entwischt!“ todesmutig daraus hervor. Dazwischen konnte man deutlich die Stimme des Baron Pinkus unterscheiden, der entrüstet gegen das Dasein von Riesen und dergleichen abergläubischen Nachtspuk, wovon die Streifwachen gefabelt, im Namen der Aufklärung protestierte. Jetzt aber erblickten sie den Rüpel, den sie anfangs für einen knorrigen Baumstamm angesehen hatten, und hielten plötzlich an. Niemand wagte sich zu regen, es war so still, daß man fast die Gedanken hören konnte; überall nichts als ein irres Flüstern mit den Augen, todbleiche Gesichter und fliegende Röte dazwischen, kurz, alle Symptome einer allgemeinen Verschwindsucht. Bei Pinkus endlich kam sie zum Ausbruch. Erst ganz leise mit langen, langen Schritten, den Kopf noch immer zurückgewendet, dann unaufhaltsam in immer weiteren Sprüngen, daß ihm der Opferzopf hoch in der Luft nachflog, stürzte er nach dem Schlosse und die bewaffnete Macht in wildester Flucht ihm nach. Rüpel hatte eben nur noch Zeit genug, den behenden Pinkus mit ein paar gewaltigen Sätzen am Zipfel seines Zopfes zu erfassen, aber er behielt den Zopf allein in der Hand, und damit hieb er wütend rechts und links und trieb sie alle vor sich her; ja, er wäre ohne Zweifel mit ihnen zugleich in das Schloß gedrungen, wenn er nicht in der Hitze des Gefechtes an den Schwibbogen des Tores mit solcher Vehemenz mit dem Kopfe angerannt wäre, daß er unversehens rücklings zu Boden fiel, was den empfindlich Geschlagenen notdürftigen Vorsprung gewährte, sich in das Schloß zu salvieren und, ehe Rüpel sich wieder aufraffte, die eisernen Torflügel dicht vor ihm krachend zuzuwerfen.

Nun wandte sich Rüpel sehr vergnügt um, mit Magog weiteren Kriegsrat zu pflegen. Aber wie erstaunte er, als er niemand hinter sich erblickte. Vergebens ging und rief er am Rande des Waldes auf und nieder, die beiden Liebenden waren spurlos verschwunden. Die Libertas mag

sich wohl vor dem Schlachtlärme etwas tiefer in den Wald zurückgezogen haben, dachte er; er hoffte noch immer sie wiederzufinden und ging und rief von neuem immer weiter fort, worüber er aber mit dem Echo, das ihm lauter unvernünftige Antworten gab, in einen ebenso heftigen als fruchtlosen Wortwechsel geriet. Und so hatte er denn von der ganzen großen Unternehmung nichts als ein paar neue Löcher in seiner alten Wildschur gewonnen und schritt endlich voller Zorn und so eilfertig wieder in den Urwald zurück, daß wir ihm unmöglich weiter nachgehen können.

Wie aber war die Libertas so unverhofft aus ihrem Turme entkommen?
Wir haben schon früher gesehen, daß seit ihrer Gefangenschaft im Pinkusschen Schlosse und Garten die gute alte Zeit wieder repariert und neu vergoldet worden, wo sie durch ihre impertinente Einmischung etwa gelitten hatte. Alles schämte sich pflichtschuldigst der augenblicklichen Verführung und Verwilderung; in der schillernden Mittagsschwüle plätscherten die Wasserkünste wieder wie blödsinnig immerfort in endloser Einförmigkeit; die Statuen sahen die Buchsbäume, die Buchsbäume die Statuen an, und die Sonne vertrieb sich die Zeit damit, auf den Marmorplatten vor dem Schlosse glitzernde Schnörkel und Ringe zu machen; es war zum Sterben langweilig. Libertas hatte daher schon lange nachgedacht, wie sie sich befreien könnte, und sann und sann, bis endlich die Nacht der ganzen Industrie im Schloß das Handwerk gelegt und draußen die Welt ungestört wieder aufatmete. Auch der Schwan auf dem Wallgraben unter dem Turm war nun eingeschlummert, und drüben standen die Wälder im Mondschein. Da trat Libertas an das offene Fenster und sprach:

„Wie rauscht so sacht
Durch alle Wipfel
Die stille Nacht,

Hat Tal und Gipfel
Zur Ruh' gebracht.
Nur in den Bäumen
Die Nachtigall wacht
Und singt, was sie träumen
In der stillen Pracht."

Die Nachtigall aber antwortete aus dem Fliederbusche unten:

„In der stillen Pracht,
In allen frischen Büschen, Bäumen flüstert's in Träumen
Die ganze Nacht,
Denn über den mondbeglänzten Ländern
Mit langen weißen Gewändern
Ziehen die schlanken
Wolkenfrauen, wie geheime Gedanken,
Senden von den Felsenwänden herab die behenden
Frühlingsgesellen: die hellen Waldquellen,
Um's unten zu bestellen
An die duftigen Tiefen,
Die tun, als ob sie schliefen,
Und wiegen und neigen in verstelltem Schweigen
Sich doch so eigen mit Ähren und Zweigen,
Erzählen's den Winden,
Die durch die blühenden Linden,
Vorüber an den grasenden Rehen
Säuselnd über die Seen gehen,
Daß die Nixen verschlafen auftauchen
Und fragen,
Was sie so lieblich hauchen?
Ich weiß es wohl, dürft' ich nur alles, alles sagen."

Hier kam plötzlich ein Storch aus dem Gesträuch und klapperte zornig nach dem Fliederbusche hin, und die Nachtigall schwieg auf einmal. – „Was hat nur der Storch mit der Nachtigall zu so später Zeit? er ruht doch sonst

auch gern bei Nacht“, sagte Libertas zu sich selbst und wußte gar nicht, was sie davon denken sollte.

Aber die Nachtigall wußte es recht gut, und daß sie in der Nähe des Schlosses nicht so viel ausplaudern sollte; denn unter den freien Tieren des Waldes war in jener großen nächtlichen Versammlung, die Magog auf seiner Wanderschaft von ferne mit angesehen hatte, eine geheime Verschwörung gemacht worden und sollte eben in der heutigen Nacht zum Ausbruch kommen. Schon am vorigen Abend war es den Landleuten, die vor Schlafengehen noch ihre Saaten in Augenschein nahmen, sehr aufgefallen, wie da über der Au im Tale, wo die glänzenden Sommerfäden an den Gräsern hingen, so viele Schwalben emsig hin und her schweiften und mit ihren Schnäblein die Fäden aufrafften, soviel eine jede im Fluge erhaschen konnte, daß sie, als sie damit durch die Luft flogen, wie in langen silbernen Schleiern dahinzogen. Dieses feine Gespinst aber breiteten die Schwalben sodann auf einer einsamen Waldwiese im Mondschein aus; da kamen hurtig unzählige kleine Spinnen, die schon darauf gewartet, rote, braune und grüne, und drehten die Fäden fleißig zusammen und woben, damit es besser aussähe, auch etwas Mondschein darein, während die Johannisfünkchen ihnen dabei leuchteten und die Heimchen dazu sangen. Kaum aber hatten sie die letzten Maschen geknüpft, so säuselte es leise leise durch die Stille, von allen Seiten kamen Bienen, die heute Schlaf und Honig vergaßen, dicke Päckchen an ihren Füßen, die streckten und steiften mit dem Wachse das ganze Gespinst gar kunstreich zu einer langen Strickleiter. Unterdes sah man bei dem klaren Mondlicht bald da, bald dort am Waldessaume ein Reh mit den klugen Augen hervorgucken und schnell wieder im Dickicht verschwinden, denn das wachsame Wild machte die Runde, um sogleich zu warnen, wenn etwa Verrat drohte. Der getreue Storch aber, der vorher die Nachtigall wegen ihrer Plauderhaftigkeit ausgescholten, stand die ganze Zeit hindurch, nur ein paarmal wider Willen einnickend, unbe-

weglich auf einem Beine bei den Spinnen und Bienen, um auf ihr Werk aufzupassen und ohne Nachsicht jeden wegzuschnappen, der sich bei der Arbeit saumselig zeigte. Und als die Leiter fertig war, prüfte er sie bedächtig, hing sie dann an den Ast des nächsten Baumes und stieg selbst daran hinauf, um zu versuchen, ob sie fest genug, wobei er sich aber so ungeschickt und seltsam anstellte, daß die kleinen behenden Kreaturen ringsumher einigemal heimlich kichern mußten und die Heimchen neckend: „Storch, Storch, Steiner, hast so lange Beiner!" zu ihm hinüberriefen, worüber er jedesmal sehr böse wurde und mit seinem langen Schnabel nach ihnen hackte.

Als er nun aber sah, daß alles gut war, nahm er das eine Ende der luftigen Leiter in den Schnabel, flog damit zu dem Fenster der Libertas hinan und schlang es fest um das Fensterkreuz. Zu gleicher Zeit schlug die Wachtel gellend in dem nahen Kornfelde; das war das verabredete Zeichen. Da erwachten alle Waldvögel draußen, die ohnedies nicht fest geschlafen vor Freude und Erwartung, und weil die Nachtigall die ganze Nacht so laut geschmettert hatte. Die flogen nun alle nach dem Turmfenster droben, pickten an die Scheiben und sangen ganz leise:

„Frau Libertas, komm heraus!
Denn der liebe Gott hat lange
Draußen unser grünes Haus
Schon geschmückt dir zum Empfange,
Hat zur Nacht die stillen Tale
Rings mit Mondenschein bedeckt,
Und in seinem Himmelssaale
Alle Lichter angesteckt.
Horch, das rauscht so kühl herauf,
Frau Libertas, wache auf!"

Aber Libertas, die an dem heimlichen Treiben draußen längst alles gemerkt, hatte schon ihr Bündel geschnürt und betrat, die treuen Vögel freundlich grüßend, die Stricklei-

ter, und wie sie so in die Nacht hinabstieg, boten ihr die kleinen Birken, die aus den Mauerritzen des alten Turmes wuchsen, überall helfend die grünen Hände, und von unten wehte ihr der Duft der Wälder und Wiesen erfrischend entgegen. Als sie aber an den breiten Wallgraben kam, war schon der Schwan am Ufer und schwellte stolz seine Flügel wie zwei schneeweiße Segel. Da setzte sich Libertas dazwischen, und er glitt mit ihr hinüber und betrachtete voll Entzücken ihr schönes Bild, das auf dem Spiegel des Weihers neben ihm dahinschwebte. Unterdes aber hatte der Kettenhund im Hofe schon lange die Ohren gespitzt und weckte jetzt laut bellend seinen Nachbar, den boshaften Puter, der hätte bald alles verraten, er kollerte so heftig, daß er ganz rot und blau am Kragen wurde vor Zorn und Hoffart, darüber wachten auch die Gänse im Stalle auf und schrien Zeter und abermals Zeter, denn sie hatten die rechte Witterung von den heimlichen Umtrieben am Turme und fürchteten alle, wenn die Libertas entwischte, aus dem guten Futter zu kommen und zu den andern gemeinen Vögeln in die Freiheit gesetzt zu werden. Aber ihr Lärm und Ärger kam zu spät, Libertas war schon jenseits des Wallgrabens. Drüben aber stand ein Hirsch am Waldessaume und neigte die Knie und sein Geweih vor ihr bis auf den Rasen. Da schwang sie sich rasch hinauf, und fort ging es durch Nacht und Wald, und der Storch mit den andern Vögeln, um ihr das Geleit zu geben, stürzte sich hinterdrein vom Turme in die Luft, in stillen Kreisen über den mondbeglänzten Gärten, Wäldern und Seen schwebend. Die im Schlosse merkten es erst bei Tagesanbruch, wo sie, wie wir gesehen, zu ihrem Unglück auf ihre Verfolgung ausrückten. Nur die Hirten, die an den Bergeshängen bei ihren Herden wachten, hörten erstaunt den Gesang in den Lüften und die geheimnisvolle Flucht im Waldesgrund an den einsamen Weilern vorüberziehen. Und das war eben die schöne Frauengestalt auf dem Hirsch, die in derselben Nacht Rüpel und Magog auf ihrer Wanderschaft im Urwald gesehen, ohne die Libertas zu

erkennen, auf deren Befreiung sie so schlau und vorsichtig ausgezogen.
Die Amazone aber, die sie gerettet hatten, war niemand anders als die Pinkussche Silberwäscherin Marzebille, ein herzhaftes Frauenzimmer, die schon früher als Marketenderin mit den Aufklärungstruppen durch dick und dünn mit fortgeschritten und nirgends fehlte, wo es was Neues gab. Die hatte nun seit der Libertas Erscheinung eine inkurable Begeisterung erlitten und sich daher an jenem denkwürdigen Morgen kurz resolviert, aus dem Schloßdienst in die Freiheit zu entlaufen. Der Doktor Magog aber war damals vor dem unverhofften Schlachtgetümmel am Schlosse so heftig erschrocken, daß er mit seiner glücklich emanzipierten Braut, die hier alle Schliche und Wege kannte, unaufhaltsam sogleich quer durch Deutschland und übers Meer bis nach Amerika entfloh, wo er wahrscheinlich die Marzebille noch heut für die Libertas hält.
Da konnte sie denn Rüpel freilich nicht mehr errufen. Und das schadet auch nichts, denn Magog hatte schon während der feierlichen Verlobung hin und her gesonnen, auf welche Weise er den Riesen, da er ihn nun nicht mehr brauchte, wieder loswerden könnte; er dachte gar nicht daran, einen so ungeschlachten Gesellen zu seinem Haushofmeister zu machen, dessen große Familie ihm wohl bald Haus und Hof verzehrt hätte. Dafür haben ihn, gleichwie die Menschen Vogelscheuchen aufzurichten pflegen, die dankbaren Vögel in Erwägung seiner vor dem Schlosse bewiesenen Bravour als Hüter des Urwaldes angestellt, mit der einzigen Verpflichtung, von Zeit zu Zeit mit den schrecklichsten Tierfellen, Mähnen und Auerochsenhörnern sich am Rande des Waldes zu zeigen. Dort also hat der Biedermann endlich sein sicheres Brot.
Die emigrierte Urtante ist gänzlich verschollen. Von der Libertas dagegen sagt man, daß sie einstweilen bei den Elfen im Traumschlosse wohne, das aber seitdem niemand wieder aufgefunden hat.

PM Press was founded in 2007 as an independent publisher with offices in the US and UK, and a veteran staff boasting a wealth of experience in print and online publishing. Operating our own printing press enables us to print and distribute short as well as large run projects, timely texts and out of print classics.

We seek to create radical and stimulating fiction and non-fiction books, pamphlets, t-shirts, visual and audio materials to entertain, educate and inspire you. We aim to distribute these through every available channel with every available technology. Whether that means you are seeing anarchist classics at our bookfair stalls, reading our latest vegan cookbook at the café over (your third) microbrew, downloading geeky fiction e-books, or digging new music and timely videos from our website.

PM Press is always on the lookout for talented and skilled volunteers, artists, activists and writers to work with. If you have a great idea for a project or can contribute in some way, please get in touch.

PM Press . PO Box 23912 . Oakland CA 94623

www.pmpress.org